RENEWALS 458-4574

THE LAKES HANDBOOK
*Volume 2*

**Also available from Blackwell Publishing:**

*The Lakes Handbook*
*Volume 1*  Limnology and Limnetic Ecology
Edited by P.E. O'Sullivan & C.S. Reynolds

# The Lakes Handbook

## VOLUME 2
## LAKE RESTORATION AND REHABILITATION

EDITED BY

**P.E. O'Sullivan**

AND

**C.S. Reynolds**

© 2005 by Blackwell Science Ltd
a Blackwell Publishing company

BLACKWELL PUBLISHING
350 Main Street, Malden, MA 02148-5020, USA
108 Cowley Road, Oxford OX4 1JF, UK
550 Swanston Street, Carlton, Victoria 3053, Australia

The rights of P.E. O'Sullivan and C.S. Reynolds to be identified as the Authors of the Editorial Material in this Work have been asserted in accordance with the UK Copyright, Designs, and Patents Act 1988.

All rights reserved. No part of this publication may be reproduced, stored in a retrieval system, or transmitted, in any form or by any means, electronic, mechanical, photocopying, recording or otherwise, except as permitted by the UK Copyright, Designs, and Patents Act 1988, without the prior permission of the publisher.

First published 2005 by Blackwell Science Ltd

*Library of Congress Cataloging-in-Publication Data*

The lakes handbook / edited by P.E. O'Sullivan and C.S. Reynolds.
   p. cm.
Includes bibliographical references and index.
ISBN 0-632-04795-X (hardback, v.1 : alk. paper)
1. Limnology.  2. Lake ecology.  I. O'Sullivan, P.E. (Patrick E.)  II. Reynolds, Colin S.
   QH96.L29 2003
   551.48′2—dc21
              2003000139

A catalogue record for this title is available from the British Library.

Set in 9 on 11.5 pt Trump Mediaeval
by SNP Best-set Typesetter Ltd., Hong Kong
Printed and bound in the United Kingdom
by TJ International, Padstow, Cornwall

The publisher's policy is to use permanent paper from mills that operate a sustainable forestry policy, and which has been manufactured from pulp processed using acid-free and elementary chlorine-free practices. Furthermore, the publisher ensures that the text paper and cover board used have met acceptable environmental accreditation standards.

For further information on
Blackwell Publishing, visit our website:
www.blackwellpublishing.com

# Contents

List of Contributors, vii

**Part I  General Issues**

1  ON THE VALUE OF LAKES, 3
   *Patrick O'Sullivan*

2  THE ASSAULT ON THE QUALITY AND VALUE OF LAKES, 25
   *Wilhelm Ripl and Klaus-Dieter Wolter*

**Part II  Regional Studies**

3  THE NORTH AMERICAN GREAT LAKES: A LAURENTIAN GREAT LAKES FOCUS, 65
   *Marlene S. Evans*

4  LAKE WASHINGTON, 96
   *W.T. Edmondson*

5  LAKES OF NORTHERN EUROPE, 117
   *Heikki Simola and Lauri Arvola*

6  EUROPEAN ALPINE LAKES, 159
   *Martin T. Dokulil*

7  LAKE BAIKAL AND OTHER GREAT LAKES OF ASIA, 179
   *Lyudmila G. Butorina*

8  LAKES IN ARID ENVIRONMENTS, 200
   *W.D. Williams*

9  FLOODPLAIN LAKES AND RESERVOIRS IN TROPICAL AND SUBTROPICAL SOUTH AMERICA: LIMNOLOGY AND HUMAN IMPACTS, 241
   *John M. Melack*

**Part III  Human Impact on Specific Lake Types**

10  EUTROPHICATION OF SHALLOW TEMPERATE LAKES, 261
    *G.L. Phillips*

11  EUTROPHICATION OF SHALLOW TROPICAL LAKES, 279
   *Patrick L. Osborne*

12  RESERVOIRS AND OTHER ARTIFICIAL WATER BODIES, 300
   *Milan Straškraba*

**Part IV  Lake and Catchment Models**

13  THE EXPORT COEFFICIENT APPROACH TO PREDICTION OF NUTRIENT LOADINGS: ERRORS AND UNCERTAINTIES IN THE BRITISH EXPERIENCE, 331
   *Helen M. Wilson*

14  THE PHOSPHORUS LOADING CONCEPT AND THE OECD EUTROPHICATION PROGRAMME: ORIGIN, APPLICATION AND CAPABILITIES, 354
   *Walter Rast and Jeffrey A. Thornton*

15  MODELS OF LAKES AND RESERVOIRS, 386
   *Sven-Erik Jørgensen*

16  THE ASSESSMENT, MANAGEMENT AND REVERSAL OF EUTROPHICATION, 438
   *Helmut Klapper*

17  BIOMANIPULATION IN SHALLOW LAKES: CONCEPTS, CASE STUDIES AND PERSPECTIVES, 462
   *S. Harry Hosper, Marie-Louise Meijer, R.D. Gulati and Ellen van Donk*

18  RESTORING ACIDIFIED LAKES: AN OVERVIEW, 483
   *Lennart Henrikson, Atle Hindar and Ingemar Abrahamsson*

**Part V  Legal Frameworks**

19  THE FRAMEWORK FOR MANAGING LAKES IN THE USA, 501
   *Thomas Davenport*

20  NORDIC LAKES—WATER LEGISLATION WITH RESPECT TO LAKES IN FINLAND AND SWEDEN, 511
   *Marianne Lindström*

21  THE PROBLEM OF REHABILITATING LAKES AND WETLANDS IN DEVELOPING COUNTRIES: THE CASE EXAMPLE OF EAST AFRICA, 523
   *F.W.B. Bugenyi*

22  SOUTH AFRICA—TOWARDS PROTECTING OUR LAKES, 534
   *G.I. Cowan*

Index, 543

# Contributors

**Ingemar Abrahamsson** *Medins Sjö- och Åbiologi AB, Företagsvägen 2, 435 33 Mölnlycke, Sweden*

**Lauri Arvola** *Lammi Biological Station, University of Helsinki, Fl-16900 Lammi, Finland*

**F.W.B. Bugenyi** *Department of Zoology, Makerere University, P.O. Box 7062, Kampala, Uganda, and Fisheries Research Institute (FIRI), PO Box 343 Jinja, Uganda*

**Lyudmila G. Butorina** *Institute for Biology of Inland Waters, Russian Academy of Sciences, Borok, Nekouz District, Yaroslavl Region 152742, Russia*

**G.I. Cowan** *Department of Environmental Affairs and Tourism, Private Bag X447, Pretoria 0001, South Africa*

**Thomas Davenport** *EPA Region 5, WW16-J, 77 W Jackson Boulevard, Chicago, IL 60604, USA*

**Martin T. Dokulil** *Institute of Limnology, Austrian Academy of Sciences, Mondseestrasse 9, A-5310 Mondsee, Austria*

**W.T. Edmondson** late of *Department of Zoology, University of Washington, NJ-15, Seattle, Washington 981965, USA*

**Marlene S. Evans** *National Water Research Institute, 11 Innovation Boulevard, Sakatoon, Saskatchewan S7N 3H5, Canada*

**Ramesh D. Gulati** *Netherlands Institute of Ecology—Centre for Limnology, Rijksstraatweg 6, 3631 AC Nieuwersluis, The Netherlands*

**Lennart Henrikson** *WWF Sweden, Freshwater Programme, Ulriksdals Slott, SE-170 81 Solna, Sweden*

**Atle Hindar** *Norwegian Institute for Water Research, Regional Office South, Televeien 3, N-4879 Grimstad, Norway*

**S. Harry Hosper** *Institute for Inland Water Management and Wastewater Treatment (RIZA), PO Box 17, 8200 AA, Lelystad, The Netherlands*

**Sven-Erik Jørgensen** *The Danish University of Pharmaceutical Sciences, Institute A, Section of Environmental Chemistry, University Park 2, 2100 Copenhagen Ø, Denmark*

**Helmut Klapper** *Centre for Environmental Research, Leipzig-Halle Ltd, Department of Inland Water Research, Brückstraße 3a, D-39114 Magdeburg, Germany*

**Marianne Lindström** *Finnish Environment Institute, PO Box 140, 00251 Helsinki, Finland*

**Marie-Louise Meijer** *Waterboard Hunze en Aa's, P.O. Box 164, 9640 AD Veendam, The Netherlands*

**John M. Melack** *Bren School of Environmental Science and Management, University of California, Santa Barbara, CA, USA*

**Patrick L. Osborne** *International Center for Tropical Ecology, University of Missouri-St. Louis, 8001 Natural Bridge Road, St. Louis, Missouri 63121, USA*

**Patrick O'Sullivan** *School of Earth, Ocean and Environmental Sciences, University of Plymouth, Drake Circus, Plymouth PL4 4AA, UK*

**G.L. Phillips** *Centre for Risk and Forecasting, Environment Agency, Kings Meadow House, Reading RG1 8DQ, UK*

**Walter Rast** *Aquatic Station, Department of Biology, Texas State University, 601 University Drive, San Marcos, TX 78666, USA*

**Wilhelm Ripl** *Technische Universität Berlin, Fachgebiet Limnologie, Hellriegelstrasse 6, D-14195 Berlin, Germany*

**Heikki Simola** *Karelian Institute, University of Joensuu, PO Box 111, Fl-80101 Joensuu, Finland*

**Milan Straškraba** *late of The Academy of Sciences of the Czech Republic and the University of South Bohemia, České Budějovice, Czech Republic.*

**Jeffrey A. Thornton** *Southeastern Wisconsin Regional Planning Commission, PO Box 1607, Waukesha, Wisconsin 53187, USA*

**Ellen Van Donk** *Netherlands Institute of Ecology – Centre for Limnology, Rijksstraatweg 6, 3631 AC Nieuwersluis, The Netherlands*

**W.D. Williams** *late of Department of Environmental Biology, University of Adelaide, Adelaide, SA 5005, Australia*

**Helen M. Wilson** *School of Earth, Ocean and Environmental Sciences, University of Plymouth, Drake Circus, Plymouth PL4 4AA, UK*

**Klaus-Dieter Wolter** *Technische Universität Berlin, Fachgebiet Limnologie, Department of Limnology, Hellriegelstrasse 6, D-14195 Berlin, Germany*

# *Part I* General Issues

# 1 On the Value of Lakes

PATRICK O'SULLIVAN

## 1.1 INTRODUCTION

In Volume 1 of this book, we examined:
1 lake physical and chemical processes (the nature and origin of lakes themselves, the relationship between lakes and their surrounding water table and with their catchments, lake hydrodynamics and sedimentation);
2 limnetic ecology (the nature and role of the major classes of organisms found in freshwater lakes—phytoplankton, zooplankton, macrobenthos, pelagic microbes and fish—as well as whole lake communities).
In this volume, we discuss the general impact of human societies upon lakes, as well as that experienced by lakes of selected parts of the Earth—North America (and Lake Washington in particular), the Nordic and Alpine regions of Europe, Lake Baikal (by volume by far the world's largest lake), the arid zone in general, and Latin America. We then go on to examine the problems created by human use of various different kinds of lake—shallow bodies of both the temperate zone and the tropics, reservoirs and other artificial bodies.

The volume then continues with an examination of measures developed over the past 30 years in order to combat eutrophication (the most widespread 'environmental problem' created by human impact on lakes, on a catchment scale), especially the use of catchment models to predict nutrient loadings from the land upon lakes, and of steady state and dynamic models better to understand the problem itself. This is then followed by a discussion of the general measures currently available to reverse eutrophication, and of one of these in particular (biomanipulation), and also those employed in order to restore acidified lakes. Finally, we give various authors the opportunity to describe current and recent attempts to combat human impact on lakes, and to restore them to some state resembling stability, in various countries or regions (North America, Nordic Europe, East Africa, South Africa), in the hope that these experiences may prove useful, if not inspirational, to colleagues elsewhere.

In all of this, as also in Volume 1, the **value** of lakes—their immense scientific interest, their importance as threatened habitats, and for conservation of endangered species, their extreme importance to the human community as a resource—is implicitly accepted as a 'given', in that to us, as limnologists, the value and importance of lakes is so overwhelmingly and blindingly obvious as to need no further amplification. However, if we, as scientists, are to be accountable to the society which pays our salaries, and which funds our research (a function which is today increasingly demanded of us), and if we are to explain the value of what we do to a wider society which has never been exposed to the fascinations of limnetic ecology, we need to be able to state explicitly:
1 just what it is that we find so valuable (and indeed marvellous) about these beautiful but sometimes tantalising bodies of water;
2 just what it is we are restoring, when we propose that we restore lakes to some more desirable condition, and to spell this out in terms which involve using language with which we ourselves, as scientists, may be less comfortable.

This may be because

*'Science, because of its desire for objectivity, is inadequate to teach us **all** we need to know about valuing nature. Yet value in nature, like value in human life, is something we can see and experience' (Rolston 1997, p. 61, my emphasis).*

Paradoxically, in attempting to do this, we may find ourselves reaching out beyond the laboratory, into the wider community, and enlisting the support of the growing number of our fellow citizens who may lack sophisticated limnological knowledge, but who share our concern for nature in general, and hence for lakes.

## 1.2  LAKES AS A RESOURCE

The human 'assault' upon lakes is discussed in detail in the next chapter. Here, in order to begin discussion of lakes as a resource in general, Table 1.1 depicts several of the major human uses of lakes, **as related to water quality**. It should be noted that the terms oligo- and eutrophic are used here according to their original meanings (i.e. in order to signify lake nutrient **status** or concentration [respectively, poorly and well-**nourished**; Hutchinson 1969, 1973; Rohde 1969; Edmondson 1991; O'Sullivan 1995], and not lake biological productivity). Thus, for drinking water, oligotrophic waters, which are normally, by definition, cooler and more oxygenated, and therefore potentially contain fewer pathogenic microorganisms, and fewer plankton, and thus require less treatment, therefore incurring fewer costs, are preferred. For bathing, mesotrophic waters are preferable to eutrophic, if not required, although it is sometimes possible to share the latter with blooms of benign phytoplankton, provided other distractions are available (e.g. the Wansee in August).

Non-bathing water sports are tolerable in eutrophic waters, but other recreational uses, particularly those where the aesthetic experience of lakes is as important, or more important, than their direct use, surely demand oligotrophic or mesotrophic waters, with their greater clarity, and general absence of water blooms and other nuisances. Fish culture (in the developed world, at least) depends on the species involved, with salmonids preferring those oligotrophic waters originally fished (in Britain, anyway) only by 'gentlemen'. Cyprinids, however, are tolerant of lowland, eutrophic waters, usually known (in that country) as 'coarse' fisheries, with all that the term implies regarding both nutrient and social status. The carp ponds of the Třeboň district of South Bohemia (Czech Republic) represent an early development of what is often nowadays thought of as a new idea (i.e. **permaculture**; Mollison 1988), but

Table 1.1  Lake and reservoir use according to trophic status (after Bernhardt 1981)

| Use | Required/preferred water quality | Tolerable water quality |
| --- | --- | --- |
| Drinking water | Oligotrophic | Mesotrophic |
| Bathing | Mesotrophic | Slightly eutrophic |
| Recreation (non-bathing) | Oligotrophic | Mesotrophic |
| Water sports (non-bathing) | Mesotrophic | Eutrophic |
| Fish culture: | | |
|    salmonids | Oligotrophic | Mesotrophic |
|    cyprinids | Eutrophic | |
| Commercial fisheries | Mesotrophic, eutrophic | Eutrophic |
| Landscaping | Mesotrophic | Eutrophic |
| Irrigation | Eutrophic | Strongly eutrophic |
| Industrial processes | Mesotrophic | Eutrophic |
| Transport | Mesotrophic | Highly eutrophic |
| Energy (generation) | Oligotrophic | Mesotrophic |
| Energy (cooling) | Mesotrophic | Eutrophic |

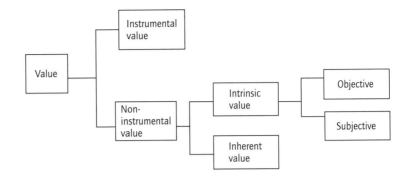

Fig. 1.1 Types of value in nature, according to sources cited in the text.

which in fact dates from the sixteenth century AD (Květ 1992).

Commercial fisheries, of necessity, rely on lakes of greater productivity, with nutrient status in the mesotrophic/eutrophic range (Sarvala et al. 2003). These are especially important in tropical countries, where other forms of protein are often expensive, and where, unfortunately, great damage to lakes is currently being done, both by eutrophication (Ogutu-Ohwayo et al. 1997), and by climate change (O'Reilly et al. 2003; Verburg et al. 2003). Landscaping is preferably carried out using mesotrophic waters, which are less valuable for drinking than the oligotrophic variety, but do not contain sufficient nutrients to support the growth of large, unsightly crops of macrophytes, filamentous algae or phytoplankton which would negate the purpose. Irrigation is one of the few economic activities which require water rich in nutrients, whilst many industrial processes demand only waters of moderate quality. Industry and transport are, of course, amongst the major factors which contribute to deterioration of water quality. Energy **generation** (hydropower) is preferable with oligotrophic waters, whereas cooling waters are usually abstracted from, and returned to, eutrophic lowland rivers, close to centres of population.

## 1.3 TYPES OF VALUE

### 1.3.1 Instrumental value

The uses to which lakes are put by human beings are therefore examples of their **instrumental value** (Figure 1.1), that is, in this case, their value **to humans**, either as the instrument of human needs or desires, or in some other way which contributes, directly or indirectly, to human welfare or human satisfaction. Instrumental value is therefore the value of an **item**\* (in this case, a lake) to some other entity (in this case, and mainly, to humans), and so does not exist independently of that entity (whom we should call the **valuer**). Crucially (for the purposes of later argument), the instrumental value of an item may be reduced (e.g. if the quality of the water in a particular lake is allowed to deteriorate beyond a point where significant use is compromised), but may also be increased (e.g. if the quality of the water in the lake is restored, so that significant use may be resumed, or even enhanced). The instrumental value of lakes to much of humanity, as discussed above, is largely economic. As pointed out in the Introduction to Volume 1 (Reynolds 2003), lakes play a crucial role in the economies of many regions. They also contribute substantially to those biospheric 'services' which, it was recently shown, provide us with the equivalent of two to three times the gross product of the world human economy (Costanza et al. 1997).

---

\* Environmental philosophers, a small sample of whose work we will shortly review, sometimes differentiate between natural 'objects' (plants, animals, bacteria, rocks), and natural 'items', a category which contains objects, but which also includes collections of objects arranged together as systems (e.g. a forest, lake or glacier). As most of what I have to say deals mainly with lakes as ecosystems, and with nature as a whole, I have chosen to use 'items' rather than 'objects' throughout, except where the latter is clearly signified.

Instrumental value need not be solely economic, however, and the recreational aspects of lakes, and their capacity to provide aesthetic experience, have already been mentioned. To go a stage further, we may say, that in common with other parts of nature in general, beyond their merely economic contribution to human welfare, lakes possess certain wider kinds of value, which Grey (1979) lists as part of his discussion of the general value (to humans) of wilderness.* Thus, Grey defines four types of wider value (of wilderness), three of which, as scientists, we would, presumably, easily recognise and subscribe to, and which, as limnologists, we could readily apply to lakes, the first of which has already been mentioned. They are (in a different order from that used by Grey):

**1 the 'gymnasium' argument.** This regards the preservation of wilderness in general, and thus lakes in particular, as important for athletic or recreational activities.

**2 the 'laboratory' argument.** Wilderness areas (and hence lakes) provide vital subject matter for scientific inquiry which furnishes us with an understanding of the intricate interdependencies of biological systems, their modes of change and development, their energy pathways and nutrient cycles, and the source of their stability. If we are to understand our own dependency on the rest of nature, we require natural systems as a norm, to inform us of ecological principles. Much of the subject matter of Volume 1 of this book falls into this category.

**3 the 'silo' argument.** One excellent reason for preserving areas of nature (including lakes) is that we thereby conserve genetic diversity (Sarvala et al. 2003). It is certainly prudent to maintain this as backup in case something should go wrong with the highly simplified biological systems which, in general, support human subsistence. Further, there is the related point that there is no way of anticipating future human needs, or the undiscovered applications of apparently useless organisms, some of which might turn out to be, for example, the source of some pharmacologically valuable drug. This might be called, perhaps, **the 'rare herb' argument**, and provides another persuasive (instrumental) justification for the preservation of wilderness, and hence (at least some) lakes.

**4 the 'cathedral' view.** Wilderness areas in general, and lakes in particular, provide a vital opportunity for human spiritual revival, moral regeneration, and aesthetic delight. Enjoyment of wilderness is often compared in this respect with religious or mystical experience. Preservation of wilderness areas for those who subscribe to this view is essential for human well-being, and its destruction conceived of as something akin to an act of vandalism, perhaps comparable to that of an important human edifice, such as the Great Wall, Angkor Wat, the Taj Mahal, the Pyramids, Stonehenge, Palenque or Macchu Picchu.

The 'cathedral' view of wilderness is, in fact, a fairly recent innovation, being largely introduced by the Romantics and Transcendentalists of the late eighteenth and nineteenth centuries, most of whom rejected the ethos of the Scientific Revolution and of the Enlightenment (Pepper 1996).† Nevertheless, as later writers (e.g. Leopold 1966) have shown, nature can also be marvelled at from a scientific as well as a philosophical or religious point of view, an attitude which may be **increased** by ecological understanding.

Lakes therefore clearly possess not only considerable economic, but also great recreational, scientific, spiritual and aesthetic value, as well as genetic potential, and as such, according to the instrumental argument, should, like other valuable resources, be protected and preserved, the better to promote not only the conservation of other species, and of the diversity of nature in general, but also the continued welfare of our own. However, some environmentalists find a definition of

---

* Although I have used 'wilderness' arguments throughout this article, it should not be construed that I am unaware of their past, present (e.g. the current removal of San people from the Kalahari 'game' reserve) and potential detrimental effects on indigenous peoples (Callicott & Nelson 1998; Woods 2001), or of the strong anti-human sentiments expressed in the past by some who advocate them (Devall & Sessions 1983).

† Not always completely, however, as witnessed by Henry David Thoreau's limnological studies of Walden Pond (Deevey 1942).

value which conceives of nature as being of value only to humans, and therefore to possess no value of its own, too limiting. Such definitions, they believe, offer only weak protection for nature, in that if nature possesses value only to us, then ultimately we will feel justified in using it for our own purposes, and in ignoring other aspects of the value of nature. Thus, Fox (1993, p. 100) writes

> '(if) the nonhuman world is only considered to be instrumentally valuable, then people are permitted to use and otherwise interfere with any aspect of it for whatever reasons they wish (i.e. no justification for interference is required). If anyone objects to such interference then, within this framework of reference, the onus is clearly on the person who objects to justify why it is more useful to humans to leave that aspect of the nonhuman world alone' (quoted in Callicott 1995, p. 4).

Similarly,

> 'A model in which nature (possesses) no value apart from human preferences will imply different conduct from one where nature projects fundamental value' (Rolston 1998, p. 144).

There is also the 'rare butterfly' example, whereby, as species become increasingly rare, their economic value correspondingly increases* (Godfrey-Smith 1980; Rolston 1998), and the point that various human uses of nature may be conflicting (Grey 1979), and may also 'humanise' nature, taming it, and turning it into some kind of theme park.† Environmental philosophers of a more biocentric, or even ecocentric persuasion, therefore propose instead that we should regard nature (and hence lakes), as possessing some kind of value which exists independently of human beings, or of their use of nature (its 'non-instrumental' value,

---

* See, for example, reports of increased commercial fishing of previously little known stocks of the Patagonian toothfish (*Dissostichus eleginoides*); *The Guardian*, London, 19 August 2003, p. 11.
† A phenomenon whose origin is not as recent as one might think (see Orwell 1946).

Figure 1.1), which therefore reflects much more accurately, the concept of nature as an entity possessing value of its own.

### 1.3.2 Intrinsic value

One such kind of value would be **intrinsic value** (Naess 1973). Unfortunately, this appears to be a concept about which philosophers themselves find it difficult to agree, and so, as a subject for non-philosophers, it represents a linguistic and intellectual minefield. However, as indicated earlier, it is at least arguable that, if scientists are to convey to the public at large what they feel about the importance of protecting nature, they first need to try to deal with such concepts. Of course, as this is my first (published) attempt, I may not have been as successful as I might have wished.

Various criteria are used to define the intrinsic value of natural items, not all of which apply, as we shall see, to every kind. Natural items are said, generally, to possess intrinsic value:

1 because they are able to exercise **preferences**, or are 'the subject of a life'. However, these criteria are used mainly by writers who see value in nature as confined mostly to animals, or even to higher animals (e.g. Singer 1976, 2001; Clark 1979; Regan 1984), and so, whilst they may clearly be important in that context, they are not of much relevance here.

2 owing to the realisation of the **interests**, or the (Aristotelian) **'good'** or well-being, of 'bearers of moral standing' (Attfield 1991, 1994). Again, this kind of criterion is used mainly by those writers (such as Attfield) who believe that value in nature is confined to individual organisms, as only these can be said to possess moral standing. It cannot therefore be said to be a property of species, of populations, of ecosystems, or of nature in general, and therefore will not be applicable to lakes.

3 when they are valued 'for their own sake', 'as an end in themselves' (Elliott 1980; O'Neill 1993; Callicott 1995; Rolston 1998), or for what they are 'in themselves' (Rolston 1980, 1998; Routley & Routley 1980; Des Jardins 2001), or when they possess 'a good of their own' (Taylor 1986; Rolston

1994a, 1998). It is also a controversial issue amongst philosophers as to whether non-sentient beings, or, more widely, non-animate nature in general, can ever possess a 'good' of their/its own (Rolston 1994a), so that, again, this argument will not be not taken further here.

**4** when they are valuable solely in virtue of their intrinsic ('non-relational') properties (O'Neill 1993), or when they possess value which exists independently of the valuation of valuers (Godfrey-Smith 1980; O'Neill 1993), 'belonging to its essential nature or contribution' (Callicott 1995, p. 5), and therefore 'without (any) contributory human reference' (Rolston 1980, p. 158).

This last criterion (which is actually two separate criteria; O'Neill 1993, 2001) is also controversial, in that (again), since the Enlightenment (and especially since Descartes, Hume, and Kant), it has been a fundamental tenet of Western philosophy that 'objective' value of this kind cannot exist, independently of the valuation of valuers. This is because, unlike mass, volume or density, Enlightenment science does not conceive of value as an objective property (as neither does Enlightenment philosophy; Elliot 1980), but one which, instead, can only be attributed to other items by rational (i.e. human) beings (Midgley 1980), or by what Callicott (1995, p. 5) terms 'the intentional act of a Cartesian subject respecting an object'.

### 1.3.3 Objective and subjective intrinsic value

Essentially, then, environmental philosophers distinguish between two types of intrinsic value, namely,
**1** 'objective' intrinsic value, where items are **valuable** 'in themselves' (Routley & Routley 1980, p. 152);
**2** 'subjective' intrinsic value, where items are **valued** 'when an interest is taken in (them)' (Elliott 1980, p. 136).
The first is therefore the kind of value which exists entirely independently of human beings, and which requires no valuers in order for it to exist. Thus, in respect of lakes, we could say that, as natural items, lakes possess intrinsic value 'in themselves', as lakes, and that this intrinsic value exists beyond any value we ourselves may place upon them as resources of any kind, and even of their aesthetic or spiritual value to us.

Unlike instrumental value, intrinsic value of this kind surely cannot be increased, in that it exists as an attribute of the item itself, and therefore both independently of ourselves, **and of the item itself**. It is therefore likely that it also cannot be **reduced**, in that any damage to the item, or even modification of it, leads not to the reduction of its intrinsic value, but rather its destruction, in the form of its integrity. For example, some North American 'deep ecologists' (e.g. Devall & Sessions 1983) believe that wilderness, which for them clearly possesses intrinsic value, is somehow 'tainted' by **any** human encroachment.

Some radical philosophers (e.g. Naess 1973) have therefore adopted 'objective' intrinsic value as a main criterion for protecting nature, in that it is essentially the only one in which value is held to exist in nature itself, entirely separately from any valuation which human beings (or any other potential valuer) may place upon it. In this way, it is hoped that the problem of nature being vulnerable to the changing needs and interests of human beings (Godfrey-Smith 1980; Des Jardins 2001) may be avoided. Thus Fox (1993, p. 101) continues

> 'If, however, the nonhuman world is considered to be intrinsically valuable, then the onus shifts to the person who wants to interfere with it to justify why they should be allowed to do so' (quoted in Callicott 1995, p. 5).

Intrinsic value of this kind therefore potentially offers much stronger protection for nature than instrumental value, in that it opposes any modification of natural items **beyond those necessary in order to satisfy vital needs** (Naess 1973, my emphasis).*

---

* This sanction is, indeed, so strong that it would, in effect, probably preclude all exploitation of nature for profit (Benton 1993)—a conclusion normally studiously ignored by most of the biocentric thinkers who advocate it (although **not** Naess (1973) or Routley & Routley (1980)).

However, it is also arguable that, as a philosophical construct, intrinsic value may not actually exist, outside the minds of human beings (see below). Conversely, there is the question that, if it does exist independently of us, why we should care about it at all (O'Neill 2001). There is also the further question as to why it should be reserved for natural items, in that many human artefacts (e.g. the Aztec calendar, the Mona Lisa, *Die Zauberflöte*) surely possess great (iconic?) value which is intrinsic to them, and which is clearly beyond any monetary value, or even instrumental value of the cathedral variety (see above, section 1.3.1, and also below, section 1.3.4).

Beyond this, there is the point that 'objective' intrinsic value clearly represents something of a moral absolute, in that either one subscribes to such a concept, or not. Whilst some philosophers may be comfortable with moral absolutes (though not all, as we shall shortly see), as scientists, we may find them unattractive, in that they cannot be easily falsified, and to use such arguments against those who adhere in all things to 'the bottom line', is often unproductive.

Others therefore suggest that whilst we can subscribe to the idea that intrinsic value in nature may exist, rather than existing independently of human beings, and therefore being 'discovered' by them, it is in fact 'generated' by human valuation of nature (Rolston 1994a), i.e. it is 'subjective' intrinsic value, 'allocated' rather than 'recognised' (Des Jardins 2001). This idea removes the problem of value only being generated 'by an intentional act of a subject' (Callicott 1995, pp. 5–6), but would still leave human beings as the ultimate judge of whether such value, in nature, actually exists. Basically, then, the problem with 'objective' intrinsic value is that it may not exist, whereas 'subjective' intrinsic value may not be strong enough to protect nature.

### 1.3.4 Inherent value

Some environmental philosophers (e.g. Frankena 1979; Taylor 1986, 1998; Attfield 1991) distinguish between intrinsic value, and **inherent** value (Figure 1.1). By this they mean '(the) ability (of an item) to contribute to human life' (Attfield 1991, p. 152) 'simply because it (possesses) beauty, historical importance, or cultural significance' (Taylor 1986, pp. 73–74 [quoted in Callicott 1995, p. 10]). Taylor (1986) appears to reserve this kind of value for human artefacts (see above), whereas Attfield (1991) also extends it to natural items, including rocks, rivers, ecosystems and wilderness* (i.e. to natural **systems**). We could therefore notionally apply this kind of value to lakes (Plate 1.1).

Grey (1979, p. 2) believes that 'Insofar as the "cathedral" view holds that value in nature derives solely from human satisfactions gained from its contemplation, it is clearly an instrumentalist attitude. It does, however, frequently approach an **intrinsic** attitude, insofar as the feeling arises that there is importance in the fact that it is there to be contemplated, **whether or not anyone actually takes advantage of this fact**' (my emphasis). However, what wilderness may also possess, in this context, is a value to humans which is neither intrinsic nor instrumental, but which is still clearly 'non-instrumental' (i.e. inherent value *sensu* Attfield 1991). This is also a type of value which can be decreased, in that the value of the item could easily be diminished by human (or other) agency (e.g. a 'natural' disaster), but it could conceivably also be **increased**, by careful management or other kinds of 'encouragement'. It still leaves us with 'human chauvinism' (Routley & Routley 1980) as the ultimate sanction as to whether natural systems possess anything other than instrumental value, however, and so may also offer only relatively weak protection.

---

* Confusingly, intrinsic value, 'intrinsic worth' (Godfrey-Smith 1980) and **inherent value**, are often used in the literature as synonyms (e.g. Regan 1981), which, on consulting a dictionary (Callicott 1995), they appear to be. Even more confusingly (I am afraid), Taylor (1986) uses 'inherent worth' to signify 'objective' intrinsic value as used here, and 'intrinsic value' to describe 'an event or condition in human lives which they consider enjoyable 'in and of itself' (Callicott 1995, p. 10). Similarly, Rolston (1980, p. 158) recognises value 'which may be found in human experiences which are **enjoyable in themselves**, not needing further instrumental reference' (my emphasis). Armstrong & Botzler (1993) suggest that we should use 'intrinsic value' to mean value which is independent of valuers, and 'inherent value' for the kind of non-instrumental value which is not.

Kolivaara. Eero Järnefeltin piirros Uudelle Kuvalehdelle 1893. Museovirasto, Helsinki.

**Plate 1.1** Lithograph of the lake Pielinen (Pielisjärvi), North Karelia, eastern Finland (1893), from the Hill of Koli (347 m), by Eero Järnefelt, brother-in-law of the composer Jan Sibelius. Pielinen lies in an area rich in inherent value (as defined in this chapter), both geo-ecological, historical, anthropological and aesthetic, which provided inspiration for many Finnish National Romantic artists at the turn of the last century. During the early nineteenth century, poems of *Kalevala*, the Finnish National epic, were collected in this part of Finland. Since 1974, interdisciplinary studies of the palaeoecology and anthropology of the area have been conducted by the Karelian Institute, University of Joensuu, partly established on the initiative of the Finnish (palaeo)limnologist Jouko Meriläinen, who noted that records of swidden, the traditional Finnish method of cultivating the forests, were preserved not only in paintings by Järnefelt (e.g. *Under the Yoke (Burning the Brushwood)* (1893), and *Autumn Landscape of Lake Pielisjärvi* (1899), but also in the varved (annually laminated) sediments of many of the lakes of this region. The Koli National Park, presently extended to 3000 ha, was established in 1991 in order to include some of the most valuable local landscapes. (Copyright: Museovirasto (National Board of Antiquities), Helsinki, Finland. With thanks to Mr Kevin Given of the School of Architecture, University of Plymouth, for help with this illustration.)

### 1.3.5 Systemic value

The ecocentric philosopher Holmes Rolston III (1980; 1988; 1994a, b; 1997; 1998) has extensively explored the possibilities of developing ideas of other kinds of value in nature, especially as applied to species, to ecosystems, and to nature itself (as opposed to its individual components, see above). For example, he suggests that it is difficult to dissociate the idea of value from natural selection (Rolston 1994a), and that, as a product of this process, species (as opposed to the individual organisms of which they are composed; Attfield 1991) may themselves possess (intrinsic) value (Rolston 1980). Thus,

> 'We find no reason to say that value is an irreducible emergent at the human (or upper animal) level. We reallocate value across the whole continuum. It increases in the emergent climax, but it is continuously present in the composing precedents' (Rolston 1980, p. 157).

Similarly,

> 'There is value (in nature) wherever there is positive creativity' (Rolston 1997, p. 62).

Rolston (1998) also suggests that organisms possess value not just in themselves, or as the product of natural selection, but as part of the ecosystem, through which information flows, and that

> 'Every intrinsic value (is connected to) leading and trailing "ands" pointing to (other sources of value) from which it comes, and towards which it moves' (Rolston 1980, p. 159).

> 'Systemically, value . . . fans out from the individual to its role and matrix. Things . . . face outward and co-fit into broader nature' (Rolston 1998, p. 143).

Yet another way of expressing this would be to say that natural items possess value as 'knots in the biospheric net' (Naess 1973). Thus 'value seeps out into the system' (Rolston 1998, p. 143).

The existence of value in nature produced by the emergent, creative properties of ecological systems undergoing natural selection, resonates strongly with the concept of evolution developed by studies of complex systems (Goodwin 1994), and of 'Dynamical Systems Theory' in general (Capra 1997), although it would undoubtedly be rejected by NeoDarwinists. It is also supported by Warren (1980), who writes of the value of natural items **as parts of the natural whole**, and that

> 'It is impossible to determine the value of an organism simply by considering its individual moral (standing): we must also consider its relationships to other parts of the system' (Warren 1980, pp. 125–126).

It is also (to me, anyway) reminiscent of a much more famous quote from the (earlier) environmental literature, with which some readers, at least, may be more familiar.

> 'One basic weakness in a conservation system based wholly on economic motives is that most members of the land community have no economic value. Of the 22,000 higher plants and animals native to Wisconsin, it is doubtful whether more than 5% can be sold, fed, eaten, or otherwise put to economic use. Yet these creatures are members of the biotic community, and if (as I believe) its stability depends on its integrity, they are entitled to continuance.' (Aldo Leopold, A Sand County Almanac, with essays on conservation from Round River. Oxford University Press, Oxford, 1966, p. 221).

Finally, Rolston (1994a, 1998) has also developed the concept of the **systemic** value of ecosystems, by which he means

> 'a spontaneous order (which) envelops and produces the richness, beauty, integrity and dynamic stability of the component parts' (Rolston 1994a, p. 23).

As well as organisms, species and ecosystems, for Rolston, nature as a whole possesses emergent properties which confer upon it, at the ecosystem level and beyond, value which is both intrinsic and completely independent of any which we, as humans, may place upon it. Thus creativity is both the impetus and the outcome of the evolutionary play (Hutchinson 1965).

These ideas are more fully explored in other, lengthier publications (Rolston 1988, 1994b), neither of which was I able to consult during preparation of this article. However, there are certain problems with the 'value' approach, which we should now therefore discuss.

### 1.3.6 Problems with intrinsic value

The first of these is that intrinsic value, despite the above fairly lengthy discussion, may not actually exist, but may instead be merely a human construct. Thus Rolston (1994a) himself writes that

> 'The **attributes** under consideration are objectively there before humans come, but the **attribution** of value is subjective. The object . . . affects the subject, who . . . translates this as value' (Rolston 1994a, p. 15, original emphasis).

Similarly,

> 'Even if we somehow manage to value wild nature per se, without making any utilitarian use of it, perhaps this valuing project will prove to be a human construction. Such value will have been projected onto nature, constituted by us and our set of social forces; other peoples in other cultures might not share our views' (Rolston 1997, pp. 40–41).

In fact, according to several of the writers cited here (e.g. Callicott 1995), it is the case that intrinsic value **must** actually exist somewhere, owing to the very existence of instrumental value. This is because those items which possess only instrumental value, do not exist 'as ends in themselves', but only as 'means' to other ends, whereas, apparently, the existence of means implies the existence of ends. Items which do exist as 'ends in themselves' therefore possess intrinsic value, because they also possess 'a good of their own' (see section 1.3.3). This 'teleological proof' of the existence of intrinsic value apparently dates back to the *Nichomachaean Ethics* of Aristotle (Callicott 1995), but, as already mentioned, since the Enlightenment, Western philosophy has rejected the existence of intrinsic value in all but rational beings (i.e. [until recently, anyway] sane, adult, white, male, property-owning human beings).

Enlightenment science, and the atomistic modern economics based upon it, also find it difficult to accept teleological concepts, which means that as well as amongst some philosophers (e.g. Attfield 1991), the concept of the intrinsic value of nature may be difficult to accept on the part of scientists. Instead, they may reject objectivism, and fall back upon the subjective view, which is that value only exists in natural items insofar as there are human valuers present. This is difficult to argue against. For example

> 'Resolute subjectivists cannot, however, be defeated by argument. . . . One can always hang on to the claim that value, like a tickle, or a remorse, must be felt to be there. It is impossible by argument to dislodge anyone firmly entrenched in this belief. That there is retreat to definition, is difficult to expose, because they seem to cling too closely to inner experience. . . . At this point, the discussion can go no further' (Rolston 1980, p. 157).*

> '. . . there is no way of rationally persuading someone to adopt a new ethic or new values . . . if someone is 'value-blind' to the intrinsic worth (sic) of natural systems, I could not expect (them) to manifest anything but indifference to their destruction in the interests of what (they) took to be matters of importance' (Godfrey-Smith 1980, p. 46).

Finally,

> '. . . persons with different moral intuitions belong to a different moral world. Value systems, more clearly than empirical theories, may simply be incommensurable. Within our society . . . we

---

* I am afraid I cannot help find it amusing that those who, over the past 30 years, have repeatedly told me that I do not live 'in the real world', have demanded of me nothing but 'facts', and who (mainly since 1979), have told me that I, and the rest of humanity must, from now on, live by 'the bottom line', should be correctly called, in this context, 'subjectivists'.

*(encounter) competing moral systems; and the competing moral injunctions are concerned to promote the welfare of human beings, treating the welfare of anything else with moral indifference.' (Godfrey-Smith 1980, p. 45).*

## 1.4 NATURALNESS

At the risk of taking the long route to come full circle, I therefore conclude that, in order to protect nature (and hence lakes) from further depredation, and to identify reasons why they should be rehabilitated, it may be necessary to find a criterion (or some criteria) stronger than instrumental value (or even inherent value), which can, however, be accepted by 'resolute subjectivists'. It (or they) will therefore need to be empirical, quantifiable, and free of any connotations of non-falsifiability. One such concept, I believe, is **naturalness**.

This is one of a number of criteria used historically in order to evaluate sites and areas for nature conservation (Table 1.2). Of these, area and population size are clearly 'objective' criteria, in that they can be quantified independently. They can also be described as 'non-relational properties' (O'Neill 1993, p. 16), in that unlike richness, rarity, representativity, potential value, uniqueness, and all subsequent criteria in Table 1.2, they can be determined 'independently of the existence or non-existence of other objects' (weak interpretation), or 'without reference to other objects' (strong interpretation). Naturalness, however, at least of an ecosystem, is possibly also a non-relational property, in that it can be assessed, and even quantified, independently of other natural items (see below).

Previously employed by Ratcliffe (1977) in order to designate nature reserves, and also currently in several methods for assessing the conservation value of rivers\* (Boon & Howell 1997), the concept of naturalness has been extensively explored by Peterken (1981, 1996). His key ideas re-

---

\* Some of which recognise the importance of intrinsic value, and of 'non-instrumental utility' (e.g. O'Keefe 1997).

**Table 1.2** Criteria used internationally for nature reserve evaluation (after Usher 1997)

---
Area (or size, extent)
Population size
Richness (of habitat and/or species)
Naturalness, rarity (of habitat and/or species)
Representativity ('representativeness', 'typicalness')
Ecological fragility, position in ecological unit, potential value, uniqueness
Threat of human influence
Amenity, scientific or educational value
Recorded history
Archaeological/ethnographic interest, availability, importance for migratory, intrinsic appeal, management factors, replaceability, gene bank, successional stage, wildlife reservoir potential
---

garding the concepts of original-, past-, present-, future- and potential-naturalness, as modified in order to apply to lakes, are outlined (accurately, I hope) in Table 1.3.

Naturalness refers to an ecosystem 'unmodified by human influence' (Ratcliffe 1977, p. 7), a concept which is difficult to apply (Birks 1996 [cited in Battarbee 1997]), owing to the problem of judging accurately the degree of modification it has experienced. Original naturalness and past naturalness are clearly linked, in that they are based on features of the present ecosystem which are descended from the period before human impact. Few existing features of contemporary ecosystems can be described as 'original natural', however, as almost all parts of the Earth have been modified from their original state by some agency or other, either natural or human (Peterken 1981). Given that present and potential naturalness are hypothetical states, the two forms which actually exist are therefore past naturalness (in lakes, the 'memory' of lake **ontogeny** ['development thorough time'; Deevey 1984] to date), and future naturalness, the capacity for the present lake to follow its own future ontogeny undisturbed, in the absence of further human impact.

Naturalness is therefore precise as a concept, but imprecise as a descriptor of particular ecosystems (Peterken 1996). Rather than there being a single state of naturalness, different states may ac-

Table 1.3  Types of naturalness as applied to lakes (adapted from Peterken 1996)

| Type | Characteristics |
| --- | --- |
| Original naturalness | The physical, chemical and biological state in which lakes in any region existed before humans became a significant ecological and biogeochemical factor (i.e. mostly before the arrival of agriculture). Sometimes referred to as 'pristine' (but not *sensu* Moss *et al.* 1997) |
| Present naturalness | The physical, chemical and biological state in which lakes in any region would now exist had humans not become a significant ecological factor. Sometimes also referred to as 'pristine' (but not *sensu* Moss *et al.* 1997). As present conditions in many areas are quite different from those of pre-agricultural times, present naturalness (a hypothetical present state) differs from original naturalness (a previous but now non-existent condition) |
| Past naturalness | Those physical, chemical and biological characteristics of present day lakes (e.g. 'relict' species, population structure) inherited from original natural times. May be thought of as surviving 'memory' of past processes, states and conditions. Therefore combines elements of original and present naturalness |
| Potential naturalness | The hypothetical physical, chemical and biological condition into which a lake would develop if<br>• human influence was completely removed<br>• the resulting successional adjustment took place instantly<br>Thus, potential naturalness expresses the potential of the present system to revert (instantly) to natural conditions |
| Future naturalness | The physical, chemical and biological state into which a lake would eventually develop over time, were significant human influence to be removed. This concept takes account of possible future climatic, edaphic and other successional changes, as well as potential extinctions, introductions and colonisation, and current past naturalness. Not a return to original naturalness |

tually be applicable to each component of the system. Thus we might imagine lake **physics** (except perhaps the heat budget, or where **meromixis** develops) being little disturbed by all but the most radical of human impacts, whereas lake **chemistry** will almost certainly be changed (Stumm 2003), but probably at least partially recover, once the impact is reduced or removed (Edmondson 1991). In terms of lake **biology**, the effects may be the most extensive, and also the most prolonged (e.g. the 'memory' frequently attached to fish populations in eutrophicated lakes). In employing this concept, there is therefore a need to identify those features of past naturalness descended from original naturalness, and which contribute to future naturalness, and perhaps express these for lakes in general along a continuum similar to that used by Peterken (1996) for forests and woods (Table 1.4).

For the north temperate zone, this list looks remarkably similar to that of Moss (1988), who related mean annual total phosphorus (TP) concentration (in $\mu g L^{-1}$ TP $yr^{-1}$) to degree of human impact (i.e. broadly, to past naturalness), although the similarity breaks down in the tropics, where lakes are characteristically much richer in phosphorus. Suggested characteristic mean annual phosphorus concentrations are therefore added to Table 1.4. for certain lake categories only.* Lakes possessing the greatest degree of naturalness are therefore those which are most free from human influence (Margules & Usher 1981) except that of traditional cultures (Smith & Theberge 1986), in which natural ecological regimes are present, in which natural processes operate to the greatest extent unmodified by human impact, and in which natural lake ontogeny has therefore been most closely followed.

As discussed by Stumm (2003), those catchments with the least human impact generally produce the most pristine waters, and vice versa. However, lakes also change according to natural factors, and those processes most often affecting lakes during their ontogeny, both natural and artificial, are listed in Table 1.5. Thus we can see that succession, infilling, paludification and siltation

---

* A similar scale could perhaps be developed for tropical lakes, but using mean annual total nitrogen concentration.

Table 1.4 Categories of forests and woods, and notional categories of lakes, exhibiting different kinds of naturalness (after Peterken 1996; Moss 1988)

| Forests/woods | Lakes |
|---|---|
| Virgin forest | 'Pristine' lakes possessing many features descended from original-naturalness ($1-10\,\mu g\,L^{-1}\,TP\,yr^{-1}$) |
| Primary, ancient near-natural forests and woodlands* | Near-natural lakes possessing considerable past naturalness, only once or recently 'disturbed', and probably only in a limited sense, and/or receiving only limited human influence (e.g. from traditional human cultures) ($10-20\,\mu g\,L^{-1}\,TP\,yr^{-1}$) |
| Primary, ancient near-natural and semi-natural forests and woodlands | Lakes still possessing considerable past naturalness, but experiencing frequent significant or prolonged human influence ($20-100\,\mu g\,L^{-1}\,TP\,yr^{-1}$) |
| Secondary, disturbed primary, semi-natural woodland | Lakes with substantial recent and/or prolonged human impact ($100-1000\,\mu g\,L^{-1}\,TP\,yr^{-1}$) |
| Secondary, semi-natural woodland | Lakes of recent natural origin |
| Plantations | Reservoirs |

\* For definitions of virgin, primary, secondary, ancient and semi-natural, and their relationship to North American 'old-growth', and/or European *Urwald/prales*, see Peterken (1981 1996).

(under natural catchment regime) are part of natural lake ontogeny, whereas eutrophication* (*pace* most of the North American literature; O'Sullivan 1995) and acidification, except in certain rather particular circumstances (i.e. catchment regimes characterised by nutrient mobilisation, or by prolonged podsolisation), are not. Those lakes which have been affected by the least number of artificial processes, and which have most closely followed the natural ontogenic pathway peculiar to their own particular locality, are clearly those which have received the least human impact, and those which therefore possess the greatest (past) naturalness.

As mentioned earlier, one problem with the naturalness approach is identifying the degree of past human impact. Interestingly, environmental philosophers such as Rolston (1997, 1998) have also highlighted this problem, in that 'unlike higher animals, ecosystems have no experiences' (Rolston 1998, p. 138), and that

> 'Often the problem of scale becomes that of time, which makes much invisible to our myopic eyes. We cannot see mountains move, or the hydrological cycle, or species evolve, though sometimes one scale zooms into another. Water flows, mountains quake, rarely: and we can see incremental differences between parents and offspring. We can see occasions of mutualism and of competition, though we have no estimates of their force. We can examine the fossil record and conclude that there was a Permian period and a catastrophic extinction at the end of it' (Rolston 1997, p. 52).

However, as pointed out in Volume 1 (O'Sullivan 2003), lakes offer a unique framework for studying the above problem, in that they are one of the few ecosystems which continuously accumulate and store a record of their own ontogeny, in the form of information stored in their sediments. What is more, this information precisely bridges the gap identified by Rolston (i.e. that between geological, 'deep', evolutionary time, and the everyday, observational and instrumental record). Two main

---

\* Except, of course, for that (surely, by now) unfortunate term 'morphometric eutrophication' (Deevey 1955, 1984), which is clearly a misnomer for that part of the succession which involves reduction of lake volume as a result of infilling, deoxygenation of the sediment water interface, release of buried phosphorus from the sediments, and a speeding up of internal nutrient cycling, but **not** an overall increase in nutrient **loading** above and beyond background (Edmondson 1991), and which is therefore not, *sensu stricto*, eutrophication.

**Table 1.5** Processes characterising lake ontogeny (after Deevey 1984; Edmondson 1991; Oldfield 1983; O'Sullivan 1995)

| Process | Characteristics |
| --- | --- |
| Succession | Internally driven process |
| | Community controlled |
| | Involves change in community structure, and increasing biomass and productivity (the 'sigmoid phase') up to 'trophic equilibrium' (Deevey 1984) |
| | No increase in productivity thereafter unless external nutrient loadings increase |
| | Inorganic nutrients converted to biomass (i.e. leads to overall **decline** in lake inorganic nutrient concentrations) |
| | No overall increase in total nutrient stock |
| | Process of 'self organisation' leading to 'self stabilisation' (Oldfield 1983) |
| | In 'large' lakes, stable state may persist indefinitely |
| | In 'small' lakes, leads eventually to infilling (see below) and 'death' of lake ('senescence', dystrophy) |
| | 'Natural process' |
| Infilling | Internally driven process |
| | Community controlled |
| | Mainly affects 'small' lakes of north temperate zone (Edmondson 1991) |
| | Involves infilling with autochthonous organic material leading eventually to infilling and 'obliteration' of the hypolimnion (Deevey 1955), and thence to ombrogenous peat formation or terrestrial habitat (hydroseral succession; Walker 1970) and 'death' of lake ('senescence', dystrophy) |
| | No increase in productivity, or overall increase in total nutrient stock necessarily involved, and yet (confusingly) sometimes termed 'morphometric eutrophication' (Deevey 1984) |
| | 'Self organisation' leading to 'self stabilisation' (Oldfield 1983) |
| | 'Natural process' |
| Paludification | Related to influx of allochthonous humic organic matter (Deevey 1984) |
| | Confined mostly to northern hemisphere peatlands |
| | External forcing |
| | Leads to decline in productivity (owing to lack of nutrients and increased chemical demand for oxygen), and eventually to dystrophy |
| | Forced self organisation to new stable state (Oldfield 1983) |
| | Maintained by external factors (existence of peatlands in catchment) |
| Siltation | Related to influx of allochthonous mineral matter |
| | Confined mostly to grassland, semi-arid, or arid zone (Chapter 8), or to agricultural regions |
| | External forcing |
| | May lead to increase in lake nutrient concentration owing to inwash of soil material (**edaphic** eutrophication; Deevey 1984) |
| | Forced self organisation to new stable state (Oldfield 1983) |
| | Maintained by external factors (existence of unstable soils, or farmland in catchment) |
| | May therefore be 'natural' or artificial process |
| Eutrophication | Influx of allochthonous nutrients and/or non-humic organic matter 'above and beyond natural background' (Edmondson 1991) |
| | External forcing |
| | Increase in the proportion of lake nutrients present in the inorganic form |
| | Leads to increased biomass and productivity, and to changes in community structure |
| | Leads to instability so long as external forcing continues |
| | Attempted 'self organisation' to a new stable state (Oldfield 1983) |
| | May be 'natural process' (edaphic eutrophication, see above) but more often artificial (**cultural** eutrophication; Likens 1972) |

Table 1.5  *Continued*

| Process | Characteristics |
|---|---|
| Acidification | Influx of allochthonous hydrogen ions associated with anthropogenic emissions of $NO_x$ and $SO_4^-$ |
| | External forcing |
| | Increase in proportion of major ions present as $H^+$, reduction in bases ($Na^+, K^+, Ca^{2+}, Mg^{2+}$) |
| | Leads to reduced biomass and productivity, and to changes in community structure |
| | Leads to instability so long as external forcing continues |
| | 'Self organisation' to a new stable state (Oldfield 1983) |
| | May be 'natural process' ('edaphic oligotrophication'; Pennington 1991) but more often anthropogenic |
| Other forms of pollution (heavy metals, hydrocarbons, persistent organic compounds) | Influx of allochthonous material mainly associated with anthropogenic emissions |
| | External forcing |
| | May lead to reduced biomass and productivity, and to changes in community structure owing to toxicological effects |
| | Leads to instability so long as external forcing continues |
| | 'Self organisation' to a new stable state (Oldfield 1983) |
| | May be 'natural process' (e.g. due to influx of natural materials) but overwhelmingly anthropogenic (Stumm 2003) |

approaches have been adopted, namely 'hindcasting' of export coefficient models (Moss *et al.* 1997; see also Wilson, Chapter 15), and, as intimated above, palaeolimnology.

## 1.5 LIMNOLOGY, PALAEOLIMNOLOGY AND NATURALNESS

### 1.5.1 'Hindcasting'

Moss *et al.* (1997) describe a 'state changed' (or even 'value changed', [sic], p. 124) approach to assessing change in lakes from an original, 'baseline' state. The objective is to identify a condition to which lakes may be restored, using 'hindcasting' of current nutrient loadings based on export coefficients, and historical data on land use, livestock density and human population. Restoration to pre-settlement conditions (i.e. to a state of **original naturalness**) is considered unrealistic, whereas a pragmatic approach, involving standards for drinking water or industrial abstraction (i.e. applying standards which take into account **instrumental value** only), might not give sufficient protection. Instead, a third (functional) approach is adopted, whereby a state is identified 'which reflects the highest possible quality, consonant with (the) maintenance of current (human) populations and or agricultural use of the catchment' (p. 126). The authors consider that this approach is likely to generate the maximum conservation value (habitat and species diversity), to preserve functional values (fisheries, natural flood storage, traditional uses), and to maintain high amenity. In effect, what this approach will maximise, is **future naturalness**.

For the UK, Moss *et al.* (1997) choose AD 1931 as their 'baseline' state, on the grounds that this date clearly precedes the intensification of British agriculture which took place mainly after 1940. It also mostly pre-dates the widespread use of synthetic nitrate fertilisers, a point which is obviously not coincidental. Restoration of nutrient loadings to values characteristic of this period would also not be expensive.

Whether this would actually restore some UK lakes to a stable condition, and therefore maximise future naturalness, is an interesting point, however, as there is some evidence (developed both from hindcasting, and from palaeolimnological studies) that eutrophication (for example) of certain bodies began as early as c. AD 1900, with the final connection of many outlying regions to the railway network, and hence the national market in agricultural produce (O'Sullivan 1992, 1998). Similarly, the commercialisation of agriculture in England and Wales (i.e. production for more than a limited, mainly static, local market) was a process which took several centuries (Overton 1996). Nutrient loadings from agriculture may therefore have begun to rise in some localities as early as AD 1500 (Wharton et al. 1993).

Nevertheless, out of a sample of 90 lakes in England and Wales, 75% had suffered more than 50% change, with only 4.4% being insignificantly perturbed (Moss et al. 1997). The state-changed method also identified as significantly perturbed a subset of upland lakes in unproductive catchments which the OECD spatial-state approach (see Rast, Chapter 14) would classify as oligotrophic.

### 1.5.2  *Palaeolimnology*

A couple of years ago, when I first thought of applying palaeolimnology to George Peterken's idea of naturalness, I was quite pleased with myself. After all, it is just the kind of interdisciplinary connection I have spent my career trying to make. However, not for the first time (and probably not the last), when I consulted the literature, I found that my old colleague Rick Battarbee (1997) had been there before me. Battarbee describes an ingenious use of palaeolimnology, historical and instrumental records to develop a naturalness index for lakes (Figure 1.2), based on reconstructed habitat change ('disturbance') versus critical loadings of pH, nutrients, or other inputs (e.g. heavy metals, persistent organic pollutants (POPs), radionuclides). Again, the objective is to assess and quantify the degree to which a lake has departed from a **natural** state (i.e. as defined here, its **past** naturalness), and to identify ways in which best to restore it (i.e. to maximise **future** naturalness).

In Figure 1.2, the horizontal axis can represent any chemical change which leads to biological change, as defined above in terms of variations in critical load (Vollenweider 1975). On the vertical axis is plotted 'habitat disturbance', a variable which is more difficult to define, but which might

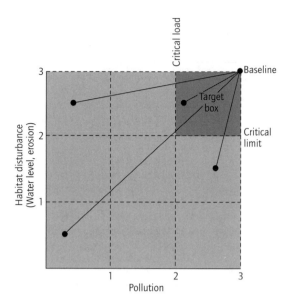

Fig. 1.2  Naturalness matrix for lakes (from Battarbee 1997). Vectors represent hypothetical lake time trajectories. For other explanation, see text.

include physical factors such as shoreline and water level changes, and inwash of catchment soils and/or other eroded material, and biological change, such as introduction of exotic species, or overfishing. Some of these will have had no discernible effect on lake ecology, whilst others will have exceeded some 'critical limit', and therefore produced changes in species composition and/or abundance. Ideally, a third axis should represent direct biological influences, such as introduction of exotic species, stocking of lakes for angling, or commercial fishing. For simplicity, however, these are included on the vertical axis.

Baseline conditions (original naturalness) are reconstructed using palaeolimnological methods, especially the 'transfer function' approach (Battarbee et al. 2001; O'Sullivan 2003). To date, these have mainly been developed for organisms characteristic of lower trophic levels (e.g. diatoms, chrysophytes, cladocera, chironomid midges, ostracods; O'Sullivan 2003). Direct (and indirect) reconstructions of changes at higher trophic levels, or of community structure

(Gregory-Eaves et al. 2003; Leavitt et al. 1994; Schindler et al. 2001), are also being developed, however. The severity of change over baseline, in terms of biological consequences, is (provisionally) expressed using **species turnover** (Gauch 1982; cited in Battarbee 1997).

The pristine or baseline state of the lake is then one which ideally combines absence of pollution and habitat disturbance with maximisation of diversity of indigenous taxa. Desirable future chemical and biological conditions are defined in terms of the target box, in which lake ecology approaches 'pristine' (i.e. original naturalness). Lakes may then be classified into three broad categories, one of which is subdivided, namely:

**1 Undisturbed lakes** (Category 3 in Figure 1.2), including all completely undisturbed sites, or where known physical change has not caused any (biological) impact upon the lake (i.e. where no critical limit has been exceeded). These lakes are then subdivided into Category 3A (Pristine—i.e. completely unaffected by human impact), and 3B (Lakes in which known physical change has not produced a biological response). The first group (in the UK) includes only a few lakes in remote mountain areas, but the second contains some major lakes (e.g. Loch Ness).* Both of these subcategories therefore contain lakes which possess both considerable past naturalness, and substantial memory of original naturalness (compare Table 1.4).

**2 Lakes in which catchment or other change has led to variations in composition and/or abundance of lake biota** (Category 2, Figure 1.2). Such events include water level changes and reworking of marginal sediments into the profundal, and accelerated soil erosion, as well as influx of exotic species, or commercial fishing. Such lakes, whilst clearly no longer pristine, may still possess substantial past naturalness, however, and also significant future naturalness (Table 1.4).

**3 Lakes in which major impact has led to significant changes in lake ecology** (Category 1, Figure 1.2). In such lakes, not only have there been changes in community structure, but also in the ecological functioning of the ecosystem. These lakes therefore possess little past or future naturalness, in that their condition has departed furthest from the course of natural ontogeny.

Figure 1.3 illustrates how the naturalness matrix can be applied to a set of seven UK lakes, four affected by acidification, and three by eutrophication. Distance from the target box depicts severity of change over baseline, and the degree to which loadings would need to be reduced in order to restore these lakes to that condition. In Figure 1.3c, biological effects of pollution, in the form of species turnover, are also shown, whilst in Figure 1.3d, scores on both axes are neatly combined to generate a state-changed classification scheme, in which 10 indicates pristine (3A above), 9 Category 3B, 4–6 perhaps Category 2, and 1–4 Category 1. Hence the baseline sediment record can be used to define some of the key biological characteristics of the restored lake. Whilst values on the $y$ axis are subjective, those on the $x$ axis are objective, and therefore offer the possibility of a quantitative approach to the 'reconstruction' of past naturalness, or more correctly, the enhancement of future naturalness. Battarbee's (1997) matrix is therefore potentially a more than useful addition to lake management.†

## 1.6 CONCLUSIONS—WHAT EXACTLY ARE WE RESTORING?

To paraphrase Grey (1979), suppose, for example, that a lake in a remote region of no economic importance was found to be so precariously balanced that *any* human intervention or contact would inevitably bring about its destruction. Those who maintained that the lake should, nevertheless, be

---

* Here, Battarbee (1997) cites building of the Caledonian Canal through the Loch during the early nineteenth century as one such event, but it now seems that the earlier 'Clearance' (i.e. eviction) of native highlanders from Glen Moriston, a valley which drains into the northern side of the Loch, did produce some effects upon its ecosystem which are detectable via palaeolimnology (O'Sullivan et al. 2000 a, b).

† Although not 'ecocentric' lake management (Battarbee 1997, p. 157), either as defined here, or, as originally, by O'Riordan (1981).

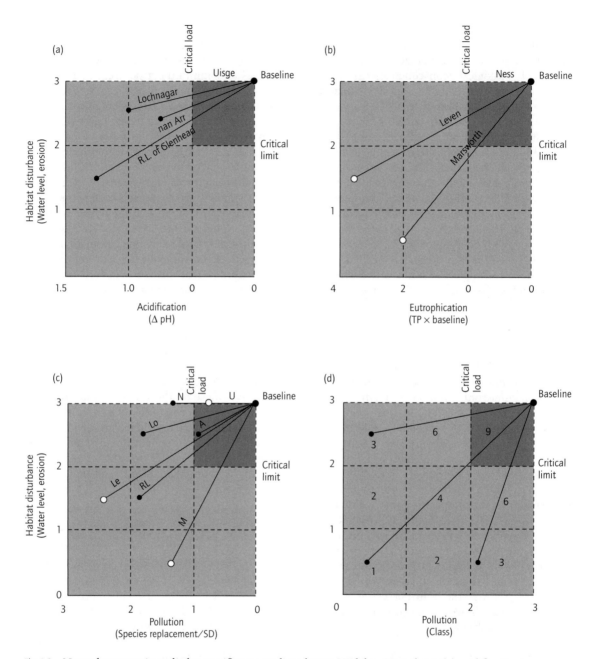

**Fig. 1.3** Naturalness matrix applied to specific case studies of seven UK lakes. Note that in (a), acidification is expressed in pH units, whereas in (b), eutrophication is cited as multiples of baseline TP concentration. For other explanation, see text.

preserved, unexperienced and unenjoyed by humans, would certainly be ascribing to it some kind of value, independent of any use to which human beings might, or could put it. Those who think it should not be preserved intact, but should not be destroyed, also presuppose that it possesses some similar kind of value. Those who think that if it was destroyed, then 'something' would be missing, must surely subscribe to the idea that this 'something' is clearly greater than the lake's economic value, or even its wider instrumental and/or inherent value (as defined above).

We could, of course, try to ague that what would be removed is the lake's inherent value (its capacity to please **us**) or even its intrinsic value (as variously defined above), but an alternative approach is to say that what is really being destroyed is its **naturalness**, a property which exists along a continuum, and which (as we have seen) can therefore be both 'hindcast', reconstructed, and quantified. The goal of lake rehabilitation would then be not to enhance the lake's instrumental value, or even its inherent value, nor even yet its intrinsic value (always assuming that this **can** be increased, which is at least questionable).

Instead, what we would be trying to enhance is its **future** naturalness (that is, its capacity to return, unaided, to its natural ontogenic pathway), by restoring the present condition of the lake to a state most closely resembling its original naturalness. In order to achieve this goal, we would need to identify those aspects of the present lake ecosystem which are most directly descended from its past naturalness, and those which stem from subsequent perturbation, using hindcasting and palaeolimnology, as outlined above. As the lake will then continue to change, under the impact of both internal and external natural factors, we cannot predict this future state, except to say that it will not be the same as its present state, or its past natural state, nor its original natural condition.

Interestingly (again!), a few environmental philosophers have identified naturalness as a source of value in nature. For example, Elliott (1994, p. 36) believes that it is the 'otherness' of nature which underwrites its value, and that this naturalness is achieved without intentional design. Similarly, Routley & Routley (1980) write

> 'Value is not subjective, but neither is it an objective feature entirely detached from nature' (p. 154). 'In simple terms, objective value is 'located' in (nature) entirely independent of valuers' (p. 151),

where it resides in

> 'diversity of systems and creatures, naturalness (sic), integrity of systems, stability of systems, harmony of systems' (p. 170)

Thus it would appear that naturalness, as modified by the palaeolimnological approach, offers us what we have so far lacked, that is, **a quantifiable measure of environmental value** (as applied to lakes, at any rate). If we also take into account that lake sediments collect information not only from lakes themselves, but from their catchment/watershed, and from their wider region (their 'airshed'; Likens 1979), we may eventually be able to apply it to ecosystems in general, or at least, wherever there are lakes.

## 1.7   REFERENCES

Armstrong, S.J. & Botzler, R.G. (1993) *Environmental Ethics. Divergence and Convergence*. McGraw Hill, New York, 567 pp.

Attfield, R. (1991) *The Ethics of the Environmental Concern* (2nd edition). Basil Blackwell, Oxford, 249 pp.

Attfield, R. (1994) Rehabilitating nature and making nature inhabitable. In: Attfield, R. & Belsey, A. (eds), *Philosophy and the Natural Environment*. Cambridge University Press, Cambridge, 45–58.

Battarbee, R.W. (1997) Freshwater quality, naturalness and palaeolimnology. In: Boon, P.J. & Howell, D.L. (eds), *Freshwater Quality: Defining the Indefinable*. Scottish Natural Heritage, The Stationery Office, Edinburgh, 155–71.

Battarbee, R.W., Jones, V.J., Flower, R.J., *et al.* (2001) Diatoms. In: Smol, J.P., Birks, H.J.B. & Last, W.M. (eds), *Tracking Environmental Change using Lake Sediments*, Vol. 3, *Terrestrial, Algal and Siliceous Indicators*. Kluwer, Dordrecht, 155–202.

Benton, T. (1993) *Natural Relations: Ecology, Animal Rights and Social Justice*. Verso, London, 246 pp.

Bernhardt, H. (1981) Reducing nutrient inflows. In: Rast, W. & Kerekes, J.J. (compilers) *Proceedings of the International Workshop on the Control of Eutrophication*. International Institute for Applied Systems Analysis, Luxemberg and Austria, 12–15 October, 43–51.

Boon, P.J. & Howell, D.L. (eds) (1997) *Freshwater Quality: Defining the Indefinable*. Scottish Natural Heritage, The Stationery Office, Edinburgh, 545 pp.

Callicott, J.B. (1995) Intrinsic value in nature: a metaethical analysis. *Electronic Journal of Analytic Philosophy*, **3**, 14 pp. (Accessed at http://ejap.louisiana.edu/EJAP/1995.spring/callicott.1995.spring.html)

Callicott, J.B. & Nelson, M.P. (eds) (1998) *The Great New Wilderness Debate*. University of Georgia Press, Athens, GA, 696 pp.

Capra, F. (1997) *The Web of Life. a New Synthesis of Mind and Matter*. Flamingo, London, 320 pp.

Clark, S.R.L. (1979) *The Moral Status of Animals*. Clarendon Press, Oxford, 179 pp.

Costanza, R., d'Arge, R., de Groot, R., *et al.* (1997) The value of the world's ecosystem services and natural capital. *Nature*, **387**, 253–60.

Deevey, E.S., Jr. (1942) A re-examination of Thoreau's 'Walden'. *Quarterly Review of Biology*, **17**, 1–10.

Deevey, E.S., Jr. (1955) The obliteration of the hypolimnion. *Memorias dell'Istituto Italiano di Idrobiologia* Supplement **8**, 9-38.

Deevey, E.S., Jr. (1984) Stress, strain, and stability of lacustrine ecosystems. In: Haworth, E.Y. & Lund, J.W.G. (eds), *Lake Sediments and Environmental History*. Leicester University Press, Leicester, 203–29.

Des Jardins, J.R. (2001) *Environmental Ethics. an Introduction to Environmental philosophy* (3rd edition). Wadsworth, Belmont, CA, 277 pp.

Devall, B. & Sessions, G. (1983) *Deep Ecology: Living as if Nature Mattered*. Peregrine Smith, Salt Lake City, UT, 229 pp.

Edmondson, W.T. (1991) *The Uses of Ecology: Lake Washington and Beyond*. University of Seattle Press, Seattle, WA, 329 pp.

Elliott, R. (1980) Why preserve species? In: Mannison, D.S., McRobbie, M.A. & Routley, R. (1980) *Environmental Philosophy*. Department of Philosophy, Research School of Social Sciences, Monograph Series, No. 2. Australian National University, Canberra, 8–29.

Elliott, R. (1994) Ecology and the ethics of environmental restoration. In: Attfield, R. & Belsey, A. (eds), *Philosophy and the Natural Environment*. Cambridge University Press, Cambridge, 31–44.

Fox, W. (1993) What does the recognition of intrinsic value entail? *Trumpeter* 10. As quoted in Callicott (1995). (Accessed at http://ejap.louisiana.edu/EJAP/1995.spring/callicott.1995.spring.html)

Frankena, W. (1979) Ethics and the environment. In: Goodpaster, K.E. & Sayre, R.M. (eds), *Ethics and Problems of the 21st Century*. University of Notre Dame Press, London, 3–20.

Godfrey-Smith, W. (1980) The rights of non-humans and intrinsic values. In: Mannison, D.S., McRobbie, M.A. & Routley, R. (eds), *Environmental Philosophy*. Department of Philosophy, Research School of Social Sciences, Monograph Series, No. 2. Australian National University, Canberra, 30–47.

Goodwin, B. (1994) *How the Leopard Changed its Spots. The Evolution of Complexity*. Phoenix, London, 233 pp.

Gregory-Eaves, I., Smol, J.P., Douglas, M.S.V. & Finney, B.P. (2003) Diatoms and sockeye salmon (*Onchorhyncus nerka*) population dynamics: reconstructions of salmon-derived nutrients over the past 2,200 years in two lakes from Kodiak Island, Alaska. *Journal of Paleolimnology*, **30**, 35–53.

Grey, W. (1979) The value of wilderness. *Environmental Ethics*, **1**, 309–19. (Accessed at http://www.uq.edu.au/~pdwgrey/pubs/vow.html)

Hutchinson, G.E. (1965) *The Ecological Theater and the Evolutionary Play*. Yale University Press, New Haven, CT, 139 pp.

Hutchinson, G.E. (1969) Eutrophication, past and present. In: Röhlich, G. (ed.), *Eutrophication: Causes, Consequences and Correctives*. National Academy of Sciences, Washington, DC, 17–26.

Hutchinson, G.E. (1973) Eutrophication. *American Scientist*, **61**, 269–79.

Květ, J. (1992) Wetlands of the Třeboň Biosphere reserve—an overview. In: Finlayson, C.M. (ed.), *Integrated Management and Conservation of Wetlands in Agricultural and Forested Landscapes*. Proceedings of a Workshop, Třeboň, Czechoslovakia, 25–31 March. IWWRB Special Publication No. 22, International Waterfowl and Wetlands Research Bureau, Slimbridge, 11–14.

Leavitt, P.R., Sanford, R.R., Carpenter, S.J. & Kitchell, J.F. (1994) An annual fossil record of production, planktivory and piscivory during whole-lake manipulation. *Journal of Paleolimnology*, **11**, 133–49.

Leopold, A. (1966) *A Sand County Almanac, with Essays on Conservation from Round River.* Oxford University Press, Oxford, 269 pp.

Likens, G.E. (1972) Eutrophication and aquatic ecosystems. In: Likens, G.E. (ed.), *Nutrients and Eutrophication. American Society for Limnology and Oceanography Special Publication*, **1**, 3–13.

Likens, G.E. (1979) The role of watershed and airshed in lake metabolism. *Archiv für Hydrobiologie, Beihefte Ergebnisse der Limnologie*, **13**, 195–211.

Margules, C. & Usher, M.B. (1981) Criteria used in assessing wildlife potential: a review. *Biological Conservation*, **21**, 79–109.

Midgley, M. (1980) Duties concerning islands. In: Elliott, R. & Gare, A. *Environmental Philosophy.* Australian National University Press, Canberra, 166–81.

Mollison, B. (1988) *Permaculture. a Designers Manual.* Tagari Publications, Tyalgum, NSW, 576 pp.

Moss, B. (1988) *The Ecology of Fresh Waters: Man & Medium* (2nd edition). Blackwell, Oxford, 417 pp.

Moss, B., Johnes, P.J. & Phillips, G. (1997) New approaches to monitoring and classifying standing waters. In: Boon, P.J. & Howell, D.L. (eds), *Freshwater Quality: Defining the Indefinable.* Scottish Natural Heritage, The Stationery Office, Edinburgh, 112–33.

Naess, A. (1973) The shallow and the deep, long-range ecology movements. *Inquiry* **16**, 95–100.

Ogutu-Ohwayo, R., Hecky, R.E.F., Cohen, A.S. & Kaufmann, L. (1997) Human impacts of the African Great Lakes. *Environmental Biology of Fishes*, **50**, 1176–31.

O'Keefe, B. (1997) Methods of assessing conservation status for natural fresh waters in the Southern Hemisphere. In: Boon, P.J. & Howell, D.L. (eds), *Freshwater Quality: Defining the Indefinable.* Scottish Natural Heritage, The Stationery Office, Edinburgh, 369–86.

Oldfield, F. (1983) Man's impact on the environment: some perspectives. *Geography*, **68**, 245–56.

O'Neill, J. (1993) *Ecology, Policy and Politics. Human Well-being and the Natural World.* Routledge, London, 227 pp.

O'Neill, J. (2001) Meta-ethics. In: Jamieson, D. (ed.), *A Companion to Environmental Philosophy.* Blackwell Publications, Oxford, 163–76.

O'Reilly, C.M., Alin, S.R., Plisnier, P.-D., Cohen, A.S. & McKee, B.A. (2003) Climate change decreases aquatic ecosystem productivity of Lake Tanganyika, Africa. *Nature* **424**, 766-768.

O'Riordan, T. (1981) *Environmentalism* (2nd edition). Pion Press, London, 409 pp.

Orwell, G. (1946) Pleasure spots. *Tribune*, January 11, 1946. Reprinted in Orwell, S. & Angus, I. (eds), *The Collected Essays, Journalism and Letters of George Orwell, Volume 4, In Front of Your Nose, 1945–1950.* Secker and Warburg, London, 555 pp. (Accessed at http://etext.library.adelaide.edu.au/o/o79e/index.html)

O'Sullivan, P.E. (1992) The eutrophication of shallow coastal lakes in Southwest England—understanding, and recommendations for restoration, based on palaeolimnology, historical records, and the modelling of changing phosphorus loads. *Hydrobiologia* **243/244**, 421–34.

O'Sullivan, P.E. (1995) Eutrophication—a review. *International Journal of Environmental Studies* **47**, 173–95.

O'Sullivan, P.E. (1998) Cows, pigs, war, but (so far as is known) no witches—the historical ecology of the Lower Ley, Slapton, Devon, and its catchment, 1840 to the present. In: Blacksell, M., Matthews, J.M. & Sims, P.C. (eds), *Environmental Management and Change in Plymouth and the South West. Essays in Honour of Dr John C. Goodridge.* University of Plymouth, Plymouth, 53–72.

O'Sullivan, P.E. (2003) Palaeolimnology. In: O'Sullivan, P.E. & Reynolds, C.S. (eds), *The Lakes Handbook.*, Vol. 1, *Limnology and Limnetic Ecology.* Blackwell, Oxford, 609–66.

O'Sullivan, P.E., Cooper, M.C., Shine, A.J., et al. (2000a) Long term response of Loch Ness, Scotland, to changes in inputs from its catchment. *Verhandlungen Internationale Vereinigung Limnologie*, **27**, 2307–11.

O'Sullivan, P.E., Cooper, M.C., Henon, D.N., et al. (2000b) Anthropogenic/climate interactions recorded in the sediments of Loch Ness, Scotland. *Terra Nostra* **2000/7**, 90–5.

Overton, M. (1996) *The Agricultural Revolution in England. The Transformation of the Agrarian Economy, 1500–1850.* Cambridge University Press, Cambridge, 257 pp.

Pennington, W. (1991) Palaeolimnology in the English Lakes—some questions and answers over fifty years. *Hydrobiologia*, **214**, 9-24.

Pepper, D. (1996) *Modern Environmentalism: an Introduction.* Routledge, London, 376 pp.

Peterken, G.F. (1981) *Woodland Conservation and Management.* Chapman & Hall, London, 328 pp.

Peterken, G.F. (1996) *Natural Woodland: Ecology and*

*Conservation in North Temperate Regions.* Cambridge University Press, Cambridge, 522 pp.

Ratcliffe, D.A. (ed.) (1977) *A Nature Conservation Review*, Vol. 1. Cambridge University Press, Cambridge, 401 pp.

Regan, T. (1981) The nature and possibility of an environmental ethic. *Environmental Ethics*, **3**, 16–31.

Regan, T. (1984) *The Case for Animal Rights.* Routledge & Kegan Paul, London, 425 pp.

Reynolds, C.S. (2003) Lakes, limnology and limnetic ecology: towards a new synthesis. In: O'Sullivan, P.E. & Reynolds, C.S. (eds), *The Lakes Handbook.*, Vol. 1, *Limnology and Limnetic Ecology.* Blackwell, Oxford, 1–7.

Rohde, W. (1969) Crystallisation of eutrophication concepts in Northern Europe. In: Rohlich, G. (ed) *Eutrophication: Causes, Consequences and Correctives.* National Academy of Sciences, Washington, DC, 50–64.

Rolston, H., III (1980) Are values in nature subjective or objective? In: Elliott, R. & Gare, A. (eds), *Environmental Philosophy.* Australian National University Press, Canberra, 135–65.

Rolston, H., III (1988). *Environmental Ethics: Duties to and Values in the Natural World.* Temple University Press, Philadelphia, 391 pp.

Rolston, H., III (1994a) Value in nature and the nature of value. In: Attfield, R. & Belsey, A. (eds), *Philosophy and the Natural Environment.* Cambridge University Press, Cambridge, 13–30.

Rolston, H., III (1994b). *Conserving Natural Value.* Columbia University Press, New York, 259 pp.

Rolston, H., III (1997) Nature for real: is nature a social construct? In: Chapple, T.D.J. (ed.), *The Philosophy of the Environment.* Edinburgh University Press, Edinburgh, 38–64.

Rolston, H., III (1998) Challenges in environmental ethics. In: Zimmermann, M. (ed.), *Environmental Philosophy. From Animal Rights to Radical Ecology* (2nd edition). Prentice Hall, New Jersey, 125–44.

Routley, R. & Routley, V. (1980) Human chauvinism and environmental ethics. In: Mannison, D.S., McRobbie, M.A. & Routley, R. (1980) *Environmental Philosophy.* Department of Philosophy, Research School of Social Sciences, Monograph Series, No. 2. Australian National University, Canberra, 96–169.

Sarvala, J., Rask, M. & Karjalainen, J. (2003) Fish community ecology. In: O'Sullivan, P.E. & Reynolds, C.S. (eds), *The Lakes Handbook.*, Vol. 1, *Limnology and Limnetic Ecology.* Blackwell, Oxford, 528–82.

Schindler, D.E., Knapp, R.A. & Leavitt, P.R. (2001) Alteration of nutrient cycles and algal production resulting from fish introductions into mountain lakes. *Ecosystems* **4**, 308–21.

Singer, P. (1976) *Animal Liberation: Towards an End to Man's Inhumanity to Animals.* Jonathan Cape, London, 303 pp.

Singer, P. (2001) Animals. In: Jamieson, D. (ed.), *A Companion to Environmental Philosophy.* Blackwell Publications, Oxford, 416–25.

Smith, P.G.R. & Theberge, J.B. (1986) A review of criteria for evaluating natural areas. *Environmental Management*, **10**, 715–34.

Stumm, W. (2003) Chemical processes regulating the composition of lake waters. In: O'Sullivan, P.E. & Reynolds, C.S. (eds), *The Lakes Handbook.*, Vol. 1, *Limnology and Limnetic Ecology.* Blackwell, Oxford, 79–106.

Taylor, P.W. (1986) *Respect for Nature: a Theory of Environmental Ethics.* Princeton University Press, Princeton, 329 pp.

Taylor, P.W. (1998) The ethics of respect for nature. In: Zimmermann, M. (ed.), *Environmental Philosophy. from Animal Rights to Radical Ecology* (2nd edition). Prentice Hall, New Jersey, 71–86.

Usher, M.B. (1997) Principles of nature conservation evaluation. In: Boon, P.J. & Howell, D.L. (eds), *Freshwater Quality: Defining the Indefinable.* Scottish Natural Heritage, The Stationery Office, Edinburgh, 199–214.

Verburg, P., Hecky, R.E.F. & Kling, H. (2003) Ecological consequences of a century of warming in Lake Tanganyika. *Science*, **301**, 505–7.

Vollenweider, R.A. (1975) Input–output models with special reference to the phosphorus-loading concept. *Schweizerische Zeitschrift für Hydrologie*, **37**, 58–83.

Walker, D. (1970) Direction and rate in some British postglacial hydroseres. In: Walker, D. & West, R.G. (eds), *Studies in the Vegetational History of the British Isles.* Cambridge University Press, Cambridge, 117–39.

Warren, M.-A. (1980) The rights of the non-human world. In: Elliott, R. & Gare, A. (eds), *Environmental Philosophy.* Australian National University Press, Canberra, 109–34.

Wharton, B.-A., Farr, K.M., & O'Sullivan, P.E. (1993) The Late Holocene history of the Shropshire–Cheshire meres, UK. *Verhandlungen Internationale Vereinigung Limnologie*, **25**, 1075–8.

Woods, M. (2001) Wilderness. In: Jamieson, D. (ed.), *A Companion to Environmental Philosophy.* Blackwell, Oxford, 349–61.

# 2 The Assault on the Quality and Value of Lakes

WILHELM RIPL AND KLAUS-DIETER WOLTER

## 2.1 INTRODUCTION

### 2.1.1 Natural functionality

In natural ecosystems, unaffected by humankind, any development or succession always occurs in the direction of almost completely closed water and matter cycles. In terms of processes, this development depends on the property of ecosystems to dissipate energy, i.e. through the agency of water, the attenuation of the daily solar-energy pulse towards a mean temperature at a given place. In this context, the most important dissipative processes are the evaporation and condensation of water (the physical dissipative-processor property of water). In unhindered system development, energy can be turned over in cycles, by way of the ever more-frequent and smaller scale cycling of water becoming more completely closed (i.e. at or near to the same place). Together with this, matter will be recycled increasingly at smaller and smaller orders of scale. Matter is therefore conserved, and the long-term stability of the system will thus be increased (a process of self-optimisation).

The more that the day–night temperature amplitude is dampened by the dissipative processes of water evaporation and condensation, the higher the overall spatiotemporal temperature equalisation. This can also be expressed at the level of the land–water ecosystem, through its high thermal efficiency. Greater thermal efficiency is always coupled to higher matter-cycling efficiency (i.e. lower irreversible matter losses). Thus, the natural functionality of ecosystems can be defined by means of their efficient dissipation of energy. Ecosystems with natural functionality exhibit a high thermal and matter-cycling efficiency. Such ecosystems possess high stability and a high level of sustainability.

### 2.1.2 Matter losses and ageing

Whatever the prevailing conditions may be, all lakes are subject to a process of natural ageing. This process is evident in the advancing terrestrialisation of a lake, as it accumulates sediment. If the terrestrialisation is slow, we may regard the lake as being intact and oligotrophic. In marked contrast, a rapid rate of terrestrialisation indicates an unstable aquatic system. This process of 'rapid lake ageing' is known in German as 'rasante Seenalterung' (Ohle 1953a). Accelerated sedimentation is a sign of excessive material loading, mostly the consequence of high nutrient and mineral loads carried in the drainage water from the lake's catchment. Thus high matter losses from the land lead to high rates of material deposition in lakes and, eventually, to the global sink—the sea.

Highly accelerated rates of lake sedimentation and corresponding high matter losses from the catchment must therefore be considered as a single coherent phenomenon—that of an unstable land–water ecosystem. Through the spatial transport of organic and inorganic material—matter transport processes—the catchment area and its corresponding surface waters experience a coupled ageing process; its speed is determined by the rate of irreversible material loss from the land (i.e. the whole catchment). In the same way, sustainability, i.e. the permanence of the system, may also be described in terms of the rates of matter loss (cf. Fritz *et al.* 1995; Ripl 1995b). The land–water ecosystem may be stable, or sustainable over a longer period, only if terrestrial matter stays largely in place. Such systems are characterised by short, localised matter cycles (i.e. recycling) which minimise matter losses to receiving surface waters. The higher the proportion of short-cycled material, then the greater is the system's efficiency in using available materials, and the more sustainable is the land–water ecosystem.

Compared with pre-industrial periods, and especially the time of the Atlantic Period (c.8000–5000 yr BP), northern Europe's ecosystems currently exhibit highly elevated rates of matter losses, accelerated ageing and lowered sustainability (Ripl 1995a). This gradual but ever-increasing material inefficiency of the landscape is clearly visible in surface waters and in the increasingly overloaded metabolism of aquatic systems. This is expressed by what we now call ontogeny of lakes: on the land, material losses lead to site deterioration and vegetation loss; in the lake (especially its littoral zone), the deposition of shed materials and leached mineral ions sponsor a different kind of vegetation in the water.

### 2.1.3 Holistic concept

The interrelation of the terrestrial and aquatic metabolism demands a holistic approach to studying the land–water ecosystem. However, such holistic approaches confront the distinctly sectoral characteristics of present-day ecological science—with its separation into terrestrial ecology and limnology. The real task of ecology, to support the understanding of ecological processes in their spatiotemporal coupling, is not assisted by the false interdisciplinary separation of land and water studies. Owing to poor understanding at the system-process level, many environmental protection measures have been developed through 'sectoral', piecemeal science. Mostly, these have not led to major successes. The 'dry energetics' of, for example, carbon dioxide, ozone, chlorofluorocarbons (CFCs), organic and inorganic toxins in soils, do not reveal a full appreciation of the entirety of energy dissipative processes involved in 'global change'. The harmful effects of these substances are not generally manifest until the landscape becomes desiccated and the dissipation of the daily solar-energy pulse by evaporation/condensation (as well as production/respiration) is already impaired.

Owing to the different points of view of various sectors, a coherent assessment of landscape and surface waters is sadly missing. Such separate sectoral approaches do not permit the development of appropriate ecosystem management strategies, aimed towards long-term stabilisation. Instead, the sectoral approach, which has prevailed over the past 100 years, has brought a dramatic increase in the productivity of systems with, at the same time, an acceleration in irreversible matter losses for ourselves (Ripl et al. 1996).

In order to characterise the land–water system, four theses may be presented.

**1** Vegetation at a given location is finite. It is limited by the pool of mineral resources, by the water regime (which exhibits, to a certain degree, feedback to the vegetation structure) and by the losses of mineral ions in the water flow from the site.

**2** The uptake of nutrients and minerals by vegetation is an active process, the energy for which must be provided by photosynthesis. In this sense, vegetation is energy-limited.

**3** The transition of the land–water system from a more or less natural state to one characterised by strong human influence has been accomplished at the cost of matter losses, which have increased by a factor of about 100. Hence, an increasing proportion of landscape exhibits a danger of desertification.

**4** The loss of minerals and nutrients from soils is manifest as an assault on lakes.

If matter losses are to be reduced, it is necessary to apply a holistic consideration of all those processes controlled by water within the land–water ecosystem. The energy–transport–reaction (ETR) conceptual model, described below, provides one such holistic approach. It addresses all the central water transfers of the system, most particularly the key physical (evaporation–condensation), chemical (dissolution and precipitation of salts) and biological (production–respiration) processes. The degree of spatiotemporal coupling between these processes determines the stability of the ecosystem, regulating losses of matter and the rate of ageing.

Essentially, description of these interrelated connections facilitates a functional understanding of the ecosystem and, in turn, identifies the measures needed for improved system efficiency and stability.

### 2.1.4  The purpose of this chapter

In respect of lakes, three attributes may be considered.

**1** Rivers and lakes may be viewed as dynamic indicators of landscape processes. The dynamic changes occurring in rivers and lakes cannot be followed by static indicators, e.g. nutrient limit concentrations, but rather by an assessment of time-series changes in their natural functionality.
**2** Lakes possess their own life-span, which stems from the dynamics of their metabolism and transport processes.
**3** Lakes are important because of their utility to humans.

In the present contribution, the assault of humankind on the quality and value of lakes, through accelerated external matter loads and the resulting internal damage to lake systems, will be demonstrated. This will be carried out using an ecosystem model of the processes involving matter both within and outside lakes. In this model, the importance of boundary exchanges at the liquid–solid interfaces to the efficient metabolism in the littoral and profundal zones is emphasised, along with the roles of low sedimentation and rapid, local cycling of matter.

In this scheme, the assault on lakes is not categorised by differentiated 'syndromes' (e.g. eutrophication, acidification, ecotoxicity or other human impacts). The anthropogenic assault on surface waters has come about through, above all else, the destabilisation of terrestrial metabolism. So, our approach seeks an understanding at the level of processes: how anthropogenically derived pulses of disturbance within the catchment bring the destabilising effects to the receiving water. Particular attention is given to the 'opening-up' of previously closed and virtually loss-free matter cycles, and the subsequent dissipation through physical, chemical and biological processes.

The point has already been made that the ineffectiveness of many attempts to restore the quality and integrity of lakes lies in the separation of treatments for the terrrestrial and aquatic components. A further aim of the present chapter is to promote the sustainable management of land–water ecosystems.

Thus, this contribution is not intended merely as an overview of contemporary state-of-the-art technical practices but, rather, the basis for future approaches. These will necessarily invoke efforts to restore sustainable system efficiency and to reduce losses from land–water ecosystems.

## 2.2  THE ENERGY–TRANSPORT–REACTION (ETR) MODEL

A general ecological model has been developed by Ripl and his co-workers (Ripl 1992, 1995a) which measures ecosystem function in terms of the dissipation of the daily, seasonally modulated solar energy pulse by the most important dynamic agent on the planet, namely water. At that time, no other model of the spatiotemporal development of ecosystems was known to the present authors and, to us, the concept of measuring ecosystem metabolism as a function of energy dissipation controlled through the medium of water remains attractive. These processes include the dissolution and precipitation of salts (=formation and neutralisation of protons), the splitting of the water dipole and its reassembly (=production and respiration of organisms participating in the carbon cycle), and the evaporation and condensation of water, which influences its discharge and transport capacity.

The energy–transport–reaction (ETR) model is based on a deductive concept. It supposes that energy dissipation, mainly accomplished by way of this dynamic agent water, produces equilibrium tendencies for temperature, precipitation, runoff and dependent chemical processes, at least over finite and estimable space and time dimensions. The model attempts to characterise the energy potential $(E)$ which the system dissipates in space and time. The incoming energy comes mainly from the sun, whence it is pulsed on a daily basis and also varies on an annual cycle. Dissipation of this energy potential leads to either a more or a less widespread transport of matter ($T$). Only by means of this transport can energy potentials be maintained. In this far-from-equilibrium system, energy dissipation and the transport of matter generate, to a greater or lesser extent, a reaction ($R$), which is to create a self-optimising dissipative network, the components of the biocoenosis and its delimited environment.

The ETR model is based on the laws of thermodynamics which determine the direction of energy dispersal as heat. Here, this entropy is regarded as being mediated mainly in the matter transported with the water flow, lost irreversibly from catchments. In this way, the model also accords with Prigogine's (1988) concept of self-organising dissipative structures. Further, the model incorporates the important role that boundary layers play in mediating processes, as well as an index of thermal and chemical efficiency in the selection of structures at different spatial scales.

Whilst the analogy with energetics is clear, the ETR approach differs in several respects from the energy model of Odum (1971), which adopts terms of energy flows in thermal equivalents. In contrast, the ETR model draws on the energy of conversions of water between its forms ('wet energetics'). For instance, about 80% of the daily energy pulse from the sun is dissipated in the evaporation and condensation of water. Ignoring all means of energy turnover, other than carbon flow, gives an inadequate reflection of the total energetics of ecosystems.

In the following paragraphs, the ETR model is described through its three components in turn — energy, transport and reaction. The last of these includes some more general conclusions about the system, its functionality, and the description of its index (or coefficient) of efficiency.

### 2.2.1 *The energetics of landscape*

Before we begin to assess the modern landscape, let us consider the energetics of a natural landscape, free to optimise processes, unhindered by the assault from humankind of the past 2000 years or so.

Water transformations provide several processor properties (Fig. 2.1). First, and most important, water dissipates the daily solar-energy pulse in space, as well as in time. Any physical structure which partitions energy into various dissipative pathways may be defined as an energy processor. The evaporation and condensation of water using the enthalpy leap between liquid water and water vapour or ice, respectively, mainly dissipate energy potentials, and constitute a physical dissipative processor. Its function leads to an almost perfect water cycle, with almost no irreversible flow of matter; thus, it is close to constituting a thermodynamic Carnot cycle, with an extremely low entropy gain. When precipitation occurs over the oceans, water currents are generated. When precipitation falls over a continent, then either streams form on the land surface, leading to a direct, coherent, topographically determined water transport, or else it is evaporated back to the atmosphere in order to create a short-circuited, near-closed water cycle.

Water expresses other important processor properties. The chemical processor property of water invokes the molecular dipole and the molecular dissociation of water. This allows water to behave as a weak acid, with a proton density of $10^{-7}$ mol L$^{-1}$ (at 20°C). Because of this dissociation, chemical reactions are enabled at the interface with solid matter, leading to the dissolution of various ionic lattices into dissolved ionic charges. Water currents at the solid–liquid interface increase the dissociation, for the water dipole is deformed in much the same way as it is at elevated temperatures. When water currents are retarded at the interface, then the dipole becomes less reactive and ions are deposited again. The game of interaction between the solid and liquid phases, dynamically in space and time, optimises the course and distribution of processes. However, reactions are only possible if the chemical potentials are maintained by a steady water flow. The absence of some means of water transport leads to the breakdown of reaction potentials: under the conditions of thermodynamic equilibrium, processes cease.

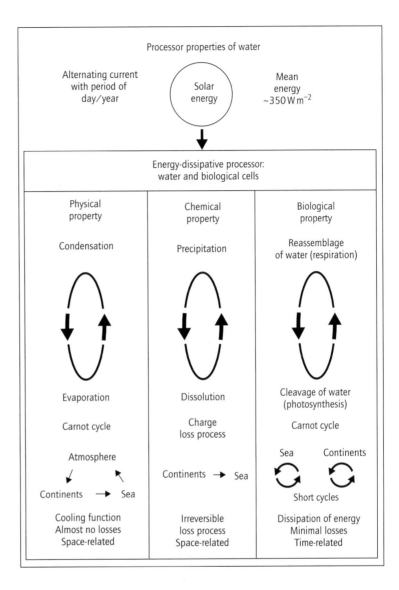

Fig. 2.1 Processor properties of water. (Redrawn from Ripl 1992.)

This chemical processor property of water is connected with the irreversible charge flow of dissolved mineral ions to the sea. If we suppose the composition of seawater to be constant, then the dissolved matter input by rivers would need to crystallise and precipitate as ocean sediment. The frequency of this process is so low that it would take something of the order of a billion years until the cycle concluded by the rising of the sea floor into a new continental mass and reversal of the mineral losses. In human timescales, however, the charge flow of mineral ions to the sea appears irreversible.

Then there are the optimised water structures that we know as autotrophs. They contain highly structured organic molecules which absorb and channel light energy in such a way that the water dipole is broken in two (the condition at pH 0).

The hydrogen thus liberated is able to reduce carbon dioxide to carbohydrate radicals and, finally, sugars or cellulose, whilst oxygen is liberated to the environment. The relevant energy-dissipative optimised water structures characterise all primary producers. In the opposite reaction, known as respiration, water molecules are reassembled in order to yield energy for the coupled processes of organisms (through the action of catalysts known as enzymes). This is a powerful, biological dissipative-processor property of water.

As these dissipative processes keep matter cycling within its own place, the losses of dissolved matter from a given site remain small. With short, almost closed, evaporative water cycles and an adequate reserve of stored energy in the form of degradable organic matter, nutrients and minerals, organisms may arrange themselves into communities and biocoenotic structures which may collectively achieve a homeostasis (dynamic stability) within the spatial domain. In this way, sustainable, functionally optimised structures (called organisms) are seen to be dissipating and damping the pulses in solar energy income and counteract the irreversible losses of matter.

These three processor properties of water (physical, chemical and biological) are mutually coupled in a recursive way, which means that the change of enthalpy stemming from chemical or biological reactions acts on the physical processor property. Whilst one area is being cooled by evapotranspiration, another is being heated by condensation of water some moment later. This thermostatic function of the landscape may be monitored. How well does the mainly cooling process of the evaporation–condensation cycle function? The temporal and spatial distribution of thermostatic dissipative landscape function may be tracked by the multi-temporal evaluation of thermal information collected, for example, by Landsat satellite (channel 6), on a region-by-region basis (Ripl *et al.* 1996). Similarly, the irreversible charge flow (i.e. a measure of the mineral ions lost from an area) may be calculated from the pattern of hydrological runoff, (say) over 1 year, multiplied by the pattern of mineral ion concentrations discharged from a given catchment.

### 2.2.2 The energy dynamics of lakes

The 'driving force' for each of these processes is a given energy potential. Any dynamic in a system emerges from the attenuation of the given energy pulse towards some mean equilibrium level free from potentials (see Fig. 2.2). In a lake, the dissipation of the diel pulse of solar energy is the driving force for all processes. One of these dissipative processes is the biological fixation of carbon where organic substances are produced through the splitting of dipole water molecules. In turn, the organic substances provide the source of energy for subsequent processes. In the course of the use and re-oxidation of organic substances, albeit spatially shifted and temporally delayed, water molecules are rebuilt (respiration). Both photosynthesis and respiration are organismically catalysed, providing the driving force for the metabolism of rivers and lakes.

The pH of lakes, like the redox potential, is similarly regulated through the recursive coupling of the three dissipative processes. The pH and redox potential, in turn, control the subordinate processes of dissolution, distribution (transport) and reaction. The conventional wisdom of considering the pH and redox conditions as the overriding control variables for metabolic processes within a lake—the simple application of 'cause and effect'—is not very helpful to the view of naturally dissipative and self-organising systems. For a deeper under-

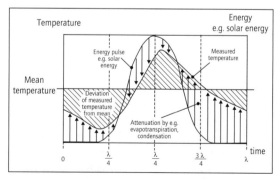

Fig. 2.2 Dampening of the temperature pulse towards a potential-free mean equilibrium. (Redrawn from Ripl *et al.* 1995.)

standing of processes in nature, at least in energy terms, the linked reactions which are most apparent need not be of major significance. What must be considered first are the mechanisms for the renewal and transport of energy potential, i.e. the rules for the spatial and temporal distribution of processes.

This coherence of energy and matter-transport processes is put together in a deductive way within the ETR model. The attempt is made to see how the energy source and the necessary transport dynamics are provided for each process and how, in turn, these energy dynamics provide the prerequisites and conditions for the various reactions between materials. From insights thus gained, it becomes more possible to define those 'probability spaces' in which aquatic processes are mainly distributed and in which metabolism takes place.

Of particular significance to the retention of matter in surface waters is the boundary layer between liquid and solid phases (such as the littoral zone and the periphyton). This is because the energy potentials here are most effective and the process density is highest. Plankton also functions according to these principles, although at smaller spatial scales. The main difference with plankton is that the liquid–solid boundary functioning processes are much more finely dispersed.

At solid–liquid boundaries, the speed of chemical reactions depends upon the partition of energy into thermal energy and the dissociation energy of water at a given temperature of the liquid–solid phases. The likelihood of water dissociating into protons and hydroxide ions cannot, statistically speaking, be assumed to be the same at the boundary layer as in the body of a large mass of water. Proton activity may vary substantially over short distances. Near a boundary layer, the probability of meeting a proton (as indicated by pH) might be somewhat enhanced. In boundary layers of macroscopic proportions, these complex interrelations apply to the interiors of individual cells. At the surfaces of internal membranes (e.g. of thylakoids, mitochondria and chloroplasts), the splitting of water and chemical synthesis occur with a high degree of efficiency.

At many solid–liquid boundaries, different charge densities exist at the surface layer itself and the local dissociation of water can be increased accordingly. Such differences can be due to various mechanical properties of different sized molecules in the solid-phase boundary, or because of differences in heat capacity between solid molecules and liquid water. Other factors can be the coherence of water molecules in their oscillations or the accelerations and decelerations which occur in clusters of water molecules. The increased dissociation of water can lead to pH being lowered locally at the boundary interface. An increased degree of water dissociation provides a form of chemical energy which will be dissipated according to the rules of energy distribution within the ETR model.

The pH and redox potential may also be influenced by the various functional groups of organic molecules present at solid surfaces or at the membranes of organisms. There may be a higher kinetic movement of water at such surfaces, giving a higher local reactivity. These mechanisms play a role in, for example, primary production and the related splitting of water (in hydrogen formation) at the surface of biological membranes. In this way, they contribute to all dissolution and metabolism processes. This being so, the preference of biological processes for boundary layers owes its efficacy to the better control of chemical reactivity at the molecular level.

As energy processors, organisms can be considered as playing a 'catalytic' role in these processes. In terms of energy potentials, however, organisms do not occupy a driving role. Photosynthesis and respiration, as well as the accumulation of matter (both organic and inorganic), are particularly developed at phase boundaries, because it is here that these processes are sorted most effectively, as, for example, by adsorption. The structure of surfaces also minimises the openness of processes, so helping the closure of cycles. Phase boundaries possess higher stability and higher processing efficiency, which are the main reasons for hosting relatively high process densities.

### 2.2.3 Matter transport

In the ETR model, energy potentials and their

spatiotemporal distribution through the various transport mechanisms and resulting reactions influence the dynamics of both natural and anthropogenically influenced systems. With respect to lakes, the various spatial scales of stratification and convective transport place a further factor influencing the many processes involved, from the microscale, molecular dimensions to the large, basin-scale events. Between these extremes, a spectrum of mixing processes affects the overall structuring of processes and the distribution of organisms. The range of phenomena covers molecular diffusion, the formation of waves and wave patterns, Langmuir circulations, Hadley cells, slicks, seiches and thermal stratification. Slow, large-scale transport is also possible in, for instance, the hypolimnetic circulation. In addition, there are also transport phenomena in the slowly circulating hypolimnion, mineral salt separation during ice formation and ice pressing. On the other hand, water movement in the wind-exposed and scarcely stratified epilimnion offers conspicuous instances of mechanical work and the level of influence on transport processes. The scale of current velocities permits a wide range of individual velocity gradients, each of which represents another potential, driving processes at a range of levels. In addition, the orientation of the lake in relation to the wind ('wind fetch') will influence patterns of water circulation and stratification, and the limits to the extent of transport.

Information concerning transport and stratification is needed in order to explain dispersive processes, the resultant distributions of organisms and the linkages to other processes. Mechanisms which transport matter away overcome the limitation of space, whereas a build-up of matter gives negative feedback to organisms and processes. However, the removal of reaction products (biomass) counters the feedback effects. In locations where transport is severely limited, such as sediments several metres thick, processes can scarcely proceed.

The various waters flowing into a lake are also instrumental in its flushing. Annual variations in influx and temperature of the inflowing water conform to patterns which may interact with the morphometry of the lake. Such influences lead to obvious dynamic structuring in the littoral and, to an extent, in the pelagic zone, and even on sediment formation in the profundal zone. The quality and quantity of inflowing waters exert further important influences on the lake.

Knowledge of external and internal transport mechanisms is important when estimating the scale of processes in a lake, and when considering their management. In highly productive lakes, for instance, being able to estimate the transport of electron acceptors (particularly of oxygen and nitrate) is essential to the anticipation of the metabolism within the hypolimnion and at the sediment–water boundary layer.

### 2.2.4 Reaction—the self-organisation of stabilising structures

#### 2.2.4.1 The dissipative ecological unit (DEU)

In the ETR model, organisms are treated as energy processors, which, together with and by means of water, help to dissipate the sun's daily energy pulse, converting it into a more or less even, more uniform flow of lower density. Useful, harmonically modulated energy emerges as a mean temperature in time and space, lacking further potential for directed (non-random) processes. The physical, chemical, biological energy-processor properties of water permit the assembly of the five essential components of a simple, functional ecosystem and delimited in time and space as a combined energy-processor. This ensemble we refer to as the 'dissipative ecological unit' (DEU). We might compare them to an electronic circuit made up from several components, and likewise ascribe to them a certain coefficient of efficiency.

The five necessary, functionally-defined components (Fig. 2.3) are as follows.
1 The primary producers fulfil a double function. As process carriers, they store energy, producing the carbohydrates and biomass which provide an energy source to other components, removed from the daily solar pulse. As process controllers, they regulate water levels, redox conditions and miner-

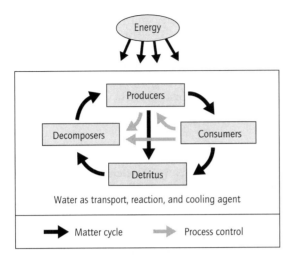

**Fig. 2.3** Dissipative ecological unit (DEU) (Ripl et al. 1995).

alisation: land plants move (literally, 'pump') water through the organism and release it by evapotranspiration; aquatic algae move water around their membranes.

2 The detritus: on land, the raw humus layer and active soil hold a pool of nutrients, minerals and chemical energy. In water, detritus comprises all non-living organic matter, in both dissolved and particulate forms (including colloids).

3 The decomposers: mainly bacteria and fungi, as process carriers responsible for the mineralisation of the detrital component. In soil, decomposers are activated by the passage of water but stop when water flow ceases, owing to the negative feedback caused by product inhibition. Decomposers might clog interstitial spaces within the humic layer, owing to their rapid reproduction, and so increase resistance against pumping, where they are not otherwise regulated by the food chain. In water, attached and free-living bacterial decomposers are strongly controlled by bacterivory.

4 The consumers: the whole food chain is a process controller, which averts space limitation by removing, and using, the autotrophic and heterotrophic process carriers as food, so contributing to the efficiency of energy dissipation.

5 Finally, water acts as the medium for energy dissipation and cooling, transport and reaction.

The DEU is a structure which manifests self-organisation. It shows, for a given set of energy and environmental conditions, a developmental progress in the direction of reduced openness of matter flow. A DEU acquires a spatially and temporally delimited structure which is more or less sustainable.

#### 2.2.4.2 Efficiency

Through the interaction between the sun and Earth's surface, energy has been used to form the landscape, through the dynamic medium of the water cycle and its subordinate atmospheric processes. Solar energy reaches Earth as a precisely structured pulse of alternating current, with a particular frequency and amplitude and thus containing information. This oscillating energy current supplies and propagates processes, simultaneously acting as a selective filter for the efficiency for all subsequent processes. By dissipating the energy pulse in the least possible space and maximising gross production to the point where net production falls to zero, the biocoenotic structure of the DEU is optimised. Consequences of the optimisation are that energy potentials are deployed mainly to drive cyclic, almost loss-free, coupled processes. Irreversible matter losses are thus minimised and the sustainability of the system is increased. Greater efficiency also imparts long-term stability.

Evolution and ecological succession are also energy-driven processes. Within a given delimited area, such as a lake catchment, with a coherent water-transport system operating on an area-dependent timescale, system development and evolution provide a strong feedback in system efficiency. Within set spatiotemporal boundaries, system efficiency ($E$) may be defined as the total turnover of matter (e.g. gross production, $P$), minus the total irreversible losses of material, $L$, divided by the proportion of the matter turned over:

$$E = (P - L)/P$$

where $E$ is efficiency, $P$ is gross productivity and $L$ is irreversible losses (base cations); all parameters are expressed as proton equivalents, i.e. 2 $H^+$ for 1 C, 1 $Ca^{2+}$ and 1 $Mg^{2+}$, and 1 $H^+$ for 1 $Na^+$ and $K^+$.

Efficiency is a measure of the openness (or relative closure) of the system under consideration. Clearly, definition of the system boundaries is of fundamental importance. Only where a maximum degree of spatiotemporal unity already exists does it make sense to delimit in time and space. Appropriate system boundaries for consideration of cycling processes are catchments and sub-catchments, because of the low material exchange across their boundaries. Moreover, it is often fairly straightforward to calculate losses from a catchment at a single monitoring point on the effluent river. A coefficient of efficiency can hardly be assigned sensibly to an individual organism, because of the relatively high level of material exchanges with its environment. Of greater relevance is the output from the networked community–organism structure, revealing the extent of matter cycling within a certain period of time.

For the kind of delimited systems being considered here, efficiency becomes a powerful criterion of selection. Inefficient (non-sustainable) structures become replaced by efficient (sustainable) structures with improved functions, depending on the scale of irreversible losses in relation to the pool of net resources and productivity. In this way, efficiency structures the distribution and coupling of parallel and sequential process, over various spatial scales, which also means that efficiency controls self-organisation.

In addition to this chemical measure of efficiency, a thermal measure for delimited systems has also been described (Ripl *et al.* 1996; Hildmann 1999). Thermal efficiency is high when temperature differences within the landscape are generally low. This is achieved through the temperature compensation provided by the evapotranspiration of vegetation. Thermal efficiency is low when the daily energy increases the temperature differences within the landscape. Over larger space and timescales, the local water cycle and the retention of matter make for close correlation of chemical and thermal efficiency.

Efficiency, stability and sustainability of systems may all be estimated from the comparison of cyclic processes against random processes. It is randomness which contributes to irreversible losses; in the case of catchments or subcatchments, losses are estimable in energetic units, or proton flow.

### 2.2.4.3 $r$ and $K$ strategies

When two relatively inefficient processes are linked together, the overall efficiency may be improved. If structures are appropriately coupled, the relatively high entropy of one can be lowered by it directly feeding the other. Entropy, in this case, is represented by irreversible charge losses, and process linkage leads to improved retention and increased sustainability. Space limitation imposes a negative feedback on the rate of processing. This mechanism underpins the transition from an individual reproductive ($r$) strategy of an individual into the survival ($K$) strategy of the community, allocating species which gain biocoenotic efficiency and which minimise the losses from the delimited area. Phases of $r$ and $K$ strategy are attained through different assemblages of organisms. During an $r$-strategy phase, the primary producers control the system (bottom-up control), whereas in a phase of $K$ strategy, the consumers are most important to system behaviour (top-down control). The controls are summarised in Table 2.1.

### 2.2.4.4 Lowering of energy flow density

The process of system self-optimisation relies on highly structured energy pulses, stable in fre-

Table 2.1 Process control during key phases of ecosystem development

| Organismic growth strategy | $r$ | $K$ |
|---|---|---|
| Phase limitation | Growth, time | Space, resource |
| System controller | Primary producers | Consumers |
| Process control | Bottom-up | Top-down |

quency, modulation and amplitude. These pulses serve as the primary clock and main time limitation for the optimisation process, regulating parallel processes and their sequential time courses. Any form of additional energy supplied randomly will tend to deteriorate already-optimised dissipative structures (such as the DEU), decrease their stability and reopen the channels for irreversible losses. The goal function for the ecosystem is thus to achieve maximum dissipation of the energy pulse, by lowering the density of energy flow, in the least area. This goal is achieved at maximum gross productivity and zero net productivity, when the metabolism of the community is maximised for the longest time period within the given space. Such a system would exhibit a minimum of large-scale atmospheric processes (affecting the upper atmosphere), owing to the lack of sufficient energy potential, and exert homeostatic conditions.

### 2.2.4.5 Self-organisation

From the standpoint of the theory of dissipative structures (Prigogine 1988), aquatic ecosystems build by moving further and further from thermodynamic equilibrium, structuring the energy supply within a limited space. In this view, even the organisms become thermodynamically necessary to the optimised energy dynamics of the limited space. Given the principles of energy dissipation and under the selective forces of efficiency, the structure of the whole ecosystem comprises the dynamic medium of water and its constituent organisms (Ripl & Hildmann 1997).

A process-related consideration of energy dissipation in rivers and lakes is necessary at several different organisational levels and scales. These are self-similar, fractal series and they include: water clusters, functional organic molecules, cells, tissue, coenobia, multicellular organisms, the dissipative ecological unit (DEU) and the system. The self-similarity, or fractal nature of these levels, does not refer to their form or structure but, rather, to their function in dissipating supplied energy in a structured way, in a limited space and conforming to the principle of efficiency. In this context, material limits (e.g. limitation of nutrients) are to be understood as being space-determined.

These structural levels are also self-organising. Each conforms to the criterion of 'least openness', with respect to matter, and in accord with the principle of efficiency selection in favour of maximum matter recycling. The delimitation of individual structures is of special significance; irreversible flow of matter across the level boundary is minimised (and so defining the limits of the fractal dissipative structure). At the next level up, further minimisation of loss occurs; system irreversibility is hierarchically organised.

Understanding ecosystems as energy-dissipative, efficiency-increasing systems is a prerequisite for their sustainable management. A local and temporary reduction of system efficiency can be understood as disturbance. If the frequency of disturbance should be greater than that which allows the system to recover, then the impact is called damage: there is a degree of system destruction and of reduced efficiency. If the disturbance frequency should be lower than the level required for recovery and if it is at all regular, then the disturbances collectively constitute a change in the initial conditions for the self-organisation of the system.

Individual organisms or species are able to accomplish matter cycles. It is the community network — the web — involving all five components of the dissipative ecological unit (DEU) that makes such circulation possible. The very act of producers, consumers and decomposers living together helps close nutrient cycles. The highest stability is shown by that structure which, under dynamic conditions, forms the most extensive matter-cycling structure.

The energy in natural waters is neither isotropically nor homogeneously distributed. Each individual location will therefore manifest different outcomes in terms of closed structures. For example, at places receiving high light income, a higher diversity of functional components is necessary for the efficiency of the DEU than if the site received much less light. All else being equal, a less useful energy potential gives lower production opportunities. Similarly, at a site with higher nutrient availability (but otherwise identical energy conditions), organisms with the fast reproduction and growth rates will be more successful than at a site with lower nutrient availability, at least until the physical space becomes filled with these organisms. Not until the space is filled will the structural potential for a succession be realised and alternative organisms with more $K$-like strategic adaptations be selected.

Foremost in the structuring of a dissipative ecological unit (DEU) are the conditions given by the distribution of energy, the daily solar energy pulse (oscillation) and the allocation of resources within the given space. The more differentiated (in distribution, etc.) these structuring elements appear in a biotope, the higher the number of species. In turn, the more varied will be the detritus structure in respect to chemical and physical form and to the throughflow of water. Each new 'invading' organism entering the DEU will either assist in the closure of matter cycles or will help their opening. If the latter, then higher matter losses (in the case of the lake, to the sediment or to the outflow) bring a reduction in efficiency and sustainability.

Conversely, positive selection of a new organism increases the efficiency, sustainability and dynamism of the DEU. It may be deduced that selection is determined not by the individual organism but by the current efficiency of the extant DEU. The probability of ecological succession in a system with relatively high efficiency is smaller than in one of lower overall efficiency. Thus, a DEU or ecosystem, optimised in this way, should support a higher number of species than one of lower efficiency. However, the converse conclusion—that high species number reflects a more stable ecosystem—is not correct, for it also can be the product of ongoing disturbance. Take, for example, the case of a dry meadow, a habitat requiring frequent management input (i.e. system disturbance) in order to maintain present structure and species composition. High species number and high successional probability reveal a reduced efficiency, relative to the optimised condition, offering little long-term possibility of maintenance of such high species numbers ('diversity'), without the continued intervention of management by organisms and ourselves.

Present efforts to maintain high species diversity by way of nature conservation are not working because the approaches do not allow self-organisation and selection to raise ecosystem efficiency and improve its functioning. The development of optimised high efficiency generally requires longer time periods (of the order of 10–100 yr) than the conservation management cycle usually allows.

The process-oriented approach advocated here accepts that, to a large extent, succession is the outcome of an interaction between water, organisms and the abiotic components. With a founding concept of a system for dissipating and attenuating pulsed inputs of solar energy, through movement (warmth, turbulence, current and water flow) and the distribution of structuring potential, the function of lake ecosystems and the scope for their deterioration and improvement can be quantified and measured. Of decisive significance is the distribution of the (light and chemical-bound) energy which is supplied and the reaction products that are removed.

## 2.3 HISTORICAL DEVELOPMENT OF LAKES AND CATCHMENTS

### 2.3.1 *Origin of lakes and stabilisation of natural landscape function*

#### 2.3.1.1 Origin of lakes in a hydromorphic landscape

Lakes are born, grow old and pass away within geological timescales. Lake basins originate in many ways—including tectonic movement, vulcanism, landslides, solution of mineral salts, through erosion, through the activities of higher animals (e.g. beaver, human beings) and even by the impacts of meteorites (Hutchinson 1957; see also Volume 1, Chapter 2). Through the agency of the hydrological cycle, it is the interactions between water and bedrock which are most often involved in lake genesis. This is particularly true of those lakes which were formed in the wake of the last glaciation and of those which now occupy tectonic basins, such as volcanic craters and geological rift valleys which are actively fed by present precipitation.

These basins evolve after formation, the extent of subsequent sedimentation and terrestrialisation depending upon their depth, shape and vegetational encroachment.

However, as a consequence of the recent explosion in anthropogenic use of alternative sources of energy (mostly as fossil and nuclear) and the powerful mechanical means for engineering the landscape (mineral extraction, reservoir construction, flood control and irrigation) the direction and rate of lake development has changed.

### 2.3.1.2 Stabilisation of natural landscape function

In the temperate zone, lake development has been closely connected with the development of vegetation cover in the catchment. For example, the initial recolonisation by vegetation of the land surface exposed by the retreating glaciers is most likely to have begun in the land–water ecotone. From the lake edge, vegetation expanded higher up the catchment and, in the opposite direction, into the euphotic zone of the lake. Vegetation development brought certain repercussions to the local water regime, initiating an optimisation process and the accumulation of a water-storing humus layer. The improved water capacity of soils led to the increasing re-routing of solar energy through localised dissipative evaporation–condensation cycles. In the context of the landscape, the development of lakes became increasingly subject to the self-optimising creation of short-distance cycles of matter and water in the catchment area. Temperature fluctuations became increasingly attenuated as a consequence of the enhanced intensity of local cycles of evaporation and condensation, and water in the landscape was subject to a more frequent and spatially smaller circulation. Runoff and throughflow of groundwater flow became slower and fluctuations became increasingly damped, to the extent that evapotranspiration and dew formation became the main processes contributing to the system of environmental temperature compensation and equalisation.

Such development of the water regime of landscapes is reflected in other natural patterns. These include (i) the decreased amplitude of daily and seasonal temperature variations and (ii) the diminished and damped amplitude of river discharges. Short-distance water circulation also steadies the water flow during the summer, so that seasonal extremes of river and lake water discharge (baseflow, flood) are more nearly equalised. During the colder seasons, when the relative proportion of precipitation discharged into rivers compared with that evaporated increases, potential flood volumes are offset by the enhanced water-logging capacity of the soil.

The chemical composition of soil water which reaches streams and rivers may also be viewed within the functional context of the vegetation–water–ecosystem development. In the self-optimising landscape, the increasing organic enrichment of soil surface layers (humus, peat), the enhanced evaporation leads to a generally more horizontal reorientation of soil water flow. The consequent reduction in the vertical percolation of water through increasingly water-saturated soils brings greater stress on the oxygen content and oxidative capacity, leading to the reduction of nitrate and sulphate in soil water. Iron, calcium and aluminium ions, as well as bicarbonate, are rather more characteristic of the chemical solutes in soil water.

Typical of the natural environment are spatial and temporal patterns—the determination and interpretation of the frequency and amplitude of these patterns facilitate statements concerning the stability of ecosystems. In natural systems, the frequency of processes are brought into phase ('tuned') with the periodicity of the daily, though annually modulated, input of solar energy. The amplitude of process intensity is already damped down short in space and time, which means that process intensity is more or less equalised.

### 2.3.1.3 Vegetation in the climax stage

The post-glacial development of vegetation, up to a fully developed 'climax' community stage, may be reconstructed with the help of pollen analysis of ancient lake sediments. From the various pollen fractions represented and a knowledge of the ecological requirements of the tree, herb and grass taxa concerned, the past climatic and edaphic conditions may be traced. This reconstructed soil development and past climates changes are reflected in the lake deposits themselves (Digerfeldt 1972). During the Late-glacial period, following the retreat of the ice sheet from the surroundings of Lake Trummen (Sweden), the unleached, neutral to basic mineral soil was inhabited by herbs and grasses, later followed by birch and pine (*Betula* and *Pinus*). With the onset of the post-glacial Pre-Boreal period, which began about 10,300 years BP, the herbs and grasses were largely replaced by trees. During the next 6500 years, vegetation developed from open pioneer to dense deciduous forest. Through this succession, the soil was successively depleted by the net primary production of the forest. However, deposition rates of base cations and nutrients in the sediments of Lake Trummen gradually diminished from the beginning of the Pre-Boreal period, to a minimum at the onset of the Atlantic period (Digerfeldt 1972; see also section 2.4.4).

The input of base cations to the lake, together with the transport of nitrogen and phosphorus, appears to be autocorrelated. In parallel with the lowering of matter in the runoff from the catchment, the lake became increasingly oligotrophic ('post-glacial oligotrophication': Digerfeldt 1972). Phosphates would have co-precipitated with aluminium, iron or calcium ions; following mineralisation, any excess of nitrogen compounds would have been denitrified in the accumulating wet peat layers and recirculated to the atmosphere. Accordingly, relatively little residual nutrient would have been shed to the lake.

The ecological structuring of lakes is necessarily viewed holistically, as a section of energy-dissipative temperature compensation processes over the entire catchment. With increasing terrestrial dissipation of energy, runoff and other hydrological processes, including the leaching of minerals, are diminished in amplitude and the material processes in lakes become severely resource-constrained. Rivers and lakes are moved from being dynamic and very open systems, towards a maximum of structural stability and diversity of form and minimised rates of terrestrialisation and sediment formation.

With succession directed towards stable terrestrial biocoenoses, maintaining short-distance matter cycles, mosaic-cycle structures emerge (cf. Remmert 1991). Such cycles are well exemplified in peat bogs (see section 2.3.1.5). As vegetation comes increasingly to control water and matter cycles, the influence of external factors, such as macroclimate or geology, diminishes. External processes are internalised. Differences in landscape potentials (habitat differences) diminish and the dynamics of individual habitats become more equalised. Once again, the tendencies to closed material cycles at every fractal level, balanced energy flow and increased stability and sustainability are plainly evident.

### 2.3.1.4 Climatic development

Vegetation began to inherit the land once the postglacial conditions allowed. The rate of development of vegetation would have been greatly influenced by the general weather conditions, pervading every aspect of habitat (from geological substrate to developing soil). The macroclimate was the decisive controlling agency. However, as vegetation became well established, the influence of dissipative evaporation and condensation on microclimate increased accordingly. The spatially wide, long-distance movement of water between the oceans and land, which characterised the early stage of landscape development, constitutes a 'long-wave' or 'low-frequency' cycle. The sparseness of the early vegetation cover would have provided little landscape cooling and little to counter irregular precipitation events. At the same time, the ratio of precipitation percolating downwards through soils would have been high in relation to evaporation, with consequent high losses of leached mineral ions.

As the spatiotemporal separation of evapotranspiration from condensation is steadily reduced as the cover of vegetation extends and thickens and the temperature regime is increasingly moderated, decomposition of matter becomes more closely coupled to production, and the uptake of nutrients becomes more closely linked to the rate of remineralisation. As the vegetation becomes a more

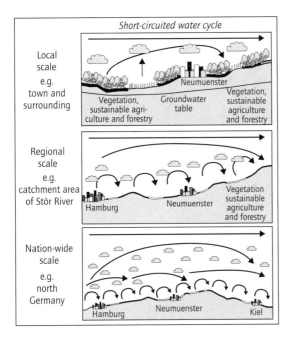

**Fig. 2.4** Short-circuited or short-distance water cycles. (Redrawn from Ripl et al. 1995.)

fully developed part of the landscape, water cycles become spatiotemporally shorter (Fig.2.4). Short-circuited cycles of water and associated matter thus counter losses through export.

During the advance of vegetation succession over large areas of the landscape, the worst aspects of a macroclimate may be softened. The extent of 'rare' extreme weather events, such as severe rainstorms or periods of long drought, can be attenuated and, hence, ameliorated by the energy-dissipative evaporation–condensation process of extensive vegetation cover. Similarly, in a landscape with a more equalised temperature regime, the probability of extreme macroclimatic events is diminished. The gradual transition from a long-wave water cycle to the more frequent, short-distance water cycle, which is more within organismic control through truncated matter cycles, is a further instance of a large-scale external process (in this case, global water circulation) becoming increasingly internalised.

### 2.3.1.5 Formation and terrestrialisation of bogs

In many of the lowland areas affected by the last glaciation, many of the shallow water-filled hollows left have become wholly covered by vegetation. As decomposition processes in standing waters are limited by the availability of suitable electron acceptors (oxidising agents) such as oxygen, nitrate and sulphate, incompletely decayed organic matter accumulated in these basins is gradually filling them in.

The fens and mires thus formed represent developmental stages in the terrestrialisation process. The net accumulation of matter, being first fixed in biomass and then accumulated in the abundant necromass that is peat, minimises its irreversible loss from the site of its production. The formation of this type of bog took place against a background of landscape still characterised by long-distance matter cycles. With the advancing efficiency of dissipative vegetation structures with better catchment storage and reduced transport of nutrients and bases, the tendency of peat development was towards more ombrotrophic peat bogs, which derive their water directly from rainfall and receive their nutrient input mainly from the atmosphere. The growth of ombrotrophic bogs is directed by the poverty of bases and is limited by the small nutrient inputs from wet and dry air deposition.

Nevertheless, peat bogs provide especially vivid examples of structures which function with markedly low losses. Matter decomposition and production are minimised and intimately coupled, as is illustrated particularly by the cyclical changes which involve hummocks of bog moss (*Sphagnum*) and the intemediate wet hollows (Fig. 2.5). In these 'hummock-and-hollow' systems, seasonally fluctuating water levels incur alternating wet and dry phases, with alternately improved supplies of water and oxygen. The alternations are most developed in the relatively elevated hummocks, where the fastest rate of organic-matter mineralisation occurs. The nutrients drain to the ecotone between hummock and wet hollow. At the ecotone, biomass production continues rela-

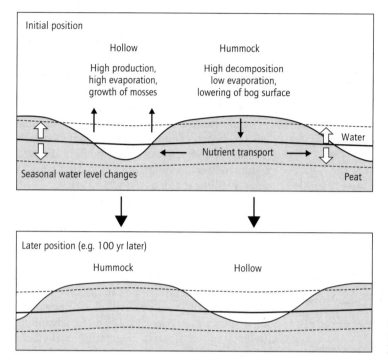

**Fig. 2.5** Functional description of the mosaic cycle of hummocks and hollows in a raised bog.

tively well, as it benefits from the nutrients mineralised in the adjacent hummock. Thus, in relation to the ecotone, the surface of the hummock is gradually lowering, whilst the surface of the ecotone, and of the hollow pools, is steadily rising. The growth and increase in height of the newly forming hummock cease at a critical threshold — where the supply of nutrients and water for the plants becomes energetically more costly to the plants. As a result, cyclical changes in the spatial locations of hummock and hollows occur through time. Depending on the overall growth rate of the peat bog, these cycles possess frequencies of decades to centuries.

#### 2.3.1.6 Palaeoecology and palaeolimnology

We have argued from first principles that the historical development of a lake is closely coupled to the flow of matter from the surrounding landscape. Evidence that this is so is available from the palaeoecological investigation of lake sediments.

From the many investigations of the composition and deposition rates of lake sediments, such as that of Lake Trummen (Digerfeldt 1972) and highlighted in this chapter, it is safe to estimate that the flow of cations from catchments to lakes was some two to three times higher at the end of the last ice age and nutrient flows were about two to ten times higher than the equivalent rates in the Atlantic period, of about 2000–3000 years later (see also section 3.4). The low sedimentation rates of the Atlantic period show that, at this time, closed matter cycles were extensively well-developed and that the associated lakes were intact.

### 2.3.2 Human settlement

Stone Age forager (hunter–gatherer) peoples hardly affected the landscape–water regime. Disturbances were first experienced only after humans began to settle down, to clear forest by fire and to impose agriculture on the landscape, mainly within the past 5000 years. The initial small uses made of forest clearings for pasture were still not a problem but, through the inexorable increase in land use, involving ever larger areas, the wholesale clearance of forest for agriculture completely changed landscapes. The measures, and their effects, were not uniform, but were differentiated in a non-linear way. However, the interventions became so profound that they became kinetically more rapid than the efficiency-increasing self-optimisation processes at work in the ecosystems of the natural landscape as a whole.

As the vegetation cover was removed, so its important dissipative structure and cooling function was undermined. There followed reductions in topsoil and its rich resource of organic matter and, in consequence, the water-retaining capacity was also weakened. It began to be difficult for plants to survive extended drought periods without marked reductions in their transpiration rates. The fluctuations in soil-moisture content and temperature became amplified. More mineralisation was favoured and the transport of mineral ions in water flow became easier. In these ways, the previously short-distance, short-circuited water and matter cycles became increasingly opened up.

Leaching from soils was further increased through the raised net productivity of arable land. In order to compensate for the uptake of base cations and nutrients removed with the harvest, plants release increased quantities of protons into the soil. This increased proton flow accelerates the dissolution of acid-soluble soil constituents and, subsequently, their flushing from the land to the sea.

However, by missing out on the possibility of wide-scale drainage schemes, matter losses and the loads transported to fresh waters still rose only slightly, in comparison to the events of the industrial period, mainly because the landscape remained relatively moist. Increased intervention in land drainage began about 700 years ago, with the introduction of the first mill dams. The effects became much stronger in the past 300 years or so, once the capability to manipulate lake levels, to drain swamps and bogs and to divert rivers, all in the interests of expanding and improving agricultural land, was developed. As a rough estimate, deduced from sediment cores extracted from eutrophicated lakes, the losses of base cations during the period of human settlement have increased by a factor of two to five. Over the same period, the losses of nutrients are between five and 100 times greater than they were under the climactic forest conditions of the Atlantic period.

### 2.3.3 Historical development of the assault on lakes and their catchments

Anthropogenic interventions before industrialisation (before approximately AD 1850) mainly comprised forest clearance and land management without strong repercussions on the wide-range soil-water balance. The sustainability of the ecosystems, measured as a relation of material cycling to material losses, was only marginally lowered under this management.

Substantially greater damage to the landscape and to waters has followed industrialisation. Irreversible material losses to water have been drastically increased through the intensified application of accessory (mostly fossil and nuclear) energy to large areas. The resultant destabilisation of the landscape can be divided into two phases, one spanning the beginning of the industrialisation to approximately 1950, the other from 1950 until today.

#### 2.3.3.1 Phase I: beginning of industrialisation (c.1850) until 1950

The technological development and the use of accessory energy after the beginning of industrialisation both demanded and led to a shift of the human population from rural areas to cities. Costs of transport fell with technical developments, so that an even greater range of goods and products, such as food, building material and manufactured goods, could be supplied to or exported from developing cities. With the increasing volume of transport and the expansive growth of cities, material cycles were opened ever wider. Biological production, consumption and biodecomposition became increasingly separated, in space and, with the development of industrial waste management, in time as well.

Destabilisation of landscape, however, was initiated through direct interventions into its water balance, with the engineering of water courses and the impacts of sewage treatment and the mining industry. The water demands of the fast growing urban population were met by groundwater pumping and the apparently inexhaustible filtration capacity of the soil. This dehydration of the landscape led to increased fluctuations in the soil water table and soil temperature (Ripl et al. 1996; Hildmann 1999). The accelerated mineralisation of organic substances in the soil was facilitated; increased leachate seeped first into the groundwater and then on into the open waters of lakes and rivers. The introduction of water-borne systems for the transport and disposal of sewage from households transformed the scale of urban water exchanges as well as contributing to the degradation of the quality of the receiving waters in a way which did not happen with cess pits or septic tanks (Kluge & Schramm 1995).

#### 2.3.3.2 Phase II: industrialisation from 1950 until today

Irreversible material losses have increased enormously since 1950 and, with them, the stability and sustainability of ecosystems has been correspondingly reduced. The increase of agricultural production ('the green revolution') since the 1950s is a substantial causative factor. Productivity was impressively increased through the enlargement of fields (redistribution of farmland), the installation of monocultures (single crop farming), the use of inorganic fertilisers, the increased use of plant protective agents and seed preservatives. Ameliorative measures such as land improvement, soil drainage and deep ploughing were, however, the main agents for degradation in terrestrial ecosystems. Leaching processes in the soil, initiated in the preceding first phase of industrialisation, have been substantially increased and over huge areas. At the same time, material losses have been augmented through the routine operation of point discharges to water courses (purification plants, industrial disposal). The transport of mineral salts, heavy metals, organic pollutants, biodegradable organic substances and, inevitably, nutrients is greatly enhanced through this medium.

The increased movement of traffic, both public and private, in and out of cities is a further factor in increasing wash out of matter into surface waters. The consequences of increased use of lakes for recreation have come not just from the physical damage to bank areas but from the introduction of alien plants and animals (e.g. Canadian waterweed, *Elodea canadensis*; grass carp, *Ctenopharyngodon idella*; zebra mussel, *Dreissena polymorpha*). The destabilisation of socialised coenoses which these invasions caused has coincided with a disturbance of the material cycles in waters. This is demonstrated with particular clarity by experiences with the introduction of the grass carp, which entailed not just intensified development of phytoplankton (Kucklentz & Hamm 1988) but increased sedimentation rates and accelerated lake infilling.

Waste-water disposal was never integrated into a sustainable landscape function. Increase in the urban connection to purification plants and the gradual improvement of the purification technique do not increase the sustainability of treatment. The useful matter available in waste water (nutrients and mineral matter) is still either mineralised, released as gases into the atmosphere, or exported in solution to receiving waters or is precipitated from the water. The nutrient-containing sludges of sewage treatment are more often treated as hazardous or special waste because of its contamination with trace matter. Thus life-cycle management of nutrients and utilisable matter is hardly practised at all.

## 2.4 NATURAL FUNCTIONALITY AND THE ASSAULT ON LAKES AND THEIR CATCHMENTS

In this section, natural processes in lakes and the damage inflicted by anthropogenic assault are considered in the context of the ETR model. Since energetic interactions become effective mainly at interfaces, the importance of the liquid–solid boundary layer for metabolism with little loss and high efficiency is stressed. Thus, the littoral and benthic zones are of particular importance in lake function.

Because the ETR model treats sustainability in terms of chemical efficiency (cf. section 2.2), damage to ecosystem function can be considered by reference to the same measures: energetic potential, transport at various fractal levels and spatial scales, the dissipation of energy by processor properties of water, their spatiotemporal coupling and self-organisation.

Functionally intact oligotrophic lakes are considered to be indicative of a sustainable, functioning landscape. Functional integrity is, of course, an outcome of the sum of the metabolic processes in the lake and not some ideal of a classification system of water quality. Here, we make the case only for intact shallow lakes, defined according to Wetzel's (1983) relative-depth criterion, that the maximum depth is <4% of the median lake diameter. Typical examples are found in the areas of the northern temperate zone which were glaciated during the last Ice Age.

The intact lake is distinguished by receiving only minimal material loads from its catchment area and by supporting almost closed matter cycles within the aquatic part of the ecosystem. This control of the lake by organismic metabolism is manifest in a low, almost negligible, rate of sedimentation. In contrast to the situation of intact lakes, the poorly functioning catchments which have become wide-open, deliver substantial material loads to lakes, causing considerable metabolic disturbance. Organismic regulation of the metabolic processes in the lake becomes overburdened. Thus disturbed, the lake loses control of its matter-loss function and shows increasingly the symptoms of rapid ageing.

Intact lakes, along with their closely coupled intact landscapes, are particularly valuable for species diversity (Ripl 1995c). Permanent, mixed-forest vegetation, abundant litter and humus, efficient water retention and temperature regulation all make for smooth ecotones and stable habitat for appropriate species. A highly diversified floristic and faunistic zonation is favoured by these low moisture gradients, allowing the strengthening of the network of organisms and the improvement of efficiency. At the zone boundaries, where gradients of turnover are more marked, successional processes contribute to optimisation through enhanced efficiency. Examples of such structures are coral reefs and periphyton communities. High species diversity and organismic abundance are found at higher levels of the food web, where persistence of rarer species is a consequence of habitat complexity.

### 2.4.1 Disturbance pulses in the ecosystem as a cause of damage to lakes

Transport of dissolved and particulate material to lakes occurs mainly through the medium of water flow. Thus, besides the vegetation and energetic relationships of the catchment, the delivery of material loadings to the lake is sensitive to the hydrograph. Since matter loss from the catchment is primarily a surface process, the ratio of catchment area to lake surface is a key quantity in assessment of lake behaviour (Ohle 1971). For the lake itself, the notional water-retention time is the most relevant to ecological processes, although actual exchanges may be subject to short-circuiting flows. These may be temporal as well as spatial, with the variability in flows from increasingly degraded catchments widening to constitute distinct pulses of water flow and material loads.

In self-organising systems, processes and structures are *not* random, although they do damp out regularly recurrent signals (such as diel temperature variation and annual variations in discharge). The system reacts to and accommodates according to the patterns established, so that the outputs tend to a steady state, reflecting the average with a decreased variance. The existing patterns also govern the survival and selection of organisms and they, in turn, increase the damping, dissipative characteristics of the system, thus consolidating the structure.

So, when the amplitude of variability is increased, or its periodicity becomes more stochastic, the likelihood of usurping the steady state is compromised. Pulses become system disturbances when the biological tolerances of organisms of the present assemblage are no longer able to tolerate or adapt to the extremes of the imposed variability. When they fail, there are mortalities, perhaps of disastrous proportions: mass kills of fish as a result of abruptly lowered concentrations of dissolved oxygen provide a graphic example. At a larger scale, increased pulsing of the material losses from the land to the water, as a consequence of anthropogenic management of the landscape, strain and, ultimately, reset the natural forces for optimisation. Such destabilisations affect catchments and receiving waters alike. The symptoms of destabilisation in lakes are manifest in accelerated rates of sedimentation.

### 2.4.2 Loading pathways to lakes

The following pathways for materials to lakes are recognised.
- Atmospheric influx. Extent depends on the temperature damping (lowering of energy flux density) in the surrounding landscape. Spatial temperature differences are minimised under optimal temperature attenuation by evaporation and dew formation. Compensatory movements of air, which take place between the drier (uncooled) and the moister (cooler) areas, are generally low.
- Loading of lakes with fine, airborne dust is slight in cooled landscapes.
- Loading via inflow. Both the inflow and their attendant material charge depend on the attenuation characteristics of the surrounding landscape with regard to average and variance. The type of materials carried into the lakes also relates directly to the attenuation.
- Loading via bank erosion and groundwater inflow or interflow. Influenced by elevation and material composition, loadings to lakes reflect the attenuation characteristics and landscape dissipation of the energy flux density. The condition of the bank greatly affects loading to the lake through its vegetation and retentive capacity.
- Direct loading of the lakes by, for example, accidents or material dumping. This loading usually occurs irregularly and therefore exhibits no recognisable temporal pattern. Usually it is classified as being quantitatively more or less important to the material loadings from groundwater and intermediate discharge.

Two substantial sources of loading to lakes are generally recognised, conforming by analogy to the above pathways: external point sources, where materials enter the system at discrete locations; and diffuse sources, where material entry occurs as a background over a wide area of landscape.

### 2.4.3 Atmosphere and climate

#### 2.4.3.1 Water cycle

The atmospheric water circulation was changed extensively by the removal of large areas of forest vegetation and by far-reaching drainage operations which increased the amount and rate of rainfall runoff. The elimination of evaporating structures in the landscape and cities, the reduced water retention capacity of topsoil and the long-range fall in water tables contribute to shortages of water at the soil surface in summer and to diminished rates of evaporation. There are several dependent reactions. During the day, the land surface loses its cooling function owing to lack of evaporation. On dry, vegetation-free surfaces (e.g. cities, uncovered areas of arable land), surface temperature might vary by more than 10°C from structures evaporating water (e.g. forests). The global distribution of evaporative structures is evident from temperature-sensitive remote sensing and has been used in the assessment of the thermal efficiency of landscapes (Ripl et al. 1995, 1996).

Reduced evaporation ashore and accelerated discharges to the sea might also influence the runoff to coastal and shelf waters. In the former West Germany, a discharge yield equivalent to 313 mm yr$^{-1}$ from a mean precipitation of 837 mm yr$^{-1}$ establishes an evapotranspiration of 524 mm yr$^{-1}$ (Keller 1979). This runoff coefficient is 37%. After corrections for an underestimation of precipitation of 15–20% on average, the runoff coefficient is reduced to c.32%. From this, it follows that about two-thirds of the precipitation on the surface of the former West Germany re-evaporates again. Similar conditions can be diagnosed on the global

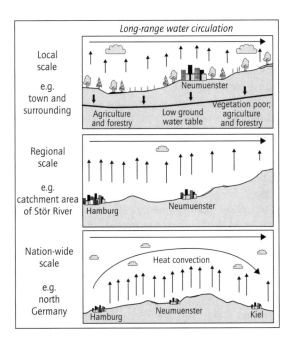

**Fig. 2.6** Long-range water circulation (cf. short-circuited water circulation, Fig. 2.4 in section 2.3.1.4). (Redrawn from Ripl et al. 1995.)

scale for mainland areas with surface discharge. The mean global precipitation is estimated to be 924 mm yr$^{-1}$, mean evaporation at 529 mm yr$^{-1}$ and runoff at 395 mm yr$^{-1}$ (Dyck & Peschke 1989); runoff coefficient is c. 43%.

Under modern conditions of superficial landscape desiccation, the short-circuited water cycle of evaporation and condensation is anthropogenically disturbed and more spatiotemporally separated than under natural conditions (Fig. 2.6). This damage to the hydrological cycle affects climate (failure of cooling function in sunshine), dust generation and temperature-regulated mineralisation and loss processes in the soil.

#### 2.4.3.2 Pollutant emissions from heating surfaces

In catchment areas with soil degradation the maximum depth of the groundwater table and water drainage in the soils have fallen below the root zone of trees. Thus limited, a lower rate of evapotranspiration permits an overheating of the landscape, particularly in summer. Surfaces with reduced temperature damping exhibit a strong emission of intermediate reaction gases (nitrogen oxides, partly methane, ozone formed near the surface). By compensatory movements of air from warmer (drier) to cooler (moister) areas, the damper areas are increasingly loaded with fine dust on which organic and inorganic pollutants can be adsorbed in high concentrations. In this way, even pollutant-contaminated surface sediments occur in areas well beyond their industrial sources. A useful example comes from Pechsee, in Berlin, Germany.

Pechsee is in the centre of a protected area, inaccessible to the public. It is situated in a kettle-like valley completely wooded with pines, oaks and birches, and is surrounded by a mire with typical vegetation. The lake is 0.8 ha in surface area, with a catchment of approximately 4 ha and no superficial inflow. It is situated in the Grunewald, a large, dense recovery forest. It is more than 800 m from the nearest highway. Nevertheless, its surface sediments are enriched by heavy metals to remarkable concentrations: Zn 1100, Pb 420, Cu 120, Cr 22 and Cd 6.0 $\mu$g g$^{-1}$ dry weight (means of 11 to 12 samples in each case; unpublished data of Ripl et al.).

### 2.4.4 Catchment area and landscape

#### 2.4.4.1 Spatial distribution of land use

Present ageing of the landscape and its lakes no longer relates to their pristine development. Rather, its development bears the characteristics of randomness with a marked lack of capacity to dissipate the precisely structured natural energy pulse. Anthropogenic management of land and waters is not adapted to natural phases but, rather, to technical and economic requirements. Vegetation systems, optimised over long periods of water balance, are extensively biodegraded, with separated functions and open matter cycles.

Modern land evaluation is based on its capacity for short-term economic exploitation, which, ironically, is the main driver for the destruction of its landscape functionality. Such use of land opposes the sustainable functioning of landscapes. Use as arable land leads directly to high material leakage, usually via a drainage system feeding streams and rivers. With such disregard for landscape matter flows, the ageing of the landscape and its transformation to desert is not accelerated linearly. Full landscape functionality is possible only with an intact landscape water balance, with its flattened regional moisture extending to and over its ecotones with surface waters. From an ecological perspective, the present distribution of land use is not coincidental and cannot be adapted. Monetary valuations which ignore ecological functionality must, sooner or later, damage the landscape and its rivers and lakes.

An example of this is the use of lake or river banks, which are important matter retention areas of a landscape. Yet they are important to recreation and are often overused. Redevelopment of the retention function of these areas, through which sustainability of the adjacent catchment area might be increased, would require generously dimensioned protected zones of 100 or 200 m along the banks.

### 2.4.4.2 Interference with the landscape water balance

High uniformity of runoff discharge and strong attenuation of flood water is said to characterise intact, perennially vegetated catchment areas (Dyck & Peschke 1989; Risser 1990). Uniform discharge is not a feature of damaged landscapes. Material and nutrient loadings from natural landscapes are lowered during summer, owing to filtration of flow by vegetation and to the short-circuited water cycle. In contrast, degraded landscapes may not maintain such nutrient-poor discharges.

The consequent strain on landscapes and lakes can be classed as a disturbance to the water cycle or as a malfunction of energy dissipation. These disturbances are initiated as hydrological interventions on the catchment scale. They include:
- drainage of arable land and forested areas;
- drainage of swamps, riparian forests, forests and wetlands with a network of ditches;
- groundwater abstraction of drinking water;
- waterborne sewage system and central purification plants;
- river straightening, deepening and hydraulic engineering of rivers;
- regulation of water level in lakes, artificial lakes, navigable rivers and channels;
- open-cast mining with associated lowering of groundwater level.

Lakes also act upon their own hydrological regime. The intact lake damps the fluctuations in lake level and the discharge to its outflows. Whilst the lake maintains a more regular water level through the year, the amplitude of successive wet–dry phases in the surrounding soil is also kept low, as is the leaching of minerals. Water in the ground but above the water table, not usually considered as 'ground water' but regarded as 'process water', is involved in a high intensity of processes (Fig. 2.7). Fluctuations in level increase the intensity of these processes, leading to oscillations between aerobic (aerated) and anaerobic conditions (saturated soil). Microbial degradation processes are strongly favoured by these alternating wet–dry phases. Thus, the 'pumping movement' of water level in the soil leads to a better supply of reactants (e.g. electron acceptors) than is found at steady water levels and, in consequence, accelerated loss of the products of bacterial metabolism.

The process intensity of water in the soil is clearly weaker in a low-loss, fully functioning ecosystem. The groundwater is characterised by a slow, relatively steady discharge (contributing the base-water flow of rivers) with quite small fluctuations of the water level. Moreover, the superficial organic enhancement of an unimpaired landscape favours discharge close to the soil surface. During the passage of water through wet, near-surface soil layers, its oxygen content is consumed rapidly. Oxidant anions (nitrates, sulphates) are used in the anaerobic oxidation of carbon. Carbon dioxide is a product of oxidation and hydroxyl ions are released as a consequence of the reduction of anions. The latter lead to an increase of pH, encouraging the

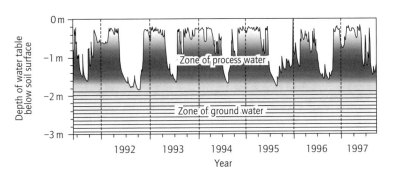

**Fig. 2.7** Functional description of process water and groundwater. Differentiation of a dynamic zone of 'process water' and a non-dynamic zone of 'true groundwater' by the periodically fluctuating water level in the soil (Himmelreich, Stör catchment area, Germany).

immobilisation of base cations. With the development of water-storing soil structures, oxidising, acidifying and base-dissolving processes are decreased and matter retention (denitrification, desulphurisation) is increased.

Today, groundwater levels generally stand lower in the central European landscape than they did in recent (historical) times. Former higher levels were due to the lack of artificial drainage of the unmodified natural morphology of running waters. Small level fluctuations occur in semi-natural streams and rivers with extensive floodplains. In such undisturbed landscapes, a greater proportion of the precipitation was processed in the litter or humus. With floodplain discharge much nearer to the surface than is the case today, the proportion of groundwater discharged to rivers and lakes was smaller and, with it, the proportion of electrolyte-poor, superficial discharge owing to precipitation was clearly higher than is now the case.

Modern modifications to rivers include straightening, deepening and engineering of banks. A rapid discharge of water is favoured in wall-sheeted or narrowed channels. As a function of the precipitation, which is erratic in the anthropogenically deformed landscape, discharges and water levels vary conspicuously. Variation is apparent also in the adjacent riparian zone as a consequence of fluctuating soil-water level. Enhanced discharges modify drainage effects, including the rapid runoff response to precipitation. Altered drainage morphology raises the proportion of process water in the soil (dynamisation of soil water), with accelerated losses of water-conserving soil components (e.g. clay translocation, respiration of organic substance).

When water can no longer be retained near the soil surface and confined to local evaporation–condensation cycles, the formerly attenuated discharge pattern is replaced by a steepening hydrograph response. This leads to increased loadings to receiving waters. Especially in autumn, when the proportion of the precipitation appearing as runoff relative to evaporation is increased, materials set free by mineralisation (nitrate, sulphates, base cations) are readily transferred to surface waters and on a greater scale. Conditions of discharge generation in a natural and an anthropogenically modified landscape are illustrated in Figs 2.8 and 2.9, respectively.

### 2.4.4.3 Soil and agriculture

Modern processes of material fixation, loss and leaching mechanisms must be considered in the context of water balance. Material losses under intensive agriculture, forestry and extensive groundwater abstractions are high. Soils are converted from the hydrologically balanced interface between geological subsoil and vegetation with an optimised water balance to an industrial medium for maximised agricultural production. Prior to industrialisation, any increase in agricultural productivity was gained at the expense of drainage and oxidation of nutrient-rich, organic soil stores. Whilst organic accumulation was favoured

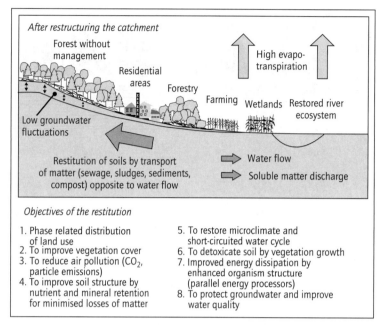

**Fig. 2.8** Landscape processes. Former near-surface discharge from the landscape. (Redrawn from Ripl et al. 1995.)

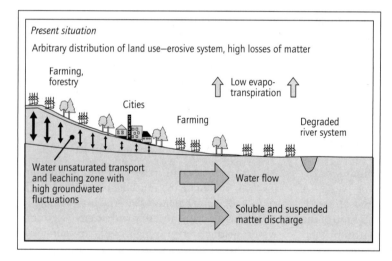

**Fig. 2.9** Landscape processes. Present discharge from the landscape characterised by groundwater. (Redrawn from Ripl et al. 1995.)

in former times, today it has been reversed. In around 100 years, large areas of land have experienced mineralisation and washout, followed by the use of fertilisers and irrigation, supplied by the input of accessory (mostly fossil and nuclear) energy.

Effects on the lime–carbonic acid equilibrium and the solubility of bases, sulphur, nitrogen and phosphorus parallel the oxidation of organic materials. The formation of strong (e.g. nitric, sulphuric) acids is no longer countered by bases or high water tables. At the same time, enhanced

Table 2.2  Sulphate and chloride concentrations in the River Spree in the Berlin region

|  |  | 1890–1895 | 1926 | 1982–1993 |
|---|---|---|---|---|
| Sulphate (mg L$^{-1}$) | Single value |  | 22 |  |
|  | Mean | 20 |  | 160 |
|  | Median | 17 |  | 159 |
| Chloride (mg L$^{-1}$) | Single value |  | 25 |  |
|  | Mean | 30 |  | 56 |
|  | Median | 28 |  | 54 |
| Number $n$ |  | 9 |  | 163 |
| Source |  | N.N. (1895) | Höll (1928) | Ripl et al. (unpublished) |

respiration increases the partial pressure of carbon dioxide in soil. In consequence, the equilibrium conditions for base solution of bases and phosphate are moved upwards.

Measurements from the Berlin region demonstrate accelerated landscape losses of anions. Investigations show (see Table 2.2) that sulphate concentration in the River Spree has increased eightfold and chloride by a factor of nearly two, since the end of the nineteenth century. In addition, phosphorus concentrations of up to half a milligram (e.g. Kloos 1986) were found in the groundwater.

In Germany, the landscape water balance is disturbed substantially by the open-cast mining of brown coal (lignite). The soil is heavily leached owing to a severe fall in soil water levels, leading to steppe formation, as well as degradation of the Spree River catchment in Lower Lusatia. Strong flows of material in the catchment area have occurred with the oxidation of pyrite-containing mine tailings following the lowering of groundwater levels. The sulphuric acid formed from pyrite is especially effective in leaching calcium and other bases from the soil, and accounts for the strong increase in sulphate concentrations in the River Spree. In the Havel River above Berlin, whose catchment area is characterised by agriculture and forestry and with few larger settlements, solute concentrations of up to 85 mg L$^{-1}$ sulphate and 37 mg L$^{-1}$ chloride have been encountered between 1984 and 1993 (Ripl et al., unpublished).

In other German rivers, the carriage of base cations reflects the dynamisation of soil processes.

Table 2.3  Outline of the minimum and maximum yearly transport of base cations (Ca, Mg and K) of the river basins in Germany (at Danube without potassium). List after Ripl et al. (1996)

|  | Minimum annual removal (kg ha$^{-1}$ yr$^{-1}$) | Maximum annual removal (kg ha$^{-1}$ yr$^{-1}$) |
|---|---|---|
| Elbe | 150 | 193 |
| Rhine | 205 | 437 |
| Weser | 157 | 437 |
| Danube | 255 | 476 |
| Ems | 101 | 342 |
| Vechte | 138 | 458 |
| Areal mean | 185 | 367 |

The combined losses of calcium, magnesium and potassium from the river basins of the Elbe, Rhine, Weser, Danube, Ems and Vechte are equivalent to between 185 and 367 kg ha$^{-1}$ yr$^{-1}$ (see Table 2.3). Of this, about 78% by mass is calcium, 14% magnesium and 8% potassium (Ripl et al. 1996). This loss is not compensated by application of conventional N–P–K agricultural fertilisers.

It must be emphasised that it is not only urban agglomerations which experience high matter losses with runoff, but also their supplying hinterlands. Whilst the electrical conductivity of waters discharged from optimised landscapes may fall in the range 10–40 $\mu$S cm$^{-1}$, they rise to some 600–700 $\mu$S cm$^{-1}$ in areas of intensive agriculture; local peak values may exceed 1000 $\mu$S cm$^{-1}$ (Ripl et al.,

Table 2.4 Material losses of base cations and nutrients during the Holocene (recalculated after Ripl & Feibicke 1992)

| Phase and age | Development of reactivity and concentration of charges in recipient waters | | Loss factor of bases | Loss factor of nutrients |
|---|---|---|---|---|
| | pH | Conductivity ($\mu S\,cm^{-1}$) | | |
| Phase 1: pioneer vegetation 12,000–9000 BC | 6.5–8 | 20–200 | 2–3 | 2–10 |
| Phase 2: development to climax conditions, 9000 to c. 1500 BC | 6.5–8 | 10–40 | 1 | 1 |
| Phase 3: anthropogenic development, c. 1500 BC to AD 1850 | 4–9 | 30–300 | 2–5 | 5–100 |
| Phase 4: desertification, AD 1850–2000 | 3–11 | 50–1000 | 5–100 | 10–1000 |
| Phase 5: restitution before the next glaciation? After AD 2000 | 4–10 | Decreasing | Low | Decreasing |

unpublished). Ripl (1995a) proposed that base cations are being leached from central Europe roughly 100 times more rapidly than they were at the Atlantic climax, 7000 years ago, and that the equivalent multiplier for nutrients is nearer 1000 (see also Table 2.4).

When consideration is given to the scale of losses in relation to the longevity of greatly increased agricultural productivity and the shortened opportunity for useful life of the land, serious concerns about the sustainability of human exploitation must arise.

### 2.4.4.4 Acidification

Acidification of lakes and forest soils was originally supposed to be due to precipitation. Rainwater possesses a natural proton concentration, typically corresponding to a pH of 4.5. Acid precipitation plays a major interactive role with exposed surfaces, a process which has been affected by changes in water balance. Well-known examples of areas affected include the Krusne Hory, or Erzgebirge, separating the Czech Republic and Germany, as well as those areas which have been subject to high industrial emissions of sulphur dioxide. However, the extent of acid damage and its proximal causes are evaluated separately in many instances (cf. Kandler 1994; Ellenberg 1995).

Matter flow charts show that the proton and cation quantities carried in precipitation did not correspond at all well to the quantities delivered into rivers and lakes, via soils. Rather, discharges of these materials were increased, mainly as a consequence of the soil acidification owing to the high net productivity and forestry use (Ripl et al. 1996; cf. also Alexander & Cresser 1995). Accordingly, receiving waters also experienced increasingly acidic hydraulic loads. With the impoverishment of cations from the landscape, vegetation is forced to consume more energy (photosynthetic product) in the gain of mineral matter. Processes in the root biomass rely on small amounts of nutrient and a supply of bases for their reduction. With nitrogen and sulphur figuring in these exchanges, the formation of strong acids (nitric and sulphuric acid) contributes to the loss of bases. Acid production modifies the soil environment and contributes to material wash-out.

The temporary local decrease of pH also dissolves metals in soil, favouring further protolysing reactions (Stumm & Morgan 1981), for instance:

$$Fe^{3+} + 3H_2O \Rightarrow Fe(OH)_3(s) + 3H^+$$

Disturbances by acidification are observed in many smaller forest lakes in the spring months if dissolved humic matter, protons and metal ions are carried into receiving lakes. These materials affect the feeding of fish and interfere with fish reproduction (spawning), especially in the riparian zone. Other damage may include poor development of plankton and periphyton and disruption of the age-class structure of animal populations. Gradually, acid-sensitive fish types have disappeared completely from the lakes affected, or the age structure has changed in favour of older, predatory survivors, whilst the crustacean basis of the food chain for the younger planktivorous fish is eroded. Increasingly, the visible symptoms of lake acidification are recognised to be attributable to damage to catchments.

Despite the growing problems associated with landscape management, measures which favour the loss of base cations continue to be followed. Nitrogen loading, a primary factor, is still being increased by precipitation (Ellenberg 1995; Flaig 1996). This favours increased production generally but particularly in forestry. However, as the trees continue to demand a greater supply of bases, the weathering processes in the root zone must respond to the lowering of soil pH.

### 2.4.5 Running waters

#### 2.4.5.1 Boundary layer between atmosphere and water

The interface between water and the atmosphere is important for the mass transfer of gases which participate in metabolism and are more or less soluble in water. As gas exchange is a function of surface area, engineering variations in the surface/volume ratio of water bodies, such as by deepening or straightening of watercourses, distinctly affect their gas metabolism. Effects on the metabolism of water bodies because of the absence or surplus of oxygen, or of carbon dioxide, and of reduced gases such as methane and hydrogen sulphide, are far-reaching.

#### 2.4.5.2 Damage to river morphology

Engineering measures such as deepening, straightening or bed regrading are generally applied to

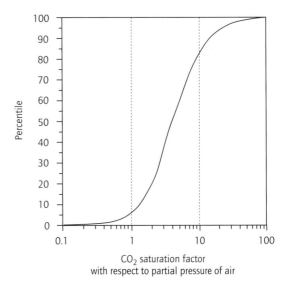

Fig. 2.10 Distribution of the carbon dioxide saturation factor in the waters of Berlin. Sample depth 0–1 m, 1980–1995, $n = 8178$. (Data Department of Limnology, Technische Universität Berlin.)

increase flow and transport capacity. In so doing, important material-retention properties, backwater coenoses and river-bed structures have been obliterated. Instead, long-range translocations of solid and dissolved materials are encouraged, along with strongly fluctuating discharge patterns, with resultant increased material loads to receiving waters downstream.

#### 2.4.5.3 Changes to the carbonic acid equilibrium

For much of their length, running waters are characterised by oxidative processes: they are, on net balance, heterotrophic. They exhibit almost permanent supersaturation with respect to carbonic acid. In waters of the Berlin region, a carbonic acid saturation factor of the water of less than 1, testifying to a carbon-dioxide content balanced to the atmosphere, lies below 6% in 8000 samples. The median readings show that the factor of supersaturation with carbonic acid is around 4, i.e. 400% (Fig. 2.10). Causes include the accelerated carbonate leaching of catchment soils, the organic load to water and the small amount of carbon-dioxide consuming load of organic material into waters and, above all, a weakened rate of exchange of carbon dioxide between water and atmosphere. Channel engineering has greatly reduced the boundary layer between the surface of waters and the atmosphere and, with it, reduced the width–depth variance of the reworked channels. Water velocity is also accelerated. Free exchange of the carbonic acid partial pressure with the atmosphere is impeded. The precipitation of carbonate and silicic acid, which could occur under an atmospherically equilibrated carbonic acid concentration, is also reduced as a consequence.

### 2.4.6 The riparian ecotone, between land and water

#### 2.4.6.1 The functional view of littoral structures

In the intact lake, the highest process dynamics are seen in the littoral zone. Compared with the pelagic zone, water turnover in the littoral zone is encouraged by turbulence and matter turnover is more short-circuited by high densities of macrophytes and periphyton. The littoral zone provides more benign environments with respect to exposure, light, oxygen supply and nutrient availability than does the body of the lake. The favourable conditions select in favour of the establishment of many kinds of organisms, of rapid rates of material turnover and for high rates of community metabolism. Here, periphyton ('Aufwuchs'), characteristic of the littoral zone, settle on the surfaces of living plants or dead substrate, and soon become a main autotrophic constituent. Under oligotrophic boundary conditions, periphytic biomass may be small but it is pivotal in coupling autotrophic and heterotrophic processes and in developing short-circuited, low-loss cycling, with high chemical efficiency.

The component producers (e.g. reed), detritus (organic deposits), decomposers (bacteria), a consumer network and water (as the medium of transport and reaction) comprise a viable, dissipative ecological unit (DEU) in the littoral zone (cf. section 2.2.4.1). Once again, we see the work of a self-optimising structure, dissipating the solar power effective in the day–night pulse and progress towards the closure of matter cycles within the habitat.

The macrophytes are of special importance to the self-optimisation of the littoral DEU in littoral zones. They maximise the living space and bring close spatial and temporal coupling of substance decomposition and production, with feedback to the water and matter balance in their location. They increase the water retention capacity of the soil in the adjacent banks above the mean water line, through organic deposits and microclimatic sheltering. With increasing soil moisture, mineralisation processes are frequently restricted to the vicinity of the root zone, but where, through plant transpiration, drier phases may well arise. Matter cycling is locally restricted and losses are small, so materials are conserved and stored.

Detritus is likewise enriched on the water side of the littoral interface. By positive feedback, the water becomes shallower and amenable to the establishment of further macrophytes. Thus, the plants increasingly affect the material processes within the rooted zone. They mediate the water transport and oxygen supply in the former anoxic sediment through transpiration. Mineralisation near the roots is favoured and the nutrient supply to support herbal production is encouraged.

Macrophytic development influences the production and distribution of littoral detritus, equalising water and matter balances and generally flattening the ecotone. Diversity and community stability also benefit from this development: as environmental gradients are smoothed and habitat variability occurs on ever finer (niche-level) scales, so the strength of the network is compounded (Ripl 1995c).

Structural robustness is also important as a defence against mechanical erosion. Separation of the littoral from the profundal zone is frequently marked by the limit of accumulation of fine sediment and the boundary between soft sediments and the more minerogenic deposits of the exposed areas of the littoral zone. In lakes where the matter loading is within the natural control range of the aquatic biocoenoses, the effect of light on the soft deposits may permit the development of submerged aquatic plants. However, a significant underwater vegetation of *Lobelia dortmanna*, *Littorella uniflora* (shoreweed) and *Isoëtes lacustris* (common quillwort) is to be found only in oligotrophic soft waters, whilst, in carbonate-rich, marl-forming lakes, the main plants may be Characeae (see also Volume 1, Chapter 11). Submerged vegetation also resists erosion of the riparian zone by the absorption of wave energy and by binding and stabilising substrata through the root system (Wöbbecke & Ripl 1990). Moreover, this stability favours the more littoral sedimentation and the centripetal expansion of the entire littoral biocoenosis (Lundqvist 1927). The littoral zone functions as a self-optimising, dissipative unit, yet it is dynamic in terms of structural renewal, consistently refreshing the characteristics of the littoral dissipative ecological unit.

#### 2.4.6.2 Temporal changes to reedbeds — spread and decline

The composition and robustness of littoral structures and their durability are influenced decisively by the nutrient supply, as a space-limited function. The littoral development of rivers and lakes is just another facet of the material balancing function of the surrounding catchment. The linkages may be illustrated through descriptive models for the spread and decline of reed, based on work of Ripl *et al.* (1994a,b,c).

Colonisation of a river or lake bank with reed starts at the land–water ecotone. Nutrient concentrations, exposure to wind and waves and the dampness gradient work selectively to influence the rooting for torn-off rhizome fragments, stems or seeds. Thereafter, macrophytic development may proceed in accord within the rules of self-optimisation. The early development of macrophytes exemplifies the assembly of littoral structures functioning with low loss. This is noted first in the visible structure. A strong spatial connection between progressive zones (areas with vegetation increase) and regressive zones (areas with vegetation decline) links the various developmental stages, with optimal use of mineralised matter but a rather light development of the standing stock. A further feature of a low-loss function arises through the surface/circumference ratio of the lake, which progressively shortens the outer boundary of the reed bed. Normally, the outside edge is the site of the greatest material losses. On this, the landward side of the reed bed, losses are associated with the influence of terrestrial plants, as opposed to that of macrophytes on the water and the material balance of the inner margin, in decreasing water depth and motion. In this way, dense, closed reed beds contrast with the operation of sparser beds where losses are less severe and there is thus greater stability.

As water and matter cycles become increasingly closed with the unhindered development of the catchment, then more low-loss functioning communities are favoured in the receiving waters. At the same time, reed in the littoral zone gives way to space limitation and shading by the developing vegetation of the riparian zone (Ripl *et al.* 1994a) and may even be replaced by a new, more tolerant species. The main species within the littoral DEU must be compatible with the optimisation process of the adjacent unit of catchment. Only the types best-adapted to the changing boundary conditions will persist.

In this way, the spread and the decline of reeds are to be regarded as phases within the energy-dissipative, self-optimisation process of landscapes. The rules for the expansion of reed comply with the requirement of increased energetic efficiency. It follows that in order to explain the dynamic, spatiotemporal view of the system a knowledge of the ecological criteria for the occurrence and survival of organisms is required. On the other hand, knowledge of the autoecological habitat requirements of individual organisms is not enough to predict precisely in which systems and processes they will participate.

#### 2.4.6.3 Damage to the functionality of the littoral zone

What may we then infer about damage to riparian

zones? Impairment is associated with weakening of material retention ashore and the higher strain placed on metabolism within the lake itself (Odum 1990). Destruction is correlated with modifications to the processes governing plankton development and sediment processes, concomitant with a decline in overall efficiency. Damage to the bank structure favours a shift in the spectrum of fish types towards planktivorous species. This shift is almost always accompanied by an increasing development of the planktonic biocoenosis and deterioration of the oxygen budget of the profundal zone. Functionally, the nutrient (space) limitation on planktonic production is alleviated whilst the additional biomass which may be produced raises the export from the pelagic zone, and the chemical efficiency of the lake is diminished.

### 2.4.6.4 Regulation of the water level of lakes

Usually, the surface level of a lake corresponds to the soil-water level in the adjacent catchment. Many lakes today possess regulated water levels, in order to secure flood protection and different water relations in the adjacent managed areas of land. The same applies to artificial lakes which, for the most part, are the products of interference with the landscape and/or its water balance. Lakes in former gravel pits or pits left by open-cast mining and reservoirs may be included here. The adjacent soil-water balance is affected negatively by the regulation of the water level. Enlargement of the area subject to wet–dry alternation leads to increased leaching of bases and nutrients in the seepage water (as described in Section 2.4.4).

Constant water levels secured through dam regulation may also negatively affect the riparian zone. Since, over a period of time, waves mostly attack the same parts of the shoreline, locally intensified erosion may be forced. If it is reinforced, then there may be impacts on the totality of metabolic processes in the water body.

### 2.4.6.5 Disturbance and removal of retentive structures

Often, the riparian vegetation along the landward side of banks of rivers and lakes was removed to a large extent during the phase of industrialisation. Accordingly, the ability of these belts to retain materials washed from higher ground in the adjacent catchment area has been diminished. Moreover, the habitat for periphyton was sacrificed. Both landward-side vegetation and periphyton support the development of material short-circuits countering the material losses by erosion and profundal sedimentation. The impaired function of littoral vegetation also weakens the capacity to intercept subsurface flow (interflow) and the materials washed out from catchments, which may now be transported to the sea.

### 2.4.6.6 The reed decline

Substantial decreases in reed cover in the littoral zone of lakes in central Europe have been reported during the second half of the twentieth century. Ripl *et al.* (1994a,b,c) investigated the example of Havel River, in Berlin. During the phase of reed expansion, processes are characterised by strengthening internal cycling. Contrariwise, the phase of reed-stand decrease is marked by changes in the external influences of water quality and currents. Production and mineralisation become increasingly decoupled. The attendant material losses and water flows favour the establishment of alternative, better adapted plant structures. Whilst the spread of reed through nutrient limitation is a particularly slow process, the retreat often occurs in substantially shorter periods. Successors on the landward side include communities of sedges whilst, on the water side, nitrophilous filamentous algae and phytoplankton develop, symptomatic of eutrophication.

The water level of the Havel River is maintained for navigation through the use of weirs and sluices. Discharge fluctuations at constant water level are manifest in velocity variations, which tend to attack and erode at the same littoral sites. The effect is magnified by the wash of passing ships. The pressure on the littoral zone and simultaneous increases in nutrient loads have favoured the development of algal blooms and the almost complete loss of submerged vegetation.

Natural hydrological patterns in the Havel River are disturbed by expansion of well networks for groundwater abstraction. In their vicinity, organic substances are mineralised and cohesive minerals (silts, clays) dry and crumble, allowing increased vertical seepage. These areas experience a diminution of water-retentive capacity. This mechanism has contributed to reed decline in several locations (Markstein & Sukopp 1980; Wöbbecke & Ripl 1990; Wöbbecke 1993).

Apart from these obviously recent findings, ancient instances of reed decline are known. In about AD 800, the littoral zone of the Havel abutted an intact riparian forest, with heavy organic debris. In this condition, water drainage was close to the surface and carried lower concentrations of solute. Reed propagation was restricted by shading and by low nutrient availability. Following the German settlement of the Berlin region, approximately 800 years ago, the riparian forest was removed and the banks of the waters were used as meadows and pastures. The litter layer gradually retreated uphill in response to intensified use of the land. The consequence was the deeper percolation of rainwater and an increased nutrient load into Havel River. Given the enhanced light and nutrient conditions, reed spread strongly. With regional industrialisation and intensified use of the river, the decrease of macrophytes and the increase of algal blooms became inevitable. Reed growth may have been at first encouraged by the eutrophication but the connection between production and decomposition was already broken.

The cycle of propagation and decline of reeds cannot be divorced from the overarching changes to the water and material balances at the catchment level. Moreover, small, intermediate hydrological modifications, such as more sensitive river regulation (mud polders, public bathing areas), scarcely off-set the long-range trends of human interventions on landscape water balances.

### 2.4.7 Disturbances to internal processes in lakes

Internal disturbances refer to impacts which lower the efficiency of material balance within the body of a lake. Normally, the balance reflects the chemical efficiency (section 4.2.3) of matter retention versus loss. Reduced efficiency may arise through enforced adjustments in the flow and relative enrichment of the groundwater (sections 2.4.4, 2.4.6). Other perturbations of the biotic communities are caused by interventions to the littoral vegetation or to the stocks of fish. Internal disturbances generally coincide with altered external circumstances, especially changes relating to increased rates of material inwash (cf. section 2.4.6).

#### 2.4.7.1 Disturbances of stratification and hydraulics in lakes

Many interventions in lakes produce intended or unintentional effects on hydraulic conditions. Aeration and artificial circulation are used to override stratification in lakes and reservoirs or extend mixing, principally to transport of oxygen and other electron acceptors to the hypolimnion or bottom sediments in order to stabilise the redox conditions. Metabolism at the sediment surface may be enhanced on a temporary or medium-term basis.

However, under certain circumstances, the effects of such measures can be harmful. For instance, easily degradable organic substances may exhaust the acceptor capacity to below the threshold of denitrification, leading to the release of iron, manganese and phosphorus into the water.

Decomposition processes which locally exceed re-aeration can also produce some undesirable results. Substantial quantities of reduced sulphur were formed and deposited in the discharge water when hypolimnetic water was withdrawn by an application of an Olszewski tube (Olszewski 1961) from the Piburger See in Austria. The quantities of sulphur involved would hardly have developed to this extent under stagnating conditions in the lake, invoking the explanation of limitation of the sulphate supply in the discharge and the negative feedback of further hydrogen sulphide formation by sulphur-reducing bacteria. The rate of sulphur reduction is still from time to time increased by hypolimnetic withdrawals (Pechlaner 1979).

Analogous increases in hydrogen-sulphide formation were observed recently in Ploetzensee (Berlin), after an external plant for phosphate elimination had dynamised hydraulic conditions in the lake by withdrawal and return of water into the lake.

### 2.4.7.2 Eutrophication

The problem of eutrophication has received substantial attention in the limnological literature. Eutrophication generally refers to the increase of nutrient loading and production in the pelagic zone (Rohlich 1969). The process is usually explained in terms of increased phosphorus loadings to water. The basis for the well-known quantitative explanation was developed by Vollenweider (see e.g. Vollenweider 1968) from a statistical analysis of data collected mostly from larger, deeper temperate lakes. A primary causal connection was inferred between eutrophication and the point sources of phosphorus, most particularly those from plants purifying sewage to secondary standards (full removal of organic content and fully mineralised effluent). The result of this work and the work based on it created the demand for the provision of wastewater treatment which eliminated the phosphate content. Such tertiary treatment was quickly implemented in the highly developed, industrialised countries. Today, chemical purification in Scandinavia regularly achieves effluent concentrations far below $100\,\mu g\,L^{-1}$ P.

Against this, the original objective of restoring to lakes the previous oligotrophic conditions has not been fulfilled. Diffuse sources of phosphorus, together with that released from the lake sediment, have frustrated the success of reducing external point loads.

That the limiting role of nitrogen is less important practically for, as was shown by Schindler et al. (1973), the ability of micro-organisms to fix nitrogen moves the constraint back to phosphorus. Carbon is not regarded as a factor limiting the attainable biomass but its supply can limit rates of primary production.

### 2.4.7.3 Oxygen in the hypolimnion

The dynamics of the supply of oxygen in parts of the system isolated from the main movements of water, including especially the lake hypolimnion, play a large part in the metabolic dynamics of the system as a whole. Feedbacks between the benthic communities and the plankton, including the redox-mediated releases of nutrients to plankton production, are well recognised. The high sensitivity of these processes is such that small changes in the oxygen budget may affect the metabolism of the entire lake.

### 2.4.7.4 Benthic processes

The sediments of a lake comprise allochthonous (material carried in via inflows or the banks) and autochthonous (produced within the lake) components. The latter may comprise inorganic (biogenic or abiogenic marl precipitate, siliceous diatom frustules and debris) and organic (fine and coarse detritus, including the remains of littoral vegetation) fractions.

The energy content of organic deposits is compressed and its value as a substrate for bacterial processes is substantially confined to the sediment–water boundary layer. Depending on the supply of oxygen and other electron acceptors, bacteria can increase rapidly and supply the benthic food chain of heterotrophs: ciliates, rotifers, tubificids, midge larvae and fishes.

Whilst the food supply at the sediment surface is hardly a limiting resource in mesotrophic to eutrophic lakes, the oxygen conditions are of crucial importance (see also Volume 1, Chapter 12). Some red-coloured organisms provided with haemoglobin (tubificids, chironomids) can survive microaerobic conditions if hydrogen sulphide which is not free results from sulphate reduction. The boundary layer becomes otherwise closed to the activities of higher organisms, including fish.

The extent of oxygen depletion in the sediments is influenced by water movement and the renewal of suitable electron acceptors in relation to the degradability of the organic deposit. Even so, the rate of the transport-limited supply of electron

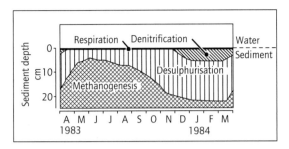

Fig. 2.11 Respiration processes in the anaerobic sapropel of the Schlei estuary.

acceptors frequently falls far behind the oxidative demand represented by the organic content of the deposited material. The accumulation of the reduced organic substance and the use of oxygen, nitrate and sulphate at the sediment surface leads to a characteristic sequence of the bacterial reduction processes beneath it. Seasonal variability of the gradients of electron acceptors in the sediment interstitial water of the flat, highly eutrophic Schlei estuary (Schleswig-Holstein, Germany) during 1983–84 is illustrated in Fig. 2.11. The data (from Ripl 1978; Ripl & Lindmark 1978) illustrate the rapid exhaustion of oxygen from the upper few millimetres of sediment and the subsurface distribution of responses of chemoautotrophic and heterotrophic reduction of nitrate and sulphate. In the winter months, denitrification characterises the upper 5 cm of brown-coloured sediment but, by spring, after nitrate is completely exhausted, it is succeeded by desulphurisation (sulphur reduction) to within a few millimetres of the sediment surface. The sulphide ions thus discharged precipitate with reduced iron to colour the sediment black. Beneath this, only the carbon fermentation by methanogens is possible.

#### 2.4.7.5 Phosphorus release from the sediment

The phosphorus-binding properties of sediments are sensitive to sediment metabolism, and relative dissociation within the sedimentary boundary layer corresponds to the distribution of the reduction potential. The composition and structure of the sediment layer are thus crucial to the more-or-less delayed recovery responses to managed reductions in the phosphorus loads delivered to lakes. The mainspring of sedimentary releases of phosphorus from the sediment to the water is the availability to bacteria of the easily degradable, organically bound energy in the sediments. More recalcitrant organic substances, which are deposited in oligotrophic lakes at a rate well below 1 mm yr$^{-1}$, do not offer a high potential for oxygen-consuming bacterial processes, much less denitrification and desulphurisation. Little remobilisation of phosphorus occurs and the rapid lake-ageing ('rasante Seenalterung'), described by Ohle (1953a, 1955), is avoided. It should be noted, however, that the return of phosphorus to the pelagic zone is caused not by seasonal oxygen deficiency (Mortimer 1941–1942) so much as by sulphate reduction (Ohle 1954). After the exhaustion of oxygen and reduction of nitrate at the sediment–water contact, desulphurisation produces hydrogen sulphide, which reduces trivalent (oxidised) iron and precipitates it as bivalent iron sulphide. During this process, the phosphorus which is bound to oxidised iron is set free into the sediment pore water and may be released into the water column where there may remain a relative deficiency of iron-based chemical binding sites (Ripl 1978; Ripl & Lindmark 1979).

Prior to industrialisation, sulphate concentrations were rather less than 20 mg L$^{-1}$, often <10 mg L$^{-1}$, in the waters around Berlin (Höll 1928; section 4.4 and Table 2.2). Even then, Ohle (1953b) pointed out that soil drainage and deep ploughing were contributing to an increase of sulphate concentrations in drainage waters. Until then, sulphate reduction in lakes had been constrained by low, pre-industrial sulphate concentrations. The remobilisation of phosphorus from lake sediments was also slight because of the low intensity of sulphide production.

### 2.4.8 Utilisation of lakes

In addition to the major damage to lakes inflicted by hydrological regulation, altered bank filtration and use as tributaries for accelerated material transport, some consequences of the direct anthropogenic use of lakes may be enumerated.

#### 2.4.8.1 Bathing waters

Use of lakes for recreational bathing causes perturbation in several respects, including the direct introduction of nutrients (e.g. by urine, sun protective agents) and deliberate and incidental modification to bank structures and accessible sediments. In small, relatively shallow lakes, these disturbances may be correspondingly more significant.

#### 2.4.8.2 Fisheries

Damage occurs through the management of lakes as fisheries, especially if particular species (white fish, predators) are stocked or are fished preferentially. These changes may well affect the structure and balance of the supportive food web, leading to overexploitation of zooplankton and simultaneous increases in the phytoplankton biomass and sedimentation losses (Hrbáček 1994). Autotrophic components thrive in relation to heterotrophic biomass. The erstwhile coupling of production and respiration is separated in space and time. Lower chemical efficiency is revealed by the fact that phases of production and respiration increase in amplitude and that dissolved oxygen concentrations vary more conspicuously.

#### 2.4.8.3 Navigation

The additional erosive effects of primary waves on the littoral vegetation and bank structure are well-known in lakes subject to a high frequency of passage by passenger and freight ships and fast sport boats.

### 2.5 SUMMARY AND CONCLUSIONS

Anthropogenic disturbances have been evaluated in relation to the natural function of landscape and waters. The principal damage arises as a consequence of opening of the otherwise closed material cycles of landscapes without the simultaneous creation of spatially coupled sinks. Natural sinks are frequently overloaded: the example of sewage phosphorus is a classic case in point: phosphorus is diluted with mineral-rich drinking water, it is transported to purification plants for mineralisation and it is discharged to the environment as an inorganic solute. Until recently, sewage purification was not designed to capture much phosphorus or any other of the dissolved bases. The natural sink for these nutrients and mineral materials, the soil, has been by-passed. They are simply lost to the sea. The coupling of sources to sinks is broken within the management time frame and the cycle of ecosystem self-organisation is substantially subverted.

The sectored approach of science and the tendency to model processes in isolated subcompartments are not well-adapted to deal holistically with the broad, general system-spreading implications of environmental exploitation. The water-mediated, energetic connections between individual landscape compartments just cannot be reasonably analysed or realistically addressed without breaking from the traditional separation of scientific disciplines. In consequence, the impacts of the water cycle and the vastness of the material transport on which it impinges are still poorly understood, especially in respect of the underestimation of the relevance of the cooling function of evaporation from vegetation in the context of the 'climate change' debate.

The damage to landscape and waters is manifest on fractally-divergent spatiotemporal scales. Without a matching concept to address their impacts, any chance of recovering fully sustainable systems seems remote.

### 2.6 ACKNOWLEDGEMENTS

The team of the Department of Limnology at Technische Universität Berlin (C. Hildmann, T. Janssen, I. Otto and others) have co-developed the ETR model. S. Heller accomplished data processing of the water-chemical analyses. A part of the English text was improved linguistically by S. Ridgill.

## 2.7 REFERENCES

Alexander, C.E. & Cresser, M.S. (1995) An assessment of the possible impact of expansion of native woodland cover on the chemistry of Scottish fresh-waters. *Forest Ecology and Management*, **73**(1–3), 1–27.

Digerfeldt, G. (1972) The post-glacial development of Lake Trummen. Regional vegetation history, water level changes and paleolimnology. *Folia Limnologica Scandinavica*, **16**, 1–104.

Dyck, S. & Peschke, G. (1989) *Grundlagen der Hydrologie*. VEB Verlag für Bauwesen, Berlin, 408 pp.

Ellenberg, H. (1995) Allgemeines Waldsterben—ein Konstrukt? *Naturwissenschaftliche Rundschau*, **48**(3), 93–6.

Flaig, H. (1996) Zuviel Stickstoff—ein Kreislauf in der Krise. *Akademie für Technikfolgenabschätzung, TA-Informationen*, **2/96**, 2–8.

Fritz, P., Huber, J. & Levi, H.W. (1995) Das Konzept der nachhaltigen Entwicklung als neue Etappe der Suche nach einem umweltverträglichen Entwicklungsmodell der modernen Gesellschaft. Einleitung und Überblick zu den Beiträgen des Bandes. In: Fritz, P., Huber, J. & Levi, H.W. (eds), *Nachhaltigkeit in naturwissenschaftlicher und sozialwissenschaftlicher Perspektive*. Hirzel, Stuttgart, 7–16.

Hildmann, C. (1999) *Temperaturen in Zönosen als Indikatoren zur Prozeßanalyse und zur Bestimmung des Wirkungsgrades: Energiedissipation und beschleunigte Alterung der Landschaft*. Dissertation, Technische Universität Berlin, Fachbereich Umwelt und Gesellschaft, Berlin.

Höll, K. (1928) Ökologie der Peridineen. Studie über den Einfluß chemischer und physikalischer Faktoren auf die Verbreitung der Dinoflagellaten im Süßwasser. *Pflanzenforschung, Jena*, **11**, 105 pp.

Hrbáček, J. (1994) Food web relations. In: Eiseltová, M. & Biggs, J. (eds.), *Restoration of Lake Ecosystems—a Holistic Approach*. IWRB Publication 32, International Waterfowl and Wetlands Research Bureau, Slimbridge, 44–58.

Hutchinson, G.E. (1957) *A Treatise on Limnology*, Vol. I, *Geography, Physics and Chemistry*. Wiley, New York, 1016 pp.

Kandler, O. (1994) Vierzehn Jahre Waldschadensdiskussion. Szenarien und Fakten. *Naturwissenschaftliche Rundschau*, **47**(11), 419–30.

Keller, R. (1979) *Hydrologischer Atlas der Bundesrepublik Deutschland*. De Haar, H. von U., Keller, R., Liebscher, H.-J., Richter, W. & Schirmer, H. Deutsche Forschungsgemeinschaft, Boldt, Boppard, 365 pp. + annex.

Kloos, R. (1986) *Das Grundwasser in Berlin—Bedeutung, Probleme, Sanierungskonzeption—Besondere Mitteilungen zum Gewässerkundlichen*. Jahresbericht des Landes Berlin. Senator für Stadtentwicklung und Umweltschutz, Berlin, 165 pp.

Kluge, T & Schramm, E. (1995) Wasser als Problem—Wasser als Politik. Eine Chronologie der Wasserdebatte in Deutschland. In: Altner, G., Mettler-Meibom, B., Simonis, U.E., Weizsäcker, E.U.v.(eds), *Jahrbuch Ökologie 1995*. Beck, München, 226–39.

Kucklentz, V. & Hamm, A. (1988) *Möglichkeiten und Erfolgsaussichten der Seenrestaurierung*. Bayerische Landesanstalt für Wasserforschung, München, 212 pp.

Lundqvist, G. (1927) Bodenablagerungen und Entwicklunstypen der Seen. *Die Binnengewässer*, **2**, 1–124.

Markstein, B. & Sukopp, H. (1980) Die Ufervegetation der Berliner Havel 1962–77. *Garten und Landschaft*, **1**, 30–6.

Mortimer, C.H. (1941–1942) The exchange of dissolved substances between mud and water in lakes. *Journal of Ecology*, **29**, 280–329; **30**, 147–201.

N.N. (1895) Graphische Zusammenstellung die Ergebnisse von Spreewasseruntersuchungen (1890–95). Tiefbauamt Charlottenburg, Abteilung Plankammer. Bl. I und Blatt II.

Odum, E.P. (1971) *Fundamentals of Ecology*. Saunders, Philadelphia.

Odum, W.E. (1990) Internal processes influencing the maintenance of ecotones: do they exist? In: Naiman, R.J. & Décamps, H. (eds), *The Ecology and Management of Aquatic Terrestrial Ecotones*. MAB Series 4, UNESCO, Paris, 91–102.

Ohle, W. (1953a) Der Vorgang rasanter Seenalterung in Holstein. *Naturwissenschaften*, **40**, 153–62.

Ohle, W. (1953b) Sulfatanreicherung der Fließgewässer und Seen infolge von Bodenmeliorationen. Berichte der Limnologischen Flußstation Freudenthal. *Außenstelle der Hydrobiologischen Anstalt der Max-Planck-Gesellschaft, Hannoversch-Münden*, **4**, 40–3.

Ohle, W. (1954) Sulfat als 'Katalysator' des limnischen Stoffkreislaufes. *Vom Wasser*, **21**, 13–32.

Ohle, W. (1955) Die Ursachen der rasanten Seeneutrophierung. *Verhandlungen der internationale Vereinigung für theoretische und angewandte Limnologie*, **12**, 373–82.

Ohle, W. (1971) Gewässer und Umgebung als ökologische Einheit in ihrer Bedeutung für die Gewässereutrophierung. *Wasser—Abwasser (Aachen)*, **1971**(2), 437–56.

Olszewski, P. (1961) Versuch einer Ableitung des hypolimnischen Wassers aus einem See. *Verhandlungen der internationale Vereinigung für theoretische und angewandte Limnologie*, **14**, 855–61.

Pechlaner, R. (1979) Response of the eutrophied Piburger See to reduced external loading and removal of monimolimnic water. *Archiv für Hydrobiologie Beihefte. Ergebnisse der Limnologie*, **13**, 293–305.

Prigogine, I. (1988) *Vom Sein zum Werden. Zeit und Komplexität in den Naturwissenschaften*. Piper, München, 304 pp.

Remmert, H. (ed.) (1991) *The Mosaic–Cycle Concept of Ecosystems*. Ecological Studies, No. 85, Springer-Verlag, Berlin, 168 pp.

Ripl, W. (1978) *Oxidation of Lake Sediments with Nitrate—a Restoration Method for Former Recipients*. Institute of Limnology, University of Lund, 151 pp.

Ripl, W. (1992) Management of water cycle: An approach to Urban Ecology. *Water Pollution Research Journal of Canada*, **27**, 221–37.

Ripl, W. (1995a) Management of water cycle and energy flow for ecosystem control: the energy-transport-reaction (ETR) model. *Ecological Modelling*, **78**, 61–76.

Ripl W. (1995b) Nachhaltige Bewirtschaftung von Ökosystemen aus wasserwirtschaftlicher Sicht. In: Fritz, P., Huber, J. & Levi, H.W. (eds), *Nachhaltigkeit in naturwissenschaftlicher und sozialwissenschaftlicher Perspektive*. Hirzel, Stuttgart, 69–80.

Ripl, W. (1995c) Der landschaftliche Wirkungsgrad als Maß für die Nachhaltigkeit. In: Backhaus, R. & Grunwald, A. (eds), *Umwelt und Fernerkundung: Was leisten integrierte Geo-Daten für die Entwicklung und Umsetzung von Umweltstrategien)*, Wichmann, Heidelberg, 40–52.

Ripl, W. & Feibicke, M. (1992) Nitrogen metabolism in ecosystems—a new approach. *Internationale Revue der gesamten Hydrobiologie*, **77**, 5–27.

Ripl, W. & Hildmann, C. (1997) Ökosysteme als thermodynamische Notwendigkeit. In: Fränzle, O., Müller, F. & Schröder, W. (eds), *Handbuch der Umweltwissenschaften. Grundlagen und Anwendungen der Ökosystemforschung. Grundwerk*. Ecomed, Landsberg am Lech (Loseblattsammlung), 12 pp.

Ripl, W. & Lindmark, G. (1978) Ecosystem control by nitrogen metabolism in sediment. *Vatten*, **34**, 135–44.

Ripl, W. & Lindmark, G. (1979) The impact of algae and nutrient composition on sediment exchange dynamics. *Archiv für Hydrobiologie*, **86**, 45–65.

Ripl, W., Hildmann, C. & Janssen, T. (1994a) Das Röhricht—ein itegrativer System-Ansatz auf der Grundlage des ETR-Modells. In: Ripl, W., Szabo, I., Szeglet, P., Balogh, M., Balogh, A., Marialigeti, K., Hildmann, C., Janssen, T. & Markwitz, M., *Das Röhricht—modellhafte Vorstellungen und Beispiele aus Ungarn (Balaton, Velenci-See) und Deutschland (Berlin, Unterhavel). Abschlußbericht*, Im Auftrag der Senatsverwaltung für Stadtentwicklung und Umweltschutz, Berlin. Technische Universität, Berlin, 5–14.

Ripl, W., Hildmann, C., Markwitz, M. & Janssen, T. (1994b) Das Beispiel der Unterhavel (Berlin, Deutschland). In: Ripl, W., Szabo, I., Szeglet, P., Balogh, M., Balogh, A., Marialigeti, K., Hildmann, C., Janssen, T. & Markwitz M., *Das Röhricht—modellhafte Vorstellungen und Beispiele aus Ungarn (Balaton, Velenci-See) und Deutschland (Berlin, Unterhavel). Abschlußbericht*, Im Auftrag der Senatsverwaltung für Stadtentwicklung und Umweltschutz, Berlin. Technische Universität, Berlin, 119–61.

Ripl, W., Hildmann, C. & Janssen, T. (1994c) Schlußfolgerungen. In: Ripl, W., Szabo, I., Szeglet, P., Balogh, M., Balogh, A., Marialigeti, K., Hildmann, C., Janssen, T. & Markwitz M., *Das Röhricht—modellhafte Vorstellungen und Beispiele aus Ungarn (Balaton, Velenci-See) und Deutschland (Berlin, Unterhavel). Abschlußbericht*, Im Auftrag der Senatsverwaltung für Stadtentwicklung und Umweltschutz, Berlin. Technische Universität, Berlin, 163–5.

Ripl, W., Hildmann, C., Janssen, T., Gerlach, I., Heller, S. & Ridgill, S. (1995) Sustainable redevelopment of a river and its catchment—the Stör River Project. In: Eiseltová, M. & Biggs, J. (eds.), *Restoration of Lake Ecosystems—a Holistic Approach*. IWRB Publication 32, International Waterfowl and Wetlands Research Bureau, Slimbridge, 76–112,

Ripl, W., Trillitzsch, F., Backhaus, R., Blume, H.-P., Widmoser, P., Janssen, T., Hildmann, C. & Otto, I. (1996) *Entwicklung eines Land-Gewässer Bewirtschaftungskonzeptes zur Senkung von Stoffverlusten an Gewässer (Stör-Projekt I und II). Endbericht*. Bundesministerium für Bildung, Wissenschaft, Forschung und Technologie (BMBF), Landesamt für Wasserhaushalt und Küsten Schleswig-Holstein, Kiel, Technische Universität Berlin, Fachgebiet Limnologie. 203 pp. + annex.

Risser, P.G. (1990) The ecological importance of land–water ecotones. In: Naiman, R.J. & Décamps,

H. (eds), *The Ecology and Management of Aquatic Terrestrial Ecotones*. MAB Series 4, UNESCO, Paris, 7–21.

Rohlich, G.A. (ed.) (1969) *Eutrophication; Causes, Consequences, Correctives*. National Academy of Sciences, Washington, 661 pp.

Schindler, D.W., Kling, H., Schmidt, R.V., Prokopowich, J., Frost, V.E., Reid, R.A. & Capel, M. (1973) Eutrophication of Lake 227 by addition of phosphate and nitrate: the second, third, and fourth years of enrichment, 1970, 1971 and 1972. *Journal of the Fisheries Research Board of Canada*, **30**, 1415–60.

Stumm, W. & Morgan, J.J. (1981) *Aquatic Chemistry. An Introduction Emphasizing Chemical Equilibria in Natural Waters*. Wiley, New York, 780 pp.

Vollenweider, R.A. (1968) *Scientific Fundamentals of the Eutrophication of Lakes and Fflowing Waters, with Particular Reference to Nitrogen and Phosphorus as Factors in Eutrophication*. Organisation for Economic Co-operation and Development (OECD), Paris, 159 pp.

Wetzel, R.G. (1983) *Limnology*. Saunders, Philadelphia, 837 pp.

Wöbbecke, K. (1993) *Der Einfluß der Trinkwassergewinnung durch Uferfiltration auf den Röhrichtbestand der Berliner Unterhavel*. Inauguraldissertation Fachbereich Landschaftsentwicklung der Technischen Universität, Berlin.

Wöbbecke, K. & Ripl, W. (1990) Untersuchungen zum Röhrichtrückgang an der Berliner Havel. In: Sukopp, H. & Krauss, M. (eds), *Ökologie, Gefährdung und Schutz von Röhrichtpflanzen. Landschaftsentwicklung und Umweltforschung*. Technische Universität, Berlin, 94–102.

# *Part II*  Regional Studies

# 3 The North American Great Lakes: a Laurentian Great Lakes Focus

MARLENE S. EVANS

## 3.1 INTRODUCTION

The North American Great Lakes (Fig. 3.1) consist of a chain of large lakes carved out of the North American landscape by a series of glaciations, the last of which ended some 7500–12,000 years ago (Patalas 1975). The most familiar of these are the Laurentian Great Lakes (i.e. lakes Ontario, Erie, Huron, Michigan and Superior). These five possess a collective volume of 22,684 km$^3$, accounting for c.18.2% of the world's volume of freshwater lakes (Table 3.1; Wetzel 1975). Lake Winnipeg, Lake Winnipegosis and Lake Manitoba (Table 3.2) are large but shallow lakes situated in the boreal region of Manitoba (Herdendorf 1982). To the northwest are Lake Athabasca, a moderately deep lake, and Great Slave Lake, a deep lake (both also located in the boreal region), and Great Bear Lake, a large, deep lake in the Canadian subarctic (see also Table 3.2). Collectively, these North American Great Lakes account for 22.1% of the world's supply volume of freshwater lakes, and the major volume of Canada's economically useable surface freshwaters.

Drainage patterns in the North American Great Lakes watershed are complex (Fig. 3.1). The Laurentian Great Lakes proper drain an area of 521,830 km$^2$ (Table 3.1), and discharge via the St Lawrence River into the Atlantic Ocean. Lakes Winnipeg, Manitoba and Winnipegosis receive most of their waters from rivers draining the Prairies, and discharge via the Nelson River into the cold waters of Hudson Bay; their total drainage area is c.980,000 km$^2$ (http://climatechangeconnection.org/pagesl-1ake_winnipeg.html 2004). Lake Athabasca is fed mainly by the Athabasca River (drainage basin 255,000 km$^2$) which originates in the mountainous regions of British Columbia, although some water also enters the western end of the lake via the Peace River inflow (drainage basin 307,000 km$^2$) and from the east from the Fond du Lac River. Lake Athabasca drains north via the Slave River into the Great Slave Lake.

Great Slave Lake, with a total watershed of 983,040 km$^2$, receives approximately 87% of its waters from Slave River (Rawson 1950). In turn, it forms the headwaters of the Mackenzie River which flows north into the Arctic Ocean. Great Bear Lake, which flows into the Mackenzie River, drains a relatively small watershed of only 145,780 km$^2$ (Johnson 1975a).

Direct anthropogenic influences upon the North American Great Lakes vary markedly, from the very strong impact upon Lake Ontario, to the near pristine conditions of Great Bear Lake. Geographical gradients in anthropogenic impacts are, in large measure, based on two considerations: climate and river drainage patterns. Both have influenced the degree to which the immediate watersheds of these lakes have been developed, particularly during early European settlement, and both continue to be important, with much of the boreal and subarctic region characterised by low population density, and natural-resource-driven economies (i.e. forestry, mining and commercial fishing; Evans 2000). In contrast, in the Lake Ontario, Erie and Michigan watersheds, population density is great, and economies are largely driven by manufacturing, trade and community services (Table 3.1; Environment Canada 1995).

In this chapter, a broad overview is provided of the history of significant anthropogenic activities affecting the North American Great Lakes. The chapter focuses on the Laurentian Great Lakes, the most developed, most heavily influenced and the most closely investigated of the North

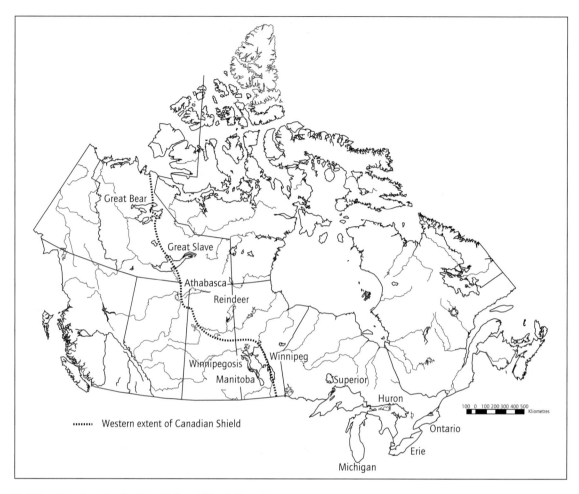

Fig. 3.1 Map showing the Great Lakes of North America.

American Great Lakes. However, brief overviews are also provided of the prairie, boreal and subarctic great lakes. Economic development in these regions is accelerating, and may possibly repeat some aspects of the history of environmental degradation of the Laurentian Great Lakes. Moreover, organic contaminants such as PCBs and toxaphene are widespread in northern ecosystems through long-range atmospheric transport, reaching high concentrations in predatory fish and marine mammals (Jensen *et al.* 1997).

## 3.2  EUROPEAN COLONISATION

Prior to European contact, what is now Canada was inhabited by slightly more than 500,000 people, with 100,000–150,000 living around the eastern Great Lakes and St Lawrence valley, and 50,000–100,000 in the grassland–parkland areas of the Western Interior (Ray 1996). People lived as gatherers, hunters and fishers. Whilst Vikings had initially made contact along the northeast coast of North America as early as the end of the tenth cen-

Table 3.1  Physical features and population for the Laurentian Great Lakes. (Modified from Environment Canada 1995)

|  | Superior | Michigan | Huron | Erie | Ontario |
|---|---|---|---|---|---|
| Elevation above sea level (m) | 183 | 176 | 176 | 173 | 74 |
| Length (km) | 563 | 494 | 332 | 388 | 311 |
| Width (km) | 257 | 190 | 245 | 92 | 85 |
| Average depth (m) | 147 | 85 | 59 | 19 | 86 |
| Maximum depth (m) | 405 | 281 | 229 | 64 | 244 |
| Volume (km$^3$) | 12,100 | 4920 | 3540 | 484 | 1640 |
| Lake area (km$^2$) | 82,100 | 57,800 | 59,600 | 25,700 | 18,960 |
| Drainage area (km$^2$) | 127,700 | 118,000 | 134,100 | 78,000 | 64,030 |
| Total area (km$^2$) | 209,800 | 175,800 | 193,700 | 103,700 | 82,990 |
| Land : water area | 1.56 : 1 | 2.04 : 1 | 2.25 : 1 | 3.04 : 1 | 3.38 : 1 |
| Shoreline length* (km) | 4385 | 2633 | 6157 | 1402 | 1175 |
| Shoreline development† | 4.32 | 3.12 | 7.11 | 2.47 | 2.41 |
| Retention time (years) | 191 | 99 | 22 | 2.6 | 6 |
| Population: USA (1980) | 425,548 | 10,057,026 | 1,502,687 | 10,017,530 | 2,704,284 |
| Canada (1981) | 181,573 | – | 1,191,467 | 1,664,639 | 5,446,611 |
| Totals | 607,121 | 10,057,026 | 2,694,154 | 11,682,169 | 8,150,895 |
| Population density (people km$^{-2}$) | 4.8 | 85.2 | 20.1 | 149.8 | 127.3 |

\* Includes islands.
† The ratio of the length of the shoreline to the circumference of a circle of area equal to that of the lake (see Wetzel 1975).

Table 3.2  The prairie, boreal and subarctic North American Great Lakes—areas and volumes

|  | Area (km$^2$) | Volume (km$^3$) |
|---|---|---|
| Lake Winnipeg | 24,387 | 371 |
| Lake Winnipegosis | 5375 | 16 |
| Lake Manitoba | 4625 | 17 |
| Lake Athabasca | 7935 | 110 |
| Great Slave Lake | 28,568 | 2088 |
| Great Bear Lake | 31,326 | 2292 |

tury, it was not until 1497, when Cabot reached the Newfoundland (or Labrador) coast, that European influence became significant. Like many other fifteenth century European global explorations, this was based on attempts to find a rapid marine route to rich Asian sources of precious commodities. The current routes, across the centre of Asia, or around the tip of Africa, were long and dangerous.

European exploration of the North American continent accelerated throughout the sixteenth and seventeenth centuries, as explorers travelled along major waterways continuing to seek alternative routes to Asia (Ray 1996). As this occurred, the rich natural resources of the continent were encountered, promoting a new wave of European travel. By the 1530s, Europeans had established a growing cod and whaling industry in the Gulf of St Lawrence region, and had began to purchase increasing numbers of furs and skins from indigenous peoples. In 1534, Jacques Cartier travelled up the St Lawrence as far as what is now Montreal. The following year, he set up winter camp in the area of Quebec City. These explorations, and developing economic ventures, continued through the sixteenth century, with North American trade becoming increasingly of greater importance than the original goal of seeking alternative trade routes to Asia. By the early seventeenth century, a major fur-trade route had been established from the St Lawrence River, via the Ottawa–Nippissing River systems, to Georgian Bay, Lake Huron.

As trade networks became established, these routes served other interests. Missionary activity began intensifying in the 1600s, with missionaries also playing a major role in the early exploration of the North American interior. By the early eighteenth century, trade routes extended west through the Great Lakes to Lake-of-the-Woods in Manitoba. By the 1760s, routes had been established through Lake Winnipeg, north to York Factory on Hudson's Bay, and east to Cumberland House. By 1810, they had been developed to Lake Athabasca, north along the Slave River to the Great Slave Lake, and along the southern reaches of the Mackenzie River. Routes were not established along the entire Mackenzie River, north to the Arctic Ocean, until after 1810, however (Ray 1996).

Early exploration of the North American interior was financed largely by European government, business and church interests. This changed during the seventeenth century, as small groups of people migrated to North America settling close to the coasts and major waterways. Between 1792 and 1800, the population of southern Ontario and Quebec increased from 20,000 to 60,000. Forests were cleared and agricultural activity increased. By the mid-nineteenth century, approximately 1.5 million people were living in Michigan, Wisconsin, southern Ontario and Quebec alone (Environment Canada 1995). The total population for the Laurentian Great Lakes watershed at the turn of the twentieth century was c.11 million (Fig. 3.2).

Population continued to swell in the Lake Ontario, Erie and Huron watersheds over the next 100 years, stimulated by the overall favourable features of much of the region for agriculture. Later discoveries of iron in the Lake Superior drainage basin, coal in the Lake Erie and Ontario basins, and of limestone in the lakes Erie, Huron and Michigan basins, fuelled steel manufacturing and other associated industries. Development also was facilitated by an abundant supply of lake water for transportation, and for industrial and urban uses. By 1980–1981, there were over 36 million people living in the Great Lakes drainage basins, with those of lakes Michigan, Erie and Ontario containing the greatest density (Table 3.1). In contrast, the value for the Lake Superior watershed was 5.8 people km$^{-2}$, a total population of only 740,000.

Population numbers and density decline along

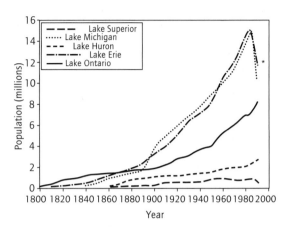

**Fig. 3.2** Population trends (in millions) in the Laurentian Great Lakes drainage basins. * Population data for 1990–1991 were collected on different watershed boundaries and are not directly comparable with data from previous years. (From Beeton (1969), Environment Canada (1987) and Environment Canada and the U. S. Environmental Protection Agency (2001).)

the northwestern side of the North American Great Lakes. The Red River Valley, south of Lake Winnipeg, did not begin to become settled until the early 1880s (Ray 1996). The economy was based largely on agriculture. Population remained low throughout the twentieth century, with the province of Manitoba containing a reported population of 1.1 million in the early 1990s, i.e. not appreciably greater than that of the Lake Superior basin. Winnipeg, founded in 1812, contained a population of 617,000 in 1991 (Marsh 1988).

The Lake Athabasca region is also thinly populated. During the mid-nineteenth century, the population of Fort Chipewyan, the major community, was approximately 920. Uranium City was created during the 1950s, but mining operations were shut down during the early 1980s (Mitchell & Prepas 1990). Less hospitable climates for agriculture, the lack of readily accessible mineral resources and difficult trade routes along relatively small and shallow northern flowing rivers also limited development. These circumstances are now changing as technology is allowing the vast oil sands reserves to be extracted from along the northern Athabasca River, and diamonds from north of the Great Slave Lake (Evans 2000). Improved air transport and winter roads allow even the most remote community to receive a steady supply of shipped foods, and provide freedom from the time demands of subsistence hunting and fishing. Other types of economy are now feasible.

The Great Slave Lake basin is the most populated of those of the northern great lakes, with approximately 17,000 people living in Yellowknife. Although gold was discovered as early as 1896, it was not until the early 1930s that this began to be exploited (Lopatka *et al.* 1990). Yellowknife was founded in 1935, and is the territorial capital of the Northwest Territories. Hay River, a town of approximately 3000 people, is a commercial fishing port on Great Slave Lake, and an important port connecting to the Mackenzie River. Its importance as a shipping port has declined in recent years, as oil exploration activity in the Mackenzie Basin has diminished. Hay River is connected to southern Canada by highway and railroad.

The Great Bear Lake region is thinly populated, with Deline (Fort Franklin) the primary population centre, of approximately 600 people. Port Radium was developed during the early 1930s, following the discovery of pitchblende (a uranium ore) at Great Bear Lake, oil and gas at Normal Wells (on the Mackenzie River), and technological improvements in transportation which enabled mining operations to become commercially feasible. The mine is no longer operational. Access to Great Bear Lake is by air, boat and winter road. Trapping, hunting, fishing and tourism are the major economic activities, and there are scattered fishing lodges around the lake.

## 3.3 THE NORTH AMERICAN GREAT LAKES BEFORE THE COLONIAL PERIOD

Prior to the colonial period, the watersheds of the North American Great Lakes were largely forested, and the lakes were, in large measure, inherently oligotrophic. However, some, such as Lake Winnipeg and Lake Athabasca, were fed by rivers which flowed north through prairie grasslands. Chapra & Robertson (1977) estimated that most regions of the Laurentian Great Lakes were naturally oligotrophic, i.e. with total phosphorus concentrations <$10\,\mu g\,L^{-1}$. Only shallow embayments, such as Saginaw Bay (Lake Huron) and lower Green Bay (Lake Michigan), were naturally mesotrophic, with concentrations averaging $11\,\mu g\,L^{-1}$ and $15\,\mu g\,L^{-1}$, respectively. An average concentration of $20\,\mu g\,L^{-1}$ was established as the boundary separating mesotrophic from eutrophic conditions.

Whilst similar modelling studies have not been conducted for the prairie, boreal and subarctic Great Lakes (Table 3.2), these systems are unproductive, primarily because of climate (i.e. the growing season becomes progressively shorter with increasing latitude). Some, such as Great Slave and Great Bear Lake, are also unproductive because they are deep. For example, the mean depth of the West Basin of Great Slave Lake is 41 m, whilst that of the East Arm is 185 m; the mean depth of Great Bear Lake is 62 m (Evans 2000).

However, total phosphorus concentrations are not low in any of these more northern Great Lakes, and are generally characteristic of mesotrophic conditions. Thus whilst they are unproductive, they are not, in the strict sense, oligotrophic. Concentrations in Lake Athabasca are $9-14\,\mu g\,L^{-1}$, whilst those of Great Slave Lake range from 8 to $15\,\mu g\,L^{-1}$ (Fee *et al.* 1985; Mitchell & Prepas 1990). Such high concentrations are, in part, a function of the large watersheds of these lakes, which give rise to significant river-borne loadings (Brunskill *et al.* 1975). In addition, phosphorus may be efficiently retained in the water column by physical processes such as strong vertical mixing, and various efficiencies in trophic structure. Great Bear Lake, with its small watershed and deep waters, contains the lowest concentrations, of $5\,\mu g\,L^{-1}$ (Evans, unpublished data).

The North American Great Lakes originally supported a diverse assemblage of organisms which were generally characteristic of oligotrophic environments. Great Bear Lake, the northernmost lake, continues to support a relatively low diversity of zooplankton and fish (Patalas 1975; Johnson 1975b). General information on fish, invertebrate and algal populations in Great Slave Lake, Lake Athabasca and Lake Winnipeg appear in Rawson (1951, 1953, 1956), Mitchell & Prepas (1990), Patalas & Salki (1992), Flannagan et al. (1994), Kling (1996) and Franzin et al. (1996). With the exception of Lake Winnipeg, there are no known introductions or losses of species from these lakes in recent decades.

Limnological investigations of the Laurentian Great Lakes began during the late nineteenth century, and continued, to varying degrees of intensity, into the mid-twentieth century (Welch 1935; Beeton & Chandler 1963). However, by the time these studies were conducted, many of these lakes were already seriously perturbed. Thus, in order to reconstruct the major features of these lakes before colonial impact, forensic scientific approaches must be used.

Relatively comprehensive qualitative information exists on fish composition in the Great Lakes prior to the mid-twentieth century, when environmental degradation was at its most severe. Fish were important in the early economy of the Great Lakes region and thus, unlike phytoplankton and invertebrates, are relatively well described. Thus, fisheries biologists have determined that the Laurentian Great Lakes historically supported some 15 to 25 families of fish, representing 67–114 species (Christie 1974). The open waters were populated by two large piscivores, the lake trout (*Salvelinus namaycush*) and the burbot (*Lota lota*); Atlantic salmon (*Salmo salar*) were also present in Lake Ontario. The subfamily Coregoniae was diverse, with between three and ten species inhabiting each of the Great Lakes, both planktivores and benthivores, thus occupying different regions of the water column. Nearshore areas also contained a diverse fish fauna. Overall, food web structure is believed to have been complex and efficient (Christie 1974).

Two approaches have been used in order to reconstruct plankton and benthic invertebrate community structure prior to strong European influence. Palaeolimnological studies have provided inferences of past phytoplankton and benthic invertebrate community structure. For example, Stoermer et al. (1985, 1993, 1996) reconstructed the major features of diatom assemblages prior to European colonisation, and demonstrated the early signs of environmental degradation which in many areas accompanied this process. Similarly, information on benthic communities prior to and following the colonial period can be found in Warwick (1980), Delorme (1982) and Reynoldson & Hamilton (1993). Researchers also have investigated plankton and benthic community structure and standing stocks in relatively pristine areas of the Great Lakes (e.g. most of lakes Superior, Huron and northern Lake Michigan), and then used this information in order to reconstruct the same properties for Lake Erie, Lake Ontario and southern Lake Michigan, prior to environmental degradation. Examples of this approach can be found in the synthesis publications of Cook & Johnson (1974), Watson (1974) and Vollenweider et al. (1974).

## 3.4 EARLY EVIDENCE OF THE DEGRADATION OF THE LAURENTIAN GREAT LAKES

Anthropogenic activities began affecting some regions of the Great Lakes as early as the eighteenth century. Early palaeolimnological research determined that diatom production rates in Lake Ontario peaked from 1820 to 1850, at about 1880 in Lake Erie, and at about 1970 in Grand Traverse Bay of Lake Michigan (Schelske et al. 1983). Later studies showed that the Bay of Quinte (Lake Ontario) was affected by French settlement as early as 1699–1784 (Schelske 1991). Additional research (Stoermer et al. 1985; Schelske & Hodell 1991; Stoermer et al. 1996) provided further evidence that, by the late 1880s, clear-cutting activities in the Great Lakes drainage basins had produced detectable effects on lake productivity and algal community structure.

With increased algal production, silica was increasingly depleted from the water column, and algal composition shifted to mesotrophic and eutrophic species with lower silica requirements. However, these changes, and those in standing stocks, were not evident to researchers conducting baseline limnological studies during the late 1880s and early 1900s. Instead, changes in the fishery, as evidenced from reduced catches of prized species, were evident by the late nineteenth century, and prompted much of the early limnological research (see Reighard 1894; Ward 1896).

By the 1960s, it was evident that some of the Laurentian Great Lakes were showing signs of eutrophication (Beeton 1964). Long-term monitoring programmes at water intakes, which began during the 1920s (Damann 1960; Davis 1964; Schenk & Thompson 1965), provided the earliest evidence that phytoplankton populations were increasing in abundance in the nearshore waters of lakes Michigan, Ontario and Erie. Schenk & Thompson (1965), for example, reported a near doubling of phytoplankton abundance at the Toronto Island Filtration Plant between 1923 and 1954.

Beeton (1969) summarised many of the changes which were then being observed in the Laurentian Great Lakes. These included: increasing concentrations of total dissolved solids, sulphate and chloride in nearshore waters; increased algal abundance in water intakes; enhanced growth of the green alga *Cladophora*; changes in benthic abundance and composition, and in zooplankton composition, towards more mesotrophic and eutrophic characteristics. Changes in commercial fish catches also were highlighted.

Particular concerns were raised in the case of Lake Erie, because of the extensive oxygen depletion which occurred in the summer hypolimnion, and the near extinction of the mayfly *Hexagenia* from the western basin. Beeton (1969) linked the degree to which each of the Great Lakes was being perturbed with population and economic growth occurring in its drainage basin, i.e. lakes Erie, Ontario and Michigan were the most perturbed. Regional influences were also recognised. In Lake Michigan, the most severe pollution occurred in the southern tip of the lake and in Green Bay, whereas in Lake Huron it was confined to Saginaw Bay. The western basin of Lake Erie was much more strongly influenced by increasing pollution than the eastern basin.

In response to public and scientific concerns, extensive and intensive limnological surveys of the Great Lakes were conducted during the early 1970s by Canadian and American scientists. On the basis of nutrient and algal studies, Lake Superior and the offshore region of Lake Huron were considered oligotrophic, and the inshore region of Lake Huron, the offshore regions of Lake Michigan and Ontario, and the east basin of Lake Erie, mesotrophic. Inshore region of lakes Ontario and Michigan and the central basin of Lake Erie secondarily (periodically) were classified as eutrophic, whereas Saginaw Bay, the Bay of Quinte, Green Bay and the western basin of Lake Erie were deemed highly eutrophic (Dobson et al. 1974; Vollenweider et al. 1974). Furthermore, Chapra & Robertson (1977) estimated that, in Lake Michigan, average total phosphorus concentrations had more than doubled from natural estimated values of $3.6\,\mu g\,L^{-1}$ to $10.4\,\mu g\,L^{-1}$, in Lake Ontario from $4.6\,\mu g\,L^{-1}$ to $23.6\,\mu g\,L^{-1}$, and in western Lake Erie from $6.1\,\mu g\,L^{-1}$ to $39.6\,\mu g\,L^{-1}$. Although the main body of Lake Huron remained oligotrophic, average concentrations in Saginaw Bay had increased from 11.1 to $37.1\,\mu g\,L^{-1}$. Concentrations in Green Bay had increased from $14.6\,\mu g\,L^{-1}$ to $39.6\,\mu g\,L^{-1}$.

Phytoplankton, zooplankton and benthic studies conducted during the early 1970s also provided clear evidence that increasing total phosphorus concentrations had significantly changed community structure and standing stocks (Cook & Johnson 1974; Vollenweider et al. 1974; Watson 1974; Watson & Carpenter 1974). Low oxygen concentrations in the central basin of Lake Erie also had become of concern (Dobson & Gilbertson 1971).

Whereas degradation of the open waters of the Great Lakes did not become generally evident until the 1960s, that of harbours, bays, tributaries and some inshore regions was recognised much earlier (Beeton 1969). Much early influence occurred with logging operations during the late nineteenth century, which led to significant sawdust pollution (Berst & Spangler 1973; Wells & Mclain 1973; Lawrie 1978; Kelso et al. 1996). General environmental degradation by industrial pollution was evident in some areas as early as the 1920s (Damman 1960). However, the importance of habitat loss through the draining of coastal wetlands, and by the modification of tributaries in order to accommodate industrial and urban development, was recognised only more recently (Whillans 1979; Koonce et al. 1996).

Commercial fish records for the Great Lakes (Fig. 3.3) begin in 1867 (Baldwin et al. 2002). By the 1870s, researchers were noting declines in catches of various species (Beeton 1969). These changes promoted some of the earliest baseline limnological research studies of the Great Lakes, notably Reighard's study of Lake St Clair (1894) and Ward's investigation of the Traverse Bay region of Lake Michigan (1896). Whitefish stocking programmes were implemented, but met with little success, in that the primary causes of fish decline (overfishing, impact of the sea lamprey, habitat destruction) were not addressed.

Commercial fishing continued into the twentieth century, with the total commercial harvest (Fig. 3.3) remaining relatively constant (Kelso et al. 1996). As standing stocks of various species declined, fishing effort intensified. Thus, although total catches remained somewhat constant over time, as various species were overharvested, major changes in catch composition occurred.

Atlantic salmon, which had been present in Lake Ontario during the early nineteenth century, had almost disappeared by 1880 (Beeton 1969). During the 1950s, lake trout catches crashed in lakes Michigan, Huron, Ontario and Superior, although the timing and severity of the collapse varied between lakes (Christie 1974; Hartman 1988). Overfishing also affected numbers of deepwater cisco, with the two largest species (*Coregonus nigripinnis* and *C. johannaeo*) the most strongly affected. Populations of blue pike/walleye (*Stizostedion vitreum glaucum*), a subspecies of yellow walleye (*S. vitreum glaucum*), crashed in Lake Erie during the 1960s, and are now believed to be extinct.

Fishing pressures were particularly intense in Lake Ontario. By the mid-1940s, of the seven coregonid species originally inhabiting the lake, only the bloater (*C. hoyi*) was left (Christie 1974). Deepwater sculpin (*Myoxocephalus thompsoni*) became extinct in the lake during the early 1950s, for reasons which are still not well understood (Brandt 1986). Lake Ontario continues to possess the most perturbed fish population.

Other factors exacerbated overfishing pressures on Great Lakes fish communities (Christie 1974; Hartman 1988). In some regions of the Great Lakes, important nearshore spawning and nursery habitat were lost, thus affecting fish recruitment. Sturgeon (*Acipenser fulvescens*) were particularly vulnerable to habitat degradation. Building of the Erie Canal (1816–1825), opening of the Welland Canal (1829) and its expansion (1932), construction of the Sault locks (1855) and their continuing expansion (to 1963), and construction of the St Lawrence Seaway (1959) all facilitated rapid dispersal of sea lamprey and exotic fish species through the Great Lakes (Hatcher & Walter 1963).

The sea lamprey (*Petromyzon marinus*), a parasite, was also crucially important in decimating Great Lakes predatory fish populations. These fish were known to occur in Lake Ontario as early as the 1830s, and may have been a glacial relict inhabiting the lake. Sea lampreys spread rapidly through the Great Lakes during the early twentieth century (Christie 1974), being first reported in Lake Erie in 1921, Lake Huron in 1932, Lake Michigan in 1936 and Lake Superior in 1946. Dispersal is believed to have been facilitated by reduced numbers of predatory lake trout and burbot, which enabled sea lampreys to increase rapidly in numbers, and then disperse throughout the Great Lakes. As numbers increased, lampreys became an increasingly important factor adversely affecting lake trout, burbot and other fish. Thus, the crash in lake trout and burbot populations has been related to a combination of overfishing and increased predation pressures from the sea lamprey. The relative importance of these two factors remains a subject of some debate amongst fishery biologists.

Rainbow smelt (*Osmerus mordax*) were introduced into Crystal Lake, Michigan, in 1912, and first collected in Lake Michigan in 1923. By 1936, they had been observed in all five Great Lakes (Christie 1974). Smelt populations proliferated rapidly in Lake Michigan and Huron during the 1930s, with a commercial fishery established in Lake Michigan by 1931. They became abundant in Lake Ontario and Erie during the late 1940s, whereas in Lake Superior the population grew

# The North American Great Lakes: a Laurentian Great Lakes Focus

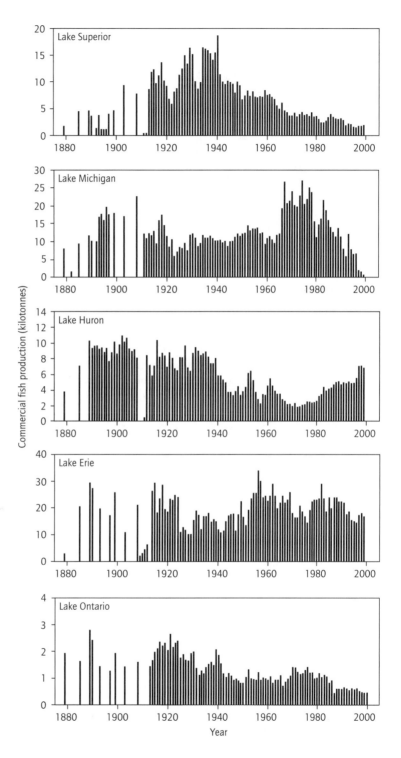

Fig. 3.3  Long-term trends in commercial fish catches (in metric tons/year) for the Laurentian Great lakes. (From Baldwin *et al.* (2002).)

gradually over the 1950s and early 1960s. Smelt remain a major component of the modern Great Lakes fish community.

The origin of the alewife (*Alosa pseudoharengus*), another exotic species to the Great Lakes, is uncertain. It may have been accidentally introduced into the lake in the 1870s or migrated earlier into the lake through the Erie Barge Canal system. It is known that alewife populations were present in Lake Ontario by the late 1880s, and the fish may also have been present in Lake Ontario for many thousands of years as a glacial relict. It was first observed in Lake Erie in 1931, Lake Huron in 1933, Lake Michigan in 1949 and Lake Superior in 1954 (Christie 1974). In 1957, alewife entered the commercial catch for Lake Michigan (Baldwin *et al.* 2002).

The rapid rate at which exotic species such as alewife and smelt were able to populate the Great Lakes is believed to be due, in large measure, to intense commercial fishing pressures, which removed the large predatory lake trout (Christie 1974; Hartman 1988). With this weakening in predatory pressures, alewife and smelt were able to proliferate rapidly through the Great Lakes, where they remain to this day.

The alewife is a size-selective planktivore; increased abundance is believed to have affected compositional changes in zooplankton assemblages in lakes Michigan, Ontario and Erie (Wells 1970; McNaught & Buzzard 1973). In later years, it was hypothesised that changes in size-selective predation pressure on the zooplankton assemblage were a major factor affecting long-term increase in algal abundance in Lake Michigan (Brooks *et al.* 1984; Edgington 1984).

## 3.5 PHOSPHORUS CONTROL PROGRAMMES IN THE LAURENTIAN GREAT LAKES

By the early 1970s, researchers had established that the Laurentian Great Lakes represented, to varying degrees, perturbed ecosystems. In 1972, the Great Lakes Water Quality Agreement (GLWQA), with the chief objective that phosphorus concentrations should be reduced to no more than $1\,mg\,L^{-1}$ in discharges from large sewage plants emptying into lakes Erie and Ontario, was signed between Canada and the USA. New limits also were set on industry (Environment Canada 1995). The 1972 GLWQA also called for research, monitoring and surveillance studies on the Great Lakes, in order to investigate problems, develop solutions and to assess the effectiveness of remedial actions.

Chapra's (1977) total phosphorus model for the Great Lakes established historical trends of increased phosphorus loadings to the lakes (Fig. 3.4). His study showed that Lake Erie had experienced the greatest increase in loading, followed by lakes Ontario and Michigan. For example, loadings on Lake Erie began to increase during the mid-twentieth century, as populations increased from c.0.25 to 1.25 million (equivalent to 3.2 to 16.0 people $km^{-2}$). Lakes Huron and Superior, whose basins contained smaller populations, received the least increases in loadings. Initially, increased loading was associated with land runoff related to clear-cutting, and to nearshore habitat disruptions. However, during the twentieth century, with increase in population size, and the concomitant expansion of sewer systems delivering untreated sewage to the lakes, human wastes became increasingly more important. Total phosphorus loadings to Lake Michigan decreased during the early twentieth century, with the diversion of the Chicago River away from Lake Michigan, and of sewage into the Chicago Sanitary and Ship Canal. These systems then flowed into the Illinois and Mississippi Rivers (Hatcher & Walter 1963). High phosphate detergents became of increasing importance as a phosphorus source during the 1940s.

In 1978, a further Great Lakes Water Quality Agreement (GLWQA) was signed between Canada and the USA, revising the 1972 Agreement. The 1978 GLWQA called for implementation of several measures to correct progressive eutrophication of the Great Lakes. Targets were set for reductions in total phosphorus loadings from 1976 estimates. These ranged from a reduction of 3000 to 2800 metric tonnes per year ($t\,yr^{-1}$) for the main body of Lake Huron, to 20,000 to 11,000 $t\,yr^{-1}$ for Lake Erie.

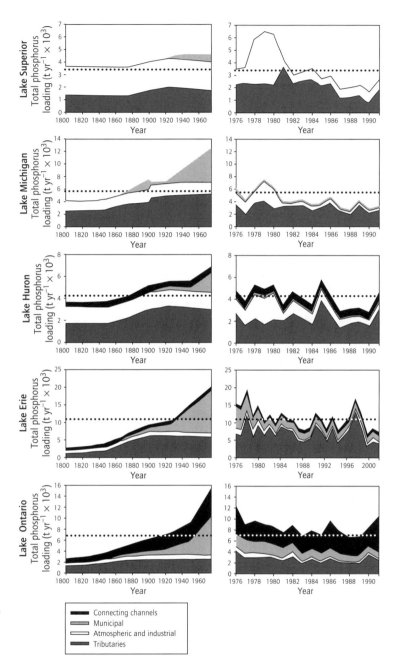

Fig. 3.4 Estimates of total phosphorus loading rates to the Laurentian Great Lakes, including sources. (Redrawn from Chapra (1977) and from Neilson et al. (1995). Lake Erie data for 1991–2001 are from Dolan & McGunagle (in preparation).)

Targets for Lake Huron were later changed to 4300 t yr$^{-1}$ in the renegotiated 1978 Agreement (Neilson et al. 1995). Loadings to Saginaw Bay were to be reduced from 870 to 440 t yr$^{-1}$.

Target loadings were not set at that time for other embayments such as the Bay of Quinte and Green Bay. The Agreement called for construction of municipal waste treatment plants in order to help meet these goals. Furthermore, municipal waste treatment plants discharging more than one million gallons (>4.5 million litres) per day were required to achieve total phosphorus concentrations of $1.0\,mg\,L^{-1}$ in the basins of lakes Superior, Michigan and Huron, and $0.5\,mg\,L^{-1}$ in those of lakes Erie and Ontario. Phosphorus concentrations in household detergents were reduced to 0.5% where necessary, in order to meet loading allocations. Finally, the 1978 GLWQA called for the restoration of year-round aerobic conditions in the bottom waters of central Lake Erie.

In the USA, phosphorus control strategies focused on municipal sewage and industrial wastes. However, between 1973 and 1979, the states of New York, Indiana, Wisconsin and Minnesota enacted legislation to limit phosphorus in detergents to no more than 0.5% by weight (Great Lakes Water Quality Board 1985). Canada had, by 1973, enacted legislation which limited phosphorus concentrations in detergents to 2.2% by weight. Since the implementation of these remedial plans, there have been significant reductions in point-source loadings to the Great Lakes, primarily through improvements in municipal waste treatment plant operations. Diffuse sources have been more difficult to control but various programmes have been implemented in order to reduce those loadings.

In Lake Superior, target loadings under the new phosphorus abatement programmes were rapidly met. By the mid-1980s, loadings were similar to or lower than Chapra's (1977) historical loading estimates (Neilson et al. 1995; Fig. 3.4). This suggests that Chapra's loading estimates for the nineteenth century were too high, especially for atmospheric inputs. Spring total phosphorus concentrations in the open waters of Lake Superior for the period 1983–1992 lay below the $5\,\mu g\,L^{-1}$ established for this lake (Neilson et al. 1995). High concentrations occur only in the western arm of the lake, in the Duluth–Superior Harbour region (Neilson et al. 1995). Chapra predicted that, with the implementation of phosphorus abatement programmes, Lake Superior would continue to remain oligotrophic to the end of the twentieth century, as has been observed. As with many of the Great Lakes, nitrate concentrations have increased more than threefold since the early twentieth century; the atmosphere, followed by tributaries, is the major source (Matheson & Munawar 1978; Thompson 1978; Weiler 1978).

Target loadings for Lake Michigan under the phosphorus abatement programmes were also rapidly met. By the mid-1980s, loadings had declined by approximately 50% over the period 1974–1990, and are now similar to Chapra's (1977) estimates for the early nineteenth century (Neilson et al. 1995; Fig. 3.4). During 1983–1992, spring total phosphorus concentrations in the open waters of Lake Michigan were below the $7\,\mu g\,L^{-1}$ guideline (Neilson et al. 1978). These observations confirm the Chapra model prediction for Lake Michigan (1977), i.e. that with implementation of remedial actions, waters would return to oligotrophic conditions during the early 1980s.

Algal standing stocks in inshore waters, as determined by intake sampling at the Chicago Central Water Filtration Plant, declined markedly from the high densities observed during the mid-1950s (Danforth & Ginsburg 1980), although changes were less obvious in offshore waters (Johengen et al. 1994). Moreover, during the 1970s and 1980s, there was no response to declining phosphorus loading either by the benthos (Nalepa 1987) or the zooplankton (Scavia et al. 1986; Makarewicz et al. 1995). Green Bay experienced major reductions in phosphorus loadings between 1973–1977 and 1978–1991; total phosphorus and chlorophyll concentrations also declined (Millard & Sager 1994).

Target loadings for Lake Huron of $5400\,t\,yr^{-1}$ were rapidly met (Fig. 3.4; Neilson et al. 1995). Part of this attainment was because atmospheric loadings were reduced, in addition to tributary inputs. Most of the cultural eutrophication of Lake Huron occurred in inner Saginaw Bay, but remedial actions implemented in the bay have been successful. Total phosphorus loadings from the Saginaw River declined from 1044t in 1974 to 472t in 1980 (Bierman et al. 1984). Total phosphorus and chlorophyll concentrations in the bay declined over the same period. Spring total phosphorus concentrations in the open waters of Lake Huron generally have remained below $5\,\mu g\,L^{-1}$ (Neilson et al. 1995). Chapra (1977) predicted that, given recommended reductions in loading, the open waters of Lake Huron would continue to remain oligotrophic to the year 2000, as has been observed. Significant enhanced nutrient loadings persist only in Saginaw Bay, and along the Ontario shore of southern Lake Huron (Neilson et al. 1995).

Total phosphorus target loadings for Lake Erie were reached during the mid-1980s, and generally these have been maintained (Neilson et al. 1995; Fig. 3.4). Lake Erie is the only lake for which total phosphorus loadings have continued to be estimated since 1991 (Dolan & McGunagle, in preparation). Most of the decrease loading to Lake Erie was accomplished via improved treatment of municipal waste, particularly that emanating from Detroit, Cleveland, and Toledo (Fraser 1987; Dolan 1993). Chapra (1977) predicted that concentrations in the West Basin would decline rapidly (from c.50 $\mu g\,L^{-1}$ during the late 1960s, to c.30 $\mu g\,L^{-1}$ during the 1980s), and then increase slightly. Concentrations in the Central Basin would decline from c.30 $\mu g\,L^{-1}$ to 15 $\mu g\,L^{-1}$, and in the East Basin from c.25 $\mu g\,L^{-1}$ to c.12 $\mu g\,L^{-1}$. Thus, the Western Basin would remain eutrophic, and the Central and Eastern basins mesotrophic, to the end of the twentieth century.

Spring total phosphorus concentrations have declined in all three basins since the 1970s, and are now approaching the $10\,\mu g\,L^{-1}$ guideline for Lake Erie (Neilson et al. 1995). However, summer concentrations in the surface waters have changed little since the early 1970s (Charlton et al. 1993). These authors hypothesised that strong nutrient regeneration from the sediments was delaying the response of summer surface waters to declining loading. Nevertheless, there is some evidence that zooplankton and phytoplankton populations responded to declining loadings during the 1980s (Makarewicz 1993a, b). Mayflies and oligochaete worms also showed some signs of recovery in the western basin, related to reduced nutrient and contaminant loading (Schloesser et al. 1995; Krieger et al. 1996). However, mussel abundance continued to decline from 1961 to 1982 (Nalepa et al. 1991).

Lake Erie is an interesting ecosystem, which continues to challenge the limnologist. During the 1960s, the lake was reported as being dead; *Hexagenia* abundance had been severely reduced in the western basin, and anoxic conditions in the central basin were of major concern. The 1978 GLWQA (Annex 4) called for 'restoration of year-round aerobic conditions in the bottom waters of the central basin'. Research studies conducted in subsequent years debated whether eutrophication of Lake Erie promoted an increase in the summer oxygen depletion rate of the central basin, or whether this is a natural feature of the central basin. These debates (Charlton 1980; Rosa 1987; Rosa & Burns 1987; Bertram 1993; Charlton et al. 1993) continue to this day.

Palaeolimnological studies clearly have demonstrated that the central basin of Lake Erie was subject to periods of low oxygen concentrations, which probably were climate driven, prior to European colonisation (Delorme 1982; Reynoldson & Hamilton 1993). The most recent evidence suggests that summer oxygen depletion rates are declining with reduced total phosphorus loading to Lake Erie, although weather, particularly long, warm periods, may mask this trend (Neilson et al. 1995). However, since target loadings are more than three times greater than historical loadings (Fig. 3.4), and spring total phosphorus guidelines are $10\,\mu g\,L^{-1}$ versus historical estimates of $<5\,\mu g\,L^{-1}$, the lake is being maintained in a mesotrophic condition, rather than being returned to its historical oligotrophic state. Anoxis is probably more intense than in colonial times.

Target loadings for Lake Ontario were approached during the late 1980s, with inflows from Lake Erie now its primary phosphorus source (Neilson et al. 1995; Fig. 3.4). Nevertheless, target loadings are three times greater than estimated loadings for the early nineteenth century, and thus will not result in restoration of the open waters of the Lake to pristine conditions. Chapra (1977) estimated that Lake Ontario, which was then eutrophic, would become mesotrophic during the late 1970s, with mesotrophy increasing slightly through the remainder of the twentieth century. Lake Ontario spring total phosphorus concentrations declined from 20 to 25 $\mu$g L$^{-1}$ during the early 1970s, to 10 $\mu$g L$^{-1}$, its target concentration, during the mid-1980s, with values changing little since (Stevens & Neilson 1987; Johengen et al. 1994; Neilson et al. 1995; Nicholls et al. 2001).

On the basis of Chapra & Robertson's (1977) study, total phosphorus concentrations in Lake Ontario are about twice as great as in pre-colonial times. Primary production rates have decreased, as has algal biomass (but not chlorophyll concentration); some eutrophic species have become rare (Gray 1987; Millard et al. 1996). Nitrogen/phosphorus ratios have increased, and may have caused some of the observed changes in algal composition. Epilimnetic zooplankton standing stocks also have declined (Mills et al. 2003).

Although total phosphorus loadings to Lake Ontario have declined markedly, there has been little evidence of recovery in zooplankton standing stocks and community structure towards those characteristics exhibited in less productive lakes such as Michigan. Intense planktivory by the abundant alewife population appears to be exerting a strong influence on zooplankton community structure, with small-bodied zooplankton continuing to characterise the assemblage (Johannsson et al. 1991; Mazumder et al. 1992). The Bay of Quinte also has exhibited a strong reduction in phosphorus loadings since 1977; total phosphorus and chlorophyll were c.41–44% lower for the period 1973–77 (post-phosphorus control) than for 1972–1977 (pre-control; Minns et al. 1986; Millard & Sager 1994). However, despite reduced phosphorus loadings and total phosphorus concentrations, phytoplankton biomass and composition in 1984–85 approached those of the pre-control period (Nicholls & Hurley 1989). This increase has been related to other, parallel changes in the trophic structure of the Bay of Quinte ecosystem.

During the 1980s, benthic populations showed signs of recovery in some nearshore regions of Lake Ontario, but declined in others and in the offshore region (Nalepa 1991). Nalepa hypothesised that these declines may be related to sublethal concentrations of contaminants in the sediments. In more recent years, amphipod and sphaerid populations have declined, as dreissenid mussels have increased in abundance (Mills et al. 2003). Overall, rehabilitation of the Lake Ontario ecosystem has proceeded at a slower rate than that of Lake Michigan (Hartig et al. 1991).

In summary, target total phosphorus loadings have been reached for all five Laurentian Great Lakes, with Superior, Michigan and Huron being managed to maintain oligotrophic conditions. Current loadings are similar to, or slightly higher (Huron) than, those estimated for the early nineteenth century. Target loadings have also been reduced for lakes Erie and Ontario, but only to values required to maintain these lakes at mesotrophic conditions, rather than their early nineteenth century oligotrophic state. Phosphorus loadings from Lake Erie tributary inputs remain high, affecting not only Lake Erie, but also Lake Ontario downstream. Spring total phosphorus concentrations have shown little evidence of continuing to decline in lakes Superior, Huron, Erie and Ontario waters over the 1990s. Summer water temperatures and winds, El-Niño events and dreissenid mussel invasions are now accounting for much of the interannual variability in lake conditions (Nicholls et al. 2001).

## 3.6 FISH MANAGEMENT PROGRAMMES IN THE LAURENTIAN GREAT LAKES

By the mid-1960s, it was evident that fish populations in the Great Lakes were highly perturbed – by intense overfishing pressures, sea lamprey predation and the invasion of exotic species. The commercial fishery was in danger of collapsing, and a public nuisance was created by massive dieoffs of alewife. Restoration of the fishery was based on a three-pronged approach: improved regulation of the commercial fishery; the development of a sea lamprey control programme; and salmonid stocking programmes.

### 3.6.1 The changing nature of the commercial fishery

Commercial fishing on the Great Lakes remains highly regulated. Total commercial catches during the 1970s were not appreciably different from those of the late 1880s (Fig. 3.3). However, catch composition, and hence commercial value, has changed markedly over the decades (Kelso et al. 1996). Moreover, although once widespread throughout the Great Lakes, the commercial fishery has become very limited in terms of its location (Environment Canada 1995).

The overall strategy for managing the commercial fishery has been to limit or to ban the harvesting of most of the species which were commercially important during the late 1880s. Thus, although large numbers of lake trout are stocked to the Great Lakes each year (apart from Lake Erie), they are no-longer harvested commercially, except in limited areas of Lake Superior (Environment Canada 1995). Fishing for lake whitefish (*Coregonus clupeaformis*) is limited to Lake Superior, and to parts of Lake Michigan and Huron. Lake herring (*Coregonus artedii*) continue to be harvested in Lake Superior, but not in the other Great Lakes. Walleye, once commercially important in the Lake Erie, are now managed primarily for sport fishery (Hartman 1988).

Conversely, exotic species have become more important. Alewife are a major component of the Lake Michigan fishery, whereas smelt are significant in Lake Superior and in Lake Erie (Hartman 1988). Overall, historical changes in the commercial fishery for the five Great Lakes are as follows.

The Lake Superior commercial fishery is the healthiest. It remains the only Great Lake where lake trout continue to be harvested commercially, although in restricted areas. Whitefish also continue to be harvested commercially. Catches increased from the 1880s, to reach a peak during the 1940s, but declined through to the early 2000s (Fig. 3.3). In 1977, the total catch was 3812 t, composed mainly of smelt, lake herring and chub (deepwater cisco), followed by lake whitefish and lake trout. Although total phosphorus loadings declined steadily from the late 1970s to the late 1980s, commercial catches did not decline appreciably during this period. Catches during the 1990s have been lower than in earlier decades; in 1999 it was 1990 t, mainly lake whitefish, lake herring and lake trout (Baldwin et al. 2002).

*The State of the Great Lakes 2001 Report* (SOE) notes that catch per unit effort has increased for lake whitefish, is declining for lake herring and that the chub fishery is almost non-existent (Environment Canada and U.S. Environmental Protection Agency 2001). Bronte et al. (2003), reporting on changes in the Lake Superior fish community over the past 30 years, note many signs of recovery, including decreasing smelt populations and recovering lake trout, lake whitefish, lake herring and other native species. Kitchell et al. (2000) model future changes in the Lake Superior fish community, and conclude that lake trout populations cannot be restored to pre-fishery and pre-lamprey numbers, in part because of the presence of exotic species. Further information on the long-term changes in the commercial fishery appear in Lawrie (1978) and Hartman (1988), and MacCallum & Selgeby (1987) discuss some of the management programmes established in order to protect the Lake Superior fishery.

Catches in Lake Michigan remained relatively constant throughout the twentieth century, and during the 1970s were equal to or greater than those observed at the beginning of the century (Fig. 3.3). However, whilst lake trout composed most of the early catches, yields plummeted during the 1940s. Whitefish, also important in the early commercial fishery, continued to be harvested in various regions of Lake Michigan until the mid-1990s, when catches collapsed (Baldwin et al. 2002). Lake herring and chub harvests also declined, and from the mid-1960s through to the end of the 1980s were replaced by alewife (Hartman 1988; Baldwin et al. 2002). The 1977 total commercial catch was 25,215 t, with alewife accounting for 87.1% by weight. Alewife catches declined gradually during the 1980s, to 4860 t in 1990, and then plummeted to near zero. In 1999, the total Lake Michigan commercial fishery catch was only 990 t, made up mostly by chub (940 t), followed by round whitefish and alewife (Baldwin et al. 2002).

The decline in the Lake Michigan catch over the 1980s and 1990s is partially related to decreased phosphorus loadings, which eventually led to a reduction in oligochaete and sphaeriid abundance (Nalepa et al. 1998). Of greater importance is the invasion of several exotic species, including dreissenid mussels, which produced a severe decline in abundance of the amphipod *Diaporeia*. Amphipods are a major food source for a variety of fish species; their declining abundance led to lowered body condition and growth of whitefish, and presumably of other species (Pothoven et al. 2001).

As in lakes Superior and Michigan, early commercial fish catches in Lake Huron were composed mainly of lake trout, whitefish and lake herring. Catches declined throughout the late nineteenth century, increased during the 1920s, and have been declining since the late 1930s (Fig. 3.3). Total catch in 1977 was 2707 t, mainly whitefish, carp (*Cyprinus carpio*) and yellow perch (*Perca flavescens*). Lake herring, which were important in the early fishery, were insignificant in the 1977 catch (Hartman 1988; Baldwin et al. 2002). Most lake herring were harvested in Saginaw Bay, where the population crashed in 1955. This collapse has been related to declining water quality of the Bay, i.e. to nutrient and/or contaminant loading. Although, by 1980, phosphorus loading to Saginaw Bay had markedly declined (Bierman et al. 1984), by 1983, lake herring populations had not recovered (Hartman 1988).

Lake-wide catches increased throughout the late 1970s to 1999, with whitefish accounting for most of the increase. The 1997 catch was 6370 t, mainly whitefish (85.7%), followed by lake trout (3.5%) and introduced Pacific salmon (3.9%). Despite this increase in yield there are continuing concerns with: (i) the loss of nearshore habitat and coastal wetlands; (ii) continued development along the lake shore line; and (iii) fish farms (for rainbow trout) in the Ontario waters of the lake. Sea lamprey populations remain high in the northern lake (Environment Canada and U.S. Environmental Protection Agency 2001).

Lake Erie supports the largest commercial fishery of the five Great Lakes. Early catches were composed mainly of lake whitefish and lake herring. Catches remained relatively constant between the 1920s and the early 1980s (Fig. 3.3) although their composition changed (Hartman 1988; Baldwin et al. 2002). Lake herring harvests plummeted during the early 1920s, and those of whitefish by the early 1950s. The 1977 catch was 20,700 t, mostly yellow perch and smelt. Walleye are also harvested commercially, but this fishery is highly regulated, mainly for sport fishery (Hartman 1988). Hatch et al. (1987) discuss the recovery of the walleye population in western Lake Erie, including management strategies. The 1997 catch, at 17,320 t, is not appreciably different from that of the 1920s, but was made up of smelt (34.4%), walleye (28.0%) and yellow perch (6.8%).

Commercial catches in Lake Ontario are the lowest of those in all five Great Lakes. During the late 1880s, they were lower even than in Lake Superior (Fig. 3.3). Lake trout, sturgeon, whitefish and lake herring made up early catches (Hartman 1988; Baldwin et al. 2002), which then declined sharply during the late 1800s, increased, and then declined again during the 1920s–1930s. Catches increased, and then declined again during the 1940s, and remained relatively constant into the early 1980s. Yellow perch, white perch (*Morone americana*), American eel (*Anguilla rostrata*) and catfish (*Ictalurus* spp.) made up the majority of the 1977 catch of 1222 t.

This harvest was 2.5 times lower than that of 1879, despite higher phosphorus loadings. Catches declined throughout the 1980s, and remained low during the 1990s. In 1999, the catch was only 460 t, composed mainly of whitefish, yellow perch (30.4% each) and bullhead catfish (21.7%). Alewife and smelt continue to make up prey fish populations and may be preventing recovery of more valued species such as whitefish, lake herring and lake trout (Environment Canada and U.S. Environmental Protection Agency 2001).

### 3.6.2 Sea lamprey control

The International Great Lakes Fishery Commission was formed in 1955, with the primary objective of finding the means, and co-ordinating the effort, to control sea lamprey populations in the

Great Lakes (Christie 1974). Eventually, the compound 3-trifluoromethyl-4-nitrophenol (TFM) was developed for application as a lampricide in sea lamprey spawning streams. It has been successfully used on all the Great Lakes where sea lamprey populations are a problem, i.e. all those except Erie, where sea lamprey habitat requirements are not well met. The sea lamprey programme continues because, although it has been successful, it has been impossible to eradicate the species completely. Not every stream is amenable to successful lampricide application. Moreover, the St Mary's River, the outflow from Lake Superior, remains an important spawning habitat and is not readily treated with TFM. Sea lamprey populations remain a regulatory issue into the early 2000s, particularly for Lake Huron (Environment Canada and U.S. Environmental Protection Agency 2001).

### 3.6.3 Stocking programmes

The third strategy for restoring the Great Lakes fish community has been the implementation of stocking programmes. Limited stocking began as early as 1874, with hatcheries being established at various sites around the Great Lakes. These early hatcheries were designed to rear whitefish, in an effort to help restore the declining fishery (Beeton & Chandler 1963). Various fish species were introduced into the Great Lakes during the late nineteenth century, in order to improve the sport fishery. For example, rainbow trout (*Salmo gairdneri*) were deliberately introduced into Lake Superior in 1895, and brown trout (*S. trutta*) around 1900 (Lawrie 1978). Histories of other fish introductions appear in Mills *et al.* (1993). Accidental introductions include alewife, smelt and carp. However, it was not until the late 1950s that a systematic stocking programme was developed in an attempt to restore the predatory lake trout population to the Great Lakes. Stocking programmes also were implemented in order to develop a salmonid-based sport fishery for the Great Lakes.

Lake trout were first stocked in Lake Superior in 1958, when a total of 987,000 fish were planted. During the mid-1960s, these were supplemented with coho and chinook salmon (*Oncorhynchus kitsutch, O. tshawytscha*). Rainbow trout, brown trout, brook trout (*S. fontinalis*) and splake (a lake trout × brook trout hybrid) also were introduced. Limited numbers of Atlantic salmon were stocked on a sporadic basis (MacCallum & Selgeby 1987). Pink salmon (*O. gorbuscha*) were accidentally introduced in 1956 (Lawrie 1978). By the early 1980s, there was strong evidence that offshore populations of lake trout were at or near their pre-sea-lamprey abundance, inshore stocks were reproducing, lake whitefish populations had recovered and lake herring populations were near recovery. Smelt, which had become important in the commercial fishery, had recovered from a major population decline during the early 1980s.

Lake trout stocking programmes in Lake Michigan began in 1965, and continue to date. Stocking has continued for more than 35 years, primarily because stocked populations of lake trout are not self-maintaining. In addition, there is a significant coho, chinook, rainbow and brown trout stocking programme (Eck & Wells 1987; Hartig *et al.* 1991).

Alewife populations began declining in Lake Michigan during the 1970s and early 1980s, apparently in response to increased lake trout stocking (Jude & Tesar 1985; Scavia *et al.* 1986). However, Eck & Wells (1987) argued that the sharp decline in alewife numbers during the early 1980s was due to poor recruitment over the period 1976–1982, a run of years characterised by relatively cold winters. As alewife numbers declined, bloater, rainbow smelt, yellow perch and deepwater sculpin increased, reflecting a strong recovery from the alewife population explosion of the 1960s.

Such top-down changes in the Lake Michigan ecosystem apparently then began to affect the zooplankton community. During the early 1980s, as alewife numbers declined, the large-bodied *Daphnia pulicaria*, a cladoceran with high filtering rates, became abundant in offshore waters. Water clarity in these waters improved markedly during 1982 and 1983. Researchers hypothesised that improved clarity was a result of top-down control of algal biomass (Scavia *et al.* 1986).

Later studies have re-examined some aspects of this research, and have questioned some of the interpretations (Lehman 1988; Evans 1996). In contrast to the offshore region, where zooplankton biomass and abundance remained relatively constant throughout the 1970s and early 1980s, zooplankton standing stocks in the inshore region plummeted (Evans 1986). This decline was associated with the yellow perch population explosion. There were no noticeable effects on water clarity during this period. Fish predation continued to be a primary factor affecting changes in zooplankton community structure in Lake Michigan throughout the 1990s (Makarewicz et al. 1995).

Lake Huron also is stocked with various salmonids, and supports a significant sports fishery based on these and other native species (Environment Canada 1995). Salmonid stocking programmes have had no reported effects on Lake Huron food webs. Lake trout were never important in the Lake Erie ecosystem. However, the lake is managed for other stocked salmonids (Kelso et al. 1996).

Lake trout and salmonid stocking programmes were implemented in Lake Ontario in 1968, and continue to date (O'Gorman et al. 1987; Hartig et al. 1991; Mills et al. 2003). As in Lake Michigan, the principal species released are lake trout, chinook and coho salmon, and brown and rainbow trout. Total salmonid plantings exceeded eight million in 1984, versus over 16 million in Lake Michigan during the same year (Eck & Wells 1987; O'Gorman et al. 1987).

In 1993, stocking was reduced because of concerns with salmonid predator demand versus forage fish prey supply; numbers are now maintained between 4 and 5.5 million annually (Mills et al. 2003). These programmes, along with other restorative measures, were instrumental in improving the status of the Lake Ontario fish community, and other components of the ecosystem (Christie et al. 1987). However, unlike in Lake Michigan, these stocking programmes have not resulted in major declines in alewife abundance, nor in concurrent increases in smelt, deepwater sculpin and bloater numbers. There are several reasons why this has not occurred.

Unlike in Lake Michigan, bloaters and deepwater sculpin are absent from the offshore waters of Lake Ontario; lake herring exist only as a remnant population. Thus, the Lake Ontario offshore fish community is considerably more perturbed than that of Lake Michigan. Interactions between alewife, smelt and salmoninds are complex in Lake Ontario, and limit the effectiveness of salmonind stocking programmes in controlling alewife standing stocks (O'Gorman et al. 1987). When adult alewife are abundant, they feed, along with young-of-the-year (YOY) fish, on the limited zooplankton resources, which can reduce the latter's growth and survivorship rates.

Adult alewife may interact more strongly with YOY fish in Lake Ontario than in Lake Michigan, where zooplankton tend to be larger, and amphipods more abundant. Reduced YOY growth and survivorship, in turn, adversely affects young salmonines. Because total phosphorus loadings to Lake Ontario are being reduced, alewife resource-limitation may become even more significant as primary and zooplankton production rates continue to decline.

One proposed strategy for better managing alewife populations in Lake Ontario is to increase salmonine stocking rates. This would depress alewife abundance, and favour more desirable fish species. However, alewife populations in Lake Ontario are subject to periodic die-offs following severe winters. When these occur, stocked salmonids may become prey-limited. Ultimately, this can negatively affect salmonid growth and survivorship rates. Jones et al. (1993) recommended further research in order better to assess the balance between reducing total phosphorus loading and salmonid stocking to Lake Ontario. Further discussion of food web changes in Lake Ontario (and the Bay of Quinte) can be found in Christie et al. (1987), Minns et al. (1986) and Nicholls & Hurley (1989).

Various actions in an effort to restore the Great Lakes fishery continue. Despite more than 30 years of stocking effort, lake trout populations still are not self-reproducing in most regions of the Great Lakes. A variety of causal factors have been implicated, including (i) poor homing behaviour to historic spawning reefs, (ii) the presence of high concentrations of organic contaminants in the eggs, (iii) thiamine deficiencies owing to parental diet of alewife, (iv) effects of sea lampreys and (v) differences in the genetic composition of hatchery-reared fish versus native (wild) stock, which are better adapted to the Great Lakes environment (Environment Canada and U.S. Environmental Protection Agency 2001).

Loftus *et al.* (1987) predicted that the various remedial actions would meet with success, and that, by the year 2020, fish yields and composition would improve in most of the Great Lakes. They estimated that in Lake Superior yields would increase from $0.38\,kg\,ha^{-1}\,yr^{-1}$ during 1979–1983 to $0.94\,kg\,ha^{-1}\,yr^{-1}$ by the year 2020. Coregonines and salmonines would make up most of the catch. Lake Huron harvests would increase from 0.56 to $1.36\,kg\,ha^{-1}\,yr^{-1}$ by 2020, mostly salmonines and coregonines. In contrast, yields in Lake Michigan would decline from 2.76 to $1.64\,kg\,ha^{-1}\,yr^{-1}$ by 2020, in association with reduced alewife standing stocks. Salmonines and coregonines would be the main species caught.

Catches in Lake Erie were predicted to decline from 9.60 to $7.43\,kg\,ha^{-1}\,yr^{-1}$, and to be composed mainly of rainbow smelt and percids. Lake Ontario yield was predicted to increase from 0.55 to $0.88\,kg\,ha^{-1}\,yr^{-1}$, with salmonines and percids constituting the majority of the catch. However, recent invasions of the Great Lakes by exotic species cast some doubt on these predictions.

## 3.7 RECENT INVASIONS OF THE LAURENTIAN GREAT LAKES BY EXOTIC INVERTEBRATES

There is a long history of introduction of exotic species to the Laurentian Great Lakes, both deliberate and accidental (Mills *et al.* 1993). As already noted, exotic fish species such as alewife and smelt have produced major impacts on fish communities. Alewife populations have also substantially affected zooplankton assemblages in lakes Michigan and Ontario, and, as a consequence, other features of the ecosystem. However, it was not until the 1980s that invertebrate introductions began to emerge as an issue of significant environmental concern.

Most (c.70%) of these species are endemic to Ponto-Caspian basins in eastern Europe (see Löffler 2003, Volume 1, Chapter 2), and are transported to the Great Lakes along with ballast water in ocean-going ships (Reid & Orlova 2002). These species are tolerant of a wide range of environmental conditions, allowing them to survive transoceanic transport, and successfully to invade the Great Lakes. Although the ecological consequences of these invasions for most species are unknown, some, such as dreissenid mussels, have severely affected several Great Lakes.

During the 1980s, the spiny waterflea *Bythotrephes longinmanus* (formerly *B. cederstroem*) became established in all five Great Lakes (Mills *et al.* 1993). This cladoceran is a predator, feeding on rotifers and small zooplankton, including immature *Daphnia*. The species has since become assimilated into all Great Lakes' food webs. Although concerns were expressed that *B. cederstroemi* would depress *Daphnia* abundance in Lake Michigan, and adversely effect zooplanktivorous fish populations, pronounced effects on zooplankton communities have not been observed (Lehman & Caceres 1993; Makarewicz *et al.* 1995). Overall, however, the impact of *B. cederstroemi* on Great Lakes ecosystems has been moderate.

The exotic cladoceran *Cercopagis pengoi* was discovered in Lake Ontario in 1998 and, like *B. cederstroemi*, may have adversely affected small-bodied zooplankton (MacIsaac *et al.* 1999). Because both species possess a long spine, they are not readily consumed by fish. Thus their presence may lead to less efficient trophic transfer from zooplankton to fish.

Zebra mussels (*Dreissena polymorpha*) were first discovered in North America in Lake St Clair during June 1988, and have since become widespread, inhabiting the Hudson River, the upper Mississippi and the Susquehanna drainage basins (Mills *et al.* 1993). A related species, *D. bugensis* (the quagga mussel), also has become widespread (Dermott & Munawar 1993). Dreissenids have produced major negative economic impacts wherever they have become well established, forming dense aggregations on a wide variety of substrates and, in the process, clogging water intakes, encrusting boat hulls and other structures. Preventative and restorative measures have been very costly.

In certain areas, dreissenids, which possess high filtering rates, have produced marked improvements in water quality. In regions where they are abundant, they can filter a major fraction of the water column in a day. Such areas tend to be shallow, rocky and productive, such as Saginaw Bay and parts of Lake Erie. In Saginaw Bay, the zebra mussel invasion in the early 1990s led to a major reduction in total phosphorus and chlorophyll concentrations in the inner bay, and water clarity and macrophyte growth improved (Nalepa & Fahnensteil 1995). Because dreissenids are efficient filter-feeders, they can replace other benthic invertebrates and depress zooplankton abundance.

With the dreissenid population expansion, water clarity also improved in western Lake Erie. Diatom, zooplankton and unionid bivalve abundance declined during the same period (Beeton & Hageman 1992; Holland 1993; Schloesser & Nalepa 1994). In contrast, total phosphorus concentrations did not change appreciably, and dissolved nutrient concentrations increased (Holland et al. 1995).

Dreissenids also adversely affected Lake Ontario during the 1990s (Mills et al. 2003). Populations of sphaerids (fingernail clams) collapsed, as did those of amphipods; zooplankton production declined as the burgeoning mussel population consumed an increasingly large fraction of primary production. Dreissenids also affected the benthos and fish communities of Lake Michigan (Nalepa et al. 1998; Pothoven et al. 2001). Their presence in lakes Erie, Ontario and Michigan continues to be of concern, but their impact has been less pronounced in lakes Superior and Huron.

## 3.8 TOXIC SUBSTANCES IN THE LAURENTIAN GREAT LAKES

Toxic substances became of increasing concern in the Laurentian Great Lakes during the 1970s, as an increasingly wide variety and concentration of contaminants entered from various industrial and agricultural activities in their watersheds. Lakes Ontario, Erie and Michigan were particularly affected. By 1978, the GLWQA contained specific objectives for 12 organic and 11 persistent toxic substances, and four non-persistent toxic organic compounds.

In general, inorganic compounds tended to be of greatest concern in inshore areas, where large amounts of these compounds were released by industrial activities, whereas organic contaminants were more widespread. Many of these inshore regions were later classified as Areas of Concern (AOC), meriting specialised remedial actions. During the 1970s, the commercial fishery was closed in Thunder Bay (Lake Superior) and in western Lake Erie, because of high mercury concentrations in predatory fish (MacCallum & Selgeby 1987; Hartman 1988; Koonce et al. 1996). Such closures were instrumental in enabling these fish populations to recover from overexploitation.

Most concern regarding toxic substances in the Great Lakes has focused on persistent organochlorine contaminants, particularly PCBs, DDT and its metabolites, and toxaphene. These compounds are strongly biomagnified by the biota, reaching their highest concentrations in top predators such as walleye and lake trout (DeVault et al. 1996). Contaminants have reached sufficiently high concentrations in these fish for consumptive advisories to be issued.

Use of DDT was banned in 1970, dieldrin in 1974 and PCB in 1986 (Hesselberg et al. 1990). In the most contaminated areas of the Great Lakes, the primary source of these compounds was runoff and discharges from industrial and agricultural areas; however, the atmosphere also was a major source of these contaminants, which are highly volatile. This was clearly demonstrated in Swain's (1978) investigation of organochlorine contaminants in lake trout collected from Siskiwit Lake, located on Isle Royale in Lake Superior. Fish inhabiting this island lake contained twice the PCB and nearly ten times the $p,p$-DDE concentration of lake trout collected in Lake Superior. Thus, even in the absence of localised sources, significant quantities of persistent organic contaminants enter the

Great Lakes via atmospheric routes, with predaceous fish attaining particularly high concentrations.

Contaminant concentrations have been declining in Great Lakes fish since the implementation of production and use bans (Hesselberg et al. 1990; Borgmann & Whittle 1992; Suns et al. 1993; Huestis et al. 1996; DeVault et al. 1996). Nevertheless, mean PCB concentrations exceed the $1.9\,\mu g\,g^{-1}$ consumption guideline for lake trout from lakes Michigan and Ontario (DeVault et al. 1996; Fig. 3.5). Concentrations in Lake Huron have declined from over $3\,\mu g\,g^{-1}$ to $>1$–$1.6\,\mu g\,g^{-1}$. Mean values in Lake Superior during the early 1990s were in the range $0.2$–$0.5\,\mu g\,g^{-1}$.

In 1992, mean PCB concentrations in Lake Erie walleye were $2.2\,\mu g\,g\,m^{-1}$, and thus above the range of recommended consumption guidelines. Concentrations of DDT have also fallen dramatically, with the slowest decline in Lake Ontario (Fig. 3.5). In recent years, the rate of overall contaminant decline has decreased, for reasons which are not entirely well documented, but which are related to contaminant recycling within the lakes, continued loadings from their watersheds and long-range transport.

The presence of persistent organic contaminants has posed a variety of concerns to the Great Lakes ecosystem, in addition to consumption advisories. Palaeolimnological studies by Warwick (1980) demonstrated an increased incidence of deformities in chironomid head capsules coincident with increased industrial activity in the Bay of Quinte. Gilbertson (1988) presented compelling arguments for impaired bird and mammal health, and the presence of persistent organic contaminants in their environment during the 1960s and 1970s. Researchers continue to argue that persistent organic contaminants are occurring in sufficiently high concentrations in Great Lakes biota to affect the health of fish-eating birds (Fox 1993), salmonids (Leatherland 1993), and lake trout eggs (Mac et al. 1993). There also is evidence of contaminant effects on the cognitive functioning of children born to significant consumers of Great Lakes fish (Jacobson & Jacobson 1993).

## 3.9 AREAS OF CONCERN FOR THE LAURENTIAN GREAT LAKES

Although much progress has been made in improving the environmental status of the open waters of the Great Lakes, a number of regions continue to exhibit serious environmental degradation. These have been designated Areas of Concern (AOC) by the Great Lakes Water Quality Board of the International Joint Commission. Such areas are harbours, river mouths and connecting channels, which have been severely affected by a variety of anthropogenic activities (Environment Canada 1995). Five AOCs exist for the Lake Michigan watershed, six for Lake Huron, eight each for lakes Erie and Ontario, and five for the connecting channels, e.g. the St Mary's River connecting lakes Superior and Huron (Fig. 3.6). Most AOCs are contaminated with heavy metals and toxic organic compounds, primarily from municipal industrial point sources (Table 3.3). As a consequence, sediments are contaminated, biological impacts have been observed and fish consumption advisories issued.

All five connecting channels, all eight of the AOCs in Lake Erie, seven in Lake Ontario, five in Lake Michigan and six in Lake Superior are contaminated by conventional pollutants, e.g. by nutrients, bacteria and oxygen-consuming materials. These pollutants enter from non-point and municipal and industrial point sources. There also are several hazardous waste sites and waste discharge sites of concern within the Great Lakes watershed (Fig. 3.6). There is a particularly large concentration of hazardous waste sites in the Niagara River region. Remedial action plans have been developed for several AOCs. These involve a variety of jurisdictions, partnerships and interest groups. Successes and concerns regarding these plans have recently been discussed in Mackenzie (1993), the Great Lakes Water Quality Board (1997) and Kellogg (1997).

## 3.10 CONCLUDING REMARKS

The North American Great Lakes are one of

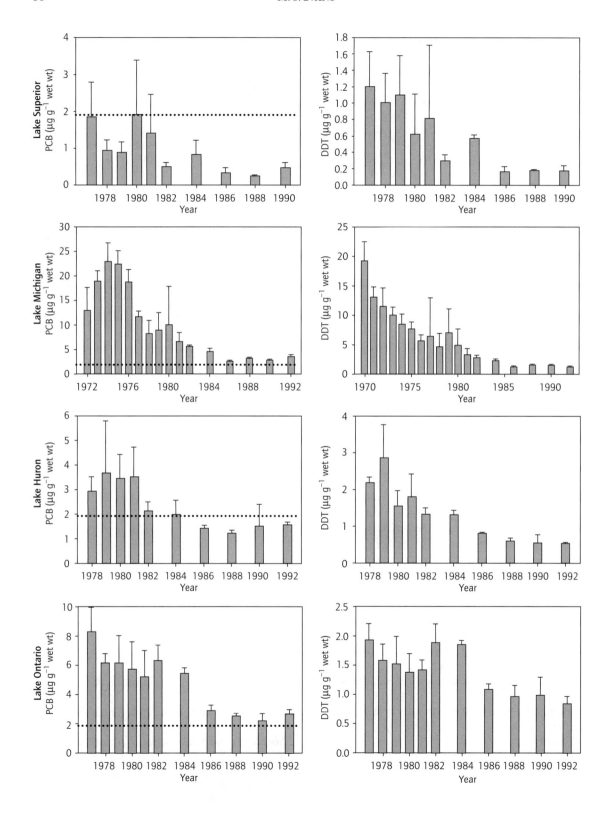

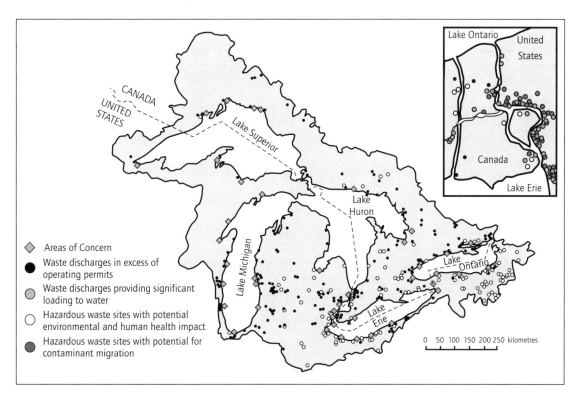

**Fig. 3.6** Map showing areas of concern for the Laurentian Great Lakes drainage basin. Also shown are waste sites and sites where waste discharges are of concern. (Redrawn from Environment Canada (1987).)

**Fig. 3.5** Long-term trends in PCB and DDT concentrations (mean wet weight and 95% confidence interval) in lake trout from lakes Superior, Michigan, Huron and Ontario. Data are from DeVault et al. (1996). Also shown is the upper consumption guideline for PCBs. It is recommended that fish containing PCB concentrations in excess of this guideline (1.9 $\mu$g g$^{-1}$ wet wt) should not be eaten on a regular basis.

Canada's most precious natural resources. Of these lakes, the Laurentian Great Lakes are the most utilised and, as a consequence, the most perturbed. Over the past 100 years, the lakes have been subject to a wide variety of perturbations: habitat loss, overfishing, increased nutrient loading, colonisation by exotic species and increased contaminant loading. By the late 1960s, it had become broadly apparent that the lakes were being rapidly and adversely affected by anthropogenic activities. A wide variety of remedial actions were implemented in order to restore the lakes: reductions in phosphorus and contaminant loadings, limits to the commercial fishery, sea lamprey control and salmonid stocking programmes, to mention a few.

These programmes have met with many successes. Phosphorus loadings to the Great Lakes have diminished and, for lakes Superior, Huron and Michigan, appear to be reaching historically low values. Conversely, although total phosphorus loadings to lakes Erie and Ontario have been significantly reduced, current and target loadings will ensure that these lakes will remain mesotrophic, rather than returning to their historical oligotrophic state.

**Table 3.3** Areas of concern, types of environmental problems and sources of pollution for the Laurentian Great Lakes. (Modified from Environment Canada 1995)

| Lake basin | Area of concern | Types of problems* | | | | | Source of pollution† | | | | |
|---|---|---|---|---|---|---|---|---|---|---|---|
| | | 1 | 2 | 3 | 4 | 5 | 6 | 7 | 8 | 9 | 10 |
| Lake Superior | Peninsula Harbour | X | X | X | | X | | | X | | X |
| | Jackfish Bay | X | X | X | | X | | | X | | X |
| | Nipigon Bay | X | X | X | | X | | | X | | X |
| | Thunder Bay | X | X | X | | X | X | | X | | X |
| | St. Louis River | X | X | X | | X | | | | | X |
| | Torch Lake | X | X | X | | X | | | X | | X |
| | Deer Lake–Carp Creek–Carp River | | X | X | | X | | | | | |
| Lake Michigan | Manistique River | | X | X | | X | | | X | | X |
| | Menominee River | | X | X | | | | | | | X |
| | Fox River/Southern Green Bay | X | X | X | | X | | | X | | X |
| | Sheboygan | | X | X | | X | | | | | X |
| | Milwaukee Estuary | X | X | X | | X | X | X | X | | X |
| | Waukegan Harbor | | X | X | | X | | | | | X |
| | Grand Calumet River/Indiana Harbour Canal | X | X | X | | X | X | X | X | | X |
| | Kalamazoo River | | X | X | | X | | | | | |
| | Muskegon Lake | | | | | | | | | | |
| | White Lake | | X | | | | | | X | | X |
| Lake Huron | Saginaw River/Saginaw Bay | | X | X | X | X | | X | X | | X |
| | Collingwood Harbour | | X | X | X | X | | | X | | X |
| | Penetang Bay to Sturgeon Bay | | | | X | | | X | X | | |
| | Spanish River Mouth | | | X | | | | | X | | X |
| Lake Erie | Clinton River | X | X | X | | X | | X | X | | X |
| | Rouge River | X | X | X | | X | | X | X | | X |
| | Raisin River | X | X | X | | X | | X | X | | X |
| | Maumee River | X | X | X | X | X | | X | X | | X |
| | Black River | X | X | X | X | X | X | X | X | X | X |
| | Cuyahoga River | X | X | X | | X | | X | X | | X |
| | Ashtabula River | X | X | X | | X | | X | X | | X |
| | Wheatley Harbour | X | X | X | X | X | | | X | | X |
| Lake Ontario | Buffalo River | X | X | X | | X | | | X | X | X |
| | Eighteen Mile Creek | X | X | X | | | | X | X | | X |
| | Rochester Embayment | X | X | X | | X | | X | X | | X |
| | Oswego River | X | X | X | | X | | X | X | X | X |
| | Bay of Quinte | X | X | | X | | | X | X | | |
| | Port Hope | | X | | | | | | | X | X |
| | Toronto Waterfront | X | X | X | | X | X | X | X | | X |
| | Hamilton Harbour | X | X | X | X | X | | X | X | | X |
| Connecting channels | St. Marys River | X | X | X | | X | | X | X | | X |
| | St. Clair River | X | X | X | | X | X | | X | | X |
| | Detroit River | X | X | X | X | X | X | X | X | | X |
| | Niagara River | X | X | X | | X | | X | X | X | X |
| | St. Lawrence River | X | X | X | | X | X | X | X | X | X |

\* 1, conventional pollutants; 2, heavy metal and toxic organics; 3, contaminated sediments; 4, eutrophication; 5, biological impacts and fish advisories; 6, beach closings.

† 7, non-point sources; 8, municipal industrial point sources; 9, waste disposal sites; 10, in-place pollutants.

Fish management programmes have been successful. Fish yield and composition for lakes Superior, Michigan and Huron are predicted to return to many of the features exhibited by fish communities during the late nineteenth century. However, fish communities in Lake Ontario and Erie have been more severely disrupted, with major species losses, and it is not known when these lakes will recover fully. Exotic species such as alewife and smelt have become well established in the Great Lakes, and are predicted to remain a major feature of fish assemblages for years to come.

Introduction of new exotic species, particularly dreissenid mussels, is disrupting the base of the food web, with effects reverberating along the food chain to highly valued commercial and sport fish species. Salmonid stocking programmes have been successful but, with limited natural reproduction, will continue for some years to come. Should lake trout populations recover fully in the Great Lakes and become self-maintaining, the desirability of continuing to stock the lake with non-indigenous salmon species will be questioned. For Lake Ontario, compromises must be developed between the objectives of further reducing total phosphorus loading to the lake, and of maintaining a strong sport fishery.

Contaminant loadings to the Laurentian Great Lakes have been reduced in the last two decades. Nevertheless, persistent compounds such as PCB, toxaphene and DDT are ubiquitous and, in recent years, concentrations have been slow to decline. It is highly likely that these substances will be of concern for several more decades to come. Remedial action plans for the Areas of Concern are proceeding well, but many Areas will require a long time in order to recover to a healthy state.

Many impacts to the Laurentian Great Lakes have been hard to predict. Colonisations by new exotic species, such as zebra mussels, quagga mussels and others, were not readily predictable. However, they have rapidly affected the Great Lakes, including various aspects of ecosystem structure and function. Exotic species are highly likely to continue being introduced into the Great Lakes system, as transportation becomes more rapid and travel more global. Climate change, including increased UV-B incidence, also is of growing concern. In recent years, researchers have become increasingly aware of the way in which small variations in climate from one year to the next can produce major impacts upon total phosphorus concentrations, deep-water oxygen concentrations and year-class strength of fish populations.

Several factors have contributed to the success of the recovery of the Laurentian Great Lakes. Canada and the USA have worked together in order to restore the lakes, through agreements such as the 1972 and 1978 GLWQAs, and through bodies such as the International Joint Commission and the International Great Lakes Fishery Commission, their boards and committees. There is a strong commitment to surveillance, and to monitoring studies of the Great Lakes. Results of these studies are reported and published regularly. New approaches are constantly being developed, e.g. acoustic surveys of fish populations.

Research programmes also are strong, with several federal agencies (e.g. the Canada Centre for Inland Waters (National Water Research Institute and Department of Fisheries and Oceans), the Great Lakes Environmental Research Laboratory (National Oceanic and Atmospheric Administration), the U. S. Biological Survey (formerly the US Fish and Wildlife Service) and the Environmental Protection Agency) devoted to improved management of the Great Lakes being located within their drainage basin. There also are a number of state and provincial agencies which maintain significant research, surveillance and monitoring programmes on the Great Lakes. University contributions to Great Lakes research are strong, through partnerships with government laboratories, and via direct support from agencies such as Natural Sciences and Engineering Research Council of Canada, Sea Grant, National Oceanic and Atmospheric Administration, the Environmental Protection Agency, and the National Science Foundation, to name but few. The International Association for Great Lakes Research, founded in 1968, is also important, sponsoring an annual conference on Great Lakes issues, and publishing the *Journal of Great Lakes Research*, a multidisciplinary journal for the further dissemination of research results. The *Canadian Journal of Fishery and Aquatic Science* (formerly the *Journal of the Fishery Research Board of Canada*) is also a vital forum for the publication of multidisciplinary studies on the Laurentian Great Lakes.

Other North American Great Lakes, however, are relatively poorly studied. Because population around these lakes is low, human impact has been minor (Evans 2000). Lake Winnipeg, however, is showing signs of eutrophication, and some exotic fish species have invaded the lake. The massive Red River flood, which occurred in spring 1997, potentially introduced new exotic taxa into the lake (Stewart *et al.* 2003). Research investigations have detected significant anthropogenic inputs of inorganic and organic pollutants to Lake Athabasca and the Great Slave Lake, although contaminant concentrations tend to be low (Evans 2000). There is also some evidence of increased nutrient loading to the Great Slave Lake. Great Bear Lake is the most poorly studied: research is likely to detect increased inorganic and organic contaminant, and increased nutrient loading to this lake.

As Canada continues to develop the North, its northern Great Lakes will be investigated in more detail. If these studies are successful, the region will be developed without exerting major perturbations on these great lakes. Thus, the northern Great Lakes may avoid much of the environmental degradation, much of it permanent, experienced by the Laurentian Great Lakes.

## 3.11 REFERENCES

Baldwin, N. A., Saalfeld, R. W., Dochoda, M. R., et al. (2002). Commercial Fish Production in the Great Lakes 1867–2000. http://www.glfc.org/databases/commercial/commerc.asp

Beeton, A. M. (1964) Evidence for the eutrophication of Lake Erie from phytoplankton records. *Limnology and Oceanography*, **9**, 275–83.

Beeton, A. M. (1969) Changes in the environment and biota of the Great Lakes. In: Rohlich, G. (ed.), *Eutrophication: Causes, Consequences and Correctives*. National Academy of Sciences, Washington, 150–87.

Beeton, A. M. & Chandler, D. C. (1963) The St. Lawrence Great Lakes. In: Frey, D. G. (ed.), *Limnology of North America*. University of Wisconsin Press, Madison, 535–58.

Beeton, A. M. & Hageman, Jr., J. (1992) Impact of *Dreissena polymorpha* on the zooplankton community of western Lake Erie. *Verhandlungen Internationalen Vereinigung Limnologie*, **25**, 2349.

Berst, A. H. & Spangler, G. R. (1973) *Lake Huron. The Ecology of the Fish Community and Man's Effects on it*. Technical Report 21, Great Lakes Fishery Commission, Ann Arbor, MI, 41 pp.

Bertram, P. (1993) Total phosphorus and dissolved oxygen trends in the central basin of Lake Erie, 1970–1991. *Journal of Great Lakes Research*, **19**, 224–36.

Bierman, V. J., Dolan, D. M. & Kasprzyk, R. (1984) Retrospective analysis of the response of Saginaw Bay, Lake Huron, to reductions in phosphorus loadings. *Environmental Science and Technology*, **18**, 23–31.

Borgmann, U. & Whittle, D. M. (1992) DDE, PCB, and mercury concentration trends in Lake Ontario rainbow smelt (*Osmerus mordax*) and slimy sculpin (*Cottus cognatus*): 1977 to 1988. *Journal of Great Lakes Research*, **18**, 298–308.

Brandt, S. B. (1986) Disappearance of the deepwater sculpin (*Myoxocephalus thompsoni*) from Lake Ontario: the keystone predator hypothesis. *Journal of Great Lakes Research*, **12**, 18–24.

Bronte, C. R., Ebener, M. P., Schreiner, D. R., et al. (2003) Fish community change in Lake Superior, 1970–2000. *Canadian Journal of Fisheries and Aquatic Science*, **60**, 1552–74.

Brooks, A. S., Warren, G. J., Boraas, M. E., et al. (1984) Long-term phytoplankton shifts in Lake Michigan: cultural eutrophication or biotic shifts. *Verhandlungen Internationalen Vereinigung Limnologie*, **22**, 452–9.

Brunskill, G. J., Campbell, P., Elliott, S., et al. (1975) Rates of transport of total phosphorus and total nitrogen in Mackenzie and Yukon River watersheds, N. W. T. and Y. T., Canada. *Verhandlungen Internationalen Vereinigung Limnologie*, **19**, 3199–203.

Chapra, S. C. (1977) Total phosphorus model for the Great Lakes. *Journal of the Environmental Engineering Division, American Society of Civil Engineers*, **103**, 147–61.

Chapra, S. C. & Robertson, A. (1977) Great Lakes eutrophication: the effects of point source control of total phosphorus. *Science*, **196**, 1448–50.

Charlton, M. N. 1980. Oxygen depletion in Lake Erie: has there been any change? *Canadian Journal of Fisheries and Aquatic Science*, **37**, 72–81.

Charlton, M. N., Milne, J. E., Booth, W. G. & Chiocchio, F. (1993) Lake Erie offshore in 1990. Restoration and resilience in the Central Basin. *Journal of Great Lakes Research*, **19**, 291–309.

Christie, W. J. (1974) Changes in the fish species composition of the Great Lakes. *Journal of the Fisheries Research Board of Canada*, **31**, 827–54.

Christie, W. J., Scott, K. A., Sly, P. G. & Strus, R. H. (1987) Recent changes in the aquatic food web of eastern Lake Ontario. *Canadian Journal of Fisheries and Aquatic Science*, **44** (Supplement 2), 37–52.

Cook, D. G. & Johnson, M. G. (1974) Benthic macroinvertebrates of the St. Lawrence Great Lakes *Journal of the Fisheries Research Board of Canada*, **31**, 763–82.

Damman, K. E. (1960) Plankton studies of Lake Michigan. II. Thirty-three years of continuous plankton and coliform data collected from Lake Michigan at Chicago, Illinois. *Transactions of the American Microscopical Society*, **79**, 397–404.

Danforth, W. F. & Ginsburg, W. (1980) Recent changes in the phytoplankton of Lake Michigan near Chicago. *Journal of Great Lakes Research*, **6**, 307–14.

Davis, C. C. (1964) Evidence for the eutrophication of Lake Erie from phytoplankton records. *Limnology and Oceanography*, **9**, 275–283.

DeLorme, L. D. (1982) Lake Erie oxygen: the prehistoric record. *Canadian Journal of Fisheries and Aquatic Science*, **39**, 1021–9.

Dermott, R. & Munawar, M. (1993) Invasion of Lake Erie offshore sediments by *Dreissena*, and its ecological implications. *Canadian Journal of Fisheries and Aquatic Science*, **50**, 2298–304.

DeVault, D. S., Hesselberg, R., Rodgers, P. W. & Feist, T. J. (1996) Contaminant trends in lake trout and walleye from the Laurentian Great Lakes. *Journal of Great Lakes Research*, **22**, 884–95.

Dobson, H. F. H. & Gilbertson, M. (1971) Oxygen depletion in the central basin of Lake Erie, 1929 to 1970. In: *Proceedings of the 14th Conference on Great Lakes Research*, International Association for Great Lakes Research, Ann Arbor, MI, 743–8.

Dobson, H. F. H., Gilbertson, M. & Sly, P. G. (1974) A summary and comparison of nutrients and related water quality in Lakes Erie, Ontario, Huron, and Superior. *Journal of the Fisheries Research Board of Canada*, **31**, 731–8.

Dolan, D. (1993) Point source loading of phosphorus to Lake Erie. *Journal of Great Lakes Research*, **19**, 212–23.

Eck, G. W. & Wells, L. (1987) Recent changes in Lake Michigan's fish community and their probable causes, with emphasis on the role of alewife (*Alosa pseudoharengus*). *Canadian Journal of Fisheries and Aquatic Science*, **44** (Supplement 2), 53–60.

Edgington, D. (1984) Great Lakes eutrophication: fish, not phosphates. In: *The Future of Great Lakes Resources, 1982–1984. Biannual Report*. Sea Grant Report WIS-SG-84-145, University of Wisconsin, Madison, WI, 25–33.

Environment Canada (1987) *The Great Lakes. An Environmental Atlas and Resource Book*. Environment Canada, Toronto, 44 pp.

Environment Canada (1995) *The Great Lakes. An Environmental Atlas and Resource Book*. Environment Canada, Toronto, 46 pp.

Environment Canada and the U. S. Environmental Protection Agency (2001) *State of the Great Lakes 2001*. EPA 905-R-01-003, U. S. Environmental Protection Agency, Burlington, Ontario and Chicago, IL, 82 pp.

Evans, M. S. (1986) Recent major declines in zooplankton populations in the inshore region of Lake Michigan: probable causes and consequences. *Canadian Journal of Fisheries and Aquatic Science*, **43**, 154–9.

Evans, M. S. (1996) Historic changes in Lake Michigan zooplankton community structure: the 1960s revisited with implications for top-down control. *Canadian Journal of Fisheries and Aquatic Science*, **49**, 1734–49.

Evans, M. S. (2000) The Large Lake Ecosystems of Northern Canada. *Aquatic Ecosystems Health and Management*, **3**, 65–79.

Fee, E. J., Stainton, M. P. & Kling, H. J. (1985) Primary production and related limnological data for some lakes of the Yellowknife, NWT area. *Canadian Technical Reports in Fisheries and Aquatic Science*, **1409**, 55 pp.

Flannagan, J. F., Cobb, D. G. & Flannagan, P. M. (1994) A review of the research on benthos of Lake Winnipeg. *Canadian Technical Reports in Fisheries and Aquatic Science*, **2261**, 1–17.

Fox, G. A. (1993) What have biomarkers told us about the effects of contaminants on the health of fish-eating birds in the Great Lakes. The theory and a literature review. *Journal of Great Lakes Research*, **19**, 722–36.

Franzin, W. G., Stewart, K. W., Heuring, L. & Hanke, G. (1996) The fish and fisheries of Lake Winnipeg. In: Todd, B. J., Lewis, C. F., Thorleifson, L. H. & Nielsen, E. (eds), *Lake Winnipeg Project: Cruise Report and Scientific Results*. Open File 3113, Geological Survey of Canada, Ottawa, 349–54.

Fraser, A. S. (1987) Tributary and point source total phosphorus loading to Lake Erie. *Journal of Great Lakes Research*, **13**, 659–66.

Gilbertson, M. (1988) Epidemics in birds and mammals caused by chemicals in the Great Lakes. In: Evans, M. S. (ed.), *Toxic Contaminants and Ecosystem Health: a Great Lakes Focus*. Advances in Environmental Science and Technology, Vol. 21. John Wiley & Sons, New York, 133–52.

Gray, I. (1987) Differences between nearshore and offshore phytoplankton communities in Lake Ontario. *Canadian Journal of Fisheries and Aquatic Science*, **44**, 2155–2163.

Great Lakes Water Quality Board (1985) *1985 Report on Great Lakes Water Quality*. International Joint Commission, Windsor, Ontario, 212 pp.

Great Lakes Water Quality Board (1997) Great Lakes Water Quality Board position on the future of Great Lakes Remedial Action Plans, September 1996. *Journal of Great Lakes Research*, **23**, 212–21

Hartman, W. L. (1988) Historical changes in the major fish resources of the Great Lakes. In: Evans, M. S. (ed.), *Toxic Contaminants and Ecosystem Health: a Great Lakes Focus*. Advances in Environmental Science and Technology, Vol. 21. John Wiley & Sons, New York, 53–76.

Hartig, J. H., Kitchell, J. F., Scavia, D. & Brandt, S. B. (1991) Rehabilitation of Lake Ontario: the role of nutrient reduction and food web dynamics. *Canadian Journal of Fisheries and Aquatic Science*, **48**, 1574–80.

Hatch, R. W., Nepszy, S. J., Muth, K. M. & Baker, C. T. (1987) Dynamics of the recovery of Western Lake Erie walleye (*Stizostedion vitreum vitreum*) stock *Canadian Journal of Fisheries and Aquatic Science*, **44**, 15–22.

Hatcher, H. & Walter, E. A. (1963) *A Pictorial History of the Great Lakes*. Bonanza Books, New York., 344 pp.

Herdendorf, C. F. (1982) Large lakes of the world. *Journal of Great Lakes Research*, **8**, 379–412.

Hesselberg, R. J., Hickey, J. P., Nortrup, D. A. & Wilford, W. A. (1990) Contaminant residues in the bloater (*Coregonus hoyi*) of Lake Michigan, 1969–1986. *Journal of Great Lakes Research*, **16**, 121–9.

Holland, R. (1993) Changes in planktonic diatoms and water transparency in Hatchery Bay, Bass Island area, western Lake Erie since the establishment of the zebra mussel *Journal of Great Lakes Research*, **19**, 617–24.

Holland, R. E., Johengen, T. H. & Beeton, A. M. (1995) Trends in nutrient concentrations in Hatchery Bay, western Lake Erie, before and after *Dreissena polymorpha*. *Canadian Journal of Fisheries and Aquatic Science*, **52**, 1202–9.

Huestis, S. Y., Servos, M. R., Whittle, D. M. & Dixon, D. G. (1996) Temporal and age-related trends in levels of polychlorinated biphenyl congeners and organochlorine contaminants in Lake Ontario lake trout (*Salvelinus* namaycush). *Journal of Great Lakes Research*, **22**, 310–30.

Jacobson, J. L. & Jacobson, S. W. (1993) A four-year follow up study of children born to consumers of Lake Michigan fish. *Journal of Great Lakes Research*, **19**, 776–83.

Jensen, J., Adare K. & Shearer, R. (1997) *Canadian Arctic Contaminants Report*. Departm ent of Indian Affairs and Northern Development, Ottawa, 459 pp.

Johannsson, O. E., Mills, E. L. & O'Gorman, R. (1991) Changes in the nearshore and offshore zooplankton communities in Lake Ontario: 1981–88. *Canadian Journal of Fisheries and Aquatic Science*, **48**, 1546–57.

Johengen, T. H., Johannsson, O. E., Pernie, G. L. & Millard, E. S. (1994) Temporal and seasonal trends in nutrient dynamics and biomass measures in Lakes Michigan and Ontario in response to phosphorus control. *Canadian Journal of Fisheries and Aquatic Science*, **51**, 2570–8.

Johnson, L. (1975a) Physical and chemical characteristics of Great Bear Lake, Northwest Territories. *Journal of the Fisheries Research Board of Canada*, **32**, 1971–6.

Johnson, L. (1975b) Distribution of fish species in Great Bear Lake, Northwest Territories, with reference to zooplankton, benthic invertebrates, and environmental conditions *Journal of the Fisheries Research Board of Canada*, **32**, 1989–2004.

Jones, M. L., Koonce, J. F. & O'Gorman, R. (1993) Sustainability of hatchery-dependent salmonine fisheries in Lake Ontario: the conflict between predator demand and prey supply. *Transactions of the American Fisheries Society* **122**, 1002–1018.

Jude, D. J. & Tesar, F. J. (1985) Recent changes in the forage fish of Lake Michigan. *Canadian Journal of Fisheries and Aquatic Science*, **42**, 1154–1157.

Kellogg, W. (1997) Lessons from RAPs: citizen participation and the ecology of the community. *Journal of Great Lakes Research*, **23**, 227–229.

Kelso, J. R. M., Steedman, R. J. & Stoddart, S. (1996) Historical causes of change in Great Lakes fish stocks and the implications for ecosystem rehabilitation. *Canadian Journal of Fisheries and Aquatic Science*, **53** (Supplement 1), 10–19.

Kitchell, J. F., Cox, S. P., Harvey, C. J., *et al.* (2000) Sustainability of the Lake Superior fish community: interactions in a food web context. *Ecosystems*, **3**, 545–60.

Kling, H. J. (1996) Fossil and modern phytoplankton from Lake Winnipeg. In: B. J. Todd, C. F. Lewis, L. H. Thorleifson & E. Nielsen. *Lake Winnipeg Project: Cruise Report and Scientific Results.* Open File 3113, Geological Survey of Canada, Ottawa, Ontario, 283–310.

Koonce, J. F., Busch, W. D. N. & Czapla, T. (1996) Restoration of Lake Erie: contribution of water quality and natural resource management. *Canadian Journal of Fisheries and Aquatic Science*, **53** (Supplement 1), 105–12.

Krieger, K. A., Schloesser, D. W., Manny, B. A., et al. (1996) Recovery of burrowing mayflies (Ephemeroptera: Ephemeridae: Hexagenia) in western Lake Erie. *Journal of Great Lakes Research*, **22**, 254–63.

Lawrie, A. H. (1978) The fish community of Lake Superior. *Journal of Great Lakes Research*, **4**, 513–49.

Leatherland, J. F. (1993) Field observations on reproductive and developmental dysfunction in introduced and native salmonids from the Great Lakes. *Journal of Great Lakes Research*, **19**, 737–51.

Lehman, J. (1988) Algal biomass unaltered by food-web changes in Lake Michigan. *Nature*, **332**, 537–8.

Lehman, J. T. & Caceres, C. E. (1993) Food-web responses to species invasion by a predatory invertebrate: *Bythotrephes* in Lake Michigan. *Limnology and Oceanography*, **38**, 879–91.

Loftus, D. H., Olver, C. H., Brown, E. H., et al. (1987) Partitioning potential fish yields from the Great Lakes. *Canadian Journal of Fisheries and Aquatic Science*, **44** (Supplement 2), 417–24.

Lopatka, S., Ross, D. & Stoesz, R. (1990) *Northwest Territories Data Book*. Outcrop Ltd., Yellowknife, 238 pp.

Löffler, H. (2003) The origin of lake basins. In: O'Sullivan, P. E. & Reynolds, C. S. (eds) *The Lakes Handbook.*, Vol. 1, *Limnology and Limnetic Ecology*. Blackwell Publishing, Oxford, 8–60.

Mac, M. J., Schwartz, T. R. Edsall, C. C. & Frank, A. M. (1993) Polychlorinated biphenyls in Great Lakes lake trout and their eggs: relations to survival and congener composition 1979–1988. *Journal of Great Lakes Research*, **19**, 752–65.

MacCallum, W. R. & Selgeby, J. H. (1987) Lake Superior revisited. *Canadian Journal of Fisheries and Aquatic Science*, **44**, 23–36.

MacIsaac, H. J., Grigorovich, I. A., Hoyle, J. S., et al. (1999) Invasion of Lake Ontario by the Ponto-Caspian predatory cladoceran *Cercopagis pengoi*. *Canadian Journal of Fisheries and Aquatic Science*, **56**, 1–5.

Mackenzie, S. H. (1993) Ecosystem management in the Great Lakes: some observations from three RAP sites. *Journal of Great Lakes Research*, **19**, 136–45.

Makarewicz, J. C. (1993a) Phytoplankton biomass and species composition in Lake Erie, 1970 to 1987. *Journal of Great Lakes Research*, **19**, 258–74.

Makarewicz, J. C. (1993b) A lakewide comparison of zooplankton biomass and its species composition in Lake Erie, 1983–1987. *Journal of Great Lakes Research*, **19**, 275–90.

Makarewicz, J. C., Bertram, P., Lewis, T. & Brown, Jr., E. H. (1995) A decade of predatory control of zooplankton species composition of Lake Michigan *Journal of Great Lakes Research*, **21**, 620–40.

Marsh, J. H. (ed.) (1988) *The Canadian Encyclopedia*, 4 Vols. Hurtig Publishers, Edmonton.

Matheson, D. H. & Munawar, M. (1978) Lake Superior basin and its development. *Journal of Great Lakes Research*, **4**, 249–63.

Mazumder, A., Lean, D. R. S., Taylor, W. D. (1992) Dominance of small filter feeding zooplankton in the Lake Ontario foodweb. *Journal of Great Lakes Research*, **18**, 456–466.

McNaught, D. C. & Buzzard, M. (1973) Changes in zooplankton populations in Lake Ontario (1939–1972). In: *Proceedings of the 16th Conference on Great Lakes Research*. International Association for Great Lakes Research, Ann Arbor, MI, 76–86.

Millard, E. S. & Sager, P. E. (1994) Comparison of phosphorus, light climate, and photosynthesis between two culturally eutrophied bays: Green Bay, Lake Michigan, and the Bay of Quinte, Lake Ontario. *Canadian Journal of Fisheries and Aquatic Science*, **51**, 2579–90.

Millard, E. S., Myles, D. M., Johannsson, O. E. & Ralph, K. M. (1996) Phytoplankton photosynthesis at two index stations in Lake Ontario 1987–92: assessment of the long-term response to phosphorus. *Canadian Journal of Fisheries and Aquatic Science*, **53**, 1112–24.

Mills, E. L., Leach, J. H., Carlton, J. T. & Secor, C. L. (1993) Exotic species in the Great Lakes: a history of biotic crises and anthropogenic introductions. *Journal of Great Lakes Research*, **19**, 1–54.

Mills, W. L., Casselman, J. M., Dermott, R., et al. (2003) Lake Ontario: food web dynamics in a changing ecosystem (1970–2000). *Canadian Journal of Fisheries and Aquatic Science*, **60**, 471–490.

Minns, C. K., Hurley, D. A. & Nicholls, K. H. (eds) (1986) Project Quinte: point-source phosphorus control and ecosystem response in the Bay of Quinte, Lake Ontario. *Canadian Special Publications in. Fisheries and Aquatic Science*, **86**, 270 pp.

Mitchell, P. & Prepas, E. (eds) (1990) *Atlas of Alberta Lakes*. The University of Alberta Press, Edmonton, 675 pp.

Nalepa, T. F. (1987) Long-term changes in the macrobenthos of southern Lake Michigan. *Canadian Journal of Fisheries and Aquatic Science*, **44**, 515–24.

Nalepa, T. F. (1991) Status and trends of the Lake Ontario macrobenthos. *Canadian Journal of Fisheries and Aquatic Science*, **48**, 1558–67.

Nalepa, T. F. & Fahnensteil, G. L. (1995) *Dreissena polymorpha* in Saginaw Bay, Lake Huron Ecosystem: overview and perspective. *Journal of Great Lakes Research*, **21**, 411–6.

Nalepa, T. F., Manny, B. A., Roth, J. C., et al. (1991) Long-term decline in freshwater mussels (Bivalvia: Uniodidae) in the western basin of Lake Erie. *Journal of Great Lakes Research*, **17**, 214–9.

Nalepa, T. F., Hartson, D. J., Fanslow, D. L., et al. (1998) Declines in benthic macroinvertebrate populations in southern Lake Michigan, 1980–1993. *Canadian Journal of Fisheries and Aquatic Science*, **55**, 2402–13.

Nicholls, K. M. & Hurley, D. A. (1989) Recent changes in the phytoplankton of the Bay of Quinte, Lake Ontario: the relative importance of fish, nutrients, and other factors. *Canadian Journal of Fisheries and Aquatic Science*, **46**, 770–9.

Nicholls, K. H., Hopkins, G. J., Standke, S. J. & Nakamoto, L. (2001) Trends in total phosphorus in Canadian near-shore waters of the Laurentian Great Lakes: 1976–1999. *Journal of Great Lakes Research*, **27**, 402–23.

Neilson, M., L'Italien, S., Glumac, V., et al. (1995) *State of the Lakes Ecosystem Conference. Background Paper. Nutrient Trends and System Response.* Environment Canada and EPA 905-R-95-015, Environmental Protection Agency, Burlington, Ontario and Chicago, IL, 20 pp. +8 figs.

O'Gorman, R., Bergtedt, R. A. & Eckert, T. H. (1987) Prey fish dynamics and salmonine predator growth in Lake Ontario, 1978–84. *Canadian Journal of Fisheries and Aquatic Science*, **44** (Supplement 2), 390–403.

Patalas, K. (1975) The crustacean plankton communities of fourteen North American great lakes. *Verhandlungen Internationalen Vereinigung Limnologie*, **19**, 504–11.

Patalas, K. & Salki, A. (1992) Crustacean plankton in Lake Winnipeg: variation in space and time as a function of lake morphology, geology, and climate. *Canadian Journal of Fisheries and Aquatic Science*, **49**, 1035–59.

Pothoven, S. A., Nalepa, T. F., Schneeberger, P. J. & Brandt, S. B. (2001) Changes in the diet and body condition of whitefish in southern Lake Michigan associated with changes in benthos. *North American Journal of Fisheries Management*, **21**, 876–83.

Rawson, D. S. (1950) The physical limnology of Great Slave Lake. *Journal of the Fisheries Research Board of Canada*, **8**, 3–66.

Rawson, D. S. (1951) Studies of the fish of Great Slave Lake. *Journal of the Fisheries Research Board of Canada*, **8**, 207–40.

Rawson, D. S. (1953) The bottom fauna of Great Slave Lake. *Journal of the Fisheries Research Board of Canada*, **10**, 486–520.

Rawson, D. S. (1956) The net plankton of Great Slave Lake. *Journal of the Fisheries Research Board of Canada*, **13**, 53–127.

Ray, A. J. (1996) *I have Lived here since the World Began.* Lester Publishing, Toronto, 398 pp.

Reid, D. F. & Orlova, M. I. (2002) Geological and evolutionary underpinnings for the success of Ponto-Caspian species invasions in the Baltic Sea and North American Great Lakes. *Canadian Journal of Fisheries and Aquatic Science*, **59**, 1144–58.

Reighard, J. E. (1894) A biological examination of Lake St. Clair. *Bulletin Michigan Fisheries Commission*, **4**, 61 pp.

Reynoldson, T. B. & Hamilton, A. L. (1993) Historic changes in populations of burrowing mayflies (*Hexagenia limbata*) from Lake Erie based on sediment tusk profiles. *Journal of Great Lakes Research*, **13**, 250–7.

Rosa, F. (1987) Lake Erie central basin total phosphorus trend analysis from 1968 to 1982. *Journal of Great Lakes Research*, **13**, 667–73.

Rosa, F. & Burns, N. M. (1987) Lake Erie central basin oxygen depletion changes from 1929–1980. *Journal of Great Lakes Research*, **13**, 684–96.

Scavia, D., Fahnenstiel, G. L., Evans, M. S., et al. (1986) Influence of salmonine predation and weather on long-term water quality trends in Lake Michigan. *Canadian Journal of Fisheries and Aquatic Science*, **43**, 435–43.

Schelske, C. L. (1991) Historical nutrient enrichment of Lake Ontario: paleolimnological evidence. *Canadian Journal of Fisheries and Aquatic Science*, **48**, 1529–38.

Schelske, C. L. & Hodell, D. A. (1991) Recent changes in productivity and climate of Lake Ontario detected by isotopic analysis of sediments. *Limnology and Oceanography*, **36**, 961–75.

Schelske, C. L., Stoermer, E. F., Conley, D. J., et al. (1983) Early eutrophication in the lower Great Lakes: new evidence from biogenic silica in sediments. *Science*, **222**, 320–22.

Schenk, C. F. & Thompson, R. E. (1965) Long-term

changes in water chemistry and abundance of plankton at a single sampling location in Lake Ontario. In: *Proceedings Eighth Conference on Great Lakes Research*. Great Lakes Research Division, The University of Michigan, Ann Arbor, MI, 197–208.

Schloesser, D. W. & Nalepa, T. F. (1994) Dramatic decline of unionid bivalves in offshore waters of western Lake Erie after infestation by the zebra mussel, Dreissena polymorpha. *Canadian Journal of Fisheries and Aquatic Science*, **51**, 2234–42.

Schloesser, D. W., Reynoldson, T. B. & Manny, B. A. (1995) Oligochaete fauna of western Lake Erie 1961 and 1982. Signs of sediment quality recovery. *Journal of Great Lakes Research*, **21**, 294–306.

Stevens, T. J. & Neilson, M. A. (1987) Response of Lake Ontario to reductions in phosphorus load, 1967–1982. *Canadian Journal of Fisheries and Aquatic Science*, **44**, 2059–68.

Stewart, A. R., Stern, G. A., Lockhart, W. L., et al. (2003) Assessing trends in organochlorine concentrations in Lake Winnipeg fish following the 1997 Red River flood. *Journal of Great Lakes Research*, **29**, 332–54.

Stoermer, E. F., Kociolek, J. P., Schelske, C. L. & Conley, D. J. (1985) Siliceous microfossil succession in the recent history of Lake Superior. *Proceedings of the Academy of Natural Sciences, Philadelphia*, **137**, 106–18.

Stoermer, E. F., Wolin, J. A. & Schelske, C. L. (1993) Paleolimnological comparison of the Laurentian Great Lakes based on diatoms. *Limnology and Oceanography*, **38**, 1311–6.

Stoermer, E. F., Emmert, G., Julius, M. L. & Schelske, C. L. (1996) Paleolimnological evidence of rapid recent change in Lake Erie's trophic status. *Canadian Journal of Fisheries and Aquatic Science*, **53**, 1451–8.

Suns, K. R., Hitchin, G. G. & Toner, D. (1993) Spatial and temporal trends of organochlorine contaminants in spottail shiners from selected sites in the Great Lakes (1975–1990). *Journal of Great Lakes Research*, **19**, 703–14.

Swain, W. R. (1978) Chlorinated organic residues in fish, water, and precipitation from the vicinity of Isle Royale, Lake Superior *Journal of Great Lakes Research*, **4**, 398–407.

Thompson, M. E. (1978) Major ion loadings to Lake Superior. *Journal of Great Lakes Research*, **4**, 361–9.

Vollenweider, R. A., Munawar, M. & Stadelmann, P. (1974) A comparative review of phytoplankton and primary production in the Laurentian Great Lakes. *Journal of the Fisheries Research Board of Canada*, **31**, 739–62.

Ward, H. B. (1896) Biological examination of Lake Michigan in the Traverse Bay region. *Bulletin Michigan Fisheries Commission*, **6**, 100 pp + 5 plates.

Warwick, W. F. (1980) Paleolimnology of the Bay of Quinte, Lake Ontario: 2,800 years of cultural influence. *Canadian Bulletin of Fisheries and Aquatic Science*, **44**, 2178–84.

Watson, N. H. F. (1974) Zooplankton of the St. Lawrence Great Lakes – species composition, distribution, and abundance. *Journal of the Fisheries Research Board of Canada*, **31**, 783–94.

Watson, N. H. F. & Carpenter, G. F. (1974) Seasonal abundance of crustacean zooplankton and net plankton biomass of Lakes Huron, Erie, and Ontario. *Journal of the Fisheries Research Board of Canada*, **31**, 309–17.

Weiler, R. R. (1978) Chemistry of Lake Superior. *Journal of Great Lakes Research*, **4**, 370–85.

Welch, P. S. (1935) *Limnology*. McGraw-Hill, New York, 471 pp.

Wells, L. (1970) Effects of alewife predation on zooplankton populations in Lake Michigan. *Limnology and Oceanography*, **15**, 556–65.

Wells, L. & Mclain, A. L. (1973) *Lake Michigan. Man's Effects on Native Fish Stocks and other Biota*. Great Technical Report 20, Lakes Fishery Commission, Ann Arbor, MI, 55 pp.

Wetzel, R. G. (1975) *Limnology*. Saunders, Toronto, 743 pp.

Whillans, T. H. (1979) Historic transformations of fish communities in three Great Lakes bays. *Journal of Great Lakes Research*, **5**, 195–215.

# 4 Lake Washington

## W.T. EDMONDSON*

### 4.1 INTRODUCTION

Lake Washington is the second largest natural lake in the state of Washington, but is not particularly distinguished among lakes of the glaciated part of the North American temperate zone, either by area, depth, or chemical or biological features. Although it can be regarded as representative of many such non-humic, softwater lakes of glacial origin, it possesses some features which have made it especially suitable for basic research. At the same time, the entire ecosystem of which the lake is part has been subjected to a series of inadvertent disturbances and intentional manipulations which have increased its value for study.

The earliest data on the biological character of Lake Washington were recorded in 1913, in an exploratory regional study (Kemmerer et al. 1924). The first full year of limnological investigation was 1933–34 (Scheffer & Robinson 1939). Since then, diverse studies have been carried out by many people (Wissmar et al. 1981; Greenberg & Sibley 1994). My associates and I have worked since 1949 with special attention to the responses of the lake to episodes of disturbance, treated as whole-lake quasi-experiments (Edmondson 1993a,b, 1994a,b). The purpose of this chapter is to summarise these studies, in the context of general limnological knowledge.

*Sadly, Tommy Edmondson died before the appearance of this chapter, about which he wrote us several characteristically dry and entertaining letters. It has been finished with the kind help of Dr Sally Abella, to whom we are most grateful. We suspect that this is the last of Tommy's published writings. If so, we are proud that it is the Lakes' Handbook which contains the final publication of a great scientist, and a wonderful man.

### 4.2 FEATURES OF LAKE WASHINGTON

Lake Washington (Fig. 4.1) is 28 km long, with a maximum width of about 5 km. A large island nearly fills the southern part (Fig. 4.1a). The bottom of the lake (Table 4.1) drops down steeply from the shorelines, creating only a small amount of shallow water area along the lake margin. In much of the lake, the sediment surface is slightly raised up in the middle, so that the deepest parts are located toward the shores, giving a slight W shape to the vertical section of the lake (Fig. 4.1b, inset; see Gould & Budinger 1958).

Lake Washington receives water from two rivers and several creeks (Fig. 4.2). The main inlet is the Cedar River, which originates in the Cascade Mountains and enters the south end of the lake. This river is also the outlet of Chester Morse Reservoir, which serves as a water supply source for the city of Seattle (formerly called Cedar Lake), at an altitude of about 470 m. Upstream, water enters Chester Morse Reservoir from small mountain streams draining largely rocky areas at altitudes up to about 1575 m. During much of the year the Cedar River carries melting snow, with relatively little dissolved matter (Table 4.2). Its influence is affected by the amount of annual snowpack, and the time of year when the snow begins to melt. In years when the snow melts early, during the summer the river carries water with a higher chemical content.

The Sammamish River flows from Lake Sammamish into the north end of Lake Washington. Both the river and Lake Sammamish receive water from lowland creeks, which carry considerably higher concentrations of dissolved material than the Cedar River (Table 4.2). Thus both main inlets

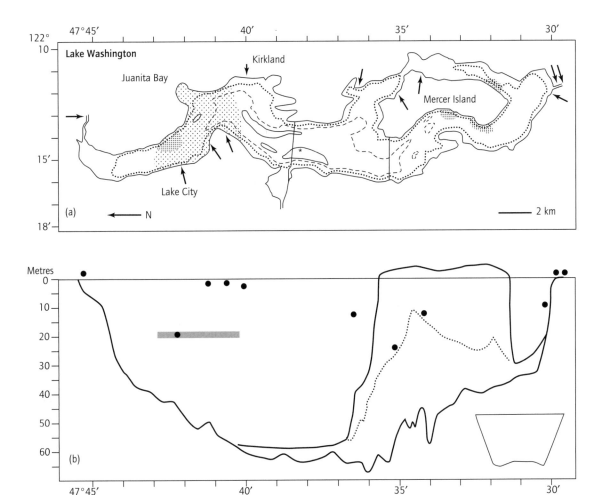

Fig. 4.1 Lake Washington. (a) Map showing features mentioned in text. Contours are shown for 10 m (dotted line), 50 m (dashed line) and 60 m (solid line). The main sampling station off Madison Park is shown by a star, and the two floating bridges by straight lines crossing the lake. Four areas of sunken forest, three by Mercer Island, and one on the east side opposite Lake City, are shown by heavy stipple. Arrows indicate the outfalls of secondary sewage treatment plants which operated between 1941 and 1968. Two emptied into the Cedar River, one into the Sammamish River. The light stipple shows the extent of spread, over six days, of tracer dye from the Lake City treatment plant (see text). (b) North–south vertical section through the deepest part off the west shore. The channel east of Mercer Island is shown by a dotted line. The line running north of Mercer Island is equidistant from the east and west shores. It shows that the depth in the middle of the lake is shallower than near the sides. Outfalls of sewage treatment plants are shown by large dots. The horizontal bar at the north end shows the extent of spread, in a thin layer, of the tracer dye shown by light stippling in (a). Vertical exaggeration 170 times. Inset: east–west vertical section at level of Madison Park sampling station to show that the deepest water is not in the middle. Vertical exaggeration 25 times. (Modified from Edmondson 1985b.)

Table 4.1 (a) Morphometric data for Lake Washington. Volumes of the slices were calculated by a trapezoidal formula. (Based on chart 6449 by the U.S. Coast and Geodetic Survey (currently NOAA 18447).) (b) Areas of land draining into waters listed (based on maps provided by the U.S. Geological Survey, Tacoma, Washington) and percentages of forested land (from Edmondson & Lehman 1981)

(a)

| Depth (m) | Area ($10^3$ m$^2$) | Depth range (m) | Volume ($10^6$ m$^3$) | Fraction of total |
|---|---|---|---|---|
| 0 | 87,615 | Total | 2885 | 1.000 |
| 5 | 80,755 | 0–10 | 805 | 0.279 |
| 10 | 72,925 | 10–20 | 673 | 0.233 |
| 15 | 67,465 | 20–bottom | 1407 | 0.488 |
| 20 | 61,585 | | | |
| 25 | 56,360 | | | |
| 30 | 50,660 | | | |
| 35 | 41,645 | | | |
| 40 | 34,770 | | | |
| 45 | 28,020 | | | |
| 50 | 20,690 | | | |
| 55 | 16,025 | | | |
| 60 | 4065 | | | |
| 65 | 20 | | | |
| 65.2 | 0 | | | |

(b)

| Water body | Drainage area (km$^2$) | Forested area (%) |
|---|---|---|
| Chester Morse Reservoir | 207 | 90 |
| Cedar River below Chester Morse | 289 | 52 |
| Lake Sammamish | 246 | 91 |
| Sammamish River below the lake | 216 | 82 |
| Lake Washington (not including Union Bay) | 316 | 53 |

to Lake Washington originate in lakes which serve as settling basins, which thus retain part of the dissolved load from the land upstream. The flushing rate of Lake Washington varies greatly from year to year depending on both snowfall in the mountains and rainfall in the lowlands. Between 1962 and 1995, the rate has varied from 0.296 lake volume per year in 1993, to 0.602 in 1972 (mean 0.425, SD 0.087). The maximum input was 1737 million m$^3$ per year, in 1972.

## 4.3 PRESENT CONDITION OF LAKE WASHINGTON

Lake Washington is mesotrophic and monomictic, with non-humic water of relatively low chemical content (Table 4.2, and Edmondson 1997). A distinct metalimnetic oxygen minimum forms annually (Shapiro 1960), but the large hypolimnion does not become anaerobic (Lehman 1988; see also fig. 13-3, Edmondson 1963), although concentrations of oxygen may become relatively low near the sediment–water interface during some years. Since 1973, the spring bloom of phytoplankton has been made up largely of diatoms; with species of *Stephanodiscus*, *Fragilaria* and *Aulacoseira* most prevalent. At other times of year, the most abundant genera (by volume) vary, with species of *Ceratium*, *Oocystis*, *Anabaena* and *Cryptomonas* commonly found (S.E.B. Abella, personal communication).

The most abundant zooplankton taxa are seven species of Cladocera, three of Copepoda and 16 of Rotifera. The most frequent are *Daphnia pulicaria*, *D. thorata*, *Bosmina longispina*, *Diaphanosoma birgei*, *Leptodiaptomus ashlandi*, *Cyclops bicuspidatus thomasi*, *Epischura nevadensis*, *Leptodora kindtii*, *Conochilus unicornis*, *C. hippocrepis*, *Kellicottia longispina* and *Keratella cochlearis*. *Synchaeta*, *Polyarthra* and several ciliate protozoa become abundant at times (Edmondson & Litt 1982, 1987).

Macrophytes are limited by the scarcity of suitable shallow bottom, but form dense growths in bays and other protected places. Major submersed genera are *Ceratophyllum*, *Elodea*, *Myriophyllum*, *Najas*, *Potamogeton* (six species), *Vallisneria* and *Zannichellia* (Patmont et al. 1981). *Myriophyllum spicatum* became a conspicuous nuisance in Union Bay during 1976, and since then has spread to all shallow, protected parts of the lake, forming dense populations. The plants support a

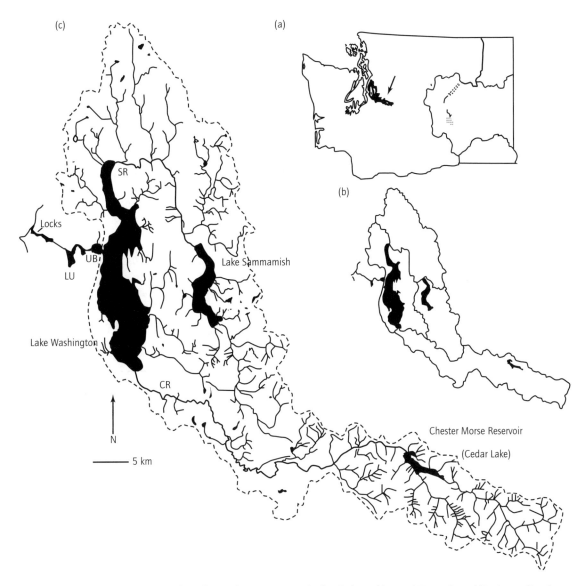

**Fig. 4.2** (a) State of Washington. The Lake Washington watershed is darkened (arrow). Puget Sound lies immediately to the west. (b) Map showing the main divisions of the Lake Washington watershed and the three largest lakes. (c) The watershed showing the rivers and small streams and lakes, based on U.S. Geological Survey quadrangles at scale 1 : 24,000, supplemented with information from U.S. Geological Survey, Tacoma, Washington. The location of the ship canal locks is shown. Abbreviations from north to south: SR, Sammamish River; UB, Union Bay; LU, Lake Union; CR, Cedar River. For details see Edmondson & Lehman (1981). For names of the small lakes, see Edmondson (1988); for names of streams, see Edmondson (1991a). (Modified from Edmondson 1988.)

Table 4.2 Chemical properties of Lake Washington, the two main inlets and two creeks. For each, data are given for the occasions of high and low alkalinity during the period in which ion analyses were made, 1989–1996. These generally correspond to high flow dominated by surface runoff or, in the Cedar River, snowmelt, and low flow dominated by base flow

| Water body | Date | Anions (meq L$^{-1}$) | | | | | Cations (meq L$^{-1}$) | | | | Conductivity (s cm$^{-1}$) |
|---|---|---|---|---|---|---|---|---|---|---|---|
| | | Alkalinity | Chloride | SO$_4$-S | NO$_3$-N | Ca | Mg | Na | K | pH | |
| Lake Washington | 8 November 1995 | 0.818 | 0.094 | 0.160 | 0.005 | 0.425 | 0.269 | 0.184 | 0.026 | 7.56 | 109 |
| | 2 February 1992 | 0.716 | 0.087 | 0.140 | 0.018 | 0.434 | 0.275 | 0.177 | 0.026 | 7.54 | 98 |
| Cedar River | 1 March 1993 | 0.970 | 0.043 | 0.067 | 0.021 | 0.356 | 0.128 | 0.099 | 0.007 | 7.61 | 73 |
| | 23 November 1990 | 0.270 | 0.039 | 0.056 | 0.046 | 0.200 | 0.082 | 0.083 | 0.010 | 7.20 | 33 |
| Sammamish River | 3 October 1994 | 1.372 | 0.129 | 0.204 | 0.027 | 0.701 | 0.658 | 0.311 | 0.048 | 7.45 | 186 |
| | 10 January 1990 | 0.470 | 0.100 | 0.173 | 0.112 | 0.322 | 0.211 | 0.173 | 0.040 | 6.82 | 89 |
| Thornton Creek | 1 August 1994 | 1.870 | 0.254 | 0.361 | 0.084 | 0.880 | 1.049 | 0.384 | 0.074 | 7.91 | 270 |
| | 23 November 1990 | 0.422 | 0.077 | 0.169 | 0.078 | 0.903 | 0.473 | 0.250 | 0.042 | 7.07 | 77 |
| Bear Creek | 20 August 1991 | 1.326 | 0.098 | 0.110 | – | 0.518 | 0.544 | 0.232 | 0.019 | 8.41 | 137 |
| | 23 November 1990 | 0.404 | 0.096 | 0.108 | 0.083 | 0.267 | 0.201 | 0.143 | 0.065 | 6.99 | 70 |

rich periphyton of bacteria, algae, protozoa and sessile rotifers and a variety of free-swimming invertebrates. The zoobenthos is composed mainly of a variety of chironomids and oligochaetes, with small numbers of other insects, molluscs and leeches (Thut 1969).

About thirty-two species of fish have been reported (T.H. Sibley, personal communication; see Eggers et al. 1978). Of particular interest is the sockeye salmon (Onchorhynchus nerka Walbaum), introduced in 1935, which became conspicuous during the mid-1960s, when several commercially valuable runs of salmon occurred. In 1989, the number leaving the lake to migrate to sea began to decrease. The reason for this change has not been determined (Fresh 1994). The surviving fish grow well during their first year of life in the lake before migrating, suggesting that the food supply is still adequate for juveniles, but there is some question, still being investigated, about the amount of food for fry. The possible effect of predatory fish is also being studied.

## 4.4 FORMATION AND DEVELOPMENT OF LAKE WASHINGTON

Lake Washington was formed at the end of the Pleistocene by the Puget lobe of the Vashon glacier, which, by 13,650 yr BP (years before present, taken as 1950), had retreated to a position just north of the present Seattle (Thorson 1980). Initially, the north end of the lake was connected with Puget Sound, and the water was brackish, as indicated by the presence in the sediments of diatoms characteristic of such water (S.E.B. Abella, personal communication). Eventually, the direct connection to Puget Sound was lost, and by about 12,000 yr BP the lake became fully fresh. An outlet subsequently developed at the south end of the lake, through the short Black River, and then on to Puget Sound by way of the Duwamish River (Fig. 4.1). The main inlet at that time was the Sammamish River. Thus it is probable that most water entering Lake Washington until modern times originated in lowland drainage from forested land (Table 4.1).

The hydrology of the lake was fundamentally changed in 1916 by human activities. A ship canal with locks, built in order to connect Lake Washington to Puget Sound, thus became the outlet of the lake instead of the Black River (Purvis 1934). The operation lowered the lake level by about 3 m, to an elevation 7 m above mean sea level. Additional water was needed to operate the locks of the ship canal, which was obtained by diverting the Cedar River into the lake through an artificial channel about 1.6 km long, dug at the southern end. Thus, both the quantity and the character of the water in Lake Washington were changed. The flushing rate was more than doubled, with the dilution potentially reducing the effect of changes in chemical input from other sources.

The motivation for building the ship canal was twofold. First, a plan was made to develop an industrial centre in Kirkland, one of the small towns on the east shore of Lake Washington. Second, the lake was to be a safe harbour for the U.S. West Coast naval fleet. The first plan was ended by a deep economic depression. The second was never implemented (Edmondson 1991b). However, there were some major economic and sociological benefits of lowering the lake level. The present real estate value of the land exposed in 1916 is many millions of dollars; about 64% of the 115 km shoreline is occupied by high-cost residential property. Commercial and recreational boat and barge traffic between Puget Sound and Lake Washington produces considerable income to the region.

Lowering of the lake level must have had some conspicuous effects, but unfortunately no limnological studies were made at the time. About 5 km$^2$ of the former bottom, 8% of the bottom area, was permanently exposed (Chrastowski 1983). This evidently caused widespread erosion and redeposition of sediment into deep water, and gave rise to a palaeolimnological time marker horizon over much of the lake, illustrated by the prominent dark band found in X-radiographs of cores (Edmondson & Allison 1970; see also Schell et al. 1983; Schell & Barnes 1986; Shapiro et al. 1971).

Little differentiation of sediments contained in core tubes is visible, but X-radiographs show considerable structure, in addition to the 1916 silt layer. Alternating light and dark bands occupy most of the top section, but are not found below the 1916 layer (Edmondson & Allison 1970). The pairs of bands can be related to seasonally varying annual events in the lake (Edmondson 1991c). At times of high flow in winter, the Cedar River delivers inorganic sediment which settles out onto the bottom of the lake. During the spring phytoplankton bloom, a layer of diatoms is deposited on top of this sediment. At the time of lowering of the lake, there was a small, temporary change in the proportions of diatom species (S.E.B. Abella, personal communication). The chironomids seemed unaffected (Wiederholm 1979).

The present condition of Lake Washington is different in important ways from conditions in both geological and recent past. The lake has been subjected to many disturbances from natural events and from human activities. Each was followed by a change in some detectable aspect of its condition (Tables 4.3 & 4.4).

## 4.5 CHANGES IN PAST CONDITIONS OF LAKE WASHINGTON

The earliest effects of disturbance are known only from palaeolimnological data. In 6800 yr BP, many of the lakes and bogs in the American Northwest and Canada received volcanic ash from the explosion of Mount Mazama that formed Crater Lake, Oregon. The eruption left a layer of ash about 5 cm thick in the sediments of Lake Washington, and this sudden influx must have had multiple effects. The lake water would have been turbid, reducing light penetration. Exchange of nutrients between sediment and water would have been reduced. Terrestrial communities and soil processes over the entire watershed would have been altered. In any

**Table 4.3** Events affecting the condition of Lake Washington. (Slightly modified from Edmondson 1994a.)

| Date | Event* |
| --- | --- |
| 13,500 yr BP | Born |
| 6800 yr BP | Volcanic ash from Mount Mazama† |
| 1100 yr BP | Major earthquake‡ |
| AD 1851 | Settlers arrive at site of future Seattle |
| c.1860 | Deforestation and land development start§ |
| 1884 | Ferry service develops |
| c.1900 | All lowland timber cut |
| | Raw sewage begins to enter |
| 1916 | Ship canal system opened¶ |
| | Lake level lowered |
| | Cedar River diverted into lake through an artificial channel |
| | Dredging begins in the artificial channel |
| 1935 | Sockeye salmon stocked in Cedar River** |
| 1936 | Diversion of raw sewage to Puget Sound completed |
| 1940 | First floating bridge across lake opens |
| | Land development accelerates on east side of lake |
| 1941 | First of ten secondary sewage treatment plants |
| 1947 | Dredging in artificial channel reduced |
| 1949–1953 | Saltwater intrusions |
| 1959–1970 | Revetments placed in upper Cedar River†† |
| 1962 | Maximum input of secondary sewage effluent |
| 1963 | Second floating bridge opens |
| | More land development on east side |
| 1963–1968 | Secondary effluent diverted |
| 1964 | Sammamish River channelised (part) |
| 1977 | Annual stocking of rainbow trout begins |
| 1990–1991 | Flood in Cedar River |
| 1995–1996 | Further flood in Cedar River |
| 1997 | Land development continues |

\* Principal sources of information: general history – Bagley (1916), McDonald (1979), Edmondson (1991b); character of Lake Washington – Edmondson (1963, 1997), Lehman (1986), Shapiro (1960); palaeolimnology – Shapiro et al. (1971), Edmondson (1974, 1991c); sewage history – Brown & Caldwell (1959), Edmondson (1961), Edmondson & Lehman (1981).
† Abella (1988) and Gould & Budinger (1958) – dates slightly different.
‡ Karlin & Abella (1992) and Edmondson (1984).
§ Davis (1973).
¶ Lake level from files of Army Corps of Engineers. Dredging information from files of King County Public Works and King County Surface Water Management. Ship Canal from chapter 20 of Bagley (1916) and from Purvis (1934).
\*\* Salmon information from Washington State Department of Fisheries and Wildlife
†† Edmondson & Abella (1988).

**Table 4.4** Selected list of changes observed in Lake Washington. The list includes responses to human influences except for the two prehistoric disturbances (Table 4.3). The lake has been affected by changes in the flow of the Cedar River which are partly managed by controlling the release of water from Chester Morse Reservoir. (Slightly modified from Edmondson 1994a.)

| Date | Change |
| --- | --- |
| 6800 yr BP | Temporary change in diatom community (Abella 1988) |
| 1100 yr BP | Areas of forested land slide into lake (Karlin & Abella 1992) |
| AD 1933–1950 | Phosphate concentration increases (Comita & Anderson 1959) |
| | Hypolimnetic oxygen content decreases (Comita & Anderson 1959) |
| 1949–1950 | Temperature anomaly (Rattray et al. 1954) |
| 1955 | *Oscillatoria* bloom (Edmondson et al. 1956) |
| 1962–1970 | Alkalinity increases (Edmondson 1990, 1991a) |
| c.1961 | Smelt, salmon begin to increase (Chigbu & Sibley 1994 a,b; Fresh 1994) |
| 1962–1967 | *Neomysis* decreases (Murtaugh 1981) |
| 1964 | Maximum input of phosphorus (Edmondson & Lehman 1981) |
| 1968 | Transparency increases |
| 1976 | *Oscillatoria* disappears (Edmondson & Litt 1982) |
| | *Daphnia* becomes abundant (Edmondson & Litt 1982) |
| | Transparency increases further (Edmondson & Litt 1982) |
| 1976–1979 | Second increase of alkalinity (Edmondson 1991a) |
| 1986–1989 | Third increase of alkalinity (Edmondson 1990, 1991a) |
| 1988 | *Aphanizomenon* pulse (Edmondson 1997) |
| | Maximum return of spawning salmon (Fresh 1994) |
| 1990 | Transitory decrease in alkalinity (Edmondson 1994a) |
| 1991–1995 | Alkalinity increases again |
| 1992 | Reduced return of salmon (Fresh 1994) |
| 1995–1996 | Another decrease in alkalinity |

case, the sedimentary record shows a considerable change in the planktonic and littoral diatoms which persisted for some time before returning to the original condition (Abella 1988; see also Edmondson 1984). Evaluation of the effect on the terrestrial community is complicated by a simultaneous change in climate (Leopold et al. 1982).

Later, in about 1100 yr BP, the Pacific Northwest was shaken by a large earthquake. Steep forested hillsides slid into Lake Washington, creating sunken forests of standing trees (Fig. 4.1a; Karlin & Abella 1992, 1996). Again, presumably the lake water became turbid. Consequences of this event include a widely distributed layer of clay in the sediment, and a transitory change in the proportions of diatom species (S.E.B. Abella, personal communication).

Interpretation of palaeolimnological data depends on modern observations of the same or similar phenomena. Effects of turbidity on the composition of phytoplankton were demonstrated by a flood of the Cedar River in early spring of 1972, at the time when the spring growth was beginning. Diatoms had begun to increase as usual, but when the transparency of the lake was reduced by the turbid flood water, they decreased sharply. *Oscillatoria*, which had been decreasing year by year, then grew to a density almost three times as great as the year before (see section on eutrophication below). This is consistent with the different light requirements of cyanobacteria and diatoms in general, but the response of the diatoms presumably may also have been affected by antibiosis by *Oscillatoria* (see Edmondson 1991c for details).

## 4.6 EFFECTS OF URBANISATION

The activities of the small resident Native American human population are not documented but probably had little effect on the lake, for they were concentrated along Puget Sound, but that changed with the arrival of European settlers in AD 1851 (Bagley 1916). Most of the known disturbances of Lake Washington in more recent times have been caused by a variety of human activities associated with urbanisation, mainly land development, and sewage disposal.

### 4.6.1 Land development

Initially, clearing of land and construction of settlements were limited to the western shoreline of Lake Washington, between the lake and Puget Sound, but by 1860, a few houses had been built near the lake. Wood was in demand for buildings in Seattle and other places; by 1900 most of the lowland timber around the lake had been cut, and the area was covered with second growth of red alder, *Alnus rubra* Bong (Davis 1973). This change in the character of the terrestrial plant community produced an important, delayed effect on the chemical condition of the lake, as will be shown below.

### 4.6.2 Sewerage

Settlements were growing along the western shore of the lake, and by 1920 drain pipes were conveying raw sewage from about 10,000 people directly to the lake. A major expansion of the Seattle sewer system diverted the sewage to a treatment plant on Puget Sound, the construction being completed in 1936. The motivation for the improvement of sewage disposal at that early time seems to have been based more on issues of public health than on the publicly perceived condition of the lake.

Meanwhile, increasing amounts of sewage were being generated by communities not served by the Seattle system. Settlement on the east side was facilitated by ferry service across the lake, beginning in 1884 (McDonald 1979). Population growth, with further land development and additional generation of sewage, was facilitated by the opening of a floating highway bridge between Seattle and Mercer Island in 1940 (earlier the island had been connected with the east shore by a bridge at the narrowest point). The increased sewage was initially accommodated by construction of secondary treatment plants in the small cities around the lake (Fig. 4.1a). The first of eleven began operating in 1941. The largest was built in 1952 in order to serve Lake City, an independent community which later was incorporated into Seattle (details in Edmondson & Lehman 1981). Four of the plants, serving 76% of the sewered population, possessed outfalls which lay deeper than 12 m (Fig. 4.1b), well below the epilimnion for most of the summer. The outfall at Lake City alone served 54% of the sewered population.

Thus a long period of land development preceded the organisation of a system for sewage disposal. Each activity affected the lake, but the effect of sewage inflow was the more conspicuous, and attracted the most public attention.

## 4.7 EUTROPHICATION AND RECOVERY

In 1950 Lake Washington produced modest amounts of phytoplankton, and Secchi disc transparency in summer was about 4 m. Cyanobacteria were present but not abundant. Small quantities of *Oscillatoria agardhii* were occasionally seen in 1950 (Comita & Anderson 1958). In 1955 the lake was in publicly acceptable condition, but growth of a conspicuous population of *Oscillatoria rubescens* in that year was a clear signal that the lake was well on its way to producing phytoplankton nuisance conditions (Edmondson et al. 1956). By 1964, the mean summer population density of phytoplankton had increased to about four times the 1950 value, and the summer transparency was 1.0 m (Fig. 4.4, and Edmondson 1993b). The algal population was then made up almost entirely by *O. agardhii*, at times composing more than 95% of phytoplankton biovolume. *Oscillatoria rubescens* appeared occasionally. There had been much experience, largely in Europe, with formerly clear lakes producing cyanobacterial blooms after enrichment with sewage. Especially striking was the number of Swiss lakes which bloomed with *O. rubescens* early in eutrophication (Hasler 1947).

Even before the bloom of *Oscillatoria*, the need for regional improvement of sewage disposal was clear, and in 1956 the mayor of Seattle appointed a committee to look into problems associated with urbanisation. As a result of an unusual public action and vote, the treatment plants dumping sewage effluent into Lake Washington were abandoned, and the effluent was diverted from the lake before serious nuisance conditions had developed. To accomplish this required action by the state legislature, and a public vote to form a Municipality of Metropolitan Seattle (Metro; Chapter 1, Edmondson 1991b), a multiple jurisdiction agency which included both cities, and adjacent unincorporated areas within King County also lying within the watershed of Lake Washington. The diversion process took five years, 1963–1968, during which lake phosphate concentration decreased significantly. Nitrate is not as enriched in sewage effluent as phosphorus, and was not greatly affected (Fig. 4.3). Diversion was followed by a conspicuous increase in transparency as a consequence of reduced abundance of phytoplankton (Fig. 4.4). *Oscillatoria* decreased, finally disappearing in 1976.

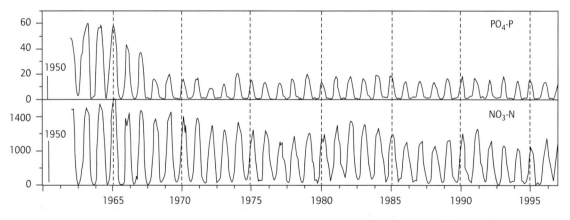

**Fig. 4.3** Long-term graphs (1962–1994) showing monthly means of concentration in surface water of phosphate-P and nitrate-N ($\mu g\,L^{-1}$). The annual ranges for 1950 are shown.

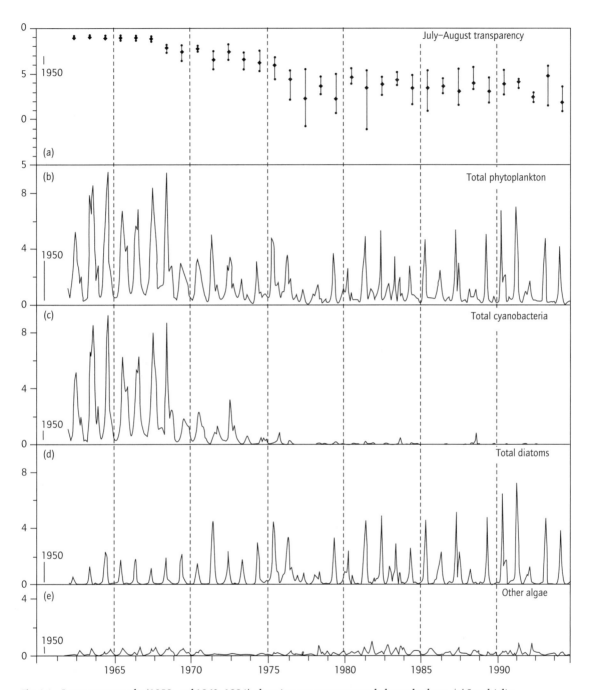

Fig. 4.4 Long-term graphs (1950, and 1962–1994), showing transparency and phytoplankton. (a) Secchi disc transparency in metres, means and range for July and August. (The maximum value ever recorded was 12.9 m on 2 June 1977.) The period during which sewage effluent was diverted is shown by a horizontal line. (b–d) Monthly means of total volume of phytoplankton and the part composed of cyanobacteria (mostly *Oscillatoria* before 1997), diatoms and all other algae (mm$^3$ L$^{-1}$). The annual range is shown for 1950. For data on earlier years see Edmondson (1993b).

The Lake City plant was the largest, and one of the last treatment plants to be diverted. During summer stratification, when the epilimnion is usually about 10m thick, its effluent flowed into the hypolimnion, and was not available to phytoplankton until mixing had deepened the mixed layer to more than 19m late in the summer. During winter, some of its phosphate would leave the lake through the outlet before the spring bloom began. Presumably, if the Lake City plant had been built with an outfall in the epilimnion, the *Oscillatoria* bloom would have occurred earlier, and been larger. The period of *Oscillatoria* abundance is well marked in the sediments, by a layer containing oscillaxanthin, a carotenoid found in only a few species of cyanobacteria (Griffiths & Edmondson 1975; Edmondson 1991c). A decrease in oscillaxanthin marks diversion of effluent from the lake.

## 4.8 CONSEQUENCES OF LAND DEVELOPMENT

A different process of chemical change in the ecosystem, begun by the early settlers of Seattle when they cleared land for housing, did not become obvious until more than 100 years later. The second growth of red alder (*Alnus rubra*), a nitrogen fixer, changed the chemical nature of the soil in a way not achieved by other trees (Cole et al. 1990; van Miegroet & Cole 1984). When topsoil is disturbed in such an area, titratable alkalinity (acid neutralising capacity) of the streams draining it may increase (Edmondson 1991a).

The spread of development has been quantified with aerial photography, by which the area of developed land in the subwatersheds of several of the creeks has been determined. For example, the area around the creek draining into Juanita Bay (Fig. 4.1a) had developed rapidly after transportation to the east side of the lake improved. Between 1965 and 1989, the land here changed from 3.8% to 72.7% urban development. In contrast, during the same interval, around the main creek draining into the Sammamish River, urban development increased from <1% to only 10.3% (S.E.B. Abella, personal communication, see also Edmondson 1991a).

The alkalinity of Lake Washington increases during spring and summer, decreasing during autumn and winter when it receives water with a lesser concentration of dissolved material, especially from the Cedar River. The concentration has varied in the long term, exhibiting a general tendency to increase, with accelerations during the late 1970s and the late 1980s (Fig. 4.5a). These changes, both annual and long term, are strongly influenced by the amount of inflow (Fig. 4.5c).

The increase in alkalinity during 1986–1988 occurred when water input was low (Fig. 4.5c). The 1988 increase attracted attention because it was accompanied by a modest outburst of *Aphanizomenon* (Edmondson 1997). Although the temporary success of *Aphanizomenon* was apparently not determined by alkalinity, it was possible to speculate that a continued increase in alkalinity might result in a proportionally greater production of cyanobacteria than of diatoms, with a consequent decrease in the nutrition of zooplankton (Edmondson 1997).

Lake Washington water is poorly buffered, and the pattern of pH variation does not parallel alkalinity (Fig. 4.5b). Seasonal changes in pH are strongly related to photosynthesis, with summer values being as high as 9.8 during the eutrophic years. Lower peak values decreased during sewage diversion. Peak values and winter minima then increased progressively until 1991.

Our opportunity to witness a possibly significant increase in cyanobacteria was temporarily thwarted by a large flood of the Cedar River, the 'Thanksgiving Flood of 1990', which took place during November 1990 to February 1991. The alkalinity of the lake in spring of 1991 was diluted to 86% of its 1988 maximum. Subsequently it increased, and in May 1994 exceeded the late summer maximum of 1988. However, the expectation of seeing a further increase in alkalinity was again thwarted by another flood in November 1995 to February 1996.

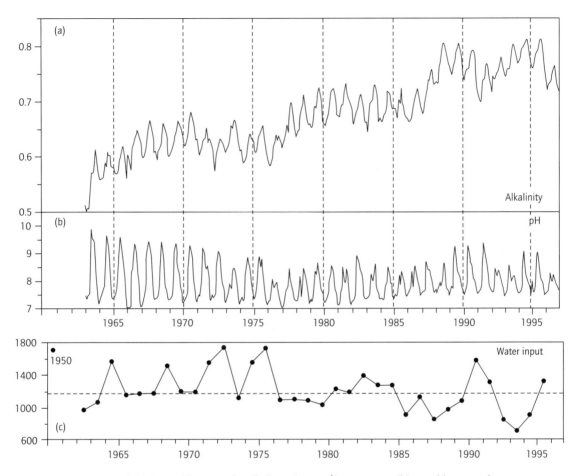

**Fig. 4.5** Long-term graphs. (a) Monthly means for alkalinity (meq L$^{-1}$), 1963–1996. (b) Monthly means for pH. (c) Annual total water input to Lake Washington 1950 and 1963–1995 (millions of m$^3$). The total includes stations on the Cedar and Sammamish Rivers gauged regularly by the U.S. Geological Survey, values calculated for other inlets by regressions, and rainfall (see Edmondson & Lehman 1981).

## 4.9 CHANGES IN ZOOPLANKTON

The response of Lake Washington to changes in sewage effluent input was informative about quantitative relationships between nutrient supply and phytoplankton, and recovery was easily predictable. Major changes took place in the zooplankton, but unlike the phytoplankton the former showed no clear response to the changes in sewage input. All species of zooplankton have gone through major changes in abundance, increases and decreases, some coordinated (Fig. 4.6). *Daphnia* had been infrequent and scarce since 1933 and was not detected in 1970–1972. It occurred in small numbers on several occasions in 1973–1975, then in 1976 became more abundant than we had seen before (Edmondson & Litt 1982). The most striking coordination shown in Fig. 4.6 is the decrease in *Diaphanosoma* after *Daphnia* became abundant (Fig. 4.6f).

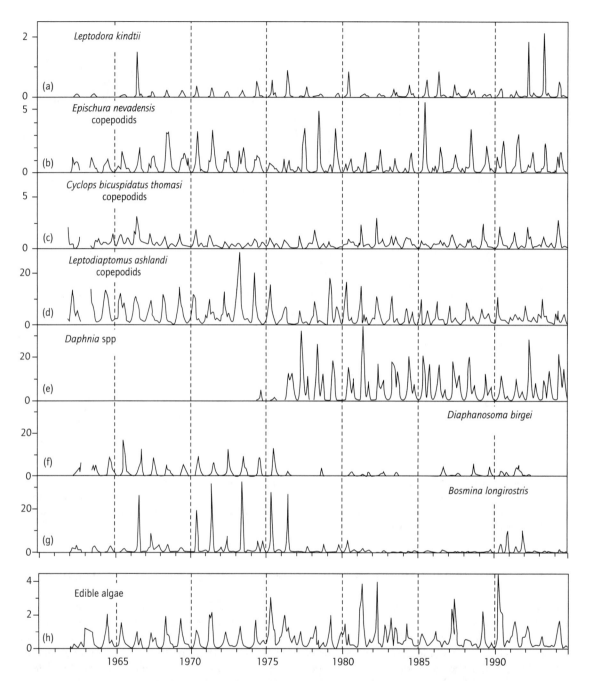

**Fig. 4.6** (a–g). Long-term graphs for major zooplankton crustaceans, 1963–1994, arranged in order of trophic status (individuals $L^{-1}$) (see text). Note the small population of *Daphnia* in 1965. The population in 1973 was too small to show at this scale. (h) Total volume of species of algae that can be eaten and digested by *Daphnia* and *Diaptomus*. It includes species eaten by *Diaphanosoma* and *Bosmina* (see Infante & Edmondson 1985).

*Bosmina* was more abundant before the advent of *Daphnia*, but not consistently, reaching only six large peaks in the fifteen years before *Daphnia* became established (Fig. 4.6g). Just before the rise of *Daphnia* (1973–1975), *Leptodiaptomus* was unusually abundant, but then decreased during the increase of *Daphnia*. The decrease at that time cannot be an effect of competition for food, because the diaptomids were producing eggs rapidly, and individuals were growing to a large size (Edmondson 1985a). Nevertheless, after 1981, the *Leptodiaptomus* population density became consistently smaller than it had been during the period 1962–1972.

These changes must be studied as a result of changes in natality and mortality following variations in availability of food, activity of predators and effects of environmental conditions. Substantial populations of algal food organisms developed after 1976, but the seasonal distribution was different because of grazing by *Daphnia* in addition to that of *Leptodiaptomus* (Fig. 4.6h). *Epischura*, which feeds on both *Bosmina* and *Daphnia* (Kerfoot 1975; Mesner 1984), has varied considerably over the years. It also feeds heavily on phytoplankton during the spring diatom bloom.

The success of *Daphnia* cannot be attributed to a simple and direct effect of the diversion of sewage, for *Daphnia* was scarce for several decades before the eutrophication episode. A probable explanation can be traced, through several steps, to efforts to control damage by floods in the Cedar River (Fig. 4.7; see Edmondson & Abella (1988) and Edmondson (1993a) for details). *Neomysis mercedis*, a predatory mysid crustacean, has been shown by experiment to specialise on *Daphnia* (Murtaugh 1981). It had been relatively abundant until the early 1960s when it decreased to about 10% of its original abundance. The decrease of *Neomysis* occurred immediately after an increase

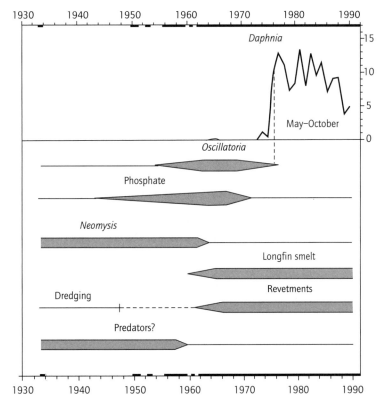

**Fig. 4.7** Schematic summary of factors related to the control of *Daphnia* in Lake Washington as described in the text. The mean abundance of *Daphnia* for May–October is shown in the top panel (individuals $L^{-1}$). Periods of abundance or activity are shown by shaded bars below. The vertical dashed guide line emphasises the relation of the increase in *Daphnia* to the decrease in *Oscillatoria*. The lowest bar takes account of the suggestion that piscivorous fish might have been involved in the control of the longfin smelt. Periods of relatively frequent sampling are shown on the top and bottom margins.

of the longfin smelt (*Spirinchus thaleichthys* Ayres), which is a specialised predator on mysids (Chigbu & Sibley 1994).

So the question shifted to conditions controlling the smelt, and relevant data were found in the files of a local governmental agency. In order to reduce damage by floods, the artificial channel of the Cedar River was deepened each summer by dredging, in order to increase channel capacity. This would have reduced the success of any fish spawning there, such as the smelt. Dredging was stopped in the mid-1960s, and the smelt subsequently increased. Coincidentally, at about the same time, construction of revetments in order to prevent erosion in the upper Cedar River evidently improved the conditions for spawning by the sockeye salmon (*Onchorhynchus nerka*), which increased at the same time as the smelt.

On the basis of limited data, it has been suggested that a decrease in a piscivorous fish (*Ptychocheilus oregonensis* Richardson, northern squawfish), shortly before the smelt increased, could have released the latter from predatory control (see Edmondson 1993b). Although this seems highly unlikely, the suggestion must be recognised.

Thus, the failure of *Daphnia* to survive before the early 1960s can be understood as the effect of predation by *Neomysis*, but that does not explain its scarcity between 1965 and 1976 when *Neomysis* was also scarce. However, that can be explained as the known inhibitory effect of *Oscillatoria* on feeding by *Daphnia*. *Oscillatoria agardhii* had been decreasing since sewage diversion. It was scarce in 1974 and 1975, and *Daphnia* was present in small numbers. *Oscillatoria* disappeared during 1976, the first year of unusual abundance of *Daphnia*. Thus, it appears that between 1933 and about 1965, *Daphnia* was excluded by *Neomysis*, and between 1965 and 1974, by *Oscillatoria*. Between 1955 and 1965, both conditions were limiting to *Daphnia*. It is also suggestive that a small population of *Daphnia* occurred in 1965 when *Oscillatoria* was less abundant than usual and *Neomysis* was beginning to decrease (Edmondson 1988).

This explanation for the change in *Daphnia* is circumstantial, but there is evidence for every step in the sequence (Edmondson 1993a). Each step could be tested experimentally. In fact, a special study of the feeding ecology of *Neomysis* was organised around the Lake Washington situation (Murtaugh 1981). The effect on *Daphnia* feeding by *Oscillatoria* was confirmed by special experiments (Infante & Abella 1985). Enough experimental work has been carried out with *Daphnia* by many investigators to be able to say that there was enough food to produce and maintain a population of *Daphnia* during most of the time it was scarce (Fig. 4.6h, Infante & Edmondson 1985; Infante & Litt 1985; A. Duncan, personal communication).

It therefore seems likely that, if *Neomysis* had not been present in Lake Washington, *Daphnia* would have thrived before eutrophication (1933–1950), and would essentially have been eliminated by *Oscillatoria* during 1955–1976. In that case, the period of *Daphnia* absence would have been the only zooplanktonic signal of the effect of sewage pollution. Not all species of *Daphnia* are equally inhibited by *Oscillatoria*, but the two which are the most common in Lake Washington are. If the dredging programme in the artificial channel had continued, presumably longfin smelt would not have increased, *Neomysis* would have continued to thrive, and *Daphnia* would have remained scarce.

## 4.10 HORIZONTAL VARIATIONS IN LAKE WASHINGTON

The descriptions of the major changes in condition of Lake Washington discussed above are based on sampling at a reference station in the midsection of the lake (Fig. 4.1), supplemented from time to time by surveys of stations in other places. Characteristically, inshore water is less transparent than offshore water, because of material suspended by waves and currents. Offshore, in water deeper than about 10 m, conditions at any one place, during much of the year, are less variable horizontally than over time. However, at some times of year, pronounced differences in conditions can develop at the north and south ends of the lake.

For example, at times of high flow of the Cedar River, low alkalinity water may be delivered faster than it can be mixed into the rest of the lake, and a gradient in alkalinity is set up between the south end and the middle. Similarly, during early spring, *Daphnia* from the Sammamish River can enter the north end of the lake before they begin to increase from their winter minimum farther south. *Daphnia* increases in Lake Sammamish much earlier than in Lake Washington, and is brought in through the Sammamish River. Spring phytoplankton growth may also begin earlier at both ends of Lake Washington than in the middle of the lake.

Despite such transitory differences, variable winds tend to transport and mix water throughout the length of the lake. Thus, in general, the same conclusions about the effect of major disturbances would have been drawn from sampling at other stations.

## 4.11 MOVEMENTS OF WATER AND CONTAINED MATERIAL IN LAKE WASHINGTON

### 4.11.1 Horizontal transport

At the height of eutrophication, an anomaly developed in the vertical distribution of phosphate, in that, after thermal stratification set in, a maximum concentration formed at about 20 m depth, and intensified during the summer (Edmondson 1972). As the outfall of the large Lake City treatment plant lies at 19.5 m deep, it seemed reasonable to check it as a source. A large quantity of rhodamine-B was put into the effluent and its path in the lake followed with fluorometry (Edmondson 1985b). The material spread in a thin layer less than 2 m thick at a depth of about 19 m (Fig. 4.1b). Within six days, it had spread 5.5 km from the outfall, and later was detected 12 km away, well south of the Madison Park station. As the phosphate maximum has not been seen since the sewage was diverted, it seems clear that Lake City had been its source.

### 4.11.2 Density flow

Processes which increase the density of water can cause flow from the surface into deep water along the bottom of a lake. As Lake Washington is somewhat deeper along the sides than in the middle (Fig. 4.1b, inset), it has been proposed that density currents are responsible for moving sediment from the steep margins out into the middle part (Gould & Budinger 1958). A clear demonstration of this phenomenon has been found in the lake. In Juanita Bay (Fig. 4.1a), on unusually cold nights, dense water masses can form. For example, on 11 January 1963, with air temperature at −6.2°C, the temperature of the water in the bay varied between 6.0° and 7.9°C. By the next day, this cold water mass had moved down the steep slope to the bottom of the lake, flowing under water at 8.1°C, and replacing it with cooler water.

This kind of flow is not limited to bays, but also occurs in shallow waters in the north part of the lake (Syck 1964). The contribution of water from the Sammamish River has not been studied, but the data suggest that it may be significant.

During much of the year, the Cedar River is colder than the surface of the lake by several degrees, and the density difference can be great enough to cause the river water to flow along the bottom rather than mixing immediately with epilimnetic water. The result is that the potential diluting effect of the Cedar River may be somewhat minimised during part of the year. The mean travel time of river water from Chester Morse Reservoir is only about five hours, so that there is not much time for it to warm up, even under full sun.

Water made dense with sediment can also flow to the lake bottom. A good example of this occurred during 1975, when there had been considerable erosion and slumping of cliffs of glacial deposits along the upper part of the Cedar River. Turbid water mixed out into the surface for about 6 km in the lake, but most of it flowed down along the bottom in a layer about 10 m thick (fig. 5, Edmondson 1991c). It was detected in water samples at the Madison Park station about 15 km from the mouth of the river, and in sediment traps nearby. The sediment deposited formed a recognisable layer in core samples.

Occasional intrusions of salt water from Puget Sound have aroused considerable public interest (Rattray *et al.* 1954). When the locks bring up a boat into the Lake Washington ship canal, some sea water is brought along with it. A drain was provided to return the salt water to Puget Sound, but under some conditions this water flows eastward along the bottom of the ship canal system. As the Puget Sound water moves toward the lake, it entrains fresh water. Occasionally it has entered Lake Washington and formed a layer of slightly saline water below 50 m, a little warmer than the water above. In each autumn of 1950–1953, by the time the water reached the lake, it contained less than 2% of the salinity of Puget Sound, which is about 28‰.

This slight increase in salinity was not enough to produce a significant biological effect, but it raised the density enough to keep the warm water from floating up, and the bottom temperature was 1.6°C, higher than the minimum in the hypolimnion. In 1988, salt water approached the western edge of Union Bay, but was prevented from entering when the flow of water through the canal away from the lake was increased by releasing extra water from Chester Morse Reservoir into the Cedar River (Fig. 4.2). This action took place in response to public anxiety about the effect of 'salt water' on the lake, which led to establishment of regulations limiting intrusion. However, if more water had entered, the thickness of the layer on the bottom would have increased, but not its salinity. Whether a thicker layer could have stabilised the lake enough to establish permanent meromixis is uncertain, but seems unlikely.

## 4.12 CONCLUDING REMARKS

One can expect any lake to respond to various disturbances, as has Lake Washington, but in different ways depending on particular features of the lake and its surroundings. The relation between those particular features and the character of the response can be informative of the mechanism by which the lake functions. Thus a dynamic comparative limnology, based on whole-lake experiments or quasi-experiments, can be a major goal of limnology. For example, in some ways Lake Washington may seem unusual. Its rapid recovery after diversion of sewage effluent surprised some limnologists (Edmondson 1991b, p. 35). That rapidity can be attributed in part to the relatively rapid flushing rate. Also, even at the height of eutrophication, the hypolimnion did not develop anaerobiosis, nor was there any consequent additional release of phosphate from the sediment; this is due in part to the relatively large volume of the hypolimnion. But such features vary among lakes. Lakes with less inflow are more vulnerable to pollution. Lakes with small hypolimnia take longer to recover because their more efficient recycling of phosphate makes them less responsive to reduction of external supplies.

During the long development of concepts of lakes as ecosystems, it became clear that, because so many conditions in lakes are generated by events in the watershed, it was productive to think of a lake and its watershed together as the effective unit of study. Much emphasis was given to one-way movement of material from the land to the lake, such as nutrients eroded from rocks and organic material from the land community. However, there is much return from the lake to the land, literally feedback. Aquatic vegetation is essential food for many species of birds and some mammals such as moose (*Alces alces* L.; Botkin *et al.* 1973). Fish are the main diet of many birds and mammals. Aerial adult chironomids and odonates, which are developed during a long growth in lakes, support large populations of birds and bats.

Some of the returns are gaseous. Hydrogen sulphide and methane, liberated by lakes as well as by marshes, are the best known, but large quantities of a variety of volatile organic compounds are also released. Some are diagnostic of the presence of particular species of phytoplankton (Jüttner *et al.* 1986; Jüttner 1987). Effects may not be limited to the immediate lake–watershed system. The presence of a large lake can affect climate. Even a small lake can have microclimatic effects on temperature and humidity.

The idea of human beings and their economies as part of ecosystems is relatively recent. Never-

**Fig. 4.8** East–west vertical section of Lake Washington at Madison Park sampling station drawn to true scale. To illustrate the scale, the Seattle Space Needle has been planted at the Madison Park sampling station. It is 185 m high. The Eiffel Tower is 300 m high.

theless, Lake Washington illustrates how human activities outside the shores of a lake can affect chemical and biological interactions within the lake that, in turn, affect human perceptions and responses. Beyond that, the human part of the ecosystem involves social and political activities which are unique to it. The changes in Lake Washington generated by human activity, which were perceived as deterioration by certain human standards, led to action by the state Legislature and by voters which affected the lake in a way that was perceived as beneficial, and caused a significant rearrangement of finances of the people living in the watershed. It is worth noting that the public vote for Metro, committing an immediate expenditure of $125,000,000 (in 1958 dollars), was taken before the lake had seriously deteriorated. It was based in part on limnological predictions of the consequences of diverting the sewage effluent or of letting it continue to increase. The success of the operation depended on the existence of a citizenry prepared to make use of such information (see Edmondson (1991b) for details).

Finally, lakes have traditionally been used as receptacles for waste which seemed to go away until the input grew larger than the lake's capacity to assimilate it. The magnitude of lakes is deceptive because their true size is not apparent. When looked at in true scale, lakes are thin sheets of water, and therefore much more highly vulnerable to the effects of human action than they appear from their surface (Fig. 4.8).

## 4.13 ACKNOWLEDGEMENTS

The long-term nature of this work, from 1958 to 1997, was made possible by a nearly continuous series of short-term grants from the National Science Foundation. Since 1986 the Andrew W. Mellon Foundation has provided strong support, which has been essential for the continuation of the work and analysis of data. Early support was given by the National Institutes of Health, the Environmental Protection Agency, the Office of Water Research and Technology, and the State of Washington Fund for Research in Biology and Medicine.

This paper is a development of Edmondson (1994a). Tables 4.3 and 4.4 were included there and are reproduced here by agreement from the North American Lake Management Society. Figure 4.2 is a modification of fig. 4.5 in Edmondson 1985b, and is reproduced here by agreement from Professor Eugenio de Fraja Frangipane, Milano, Italy.

## 4.14 REFERENCES

Abella, S.E.B. (1988) The effect of the Mazama ashfall on the planktonic diatom community of Lake Washington. *Limnology and Oceanography*, **33**, 1376–85.

Bagley, C.B. (1916) *History of Seattle from the Earliest Settlement to the Present Time*, 3 Vols. S. J. Clarke, Chicago.

Botkin, D.B., Jordan, P.A., Dominski, A.S., Lowendorf, H.S. & Hutchinson, G.E. (1973) Sodium dynamics in a northern ecosystem. *Proceedings of the National Academy of Sciences*, **70**, 2745–8.

Brown & Caldwell, Civil Chemical Engineers (1958) *Metropolitan Seattle Sewerage and Drainage Survey*. City of Seattle, Seattle. 92 pp.

Chigbu, P. & Sibley, T.H. (1994) Diet and growth of longfin smelt and juvenile sockeye salmon in Lake Washington. *Verhandlungen Internationale Vereinigung Limnologie*, **75**, 2086–91.

Chrastowski, M. (1983) Historical changes to Lake Washington and route of the Lake Washington ship canal, King County, Washington. To accompany *Water*

Resources Investigation Open File Report 81-1182. U.S. Geological Survey, Seattle, 92 pp.

Cole, D.W., Compton, J., van Miegroete, H. & Homann, P. (1990) Changes in soil properties and site production caused by red alder. *Water Air and Soil Pollution*, **54**, 231–46.

Comita, G.W. & Anderson, G.C. (1959) The seasonal development of a population of *Diaptomus ashlandi* Marsh, and related phytoplankton cycles in Lake Washington. *Limnology and Oceanography*, **4**, 37–52.

Davis, M.B. (1973) Pollen evidence of changing land use around the shores of Lake Washington. *Northwest Science*, **47**, 133–48.

Edmondson, W.T. (1961) Changes in Lake Washington following an increase in the nutrient income. *Verhandlungen Internationale Vereinigung Limnologie*, **14**, 167–75.

Edmondson, W.T. (1963) Pacific Coast and Great Basin. In: Frey, D.G. (ed.), *Limnology in North America*. University of Wisconsin Press, Madison, WI, 371–92.

Edmondson, W.T. (1972) Nutrients and phytoplankton in Lake Washington. In: Likens, G.E. (ed.), *Nutrients and Eutrophication. The Limiting Nutrient Controversy*. Special Symposia 1, American Society of Limnology and Oceanography, Waco, TX, 172–93.

Edmondson, W.T. (1974) The sedimentary record of the eutrophication of Lake Washington. *Proceedings of the National Academy of Sciences*, **71**, 5093–5.

Edmondson, W.T. (1984) Volcanic ash in lakes. *Northwest Environmental Journal*, **1**, 139–50.

Edmondson, W.T. (1985a) Reciprocal changes in abundance of *Daphnia* and *Diaptomus* in Lake Washington. *Archiv für Hydrobiologie. Ergebnisse der Limnnologie*, **21**, 475–81.

Edmondson, W.T. (1985b) Recovery of Lake Washington from eutrophication. In: Frangipane, G.F. (ed.), *Lakes Pollution and Recovery. Proceedings of the Rome Congress April 1985*. European Water Pollution Control Association, Hennef, Germany, 308–14.

Edmondson, W.T. (1988) On the modest success of *Daphnia* in Lake Washington in 1965. In: Round, F.E. (ed.), *Algae in the Aquatic Environment: Contributions in Honour of J.W.G. Lund CBE, FRS*. Biopress, Bristol, 223–43.

Edmondson, W.T. (1990) Lake Washington entered a new state in 1988. *Verhandlungen Internationale Vereinigung Limnologie*, **24**, 428–30.

Edmondson, W.T. (1991a) Responsiveness of Lake Washington to human activity in the watershed. In: *Puget Sound Research '91*. Puget Sound Water Quality Authority, Seattle, 629–38.

Edmondson, W.T. (1991b) *The Uses of Ecology: Lake Washington and Beyond*. University of Washington Press, Seattle, 329 pp.

Edmondson, W.T. (1991c) Sedimentary record of changes in the condition of Lake Washington. *Limnology and Oceanography*, **36**, 1031–44.

Edmondson, W.T. (1993a) Eutrophication effects on the food chains: long-term studies. *Memorie dell' Istituto Italiano di Idrobiologia*, **52**, 113–32.

Edmondson, W.T. (1993b) Experiments and quasi-experiments in limnology. *Bulletin of Marine Science*, **53**, 65–83.

Edmondson, W.T. (1994a) Sixty years of Lake Washington: a curriculum vitae. *Lake and Reservoir Management*, **10**, 75–84.

Edmondson, W.T. (1994b) What is Limnology? In: Margalef, R. (ed.), *Limnology Now: a Paradigm for Planetary Problems*. Elsevier, Amsterdam, 547–53.

Edmondson, W.T. (1997) *Aphanizomenon* in Lake Washington. *Archiv für Hydrobiologie (Supplementband)*, **107**, 409–46.

Edmondson, W.T. & Abella, S.E.B. (1988) Unplanned biomanipulation in Lake Washington. *Limnologica*, **19**, 73–9.

Edmondson, W.T. & Allison, D.E. (1970) Recording densitometry of X-radiographs for the study of cryptic laminations in the sediment of Lake Washington. *Limnology and Oceanography*, **15**, 138–44.

Edmondson, W.T. & Lehman, J.T. (1981) The effect of changes in the nutrient income on the condition of Lake Washington. *Limnology and Oceanography*, **26**, 1–29.

Edmondson, W.T. & Litt, A.H. (1982) *Daphnia* in Lake Washington. *Limnology and Oceanography*, **27**, 272–93.

Edmondson, W.T. & Litt, A.H. (1987) *Conochilus* in Lake Washington. *Hydrobiologia*, **147**, 157–62.

Edmondson, W.T., Anderson, G.C. & Peterson, D.R. (1956) Artificial eutrophication of Lake Washington. *Limnology and Oceanography*, **1**, 47–53.

Eggers, D.M., Bartoo, N.W., Richard, N.A., *et al.* (1978) The Lake Washington ecosystem: the perspective from the fish community and forage base. *Journal of the Fisheries Research Board of Canada*, **35**, 1553–71.

Fresh, K. (1994) Lake Washington sockeye salmon: a historical perspective: *13th International symposium of the North American Lake Management Society. Lake management and Diversity*, **9**, 148–51.

Gould, H.R. & Budinger, T.H. (1958) Control of sedimentation and bottom configuration by convection currents, Lake Washington, Washington. *Journal of Marine Research*, **17**, 183–98.

Greenberg, E.S. & Sibley, T.H. (1994) *Annotated Bibliography of the Lake Washington Drainage*. University of Washington, Seattle. 139 pp.

Griffiths, M. & Edmondson, W.T. (1975) Burial of oscillaxanthin in the sediment of Lake Washington, *Limnology and Oceanography*, **10**, 945–52.

Hasler, A.D. (1947) Eutrophication of lakes by domestic drainage. *Ecology*, **28**, 383–95.

Infante, A. & Abella, S.E.B. (1985) Inhibition of *Daphnia* by *Oscillatoria* in Lake Washington. *Limnology and Oceanography*, **30**, 1046–52.

Infante, A. & Edmondson, W.T. (1985) Edible phytoplankton and herbivorous zooplankton in Lake Washington. *Archiv für Hydrobiologie: Ergebnisse der Limnologie*, **21**, 161–71.

Infante, A. & Litt, A.H. (1985) Differences between species of *Daphnia* in the use of ten species of algae in Lake Washington. *Limnology and Oceanography*, **30**, 1053–9.

Jüttner, F. (1987) Volatile organic substances. In: Fay, P. & van Baalen, C. (eds), *Cyanobacteria*. Elsevier, Amsterdam, 453–69.

Jüttner, F., Höflacher, B. & Wurster, K. (1986) Seasonal analysis of volatile organic biogenic substances (VOBS) in freshwater phytoplankton populations dominated by *Dinobryon*, *Microcystis* and *Aphanizomenon*. *Journal of Phycology*, **22**, 169–75.

Karlin, R.E. & Abella, S.E.B. (1992) Paleoearthquakes in the Puget Sound region recorded in sediments from Lake Washington, USA. *Science*, **258**, 1617–20.

Karlin, R.E. & Abella, S.E.B. (1996) A history of Northwest earthquakes recorded in sediments from Lake Washington. *Journal of Geophysical Research*, **101B**, 6137–50.

Kemmerer, G., Bovard, J.F. & Boorman, W.R. (1924) Northwestern lakes of the United States; Biological and chemical studies with reference to possibilities in production of fish. *Fisheries Bulletin of the U S Bureau of Fisheries*, **39**, 51–140.

Kerfoot, W.C. (1975) The divergence of adjacent populations. *Ecology*, **56**, 1298–313.

Lehman, J.T. (1988) Hypolimnetic metabolism in Lake Washington: relative effects of nutrient load and food web structure on lake productivity. *Limnology and Oceanography*, **33**, 1334–47.

Leopold, E.B., Nickman, R., Hedges, J.I. & Ertel, J.R. (1982) Pollen and lignin records of late Quaternary vegetation, Lake Washington. *Science*, **218**, 1305–7.

McDonald, L. (1979) *The Lake Washington Story: a Pictorial Survey*. Superior Publishing Company, Seattle, 158 pp.

Mesner, N.O. (1984) *The feeding ecology of Epischura nevadensis in Lake Washington*. MS thesis, University of Washington, Seattle, 80 pp.

Murtaugh, P.A. (1981) Selective predation by *Neomysis mercedis* in Lake Washington. *Limnology and Oceanography*, **26**, 445–53.

Patmont, C.R., Davis, J.I. & Swartz, R.G. (1981) *Aquatic plants in selected waters of King County Distribution and Community Composition of Macrophytes*. Municipality of Metropolitan Seattle (Now King County Washington Water and Land Resources Division), Seattle. 120 pp.

Purvis, N.H. (1934) History of the Lake Washington canal. *Washington Historical Quarterly*, **25**, 114–27; 210–13.

Rattray, jr., M., Seckel, G.R. & Barnes, C.A. (1954) Salt budget in the Lake Washington ship canal. *Journal of Marine Research*, **13**, 263–75.

Scheffer, V.B. & Robinson, R.J. (1939) A limnological study of Lake Washington. *Ecological Monographs*, **9**, 95–143.

Schell, W.R. & Barnes, R.S. (1986) Environmental isotope and anthropogenic tracers of recent lake sedimentation. In: Fritz, P. & Fontes, J.Ch. (ed.), *Handbook of Isotope Geochemistry*, Vol. 2, *The Terrestrial Environment*. Elsevier, Amsterdam, 169–206.

Schell, W.R., Swanson, J.R. & Currie, L.A. (1983) Anthropogenic changes in organic carbon and trace metal input to Lake Washington. *Radiocarbon*, **25**, 621–28.

Shapiro, J. (1960) The cause of a metalimnetic minimum of dissolved oxygen. *Limnology and Oceanography*, **5**, 216–27.

Shapiro, J., Edmondson, W.T. & Allison, D.E. (1971) Changes in the chemical composition of sediments of Lake Washington, 1958–1970. *Limnology and Oceanography*, **16**, 437–52.

Syck, M.S. (1964) *Thermal convection in Lake Washington, winter 1962–1963*. MS thesis, University of Washington, 32 pp.

Thorson, R.M. (1980) Ice-sheet glaciation of the Puget Sound Lowland, Washington, during the Vashon Stade (Late Pleistocene). *Quaternary Research*, **13**, 303–21.

Thut, R.N. (1969) A study of the profundal bottom fauna of Lake Washington. *Ecological Monographs*, **39**, 79–100.

van Miegroet, H. & Cole, D.W. (1984) The impact of nitrification on soil acidification and cation leaching in a red alder ecosystem. *Journal of Environmental Quality,* **13**, 586–90.

Wiederholm, T. (1979) Chironomid remains in the recent sediments of Lake Washington. *Northwest Science,* **53**, 251–7.

Wissmar, R.C., Richey, J.E., Devol, A.H. & Eggers, D.M. (1981) Lake ecosystems of the Lake Washington drainage basins. In: Edmonds, R.L. (ed.), *Analysis of Coniferous Forest Ecosystems in the Western United States.* Douden Hutchinson & Ross, Stroudsberg, PA, 333–85.

# 5 Lakes of Northern Europe

HEIKKI SIMOLA AND LAURI ARVOLA

## 5.1 INTRODUCTION

### 5.1.1 Scope and contents

This chapter is about the northern European lakes of the Fennoscandian or Baltic Shield area of ancient crystalline bedrock, which roughly coincides with the territories of Norway, Sweden, Finland and the Russian regions of Murmansk Oblast and Karelian Republic (Fig. 5.1). A very large number of lakes are present in this area: some 200,000 of them exceed 4 ha in size; all told, their number may be over half a million. Mostly, their waters are soft and often humic, and they freeze over in winter. Even though the area is relatively sparsely populated, the lakes have been affected by various human activities and their consequences (Bernes 1993).

Following an introductory overview of monitoring studies of lake ecosystems in countries of the Fennoscandian Shield (section 5.1.2), section 5.2 provides a brief general description of the environmental setting of the area and considers the two key issues impinging on the ecosystem functioning of Shield lakes: the effects of high allochthonous organic content and of winter ice cover. Section 5.3 deals with Finland, for which we try to give a thorough overview of the various types of lake behaviours, including some restorations. Lakes in Sweden (section 5.4), Norway (section 5.5) and northwest Russia (section 5.6) are dealt with more cursorily, a few important cases and relevant overview data being included from each. Section 5.7 provides a short summary of the general trends and some assessment of the future.

### 5.1.2 Monitoring of lake environments

Regular water quality monitoring at permanent sampling sites began in all the Nordic countries during the 1960s. Earlier hydrobiological records from various lakes extend back to the latter half of the nineteenth century (e.g. Lovén 1862; Kessler 1868; Stenroos 1898) and the first physical and chemical lake studies were carried out around the turn of the century (e.g. Levander 1906a,b). There is a long tradition of regional limnological investigations in Sweden (e.g. Thunmark 1937; Lohammar 1938), in Finland (e.g. Järnefelt 1958, 1963), and in Karelia (Gerd 1947, 1956; Aleksandrov *et al.* 1959; Gordeyev *et al.* 1965). Monitoring of some basic meteorological and hydrological parameters, e.g. precipitation, air and water temperature, runoff and water level, has been conducted at a number of sites since the mid-1800s, and more systematically since the beginning of the twentieth century (Vesajoki *et al.* 1994). The longest continuous cryophenological observation series details the ice break-up dates in River Tornio since 1693 (Rosenström *et al.* 1996).

Continuous water quality monitoring was begun in the larger lakes of Finland in 1965, river monitoring having begun some years earlier. At present, some 250 lake sites are included in the basic monitoring programme conducted by the Finnish Environment Institute and regional environmental centres, in accordance with the European Environmental Agency guidelines (EEA, EUROWATERNET; Niemi *et al.* 2001). In the beginning, the factors studied were mainly physical and chemical, such as water temperature, colour, pH, conductivity, alkalinity, chemical and biological oxygen demand (COD and BOD), as well as the concentrations of phosphorus and nitrogen.

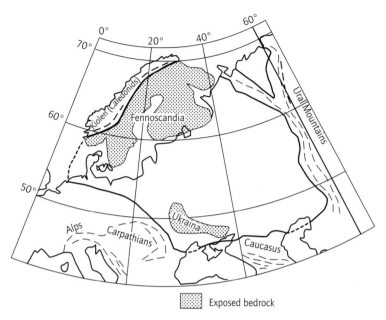

Fig. 5.1 Geographical extent of the Fennoscandian Shield, the exposed northwest part of the Fennosarmatian bedrock block, which for its southern and eastern parts is mostly covered by thick deposits of Palaeozoic and younger sedimentary rocks. (Redrawn from Simonen 1990.)

Subsequently, more determinands were included in the monitoring programme, e.g. heavy metals and toxic chemicals (e.g. PCB, DDT), and also the number of biological parameters was increased from the initial recording of only chlorophyll *a* and phytoplankton species composition and biomass. The monitoring of the large lakes in Sweden, following similar lines, began in 1965; since 1985, the programme has included 169 lakes (Persson 1996a). The sampling and analytical methods have been standardised and, since the early 1970s, the same analytical procedures have been applied in Finland, Norway and Sweden. The extensive monitoring programmes have provided basic data on water quality and enabled usability assessment for lakes and rivers (e.g. Laaksonen & Malin 1984; Vuoristo & Antikainen 1997; Niemi *et al.* 2001), but the infrequency of sampling is considered unsatisfactory for the purpose of biological monitoring (Sarvala 1992, 2003). In northwest Russia, the monitoring programmes have not been as extensive as elsewhere in the Nordic countries, but the research institutes there have conducted basic monitoring of a number of water bodies since the 1960s. These include lakes of economic importance and those affected by industrial pollution, as well as the large lakes Ladoga and Onega and the large reservoirs and regulated lakes (Balagurova *et al.* 1970; Sabylina *et al.* 1991; Timakova & Vislyanskaya 1991; Niinioja *et al.* 2000; Viljanen *et al.* 2000).

Aware of the fact that small oligotrophic lakes are the most sensitive to atmospheric pollution, the scope of monitoring was extended to such lakes in Norway and Sweden during the 1970s and in Finland in the late 1980s, leading to the establishment of networks of perennially monitored acid-sensitive lakes (e.g. Mannio & Vuorenmaa 1995; Skjelkvåle *et al.* 1997). The effects of liming have been monitored in some 160 Swedish lakes since 1983. A nation-wide survey of water quality and phytoplankton of 1250 Swedish lakes was conducted in 1972 (Rosén 1981a). The extensive surveys of lake ecosystems have produced valuable data sets of the environmental requirements of phytoplankton (Rosén 1981a,b; Willén *et al.* 1990a; Willén 1991, 1992, 2002) and macrozoobenthos (Sæther 1979; Wiederholm 1980), which are applicable to biomonitoring of lakes in a wide geographical area.

Within the framework of the Swedish national monitoring programme of environmental quality

(PMK; Programmet för övervakning av miljökvalitet), the emphasis in biological lake monitoring in Sweden was focused on a limited number of regionally representative lake ecosystems. A first stage of monitoring 26 'biological reference lakes', mostly 0.1–2 km$^2$ in area, was undertaken in the period 1989–93. The monitoring is continuing in a subset of 13 of the original selection (Persson 1996b).

In accordance with the International Convention of Long-range Transboundary Air Pollution, extensive surveys of the water quality of small lakes with low conductivity have been carried out in all the Nordic countries: in Sweden in 1985 (5700 randomly selected lakes; SNV 1986), in Norway in 1986 (1005 lakes mainly in acidification-sensitive regions; Henriksen et al. 1988) and in Finland in 1987 (a statistically representative set of 987 lakes; Forsius et al. 1990). This kind of large-scale survey was repeated in the concerted Northern European Lake Survey in 1995, which included 873 lakes in Finland, 1006 lakes in Sweden, 3075 lakes in Norway and 489 lakes in Kola and Karelia, plus a number of lakes in Denmark, Scotland and Wales (Henriksen et al. 1997).

Another international monitoring programme, established as a permanent function of the UN ECE Convention of Long-range Transboundary Air Pollution, is the Integrated Monitoring of Air Pollution Effects on Ecosystems (IM), which involves very thorough investigations of small pristine lake–watershed ecosystems (four sites in Finland, four in Sweden, two in Norway and one site in Karelia: Keskitalo & Salonen 1994; Bergström et al. 1995; Simola et al. 1991).

Palaeolimnological data on the past conditions of a lake ecosystem are often indispensable as background information for ecosystem monitoring, as most lakes have been affected by various human activities for much longer than the time span of any systematic monitoring survey (e.g. Charles et al. 1994; Smol 2002). The most detailed sedimentary records are provided by annually laminated or varved sediments, in which even seasonal patterns may be discernible (e.g. Renberg 1981; Simola 1992); most lake sediments are, however, affected by physical or biological mixing, so that a given layer represents a more or less homogenised sample of some years' or even decades' sedimentation (Håkanson & Jansson 1983).

Lake sediments are composed of both autochthonous material, produced in the lake ecosystem, and allochthonous material, entering the lake from its drainage area and the atmosphere. A multitude of physical, chemical and biological analyses are available for extracting the contained information (e.g. Haworth & Lund 1984; Berglund 1986; Last & Smol 2001a–b; Smol et al., 2001a,b). Quantitative palaeoecological reconstruction, based on sedimentary biological assemblages, is best achieved against a calibration data set of lakes in which the water-quality parameters and relevant species assemblages (e.g. surface sediment diatoms) have been analysed. Reconstruction of the water chemistry (pH, alkalinity, phosphorus concentration, etc.) proceeds through the use of transfer functions (Anderson 1993; Line et al. 1994). Whole-lake conditions are inferred by modern analogue matching (Flower et al. 1997).

Quantitative palaeolimnological methods allow detailed assessment of human impacts (eutrophication, acidification) to be determined, as well as long-term responses to events (e.g. climatic changes) in the history of the lake (e.g. Itkonen 1997). Besides retrospection, this approach allows some prognostication, particularly in cases of ecosystem recovery from assault: pre-disturbance sediment may reveal the baseline situation towards which the system may be supposed to revert (Smol 2002; for Finnish examples, see Meriläinen & Hamina 1993; Simola et al. 1996b; Miettinen et al. 2003).

A new era of water monitoring and management began in the European Union member states as well as Norway with the implementation of the Water Framework Directive (WFD; European Union 2000). The Directive emphasises the use of biological criteria in establishing a limnological typology and assessing the ecological quality of water bodies (Ruoppa & Karttunen 2002; Ruoppa et al. 2003). The final aim is to halt the quality deterioration and in fact to re-establish 'good' ecological status in most natural water bodies by the year 2015. In parallel with this the WFD should activate water pollution control by reducing emissions of harmful substances into waters.

In Finland, the WFD monitoring will be based essentially on the EUROWATERNET sampling network, but when fully implemented it will require a more extensive sampling programme than that which is presently conducted.

## 5.2 THE FENNOSCANDIAN SHIELD—ENVIRONMENT, LAKES AND LIMNOLOGY

### 5.2.1 Natural conditions

The Fennoscandian Shield consists mainly of Precambrian siliceous rocks, e.g. gneisses and granitic gneisses, formed some 1–2 billion years ago. Owing to the slow weathering rate of these acidic rocks and the scarcity of carbonates and base cations, the ground and surface waters are mostly soft and of low alkalinity. Within parts of Norway, Sweden and the Murmansk area, the bedrock is, however, overlain by Cambro-Silurian sedimentary rocks (limestones, shales, slates, conglomerates), which render a natural water quality which is less dilute and more alkaline.

Glacial landforms are widely distributed in the Shield area, which was most recently deglaciated during the past 7000 to 12,000 years (Donner 1995). Glacial action upon the hard bedrock and its mostly thin cover of loose superficial deposits created a very variable microtopography. This explains the very large number of lake basins in the entire area (Table 5.1). Most of these lakes are small to moderate in size and are generally shallow. Where the terrain is flat over wide distances, e.g. in the interior southern Finland, the drainage systems consist of chains of large lakes connected by short riverine stretches, with up to half the terrain being covered by water. The large tectonic depressions towards the margins of the Shield house the three largest freshwater lakes of Europe (Ladoga, Onega and Vänern; Table 5.2). Except for the Kiolen (Caledonid) mountain range to the north and west, the Shield area is of generally low relief. Poor drainage, in combination with the cool

Table 5.1 Numbers of lakes in different size classes in Fennoscandia. Total numbers include all lakes >0.04 km². (Data for Norway, Sweden, Finland and Murmansk Oblast from Henriksen et al. (1997); for Karelian Republic from Atlas Karelskoi ASSR (1989) and Topograficheskaya Karta Respublika Karelii (1997).)

| Country/ region | Area (km²) | | | |
|---|---|---|---|---|
| | Total | 1–10 | 10–100 | >100 |
| Norway | 38,845[a] | 2139 | 164 | 7 |
| Sweden | 60,264[b] | 3599 | 379 | 24 |
| Finland | 29,515[a] | 2164 | 276 | 47 |
| Murmansk | 15,712 | 830 | 73 | 7 |
| Karelia | 61,000[c] | 1000[b] | 85[b] | 19[b] |

[a] Total number of Norwegian lakes >0.01 km² is 134,000 (Skjelkvåle et al. 1997), and that of Finnish lakes >0.0005 km² is 187,888 (Raatikainen & Kuusisto 1990).
[b] Estimates.
[c] Lower size limit not given for lakes included in this figure.

and moist climate, has led to extensive paludification of the terrain. Peatlands compound the effects of acidic podsol soils in contributing dissolved humic substances to the surface waters; humus coloration and dystrophy are characteristic for many watercourses.

Most of the Fennoscandian Shield area belongs to the Boreal coniferous forest zone; the southernmost parts support Boreonemoral forests, in which deciduous hardwood trees also thrive. The highest, the most northerly and the most maritime coastal areas support treeless vegetation types (Table 5.3; for more details, see Ahti et al. 1968). In many respects, the natural landscape conditions of the Fennoscandian Shield resemble those of the Laurentian Shield in northeastern North America.

### 5.2.2 Humus and ice-cover as limnological constraints

#### 5.2.2.1 High humus content

Lakes of the boreal zone often receive high allochthonous input of dissolved organic matter (DOM) from their catchment areas. In northern

Table 5.2 Characteristics of the large lakes and reservoirs in northern Europe dealt with in this chapter

| Lake, country | Location | Surface area (km$^2$) | Maximum depth (m) | Volume (km$^3$) | Elevation (m a.s.l.) | Residence time (yr) | Catchment area (km$^2$) |
|---|---|---|---|---|---|---|---|
| Ladoga, Russia | 59°54'–61°47'N 29°47'–32°58'E | 17,891 | 230 | 837 | 5 | 12 | 258,000 |
| Onega, Russia | 60°55'–62°55'N 34°14'–36°30'E | 9890 | 120 | 280 | 35 | 12 | 51,450 |
| Vänern, Sweden | 58°22'–59°25'N 12°19'–14°10'E | 5648 | 106 | 153 | 44 | 9 | 41,182 |
| Saimaa, Finland | 61°04'–62°35'N 27°20'–30°00'E | 4380 | 82 | 61 | 76 | 3.5 | 61,560 |
| Vättern, Sweden | 57°46'–58°52'N 14°07'–15°00'E | 1856 | 128 | 74 | 89 | 56 | 4503 |
| Kumskoye*, Russia | 65°22'–66°25'N 30°37'–32°45'E | 1750 | 56 | 21–30 | 107–110 | 4.2 | 12,900 |
| Vigozero, Russia | 63°06'–63°55'N 34°00'–35°35'E | 1200 | 18 | 7 | 89–91 | 1.0 | 20,500 |
| Mälaren, Sweden | 59°14'–59°48'N 16°04'–18°00'E | 1140 | 61 | 13.6 | 0.3 | 2.2 | 21,460 |
| Päijänne, Finland | 61°14'–62°15'N 25°10'–26°00'E | 1100 | 98 | 16.8 | 78 | 2.7 | 25,400 |
| Inari, Finland | 68°45'–69°15'N 27°00'–28°50'E | 1050 | 96 | 15.1 | 116 | 3.4 | 13,400 |
| Segozerskoye*, Russia | 63°10'–63°30N 33°17'–34°10'E | 800 | 97 | 4.7 | 115–120 | 1.8 | 7460 |
| Knyazhegubskoye*, Russia | 66°30'–66°55'N 31°25'–32°30'E | 610 | 50 | 8.5–10.5 | 34–37 | 1.3 | 25,600 |
| Hjälmaren, Sweden | 59°08'–59°18'N 15°13'–16°19'E | 478 | 22 | 2.9 | 22 | 3.3 | 3575 |
| Lokka*, Finland | 67°50'–68°05'N 27°20'–28°40'E | 216–417 | 12 | 0.5–2.06 | 240–245 | 2.5 | 2350 |
| Mjøsa, Norway | 60°20'–61°05'N 10°40'–11°30'E | 365 | 449 | 56.2 | 120–124 | 6 | 16,420 |
| Porttipahta*, Finland | 67°55'–68°10'N 26°15'–26°50'E | 34–214 | 15 | 0.1–1.35 | 234–245 | 1.5 | 2460 |

\* Reservoirs.

Europe and particularly in some areas of Finland, Sweden and northwest Russia where peatlands cover large proportions of the land area, polyhumic lakes are quite common, especially amongst small headwater lakes with relatively long retention times. The pH is typically low in many dystrophic lakes owing to high concentrations of humic and fulvic acids. However, the organic macromolecules also provide some buffering capacity, and therefore polyhumic lakes are not as vulnerable for the atmospheric inputs of acid substances as oligohumic and oligotrophic lakes (Kortelainen et al. 1989).

Dissolved organic matter may greatly modify and regulate the physical, chemical and biological characteristics of the lake ecosystem (e.g. Sarvala

Table 5.3 Distribution (%) of the main landscape types in Fennoscandia. (Modified after Henriksen et al. (1997); Data for Karelian Republic compiled from *Atlas Karelskoi ASSR* (1989).)

| Country/region | Forest* | Heath; tundra† | Mires‡ | Cultivated land | Built-up areas | Water |
|---|---|---|---|---|---|---|
| Norway | 36 | 48 | 7 | 3 | 1 | 5 |
| Sweden | 65 | 9 | 8 | 8 | 2 | 9 |
| Finland | 75 | <1 | 5 | 7 | 3 | 10 |
| Murmansk | 39 | 33 | 26 | 0.1 | 0.1 | 8 |
| Karelia | 66 | <1 | 15 | 1 | 2 | 17 |

\* Includes forested peatlands.
† Maritime, alpine and arctic treeless areas.
‡ Treeless peatlands.

et al. 1981; Schindler et al. 1992; Arvola et al. 1996; see also Volume 1, Chapter 7). The extinction of light varies in quality and extent, depending on the type of particles and DOM present in the water. Elevated humus content affects the light climate; in mesohumic lakes of colour c. 50–100 mg Pt L$^{-1}$ red light predominates over the other wavelengths. In polyhumic lakes of colour >100 mg Pt L$^{-1}$, only the longer wavelengths penetrate any distance into the water column. In such lakes, almost all light reaching beyond the topmost few centimetres comprises the red and green wavelengths.

As a consequence of the rapid light attenuation, the thermal stratification is usually steeper in humic lakes than in oligohumic lakes where solar energy penetrates much deeper in the water column. The onset of thermal stratification in spring is very rapid and spring circulation may be incomplete in many headwater lakes possessing brownish water, small size and sheltered position (Salonen et al. 1984; Similä 1988). On the other hand, during cold nights in summer and autumn the shallow epilimnion releases a high proportion of its thermal energy to the atmosphere, so that the diel variation in temperature may be much higher in humic lakes than in oligohumic ones.

Humus affects the metabolism, dynamics and structure of the biota in many ways (Sarvala et al. 1981; Schindler et al. 1992; Arvola et al. 1996; Kankaala et al. 1996a,b). Rapid attenuation of light confines the primary production to a very shallow surface layer (e.g. Eloranta 1978; Jones & Arvola 1984), so that the euphotic zone is usually thinner even than the epilimnion. This is opposite to the situation in oligohumic lakes, where some light usually penetrates to the upper part of the hypolimnion. This will impinge upon the vertical distribution of plankton.

In polyhumic lakes, the amount of organic carbon in solution can be several orders of magnitude higher than that in living organisms (Fig. 5.2; Salonen & Hammar 1986; Münster & Chróst 1990). Even though humic substances are known to be highly refractory to microbial decomposition, DOM is potentially an important source of energy in the ecosystem (Geller 1983; Tranvik 1988; Tulonen et al. 1992; Kankaala et al. 1996a; see also Volume 1, Chapter 7). The utilisation of allochthonous DOM by microheterotrophs is strongly affected by the availability of inorganic nutrients (Benner et al. 1988; Kroer 1993). Heterotrophic production may lead to hypolimnetic oxygen depletion, particularly in small and sheltered lakes of high relative depth. In larger lakes with stronger wind mixing, the depletion of oxygen is not as pronounced. When anoxic or low-oxygen conditions prevail in the hypolimnion, steep gradients of nutrients (e.g. orthophosphate), iron and manganese compounds, pH, conductivity and water colour may develop (Jones et al. 1988). Water stratification and the complex and highly dynamic abiotic and biotic interactions involved make investigations of the system very complicated.

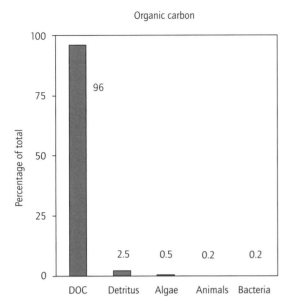

Fig. 5.2 Typical distribution of organic carbon in a polyhumic lake (Arvola et al. 1990).

#### 5.2.2.2 Ice cover

Winter conditions are especially difficult for autotrophs—macrophytes and algae—because the amount of light for photosynthesis is markedly reduced by snow and ice as well as by the low angle of incoming solar radiation. Snow and white ice also alter the spectral distribution of light penetrating into the lake water (Hawes 1985), because their absorption is highest in the red part of the visible spectrum (Roulet & Adams 1986). This shift is significant in humic lakes where red light penetrates deepest. From south to north Finland, the ice-cover period varies in the range 4 to 8 months.

Besides light, temperature is a limiting factor in winter. Water temperature is close to 0°C just beneath the ice, where the scarce illumination is greatest. This may constrain the growth rate of algae under ice, even when the light intensity might be high enough to support net photosynthesis. However, there are some algal species (e.g. *Chlamydomonas*, *Chlorogonium*, *Peridinium*, *Gymnodinium*) which are able to photosynthesise even in very cold water; some of them may produce high population densities well before the ice break (Similä 1988; Arvola & Kankaala 1989). In mid-winter, when practically no light penetrates into the water, the phytoplankton cannot photosynthesise and the energy requirements of the cells can be satisfied only by heterotrophy (mixotrophy), or by using energy stores. A survey of winter plankton by Arvola & Kankaala (1989) showed that, under mid-winter under ice and with practically no light, only very few cells possess chlorophyll in quantities sufficient to be detectable by microscopic autofluorescence.

Ice cover also prevents wind-induced water circulation and atmospheric exchange of gases. In lakes with high organic load, the lack of turbulence and gas exchange may easily lead to anoxis. Most vulnerable in this respect are small lakes, which become ice-covered earlier than large lakes. If the autumnal circulation is incomplete—a situation not uncommon in small, relatively deep lakes—the whole water column may become anoxic during the subsequent winter period. This presents a real threat to many organisms, particularly to fish. In large lakes the autumn mixing usually continues much longer than in small lakes and anoxis is likely only if the lake is heavily loaded by organic matter and nutrients or both.

## 5.3 LAKES IN FINLAND

### 5.3.1 History and forms of human impact on lake ecosystems

Until the end of the nineteenth century, the human impacts on lake ecosystems were mainly related to agricultural activities (land use, plant fibre retting (soaking), etc.) and the effects were mainly confined to small water bodies. During the twentieth century, industrialisation and growth of urban settlements led to the gradual increase of point-source pollution. As the early wood processing plants required the availability of hydropower and good connections for floating timber, many of these came to be located by the main watercourses and soon became major polluters of the downstream lakes. Industrial and urban pollution grew rapidly following the Second World War, at least until the 1970s. Since that time, construction of effective sewage treatment plants and process changes in industry have led to a marked decline of the point-source loads and improvement of water quality in the most heavily polluted areas. The effects of toxic pollutants and the internal recycling of nutrients persist in many recipient lakes. Postwar modernisation of agriculture has greatly increased its diffuse nutrient loads, which have proved to be much more difficult to control than the point-sources. Adoption of intensive forestry practices throughout the country and widespread conversion of peatlands for forestry, requiring drainage and fertilisation, has affected lakes extensively, even those in wilderness areas with little or no human habitation. The story is completed by atmospheric loading, which particularly affects the pristine headwater lakes. Besides airborne loads which are largely generated by fossil-fuel usage, other impacts of energy generation include the building of reservoirs and the regulation of levels in other lakes. Thus, the overall pattern of human assault of lake ecosystems has thoroughly changed over the past 30–40 years. Water protection measures have radically improved the situation in the most severely affected areas but, at the same time, conditions have deteriorated in areas which were previously thought to be beyond human interference (Wahlström et al. 1992, 1996; Oikari 1997).

The main types of human impact on lakes are presented in greater detail in the following sections (5.3.2–5.3.6.). Sections 5.3.7 and 5.3.8. outline some case histories of the responses of larger lakes to these various impacts.

### 5.3.2 Agriculture

The earliest cases of lake ecosystem change owing to human activities in Finland have been traced by palaeolimnological studies: even prehistoric agriculture could significantly have affected small and naturally oligotrophic lakes. Tolonen et al. (1975) report prehistoric (Late Iron Age) field erosion sediments in two lakes surrounded by clayey soils in South Finland. Numerous cases of mild lake eutrophication (diatom floral changes, increased nutrient and sediment accumulation rates) coincide with early cultivation phases revealed by pollen analysis, attesting either to arable field cultivation or to rotational slash-and-burn exploitation of forests (e.g. Miettinen et al. 2002). In areas with extensively cultivated clayey soils, e.g. in southern and southwestern Finland, the traditional agriculture caused marked eutrophication of a number of small lakes (Tolonen et al. 1990, 1994). Retting (soaking) of fibre crops (flax, hemp) in small lakes appears often to have been a significant contributor of nutrients directly into the ecosystem (e.g. Tolonen 1978, 1981; Grönlund et al. 1986). Fibre retting was noted as a cause of eutrophication when it was still practised (Hagman 1916). Since the mid-eighteenth century, rural population growth has increased the demand for arable and pasture and has led to the widespread lowering of water levels in lakes throughout the country. Anttila (1967) listed some 1500 documented lake-lowering events, mainly carried out between 1750 and 1900. Some of these were quite dramatic (Flower & Simola 1990), many lakes either disappearing completely or being permanently changed. Even fairly modest lowering, accompanied by gradual increase of external nutrient loading, appears to have triggered the onset of hypertrophic conditions with intense internal nutrient cycling in some instances. These include the lakes Enäjärvi (60°20′N, 23°40′E, lowered by about 1 m in 1928; Salonen et al. 1993) and Köyliönjärvi (61°07′N, 22°20′E, lowered in 1938–40 by about 0.7 m; Itkonen & Olander 1997); both were already eutrophic, located in agricultural areas of southern Finland. While generally shallow basins were most suitable for lowering purposes, many of them developed into very shallow eutrophic lakes with extensive macrophytic vegetation (Maristo 1941). As such, many of these have become valuable waterbird sanctuaries (Järvinen 1990); however, this is transitory pending hydroseral succession leading to their gradual terrestrialisation.

Since the 1950s, agricultural loading has accelerated into a major factor deteriorating the quality of surface waters (Kauppi 1984; Rekolainen 1989; Pietiläinen & Rekolainen 1991). In the post-war years to the 1960s, the area of arable land was increased, the drainage systems were generally improved, the use of industrial fertilisers grew rapidly and the numbers of livestock were increased. Along with an approximate doubling of the hectare yields, the nutrient loading to waters in agricultural areas also grew rapidly, as a consequence of increased soil erosion, leaching of fertiliser and careless spreading of cow manure and slurry (on snow, on frozen ground, or without subsequent tillage; Melanen *et al.* 1985; Rekolainen 1989). Tuusulanjärvi (60°25′N, 25°05′E) is an example of the many lakes in the clayey agricultural lowlands of southern Finland which were eutrophicated very early on by agriculture and which have since become hypertrophic following strongly increased agricultural loading since the 1950s (Pekkarinen 1990; Varis & Kettunen 1990). According to Rekolainen (1989), the total annual phosphorus load from Finnish agriculture is 2000–4000 t and that of nitrogen 20,000–40,000 t. As point-source loads have decreased dramatically, these figures mark agriculture as the single largest source of nutrient loads to watercourses in Finland. The particular impact of nutrient leaching is very much dependent on local and temporal conditions (topography, soils, drainage conditions, vegetation cover, season, weather). In agricultural water protection, special emphasis is nowadays placed on good practice in applying fertilisers and slurry, and on the establishment of vegetated buffer zones between cultivated fields and watercourses (Keskitalo 1990; Rekolainen 1993). This also conforms to the EU regulations, implemented in 1995, defining the environmental prerequisites for agricultural subsidies.

### 5.3.3 Impacts of forestry and peatland management on lakes

In the Boreal region (cf. Table 5.3), the influence of forest cover on both the hydrology and chemical characteristics of ground and surface waters is of paramount importance. Forest cover and peat-bogs effectively moderate hydrological extremes, such as floods, erosion events and eutrophication of rivers and lakes (see also Chapter 2, this volume).

Forest fires may transiently increase chemical erosion on catchments. Loss of forest shelter, as a consequence of fire or clear felling, may also alter the stratification pattern in small deep basins, as observed by Rask *et al.* (1993) and deduced palaeolimnologically by Korhola *et al.* (1996). During recent decades, the intensification of forestry practices (heavy soil treatment in plantations, drainage and fertilisation of peatlands for forestry) and their extensive application throughout the country have strongly affected the recipient water bodies (Kenttämies & Saukkonen 1996).

More than 60% of Finland's 10 million ha of peatland has been drained, mainly for forestry improvement. This activity peaked around 1970, when nearly 300,000 ha were ditched annually. In large areas of southern Finland, there are practically no peatlands left in natural condition (Vasander 1996). Forestry drainage may affect local hydrology by lowering the level of the groundwater table, by increasing runoff and possibly by altering watershed boundaries (Kenttämies 1981). For every 1% of a catchment drained, the annual runoff typically increases by 0.3 to 0.6%. Increased runoff is noted especially during the first years after the forestry improvement measures, but it returns gradually back to its original level during the following 15–20 years. Remedial drainage, i.e. cleaning of old ditches, again increases runoff. The same will happen also after clear felling: annual runoff might double after clearance of mature spruce forest. Moreover, the seasonal pattern of runoff may change, in favour of higher spring and summer maxima, of shorter duration and earlier occurrence.

Material leaching from forest, including particulate and dissolved organic matter as well as inorganic plant nutrients such as phosphorus, nitrogen and potassium, is of primary significance in the pollution of aquatic ecosystems. The leaching of phosphorus and nitrogen from unaffected areas is relatively low, only some 3–15 kg P km$^{-2}$ yr$^{-1}$ and 40–290 kg N km$^{-2}$ yr$^{-1}$ (Kenttämies & Saukkonen 1996). The airborne load of nitrogen may significantly add to the natural background level, particularly in the southern parts of Fennoscandia (e.g. Ruoho-Airola 1995). Leachate concentrations are usually highest during the spring and autumn peak runoff periods. Leaching from small catchment areas during the spring high flow can yield more than 50% of the total annual load in southern Finland, and greater than 60% in northern Finland.

Forest clear-felling may cause an increase in almost all leaching components. However, differences between soil types and hydrological characteristics of forests are large and all generalisations must therefore be drawn with care.

With the general intensification of forestry practices, the fertilisation of forests on both mineral and peat soils also intensified, peaking during the 1970s and early 1980s. Peatlands used for forestry have been fertilised mostly with phosphorus and potassium and forests on mineral soils with nitrogen. In some experimental peatland drainage and fertilisation areas, 0.5 to 2.5% of the applied phosphorus has been found to leach annually during the first few years. It seems that leaching continues at least for 5–10 years; in cases when PK fertiliser comprising readily soluble phosphate is used, leaching continues for up to 20 years, by which time 20% of the original application is washed out (Kenttämies & Saukkonen 1996). Fertilisation in winter, a customary practice in the early works, was found to lead to a high degree of leaching during snowmelt in spring. *Sphagnum*-peat of barren, acidic bogs possesses a very low capacity to adsorb phosphate, whereas minerotrophic peat, like mineral forest soils, possesses a higher pH and higher aluminium and iron content and, thus, binds phosphate rather better. As the total fertilised area of Finnish peatland is c. 15,000 $km^2$, it is understandable why fertilisation should have increased phosphorus concentrations in water bodies. According to Saura et al. (1995), the total phosphorus load from fertilised peatlands exceeded 300 t annually during the 1980s. Depending on the proportion of fertilised area in relation to the total catchment, the runoff phosphorus may have increased as much as five- to fifteenfold during the early years. In many cases, this is sufficient to have caused eutrophication and mass development of blue-green algae in recipient lakes (e.g. Simola et al. 1988).

Nitrogen fertilisation of mineral soils, commonly administered as ammonium nitrate or urea, has been found to increase leaching from 4 to 10% of the total during 1–2 years after the treatment (Kenttämies & Saukkonen 1996). Local conditions, especially the occurrence of heavy rains and high spring runoff, affect the process but, usually, these are of fairly short duration. However, leaching from coarse soils into the groundwater continues for longer (Ahtiainen & Huttunen 1995). Leaching of nitrogen from peatlands is much less than from mineral soils, as ammonium tends to adsorb onto peat particles. During the first year, 3–5% of the added nitrogen has been found to leach from peatland into surface drainage.

In an experimental manipulation study of small virgin-forest catchments in eastern Finland, clear-cutting of nearly 3 $km^2$ of partly paludified spruce forest increased the total phosphorus concentrations roughly fivefold, from a background of about 2.5 $\mu g L^{-1}$, throughout the first 3 years after clear-felling. Phosphate-phosphorus comprised more than half of the total. Subsequent soil treatment (scarification) perpetuated the elevated phosphorus concentrations particularly in the spring flood waters for at least another 3 years (Ahtiainen 1992). The huge, 200-fold, increases in suspended solids that continued for 3 years after the scarification (average loss, 83 $t\,km^{-2}\,yr^{-1}$) were the most damaging consequence of the treatment. Even though the clear-cutting area exceeded the present-day recommendations for good forestry practice, the experimental treatment (carried out in 1982–86) can be considered representative of the forestry activities followed extensively in eastern and northern Finland during the past three to four decades. Sedimentary records attest to forestry-related eutrophication in many formerly oligotrophic headwater lakes (Simola 1983; Rönkkö & Simola 1986; Simola et al. 1995; Meriläinen et al. 2000). Notable increases in mineral sedimentation have been observed in lakes receiving drainage from erosion-prone catchments (Sandman et al. 1990). Harmful effects are most obvious in small headwater lakes close to the impact sources, whereas, in large lakes of southern Finland, the effects of forestry are often compounded by other factors, such as agricultural and sewage effluents. Sandman *et al.* (1995) reported the paeolimnologically verified eutrophication of two fairly large wilderness lakes, Lentua (64°15′N, 29°40′E; 91 km$^2$) and Änätti (64°25′N, 29°50′E; 25 km$^2$), in Kuhmo, eastern Finland, where these other factors are negligible.

The experimental small-catchment manipulations reported by Seuna (1988) and Ahtiainen (1992) demonstrated that even fairly narrow untouched vegetation belts bordering natural drainage channels effectively prevent the leaching of particulate and dissolved nutrients from the treated forest areas into surface waters. Tossavainen (1997) reported one of the first catchment restoration projects, aimed at improving the recreational and fishery value of a lake strongly affected by consequences of forestry measures.

A particular problem associated with drainage operations is the abrupt low-pH events that may occur in areas of sulphide-rich clay mineral soils, which are particularly common in areas of former Baltic sediments (*Littorina* clays) around the Gulf of Bothnia. Groundwater fluctuation may cause oxygenation of sulphides, leading to very low pH and mobilisation of iron, aluminium and manganese in the runoff (Sevola *et al.* 1982; Palko 1994; Weppling 1997). Problems may well arise even if a small proportion (<5%) of the basin consists of drained sulphide clay. Episodic fish kills owing to acidification events have been documented from the beginning of the nineteenth century but the situation became far worse during the twentieth century, because of more effective drainage of farmland and forest. Episodic acidification may occur outside the coastal sulphide clay area: Rask *et al.* (1986) reported an acidification event in a forest lake after catchment drainage works, and this may have been attributable to the oxidation of bedrock-derived sulphide minerals.

An analysis of the occurrence of nuisance algal blooms in Finland (Kenttämies *et al.* 1995) revealed that some 7% of the reported cases could be attributed to nutrients originating from forestry practices. However, this figure might be an underestimate because of the very large number of small lakes which are potentially affected, yet which are not very important for fishing and recreation. It is probable that more complaints concern water bodies which are situated near centres of settlements and where other factors promote the algal blooms.

Blue-green algae are usually abundant in the reported algal blooms, indicating a substantial supply of phosphorus in the lake water. Another important phytoplankton taxon causing nuisance blooms is *Gonyostomum semen*, a large, flagellated, migratory alga belonging to the group of Chloromonadophyceae. It seems to be a species which favours humic lakes with a fairly high phosphorus content, at least in the hypolimnion, and the number of its reported occurrences in Fennoscandia has greatly increased since the 1980s. The reasons for its increase are, however, not fully understood (Hongve *et al.* 1988; Salonen *et al.* 1992, 2002; Lepistö *et al.* 1994; Willén 1995).

### 5.3.4 Industrial and municipal sewage loading: assault and recovery

During the past 150 years, wood processing has grown into a major industry in all of the northern European countries. Owing to the great number of pulp and paper mills, their typical siting by inland watercourses and the large outputs of process effluents, wood processing became an issue of general concern in water protection. Direct limnological impacts of other major industries, metal mining and processing and chemical production, have been less extensive. Along with industrialisation, the growth of urban settlements has been rapid. Until the 1970s, modernisation of society and increasing standards of living were accompanied by corresponding increases in municipal sewage effluents. The turning point was the development of national attitudes and policies towards protecting water. Altogether, some 600 sewage treatment plants have been built for municipalities and industry. Now, more than 76% of Finnish households are connected to sewage networks, and about 85% of all sewage water is treated by biological–chemical methods. The total phosphorus load from municipalities to waters has decreased from over 2000 t annually in the early 1970s to the present 250 t. Over the same period, organic loading (BOD) has decreased from nearly

50,000 t to some 10,000 t annually. Municipal nitrogen loads have, however, remained fairly high at some 14,000 t annually (Rosenström et al. 1996).

The history of environmental impacts of the Finnish pulp and paper industry is illustrated in Figs 5.3 and 5.4. The waste-water loading grew steadily with production increases until the energy crisis of the 1970s. This promoted the large-scale modernisation of the plants, recycling of the process chemicals and utilisation of the pulping waste for energy production. Some old mills were closed down and the outdated sulphide process had been discontinued by 1991. Further investments into process changes and waste-water purification, particularly during the late 1980s, led to a considerable decline of phosphorus, suspended solids and also organochlorines formed in the bleaching process. During the period 1989–2002, the specific organochlorine load, expressed as AOX (adsorbable organic halogens), has decreased from 2.7 to 0.16 kg t$^{-1}$ pulp bleached (Metsäteollisuus 1997, 2003). As the bulk loads have radically decreased, the focus of water quality research has shifted to ecotoxicological monitoring and risk assessment (e.g. Assmuth 1996; Oikari & Holmbom 1996).

### 5.3.5 Effects of water-level change

Hydroelectric power provides about 20% of the total electricity supply in Finland; as a versatile energy source it is widely used to reconcile the short-term and seasonal variation in energy consumption. Some 220 lakes, which, in terms of surface area, represent about one-third of total Finnish lake area, are regulated for hydropower

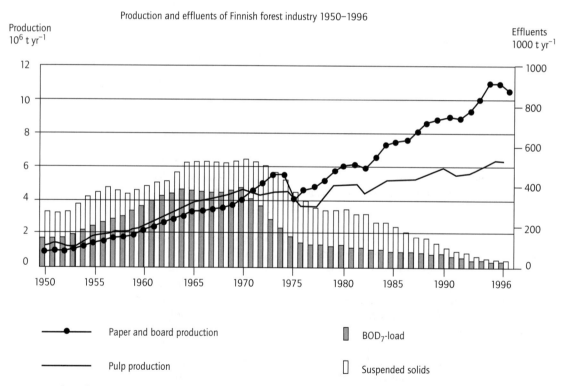

**Fig. 5.3** Pulp and paper industry production and its organic and suspended solids loading in Finland in the period 1950–1996. Organic loading expressed as biological oxygen demand (BOD). (Redrawn from Metsäteollisuus 1997.)

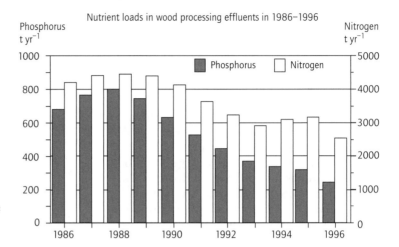

Fig. 5.4 Phosphorus and nitrogen loads in effluents of the wood-processing industry in Finland in the period 1986–1996. (Redrawn from Metsäteollisuus 1997.)

needs. The regulation amplitude in many of the large lakes in the south is, however, less than 1 m. In the typical regulation pattern, basins are filled during the open-water period and levels are lowered to a late-winter minimum. Some lakes and reservoirs are regulated in the interests of flood control (Kuusisto 1988; Alasaarela et al. 1989).

More than 60% of the hydropower is obtained from the rivers Kemijoki, Oulujoki and Iijoki in northern Finland. The two largest reservoirs of Finland (and of western Europe), Lokka and Porttipahta, were built in the upper reaches of the Kemijoki river in the late 1960s (Table 5.2). The water quality of these reservoirs has been monitored since their construction, reaching a steady state after 12–15 years (Kinnunen 1989). During the first 10 years, the late-winter oxygen conditions were worst at the extreme low-water stage. After 25 years, all the basic chemical parameters (colour, COD, total phosphorus, total nitrogen, total iron and silicon) indicated a reduction of at least 50% from the maximal concentrations observed. In Porttipahta, the reduction was slightly better than in Lokka, where about 80% of the flooded terrain was peatland. Only about one half of Porttipahta is on peat, the rest being on mineral soils.

Artificial lakes differ significantly from the natural lakes in respect of their physical, chemical and biological properties (see Chapter 12, this volume).

For instance, during the first years of their existence, the quantity of phytoplankton in both Lokka and Porttipahta was indicative of oligotrophic conditions, in spite of the high ambient concentrations of phosphorus and nitrogen (Arvola 1980; Lepistö 1995; Lepistö & Pietiläinen 1996). This was partly caused by the polyhumic water and consequently low light intensity, and partly because a major part of the nutrients were not available to phytoplankton. At this time heterotrophic and mixotrophic protists, amongst other flagellates, were abundant in the phytoplankton (Arvola 1980). At present, oxygen depletion during the late-winter, low-water phase under ice promotes nutrient release from sediments. This supports persistent blue-green algal blooms, which constitute a threat for recreation and fishery especially in Lokka reservoir (Lepistö & Pietiläinen 1996).

Hydropower regulation particularly affects the littoral zone of the impoundment but the specific consequences very much depend on the local circumstances. Erosion may be especially prominent in the early years of regulation (Alasaarela et al. 1989). Notable changes in the macrophytes, benthos and other ecosystem constituents have developed in several regulated lakes and impoundments in the Nordic countries (e.g. Nilsson 1982; Rørslett 1985).

### 5.3.6 Atmospheric deposition effects

The acidification issue in Finland was studied extensively between 1985 and 1990 through a government-funded interdisciplinary project, the Finnish Acidification Research Programme (HAPRO; Kauppi et al. 1990). A survey of statistically selected small lakes ($n = 987$) was carried out in order to quantify the impact of atmospheric loading on the acidity of lakes, which is also influenced by catchment sensitivity and the load of organic anions originating from paludified soils. The proportion found to be acidic (acid neutralising capacity $\leq 0\,\mu\text{Eq}\,\text{L}^{-1}$, as determined by Gran titration) was 12% but this was made up mainly by naturally acidic, humic lakes (75–85% of all acidic lakes; Forsius et al. 1990). Organic acids constitute the main anion fraction in all these lakes, with a median concentration of $89\,\mu\text{Eq}\,\text{L}^{-1}$. However, in the areas of high sulphur deposition in southern Finland, strong acids exceeded organic acidity in most lakes examined (Kortelainen et al. 1989; Kortelainen & Mannio 1990). Palaeolimnological investigations have revealed atmospheric acidification effects in several weakly buffered small lakes in southern Finland, either as a diatom-inferred pH decline (e.g. Tolonen & Jaakkola 1983; Simola et al. 1985; Tolonen et al. 1986) or as a loss of alkalinity (Huttunen & Turkia 1990; Huttunen et al. 1990). Notable changes owing to acidification have been observed in fish populations (Rask & Tuunainen 1990), macrozoobenthos (Meriläinen & Hynynen 1990) and macrophytic vegetation (invasion of *Sphagnum* mosses and loss of elodeids; Heitto 1990). Process modelling (Kämäri et al. 1990) showed that acidification of sensitive surface waters will continue unless significant reduction of acid deposition is achieved.

### 5.3.7 Case examples of diffuse loading effects in lakes

#### 5.3.7.1 Lappajärvi

Lappajärvi (63°10′N, 23°40′E) is a fairly large, naturally oligotrophic lake in western central Finland. Its surface area is $142\,\text{km}^2$, maximum depth 37 m, mean depth 7.5 m and drainage area $1527\,\text{km}^2$. The lake has experienced a fairly typical history of human activities affecting its condition: its water level was lowered by 1.5 m in two stages (1834, 1908) and the lake level has been regulated since 1961. Two potato-flour factories and municipal sewage have been loading the lake, but the influence on the water quality of diffuse loading from agriculture and from extensive forest drainage, peaking during the late 1960s, has been stronger and more persistent (Malve et al. 1992). During the early decades of the twentieth century, the lake was renowned for its clear water and profitable salmonid fisheries. However, the natural salmon stocks were lost when the outflow of the lake, the Ähtävänjoki, was dammed for hydropower. Prominent shore erosion commenced with the water level regulation and notable degradation of water quality has been evident since the 1960s.

Water-quality monitoring data exist from 1963 onwards and catalogues the change to mesoeutrophy. Rise of winter phosphorus concentrations from 10 to $20\,\mu\text{g}\,\text{L}^{-1}$, steady increase of BOD and COD values, greater fluctuations but a generally rising trend in water colour (up to $100\,\text{mg}\,\text{Pt}\,\text{L}^{-1}$) and, after 1976, winter anoxia in the deepest parts of the lake have all been observed (Hynynen et al. 1997). Palaeolimnological investigations (e.g. diatom and chironomid analyses) have revealed four stages in the history of ecosystem disturbance (Hynynen et al. 1997; Meriläinen et al. 2000): (i) pre-industrial period from 1900 to 1930; (ii) initial degradation from 1930s to late 1950s; (iii) period of enhanced erosion during the 1960s; (iv) pollution from 1970 onwards. During this development, the profundal chironomid fauna has changed from an oligotrophic *Micropsectra* spp.–*Stictochironomus rosencholdi*–*Heterotrissocladius subpilosus* assemblage to a eutrophic, low-oxygen assemblage characterised by *Chironomus anthracinus* gr. and *Ch. plumosus* gr. (the latter becoming most abundant during the 1980s). Benthic Quality Index values (BQI, scaling 1–5; Wiederholm 1980) have declined from 3.7–4.6 in the first stage to 1.6–2.5 in the fourth stage. Diatom-inferred pH has correspondingly increased from 5.9–6.8 in the first stage to 6.9–7.4 in the fourth, with typical assemblage changes indicating eutrophication (Meriläinen *et al.* 2000). Curtailing of the point-source loads with effective purification of the industrial and municipal effluents during the 1980s has not led to a recovery of the lake, as observed in many other cases. Nutrient concentrations have remained high and persistent algal blooming is still typical for the long autumnal turnover period. This is attributed to the continuous massive diffuse loading from the catchment and internal phosphorus recycling from the sediments (Meriläinen *et al.* 2000).

#### 5.3.7.2 Pääjärvi

Pääjärvi (61°04′N, 25°08′E) in Lammi, South Finland, has become one of the most thoroughly investigated lake ecosystems in Finland, owing to the convenient location of a biological field station on its shore (Lammi Biological Station, University of Helsinki). An extensive ecosystem study of the lake carried out in 1971–74 (Ruuhijärvi 1974; Sarvala *et al.* 1981) initiated a sustained hydrobiological research activity on the lake.

Pääjärvi, with a surface area of $13.4\,km^2$, is one of the deepest lakes in Finland with a maximum depth of 87 m and mean depth of 15 m. The catchment area ($244\,km^2$) consists mainly of forest (68%) and arable land (21%). The fields are intensively cultivated, sugar beet having been one of the main cultivated plants since the early 1970s. Sugar beet needs more fertilisers than other common crop plants and farmers have spread plenty of nutrients on their fields during the past 25–30 years.

Practically all the peat bogs (10% of the catchment area) were drained between the years 1950 and 1965 and, at the beginning of 1960s, the mean water level of the lake was lowered by 0.8 m, mainly to prevent spring flooding of the nearshore arable fields. In a multicore palaeolimnological investigation covering the recent development of the lake, Simola & Uimonen-Simola (1983) found three periods of accelerated sedimentation, of which the first two involved mainly allochthonous organic matter and the third mainly minerogenic material, respectively responding to peatland drainage and water level changes.

According to regular monitoring since the mid-1960s, both the electrical conductivity and pH have increased steadily in the lake (Fig. 5.5). The concentrations of total phosphorus and especially of total nitrogen have increased since the early 1970s, whereas hypolimnetic oxygen concentrations have decreased. Parallel long-term changes have been detected also in the biota, notably an increase in the biomass of sedentary benthos and a shift in its species composition towards the prevalence of Oligochaeta. In recent years, blue-greens have become more abundant in the lake phytoplankton (Hakala & Arvola 1994).

Although the trophic state of Pääjärvi is still oligo-mesotrophic, it is quite evident that the lake has become considerably eutrophicated during the

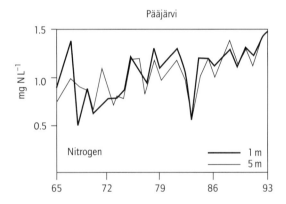

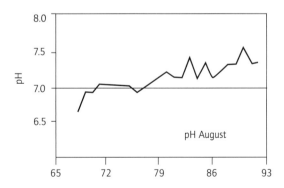

**Fig. 5.5** Epilimnetic nitrogen and summer pH in Lake Pääjärvi during the period 1965–1993.

past 20–30 years (Arvola 1991; Hakala & Arvola 1994; Itkonen & Salonen 1994). It is very clear that most of the nutrient input to this lake originates from agriculture (nitrogen also through direct deposition). Pääjärvi is thus representative of fairly large, deep lakes affected by agriculture and land-use, and where point-source loading is of only marginal importance.

The most recent results suggest, however, that the eutrophication may have weakened since the mid-1990s. Whether the observed decrease in nutrient concentrations is merely due to the rather dry weather conditions prevailing during some years, or whether it signals positive responses to the catchment management measures that have been introduced in order to protect the inflowing brooks and rivulets from nutrient leaching from the fields and farmsteads, is still very much an open question.

### 5.3.8 Point-source loading effects in large lakes

#### 5.3.8.1 Päijänne

Päijänne is the second largest lake in Finland (Table 5.2). The northern part, especially, has been severely affected by several wood processing plants. Before its rapid industrialisation and urbanisation, Päijänne was an oligotrophic lake (Itkonen & Salonen 1994; Meriläinen & Hamina 1993). According to the chironomid stratigraphy of the sediments (Meriläinen & Hamina 1993), the following developmental stages have been distinguished: (i) pre-industrial stage (the period before the Second World War); (ii) stage of increasing pollution (from c.1944 to 1973); (iii) the period of severe pollution (from 1973 to 1983); and (iv) the period of gradually improving water quality owing to water protection (c.1983 onwards). This comes out very clearly from the Benthic Quality Index values, both in actual monitoring data of profundal benthos since the early 1970s, and in the sedimentary chironomid assemblages in the northernmost basins of the lake (Meriläinen & Hamina 1993). A rather similar history of assault and initial recovery was established for Vanajavesi (61°12′N, 24°00′E), which has also been strongly affected by effluents of the wood processing industry in the town of Valkeakoski (Kansanen 1981, 1985; Kansanen & Aho 1981; Kansanen & Jaakkola 1985).

The nutrient and organic matter load originating from the local industry and communities around Päijänne has decreased drastically during the past 30 years (cf. Fig. 5.3), even though the present pulp and paper production volumes greatly exceed those of the 1960s and 1970s. Also, the water quality records indicate that the central and northern parts of the lake have recovered remarkably during the past 15 years. Transparency has increased steadily since the mid-1970s and the concentrations of lignin and total phosphorus have decreased, as have the water colour and the chemical oxygen demand.

The present condition of the lake clearly reveals the importance of paper and pulp mills and other wood-processing industry as the main sources of organic matter and nutrients. However, the phosphorus load from municipal treatment plants has decreased since 1980. In contrast, the nitrogen load from industry and communities has remained steady for the past 20 years and, at the same time, nitrogen inputs from atmosphere and agriculture have increased.

As a further indicator of improved water quality, the annual influx of head capsules of chironomids is almost twice as high in the 1983–88 sediment layer than in the 1973–79 layer. According to Meriläinen & Hamina (1993), this implies previously inhibitory effects of the high organic loads from the mill effluents, especially the organochlorine compounds, received during the 1970s (cf. Paasivirta et al. 1990). This is also supported by the observation that the profundal biomass in 1987 was twice as high as in 1984 (Meriläinen & Hynynen 1988).

Effects of the recent water quality change are not as clearly evident in phytoplankton as in the profundal benthos (A. Palomäki, University of Jyväskylä, unpublished results). In the largest basin of the lake, Tehinselkä, the biomass of phytoplankton increased substantially in the late 1970s, after which, a rapid decline, then a more gradual increase was observed. It seems that when the light conditions rapidly improved owing to increasing transparency in the end of the 1970s, diatoms could make use of the situation and their absolute biomass increased, along with their share of the total biomass. Similarly, the biomass increase since the mid-1980s was mainly accounted for by diatoms, but factors behind this development are obscure. Obviously light is one of the key factors because light penetration has improved steadily and the productive water layer has increased by some 25 to 50% within the past 20 years. At the same time, phosphorus concentration has declined. Both may support diatoms, which cannot actively regulate their position with regard to the light field in the water column, but, on the other hand, some diatoms possess very low requirements for phosphate, which may improve their ability within the phytoplankton community.

#### 5.3.8.2 Saimaa lake complex

The Saimaa is the largest lake in Finland (Table 5.2). It consists of several basins, connected by narrow straits. Natural water quality varies between basins, most notably with respect to water colour, ranging from oligo- to mesohumic. Wood-processing effluents have most strongly affected the northwestern Haukivesi basin (town of Varkaus) and the southernmost basin of Saimaa proper (several pulp and paper mills in Lappeenranta, Joutseno and Imatra). By and large, the pollution history is similar to that of Päijänne (e.g. Laine 1997; Saukkonen 1997).

A comprehensive research project, with the aim of testing and developing biological monitoring methods for large lakes, was conducted within several basins in the northern Saimaa in 1990–93. The investigation included pelagial, profundal and palaeolimnological studies. Study of ecosystem functioning in the pelagic involved extensive and co-ordinated analyses of nutrients and chlorophyll (Mononen & Niinioja 1993), primary production and respiration (Huovinen *et al.* 1993), phyto- and zooplankton and fish (Holopainen *et al.* 1993, Viljanen & Karjalainen 1993: Rahkola *et al.* 1994), as well as some of the interactions between trophic levels (Karjalainen & Viljanen 1993; Karjalainen *et al.* 1996). Synoptic basin-wide surveys revealed interesting patterning of the plankton communities (Fig. 5.6), which according to Karjalainen *et al.* (1996) can be a result of at least four possible mechanisms: (i) trophic gradient within the basin; (ii) herbivore grazing of algae; (iii) effect of algal species composition on zooplankton feeding; and (iv) regeneration and reorganisation of nutrients. Results of the studies of profundal macrobenthos and semipelagic animals, carried out in order to assess the proper sampling protocols for these groups, are presented by Veijola *et al.* (1996) and Bagge *et al.* (1996). Simola *et al.* (1996b) report the palaeolimnological studies conducted as part of the project.

The biomonitoring studies in Saimaa have been facilitated by the use of a modern, well-equipped research vessel, R/V *Muikku*. Automated data handling, including Geographical Information System (GIS) software, allow continuous monitoring of several water quality parameters, e.g. chlorophyll fluorescence, onboard the cruising ship (Holopainen *et al.* 2001).

#### 5.3.8.3 Vesijärvi

Vesijärvi, near the town of Lahti in South Finland (61°10′N, 25°30′E), is a fairly large but shallow lake (109 km$^2$, mean depth 6 m). Its drainage area is relatively small (335 km$^2$) compared with the lake area, which accounts for the lake's oligohumic character. Its water level was lowered by some 2.5 m during the eighteenth and nineteenth centuries. The present water level is 81 m a.s.l. The lake drains via Vääksy Canal (built 1871) into the southern end of Päijänne (78 m a.s.l.). At the beginning of this century, the lake was famous for its clear water — it was even called 'the largest spring in the whole country'. Agricultural runoff and, especially, municipal effluents from the rapidly growing town of Lahti changed the situation drastically. The first signs of eutrophication were recorded during the 1920s when blue-green algal blooms were identified as a cause of death of some cattle at the shore of the lake (Hindersson 1933). The eutrophication increased rapidly in the 1950s

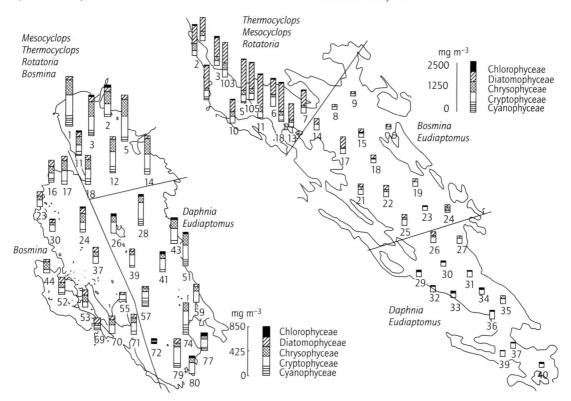

Fig. 5.6 Synoptic distribution of plankton in two basins of the Saimaa complex. On the basis of plankton composition the basins can be divided into distinct zones. The spatial patterning is partly the result of environmental gradients (e.g. nutrients) and partly a reflection of different phases in the dynamic interactions within the plankton community. Bars indicate biomass and composition of phytoplankton; the characteristic zooplankton species for each zone are also indicated. (From Karjalainen et al. 1996.)

(Keto 1982; Keto & Sammalkorpi 1988; Liukkonen et al. 1993) and massive blue-green blooms (*Planktothrix agardhii*, *Anabaena* spp. and *Microcystis* spp.) were occurring at the beginning of the 1960s. The external nutrient load reached its maximum early in the 1970s: in the most eutrophicated basin, Enonselkä, the maximum phosphorus load from municipal sewage was $2.1\,g\,m^{-2}\,yr^{-1}$, until a new treatment plant avoided the entry of sewage from Lahti in 1976. Despite the greatly reduced external loading, there was only a slight improvement in the condition of the lake: phosphorus concentrations in Enonselkä decreased from 80–90 to 40–60 $\mu g\,L^{-1}$ but remained at that level throughout the 1980s (Keto & Sammalkorpi 1988).

Ecological management and intensive research of the lake was finally initiated by the City of Lahti in 1987 with the collaboration of local and national environmental and fishery authorities, as well as of university scientists. The goals of the programme, called The Vesijärvi Project, were to eliminate the mass developments of blue-green algae, to restore the recreational value and to re-establish a sustainable commercial fishery in the lake. Restoration of the food chain by biomanipulation was suggested, using the conceptual food-web model of Keto & Sammalkorpi (1988). The study also included enclosure experiments, which indicated that a dense stock of roach (*Rutilus rutilus*) in the lake was favouring blooms of blue-greens through its bottom-feeding habit, which maintains a heavy internal phosphorus loading from the sediments (Horppila & Kairesalo 1990). Biomanipulation was preferred to direct chemical and technical methods, not least because of its lower costs.

Besides the biomanipulative treatment of Enonselkä basin, further reductions in external loading from other parts of the catchment were sought. Detailed plans were developed for 283 agricultural farms to establish buffer zones and to apply measures to cut the diffuse loads from cattle husbandry and arable cultivation. The control of internal loading in the Enonselkä basin was attempted by biomanipulation, for which fishing gear was modified to remove cyprinids preferentially. In the years of 1989–94, over 1000 t of coarse fish, mainly roach, were removed from the lake by professional fishermen using trawls and seines. In addition, local people removed a further 125 t with fykenets in the 3 years 1992–94. Altogether, more than 420 kg ha$^{-1}$ were removed from the Enonselkä basin within 5 years. After this mass cull of the cyprinids (roach and bleak, *Alburnus alburnus*), as well as smelt (*Osmerus eperlanus*), perch (*Perca fluviatilis*) and other small fish, it was estimated that the fishing effort of local fisherman was sufficient to maintain the lake in the condition achieved. A further 1.3 million yearling pike-perch (*Stizostedion lucioperca*) and other predators were stocked into the lake. The fishing treatments are described in more detail by Horppila & Peltonen (1994), Horppila & Kairesalo (1992) and Turunen *et al.* (1997).

The fish community is now structurally more diverse than 10 years ago and the roach population, especially, has diminished to less than one-third of its pre-manipulation numbers, with benefits in terms of decreased internal phosphorus loading. At the same time, the external load of phosphorus has fallen to c.0.15 g P m$^{-2}$ yr$^{-1}$, which is already under the critical load for the lake (0.3 g P m$^{-2}$ yr$^{-1}$). Since 1990, the mass occurrences of blue-greens have been less frequent, which has considerably increased the recreational value of the lake. As an unforeseen consequence of ecosystem change, Liukkonen *et al.* (1995) have observed the appearance and establishment of the brackish-water diatom *Actinocyclus normanii* f. *subsalsa* as a fairly abundant species in the lake's plankton.

The Vesijärvi project serves as a useful model for other lake restoration projects in Finland and elsewhere in north Europe, because it clearly shows that the biomanipulation approach can be applied even for large lakes. The removal of fish biomass by trawling was particularly successful (Peltonen & Horppila 1992; Horppila & Peltonen 1994; Jurvelius & Sammalkorpi 1995), and could probably be imitated elsewhere.

## 5.4 SWEDISH LAKES

### 5.4.1 Acidification and methylmercury: atmospheric deposition effects in small lake ecosystems

Lake acidification became a major environmental issue during the 1960s (Odén 1968; Almer *et al.* 1978), especially with respect to its effects on poorly buffered small lakes in areas of high sulphur and nitrogen deposition in western Sweden. Disappearance of sensitive fish species, pH decline and increase of transparency were observed in several lakes. In order to gain understanding of the mechanisms of acidification and the ecology of acidified lakes, an integrated study of Gårdsjön and its catchment (north of Gothenburg on the Swedish west coast, 58°04′N, 12°03′E) was carried out in 1979–82 (Andersson & Olsson 1985). In this case, as in many others, palaeolimnological analyses proved to be indispensable for the historical assessment of ecosystem change (Renberg & Wallin 1985; Renberg & Wik 1985).

During the 1970s, mercury came to receive special attention among environmental pollutants when considerable food-chain contamination with methyl mercury was observed in humic forest lakes, far away from any obvious industrial sources (Lindqvist *et al.* 1984; Meili 1991). In an extensive survey conducted during the early 1980s, high Hg concentrations in pike (*Esox lucius*) were observed throughout the country, except for the northernmost upland areas. The total number of lakes in which mercury in pike exceeded the concentration considered safe for human consumption (1 mg Hg kg$^{-1}$ wet weight) was estimated to be 5000–10,000 in Sweden (Lindqvist *et al.* 1984). The methylation process, which incorporates mercury into the food chain, takes place in oxygen-deficient parts of terrestrial and aquatic ecosystems. Wetlands, such as mires and paludified forest areas, possess high methylation activity and, thus, the concentration of methylated mercury and humic substances correlate positively (Meili 1991; Verta 1990).

Liming of lakes has been extensively practised in Sweden as a remedy for acidification. Henrikson & Brodin (1995) summarise over 8000 cases from the past two decades.

### 5.4.2 The four great lakes

The lakes Vänern, Vättern, Mälaren and Hjälmaren together make up about one-quarter of the total lake area in Sweden and, given the contributions of Vänern and Vättern, a much larger part of the total volume (Kvarnäs 2001). All these lakes have been variously affected by human activities. Willén (1984) presented a first comprehensive overview of the natural conditions of the four lakes, giving details of their environmental states at the time when the major detrimental influences had been curbed and the lake ecosystems were beginning to recover; their subsequent development is detailed by Willén (2001) and Wilander & Persson (2001).

#### 5.4.2.1 Mälaren

Mälaren (Table 5.2) is a complex lake, consisting of five main basins, and with a total length of about 110 km in the east–west direction. Its water level is only slightly above sea level. In fact, the lake became separated from the Baltic by isostatic land uplift during historical times. There are extensive areas of lime-rich clay deposits especially in the south and northeast parts of the catchment, which were taken into cultivation early on. The drainage area is densely populated, and the roots of its industrial history, iron and copper mining in the northwest parts of the catchment, date back to mediaeval times. The centre of political power in Sweden has been situated in the Mälaren area ever since the Vikings settled at Birka; the capital Stockholm has grown by the lake's outlet stream.

Gradual deterioration of water quality has followed the growth of human activities. As early as 1785, a traveller praised the clear and potable water of Mälaren but referred to the bad-smelling, filamentous and slimy green material which developed in shallow water during the warm season. This may be the first account of blooming of the blue-green alga *Aphanizomenon flos-aquae* in the lake (Willén *et al.* 1990b). Serious problems began, however, first during the 1950s, with the increasing municipal sewage and agricultural loading. The shallow margins of the lake near the largest population centres (Glomman at the western end, Ekoln Bay in the north, and the easternmost parts near Stockholm) developed a hypertrophic condition, whereas the open central area remained mesoeutrophic. Comprehensive limnological monitoring of the lake has been conducted since 1964 (Persson *et al.* 1990; Willén 2001). Chemical precipitation of phosphorus was introduced to municipal sewage treatment plants during 1970–75. This, and further water protection measures, e.g. sewage diversion, have led to a 95% reduction of the point-source phosphorus loading and a considerable improvement in condition of the lake. Now, some three-quarters of the phosphorus and two-thirds of the nitrogen reaching the lake come from diffuse sources, mainly agriculture. Phytoplankton biomasses have generally decreased and the patterns of seasonal phytoplankton dynamics have changed (Willén 1992). In the hypertrophic basins during the 1960s, blue-greens typically occupied the algal community from June onwards, with occasionally very high biomasses (exceeding 40 mg L$^{-1}$). During the 1980s diatoms and cryptophyceans became abundant, with blue-greens mostly of minor importance or with less persistent blooming. Corresponding changes have taken place in the zooplankton, receding from the peak biomass of the 1970s, and in the macrobenthos, showing a dramatic decline in total biomass and large shifts in the species composition (Willén et al. 1990b; Willén 2001).

Modelling of nutrient mass balances for the various sub-basins of the lake and analysis of nutrient influence on phytoplankton biomass (chlorophyll a) have established the crucial role of phosphorus as a determinant of the lake ecosystem functioning. During the 1960s, the total annual phosphorus load exceeded 1000 t. Persson et al. (1990) proposed 25 $\mu$g P L$^{-1}$ as a favourable target concentration for phosphorus in the lake. To achieve this, the total phosphorus load was required to be reduced from some 500 t yr$^{-1}$ by a further 210 t, which would be possible only by effectively overcoming the diffuse loading. The actual consequences of phosphorus reduction have greatly depended on the variable retention and release patterns of sedimentary phosphorus in the different basins (Wilander & Persson 2001).

Because of general phosphorus-limitation, excess nitrogen does not present an ecological problem in the lake. However, the high nitrogen concentrations in the outflow waters promote algal growth in the Baltic waters of the inner archipelago of Stockholm, making the control of nitrogen an important issue (Willén et al. 1990b).

The zebra mussel (*Dreissena polymorpha*) was first observed in Mälaren in 1926 and became established especially in the eastern parts of the lake (Willén et al. 1990b; Josefsson & Andersson 2001). Its impact in the Swedish lakes appears to have been much weaker than, for example, in the Lakes Erie and Huron, where the influence on the ecosystem functioning of this alien species has been profound (e.g. Fahnenstiel et al. 1995a,b; Holland 1993; see also this Volume, Chapter 3).

#### 5.4.2.2 Hjälmaren

Hjälmaren (Table 5.2) is smaller and shallower than Mälaren, into which it drains, but the lakes are rather similar in many respects. A major event in the history of Hjälmaren was the lowering of its water level by about 2 m in 1882–1886, which reduced both the lake area and the volume by nearly 30% (Håkanson 1978). Eutrophication of the lake has led to extensive development of macrophyte beds (e.g. *Phragmites*) in the shallow western and southern basins of the lake. About two-thirds of the drainage area lies to the west and southwest of the lake; the largest settlement, the town of Örebro, is at the western end. While the outlet is in the east, there is a west-to-east throughflow and, during the 1960s, diffuse and municipal loading created a strong trophic gradient across the basin, with hypertrophic conditions in the shallow western part, the Hemfjärden. The history of water quality monitoring and water protection measures of Hjälmaren is similar to that of Mälaren: the lake has been monitored since 1965, and point-source loading was drastically cut during the early 1970s. In Hjälmaren, however, the efforts have not been so successful: load reductions did decrease phosphorus concentrations in the western Hemfjärden from the very high concentrations of 300–500 $\mu$g P L$^{-1}$ (1965–76 August surface values) down to 70–150 (corresponding values in 1984–94). In the central parts, however, no significant reduction occurred and the whole lake still remains eutrophic. Internal phosphorus loading from the sediments is implicated (Persson 1996c; Wilander & Persson 2001).

#### 5.4.2.3 Vättern

Owing to its great water volume (mean depth: 40 m) and relatively small drainage area (catchment/lake ratio 2.4), Vättern (Table 5.2) is the most oligotrophic of the great lakes of Sweden. It is of great scenic and recreational value and is an important resource of drinking water to a large population. Despite domestic and industrial loading (from the town of Jönköping at the south end of the lake, and some wood-processing and metal industry), the lake has remained oligotrophic. However, during the 1960s quite alarming signs of deterioration of water quality were observed. The water was still clear but the mean Secchi-disk depths of some 17 m reported in 1888–1910 had decreased to 10 m in the 1960s, with some records as low as only 5 m. Total phosphorus concentration had increased to $15\,\mu g\,L^{-1}$ in 1967, with the increase in the estimated total-phosphorus load up from c.50 t in the 1940s to c.200 t in the mid-1960s (Willén 1984). A water conservancy plan was adopted for the lake in 1966 and measures to reduce the external loading were successfully implemented. The present phosphorus loading has been reduced to some 70 t annually and the concentration of phosphorus in the water has declined to $5-6\,\mu g\,L^{-1}$; at the same time, however, nitrogen concent-ration has gradually increased to relatively high values (total nitrogen $0.7\,mg\,L^{-1}$ in mid-1980s; Naturvårdsverket 1990). Results of the water quality monitoring of Vättern and its inflows are presented in Wilander & Willén (1994) and Wilander & Persson (2001).

### 5.4.2.4 Vänern

Vänern is the largest of the Swedish lakes; also its drainage area is quite extensive (Table 5.2). The lake consists of two large open basins, Värmlandssjön and Dalbosjön, with extensive archipelago areas along the rugged coasts: there are some 22,000 islands and islets in the lake. The natural water quality in Vänern is oligohumic (colour $25\,mg\,Pt\,L^{-1}$) and not quite as oligotrophic as that of Vättern, and the lake has been relatively more affected by industrial and domestic effluents.

Alarmingly high concentrations of mercury, other heavy metals (Zn and Cd) and xenobiotic organic compounds were detected in the lake during the late 1960s (Håkanson 1977). A chloroalkali-plant on the northern shore of the lake was the greatest polluter, with estimated emissions of some 3 t mercury each year until 1968, after which the pollution was radically diminished. Chloro-organic compounds peaked in wood-processing effluents during the mid-1960s with maximal annual AOX loads of some 2500 t. Adsorbable organic halogen loading has been effectively curbed only during the 1990s (Lindeström 1996). The present situation as regards toxic pollution is much improved: the pollutants have largely been incorporated into sediments and toxicant concentrations in the biota have decreased (Lindeström 2001).

Regular ecosystem monitoring of Vänern began in 1973, when the lake had already begun to recover from the most severe impacts. Owing to reduction of wood processing effluents, the organic content of the water has declined quite dramatically (COD from $8-10\,mg\,L^{-1}$ in mid-1970s to $3-4\,mg\,L^{-1}$ in 1993-94). Correspondingly, Secchi-disk depths have increased from 3-4 m to 5-7 m, which will considerably affect pelagic ecosystem function (Persson 1996d). Total phosphorus inputs have decreased considerably but the phosphorus concentration in the water has remained fairly low, around $10\,\mu g\,L^{-1}$, which may be explained by more efficient phosphorus recycling in the water and a reduction in net sedimentation. Overall, the phytoplankton community changed little over the period of monitoring (Willén & Wiederholm 1996). Decrease in atmospheric sulphur loading is evidenced by slightly declining sulphate concentrations and increasing alkalinity in the surface waters (Persson 1996d).

## 5.5 NORWEGIAN LAKES

### 5.5.1 Atmospheric loading impacts on dilute lakes

In Norway, there are a large number of lakes which together comprise 5.2% of the surface area of the country (Table 5.1; Skjelkvåle et al. 1997). Most of the lakes are rather small but, owing to the glacial history and mountainous terrain, many of the basins are relatively deep. Hornindalsvatn (62°00′N, 6°20′E), which is a fjord-like basin 21 km long and 514 m deep, is allegedly the deepest lake in Europe.

Surface water quality is largely determined by the widespread occurrence of minerals with low weathering rate in the bedrock and soils, and by the high precipitation and cool maritime climate. Precipitation and runoff are especially high along the southern and western coasts of Norway; high concentrations of sea salts characterise precipitation in these areas. As a whole, a very large proportion of the Norwegian lakes are oligotrophic and weakly buffered (Table 5.4). This is particularly

Table 5.4 Trophic categorisation of Fennoscandian lakes according to total phosphorus concentration, and characteristics of N:P ratio in the lakes. The data are based on water samples of a statistically representative set of altogether 5414 lakes surveyed in 1995. (Modified from Henriksen et al. 1997.)

| Country/ region | Oligotrophic ($<10\,\mu g\,P\,L^{-1}$) | | Mesotrophic ($10-35\,\mu g\,P\,L^{-1}$) | | Eutrophic ($>35\,\mu g\,P\,L^{-1}$) | | N:P ratio | |
|---|---|---|---|---|---|---|---|---|
| | Number of lakes | % | Number of lakes | % | Number of lakes | % | Median value | Percentage of lakes with N:P<12 |
| Norway | 36,242 | 92.2 | 2253 | 6.9 | 349 | 0.9 | 44 | 2.2 |
| Sweden | 35,815 | 55.8 | 15,845 | 40.8 | 1,320 | 3.4 | 39 | 2.1 |
| Finland | 11,363 | 38.5 | 15,495 | 52.5 | 2,656 | 9.0 | 31 | 2.7 |
| Murmansk | 11,768 | 74.9 | 3079 | 19.6 | 864 | 5.5 | 30 | 5.5 |

true for lakes in the treeless upland areas. In lowland lakes of forested or cultivated catchments, the organic content and nutrient concentrations tend to be somewhat higher. In most lakes, phosphorus seems to be the limiting nutrient for phytoplankton (Faafeng & Hessen 1993; Skjelkvåle et al. 1997).

Sulphur and nitrogen pollutants, transported over long distances from their sources, are heavily deposited in southern and western Norway, with a northwardly decreasing gradient. The extreme northeastern part of the country, eastern Finnmark, is also affected by industrial sulphur from the Russian Kola peninsula. Acid deposition has strongly affected an area of some $30,000\,km^2$ in southern Norway, with the loss of salmonid fishes in several watercourses. Lake acidification is demonstrable through comparisons of algological records (from 1949 and 1975; Berge 1976) and by palaeolimnological studies (e.g. Charles et al. 1989). Acidification effects include the decrease in humus content, as is inferred by sedimentary diatom assemblages in some lakes (Davis et al. 1985), and mobilisation of labile aluminium from soils into acidified waters (Hongve 1993). Even though the overall situation has improved during the past decade, through decreased industrial sulphur emissions in Europe (Fig. 5.7), the critical load is still exceeded in a large number of lakes in southern and western Norway (Skjelkvåle et al. 1997).

### 5.5.2 Mjøsa: eutrophication and recovery

Mjøsa is the largest lake in Norway (Table 5.2). It is a deep fjord-lake, the central zones of which seldom freeze during the winter. The lake is situated in the southeast lowland area of Norway, underlain by Cambro-Silurian sedimentary rocks and some limestone. Most of the drainage area is, however, on granite and gneiss bedrock. The lake itself is regulated for hydropower, as is also its main tributary flowing from the high mountain area to the northwest (Kira 1990).

The lake was gradually eutrophicated from the 1950s by increasing sewage and diffuse loading. The development culminated in the mid-1970s with the late-summer blooming of geosmin-producing *Oscillatoria bornetii* f. *tenuis* in the plankton, which jeopardised the future of the lake as a drinking-water source for about 200,000 people (Holtan, 1979; Skulberg 1988). This event initiated a nutrient-reduction programme (the Mjøsa campaign, 1977–81), which succeeded in cutting the point-source phosphorus loading by about 90%. There has been a decrease in net phytoplankton primary production in the central parts of the lake, from about $100\,g\,C\,m^{-2}\,yr^{-1}$ in 1975–76 to $20-30\,g\,C\,m^{-2}\,yr^{-1}$ since 1979, and the *Oscillatoria* blooms have ceased. Eutrophication also brought about changes in the zooplankton community (the appearance of *Daphnia cristata* and *Chydorus sphaericus*, an increase in *Polyphemus pediculus*

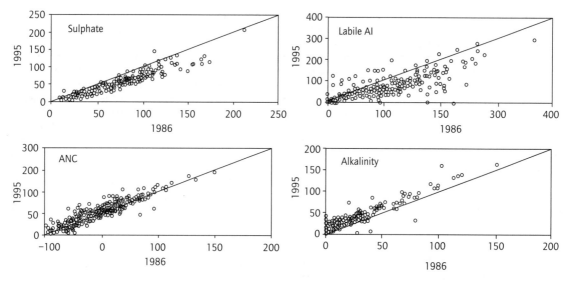

Fig. 5.7 Relaxation of acidification pressure as indicated by water chemistry parameters analysed in a statistically representative set of 485 Norwegian lakes in 1986 and 1995: ANC = acid neutralising capacity. (Redrawn from Skjelkvåle et al. 1997.)

and the disappearance of *Holopedium gibberum*), but even these changes were largely reversed with the receding eutrophy of the lake (Rognerud & Kjellberg 1990).

## 5.6 LAKES OF NORTHWEST RUSSIA

### 5.6.1 Limnology and water quality

Foundations for long-term limnological field studies in Karelia were laid at the Borodin Limnological station, Konchozero village, about 40 km north of Petrozavodsk. In 1895, a previous station had been set up by Academician I.P. Borodin in the Valdai area (between Moscow and St Petersburg), initially at Lake Bologoye and, after some years of activity there, later at Lake Seliger. In 1926, the station was moved to Karelia, where research was conducted until 1940. The studies included hydrological, chemical, hydrobiological and palaeolimnological investigations of the Karelian lakes, as well as some microbiological, ecophysiological and toxicological research (Aleksandrov et al. 1970; Veselov 1970). Following the Second World War, limnological research was continued at institutions in Petrozavodsk and St Petersburg (then Leningrad).

The Borodin station was situated on an isthmus between the lakes Konchozero and Pertozero (62°08′N, 34°00′E). We take this opportunity to point out that, contrary to somewhat obscure information given by Perfiliew (1929) and perpetuated since, the sediment of Pertozero is *not* varved as famously promulgated. Author HS cored the lake in 1991 and found the clayey sediment to be irregularly laminated but lacking in true (seasonal) varves.

Amongst the very first whole-ecosystem investigations of Fennoscandian Shield lakes were the studies of lakes Krivoe and Krugloe, conducted in 1968–69 within the International Biological Programme (IBP) framework (Alimov & Winberg 1972; Alimov et al. 1972). These two small oligotrophic lakes are close to the Kartesh Biological station near the White Sea coast, about 20 km south of the Arctic Circle (66°20′N, 33°40′E). The studies incorporated assessments of productivity of all ecosystem levels — bacteria, phytoplankton, zooplankton, benthos — into an energetic budget of the ecosystem which emphasised the role of bacterial production in ecosystem functioning. The ecosystem of the oligotrophic Lake Krasnoye, situated on the Karelian Isthmus at the southern margin of the Shield (60°34'N, 29°40'E), was similarly investigated (Andronikova et al. 1972). These studies were very advanced for their time and provided important models for lake ecosystem research elsewhere. It was most unfortunate that the 1970s' Cold War politics led to the gradual isolation of Soviet science and that the ground-breaking limnological research conducted there began to lag behind the rapid progress in research methods and instrumentation experienced in the West.

During the post-war decades, the water resources of Karelia and Kola Peninsula became extensively exploited for hydropower production. Most of the large lakes in Karelia are regulated, including Onega (since 1953), Vigozero (since 1932) and Segozero (Table 5.2). The same is true for western Kola Peninsula, e.g. Lake Imandra and Verkhnetulomskoye Reservoir serving as hydropower reservoirs. Industrial pollution of these waters gained attention only since the late 1980s. For instance, some of the lakes near the Monchegorsk smeltery, including Monche-Guba Bay of Imandra (67°55'N, 33°00'E), are very strongly affected by copper and nickel pollution (SOAER 1997, p. 103). Henriksen et al. (1997) provided water chemistry data of a statistically representative set of 435 lakes in the Kola Peninsula and some lakes in Karelia, studied primarily for the effects of atmospheric deposition. Observations on the ecotoxicological effects of heavy metal pollution in Kola lakes were presented by Yakovlev (1992) and Moiseenko et al. (1995). Freindling & Heitto (1991), Mononen & Lozovik (1994), Haimi (1997) and Holopainen et al. (2003) have provided information on water pollution in Karelia.

### 5.6.2 River Kovda drainage area

The River Kovda drainage area, bordering the Karelia and Murmansk Regions and partly extending to Kuusamo in Finland, is one of the most thoroughly regulated large catchments in Europe. Three large reservoirs on the 25,600 km$^2$ drainage system provide electric power to the industry of Murmansk Region (Fig. 5.8; data sources: Resursy 1972; Atlas Karelskoi ASSR 1989; Freindling & Heitto 1991; unpublished information provided by Anatoly Semjonov, Murmansk, and Pyotr Lozovik and Ylo Systra, Petrozavodsk). The lowermost of the reservoirs, Knyazhegubskoye (Table 5.2), was created by damming the original outlet River Kovda and raising the water level in Kovdozero by 10 m. Some smaller lakes became part of the reservoir, which discharges through Knyazhaya Guba hydropower plant (MQ 264 m$^3$ s$^{-1}$) into the White Sea. The second reservoir is Iovskoye (294 km$^2$, 72 m a.s.l.), regulated by Zarechensk hydropower plant and inundating the lakes Sushchozero, Rubozero, Sokolozero and Tumchaozero. The third and largest reservoir, Kumskoye (Table 5.2), was filled in the late 1960s: it connects two large lakes, Topozero and Pyaozero. The level of the latter was elevated by nearly 10 m, which completely submersed the old village of Oulanka at the inflow of Olanga River.

There are few published data on water quality in the reservoir system but, according to data from the mid-1970s from Knyazhegubskoye and Kumskoye, it has been good or even excellent on chemical criteria: water colour 15–25, transparency 4–5 m, BOD averages 4–5 mg O$_2$ L$^{-1}$, total phosphorus concentrations 8–10 µg L$^{-1}$. However, ecological problems arise from shore erosion and organic matter decay in the submersed forests. Especially problematical have been the winter low-water conditions: ice stranded during late-winter is disastrous for whitefish and vendace which spawn under ice in the littoral zone.

There are two remarkable lakes in the upper parts of the Kovda catchment: lakes Kitkajärvi and Paanajärvi (Fig. 5.8). Kitkajärvi is a large, ultraoligotrophic lake on the Finnish side of the border, 240 m a.s.l. (Kankaala et al. 1984, 1990). The oligotrophic Paanajärvi (Arvola et al. 1993) is a narrow fjord-like lake, 128 m deep, and famous for its scenic beauty and the rich and varied nature of its surroundings. According to plans publicised in

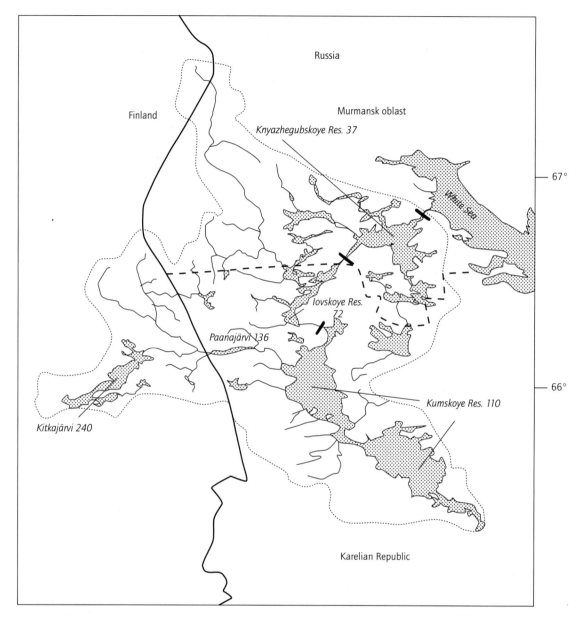

**Fig. 5.8** River Kovda drainage area, with the major lakes and reservoirs and their heights (m a.s.l.) indicated. Black bars denote hydropower stations.

1987, Paanajärvi was to become the lower reservoir of a large pumped-storage hydropower station, providing regulation power for the Murmansk industry. The planned daily regulation amplitude would have been 4 m. The plans evoked a hefty debate in both Karelia and Finland, which led to a fortunate conclusion: instead of becoming a reservoir system, the area was protected in 1992 as

Paanajärvi National Park (Koutaniemi 1993). The 1030 km² nature reserve connects with the Oulanka National Park on the Finnish side of the border. Since 1989, a joint Russian–Finnish ecological research project has been carried out in Paanajärvi and its surroundings (Arvola et al. 1993; Arvola 2000).

### 5.6.3 Lake Ladoga

Lake Ladoga (Table 5.2) is the largest lake in Europe and ranks amongst the world's top 15. Its drainage area extends to much of northwestern European Russia and eastern Finland, with three large sub-basins draining to the central lake. Of these, the drainage area of Lake Ilmen—River Volkhov (80,200 km²) is situated on sedimentary rocks of the Russian Plain, whereas those of Lake Onega—River Svir (83,200 km²) and Saimaa—River Vuoksi (66,700 km²) are mainly on the crystalline bedrock of the Fennoscandian Shield. Several smaller rivers drain a further 28,400 km².

Recent research on Lake Ladoga has been summarised in the proceedings of four International Lake Ladoga Symposia, held since 1993 (Simola et al. 1996, 1997, 2003; Peltonen et al. 2000). Existing monitoring data reveal gradual eutrophication of the lake since the 1960s, when the lake was still oligotrophic. The earliest Secchi-disk depth recordings date from 1898 to 1900, when the summer (August) values were 5.5–6 m, and 7 m visibility was observed in April (Witting 1929). According to measurements collected during 1968, Secchi-disk depth varied from 2.3 m to 3.9 (average 3.5 m) and concentrations of nutrients were relatively low: total phosphorus up to $15\,\mu g\,L^{-1}$; concentration of nitrates from 0.1 to 1.2 mg L$^{-1}$ and ammonium (NH$_4$) from 0.04 to 0.24 mg L$^{-1}$. In 1987–89, Secchi depth had declined on average to 2.2 m, the average phosphorus content of water in summer was $21\,\mu g\,L^{-1}$ in the pelagic zone and $32\,\mu g\,L^{-1}$ in the littoral area. In 1992–93, the concentrations of total phosphorus ranged from 15 to $29\,\mu g\,L^{-1}$. Eutrophication has been most clearly evident in some loaded bays and littoral areas near pollution sources (Drabkova et al. 1996).

The annual primary productivity of Lake Ladoga is reported to have increased from 14.7 g C m$^{-2}$ in 1976 to 139.5 g C m$^{-2}$ in 1985. According to the maximal daily production values measured during the 1980s, the lake could be classified as mesotrophic—some shore areas even eutrophic. In 1993, the biomass of phytoplankton varied from 0.4 to 6.6 g m$^{-3}$, the mean value being 1.38 g m$^{-3}$. During the process of eutrophication, the structure and species composition of phytoplankton changed and, in recent years, some late summer blooms of blue-green algae have been observed.

Analysis of diatom assemblages in surface sediments collected from different open-water sites in the years 1959–60, 1978–79 and 1994 confirms the process of ecosystem change, and pinpoints the period 1978–79 as a critical phase in ecosystem transition (Davydova et al. 1998). The main temporal changes in species abundances are as follows. At the time of the first sampling (1959–60), *Aulacoseira islandica* was abundant throughout the lake; in some shore areas *Aulacoseira italica* and *Asterionella formosa* were also abundant. During 1978–79, these two species and *Diatoma tenuis* were widely encountered in the plankton community, especially in the southern, shallow parts of the lake, receiving polluted waters from the industrialised River Volkhov drainage area. In the 1994 survey, the percentage proportions of *Asterionella formosa*, *Aulacoseira italica* and *Diatoma tenuis* were consistently increased at all stations, whereas the proportion of *Aulacoseira islandica* was generally lower than in the previous surveys, other than in the middle of the lake, where *A. islandica* was still a major component of the assemblage. According to a principal components analysis of the material, the main axis of variation was clearly temporal: the 1959–60 samples are clustered at low values and the 1994 samples at high values, whereas the 1978–79 samples are intermediate. Samples from the southwestern Volkhov and Svir Bay areas and from one site near the River Vuoksi (Burnaya) outlet deviate slightly from the central and northern sites. The southern part of the lake was receiving more river pollution, which was especially reflected in the increases in *Diatoma tenuis*. Later, the eutrophication was distributed more evenly throughout the pelagic zone.

Evidence of gradual ecosystem change commencing even earlier than this is provided by diatom stratigraphy and sediment chemistry in two $^{210}$Pb-dated sediment cores collected from the deep, northern part of Lake Ladoga (Kukkonen & Simola 1999). The relative increase of very small and thinly silicified frustules of *Stephanodiscus* spp. and *Cyclostephanos* spp. in the uppermost (post-1970) sediment possibly provides evidence of the transition of the diatom plankton from phosphorus-limitation to silicon-limitation, as a consequence of eutrophication.

The most recent data relating to nutrient loads suggest that the ecological condition of Ladoga may have slightly improved during the 1990s (e.g. Avinsky 2003; Avinsky & Kapustina 2003). Indeed, the external loading appears to have decreased, at least temporarily. Some outdated industrial enterprises have been shut down (pulp mills of Priozersk and Läskelä) and there have been improvements in sewage treatment by some municipalities. Agricultural loads radically decreased owing to economic recession in Russia during the early 1990s, but have since then increased again. The numbers of livestock (pigs and cattle) have nearly been halved and the use of mineral fertilisers has fallen from about 170,000 t in 1987 to 30,000 t in 1994 (Rumyantsev & Drabkova 1997).

A multidisciplinary ecological monitoring system for Lake Ladoga is now being developed, to be conducted collaboratively by administrative and scientific bodies in Russia (St Petersburg Region and Karelia) and Finland (Slepukhina et al. 1997; Rumyantsev & Drabkova 1997; Viljanen 2003). Legislation is being developed in order to safeguard the future of Lake Ladoga as an important aquatic resource (Fokin & Kudersky 1997).

### 5.6.4 Lake Onega

Lake Onega (Table 5.2) is, like Ladoga, a truly large lake with respect to its thermal properties: thermal bar development and prolonged isothermal circulation in the central water mass; prominent seiche phenomena; and geostrophic forcing in the pelagial currents. The lake ecosystem and its development have been studied quite thoroughly over several decades (Kaufman 1990). Eutrophication caused by sewage effluents is clearly evident in the western parts of the lake, notably near Petrozavodsk, and in Kondopoga Bay, whereas the eastern and central parts of the lake appear less affected (Timakova & Vislyanskaya 1991; Timakova 1991; Lozovik et al. 2000; Timakova et al. 2000). A short-core palaeolimnological survey (Davydova et al. 1993) revealed a sedimentary record very much like those observed in similarly affected large Finnish lakes (e.g. Simola et al. 1996b) with clear pollution effects of the Kondopoga pulp and paper mill complex, allegedly the largest of its kind in Europe. The future development of Onega will depend on the eventual success of planned water protection measures, which have been hampered by the economic recession of the past years.

### 5.6.5 Depopulation effects on small lakes in ceded Finnish territory

As a consequence of the 1939–44 wars, about 40,000 km$^2$ of Finnish territory on the western and northern side of Lake Ladoga was ceded to Soviet Russia. The entire Finnish population was evacuated. In the Post-war period land-use (forestry, farming, etc.) has been fairly weak or absent in large parts of this territory, which thus provides unique opportunities to investigate ecosystem recovery following relaxation of human pressure (Simola et al. 2000). Simola & Miettinen (1997) have studied palaeolimnologically the oligotrophication history of Lavijärvi, as evident in a 25-cm core of varved sediment obtained from a point 22 m deep. The sediment sequence represents the period 1935–96. There is an abrupt dimininution of mineral sedimentation, dated to 1939, obviously marking the cessation of arable field erosion. There are extensive areas of clayey soils in the drainage area, which were largely cultivated before the war but which since have been used only for cattle pasturing. As judged by the varves, mineral sediment accumulation rate fell rapidly from about 6 kg m$^{2}$ yr$^{-1}$ before 1939 to 1–1.5 kg m$^2$ yr$^{-1}$ in the post-war period. Commencing at the same horizon (21 cm upwards), a diatom species succession of about 10 years duration transforms the sedimentary assemblage from a eutrophic one (*Aulacoseira ambigua*, epiphytic *Fragilaria* spp.) into one with a much more mesotrophic character (*Cyclotella comta, Tabellaria flocculosa* v. *asterionelloides, Achnanthes minutissima*, etc.). Indeed, the stratigraphy and species succession are strikingly similar to many records of sedimentary change in Finnish forest lakes during the past decades (see section 5.3), except that in Lavijärvi the events have taken place in reversed order! This lake, as well as some other lakes in the ceded area, thus provides what is essentially a mirror image of recent environmental change in large parts of Fennoscandia.

## 5.7 SUMMARY: PRESENT SITUATION AND FUTURE TRENDS

Throughout Fennoscandia, effective load-reduction policies have considerably improved the state of waters in the vicinity of traditional pollution sources. At the same time, however, diffuse loading from both land-use and atmospheric deposition persists. Water quality may be expected still to deteriorate in former pristine or weakly affected headwaters and watercourses. The pattern is thus in a way changing from a former black-and-white situation towards one of more uniformly grey shadows. As regards the present and future impact of the diffuse atmospheric and terrestrial loads, there are great regional differences within Fennoscandia. This can be gained from the trophic categorisation (Table 5.4) and from the proportions of acidic and acid-sensitive lakes (Table 5.5; data from Henriksen *et al.* 1997). The number of oligotrophic lakes is especially large in both Norway and the Murmansk region (Kola Peninsula), as are the numbers of low-pH and low-alkalinity lakes. Hence, acid deposition will remain a major problem in these areas; the same is true for southwest Sweden.

Eutrophic conditions are encountered in Finnish lakes more commonly than elsewhere (Table 5.4), which implies that diffuse nutrient loads from terrestrial sources are critical. In Fennoscandian lakes, phosphorus concentrations exceeding the threshold of eutrophy are mainly attributable to anthropogenic impacts. However, the high phosphorus concentrations in the lakes of the Murmansk region may partly be due to the abundance of apatitic soil minerals in the region. This may also explain the relatively high number of lakes in the Murmansk region in which the critical N:P ratio of 12 is exceeded, indicating potential nitrogen limitation for phytoplankton primary production (Table 5.4).

Based on the 1995 lake survey data, Henriksen *et al.* (1997) have assessed the regional exceedance of critical loads of sulphur deposition in Fennoscandia. In the present situation, as many as 27% of Norwegian lakes and 17% of those in the Murmansk region suffer from sulphur loads exceeding the lake-specific critical loading. Corresponding figures for Sweden and Finland are both c.9%. Based on critical load exceedence models,

Table 5.5 Estimated numbers of low-pH lakes and lakes with alkalinity <20 µEq L$^{-1}$ in Fennoscandia, and their percentages of total lake numbers, according to the 1995 survey of 5414 lakes. (Modified from Henriksen *et al.* 1997.)

| Country/region | pH < 5.0 | | pH < 5.5 | | Alkalinity (<20 µEq L$^{-1}$) | |
| --- | --- | --- | --- | --- | --- | --- |
| | Number of lakes | Percentage of all lakes | Number of lakes | Percentage of all lakes | Number of lakes | Percentage of all lakes |
| Norway | 3224 | 8.3 | 7808 | 20.1 | 15,266 | 39.3 |
| Sweden | 2169 | 3.6 | 4278 | 7.1 | 8135 | 13.5 |
| Finland | 265 | 0.9 | 2066 | 7.0 | 2744 | 9.3 |
| Murmansk | 1980 | 12.6 | 2859 | 18.2 | 3016 | 19.2 |

Henriksen et al. (1994) concluded that, for most of central Fennoscandia, the acidification problem could be solved by reducing sulphur deposition but, in southern Scandinavia and parts of Kola Peninsula, the situation will remain critical if considerable nitrogen reduction cannot be achieved also.

The predicted consequences of climate warming in Fennoscandia include prolonged thaw periods and increased precipitation, especially in winter (Boer et al. 1990; Intergovernmental Panel on Climate Change 1990). According to Kivinen & Lepistö (1996), the most notable change in runoff will be in the timing of the spring high flow. With a decreasing proportion of the precipitation falling as snow, the spring flood may occur earlier but the peak runoff may drastically decrease. However, if the winter period will be wetter, with a higher groundwater table, total runoff will also increase.

The change of pattern in annual runoff may affect nutrient losses from field and forest catchments (Kallio et al. 1996). Nitrate loss may increase considerably as a result of enhanced mineralisation of organic matter under warmer conditions (Rekolainen et al. 1996). According to the results of the Finnish Research Project of Climate Change (SILMU), it appears that the limnological consequences of climate change would be greatest during the spring period, with the typical phytoplankton biomass maxima higher than today, whereas the summer biomasses would remain at the present levels. In a lake ecosystem the impacts of climate change will probably be more pronounced in the littoral zone than in the pelagial (Kankaala et al. 1996b). With access to sediment nutrients, the aquatic macrophytes can make the best of the increase in temperature. Besides, emergent vegetation also benefits from elevated $CO_2$ concentrations.

Arvola et al. (1996) and Kankaala et al. (1996a) have assessed experimentally the effects of phosphorus and allochthonous humus enrichment on the plankton community in Pääjärvi in southern Finland (see section 5.3.7). Production of both phytoplankton and bacteria was enhanced by additions of phosphorus alone and phosphorus and humic matter together; in the latter case bacterial production rose up to 59% of the total primary production in the euphotic zone. Increased loss of carbon through respiration in the several steps of the microbial food chain was observed but there was no clear response of zooplankton to the increased primary production.

Besides targets and palaeolimnological archives of human environmental impacts, the lake ecosystems also exert their own influence upon the global processes. Sediments generally act as a sink for biosphere carbon but the zero-oxygen horizon at the lake bottom—occasionally rising into the water column—is an important boundary for the reverse exchange processes between the two biospheric segments: methane ($CH_4$) formed by anaerobic metabolism and entering the atmosphere becomes an important greenhouse gas with a long half-life. Also $N_2O$ may be formed by both nitrification of ammonia and denitrification of nitrite in sediments. Investigations of $CH_4$ and $N_2O$ emissions from littoral and profundal sediments of eutrophic lakes have revealed the significance of these contributions to climate forcing (Alm et al. 1996, 1997; Silvola et al. 1996; Huttunen et al. 2003; Larmola et al. 2003).

## 5.8 ACKNOWLEDGEMENTS

We wish to thank the following persons for providing information for this text: Jukka Alm, Sari Antikainen, Pauli Haimi, Arne Henriksen, Anna-Liisa Holopainen, Ismo Holopainen, Jorma Keskitalo, Liisa Lepistö, Pyotr Lozovik, Jarmo Meriläinen, Arja Palomäki, Minna Rahkola, Sanna Saarnio, Kalevi Salonen, Ilkka Sammalkorpi, Anatoli Semyonov, Pertti Sevola, Aini Simola, Ylo Systra and Eva Willén.

## 5.9 REFERENCES

Ahti, T., Hämet-Ahti, L. & Jalas, J. (1968) Vegetation zones and their sections in northern Europe. *Annales Botanici Fennici*, **5**, 169–211.

Ahtiainen, M. (1992) The effects of forest clear-cutting

and scarification on the water quality of small brooks. *Hydrobiologia*, **243**, 465–73.

Ahtiainen, M. & Huttunen, P. (1995) Metsätaloustoimenpiteiden pitkäaikaisvaikutukset purovesien laatuun ja kuormaan. *Suomen Ympäristö*, **2**, 33–50.

Alasaarela, E., Hellsten, S. & Tikkanen, P. (1989) Ecological aspects of lake regulation in northern Finland. In: Laikari, H. (ed.), *River Basin Management, Part 5*. Pergamon Press, Oxford, 247–55.

Aleksandrov, B.M., Zytsar, N.A., Novikov, P.I., Pokrovski, V.V. & Pravdin, I.F. (eds) (1959) *Ozyora Karelii. Priroda, ryby i rybnoje khozyaistvo. Spravochnik*. Gosudarstvennoye izdatelstvo Karelskoi ASSR, Petrozavodsk.

Aleksandrov, B.M., Gordeyev, O.N. & Sokolova, V.A. (1970) Gidrobiologicheskie issledovaniya Karelii. *Uzenie Zapiski Petrozavodskogo Universiteta*, **16**(4), 1181–7.

Alimov, A.F. & Winberg, G.G. (1972) Biological productivity of two northern lakes. *Verhandlungen der internationalen Vereinigung für theoretische und angewandte Limnologie*, **18**, 65–70.

Alimov, A.F., Boullion, V.V., Finogenona, N.P., et al. (1972) Biological productivity of lakes Krivoe and Krugloe. In: Kajak, Z. & Hillbricht-Ilkowska, A. (eds), *Proceedings of the IBP-UNESCO Symposium on Productivity problems of freshwaters, Kazimierz Dolny, Poland, May 1970*. PWN Polish Scientific Publishers, Warszawa, 39–56.

Alm, J., Juutinen, S., Saarnio, S., Silvola, J., Nykänen, H. & Martikainen, P.J. (1996) Temporal and spatial variations in $CH_4$ emissions of flooded meadows and vegetated hydrolittoral. *Publications of the Academy of Finland*, **1/96**, 71–6.

Alm, J., Huttunen, J.T., Nykänen, H., Silvola, J. & Martikainen, P.J. (1997) Greenhouse gas emissions from small eutrophic lakes in Finland. In: Senesi, N. & Miano, T.M. (eds), *XIII International Symposium on Environmental Biogeochemistry*, Monopoli (Bari) Italy, Abstracts, 73.

Almer, B., Dickson, W., Ekström, C. & Hörnström, E. (1978) Sulfur pollution and the aquatic ecosystem. In: Nriagu, J. (ed.), *Sulfur in the Environment*, Vol. 2, *Ecological Impacts*, Wiley, New York, 271–311.

Anderson, N.J. (1993) Natural versus anthropogenic change in lakes: the role of the sediment record. *Trends in Ecology and Evolution*, **8**, 356–61.

Andersson, F. & Olsson, B. (eds) (1985) Lake Gårdsjön — an acid forest lake and its catchment. *Ecological Bulletins* (Stockholm), **37**, 1–336.

Andronikova, I.N., Drabkova, V.G., Kuzmenko, K.N.,
Michailova, N.F. & Stravinskaya, E.A. (1972) Biological productivity of the main communities of the Red Lake. In: Kajak, Z. & Hillbricht-Ilkowska, A. (eds), *Proceedings of the IBP-UNESCO Symposium on Productivity Problems of Freshwaters, Kazimierz Dolny, Poland, May 1970*. PWN Polish Scientific Publishers, Warszawa, 57–71.

Anttila, V. (1967) Järvenlaskuyhtiöt Suomessa. German summary: Die Seesenkungsgenossenschaften in Finnland. *Kansatieteellinen Arkisto*, **19**, 1–360.

Arvola, L. (1980) On the species composition and biomass of the phytoplankton in the Lokka reservoir, northern Finland. *Annales Botanici Fennici*, **17**, 325–5.

Arvola, L. (1991) Recent trends in the water chemistry of lake Pääjärvi. *Lammi Notes*, **18**, 1–5.

Arvola, L. (2000) Transport of organic carbon, nitrogen and phosphorus from the Finnish territory into Lake Paanajärvi. *Oulanka Reports*, **23**, 115–19.

Arvola, L. & Kankaala, P. (1989) Winter and spring variability in phyto- and bacterioplankton in lakes with different water colour. *Aqua Fennica*, **19**, 29–39.

Arvola, L., Salonen, K. & Rask, M. (1990) Chemical budgets for a small dystrophic lake in southern Finland. *Limnologica*, **20**, 243–51.

Arvola, L., Karusalmi, A. & Tulonen, T. (1993) Observations of plankton communities and primary and bacterial production of Lake Paanajärvi, a deep subarctic lake. *Oulanka Reports*, **12**, 93–107.

Arvola, L., Kankaala, P., Tulonen, T. & Ojala, A. (1996) Effects of phosphorus and allochthonous humic matter enrichment on the metabolic processes and community structure of plankton in a boreal lake (Lake Pääjärvi). *Canadian Journal of Fisheries and Aquatic Sciences*, **53**, 1646–62.

Assmuth, T. (1996) Toxicant distributions and impact models in environmntal risk analysis of waster sites. *Publications of Water and Environment Research Institute*, **20**, 1–79.

*Atlas Karelskoi ASSR* (1989) Glavnoe Upravlenie Geodesii i Kartografii Pri Sovete Ministrov SSSR, Moscow.

Avinsky, V. (2003) Lake Ladoga pelagial zone zooplankton: present state and long-term changes. *University of Joensuu, Publications of Karelian Institute*, **138**, 124–31.

Avinsky, V. & Kapustina, L. (2003) The importance of allochthonous material among the components of organic matter in Lake Ladoga. *University of Joensuu, Publications of Karelian Institute*, **138**, 132–8.

Bagge, P., Liimatainen, H.M. & Liljaniemi, P. (1996) Comparison of sampling methods for semipelagic animals in two deep basins of Lake Saimaa. *Hydrobiologia*, **322**, 293–300.

Balagurova, M.V., Freindling, V.A. & Kharkevich, N.S. (eds) (1970) *Vodnye resursy Karelii i puti ikh ispolzovaniya*. Kareliya, Petrozavodsk, 399 pp.

Benner, R., Lay, J., K'nees, E. & Hodson, R.E. (1988) Carbon conversion efficiency for bacterial growth on lignocellulose: implications for detritus-based food webs. *Limnology and Oceanography*, **33**, 1514–26.

Berge, F. (1976) Diatoms as indicators of temporal pH trends in some lakes and rivers in southern Norway. *Nova Hedwigia Beiheft*, **73**, 249–65.

Berglund, B.E. (ed.) (1986) *Handbook of Holocene Palaeoecology and Palaeohydrology*. Wiley, Chichester, 869 pp.

Bergström, I., Mäkelä, K. & Starr, M. (eds) (1995) *Integrated Monitoring Programme in Finland. First National Report, 1/1995*. Environmental Policy Department, Ministry of the Environment, Helsinki, 138 pp.

Bernes, C. (1993) *The Nordic Environment: Present State, Trends and Threats*. Nordic Council of Ministers, Copenhagen, 212 pp.

Boer, M.M., Koster, E.A. & Lundberg, H. (1990) Greenhouse impact in Fennoscandia: preliminary findings of an European workshop on the effects of climate change. *Ambio*, **19**, 2–10.

Charles, D.F., Battarbee, R.W., Renberg, I., Dam H. van & Smol, J.P. (1989) Paleoecological analysis of lake acidification trends in North America and Europe using diatoms and chrysophytes. In: Norton, S.A., Lindberg, S.E. & Page, A.L. (eds), *Acid Precipitation*, Vol. 4, *Soils, Aquatic Processes and Lake Acidification*: Springer, Berlin, 207–76.

Charles, D., Smol, J.P. & Engstrom, D. (1994) Paleolimnological approaches to biological monitoring. In: Loeb, S.L. & Spacie, A. (eds), *Biological Monitoring of Aquatic Systems*. CRC Press, Boca Raton, 233–93.

Davis, R.B., Anderson, D.S. & Berge, F. (1985) Loss of organic matter, a fundamental process in lake acidification: paleolimnological evidence. *Nature*, **316**, 436–8.

Davydova, N., Kalmykov, M., Sandman, O., Ollikainen, M. & Simola, H. (1993) Recent palaeolimnology of Kondopoga Bay, Lake Onega, reflecting pollution by a large pulp mill. *Verhandlungen der internationalen Vereinigung für theoretische und angewandte Limnologie*, **25**, 1086–90.

Davydova, N., Kukkonen, M., Simola, H. & Subetto, D. (1998) Human impact on Lake Ladoga as indicated by long-term monitoring of sedimentary diatom assemblages. *Boreal Environment Research*, **4**, 269–75.

Donner, J. (1995) *The Quaternary History of Scandinavia*. Cambridge University Press, Cambridge, 200 pp.

Drabkova, V.G., Rumyantsev, V.A., Sergeeva, L.V. & Slepukhina, T.D. (1996) Ecological problems of Lake Ladoga; cause and solutions. *Hydrobiologia*, **322**, 1–7.

Eloranta, P. (1978) Light penetration in different types of lakes in Central Finland. *Holarctic Ecology*, **1**, 362–6.

European Union (2000) *Establishing a Framework for Community Action in the Field of Water Policy*. Directive of the European Parliament and of the Council 2000/60/EC. PE/CONS 3639/1/00 REV 1, Luxembourg.

Faafeng, B. & Hessen, D. (1993) Nitrogen and phosphorus concentrations and N:P ratios in Norwegian lakes: perspectives on nutrient limitation. *Verhandlungen der internationalen Vereinigung für theoretische und angewandte Limnologie*, **25**, 465–9.

Fahnenstiel, G.L., Lang, G.A., Nalepa, T.F. & Johengen, T.H. (1995a) Effects of zebra mussel (*Dreissena polymorpha*) colonization on water quality parameters in Saginaw Bay, Lake Huron. *Journal of Great Lakes Research*, **21**, 435–8.

Fahnenstiel, G.L., Bridgeman, T.B., Lang, G.A., McCormick, M.J. & Nalepa, T.F. (1995b) Phytoplankton productivity in Saginaw Bay, Lake Huron: effects of zebra mussel (*Dreissena polymorpha*) colonization. *Journal of Great Lakes Research*, **21**, 465–75.

Flower, R. & Simola, H. (1990) Diatoms from a short sediment core from Lake Höytiäinen, eastern Finland, with special reference to a major lake level lowering in 1859 AD. In: Simola, H. (ed.), *Proceedings of the Tenth International Diatom Symposium, Joensuu, Finland, 1988*. Koeltz Scientific Books, Koenigstein, 433–41.

Flower, R., Juggins, S. & Battarbee, R.W. (1997) Matching diatom assemblages in lake sediment cores and modern surface sediment samples: the implications for lake conservation and restoration with special reference to acidified systems. *Hydrobiologia*, **344**, 27–40.

Fokin, Y. & Kudersky, L. (1997) Environmental legislation concerning nature protection in Lake Ladoga. *University of Joensuu, Publications of Karelian Institute*, **117**, 417–22.

Forsius, M., Kämäri, J., Kortelainen, P., Mannio, J., Verta, M. & Kinnunen, K. (1990) Statistical lake survey in Finland: Regional estimates of lake acidification. In: Kauppi, P., Anttila, P. & Kenttämies, K. (eds), *Acidification in Finland*. Springer-Verlag, Berlin, 759–80.

Freindling, A. & Heitto, L. (eds) (1991) *Primary Production of Inland Waters. The Second Soviet-Karelian–Finnish Symposium on Water Problems*. Volume 72,

Publications of the Water and Environment Administration, Helsinki, 1–132.

Geller, A. (1983) Degradability of dissolved organic lake water compounds in cultures of natural bacterial communities. *Archiv für Hydrobiologie*, **99**, 60–79.

Gerd, S.V. (1947) *O klassifikatsii oligotrofnykh ozer Karelii*. Izvestiya Karelsko-Finsko nauchno-issledovatelskii bazy Akademii Nauk SSSR, Petrozavodsk.

Gerd, S.V. (1956) *Opyt biolimnologicheskogo raionnirovanniya ozyor Karelii*. Trudy Karelsko-Finskogo filial Akademii Nauk SSSR 5, Petrozavodsk.

Gordeyev, O.N., Kuderskiy, L.A., Lutta, A.S., Polyanskiy, Yu, I. & Sokolova, V.A. (1965) *Fauna ozer Karelii. Becpozvonochiye*. Akademia Nauk SSSR Karelskii filial, Institut Biologii. Nauka, Petrozavodsk, 325 pp.

Grönlund, E., Simola, H. & Huttunen, P. (1986) Palaeolimnological reflections of fiber-plant retting in the sediment of a small clear-water lake. *Hydrobiologia*, **143**, 425–31.

Hagman, N. (1916) Vähäsen pellavanliotuksen vaikutuksesta kalojen ravintoon järvissämme. *Suomen Kalastuslehti*, **23**(2), 17–22.

Haimi, P. (1997) *The Development of Environment Monitoring in the Republic of Karelia, Russia*. Number 15/16, Mimeograph Series, North Karelia Regional Environment Centre, Joensuu, 1–47. (In Finnish and Russian, with English abstract.)

Hakala, I. & Arvola, L. (1994) Alarming signs of eutrophication in Lake Pääjärvi. *Lammi Notes*, **21**, 1–5.

Håkanson, L. (1977) *Sediments as indicators of contamination—investigations in the four largest Swedish lakes*. Rapport 839, Statens Naturvårdsverket, Solna.

Håkanson, L. (1978) *Hjälmaren—en naturgeografisk beskrivning*. Rapport 1079, Statens Naturvårdsverket, Solna.

Håkanson, L. & Jansson, M. (1983) *Principles of Lake Sedimentology*. Springer-Verlag, Berlin, 316 p.

Hawes, I. (1985) Light climate and phytoplankton photosynthesis in maritime Antarctic lakes. *Hydrobiologia*, **123**, 89–95.

Haworth, E.Y. & Lund, J.W.G. (eds) (1984) *Lake Sediments and Environmental History*. Leicester University Press, Leicester, 411 pp.

Heitto, L. (1990) Macrophytes in Finnish forest lakes and possible effects of airborne acidification. In: Kauppi, P., Anttila, P. & Kenttämies, K. (eds), *Acidification in Finland*. Springer-Verlag, Berlin, 963–972.

Henriksen, A., Lien, L., Traaen, T.S., Sevaldrud, I.S. & Brakke, D. (1988) Lake acidification in Norway—present and predicted chemical status. *Ambio*, **17**, 259–66.

Henriksen, A., Kämäri, J., Posch, M., Forsius, M., Wilander, A. & Moiseenko, T. (1994) Critical loads for surface waters in Scandinavia and the Kola region. In: Raitio, H. & Kilponen, T. (eds), *Critical Loads and Critical Limit Values*. Research Papers No. 513, Finnish Forest Research Institute, Helsinki, 97–108.

Henriksen, A., Skjelkvåle, B.L., Traaen, T.S., *et al.* (1997) *Results of National Lake Surveys (1995) in Finland, Norway, Sweden, Denmark, Russian Kola, Russian Karelia, Scotland and Wales*. Report No. 3645-97, Norsk Institutt for Vannforskning, Oslo.

Henrikson, L. & Brodin, Y.W. (eds) (1995) *Liming of Acidified Surface Waters—a Swedish Synthesis*. Springer-Verlag, Berlin, 458 pp.

Hindersson, R. (1933) Förgiftning av nötkreatur genom sötvattensplankton. *Finska Veterinärtidskrift*, **39**, 179–89.

Holland, R.E. (1993) Changes in planktonic diatoms and water transparency in Hatchery Bay, Bass Island area, western Lake Erie, since the establishment of the zebra mussel. *Journal of Great Lakes Research*, **19**, 617–24.

Holopainen, A.L., Huovinen, P. & Huttunen, P. (1993) Horizontal distribution of phytoplankton in two large lakes in eastern Finland. *Verhandlungen der internationalen Vereinigung für theoretische und angewandte Limnologie*, **25**, 557–62.

Holopainen, A.L., Viljanen, M. & Lempinen, R. (2001) Fluorometer and CTD monitoring of the Vuoksi drainage basin, eastern Finland. *Verhandlungen der internationalen Vereinigung für theoretische und angewandte Limnologie*, **27**, 2194–7.

Holopainen, I.J., Holopainen, A.-L., Hämäläinen, H., Rahkola-Sorsa, M., Tkatcheva, V. & Viljanen, M. (2003) Effects of mining industry waste waters on a shallow lake ecosystem in Karelia, north-west Russia. *Hydrobiologia*, **506–9**, 111–19.

Holtan, H. (1979) The Lake Mjøsa story. *Ergebnisse der Limnologie*, **13**, 242–58.

Hongve, D. (1993) Total and reactive aluminium concentrations in non-turbid Norwegian surface waters. *Verhandlungen der internationalen Vereinigung für theoretische und angewandte Limnologie*, **25**, 133–6.

Hongve, D., Løvstad, Ø. & Bjørndalen, K. (1988) *Gonyostomum semen*—a new nuisance to bathers in Norwegian lakes. *Verhandlungen der internationalen Vereinigung für theoretische und angewandte Limnologie*, **23**, 430–4.

Horppila, J. & Kairesalo, T. (1990) A fading recovery: the role of roach (*Rutilus rutilus* L.) in maintaining high algal productivity and biomass in Lake Vesijärvi, southern Finland. *Hydrobiologia*, **200/201**, 153–65.

Horppila, J. & Kairesalo, T. (1992) Impacts of bleak (*Alburnus alburnus*) and roach (*Rutilus rutilus*) on water quality, sedimentation and internal nutrient loading. *Hydrobiologia*, **243/244**, 323–31.

Horppila, J. & Peltonen, H. (1994) Optimizing sampling from trawl catches: contemporarous multistage sampling from age and length structures. *Canadian Journal of Fisheries and Aquatic Sciences*, **49**, 1555–9.

Huovinen, P., Holopainen, A.L. & Huttunen, P. (1993) Spatial variation of community respiration and primary productivity in two large lakes in eastern Finland *Verhandlungen der internationalen Vereinigung für theoretische und angewandte Limnologie*, **25**, 552–6.

Huttunen, J., Alm, J., Liikanen, A., et al. (2003) Fluxes of methane, carbon dioxide and nitrous oxide in boreal lakes and potential anthropogenic effects on the aquatic greenhouse gas emissions. *Chemosphere*, **52**, 609–21.

Huttunen, P. & Turkia, J. (1990) Estimation of palaeoalkalinity from diatom assemblages by means of CCA. In: Simola, H. (ed.), *Proceedings of the Tenth International Diatom Symposium, Joensuu, Finland, 1988.* Koeltz Scientific Books, Koenigstein, 443–50.

Huttunen, P., Kenttämies, K., Liehu, A., et al. (1990) Palaeoecological evaluation of the recent acidification of susceptible lakes in Finland. In: Kauppi, P., Anttila, P. & Kenttämies, K. (eds), *Acidification in Finland*. Springer-Verlag, Berlin, 1071–90.

Hynynen, J., Teppo, A., Palomäki, A., Meriläinen, J.J., Granberg, K. & Reinikainen, P. (1997) *Lappajärven paleolimnologinen historia: rehevöitymiseen johtanut kehitys*. (Recent environmental history of Lake Lappajärvi: A palaeolimnological study of chironomids and diatoms). Report 148, Institute of Environmental Research, University of Jyväskylä, 1–32.

Intergovernmental Panel on Climate Change (1990) *Climate Change: the IPCC Scientific Assessment*. Cambridge University Press, Cambridge, 365 pp.

Itkonen, A. (1997) Past trophic responses of Boreal Shield lakes and the Baltic Sea to geological, climatic and anthropogenic inputs as inferred from sediment geochemistry. *Annales Universitatis Turkuensis A II*, **103**, 1–128.

Itkonen, A. & Olander, H. (1997) The origin of the hypertrophic state of a shallow boreal shield lake. *Boreal Environment Research*, **2**, 183–98.

Itkonen, A. & Salonen, V.P. (1994) The response of sedimentation in three varved lacustrine sequences to air temperature, precipitation and human impact. *Journal of Paleolimnology*, **11**, 323–32.

Järnefelt, H. (1958) On the typology of the northern lakes Finland *Verhandlungen der internationalen Vereinigung für theoretische und angewandte Limnologie*, **13**, 228–35.

Järnefelt, H. (1963) Zur Limnologie einiger Gewässer Finnlands XX. Zusammenfassende Besprechung der Thermik und Chemie finnischer Seen. *Annales Zoologici Societatis Vanamo*, **25**(1), 1–51.

Järvinen, O. (1990) Finland. In: Grimmet, R.F.A. & Jones, T.A. (eds), *Important bird areas in Europe*. Technical Publication 9, International Council of Bird Preservation, Cambridge, 163–88.

Jones, R. & Arvola, L. (1984) Light penetration and some related characteristics in small forest lakes in southern Finland. *Verhandlungen der internationalen Vereinigung für theoretische und angewandte Limnologie*, **22**, 811–16.

Jones, R.I., Salonen, K. & DeHaan, H. (1988) Phosphorus transformations in the epilimnion of humic lakes: abiotic interactions between dissolved humic materials and phosphate. *Freshwater Biology*, **19**, 357–69.

Josefsson, M. & Andersson, B. (2001) The environmental consequences of alien species in the Swedish lakes Mälaren, Hjälmaren, Vänern and Vättern. *Ambio*, **30**, 514–21.

Jurvelius, J. & Sammalkorpi, I. (1995) Hydroacoustic monitoring of the distribution, density and the massremoval of pelagic fish in a eutrophic lake. *Hydrobiologia*, **316**, 34–41.

Kallio, K., Huttunen, M., Vehviläinen, B., Ekholm, P. & Laine, Y. (1996) Effect of climate change on nutrient transport from an agrigultural basin. In: Roos, J. (ed.), *The Finnish Research Programme on Climate Change*. Publication 4/96, Academy of Finland, Helsinki, 130–5.

Kämäri, J., Forsius, M. & Lepistö, A. (1990) Modeling long-term development of surface water acidification in Finland. In: Kauppi, P., Anttila, P. & Kenttämies, K. (eds), *Acidification in Finland*. Springer-Verlag, Berlin, 781–910.

Kankaala, P., Hellstén, S. & Alasaarela, E. (1984) Primary production of phytoplankton in the oligohumic Kitka lakes in northern Finland. *Aqua Fennica*, **14**, 65–78.

Kankaala, P., Vasama, A., Eskonen, K. & Hyytinen, L. (1990) Zooplankton of Lake Ala-Kitka (NE Finland) in relation to phytoplankton and predation by vendace. *Aqua Fennica*, **20**, 81–94.

Kankaala, P., Arvola, L., Tulonen, T. & Ojala, A. (1996a) Carbon budget for the pelagic food web of the euphotic zone in a boreal lake (Lake Pääjärvi). *Canadian Journal of Fisheries and Aquatic Research*, **53**, 1663–74.

Kankaala, P., Ojala, A., Tulonen, T., Haapamäki, J. &

Arvola, L. (1996b) Impact of climate change on carbon cycle in freshwater ecosystems. In: Roos, J. (ed.), *The Finnish Research Programme on Climate Change*. Publication 4/96, Academy of Finland, Helsinki, 196–201.

Kansanen, P. (1981) Effects of heavy pollution on the zoobenthos in Lake Vanajavesi, southern Finland, with special reference to meiozoobenthos. *Annales Zoologici Fennici*, **18**, 243–51.

Kansanen, P. (1985) Assessment of pollution history in recent sediments in Lake Vanajavesi, southern Finland II. Changes in the Chironomidae, Chaoboridae and Ceratopogonidae (Diptera) fauna. *Annales Zoologici Fennici*, **22**, 57–90.

Kansanen, P. & Aho, J. (1981) Changes in the macrozoobenthos asociations in the polluted Lake Vanajavesi, southern Finland, over a period of 50 years. *Annales Zoologici Fennici*, **18**, 73–101.

Kansanen, P. & Jaakkola, T. (1985) Assessment of pollution history in recent sediments in Lake Vanajavesi, southern Finland. I. Selection of representative profiles, their dating and chemostratigraphy. *Annales Zoologici Fennici*, **22**, 13–55.

Karjalainen, J. & Viljanen, M. (1993) Changes in the zooplankton community in Lake Puruvesi, Finland, in relation to the stock of vendace (*Coregonus albula* (L.). *Verhandlungen der internationalen Vereinigung für theoretische und angewandte Limnologie*, **25**, 563–6.

Karjalainen, J., Holopainen, A.L. & Huttunen, P. (1996) Spatial patterns and relationships between phytoplankton, zooplankton and water quality in the Saimaa lake system, Finland. *Hydrobiologia*, **322**, 267–76.

Kaufman, Z.S. (ed.) (1990) *Ekosistema Onezhskogo ozera i tendentsii eyo izmeneniya*. Nauka, Leningrad, 264 pp.

Kauppi, L. (1984) Contribution of agricultural loading to the deterioration of surface waters in Finland. *Publications of the Water Research Institute, National Board of Waters, Finland*, **57**, 5–11.

Kauppi, P., Anttila, P. & Kenttämies, K. (eds) (1990) *Acidification in Finland*. Springer-Verlag, Berlin, 1237 pp.

Kenttämies, K. (1981) The effects on water quality of forest drainage and fertilization in peatlands. *Publications of the Water Research Institute, National Board of Waters, Finland*, **43**, 24–31.

Kenttämies, K. & Saukkonen, S. (1996) *Metsätalous ja vesistöt, Yhteistutkimusprojektin Metsätalouden vesistöhaitat ja niiden torjunta (METVE) yhteenveto* (English summary). Maa-ja metsätalousministeriön julkaisuja, Helsinki.

Kenttämies, K., Lepistö, L. & Vilhunen, O. (1995) Metsätalouden osuus Suomessa todetuista vesistöjen levähaitoista. *Suomen Ympäristö*, **2**, 229–39.

Keskitalo, J. (1990) Occurrence of vegetated buffer zones along brooks in the catchment area of Lake Tuusulanjärvi, southern Finland. *Aqua Fennica*, **20**, 55–64.

Keskitalo, J. & Salonen, K. (1994) Manual for Integrated Monitoring. Subprogramme Hydrobiology of Lakes. *Publications of the Water and Environment Institute*, Helsinki, B, **16**, 1–41.

Kessler, K. (1868) *Materialy dlya poznaniya Onezhskago Ozera i obonezhskago kraya*. St Petersburg.

Keto, J. (1982) The recovery of L. Vesijärvi following sewage diversion. *Hydrobiologia*, **86**, 195–9.

Keto, J. & Sammalkorpi, I. (1988) A fading recovery: a conceptual model for Lake Vesijärvi management and research. *Aqua Fennica*, **18**, 193–204.

Kinnunen, K. (1989) Water quality development of the artificial lakes Lokka and Porttipahta in Finnish Lapland. *Aqua Fennica*, **19**, 11–17.

Kira, T. (ed.) (1990) *Survey of the State of World Lakes and Data Book of World Lake Environments*. International Lake Environment Committee, Otsu (Pt III, Eur 32, pp. 1–13 for Lake Mjøsa).

Kivinen, Y. & Lepistö, A. (1996) The effects of climate change on the hydrology of a forested catchment in southern Finland. In: Roos, J. (ed.), *The Finnish Research Programme on Climate Change*. Publication 4/96, Academy of Finland, Helsinki, No., 114–18.

Korhola, A., Virkanen, J., Tikkanen, M. & Blom, T. (1996) Fire-induced pH rise in a naturally acid hill-top lake, southern Finland: a palaeoecological study. *Journal of Ecology*, **84**, 257–65.

Kortelainen, P. & Mannio, J. (1990) Organic acidity in Finnish lakes. In: Kauppi, P., Anttila, P. & Kenttämies, K. (eds), *Acidification in Finland*, Springer-Verlag, Berlin, 849–63.

Kortelainen, P., Mannio, J., Forsius, M., Kämäri, J. & Verta, M. (1989) Finnish lake survey: the role of organic and anthropogenic acidity. *Water, Air and Soil Pollution*, **46**, 235–49.

Koutaniemi, L. (ed.) (1993) *Paanayarvskii natsionalnyij park*. Fond Paanayarvi-Oulanka, Kuusamo, 159 pp.

Kroer, N. (1993) Bacterial growth efficiency on natural dissolved organic matter. *Limnology and Oceanography*, **38**, 1282–90.

Kukkonen, M. & Simola, H. (1999) Stratigraphy of diatoms in two $^{210}$Pb-dated deep-water cores from Lake Ladoga: evidence of environmental change in the largest lake of Europe. *Proceedings of the 14th International Diatom Symposium, Tokyo, 1996*. Koeltz, Königstein, 427–35.

Kuusisto, E. (1988) Säännöstelyn vaikutus vesistöjen hydrologiaan. *Vesi- ja ympäristöhallituksen monistesarja*, **80**, 25–31.

Kvarnäs, H. (2001) Morphometry and hydrology of the four large lakes of Sweden. *Ambio*, **30**, 467–74.

Laaksonen, R. & Malin, V. (1984) Changes in water quality in Finnish lakes 1965–1982. *Publications of the Water Research Institute, National Board of Waters, Finland*, **57**, 52–8.

Laine, P. (1997) Water protection steps and waste water loading of pulp, paper and kraftline mills on the southern Lake Saimaa. *University of Joensuu, Publications of Karelian Institute*, **117**, 339–43.

Larmola, T., Alm, J., Juutinen, S., Martikainen, P. & Silvola, J. (2003) Ecosystem $CO_2$ exchange and plant biomass in the littoral zone of a boreal eutrophic lake. *Freshwater Biology*, **48**, 1295–310.

Last, B. & Smol, J. (eds) (2001a–b) *Tracking Environmental Changes using Lake Sediments*, Vols 1–2. Kluwer, Dordrecht.

Lepistö, L. (1995) Phytoplankton succession from 1968 to 1990 in the subarctic Lokka reservoir. *Publications of the Water and Environment Institute, Helsinki*, **13**, 64 pp.

Lepistö, L., Antikainen, S. & Kivinen, J. (1994) The occurrence of *Gonyostomum semen* (Ehr.) Diesing in Finnish lakes. *Hydrobiologia*, **273**, 1–8.

Lepistö, L. & Pietiläinen, O-P. (1996) Kasviplanktonin määrän ja koostumuksen muutokset Lokassa, Porttipahdassa ja Kemijärvessä (Changes in the quantity and composition of phytoplankton in two reservoirs, Lokka and Porttipahta, and in Lake Kemijärvi). *Suomen ympäristö*, **13**, 78 pp.

Levander, K.M. (1906a) Beiträge zur Kenntnis des Sees Valkea-Mustajärvi der Fischereiversuchsstation Evois. *Acta Societes Fauna et Flora Fennica*, **28**, 1–28.

Levander, K.M. (1906b) Beiträge zur Kenntnis des Sees Pitkäniemenjärvi der Fischereiversuchsstation Evois. *Acta Societes Fauna et Flora Fennica*, **29**, 1–15.

Lindeström, L. (1996) Metaller och stabila organiska ämnen. In: Wallin, M. (ed.), *Vänerns miljötillstånd och utveckling 1973–1994*. Rapport 4619, Statens Naturvårdsverket, Solna, 54–64.

Lindeström, L. (2001) Mercury in sediment and fish communities of Lake Vänern, Sweden: recovery from contamination. *Ambio*, **30**, 538–44.

Lindqvist, O., Jernelöv, A., Johansson, K. & Rodhe, H. (1984) *Mercury in the Swedish Environment. Global and Local Sources*. Report 1816, Swedish National Environment Protection Board, Solna, 1–110.

Line, J.M., ter Braak, C.J.F. & Birks, H.J.B. (1994) WACAL-IB version 3.3 — a computer program to reconstruct environmental variables from fossil assemblages by weighted averaging and to derive sample-specific errors of prediction. *Journal of Paleolimnology*, **10**, 147–52.

Liukkonen, M., Kairesalo, T. & Keto, J. (1993) Eutrophication and recovery of Lake Vesijärvi (South Finland): diatom frustules in varved sediments over 30-year period. *Hydrobiologia*, **269/270**, 415–26.

Liukkonen, M., Kairesalo, T., Haworth, E.Y. & Salkinoja-Salonen, M. (1995) Changes in the diatom community during the biomanipulation of Lake Vesijärvi. Paleolimnological analysis evidenced the initiation of a new species *Actinocyclus normanii* f. *subsalsa* into the lake's plankton. In: Saski, E.K. & Saarinen, T. (eds.), *Proceedings of the Second Finnish Conference of Environmental Sciences*, University of Helsinki, Helsinki, 293–6.

Lohammar, G. (1938) Wasserchemie und höhere Vegetation schwedischer Seen. *Symbolae Botanicae Uppsaliense*, **3**(1): 1–252.

Lovén, S. (1862) Till frågan om ishavsfaunans fordna utsträckning öfver en del af Nordens fastland. Öfvers. af Kungliga Vetenskapsakademins Förhandlingar, Stockholm.

Lozovik, P., Kulikova, T., Filatov, N., Sabylina, A., Polyakova, T. & Vislyanskaya, I. (2000) The present state of Lake Onega, Lake Ladoga and the Vygozero reservoir and their monitoring. *University of Joensuu, Publications of Karelian Institute*, **129**, 250–7.

Malve, O., Huttula, T. & Lehtinen, K. (1992) Modelling of eutrophication and oxygen depletion in the Lake Lappajärvi. *First Conference on Water Pollution*, Wessex Institute of Technology, Southampton, Abstracts.

Mannio, J. & Vuorenmaa, J. (1995) Regional monitoring of lake acidification in Finland. *Water, Air and Soil Pollution*, **85**, 571–6.

Maristo, L. (1941) Die Seetypen Finnlands auf floristischer und vegetationsphysiognomischer Grundlage. *Annales Botanici Societatis Zoologicae Botanicae Vanamo*, **15**(5), 1–314.

Meili, M. (1991) Mercury in forest lake ecosystems — bioavailability, bioaccumulation and biomagnification. *Water, Air and Soil Pollution*, **55**, 131–57.

Melanen, M., Jaakkola, A., Melkas, M., Ahtiainen, M. & Matinvesi, J. (1985) Leaching resulting from land application of sewage sludge and slurry. *Publications of the Water Research Institute, National Board of Waters, Finland*, **61**, 1–124.

Meriläinen, J.J. & Hamina, V. (1993) Recent environmental history of a large, originally oligotrophic lake in Finland: a paleolimnological study of chironomid remains. *Journal of Paleolimnology*, **9**, 129–40.

Meriläinen, J.J. & Hynynen, J. (1988) *Pohjois-Päijänten velvoitetarkkailu vuonna 1987. Jätevesien vaikutus pohjaeläimistöön vuonna 1987 (Zoobenthos in Northern Lake Päijänne in 1987)*. Institute for Environmental Research, University of Jyväskylä, Jyväskylä, 13 pp.

Meriläinen, J.J. & Hynynen J. (1990) Benthic invertebrates in relation to acidity in Finnish forest lakes. In: Kauppi, P., Anttila, P. & Kenttämies, K. (eds), *Acidification in Finland*, Springer-Verlag, Berlin, 1029–50.

Meriläinen, J.J., Hynynen, J., Teppo, A., Palomäki, A., Granberg, K. & Reinikainen, P. (2000) Importance of diffuse nutrient loading and lake level changes to the eutrophication of an originally oligotrophic boreal lake: a palaeolimnological diatom and chironomid analysis. *Journal of Paleolimnology*, **24**, 251–70.

Metsäteollisuus (1997) *Metsäteollisuustilastot vuodelta 1996* (Forest industry statistics for 1996). Metsäteollisuus ry, Helsinki, 64 pp.

Metsäteollisuus (2003) *Metsäteollisuustilastot vuodelta 2002* (Forest industry statistics for 2002). Metsäteollisuus ry, Helsinki, 64 pp.

Miettinen, J., Grönlund, E., Simola, H. & Huttunen, P. (2002) Palaeolimnology of Lake Pieni-Kuuppalanlampi (Kurkijoki, Karelian Republic, Russia): isolation history, lake ecosystem development and human impact. *Journal of Paleolimnology*, **27**, 29–44.

Miettinen, J., Kukkonen, M. & Simola, H. (2003) Reconstructing background concentrations of total phosphorus in lakes using sedimented diatoms. *TemaNord*, **547**, 64–7.

Moiseenko, T.I., Kudryavtseva, L.P., Rodyushkin, I.V., Dauvalter, V.A., Lukin, A.A. & Kashulin, N.A. (1995) Airborne contamination by heavy metals and aluminum in the freshwater ecosystems of the Kola subarctic region. *Science of the Total Environment*, **160/161**, 715–27.

Mononen, P. & Lozovik, P. (eds) (1994) Acidification of inland waters. *The Third Soviet-Karelian–Finnish Symposium on Water Problems*. Water and Environment Administration (Finland), Helsinki, 141 pp.

Mononen, P. & Niinioja, R. (1993) Nutrient and chlorophyll concentrations in the northern part of Lake Saimaa, Finland. *Verhandlungen der internationalen Vereinigung für theoretische und angewandte Limnologie*, **25**, 544–7.

Münster, U. & Chróst, R.J. (1990) Origin, composition and microbial utilization of dissolved organic matter. In: Overbeck, J. & Chróst, R.J. (eds), *Aquatic microbial ecology*. Brock/Springer, New York, 8–46.

Naturvårdsverket (1990) *Stora sjöar; miljösituation och förslag till åtgärder*. Rapport 3839, Naturvårdsverket, Solna, 78 pp.

Niemi, J., Heinonen, P., Mitikka, S., Vuoristo, H., Pietiläinen, O.-P., Puupponen, M. & Rönkä, S. (2001) The Finnish EUROWATERNET—The European Environment Agency's monitoring network for Finnish inland waters with information of Finnish water resources and monitoring strategies. *Finnish Environment*, **414**, 1–62.

Nilsson, C. (1982) *Biological Effects of Small Hydro Powerplants*. Report No 1593, Swedish Environmental Protection Agency, Solna. (In Swedish).

Niinioja, R., Tanskanen, A.-L., Rumyantsev, V., et al. (2000) Water management policy for large lakes. *Finnish Environment*, **414**, 1–288.

Odén, S. (1968) Nederbördens och luftens försurning. Dess förlopp och verkan i olika miljöer. *Ecological Bulletins*, **1**, 1–117.

Oikari, A. (1997) Vesistöjen pilaamisesta vesiensuojelun aikaan. *Luonnon Tutkija*, **100**(5), 13–24.

Oikari, A. & Holmbom, B. (1996) Ecotoxicological effects of process changes implemented in a pulp and paper mill: a Nordic case study. In: Servos, M., Munkitrick, K., Carey, J. & Van der Kraak, G. (eds), *Environmental Fate and Effects of Pulp and Paper Mill Effluents*, St Lucie Press, Boca Raton, FL, 613–25.

Paasivirta, J., Hakala, H., Knuutinen, J., et al. (1990). Organic chlorine compounds in lake sediments. 3. Chlorohydrocarbons, free and chemically bound chlorophenols. *Chemosphere*, **21**, 1355–70.

Palko J. (1994) Acid sulphate soils and their agricultural and environmental problems in Finland. *Acta Universitatis Ouluensis, Series C, Technica*, **75**, 1–58

Pekkarinen, M. (1990) Comprehensive survey of the hypertrophic Lake Tuusulanjärvi, nutrient loading, water quality and prospects of restoration. *Aqua Fennica*, **20**, 13–25.

Peltonen, H. & Horppila, J. (1992) Effects of mass removal on the roach stock of Lake Vesijärvi estimated with VPA within one season. *Journal of Fisheries Biology*, **40**, 293–301.

Peltonen, A., Grönlund, E. & Viljanen, M. (eds) (2000) Proceedings of the Third International Lake Ladoga

Symposium. *University of Joensuu, Publications of Kerelian Institute*, **129**, 1–519.

Perfiliew, B.W. (1929) Zur Mikrobiologie der Bodenablagerungen. *Verhandlungen der Internationalen Vereinigung für theoretische und angewandte Limnologie*, **4**, 107–26.

Persson, G. (ed.) (1996a) *Sjöar & vattendrag; årsskrift från miljöövervakningen 1995*. Institutionen för Miljöanalys, Svenska Lantbruksuniversitetet, Uppsala, 95 pp.

Persson, G. (1996b) *26 svenska referenssjöar 1989–1993. En kemisk-biologisk statusbeskrivning*. Statens Naturvårdsverk, Rapport 4552, 141 pp.

Persson, G. (ed.) (1996c) *Hjälmaren under 29 år. Undersökningar inom PMK 1965–1994*. Rapport 4535, Statens Naturvårdsverk, Solna, 74 pp.

Persson, G. (1996d) Vänerns vattenkemiska utveckling. In: Wallin M. (ed.), *Vänerns miljötillstånd och utveckling 1973–1994*. Rapport 4619, Statens Naturvårdsverk, Solna, 15–21.

Persson, G., Olsson, H. & Willén, E. (1990) *Mälarens vattenkvalitet under 20 år.1. Växtnäring: tillförsel, sjökoncentrationer och växtplanktonmängder*. Rapport 3759, Statens Naturvårdsverk, Solna, 47 pp.

Pietiläinen, O.P. & Rekolainen, S. (1991) Dissolved reactive and total phosphorus load from agricultural and forested basins to surface waters in Finland. *Aqua Fennica*, **21**, 127–36.

Raatikainen, M. & Kuusisto, E. (1990) Suomen järvien lukumäärä ja pinta-ala (Abstract: The number and surface area of the lakes in Finland). *Terra*, **102**, 97–110.

Rahkola, M., Karjalainen, J. & Viljanen, M. (1994) Evaluation of a pumping system for sampling zooplankton. *Journal of Plankton Research*, **16**, 905–10.

Rask, M. & Tuunainen, P. (1990) Acid-induced changes in fish populations in small Finnish lakes. In: Kauppi, P., Anttila, P. & Kenttämies, K. (eds), *Acidification in Finland*. Springer-Verlag, Berlin, 911–28.

Rask, M., Arvola, L. & Salonen, K. (1993) Effects of catchment deforestation and burning on the limnology of a small forest lake in southern Finland. *Verhandlungen der internationalen Vereinigung für theoretische und angewandte Limnologie*, **25**, 525–8.

Rask, M., Heinänen, A., Salonen, K., et al. (1986) The limnology of a small, naturally acidic, highly humic forest lake. *Archiv für Hydrobiologie*, **106**, 351–71.

Rekolainen, S. (1989) Phosphorus and nitrogen load from forest and agricultural areas in Finland. *Aqua Fennica*, **19**, 95–107.

Rekolainen S. (1993) Assessment and mitigation of agricultural water pollution. *Publications of the Water and Environment Institute, Helsinki*, B, **11**, 1–34.

Rekolainen, S., Granlund, K., Laine, Y., Johnsson, H. & Hoffman, M. (1996) Regionalisation of the impacts of climate change on erosion and nutrient losses from agriculture. In: Roos, J. (ed.), *The Finnish Research Programme on Climate Change*, Publication 4/96, Academy of Finland, Helsinki, 146–51.

Renberg, I. (1981) Improved methods of sampling, photographing and varve-counting of varved lake sediments. *Boreas*, **10**, 255–8.

Renberg, I. & Wallin, J.E. (1985) The history of the acidification of Lake Gårdsjön as deduced from diatoms and Sphagnum leaves in the sediment. In: Andersson, F. & Olsson, B. (eds), *Lake Gårdsjön—an Acid Forest Lake and its Catchment. Ecological Bulletins (Stockholm)*, **37**, 47–52.

Renberg, I. & Wik, M. (1985) Soot particle counting in recent lake sediments. An indirect dating method. In: Andersson, F. & Olsson, B. (eds), *Lake Gårdsjön—an acid forest lake and its catchment. Ecological Bulletins (Stockholm)*, **37**, 53–7.

Resursy (1972) *Resursy poverkhnostnykh vod SSSR, 2(3) —Karelia and North-West*. Gidrometizdat, Leningrad.

Rognerud, S. & Kjellberg, G. (1990) Long-term dynamics of the zooplankton community in Lake Mjøsa, the largest lake in Norway. *Verhandlungen der internationalen Vereinigung für theoretische und angewandte Limnologie*, **24**, 580–5.

Rönkkö, J. & Simola, H. (1986) Geological control upon the floral manifestation of eutrophication in two headwater lakes. *University of Joensuu, Publications of Karelian Institute*, **79**, 89–96.

Rørslett, B. (1985) Environmental factors and aquatic macrophyte response in regulated lakes: a statistical approach. *Aquatic Botany*, **19**, 199–220.

Rosén, G. (1981a) *Tusen sjöar. Växtplanktons miljökrav*. Naturvårdsverkets Rapport, Liber Förlag, Stockholm, 120 pp.

Rosén, G. (1981b) Phytoplankton indicators and their relations to certain chemical and physical factors. *Limnologica (Berlin)*, **13**, 263–90.

Rosenström, U., Lehtonen, M. & Muurman, J. (1996) Trends in the Finnish environment. Indicators for the 1997 OECD Environmental Performance Review of Finland. *The Finnish Environment*, **63**, 1–124.

Roulet, N.T. & Adams, W.P. (1986) Spectral distribution of light under a subarctic winter lake cover. *Hydrobiologia*, **134**, 89–95.

Rumyantsev, V. & Drabkova, V. (1997) Strategies for ecological sustainability assessment in Lake Ladoga as a basis for human impact control. *University of Joensuu, Publications of Karelian Institute*, **117**, 402–9.

Ruoho-Airola, T. (1995) Bulk deposition. In: Bergström, I., Mäkelä, K. & Starr, M. (eds). *Integrated Monitoring Programme in Finland, First National Report*. Environmental Policy Department, Ministry of Environment, Helsinki, 54 pp.

Ruoppa, M. & Karttunen, K. (eds) (2002) Typology and ecological classification of lakes and rivers. *TemaNord*, **566**, 1–136.

Ruoppa, M., Heinonen, P., Pilke, A., Rekolainen, S., Toivonen, H. & Vuoristo, H. (eds) (2003) How to assess and monitor ecological quality in freshwaters. *TemaNord*, **547**, 1–214.

Ruuhijärvi, R. (1974) A general description of the oligotrophic lake Pääjärvi, southern Finland, and the ecological studies on it. *Annales Botanici Fennici*, **11**, 95–104.

Sabylina, A.V., Basov, M.I., Harkevitch, N.S. & Mitina, I.F. (1991) Abiotic factors, plankton primary production and organic matter destruction in the basins of Karelia. *Vesi- ja ympäristöhallinnon julkaisuja A*, **72**, 81–93.

Sæther, O. (1979) Chironomid communities as water quality indicators. *Holarctic Ecology*, **2**, 65–74.

Salonen, K. & Hammar, T. (1986) On the importance of dissolved organic matter in the nutrition of zooplankton in some waters. *Oecologia*, **68**, 246–53.

Salonen, K., Arvola, L. & Rask, M. (1984) Autumnal and vernal circulation of small lakes in southern Finland. *Verhandlungen der internationalen Vereinigung für theoretische und angewandte Limnologie*, **22**, 103–7.

Salonen, K., Kankaala, P., Tulonen, T., *et al.* (1992) Planktonic food chains of a highly humic lake. II. A mesocosm experiment in summer during dominance of heterotrophic processes. *Hydrobiologia*, **229**, 143–57.

Salonen, K., Holopainen, A.-L. & Keskitalo, J. (2002) Regular high contribution of *Gonyostomum semen* to phytoplankton biomass in a small humic lake. *Verhandlungen der internationalen Vereinigung für theoretische und angewandte Limnologie*, **28**, 488–91.

Salonen, V.P., Alhonen, P., Itkonen, A. & Olander, H. (1993) The trophic history of Enäjärvi, SW Finland, with special reference to its restoration problems. *Hydrobiologia*, **268**, 147–62.

Sandman, O., Liehu, A. & Simola, H. (1990) Drainage ditch erosion history as recorded in the varved sediment of a small lake in East Finland. *Journal of Paleolimnology*, **3**, 161–9.

Sandman, O., Turkia, J. & Huttunen, P. (1995) Metsänkäsittelyn vaikutukset järvien paleolimnologisiin muutoksiin. *Suomen Ympäristö*, **2**, 213–27.

Sarvala, J. (1992) Trends in Finnish limnology during 1940–1989. *Hydrobiologia*, **243/244**, 1–19.

Sarvala, J. (2003) Long-term ecological monitoring in assessing the ecological state of lakes and guiding the management. *TemaNord*, **547**, 91–4.

Sarvala, J., Ilmavirta, V., Paasivirta, L. & Salonen, K. (1981) The ecosystem of the oligotrophic Lake Pääjärvi. 3. Secondary production and an ecological energy budget of the lake. *Verhandlungen der internationalen Vereinigung für theoretische und angewandte Limnologie*, **21**, 454–9.

Saukkonen P. (1997) Water quality in southern Lake Saimaa and River Vuoksi, as described by numerical quality indices. *University of Joensuu, Publications of Karelian Institute*, **117**, 344–9.

Saura, M., Sallantaus, T., Bilaletdin, Ä. & Frisk, T. (1995) Metsänlannotteiden huuhtoutuminen Kalliojärven valuma-alueelta. *Suomen ympäristö*, **2**, 87–104.

Schindler, D.W. Bayley, S.E. Curtis, P.J., Parker, B.R., Stainton, M.P. & Kelly, C.A. (1992) Natural and man-caused factors affecting the abundance and cycling of dissolved organic substances in precambrian shield lakes. *Hydrobiologia*, **229**, 1–21.

Seuna P. (1988) Effects of clear-cutting and forestry drainage on runoff in the Nurmes-study. In: Rantajärvi, L. (ed.) *Symposium on the Hydrology of Wetlands in Temperate and Cold Regions*, Vol. 1. Academy of Finland, Helsinki, 107–21.

Sevola, P., Hudd, R. & Hilden, M. (1982) Luontaiset mahdollisuudet tutkia happamuuden vaikutuksia vesiin Suomessa tulisi käyttää hyväksi. (Abstract: Acidification of watercourses in sulphide clay areas). *Luonnon Tutkija*, **86**, 62–4.

Silvola, J., Alm, J., Saarnio, S. & Martikainen, P.J. (1996) Greenhouse gas flux dynamics in wetlands. In: Roos, J. (ed.), *The Finnish Research Programme on Climate Change*, Publication 4/96, Academy of Finland, Helsinki, 384–94.

Similä, A. (1988) Spring development of a *Chlamydomonas* population in Lake Nimetön, a small humic lake in southern Finland. *Hydrobiologia*, **161**, 149–57.

Simola, H. (1983) Limnological effects of peatland drainage and fertilization as reflected in the varved sediment of a deep lake. *Hydrobiologia*, **106**, 43–57.

Simola, H. (1992) Structural elements in varved lake sediments. In: Saarnisto, M. & Kahra, A. (eds), *INQUA Workshop on Laminated Sediments, Lammi, Finland, June 1990*. Geological Survey of Finland, Special Papers, **14**, 5–9.

Simola, H. & Miettinen, J. (1997) Mitä tapahtui kun väestö evakuoitiin? —luonnontilan palautumisprosessi on tallentunut luovutetun Karjalan järvien sedimentteihin. Summary: Ecosystem recovery following depopulation: palaeolimnological records in former Finnish territory in Russian Karelia. *Geologi*, **49**, 86–9.

Simola, H. & Uimonen-Simola, P. (1983) Recent stratigraphy and accumulation of sediment in the deep, oligotrophic Lake Pääjärvi in South Finland. *Hydrobiologia*, **103**, 287–93.

Simola, H., Kenttämies, K. & Sandman, O. (1985) The recent pH-history of some Finnish headwater and seepage lakes, studied by means of diatom analysis of $^{210}$Pb dated sediment cores. *Aqua Fennica*, **15**, 245–55.

Simola, H., Huttunen, P., Uimonen-Simola, P., Selin, P. & Meriläinen, J.J. (1988) Effects of peatland forestry management and fuel peat mining in Lake Ilajanjärvi, east Finland—a paleolimnological study. In: Rantajärvi, L. (ed.) *Symposium on the Hydrology of Wetlands in Temperate and Cold Regions*, Vol. 1. Academy of Finland, Helsinki, 285–90.

Simola, H., Huttunen, P., Rönkkö, J. & Uimonen-Simola, P. (1991) Palaeolimnological study of an environmental monitoring area, or, Are there pristine lakes in Finland? *Hydrobiologia*, **214**, 187–90.

Simola, H., Kukkonen, M., Lahtinen, J. & Tossavainen, T. (1995) Effects of intensive forestry and peatland management on forest lake ecosystems in Finland: sedimentary records of diatom floral changes. In: Marino, D. & Montressor, M. (eds), *Proceedings of the 13th International Diatom Symposium*: Biopress, Bristol, 121–8.

Simola, H., Viljanen, M., Slepukhina, T. & Murthy, R. (eds) (1996a) *The 1st International Lake Ladoga Symposium*. Developments in Hydrobiology No. 113, Kluwer, Dordrecht, xi + 328 pp.

Simola, H., Meriläinen, J.J., Sandman, O., *et al.* (1996b) Palaeolimnological analyses as information source for large lake biomonitoring. *Hydrobiologia*, **322**, 283–92.

Simola, H., Viljanen, M. & Slepukhina, T (eds) (1997) *Proceedings of the Second International Lake Ladoga Symposium 1996*. *University of Joensuu, Publications of Karelian Institute*, **117**, 450 pp.

Simola, H., Miettinen, J., Grönlund, E. & Saksa, A. (2000) Palaeoecological reflections of rapid land-use change in the northern Lake Ladoga region. *University of Joensuu, Publications of Karelian Institute*, **129**, 504–7.

Simola, H., Terzhevik, A., Viljanen, M. & Holopainen, I. (eds) (2003) *Proceedings of the Fourth International Lake Ladoga Symposium 2002*. *University of Joensuu, Publications of Karelian Institute*, **138**, 1–592.

Simonen, A. (1990) Bedrock. *Atlas of Finland* (5th edition). National Board of Survey, Finland, Helsinki, 1–4.

Skjelkvåle, B.L., Henriksen, A., Faafeng, B., Fjeld, E., Traaen, T., Lien, L., Lydersen, E. & Buan, A.K. (1997) *Regional innsjøundersøkelse 1995*. Statlig program for forurensningsovervåkning, Rapport 677/96, Norsk Institutt for Vannforskning, Oslo.

Skulberg, O. (1988) *Blågrønnalger—vannkvalitet. Toksiner. Lukt- og smakstoffer. Nitrogenbindning.* Report 2116, Norsk Institutt for Vannforskning, Oslo.

Slepukhina, T., Frumin, G., Barbashova, M. & Barkan, L. (1997) Concepts for ecological monitoring of Lake Ladoga. *University of Joensuu, Publications of Karelian Institute*, **117**, 16–25.

Smol, J.P. (2002) *Pollution of Lakes and Rivers: a Paleoenvironmental Perspective*. Arnold, London, and Oxford University Press, New York, 280 pp.

Smol, J.P., Birks, H.J.B. & Last, W.M. (eds) (2001a) *Tracking Environmental Change using Lake Sediments*, Vol. 3, *Terrestrial, Algal and Siliceous Indicators*. Kluwer, Dordrecht, 371 pp.

Smol, J.P., Birks, H.J.B. & Last, W.M. (eds) (2001b) *Tracking Environmental Change using Lake Sediments*, Vol. 4, *Zoological Indicators*. Kluwer, Dordrecht, 217 pp.

SNV (1986) *Acid and acidified waters. Monitor 1986*. Swedish Environment Protection Board, Solna.

SOAER (1997) *Arctic Pollution Issues: a State of the Arctic Environment Report*. Arctic Monitoring and Assessment Programme, Oslo, 188 pp.

Stenroos, K.E. (1898) Das Thierleben im Nurmijärvi-See. Eine faunistisch-biologische Studie. *Acta Societatis pro Fauna et Flora Fennica*, **17**(1), 1–259.

Thunmark, S. (1937) Über die regionale Limnologie von Südschweden. *Sveriges Geologiske Undersökelse Årsbok*, **31**, 6.

Timakova, I.M. (1991) The role of bacterioplankton in the formation of organic matter in Lake Onega. *Publications of the Water and Environment Administration (Finland) A*, **72**, 7–14.

Timakova, T.M. & Vislyanskaya, I.G. (1991) Primary production in Lake Onega. *Publications of the Water and Environment Administration (Finland) A*, **72**, 73–9.

Timakova, T., Kulikova, T., Polyakova, T., Vislyanskaya, I. & Syarki, M. (2000) Effects of eutrophication on the aquatic biota of Lake Onega. *University of Joensuu, Publications of Karelian Institute*, **129**, 299–304.

Tolonen, K. & Jaakkola, T. (1983) History of lake acidification and air pollution studied by sediments in South Finland. *Annales Botanici Fennici*, **20**, 57–78.

Tolonen, K., Siiriäinen, A. & Thompson, R. (1975) Prehistoric field erosion sediment in Lake Lojärvi, S. Finland, and its palaeomagnetic dating. *Annales Botanici Fennici*, **12**, 161–4

Tolonen, K., Liukkonen, M., Harjula, R. & Pätilä, A. (1986) Acidification of small lakes in Finland by sedimentary diatom and chrysophycean remains. In: Smol, J.P., Battarbee, R. & Meriläinen, J.J. (eds), *Diatoms and Lake Acidity*. W. Junk, Den Haag, 169–99.

Tolonen, K., Ilmavirta, V., Hartikainen, H. & Suksi, J. (1990) Paleolimnological investigation of the history of eutrophication of Lake Tuusulanjärvi, southern Finland. *Aqua Fennica*, **20**, 27–41.

Tolonen, K., Ilmavirta, V., Uimonen-Simola, P. & Suksi, J. (1994) Eutrophication history of a small pelotrophic lake, Rusutjärvi, southern Finland. *Aqua Fennica*, **24**, 141–62.

Tolonen, M. (1978). Paleoecology of annualy laminated sediments in Lake Ahvenainen, S. Finland. III. Human influence on the lake development. *Annales Botanici Fennici*, **15**, 223–40.

Tolonen, M. (1981) An absolute and relative pollen analytic study of prehistoric agriculture in South Finland. *Annales Botanici Fennici*, **18**, 213–20.

*Topograficheskaya Karta Respublika Karelii* (1997) Voyenno-topograficheskoye Upravlenie Generalnogo Shtaba, Moscow.

Tossavainen, T. (1997) Nurmeksen Kuohattijärven ympäristönhoitosuunnitelma. Summary: The plan for the restoration project of Lake Kuohattijärvi, Nurmes, North Karelia. *Mimeograph Series of North Karelia Regional Environment Centre*, **14**, 1–37.

Tranvik, L.J. (1988) Availability of dissolved organic carbon for planktonic bacteria in oligotrophic lakes of differing humic content. *Microbial Ecology*, **16**, 311–22.

Tulonen, T., Salonen, K. & Arvola, L. (1992) Effect of different molecular weight fractions of dissolved organic matter on the growth of bacteria, algae and protozoa from a highly humic lake. *Hydrobiologia*, **229**, 239–52.

Turunen, T., Sammealkorpi, I. & Suuronen, P. (1997). Sutability of motorized under-ice seining in selective mass removal of coarse fish. *Fisheries Research*, **31**, 73–82.

Varis, O. & Kettunen, J. (1990) Modeling of water quality in Lake Tuusulanjärvi. *Aqua Fennica*, **20**, 43–54.

Vasander, H. (ed.) (1996) *Peatlands in Finland*. Finnish Peatland Society, Helsinki, 168 pp.

Veijola, H., Meriläinen, J.J. & Marttila, V. (1996) Sample size in the monitoring of benthic macrofauna in the profundal of lakes: evaluation of the precision of estimates. *Hydrobiologia*, **322**, 301–15.

Vesajoki, H., Tornberg, M. & Kajander, J. (1994) Documentary sources for reconstructing past weather and climate in Finland. *Archaeological Museum of Stavanger AmS-Varia*, **7**, 39–42.

Verta, M. (1990) Mercury in Finnish forest lakes and reservoirs: anthropogenic contribution to the load and accumulation in fish. *Publications of the Water and Environment Institute, Helsinki*, **6**, 1–34.

Veselov, E.A. (1970) Razvitie issledovanii po ekologicheskoi fiziologii vodnikh organizmov i vodnoi toksikologii. *Uzenie Zapiski Petrozavodskogo Universiteta*, **16**(4), 194–201.

Viljanen, M. (2003) Development and implementation of an integrated programme for environmental monitoring of Lake Ladoga (DIMPLA). *University of Joensuu, Publications of Karelian Institute*, **138**, 430–5.

Viljanen, M. & Karjalainen, J. (1993) Horizontal distribution of zooplankton in two large lakes in eastern Finland. *Verhandlungen der internationalen Vereinigung für theoretische und angewandte Limnologie*, **25**, 548–51.

Viljanen, M., Niinioja, R. Huttula, T, Filatov, N., Drabkova, V., Rumyantsev, V., Budarin, N. & Feshenko, M. (2000) Development and implementation of an intagrated programme for environmental monitoring of Lake Ladoga (DIMPLA): project summary and conclusions. *University of Joensuu, Publications of Karelian Institute*, **130**, 1–31.

Vuoristo, H. & Antikainen, S. (1997) *Water Quality in Lakes and Rivers* (Brochure). Finnish Environment Institute, Helsinki, 2 pp.

Wahlström, E., Reinikainen, T. & Hallanaro, E.L. (eds) (1992) *Ympäriston tila Suomessa*. Gaudeamus Kirja, Helsinki, 364 pp.

Wahlstrom, E., Hallanaro, E.L. & Manninen, S. (eds) (1996) *The Future of the Finnish Environment*. Edita, Finnish Environment Institute, Helsinki, 272 pp.

Weppling, K. (1997) *On the assesment of feasible liming strategies for acid sulphate waters in Finland*. Dissertationes geographicae Universitatis Tartuensis, 5, Tartu.

Wiederholm, T. (1980) Use of benthos in lake monitoring. *Journal of Water Pollution Control Federation*, **52**, 537–47.

Wilander, A. & Persson, G. (2001) Recovery from eutrophication: experiences of reduced phosphorus input to the four largest lakes in Sweden. *Ambio*, **30**, 475–85.

Wilander, A. & Willén, E. (1994) *Vättern och dess tillflöden 1971–1994*. Rapport 40, Jönköpings Län Vätternvårdsförbundets, Jönköping, 88 pp.

Willén, E. (1984) The large lakes of Sweden: Vänern,

Vättern, Mälaren and Hjälmaren. In: Taub F.B (ed.), *Lakes and Reservoirs. Ecosystems of the World*, Vol. 23, Elsevier, Amsterdam, 107–34.

Willén, E. (1991) Planktonic diatoms—an ecological review. *Algological Studies*, **62**, 69–106.

Willén, E. (1992) Planktonic green algae in an acidification gradient of nutrient-poor lakes. *Archiv für Protistenkunde*, **141**, 47–64.

Willén, E. (1995) Skadliga alger. *Nytt från Institutionen för Miljöanalys*, **2**, 3–4.

Willén, E. (2001) Four decades of research on the Swedish large lakes Mälaren, Hjälmaren, Vättern and Vänern: the significance of monitoring and remedial measures for a sustainable society. *Ambio*, **30**, 458–66.

Willén, E. (2002) Phytoplankton in water quality assessment—an indicator concept. In: Heinonen, P., Ziglio, G. & Van der Beken, A. (eds) *Hydrological and Limnological Aspects of Lake Monitoring*. Wiley, Chichester, 57–80.

Willén, E. & Wiederholm, A.M. (1996) Växtplankton i Vänern. In: Wallin, M. (ed.), *Vänerns miljötillstånd och utveckling 1973–1994*. Rapport 4619, Statens Naturvårdsverk, Solna, 35–41.

Willén, E., Hajdu, S. & Pejler, Y. (1990a) Summer phytoplankton in 73 nutrient-poor Swedish lakes. Classification, ordination and choice of long-term monitoring objects. *Limnologica (Berlin)*, **20**, 217–27.

Willén, E., Wiederholm, T. & Persson, G. (1990b) *Mälarens vattenkvalitet under 20 år. 2. Strandvegetation, plankton, bottendjur och fisk*. Rapport 3842, Statens Naturvårdsverk, Solna, 42 pp.

Witting, R. (1929) Beobachtungen im Ladoga-See in den Jahren 1898–1903. *Merentutkimuslaitoksen Julkaisu*, **60**, 1–34.

Yakovlev, V. (1992) Benthic invertebrates, zooplankton and phytoplankton communities in relation to pollution of waters in the Kola Peninsula. *Publications of the Arctic Centre, University of Lapland*, **4**, 165–7.

# 6 European Alpine Lakes

MARTIN T. DOKULIL

## 6.1 INTRODUCTION

Although the term 'alpine lake' is widely accepted and used as a descriptor for a certain type of lake found around the globe, an adequate and precise definition is not available. Alpine lakes can be considered to be water bodies filling depressions formed by glacial erosion during the Pleistocene, but not now in direct contact with a perennial ice sheet.

Hutchinson (1957) classifies the lakes of the European Alps with certain subtypes of glacial rock basin (his types 27–29). According to the classification of Timms (1992), lakes originating from glacial erosion are one of the four major groups of glacial lakes. Three of these are categories of alpine lakes, distinguished by their formative processes and their elevation: high alpine lakes, alpine lakes and pre- or sub-alpine lakes. High alpine lakes are situated at high altitude, usually above the tree line. These lakes are the ice-excavated cirque or kar lakes (Timms 1992; type 27 of Hutchinson 1957). Alpine lakes *sensu stricto* are glacial valley lakes (Timms 1992) or fjord-type lakes (type 28b of Hutchinson 1957), occupying former glacier channels. Pre- or sub-alpine lakes are formed by large glaciers descending through long valleys to relatively low elevations. Such lakes are also termed piedmont lakes (type 28c of Hutchinson 1957; Timms 1992).

All the lakes of the Alps region of Europe were formed by glaciers (Fahn 1981), which filled parallel valleys between the mountain ridges which strike northeast to southwest (Fig. 6.1): the Swiss and Austrian lakes lie to the north of this axis; the Italian lakes are to the south. In many cases, the lakes have been deepened by moraine material forming dam-like barriers at their lower ends. There also exist many smaller kettle lakes (type 38 of Hutchinson 1957), which occupy cavities in the moraine created by the decay of ice blocks detached from the glaciers during their retreat phases. The tendency of glaciers to fan out at lower elevations is also mirrored in the lakes which now occupy their valleys. Lago di Como is perhaps one of the best examples, but the effect can also be seen in Lago di Garda and Lake Constance (Bodensee).

Depending on their geographical and geological situations, the central European Alpine Lakes can be divided into the following groups (Thomas 1969):

1 Bavarian lakes (Germany)—situated on the Swabian–Bavarian plateau at about 500m a.s.l. in calcareous-rich catchments; lakes in the Salzkammergut (Austria)—mountainous lakes in an area rich in dolomite and calcium carbonate;

2 Carinthian lakes (Austria)—situated in mountains with partly crystalline geology, and many are meromictic;

3 lakes of the southern Alps (Italy and Switzerland)—catchments differ in lithology; Savoyan lakes (France)—situated in a calcareous region but hardness not extremely high;

4 lakes of the Swiss midlands (Switzerland)—mainly receiving water from crystalline regions of the Alps;

5 lakes of the higher Alps—usually low salt content; a few are meromictic; non-homogeneous group.

The lakes and rivers in the European Alps are largely influenced by the unique hydrographical situation (Baumgartner *et al.* 1983). In the mid-continental mountain ranges north of 46°N latitude, the general water budget relates first to the high absolute run-off and its magnitude in relation to precipitation (Table 6.1). Although the catchments of the tropical rivers Amazon and Zaire are three to six times greater and the run-off they

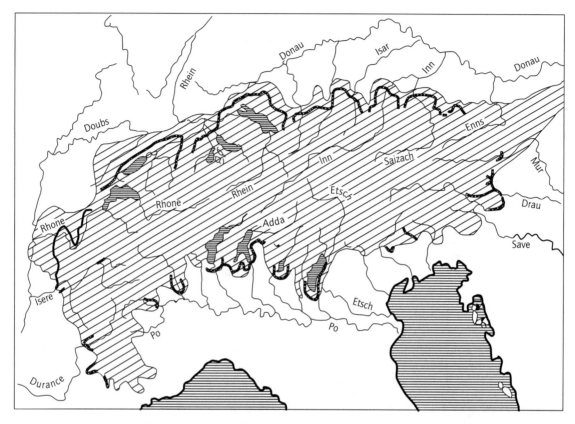

**Fig. 6.1** Maximum extent of the last glaciation of the Alps (shaded area). Thick lines represent moraines (From Pesta 1929).

**Table 6.1** Hydrography for regions north of 46°N: $D$, run-off height; $D/P$, relationship of run-off height to precipitation. (From Baumgartner *et al.* 1983.)

| Variable | Alps | St Lorenz | Colorado | Amur | Volga | Missouri |
|---|---|---|---|---|---|---|
| $D$ (cm) | 101 | 34 | 32 | 20 | 19 | 5 |
| $D/P$ (%) | 67 | 37 | 41 | 33 | 29 | 8 |

generate is five to 25 times larger than that of the Alps, the run-off yield from the Amazon is almost equal to that from the European Alps, whereas that of the Zaire River is approximately half. The annual evaporation of about 110 cm in the tropical catchments is about double that in the Alps. Precipitation in the Zaire catchment is approximately equal to that in the Alps but is about 40% higher in the Amazon catchment.

The total area covered by lakes in the Alps is estimated to be 3440 km$^2$ (Meybeck 1995). Taking the area of the region as some 40,000 km$^2$, the nominal limnic ratio is 8.6%. The total number of lakes in the size range 0.01 to 10$^5$ km$^2$ is 1357. Small lakes of less than 0.1 km$^2$ in size are most abundant (Table 6.2). Their areal density (0.025 km$^{-2}$) is typical for those of lakes on sedimentary platforms (Meybeck 1995). However, the number of lakes in the Alps is greatly increased if those smaller than 0.01 km$^2$ are considered. Table 6.3 gives an example, from the Carinthian region in Austria. Most of these small lakes are located at

**Table 6.2** Distribution of lakes in the Alps: $DL$, number km$^{-2}$; $A_0$, total lake area (km$^2$); $n$, number of lakes. Area of the Alps = 40,000 km$^2$; total lake area = 3,440 km$^2$; limnic ratio = 8.6%. (From Meybeck 1995.)

| Area class (km$^2$) | $n$ | $A_0$ | $DL$ |
| --- | --- | --- | --- |
| 0.01–0.1 | 1000 | 25 | 0.025 |
| 0.1–1 | 250 | 65 | 0.0062 |
| 1–10 | 70 | 180 | 0.00175 |
| 10–100 | 30 | 995 | 0.00075 |
| 100–1,000 | 7 | 2176 | 0.00018 |

**Table 6.3** Frequency distribution of lakes in Carinthia, Austria. (From Sampl 1976.)

| Area class (km$^2$) | Altitude <1000 m | Altitude >1000 m | Total |
| --- | --- | --- | --- |
| >10 | 3 | 0 | 3 |
| 1–10 | 4 | 0 | 4 |
| 0.1–1 | 112 | 60 | 172 |
| 0.01–0.1 | 520 | 571 | 1091 |
| Total | | | 1270 |

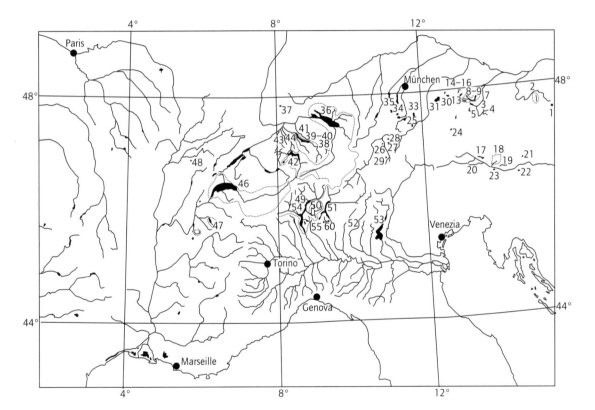

**Fig. 6.2** Map of the Alps showing the location of 60 alpine lakes. Their names and morphometric characteristics can be identified from Table 6.4.

high altitudes in the mountains. The geographical positions of several alpine lakes and their morphometric characteristics are given in Fig. 6.2 and Table 6.4.

One of the main differences between lakes in the Alps and those of other deglaciated lake regions is their much greater depths, which are the result of glacial scouring (see Table 6.4). Amongst

**Table 6.4** Morphometric characteristics of 60 alpine lakes: $Z_{max}$, maximum depth; $Z_{mean}$, mean depth; $t_w$, theoretical retention time. (Compiled from several sources.)

| Country/area | Lake name | Altitude (m, a.s.l.) | Catchment area (km²) | Lake area (km²) | Volume (×10⁶ m³) | $Z_{max}$ (m) | $Z_{mean}$ (m) | $t_w$ (yr) |
|---|---|---|---|---|---|---|---|---|
| Austria | 1 Erlaufsee | 835 | 10 | 0.58 | 12.4 | 38 | 21.2 | 1.5 |
| | 2 Lunzer Untersee | 608 | 27 | 0.68 | 13 | 33.7 | 20 | 0.3 |
| Salzkammergut | 3 Grundlsee | 709 | 125 | 4.14 | 170 | 63.8 | 32.2 | 0.92 |
| | 4 Altauseersee | 712 | 54.5 | 2.1 | 72 | 52.8 | 34.6 | 0.58 |
| | 5 Halstättersee | 508 | 646.5 | 8.58 | 557 | 125.2 | 64.9 | 0.5 |
| | 6 Wolfgangsee | 538 | 123 | 13.15 | 619 | 114 | 47.1 | 3.6 |
| | 7 Traunsee | 422 | 1417 | 25.6 | 2303 | 191 | 89.7 | 1.0 |
| | 8 Krottensee | 579 | | 0.09 | 2.2 | 46.8 | 26 | |
| | 9 Fuschlsee | 663 | 29.5 | 2.66 | 99.5 | 67.3 | 37.4 | 2.6 |
| | 10 Irrsee (Zellersee) | 533 | 27.5 | 3.47 | 53 | 32 | 15.3 | 1.7 |
| | 11 Mondsee | 481 | 247 | 14.21 | 510 | 68.3 | 36 | 1.7 |
| | 12 Attersee | 469.2 | 463.5 | 45.6 | 3944.6 | 170.6 | 84.2 | 7.0 |
| | 13 Wallersee | 505 | 110 | 6.39 | 71 | 24 | 9.4 | 2.0 |
| | 14 Obertrumersee | 550 | 45.9 | 4.8 | 86 | 35 | 14.1 | 1.3 |
| | 15 Niedertumersee | 500 | 12.5 | 3.25 | 48 | 40 | 14.8 | 3.1 |
| | 16 Grabensee | 498 | 64.8 | 1.26 | 9 | 13 | 7.2 | 0.12 |
| Carinthian Lake District | 17 Millstättersee | 588 | 276 | 13.28 | 1176.6 | 141 | 89 | 7 |
| | 18 Ossiachersee | 501 | 154.8 | 10.79 | 215.1 | 52 | 19.9 | 2 |
| | 19 Wörthersee | 439 | 164 | 19.38 | 816.3 | 85.2 | 42.1 | 9.5 |
| | 20 Weissensee | 930 | 50 | 6.53 | 238.1 | 99 | 36 | 11 |
| | 21 Längsee | 548 | | 0.76 | 9 | 21 | 11.1 | |
| | 22 Klopeinersee | 446 | 4.4 | 1.11 | 24.9 | 46 | 22.6 | 11.5 |
| | 23 Faaker See | 554 | 35.6 | 2.2 | 32.7 | 29.5 | 14.9 | 1.2 |
| Central Alps and Tyrol | 24 Zeller See | 749.5 | 54.7 | 4.55 | 178.2 | 68.4 | 39.2 | 4.1 |
| | 25 Achensee | 929 | 218.1 | 6.8 | 481 | 133 | 66.8 | 1.6 |
| | 26 Piburgersee | 913 | 2.65 | 0.13 | 1.8 | 24.6 | 13.7 | 2.7 |
| | 27 Vdr Finstertalersee | 2237 | | 0.16 | 2.3 | 28.3 | 14.8 | |
| | 28 Gossenköllesee | 2413 | 0.3 | 0.016 | | 9.9 | 4.7 | |
| | 29 Schwarzsee/Sölden | 2799 | 0.2 | 0.034 | | 18 | 10 | |
| Germany | 30 Wagingersee | | 123.5 | 8.9 | 156.7 | 27 | 15.6 | |
| | 31 Königssee | 603.3 | 137.6 | 5.3 | 511 | 189 | 93.1 | 2.3 |
| | 32 Chiemsee | 518.2 | 1388.3 | 82.2 | 2053 | 73.4 | 25.6 | 1.4 |
| | 33 Tegernsee | | 210.5 | 9.1 | 324 | 72.2 | 36.6 | |
| | 34 Starnberger See | 584.2 | 314.7 | 56.4 | 2998 | 127.8 | 53.2 | 21 |
| | 35 Ammersee | | 988.5 | 47.6 | 1775 | 82.5 | 37.8 | 2.5 |
| | 36 Bodensee | 395 | 10,900 | 539 | 48,400 | 252 | 100 | 4.7 |
| | 37 Feldsee | 1109 | 1.5 | 0.9 | 1.4 | 32 | 15.7 | 0.5 |
| Switzerland | 38 Walensee | 419 | 1061 | 24.2 | 2420 | 145 | 100 | 1.4 |
| | 39 Zürichsee (OberS.) | 408 | 1564.8 | 20.3 | 467 | 48 | 23 | 0.2 |
| | 40 Zürichsee (UnterS.) | 406 | 1740.1 | 65.1 | 3300 | 136 | 51 | 1.4 |
| | 41 Greifensee | 436 | 160 | 8.5 | 150 | 32.6 | 17.7 | 1.5 |
| | 42 Vierwaldstättersee | | 1831 | 87.8 | 12,280 | 214 | 156 | 1.8 |
| | 43 Baldeggersee | 463 | 67.9 | 5.3 | 178 | 66 | 34 | 6 |

Table 6.4  Continued

| Country/area | Lake name | Altitude (m, a.s.l.) | Catchment area (km²) | Lake area (km²) | Volume (×10⁶ m³) | $Z_{max}$ (m) | $Z_{mean}$ (m) | $t_w$ (yr) |
|---|---|---|---|---|---|---|---|---|
| | 44 Hallwilersee | 449 | 127.7 | 9.9 | 285 | 48 | 20.6 | 4.1 |
| | 45 Sempachersee | 504 | 62.6 | 14.4 | 624 | 87 | 46 | 17 |
| France | 46 Lac Léman | 372 | 7975 | 582.4 | 88,900 | 309.7 | 152.7 | 11.8 |
| | 47 Lac d'Annecy | 446.5 | 278 | 27 | 1120 | 64.7 | 41.5 | 3.8 |
| | 48 Lac Nantua | 474.5 | 50.3 | 1.4 | 40 | 42.8 | 28.4 | 0.7 |
| Italy | 49 Lago Maggiore | 193.5 | 6599 | 212.5 | 37,700 | 370 | 177 | 4 |
| | 50 Lago Lugano | 271 | 615 | | 6500 | 288 | 130 | 0.8 |
| | 51 Lago di Como | 198 | 4572 | 145.9 | 22,500 | 410 | 153 | 4.5 |
| | 52 Lago Iseo | 186 | 1842 | 61.8 | 7600 | 251 | 123 | 4.1 |
| | 53 Lago di Garda | 65 | 2350 | 370 | 49,000 | 346 | 136 | 26.6 |
| | 54 Lago die Mergozzo | 194 | | | 83 | 73 | 45.4 | 4.7 |
| | 55 Lago di Oggione | 225 | 12.7 | 3.8 | 24 | 11.3 | 6.3 | 1.6 |
| | 56 Lago die Pusiano | 260 | 90.6 | 4.9 | 69 | 24.3 | 14.0 | 0.8 |
| | 57 Lago di Montorfano | 394 | 1.7 | 0.46 | 2 | 6.8 | 4.1 | 1.1 |
| | 58 Lago di Annone | 225 | 14.5 | 1.7 | 6.8 | 10.1 | 4 | 0.4 |
| | 59 Lago del Segrino | 374 | 3.4 | 0.38 | 1.2 | 8.6 | 3.2 | 0.3 |
| | 60 Lago d'Alserio | 260 | 18.4 | 1.23 | 6.55 | 8.1 | 5.3 | 0.4 |

alpine lakes, mean depth is, on average, 0.46 of the maximum depth (Ventz 1973). This relation is actually similar to the world average of 0.47 (range 0.33–0.67; Neumann 1959) but somewhat higher than the 0.37 estimated by Herdendorf (1990) for large lakes. This indicates that the depth relation of alpine lakes fits to the global range.

The history and development of alpine lakes are reconstructed through palaeolimnological investigations (Smol 1992). Specific transfer functions, used to infer past natural and anthropogenic environmental changes, rely on the sensitivity of modern biological indicator species to gradients of variability (e.g. Anderson 1993; Wunsam & Schmidt 1995).

## 6.2  PHYSICS

Depth soundings of water temperature in lakes were initiated in Switzerland and Austria as early as 1848 (Brunner 1849; Simony 1850). Annual temperature cycles and the timing of the onset and extent of stratification and mixing conditions are highly variable among the alpine lakes, depending on their size, depth and altitude.

Epilimnetic temperatures in the larger lakes of the Salzkammergut (e.g. Traunsee, Hällstättersee), rarely exceed 20°C. Lakes in the Carinthian lake district of Austria, south of the Alps, experience comparatively higher epilimnetic water temperatures during summer, owing to the high insolation in this region and the wind-protected situation and low flushing rates of the water bodies (Sampl 1976). Most alpine lakes are of the warm-monomictic or warm-dimictic types. However, some high-alpine lakes are not ice-free for long enough to warm above 4°C. These lakes are classed as cold-monomictic. Most of the kettle lakes are meromictic (i.e. they rarely, if ever, mix completely; Findenegg 1935).

Lakes above the tree-line and small lakes freeze over every year but larger and deeper alpine and sub-alpine lakes experience irregular freezing intervals. Examples include several of the large Swiss lakes, Lake Constance and the main lakes of

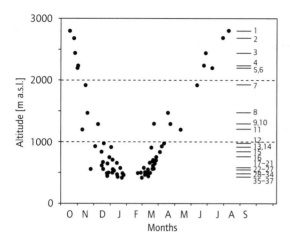

Fig. 6.3 Average beginning and end of the annual ice-cover period for several alpine lakes versus altitude: 1, Schwarzsee; 2, Barrenlesee; 3, Plenderlesee; 4, V. Finstertalersee; 5, Wangenitzsee; 6, Brettersee; 7, Giglacbsee; 8, Reschensee; 9, Brennersee; 10, Möserersee; 11, Lunzer Obersee; 12, Plansee; 13, Weißensee; 14, Achensee; 15, Erlaufsee; 16, Zellersee; 17, Grundlsee; 18, Altausseersee; 19, V. Langbathsee; 20, Fuschlsee; 21, Offensee; 22, Lunzer Untersee; 23, Millstättersee; 24, Faakersee; 25, Presseggersee; 26, Irrsee; 27, Wolfgangsee; 28, Wallersee; 29, Obertrummersee; 30, Niedertrummersee; 31, Halistättersee; 32, Ossiachersee; 33, Mondsee; 34, Attersee; 35, Klopeinersee; 36, Wörthersee; 37, Traunsee. (Modified from Eckel 1955.)

the Salzkammergut region in Austria (Hallstättersee, Traunsee and Attersee). All these lakes possess a large heat capacity and/or a high flushing rate. The last occasion when any was completely frozen was during the cold winter of 1962–3.

In general the average duration of the freezing period depends primarily on the altitude of the lake (Fig. 6.3). The duration of lake freezing is about 2 months in the lowlands, about 4 months at 1000 m, and approximately 7 months at 2000 m. At very high altitude, ice cover may persist for 11 months each year. Other factors may be important in specific lakes, such as the shading by mountains or the size and restricted depth of the lake.

According to Eckel (1955), the average duration ($D$, in days) of freezing can be calculated from the empirical equation:

$$D = 14 + 10.7A \qquad (1)$$

where $A$ is altitude in hectometres. Corresponding with ice duration, the thickness of the ice cover also increases generally with altitude, rising to 1–2 m in the high alpine lakes. However, the large, low-altitude (603.3 m), fjord-type lake, Königssee (Germany), is frequently covered by ice sufficiently thick to allow car traffic to pass across it. This is due to a comparatively low flushing rate (theoretical retention time of Königssee is 2.3 yr) and the low annual heat gain of approximately 12 kcal cm$^{-2}$ (Siebeck 1982, 1989).

Alpine lakes are usually oligotrophic, characterised by clear water and high transparency. Maximum Secchi-disc depths of 18–20 m are not uncommon (e.g. Attersee in Austria, Königssee in Germany, Vierwaldstättersee in Switzerland, Lago Maggiore in Italy). However, minimum transparency can be as low as 2–3 m during summer, mainly as a consequence of calcite precipitation. Annual average Secchi-disc depth varies between 12 m in the clearest lakes and 2 m in lakes more influenced by anthropogenic impacts (Dokulil et al. 2000). Secchi-disc depth is related to the euphotic zone by a factor of 2–2.5.

Attenuation of solar radiation with depth for several alpine lakes is shown in Fig. 6.4. Light transmission is greatest in the unproductive lakes, such as Attersee and Königssee (Dokulil 2001; Siebeck 1984), with the most penetrating component in the blue-green spectral region (500–520 nm). As the amount of organic substances in the water increases, maximum transparency progressively shifts to longer wavelengths resulting in green to olive-green water colour.

High alpine lakes are usually very clear and transparent. Secchi-disc readings of 21.5 m are reported for some of the deeper alpine lakes (Turnowsky 1976). Visibility to the lake bottom is not uncommon among the majority of the high alpine lakes, their average mean depths being about 15 m (Pesta 1929). As a consequence of high

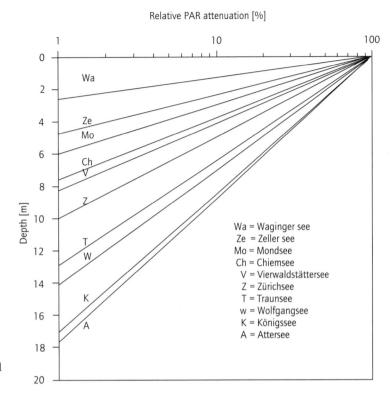

Fig. 6.4 Relative attenuation of photosynthetically active radiation (PAR) versus depth for a number of alpine lakes. (Compiled from several sources including original data.)

water clarity, penetration of ultraviolet (UV) radiation is considerable in the high alpine lakes. In Gossenköllesee, Austria, for instance, the coefficient of vertical attenuation in the 305-nm UV fraction (as measured by Sommaruga & Psenner 1997) means that ≥10% frequently reaches to the bottom at 9.9 m.

## 6.3 CHEMISTRY

The chemical composition of lake water depends largely on the geology of the watershed. The total salt content of most alpine and pre-alpine lakes is 100–200 mg L$^{-1}$. Lakes in calcareous regions are rich in bicarbonate (>90% of total salt content) and are well buffered. In many of these lakes, calcite precipitation is a regular phenomenon, forming large carbonate platforms ('Seekreide': Ruttner 1962) along the littoral zone in several cases. Photosynthesis of plankton and benthic algae withdraws nutrients and, especially, carbon dioxide and bicarbonate from surface waters, resulting in a rise of pH and subsequent oversaturation of $CaCO_3$. Carbonate then precipitates from the epilimnion in the form of crystals and aggregates. This biogenic epilimnetic decalcification affects the entire lake area during summer. According to Schneider et al. (1987), this process has been taking place throughout the 13,000 years of the post-glacial period.

Lakes in predominantly siliceous areas contain soft waters with low conductivity and alkalinity. Their total salt contents are usually less than 40 mg L$^{-1}$. These lakes are sensitive to acid precipitation.

Oxygen concentrations are more or less uniformly distributed with depth in unproductive alpine lakes (orthograde curve), whereas more productive lakes develop clinograde curves during the summer (oxygen concentrations decreasing with depth). In some cases, the deeper parts of the hypolimnion become deoxygenated towards the end of the growing season, owing to oxygen consumption during the mineralisation of sedimenting particles. Oxygen depletion may also occur in winter under ice cover, usually as a function of mean depth and trophic level, as observed elsewhere (Mathias & Barica 1979).

The orthograde and the clinograde oxygen distribution are two extremes of a continuum between the oligotrophic and eutrophic states. Deviations from these two types are not uncommon in alpine lakes (Ruttner 1962). Most commonly, either metalimnetic maxima or minima occur (positive or negative heterograde curves). Supersaturation in the metalimnion is mainly due to photosynthesis by phytoplankton in this stabilised layer (Fig. 6.5a). The filamentous cyanoprokaryote, *Planktothrix rubescens*, is very often responsible for the pronounced metalimnetic oxygen maxima which have been frequently reported from many lakes in the European Alps (Findenegg 1973). Among smaller lakes, such as Lunzer Obersee, oxygen supersaturation in the metalimnion can occur as a result of intense photosynthesis of macrophytes on the slope at the depth of metalimnion (Ruttner 1962).

Metalimnetic minima develop when the oxygen uptake in this stratum is greater than in the upper hypolimnion. The most widely accepted theory is that it represents the result of the accumulation and mineralisation of oxidisable material. However, in some instances metalimnetic oxygen deficits have developed during recovering phases from eutrophication (e.g. Ammersee) and cannot readily be explained (Kucklentz et al. 2001). Sometimes, both types of heterograde curves may be observed at different times in the same lake (Fig. 6.5b).

Permanent deoxygenation is characteristic of stagnant monimolimnia of meromictic lakes, even when productivity is quite modest (Findenegg 1934). The deep water may also be permanently devoid of oxygen and considerable variation in $O_2$ concentration may occur at lesser depths, depending on the extent and frequency of convective circulation. In particular, larger lakes can be considerably ventilated in this way, as long as no ice cover develops. Such cases have been reported from lakes in Carinthia, Austria (e.g. Wörthersee: Sampl & Schulz 1996).

## 6.4 PLANKTON

Corresponding to the large variation in size, depth, temperature regime and lake type, alpine phytoplankton assemblages differ extensively in their biomass and species composition. However, some common features may be deduced from the published observations on the lakes of the European Alps.

Annual periodicity of phytoplankton in subalpine and alpine lakes is usually characterised by a smaller spring and a larger late summer to early autumn maximum, often made up largely by diatoms (Dokulil & Skolaut 1986). As a consequence of zooplankton development and increased feeding activity, the spring peak of phytoplankton is followed in most cases by a period of low algal biomass and greater transparency, known as 'the clear water phase'. In addition to zooplankton activity, the exhaustion of soluble reactive silica and fungal parasitism contribute to the rapid decline of the spring diatom populations.

The timing and composition of the second peak

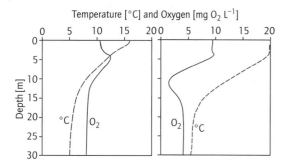

Fig. 6.5 Oxygen and temperature distribution on two occasions in Ossiachersee showing metalimnetic maximum (left panel) and metalimnetic minimum (right panel). (Modified from Ruttner 1962; after Findenegg (unpublished).)

are more variable. In Lago Maggiore, for instance, the July to August period is characterised by cyanoprokaryotes (Ruggio et al. 1993). In contrast, summer peaks in Lac Léman and Ammersee, Bavaria, are made up largely of Dinoflagellates, especially *Ceratium hirundinella* (Steinberg 1980, Pelletier et al. 1987, Revaclier et al. 1988). Similarly, *Ceratium* is a main component of the summer peaks in Lake Constance but with significant additions of larger diatoms such as *Asterionella* and *Fragilaria* (Sommer 1981). Cryptophytes, represented by *Rhodomonas lacustris* and a variety of *Cryptomonas* species, are important components in these lakes throughout the year, as they are in lakes of the Austrian Salzkammergut (Dokulil 1992). Green algae are never very important, at least for so long as the trophic level remains low (Dokulil 1991a). Other algal groups of importance are Chrysophyceae, mainly species of *Dinobryon* (Dokulil & Skolaut 1991). An example of a characteristic species sequence in an alpine lake is shown in Fig. 6.6, drawn from data from Mondsee 1982–1984. The general pattern is more or less similar every year but the species involved at certain periods can vary from year to year. Cryptophytes are important components during late autumn and winter (Dokulil 1988). In spring, there is a characteristically rapid development of diatoms, mainly *Asterionella* or *Aulacoseira*. *Planktothrix* and *Ceratium* undergo their main development during summer. Species of *Dinobryon* appear at various times each year. This is an opportunist taxon capable of rapid biomass development when a window of suitable conditions opens, such as the breakdown of a preceding population. In late autumn, diatoms usually appear again. The sequence of events briefly described here is, with modifications in species composition, representative for most of the larger mesotrophic lakes in the Alps.

The spring maximum of diatoms is larger than the autumn peak in some lakes (e.g. Traunsee), or elsewhere, in particular years (e.g. Mondsee in 1983). In ultra-oligotrophic lakes, such as Königssee in Bavaria and Attersee in Austria, only one phytoplankton maximum is observed, usually largely composed of small centric diatoms.

Interactions among components of the food web and their environment in alpine lakes have been described (e.g. Morscheid & Morscheid 2001). In general, phytoplankton is, in most near-natural cases, resource-controlled from the 'bottom up'. On the other hand, some of the lakes contain great stocks of plankton-feeding fish which effectively control the zooplankton from the 'top down'. Cascading trophic interactions are not uncommon.

Phytoplankton is usually sparse in high alpine lakes because of low productivity and poor nutrient support. As in Arctic lakes, the species assemblage consists of very small forms, particularly small centric diatoms, desmids and certain dinoflagellates such as *Peridinium*. Many of the most abundant algal species are adapted to very low light intensities and can therefore thrive even under a thick ice-cover for prolonged times. Some are mixotrophic, facultatively exploiting some organic carbon sources. During the ice-free season, maximum development of mixotrophs occurs at depth and generally peaks during July or August. If the lakes are not very deep, as is the case in Gossenköllesee ($z_m = 9.9$ m), phytoplankton abundance is greatest close to the lake bottom.

As also in the Arctic, a low level of primary production supports only a limited zooplankton fauna. In Vorderer Finstertaler See, two rotifers (*Polyarthra dolichoptera* and *Keratella hiemalis*), together with the copepod *Cyclops abyssorum tatricus*, form 90% of the zooplankton biomass. Similarly, only *Cyclops* and *Polyarthra* inhabit Gossenköllesee, where the older development stages of *Cyclops a. tatricus* live near the bottom. Their food is mainly benthic, with a large contribution from nematodes (Bretschko 1975).

## 6.5 PRIMARY PRODUCTION

Primary production of phytoplankton depends mainly on the photosynthetic rates attained per unit volume, which are limited by nutrient supply, and is under control of underwater light availability and temperature. In addition, size-related strategies and pigment distribution among phytoplankton assemblages are key factors influencing

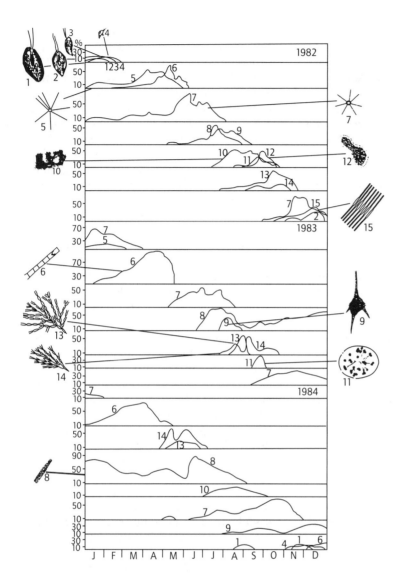

Fig. 6.6 Seasonal sequence of phytoplankton species as relative biomass contribution in Mondsee, Austria. Species are as follows: 1, *Cryptomonas* cf. *ovata*; 2, *C. pusilla*; 3, *C. marsonii*; 4, *Rhodomonas minuta* var. *nanoplanktica*; 5, *Asterionella formosa*; 6, *Aulacoseira islandica*; 7, *Tabellaria flocculosa* var. *asterionelloides*; 8, *Planktothrix rubescens*; 9, *Ceratium hirundinella*; 10, *Microcystis* spp.; 11, *Gomphosphaeria* sp.; 12, *Aphanotheca clathrata*; 13, *Dinobryon divergens*; 14, *D. sociale*; 15, *Fragilaria crotonensis*.

the shape of vertical profiles (Teubner *et al.* 2001). Several types of vertical profile are realised among the lakes considered here, depending on lake mixing, algal distribution and trophic level (Fig. 6.7). Clear oligotrophic lakes (e.g. Attersee, Lunzer Untersee, Königssee) support low community photosynthetic rates but they extend to considerable depth, giving atypical profiles. Mesotrophic lakes (e.g. Obertrumersee, Ossiachersee, Mondsee) support faster rates of photosynthesis over a truncated depth range, yielding the rather more familiar depth distributions with a maximum at a certain distance below the surface (Fig. 6.7). Two maxima appear in lakes with well-developed metalimnetic algal populations, such as Wörthersee (Fig. 6.7), a smaller one near the surface and a much larger one in the layer where, in this case, *Planktothrix rubescens* flourishes. Meromictic

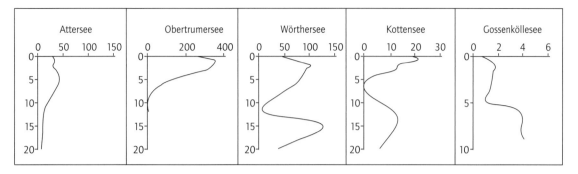

**Fig. 6.7** Depth distribution of photosynthetic rates as mg C m$^{-3}$ day$^{-1}$ for alpine lakes in Austria of different trophic conditions. (Original data except Gossenköllesee, which was redrawn from Khan 1981.)

lakes, such as Krottensee, are characterised by low epilimnetic rates of photosynthesis but exhibit a second peak of carbon-uptake rates near the oxycline (Fig. 6.7), as a consequence of the activity of chemosynthetic microorganisms.

In the high alpine lakes Gossenköllesee (Fig. 6.7), Schwarzsee and Drachensee, photosynthetic rates increase with depth corresponding with maxima of the phytoplankton biomass near the bottom (Rodhe et al. 1966; Khan 1981). Similar observations have been recorded in other high altitude lakes of the Alps (e.g. Schneider 1984). The disproportion between nutrient availability and high radiation is responsible for this distribution (Pechlaner 1971; Witt 1977). The high intensity of UV radiation might also be an influential factor. In many of the lakes where there is also a rich algal development on the well-illuminated lake bottom, production at the bottom is higher than the production in the pelagial.

Daily rates of primary productivity in the euphotic zones of several alpine lakes, ranging from ultra-oligotrophic to eutrophic, are summarised in Fig. 6.8. As would be expected, rates increase in general and become more variable along the trophic gradient represented by the lakes selected. However, in some cases, average productivity is higher than would be anticipated from the estimated trophic level.

## 6.6 BENTHOS

Submerged macrophytes such as *Myriophyllum*, *Potamogeton* and *Elodea* occur near the shoreline of most alpine lakes. Deeper regions of the littoral zone are inhabited by Characeans which are substituted by the moss *Fontinalis* at greater depth.

The benthos of the deep oligotrophic alpine lakes is usually composed largely of oligochaetes. Other groups of importance are chironomids, nematodes and, sometimes, ostracods. Chironomid larvae are the only group of insects inhabiting the deep profundal (Pesta 1929). Molluscs are quite important in some alpine lakes, e.g. Traunsee. The aquatic larvae of the phantom midges, *Chaoborus* spp., are often found in smaller or meromictic lakes where they live in the substratum during the day. During the night, they rise through the water column where they feed on zooplankton. Their predation can produce a marked effect on the other components of the zooplankton. The composition of the benthic community varies with the type of substratum, both within and between lakes. However, species richness is largely reduced when lakes become enriched with nutrients. In high alpine lakes, such as Gossenköllesee, chironomids overwhelmingly make up the benthos (Zaderer 1980). Mass appearance of trichopteran larvae in the littoral is a common phenomenon in many lakes in the high Alps (Turnowsky 1976). Only a few mollusc species occur at these altitudes,

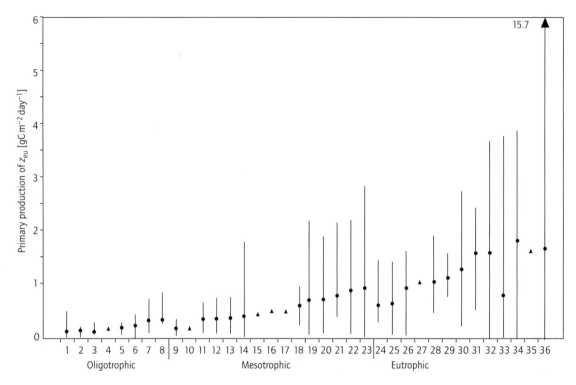

**Fig. 6.8** Average, maximum and minimum of euphotic zone production per day given as g C m$^{-2}$ day$^{-1}$ for a number of alpine lakes. Triangles indicate single observations. Lakes are arranged according to independent estimates of trophic level. Compiled from various authors, mainly from Fricker (1980), with additions of original data. Oligotrophic: 1, Finstertalersee; 2, Gossenköllesee; 3, Lunzer Obersee; 4, Traunsee; 5, Attersee; 6, Königssee; 7, Lago di Montorfano; 8, Lunzer Untersee. Mesotrophic: 9, Piburgersee; 10, Längsee; 11, Lac d' Annecy; 12, Ossiachersee; 13, Mondsee; 14, Lago Mergozzo; 15, Wolfgangsee; 16, Wörthersee; 17, Jeserzersee; 18, Lago di Annone; 19, Walensee; 20, Vierwaldstättersee; 21, Lago Maggiore; 22, Zürichsee Untersee; 23, Lac Léman. Eutrophic: 24, Lac Nantua; 25, Zürichsee Obersee; 26, Halwilersee; 27, Goggausee; 28, Lago del Segrino; 29, Lago di Pusiano; 30, Lago di Oggiono; 31, Lago d' Alserio; 32, Greifensee; 33, Bodensee; 34, Lago Lugano; 35, Obertrumer See; 36, Sempachersee.

of which the genus *Pisidium* is of greatest importance.

## 6.7 FISH

The main fish species occurring in high alpine lakes is brown trout (*Salmo trutta forma fario*), sometimes in company with bullhead (*Cottus gobio*) and minnow (*Phoxinus phoxinus*) (Steinböck 1951). They feed on the bottom of the lakes. Their diet is augmented by surface animals during the ice-free period and by zooplankton. In many cases, introduced fish grow only very slowly and remain as small-appetite forms (Steinböck 1950, Pechlaner 1966). In addition, the arctic char (*Salvelinus alpinus*) has been introduced into many high mountain lakes, where it feeds mainly on chironomid larvae (Cavalli & Chappaz 1996).

Alpine lakes with moderate water temperatures, and which are also rich in oxygen, are characterised by populations of arctic char, although various morphotypes are known, either feeding principally on zooplankton or carnivorously on other fish. In some lakes (e.g. Attersee, Lake Constance), a deep-living morphotype feeds on benthic animals.

White fish form a second main component of the fish populations in the larger alpine lakes such as Lac Léman, Königssee and Mondsee. These coregonids also appear in a variety of morphotypes, which, historically, have been classified as different species. Today all these forms are attributed to *Coregonus lavaretus* (Wanzenböck *et al.* 2002). Additional species are the lake form of brown trout (*Salmo trutta* forma *lacustris*), pike (*Esox lucius*) and some Cyprinid species. The more shallow pre-alpine lakes in which oxygen depletion may occur at depth are inhabited by a variety of Cyprinids, most commonly bream (*Abramis brama*) and roach (*Rutilus rutilus*). These distributions have been further differentiated by lake types by several authors, although not always consistently (Haempel 1930; Spindler 1995). In general, species diversity increases with the trophic status of the alpine lakes and varies inversely with altitude. Diversity is low among oligotrophic high-alpine lakes, it increases at decreasing elevations and is greatest in the richer pre-alpine lakes (Gassner *et al.* 2002).

## 6.8 HUMAN IMPACTS

Fresh-waters are influenced by anthropogenic activities world-wide (see also Chapter 2). Impacts of significant importance to alpine lakes include the catchment-oriented process of eutrophication, the transboundary problem of acidification and the global effects of climatic change.

Acidification and eutrophication share some effects in common (such as changes in species composition, extinction of sensitive species and toxicity). Restoration measures are directed towards load reduction (nutrient, acidity) in either case. Nevertheless, the sources, sites and mechanisms differ (cf. table 1 of Psenner 1994) and the reactions to the two problems are strikingly different. Although the effects of acid rain on water bodies were clearly and precisely described by Smith (1872), more than a century ago, countermeasures were not initiated before 1970. Schemes to restore lakes enriched with nutrients have been initiated within only one decade of its diagnosis as a major problem.

Global climatic changes will affect the perialpine area and its lakes in several ways (Livingstone & Dokulil 2001). Local precipitation patterns, runoff, erosion and temperature distribution will change drastically as the climate becomes warmer and drier (Dokulil *et al.* 1992). In turn, this will affect in-lake processes. Ultraviolet radiation flux in alpine regions has already increased significantly as a result of atmospheric ozone depletion (Blumthaler & Ambach 1990).

### 6.8.1 Eutrophication

Effects of increased nutrient input as a result of human activity were recognised in alpine lakes early in the nineteenth century. De Candolle (1825) described a mass appearance of *Oscillatoria (Planktothrix) rubescens* from a lake in Switzerland. A few years later this filamentous cyanoprokaryote appeared in several other Swiss lakes, such as Zürichsee and Vierwaldstättersee. According to Findenegg (1973), *Oscillatoria* first appeared in Austria around 1909, establishing a large population in the Wörthersee. The increased discharge of untreated nutrients, as a consequence of rapidly expanding tourism in the 1950s, and elevated phosphorus concentrations in the lakes helped the alga to spread to many other places throughout the alpine region (Thomas 1969; Findenegg 1973). In Mondsee, for example, the organism was first observed in 1968, bloomed in the following years and began to decline significantly after sewage works began to operate in 1973 (Dokulil 1987; see Fig. 6.9). Changes in the phytoplankton structure were discernible also in the amount and species composition of diatoms as evidenced from plankton data and subfossil records from the sediment (Dokulil 1991b; Schmidt 1991). Elevated biomass drastically decreased light transmission and, hence, Secchi-disc depth, as demonstrated by Findenegg (1972). Increased hypolimnetic mineralisation of the enhanced phytoplankton biomass led consequently to oxy-

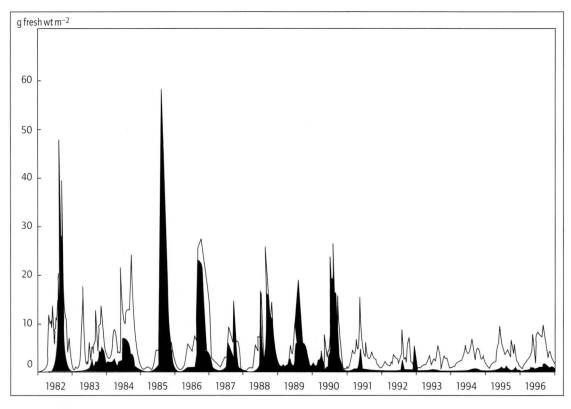

Fig. 6.9 Phytoplankton development in Mondsee, Austria from 1957 to 1996 as g fresh wt m$^{-2}$. The shaded area shows the contribution of the filamentous cyanoprokaryote *Planktothrix* (*Oscillatoria*) *rubescens*. (The 1957 to 1980 data are compiled from various sources. For identification of species consult Dokulil (1987). Data from 1982 to present are original data.)

gen depletion in the deeper parts of the lake and to complete deoxygenation of large areas of bottom sediment. Species composition and biomass of zooplankton and fish populations also changed drastically (Jagsch 1982; Nauwerck 1988). During the eutrophication, peak numbers of *Daphnia cucculata* and rotifers, especially *Keratella cochlearis*, also largely increased. The arctic char (*Salvelinus alpinus*) began to decline during the late 1950s, primarily in response to the influx of large amounts of inorganic material washed into the lakes during highway construction. Correspondingly, populations of Cyprinids and Coregonids increased. Today *Chalcalburnus chalcoides mento* (Danube bleak) makes up the majority of the fish fauna, and occupies most of the trophic relationships in Mondsee (Dokulil et al. 1990). Although most direct effects of tourism (Dokulil & Jagsch 1989) have been minimised and the lake is now in a stage of re-oligotrophication (Dokulil 1984), after effects of the eutrophic period are still recognisable. Mass-appearances of opportunistic algal populations, such as *Microcystis* or *Dinobryon*, occur at irregular intervals when conditions are optimal (Dokulil & Skolaut 1991).

Similar symptoms of eutrophication and recovery have been observed in many other alpine lakes (Thomas 1969; Ambühl 1980; Fricker 1980; Sas 1989). Reports of delayed recovery from eutrophication are mainly attributable either to significant phosphorus release from the sediment or to large residual nutrient inputs derived from diffuse sources, in most cases intensively used agricultural land.

Evaluation of restoration measures applied to eutrophicated lakes has been made possible through the development of the critical load concept by Vollenweider (1968, 1976), which has been applied successfully to many alpine lakes throughout the region.

### 6.8.2 Acidification

Lakes become acidic when the input of hydrogen ions exceeds the amount of bases produced in the watershed, by the weathering of rocks, or in the lake itself, through the reduction of acid anions, such as sulphate and nitrate. By definition, acid waters possess a pH of 5.5 or lower. Acidification occurs in siliceous catchments with low rates of weathering. Therefore, acidic lakes occur in regions of the western and eastern Central Alps, especially at high altitudes (e.g. Mosello 1984; Psenner 1989; Kenttämies & Merilehto 1990; Marchetto et al. 1995). Areas affected in Austria include lakes on granitic bedrock at altitudes above 1500 m in the provinces of Tyrol and Carinthia. In Italy, 123 of the 630 high alpine lakes surveyed were affected by or sensitive to acidification. Similarly, 37 lakes in the Canton Ticino, Switzerland, are reported to be acidified.

Under alpine conditions, the deposition of atmospheric pollutants is strongly influenced by local circumstances. However, the spatial pattern of pollutant deposition is similar to that of annual precipitation, which is, in turn, governed largely by altitude and screening by windward mountains. Therefore, precipitation maxima, which increase strongly with altitude, occur at the outer chains of the Alps. The interior of the Alps is considerably drier and altitude plays a minor role below 2000 m. Annual sums of precipitation increase sharply between 2000 and 3000 m above sea level (Kuhn 1990), whilst weathering rates decrease with increasing elevation (Zobrist & Drever 1990).

The concentrations of acids and nutrients in precipitation are therefore considerably lower in the Central Alps than they are at points along the northern and southern borders of the Alps. High alpine lakes are exposed to low or moderate acid loads. Moreover, frequent episodic deposition of atmospheric dust (mainly from the Sahara) can be sufficient to neutralise large amounts of acids (Psenner & Nickus 1986). Contrary to many other lakes in areas affected by acid deposition, such as Scandinavia and North America, high lakes in the European Alps receive considerable amounts of sulphate from the watershed. The chemical composition of 210 high alpine lakes is summarised in Table 6.5. Calcium is most abundant among the cations and sulphate is the leading anion. Nitrate proved to be quantitatively important as an acid

**Table 6.5** Ionic composition of high alpine lakes from the Alps in siliceous catchments. (From Psenner 1994; after Mosello et al. 1992.)

| Variable | Unit | Mean | Minimum | Median | Maximum |
|---|---|---|---|---|---|
| Conductivity | $\mu S_{18}\,cm^{-1}$ | 22 | 5 | 18 | 24 |
| pH |  | 6.57 | 4.64 | 6.59 | 8.51 |
| Alkalinity | $\mu Equiv\,L^{-1}$ | 95 | −23 | 51 | 1521 |
| Calcium | $\mu Equiv\,L^{-1}$ | 152 | 15 | 109 | 2495 |
| Magnesium | $\mu Equiv\,L^{-1}$ | 34 | 3 | 21 | 526 |
| Sodium | $\mu Equiv\,L^{-1}$ | 17 | 2 | 15 | 106 |
| Potassium | $\mu Equiv\,L^{-1}$ | 10 | 0 | 8 | 123 |
| Sulphate | $\mu Equiv\,L^{-1}$ | 91 | 17 | 67 | 141 |
| Nitrate | $\mu Equiv\,L^{-1}$ | 18 | 0 | 18 | 51 |
| Chloride | $\mu Equiv\,L^{-1}$ | 6 | 0 | 5 | 50 |
| Silica | $\mu mol\,L^{-1}$ | 32 | 1 | 27 | 231 |

anion, in contrast to earlier observations in Norway.

More than 20% of the lakes are below pH 6. Two-thirds of all the lakes and more than 80% of lakes in acidic catchments possessed alkalinities $<200\,\mu$Equiv L$^{-1}$ and were therefore classified as sensitive to acidification. One-third of all lakes and 50% of lakes in the acidic catchments possessed alkalinities of $<50\,\mu$Equiv L$^{-1}$ and were regarded as 'very sensitive' (Psenner 1989, 1994). Although the generally accepted hypothesis is that alkalinity in high altitude lakes is generated by chemical weathering in the watershed, in-lake production can contribute significantly (Psenner & Zapf 1990).

Effects of acidification on the biota are highly variable and group specific. Phytoplankton numbers and biomass are not correlated to pH or alkalinity but species composition and abundance ratio are clearly influenced. Diatoms and Chrysophytes prefer lakes with pH values > 6.0. Dinophyceae are most abundant in lakes with pH < 6.0. The zooplankton is characterised by small numbers of species which possess no clear pH dependency. For several sensitive species of molluscs, crustaceans and some fish, a pH of 5.0 represents a critical limit. Below pH 5.0, acidity is harmful to eggs and fry of salmonids. Below 4.0, fish can scarcely survive at all (Lenhart & Steinberg 1984; Jørgensen 1992).

Estimation of the potential impact of atmospheric depositions on sensitive systems can be achieved by evaluating critical acid deposition loads, analogous to the critical load calculations used in eutrophication studies (Marchetto et al. 1994).

### 6.8.3 Impact of ultraviolet radiation

Ozone reduction in the stratosphere, discovered over both poles, has also been reported at mid-latitudes (e.g. Madronich et al. 1995). Blumenthaler & Ambach (1990) found a trend of increasing UV-B radiation in the Swiss Alps with an additional elevation effect of between 11 and 19% per 1000 m (Blumenthaler et al. 1992, 1993). The effect is wavelength-dependent and is much greater for UV-B (280–320 nm) than for UV-A (320–400 nm). Many alpine lakes, especially at high altitudes, are therefore potentially exposed to increased fluxes of UV radiation (Williamson 1995).

Attenuation of both UV-A and UV-B radiation in temperate freshwater lakes is largely influenced by dissolved organic carbon (DOC) and, in certain cases, by phytoplankton (Kirk 1994; Scully & Lean 1994; Sommaruga & Psenner 1997). Although transmittance of photosynthetically active radiation (PAR) and UV in ice- and snow-covered high alpine lakes is reduced to extremely low values, UV penetration during the ice-free season is considerable, because of high transparency and low DOC concentrations.

Organisms living in such ecosystems have evolved a variety of strategies to cope with high UV values. These have been summarised by Karentz et al. (1994), Siebeck et al. (1994) and Sommaruga (2001). However, further increases in radiation associated with changes in spectral composition impose additional stresses. Although motile phytoflagellates and zooplankton can avoid exposure to damaging amounts of radiation through vertical migration (Rodhe et al. 1966; Rott 1988; Siebeck & Böhm 1994), periphyton on rocks in alpine lakes is particularly sensitive to UV radiation, as demonstrated by Vinebrooke & Leavitt (1996). The biologically damaging effects of UV-B radiation have been reviewed by Madronich (1994).

## 6.9 ACKNOWLEDGEMENTS

The manuscript has improved considerably through stimulating discussions with several scientific colleagues. Particularly, the help by Joseph Wanzenböck with alpine fish and by Katrin Teubner who critically read and discussed the manuscript at various stages is kindly acknowledged.

## 6.10 REFERENCES

Ambühl, H. (1980) Eutrophication of alpine lakes. *Progress in Water Technology*, **12**, 89–101.

Anderson, N.J. (1993) Natural versus anthropogenic change in lakes: The role of the sediment record. *Trends in Ecology and Evolution*, **8**, 356–61.

Baumgartner, A., Reichel, E. & Weber, G. (1983) *Der Wasserhaushalt der Alpen*. Oldenbourg Verlag, München, 343 pp.

Blumenthaler, M. & Ambach, W. (1990) Indication of increasing solar ultraviolet-B radiation flux in Alpine regions. *Science*, **248**, 206–8.

Blumenthaler, M., Ambach, W. & Huber, M. (1993) Altitude effect of solar UV radiation dependent on albedo, turbidity and solar elevation. *Meteorologische Zeitschrift*, **2**, 116–20.

Blumenthaler, M., Ambach, W. & Rehwald, W. (1992) Solar UV-A and UV-B radiation flux at two alpine stations at different altitudes. *Theoretical and Applied Climatology*, **46**, 39–44.

Bretschko, G. (1975) Annual benthic biomass distribution in a high mountain lake (Vorderer Finstertaler See, Tyrol, Austria). *Verhandlungen der internationalen Vereinigung für Limnologie*, **19**, 1279–85.

Brunner, A. (1849) Recherches sur la température du lac de Thoune. *Mémoires de la Société de Physiques et d' Historie naturel de Genève*, 1849.

Cavalli, L. & Chappaz, R. (1996) Diet, growth and reproduction of the Arctic char in a high alpine lake. *Journal of Fish Biology*, **49**, 953–64.

De Candolle, A.P. (1825) Notice sur la matière qui a colorè en rouge de lac de Morat au printemps de 1825. *Mémoires Société Physiques et d' Historie naturel de Genève*, 1825.

Dokulil, M.T. (1984) Die Reoligotrophierung des Mondsees. *Laufener Seminarbeiträge Bayerische Akademie für Naturschutz und Landschaftspflege*, **2**, 46–53.

Dokulil, M.T. (1987) Long-term occurrence of blue-green algae in Mondsee during eutrophication and after nutrient reduction, with special reference to *Oscillatoria rubescens*. *Schweizerische Zeitschrift für Hydrologie*, **49**, 378.

Dokulil, M.T. (1988) Seasonal and spatial distribution of cryptophycean species in the deep, stratifying alpine lake Mondsee and their role in the food web. *Hydrobiologia*, **161**, 185–201.

Dokulil, M.T. (1991a) Contribution of green-algae to the phytoplankton assemblage in a mesotrophic lake, Mondsee, Austria. *Archiv für Protistenkunde*, **139**, 1–4.

Dokulil, M.T. (1991b) Populationsdynamik der Phytoplankton-Diatomeen im Mondsee seit 1957. *Wasser and Abwasser*, **35**, 53–75.

Dokulil, M.T. (1992) Langzeitveränderungen des Phytoplanktons in Alpenseen. In: *DGL, Erweiterte Zusammenfassungen der Jahrestagung 1991 in Mondsee, Österreich*. Frank GmbH, München, 13–17.

Dokulil, M.T. (2001). Physikalische und chemische Umwelt. In: Dokulil, M.T., Hamm, A. & Kohl, J.-G. (eds), *Ökologie und Schutz von Seen*. Facultas, Wien, 47–67.

Dokulil, M.T. & Jagsch, A. (1989) Some aspects of the impact of tourism on Mondsee, Austria. *Symposia Biologica Hungarica*, **38**, 415–28.

Dokulil, M.T. & Skolaut, C. (1986) Succession of phytoplankton in a deep stratifying lake: Mondsee, Austria. *Hydrobiologia*, **138**, 9–24.

Dokulil, M.T. & Skolaut, C. (1991) Aspects of phytoplankton seasonal succession in Mondsee, Austria, with particular reference to the ecology of *Dinobryon* Ehrenb. *Verhandlungen der Internationale Vereinigung für theoretische und angewandte Limnologie*, **24**, 968–73.

Dokulil, M.T., Herzig, A. & Jagsch, A. (1990) Trophic relationships in the pelagic zone of Mondsee, Austria. *Hydrobiologia*, **191**, 199–212.

Dokulil, M.T., Humpesch, U.H., Pöckl, M. & Schmidt, R. (1992) Auswirkungen geänderter Klimaverhältnisse auf die Ökologie von Oberflächengewässer in Österreich. In: Kommission Reinhaltung der Luft der Österreichische Akademie des Wissenschaften (eds), *Bestandsaufnahme—Anthropogene Klimaänderungen: Mögliche Auswirkungen auf Österreich—Mögliche Massnahmen in Österreich.*, BMfWF und BMfUJF, Wien, 5.1–5.14.

Dokulil, M.T., Schwarz, K. & Jagsch, A. (2000) Die Reoligotrophierung österreichischer Seen: Sanierung, Restaurierung und Nachhaltigkeit. Ein Überblick. *Münchner Beiträge für Abwasser-, Fischerei- und Flussbiologie*, **53**, 307–21.

Eckel, O. (1955) Statistisches zur Vereisung der Ostalpenseen. *Wetter und Leben*, **3/4**, 49–57.

Fahn, H.J. (1981) Birth and decline of the lakes of the alpine foothills. *Wasserwirtschaft*, **71**, 224–8.

Findenegg, I. (1934) Zur Frage der Entstehung pseudoeutropher Schichtungs-Verhältnisse in den Seen. *Archiv für Hydrobiologie*, **27**, 621–5.

Findenegg, I. (1935) Limnologische Untersuchungen im Kärntner Seengebiet. Ein Beitrag zur Kenntnis des Stoffhaushaltes in Alpenseen. *Internationale Revue der gesamten Hydrobiologie*, **32**, 369–423.

Findenegg, I. (1972) Die Auswirkungen der Eutrophierung einiger Ostalpenseen auf die Lichttransmission ihres Wassers. *Wetter und Leben*, **24**, 110–8.

Findenegg, I. (1973) Occurrence and biological behaviour of the blue-green alga *Oscillatoria rubescens* in the Austrian alpine lakes. *Carinthia II*, **163**, 317–30. (In German.)

Fricker, H.J. (1980) *OECD-Eutrophication Programme: Regional Project—Alpine Lakes*. Swiss Federal Board for Environmental Protection, Bern, 233 pp.

Gassner, H., Wanzenböck, J. & Tischler, G. (2002) Towards a new ecological integrity assessment method for lakes using fish communities—suggestions from two Austrian pre-alpine lakes. *Freshwater Biology*, **47**, 635–652.

Haempel, O. (1930) *Fischereibiologie der Alpenseen*. Die Binnengewässer Bd. 10, Schweizerbart'sche Verlagsbuchhandlung, Stuttgart, 259 pp.

Herdendorf, C.E. (1990) Distribution of the world's large lakes. In: Tilzer, M.M. & Serruya, C. (eds), *Large Lakes. Ecological Structure and Function*. Springer-Verlag, Berlin, 1–38.

Hutchinson, G.E. (1957) *A Treatise on Limnology*, Vol. I, *Geography, Physics and Chemistry*. John Wiley & Sons, New York, 1016 pp.

Jagsch, A. (1982) Mondsee. In: *Seenreinhaltung in Österreich. Schriftenreihe Wasserwirtschaft*, **6**, 155–63.

Jørgensen, S.E. (1992) Effects of lake acidification. In: Jørgensen SE. (ed.), *Management of Lake Acidification*. Guidelines of Lake Management, Vol. 5, International Lake Environment Committee, Lake Biwa, Japan, 47–69.

Karentz, D., Bothwell, M.L., Coffin, R.B., et al. (1994) Impact of UV-B radiation on pelagic freshwater ecosystems: Report of working group on bacteria and phytoplankton. *Archiv für Hydrobiologie Beihefte (Ergebnisse der Limnologie)*, **43**, 31–69.

Kenttämies, K. & Merilehto, K. (1990) The geographical extent of acidification of lakes and streams in the ECE region. In: Johannessen, M., Mosello, R. & Barth, H. (eds), *Acidification Processes in Remote Mountain Lakes*. Air Pollution Research Report No. 20, European Economic Community, Brussels, 1–10.

Khan, M.A. (1981) Summer phytoplankton measurements of two lakes from Austrian Alps using carbon-14 technique. *Jahresbericht der Abteilung für Limnologie Innsbruck*, **7**, 192–195.

Kirk, J.T.O. (1994) Optics of UV-B radiation in natural waters. *Archiv für Hydrobiologie Beihefte (Ergebnisse der Limnologie)*, **43**, 1–16.

Kucklentz, V., Hamm, A., Jöhnk, K., et al. (2001) *Antwort bayerischer Voralpenseen auf verringerte Nährstoffzufuhr*. Informationsberichte 2/2001, Bayerisches Landesamt für Wasserwirtschaft, München, 272 pp.

Kuhn, M. (1990) Meteorology and pollution processes in the Alps. In: Johannessen, M., Mosello, R. & Barth, H. (eds), *Acidification Processes in Remote Mountain Lakes*. Air Pollution Research Report No. 20, European Economic Community, Brussels, 11–21.

Lenhart, B. & Steinberg, C. (1984) *Gewässerversauerung. Limnologie für die Praxis. Grundlagen des Gewässerschutzes*. 2. Aufl. Ecomed Verlagsgesellschaft, Landsberg, München, 62 pp.

Livingstone, D.M. & Dokulil, M.T. (2001) Eighty years of spatially coherent Austrian lake surface temperatures and their relationship to regional air temperatures and to the North Atlantic Oscillation. *Limnology & Oceanography*, **46**, 1220–7.

Madronich, S. (1994) Increases in biologically damaging UV-B radiation due to stratospheric ozone reductions: a brief review. *Archiv für Hydrobiologie Beihefte (Ergebnisse der Limnologie)*, **43**, 17–30.

Madronich, S., McKanzie, R.L., Caldwell, M. & Björn, L.O. (1995) Changes in ultraviolet radiation reaching the earth's surface. *Ambio*, **24**, 143–52.

Marchetto, A., Mosello, R., Psenner, R., et al. (1994) Evaluation of the level of acidification and the critical loads for alpine lakes. *Ambio*, **23**, 150–4.

Marchetto, A., Mosello, R., Psenner, R., et al. (1995) Factors affecting water chemistry of alpine lakes. *Aquatic Sciences*, **57**, 81–9.

Mathias, J.A. & Barica, J. (1979) Factors controlling oxygen depletion in ice-covered lakes. *Canadian Journal of Fisheries and Aquatic Sciences*, **37**, 185–94.

Meybeck, M. (1995) Global distribution of lakes. In: Lerman, A., Imboden, D.M. & Gat, J.R. (eds), *Physics and Chemistry of Lakes* (2nd edition), Springer-Verlag, Berlin, 1–35.

Morscheid, M. & Morscheid, M. (2001) Ökosystemare Zusammenhänge am Beispiel des Ammersees. In: Dokulil, M.T., Hamm, A & Kohl, J.-G. (eds), *Ökologie und Schutz von Seen*. Facultas, Wien, 385–400.

Mosello, R. (1984) Hydrochemistry of high altitude alpine lakes. *Schweizerische Zeitschrift für Hydrologie*, **46**, 86–99.

Mosello, R., Marchetto, A., Boggero, A., et al. (1992) *Quantification of the Susceptibility of Alpine Lakes to Acidification*. Final Report to the European Economic Community (Project CNR EV4V-0114, 1988-1991), Brussels, 75 pp.

Nauwerck, A. (1988) Veränderungen im Zooplankton des Mondsees 1943–1988. *Berichte des naturwissenschaftlich-medizinischen Vereins Salzburg*, **9**, 101–33.

Neumann, J. (1959) Maximum depth and average depth

of lakes. *Journal of the Fisheries Research Board of Canada*, **16**, 923–7.

Pechlaner R. (1966) Salmonideneinsätze in Hochgebirgsseen und -tümpel der Ostalpen. *Verhandlungen der Internationalen Vereinigung für Limnologie* **16**, 1182–91.

Pechlaner, R. (1971) Factors that control the production rate and biomass of phytoplankton in high-mountain lakes. *Mitteilungen der Internationalen Vereinigung für theoretische und angewandte Limnologie*, **19**, 125–45.

Pelletier, J., Balvay, G., Druart, J.C. & Revaclier, R. (1987) Evolution du Plancton du Léman, Campagne 1986. *Rapport Commission international protection des eaux Léman contre pollution, Campagne 1986*, 47–68.

Pesta, O. (1929) *Der Hochgebirgssee der Alpen*. Die Binnengewässer VIII, E.Schweizerbart, Stuttgart, 156 pp.

Psenner, R. (1989) Chemistry of high mountain lakes in siliceous catchments of the Central Eastern Alps. *Aquatic Sciences*, **51**, 108–28.

Psenner, R. (1994) Environmental impacts on freshwaters: acidification as a global problem. *The Science of the Total Environment*, **143**, 53–61.

Psenner, R. & Nickus, U. (1986) Snow chemistry of a glacier in the Central Eastern Alps (Hintereisferner, Tyrol, Austria. *Zeitschrift für Gletscherkunde und Glazialgeologie*, **22**, 1–18.

Psenner, R. & Zapf, F. (1990) High mountain lakes in the Alps: Pecularities and biology. In: Johannessen, M., Mosello, R. & Barth, H. (eds), *Acidification Processes in Remote Mountain Lakes*. Air Pollution Research Report No. 20, European Economic Community, Brussels, 22–37.

Revaclier, R., Balvay, G., Druart, J.C. & Pelletier, J. (1988) Evolution du Plancton du Léman, Campagne 1987. *Rapport Commission international protection des eaux Léman contre pollution, Campagne 1987*, 53–75.

Rodhe, W., Hobbie, J.E. & Wright, R.T. (1966) Phototrophy and heterotrophy in high mountain lakes. *Verhandlungen der Internationalen Vereinigung für theoretische und angewandte Limnologie*, **16**, 302–13.

Rott, E. (1988) Some aspects of the seasonal distribution of flagellates in mountain lakes. *Hydrobiologia*, **161**, 159–70.

Ruggio, D., Panzani, P. & Morabito, G. (1993) Indagini sul fitoplancton. Ricerche sull' evoluzione del Lago Maggiore—Aspetti limnologici. In: Comissione Internationale per la protezione delle acque italo-svizzera, *Programma quinquennale 1988–1992*. Istituto Italiano di Idrobiologia, Pallanza, 65–8.

Ruttner, F. (1962) *Grundriss der Limnologie* (3rd edition). Walter de Gruyter & Co., Berlin, 332 pp.

Sampl, H. (1976) Die Seen der Tallagen. In *Die Natur Kärntens*, Part 2. Verlag J. Heyn, Klagenfurt, 165–221.

Sampl, H. & Schulz, L. (1996) Kärntner Seenbericht 1996. *Veröffentlichungen des Kärntner Instituts für Seenforschung*, **11**, 1–166.

Sas, H. (1989) *Lake Restoration by Reduction of Nutrient Loading. Expectations, Experiences, Extrapolations*. Academia Verlag, Sankt Augustin, 497 pp.

Schmidt, R. (1991) Diatomeenanalytische Auswertung laminierter Sedimente für die Beurteilung trophischer Langzeittrends am Beispiel des Mondsees. *Wasser und Abwasser*, **35**, 109–23.

Schneider, J., Müller, J. & Stumm, M. (1987) Die sedimentgeologische Entwicklung des Attersees und des Traunsees im Spät- und Postglazial. *Mitteilungen der Kommission für Quartärforschung der Österreichischen Akademie der Wissenschaften*, **7**, 51–78.

Schneider, U. (1984) Phytoplankton und Primärproduktion in Hochgebirgsseen des Kanton Tessin (Schweiz). In: Zielonkowski, W. & Schumacher, R. (eds), *Ökologie alpiner Seen* (Laufener Seminarbeiträge 2/84). Bayerische Akademie für Naturschutz und Landschafspflege, Laufen, 6–15.

Scully, N.M. & Lean, D.R.S. (1994) The attenuation of ultraviolet radiation in temperate lakes. *Archiv für Hydrobiologie Beihefte (Ergebnisse der Limnologie)*, **43**, 135–44.

Siebeck, O. (1982) *Der Königssee. Eine limnologische Projektstudie*. Nationalpark, Berchtesgaden, 131 pp.

Siebeck, O. (1984) Die Sonderstellung des Königssees unter den Bayerischen Seen. In: Zielonkowski, W. & Schumacher, R. (eds), *Ökologie alpiner Seen* (Laufener Seminarbeiträge 2/84). Bayerische Akademie für Naturschutz und Landschafspflege, Laufen, 77–96.

Siebeck, O. (1989) Königssee—an oligotrophic lake. In: Lampert, W. & Rothaupt, K.O. (eds), *Limnology in the Federal Republic of Germany*. International Association for Theoretical and Applied Limnology, Plön, 32–6.

Siebeck, O. & Böhm, U. (1994) Challenges for an appraisal of UV-B effects upon planktonic crustaceans under natural radiation conditions with non-migrating (*Daphnia pulex obtusa*) and a migrating cladoceran (*Daphnia galeata*). *Archiv für Hydrobiologie Beihefte (Ergebnisse der Limnologie)*, **43**, 197–206.

Siebeck, O., Vail, T.L., Williamson, C.E., et al. (1994) Impact of UV-B radiation on zooplankton and fish in pelagic freshwater ecosystems. *Archiv für Hydrobiologie Beihefte (Ergebnisse der Limnologie)*, **43**, 101–4.

Simony, F. (1850) Die Seen des Salzkammergutes. *Sitzungs-Berichte der Kaiserlichen und Königlichen Akademie der Wissenschaften, Wien, mathematisch-naturwissenschaftliche Klasse*, **9**, 542–66.

Smith, R.A. (1872) *Air and Rain: the Beginnings of Chemical Climatology*. Longman Green, London, 600 pp.

Smol, J.P. (1992) Palaeolimnology: An important tool for effective ecosystem management. *Journal Aquatic Ecosystem Health*, **1**, 49–58.

Sommaruga, R. (2001) Die Rolle der UV-Strahlung in Binnengewässern. In: Dokulil, M.T., Hamm, A. & Kohl, J.-G. (eds), *Ökologie und Schutz von Seen*, Facultas, Wien, 341–59.

Sommaruga, R. & Psenner, R. (1997) Ultraviolet radiation in a high mountain lake of the Austrian Alps: Air and underwater measurements. *Photochemistry and Photobiology*, **65**, 957–63.

Sommer, U. (1981) The role of r- and K-selection in the succession of phytoplankton in Lake Constance. *Acta Oecologia*, **2**, 327–42.

Spindler, Th. (1995) *Fischfauna in Österreich*. Ökologie-Gefährdung-Bioindikation-Fischerei-Gesetzgebung Monographien, 53, BM Umwelt, Wien, 120 pp.

Steinberg, C. (1980) Nutrient enrichment in a sub-alpine lake: its degree and effects on the phytoplankton of Lake Ammersee. *Gewässer und Abwässer*, **66/67**, 175–187. (In German.)

Steinböck, O. (1950) Probleme der Ernährung und des Wachstums bei Salmoniden. *Schweizerische Fischereizeitung*, 1950(3/4), 1–7.

Steinböck, O. (1951) Die Fische der Hochgebirgsseen. *Alpenvereins-Jahrbuch*, 1951, 134–44.

Teubner, K., Sarobe, A., Vadruzzi, M. & Dokulil, M.T. (2001) $^{14}$C photosynthesis and pigment pattern of phytoplankton as size related adaptation strategies in alpine lakes. *Aquatic Sciences*, **63**, 310–25.

Thomas, E.A. (1969) The process of Eutrophication in Central European lakes. In: Rohlich, G.A. (ed), *Eutrophication. Causes, Consequences, Correctives*, National Academy of Sciences, Washington, DC, 29–49.

Timms, B.V. (1992) *Lake Geomorphology*. Gleneagles Publishers, Adelaide, 180 pp.

Turnowsky, F. (1976) Seen im Hochgebirge. In *Die Natur Kärntens*, Part. 2, Verlag J. Heyn, Klagenfurt, 223–45.

Ventz, D. (1973) Beitrag zur Hydrographie von Seen. *Acta Hydrophysica*, **17**, 307–16.

Vinebrooke, R.D. & Leavitt, P.R. (1996) Effects of ultraviolet radiation on periphyton in an alpine lake. *Limnology and Oceanography*, **41**, 1035–40.

Vollenweider, R.A. (1968) *Scientific Fundamentals of the Eutrophication of Lakes and Flowing Waters, with Particular Reference to Nitrogen and Phosphorus as Factors in Eutrophication*. Organisation of Economic Cooperation and Development, Paris, 159 pp.

Vollenweider, R.A. (1976) Advances in defining critical loading levels for phosphorus in lake eutrophication. *Memorie dell' Istituto Italiano di Idrobiologia*, **33**, 53–83.

Wanzenböck, J., Lahnsteiner, B., Hassan, Y., Hauseder, G. & Gassner, H. (2002) Ecology of European whitefish, *Coregonus lavaretus*, in two Austrian lakes in relation to fisheries management and lake productivity. In: Cowx, I. G. (ed.), *Management and Ecology of Lake and Reservoir Fisheries*. Blackwell Science, Oxford, 58–69.

Williamson, C.E. (1995) What role does UV-B radiation play in freshwater ecosystems? *Limnology and Oceanography*, **40**, 386–92.

Witt, U. (1977) Auswirkungen der künstlichen Düngung eines Hochgebirgsees (Vorderer Finstertaler See, Kühtei, Tirol). *Archiv für Hydrobiologie*, **81**, 211–32.

Wunsam, S. & Schmidt, R. (1995) A diatom-phosphorus transfer function for alpine and pre-alpine lakes. *Memorie dell' Istituto Italiano di Idrobiologia*, **53**, 85–99.

Zaderer, P. (1980) Die Chironomiden des Gossenköllesees: Artenliste, Bestand und Schlüpfraten im Jahr 1979. *Jahresbericht der Abteilung Limnologie Innsbruck*, **6**, 117–23.

Zobrist, J. & Drever, I. (1990) Weathering processes in alpine watersheds sensitive to acidification. In: Johannessen, M., Mosello, R. & Barth, H. (eds), *Acidification Processes in Remote Mountain Lakes*. Air Pollution Research Report No. 20, European Economic Community, Brussels, 149–61.

# 7 Lake Baikal and other Great Lakes of Asia

LYUDMILA G. BUTORINA

## 7.1 INTRODUCTION

Asian Great Lakes differ in area, depth, water transparency and quality. However, they are similar in experiencing threats to their integrity from changes in global climate, irregularities of the development of Earth and, primarily, from the deadly forcing of anthropogenic economic activity. This chapter presents assessments of the ecosystems of three Asian great lakes (Baikal, The Aral and Balkhash; see Table 7.1), how they are threatened, how they are changing and what might be achieved in order to reverse these trends.

## 7.2 LAKE BAIKAL

### 7.2.1 General characteristics

Lake Baikal is located in eastern Siberia, 53°00′N, 107°40′E (Fig. 7.1). It is the deepest lake in the world and it ranks eighth amongst the 22 world's largest. The lake is a unique store of fresh water, containing 20% of the world's storage and 80% of the surface fresh water of Russia. Baikal is one of the most ancient lakes in the world. It is some 20 to 25 million years old (Galaziy 1988; Solomatina 1995). It is tectonic in origin and remains in a seismically active area (Galaziy 1988; Suzyumov & Tsiproruha 1991; Solomatina 1995). About 2000 weak earthquakes are registered here annually and earthquakes of 5 to 6 points on the Richter Scale occur about once in every 10 to 12 years. Earthquakes of 7 to 8 points occur on a probability of once in 20–23 years, but those over 9 seldom occur. Lake Baikal's shores are moving apart at a rate of 2 cm each year, as its bed widens.

Geological studies enable us to make the supposition that Lake Baikal is a new ocean, in the early stages of its formation (Druyanov 1984; Galaziy 1988; Suzyumov & Tsiproruha 1991). Baikal exhibits many features of an ocean: abyssal depth, internal waves and seiches, tides, violent storms in which wave-height reaches up to 4 m, and deep-water luminescence of uncertain nature (Galaziy 1988). The bottom of Lake Baikal lies 1181 m below sea level and possesses a complex relief, with shelves, slopes, cliffs, mountain ridges, deep fractures and canyons. It is covered with unconsolidated bottom deposits from 100 m to 7.5 km thick (Galaziy 1988; Kuz'min 1994). Annual sediment deposition varies from zero to 9 mm, depending on the topography of the bottom (Agafonov 1993).

The water level of Lake Baikal, as in other great lakes, is subject to long-term and short-term variations (Galaziy 1988). Smooth long-term variations have been determined by historical cyclical fluctuations of climate and hydrology, affecting the volume of water stored in the whole lake. The amplitude of long-term water level variations in Baikal is 2.17 m, with the highest levels occurring every 65–70 years (Shimarev 1985; Galaziy 1988). Long-term variations are accompanied by interrelated changes of water temperature, composition of zooplankton and, as a result, changes in self-purifying ability of the lake. In the years when lake level is low, surface water temperatures are higher, but numbers of the main filter-feeding zooplankton, *Epischura* (Crustacea, Calanoida), decrease and the water quality becomes poorer.

Short-term variations of water level are local in character. They are determined by variations of atmospheric pressure, water surges owing to wind action, changes in temperature and density. Local water-level fluctuations may account for about 1 m (Shimarev 1985; Galaziy 1988). The amplitude of seasonal water level variations is from 0.8 to 1.0 m; that of the diurnal tidal cycle is from 2 to 3 cm.

Table 7.1  Morphometric and hydrological characteristics of lakes

| Lake | Elevation above sea level (m) | Area (km$^2$) | Maximum depth (m) | Transparency (m) | Volume (km$^3$) | Number of inflowing rivers | Salinity (mg L$^{-1}$) |
|---|---|---|---|---|---|---|---|
| Baikal | 465 | 31,500 | 1637 | 250–1200 | 23,600 | 336 | 0.1–0.2 |
| Aral | 53 | 64,500 | 68 | 0.2–25 | 1064 | 2 | 10.0–11.0 |
| Balkhash | 340 | 19,500 | 27 | 0.2–12 | 112 | 6–8 | 0.7–5.2 |

The rate of hydraulic exchange of Lake Baikal is very slow, with an average theoretical retention time of 383 years. Direct atmospheric precipitation accounts for 13 to 17% of the annual water influx, and river runoff (58.75 km$^3$) accounts for the balance (Druyanov 1984; Galaziy 1988). The River Selenga is the greatest affluent of Baikal, accounting for about 50% of the inflow. The Angara is the only riverine outflow, discharging about 60 km$^3$ annually (Galaziy 1988). The annual water balance of the lake is zero.

Baikal is a very turbulent lake. The entire water mass is subject to mixing owing to wind-generated currents and there is a perpetual vertical circulation caused by temperature variations and diffusive convection (Kozhov 1963). From the middle of December until the beginning or middle of May, the lake is covered with ice, from 1 m thick in open parts and up to 2 m thick in some sheltered areas (Galaziy 1988). Summer temperatures of surface water in open parts of the lake are in the range 14–15°C, whereas in bays it may rise to 22–23°C (Kozhov 1963; Shimarev 1985; Galaziy 1988). Diel variations of water temperature do not exceed 2°C when it is calm, but may vary by up to 10–12°C during storms. Diel temperature variations are observed only in the upper 10 to 12 m of the water column. Below that, down to the layer of temperature discontinuity at a depth of 200 to 250 m, temperature variations are seasonal. Beyond that, the water temperatures are almost constant. Near the bottom in the deepest part of the lake, the temperature varies very little (between 3.2 and 3.3°C).

Baikal contains the purest, clearest fresh water on Earth (Galaziy 1988; Solomatina 1995). Its transparency varies from 250 to 1200 m and is close to that observed in the ocean. The density of water in Lake Baikal is close to that of distilled water (Galaziy 1988). Almost half of Baikal's water (that from depths below 400 to 500 m) would satisfy the strict international standards set by the World Health Organization for high-quality drinking water. Heavy metals and chlorinated organic pollutants occur in Baikal at concentrations several orders of magnitude below the upper limits set by the same international standards (Solomatina 1995).

The water of Baikal is soft, mildly alkaline, with low mineral content. Intensive oxidation processes occur in the surface water layers. Bicarbonate and calcium are the most abundant ions. Gaseous and ionic composition, as well as the contents of organic matter, phosphorus and nitrogen, are interseasonally quite stable (Kozhov 1963; Galaziy 1988). At 96.5 mg L$^{-1}$, the mineral content of the near-surface lake water is 21 mg L$^{-1}$ less than that of the main inflows (Druyanov 1984; Galaziy 1996). Sulphates account for 5 mg L$^{-1}$, chloride 0.4 mg L$^{-1}$, dissolved organic matter about 41 mg L$^{-1}$, and particulate organic matter occurs at a concentration of 1.5 mg L$^{-1}$. The water is deficient in iodine (Galaziy 1988; Tarasova & Meshcheryakova 1992). Mineral content decreases with depth, a unique phenomenon of Baikal. It may be explained by a probable significant inflow of ultra-fresh water of mantle origin from a depth of 70–80 km. Lake Baikal is well-oxygenated at all depths. Dissolved oxygen content at the surface varies between 11.7 and 11.9 mg L$^{-1}$ and between 9.9 and 10.6 mg L$^{-1}$ at a depth of 1400 m. This is also unique. An explanative hypothesis invokes the lake's abyssal

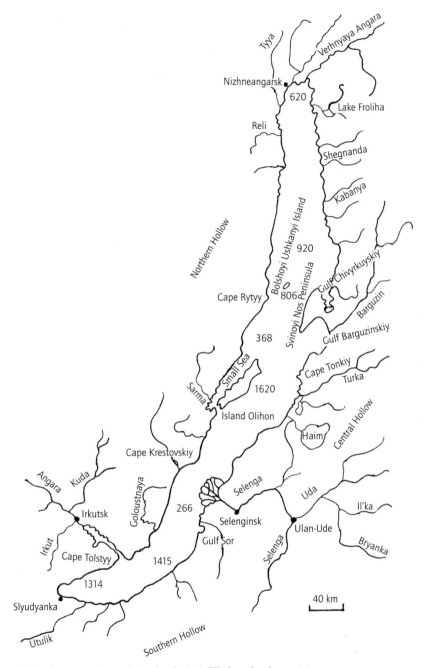

Fig. 7.1 Lake Baikal (Galaziy 1988): numbers, depths (m); filled circles, large cities.

water being sourced from Earth's mantle and the manner in which water in the lake is vertically exchanged. Near uniform distribution of oxygen determines a high ecosystem stability (Solomatina 1995).

### 7.2.2 Flora and fauna of the lake

The flora and fauna of Lake Baikal are highly diverse. The presence of some 1085 species of plant and 1550 species of animals is recorded (Kozhov 1963; Galaziy 1988). Almost 60% of the species are endemic to Baikal. Diatoms are abundant amongst vegetation. Gammarids and Baikal oilfish (*Comephorus baikalensis*) are the most numerous members of the fauna. Sponges are the most ancient inhabitants of Baikal. The evolution of the Baikal assemblage is partly due to the long isolation of the lake but it is the stable temperature and chemical conditions which have led to an increase of stenobiont, oxyphilic species (Votintsev 1992). Baikal is functionally oligotrophic (Galaziy 1988; Votintsev 1992); its productivity is comparable with that of the ocean. Phytoplankton productivity accounts for 3,925,000 t of organic matter (130 g $m^{-2}$) every year, or 90% of the total sum of annual organic matter input into the lake (Galaziy 1988; Votintsev 1992). Each year, phytoplankton production in Baikal turns over 286,000 t of nitrogen, 62,000 t of phosphorus, 300,000 t of silicon, about 1000 t of iron and it produces up to 10,500,000 t of oxygen. Average annual production of zooplankton is 5,300,000 t to maintain a biomass of 462,000 t. *Epischura* (Calanoida) accounts for 90% of the zooplankton biomass. Crustaceans of this genus form a very efficient biofilter of Lake Baikal water, processing between 500 and 1000 km³ of water annually, that is, some 10 to 15 times more than the annual water from all affluents (Galaziy 1988). Average annual production of zoobenthos is 250,000 t and its standing biomass is about 700,000 t. Annual fish production in the lake is between 190,000 and 220,000 t and its biomass is about 230,000 t. Annual commercial fish yield varies between 11,000 and 14,000 t. About half the annual non-commercial fish production in the lake is consumed by seals (*Pusa sibirica*).

The self-purifying potential of the Lake Baikal ecosystem is determined by the balance in the system of organic matter and oxygen. With an average organic matter content of about 1 mg of carbon per litre (Votintsev 1992), the total organic carbon content in the lake amounts to 23,000,000 t, with annual external input of an estimated 4,300,000 t. Complete renewal of the organic matter in the lake takes 6 years. Photosynthetic oxygen generation is one of the main sources of oxidant for the decomposition of organic matter in Baikal, around 90% (Galaziy 1988; Votintsev 1992). Photosynthesis is confined to the upper 100-m layer of the water mass, where 145 g $m^{-2}$ of organic carbon are oxidised annually. The corresponding oxygen requirement is 406 g $m^{-2}$ of oxygen. The annual respiration of *Epischura* (Crustacea, Calanoida), predatory zooplankton and pelagic fish demand 93.7, 9.2 and 12.6 g $m^{-2}$ of oxygen respectively. Thus, only 51.3 g $m^{-2}$ of oxygen remains to fulfil the self-purification potential, which, as shown, is really rather limited.

### 7.2.3 Ecological problems of Lake Baikal

Being a unique reservoir of high quality fresh water, Baikal demands a very cautious attitude and special protection. Economic exploitation of the lake began over 300 years ago (Izrael & Anihina 1991). Industrial use of the lake, with consequent chemical pollution persisting to the present day, began in 1968. According to Galaziy (1990, 1996), the main sources of present pollution are:

**1** Baikal paper and pulp plant (BPPM), releasing into the lake 240,000 m³ of purified industrial wastes and 150,000 m³ of relatively pure water each day.

**2** Inadequately purified industrial and municipal wastes of the city Ulan-Ude, over 120,000 m³ per day.

**3** Selenga paper and cardboard plant (SPCM), releasing more than 60,000 m³ of poorly purified water per day.

**4** Water and wind erosion—the soil and agricultural pollutants that wash off from 907,000 ha (9070 km²).

**5** Inappropriate agricultural land use and forest

utilisation in Baikal's catchment area, the latter resulting in timber cutting exceeding the highest permitted annual limits five to sixfold—therefore in some areas these limits already have been exceeded by two orders of magnitude.

6 Floating of about 2 million m$^3$ of timber per year.

7 Atmospheric discharges from industrial enterprises, which amount to 100 t per day, or 35,000 t per year.

8 Deterioration of soil and vegetation in the area of Baikal–Amur railroad and in the wildlife protection zone of the lake's catchment area, which led to solid matter runoff as high as 2230 g m$^{-2}$.

9 Navigation, which has led to water pollution from oil products, in some areas exceeding the highest permissible limits by a factor of 20 to 40.

10 Uncontrolled tourism, leading to heavy littering of some areas, forest cutting and fires.

The purified industrial waste waters from BPPM possess a mineral content seven to eight times more concentrated than that of the lake water (Galaziy 1996). They contain 100 times more sulphate, 275 times more chloride, 17.5 times more organic matter and two times less oxygen found in the lake water. They contain 0.28 mg L$^{-1}$ of polychlorinated biphenyls (PCBs).

The area of precipitation (fallout) of the BPPM gaseous dust wastes is 400 km$^2$. The area polluted by industrial effluents of the plant is estimated by chemical and biological indices as 30 to 35 and 8 to 16 km$^2$ respectively (Feodorov et al. 1996). The zone of pollution takes the shape of a flame tongue and spreads along the lake in a westerly direction. The area near the BPPM discharge may be divided into three zones of pollution by industrial sulphates: heavily polluted, moderately polluted and the zone transitional to the clear water. Concentrations of sulphates in these zones are 2 to 100% higher than in the central part of Baikal. In 1993, the combined area of the three zones was 93 km$^2$. The size, shape and direction of movement of the zone of elevated sulphate concentrations within the surface water layer of Baikal are spatially and temporally variable, depending upon the intensity of floods, the direction and velocity of the wind and undersurface flows, and the sulphate content of the effluent. Concentrations of sulphur-34 increase toward the open part of the lake. Suspended matter in the BPPM effluent, comprising cellulose, lignin, tarry and mineral substances, sediments over an area of about 2 km$^2$ (Izrael & Zenin 1973).

About 12,000 t of mineral substances, some 3400 t of organic substances and 135 t of suspended matter are discharged annually into Baikal from the SPCM (Galaziy 1990). In addition, over 44,000 t of aerosol and hazardous oxides are discharged into the atmosphere by this plant. Pollutants are carried with the flow of the Selenga river for 130 km to reach the opposite bank of Lake Baikal. The area affected by pollution is about 1500 km$^2$ and it is currently expanding.

Together, the two plants (BPPM, SPCM) discharge more than 100 t of hazardous substances per day into the atmosphere (Galaziy 1990). Atmospheric discharges spread up to 160 km northeastwards along the side of Baikal, up to 40 or 50 km to the west and rise up to 1.5 to 1.8 km. Under conditions of high humidity and fog, sulphur and nitrogen oxides, as well as other hazardous oxides, hydrolyse and precipitate as acid rain or snow.

Some 15,000 km$^3$ of industrial wastes have been discharged into Lake Baikal in the 22 years up to Galaziy's (1988) report. This is equivalent to over half the volume of the lake (Table 7.1). The anthropogenic impact on Baikal is increasing and spreading over the entire area of the lake (Galaziy 1990, 1996; Votintsev 1992; Feodorov et al. 1996). The mineral content of the lake water is increasing, bringing with it a rise in the osmotic pressure experienced by organisms. This may cause death by dehydration, even if the dissolved salts themselves are otherwise non-toxic. Increased concentrations of sulphate ions, chloride ions, phenols, methanol, terpene hydrocarbons, petroleum products and other industrial pollutants have caused changes to the natural microbial communities. In the area near the points of discharge, coccoid bacteria have begun to increase relative to other forms; fungi have also appeared. The number of heterotrophic bacteria in water increased 40- to 100-fold and by 10 to 80 times in the bottom deposits. The organic matter content of bottom deposits in the polluted areas has increased by up to 16%. This is sufficient to bring about an oxygen deficit, an unusual phenomenon for Baikal. Cellulose- and sulphate-reducing bacteria are now found in these areas. Thiobacteria are found in industrial effluents.

Since the 1960s, there have been considerable changes in the planktonic community of Lake Baikal. Endemic species have begun to disappear and to be substituted by cosmopolitan ones. *Nitzschia acicularis*, not normally associated with water of high purity, has become abundant in the phytoplankton, at the expense of the previously flourishing *Melosira (Aulacoseira) bajcalensis*. Small-celled algae have been replacing larger celled species. In the zooplankton, the characteristic *Cyclops*, the main food of fish in the open water, has declined in relative abundance. By 1996, the biomass of zooplankton during the autumn peak of its development fell to half what it had been in the period 1961–1965. As a direct result of increased mortality rates, the numbers and biomass of zooplankton decreased particularly in the area of discharge of the BPPM effluents. Since 1966–1967, the growth rate, fecundity and fatness of fish and ringed seal decreased sharply. The general physiological condition of fish and, especially, of the Baikal cisco (*Coregonus baicalensis*) deteriorated. In the 1950s, the Baikal cisco would commence spawning at the age of 4 years, when they weighed about 500 g. Presently, the fish does not spawn until it is 7 or 8 years old, with a weight of only 180 to 200 g. The immune systems of the Lake Baikal organisms have also been weakened as a result of poisoning. This has led to increases in disease and mortality. In 1987–1988, 10% of the ringed seal population died. The mortality of bullheads (*Cottus gobio*) is 20 specimens per square kilometre of lake margin. Mutations are observed amongst valuable fish species in areas affected by industrial effluent discharges.

Yet another symptom of pollution is well-expressed in Baikal: accumulation of hazardous compounds in the food chain (Galaziy 1988, 1996; Votintsev 1992). Concentrations of hazardous chemicals in lake organisms exceed the equivalent lake concentrations by two to six orders of magnitude, reaching pathogenic or lethal doses. For example, concentration of PCBs in the lipids of ringed seals may be as high as 38 to $64\mu g g^{-1}$ and dioxins equivalent to $93 pg g^{-1}$ of lipid have been reported.

Changes to terrestrial ecosystems are occurring as a result of atmospheric discharges (Galaziy 1988, 1990, 1996). Conifer and, especially, Siberian-fir forests have been dramatically affected by gaseous dust discharges of paper mills and other coal-burning enterprises. Mass drying of trees has occurred over an area of $2500 km^2$ and approximately $400 km^2$ of these forests have been completely lost. Amongst other tree species, there has been a decrease in the rate of wood growth and fruiting. A great number of seeds of pine and birch appear to be non-fertile. Plants growing from seeds affected by pollution exhibit less energy for growth and fruiting. This may lead to reduction of biological productivity of forests in future.

### 7.2.4 Measures to protect Lake Baikal

Various measures have been adopted in order to prevent the further pollution of Baikal (Izrael & Zenin 1973; Anohin *et al.* 1984; Galaziy 1990; Shutov 1995). A three-stage purification of wastes, including mechanical, chemical and biological purification, has been introduced at BPPM. Waste waters are intensively aerated for 4 to 6 hours and then discharged to the lake 160 m from the shore at a depth of 40 m. The quality of water discharged into the lake is checked by analysis of water sampled three to four times a day. A research centre for controlling water quality in the lake areas adjacent to the area of waste water discharge has also been established at BPPM.

Stricter standards of permissible concentrations of chemicals in purified waste water were introduced especially for Baikal. The scientific institutions of the Baikal region conduct hydrological, hydrochemical and hydrobiological surveys of the whole water area of the lake two or three times a year. Since 1971, construction of new industrial enterprises and wood-cutting along Baikal shores and inflowing rivers have been prohibited.

Timber rafting has been restricted. A special law for the protection of Lake Baikal was issued by the Russian Government; unfortunately, the Russian Government was unable to provide enough financial resources to implement this law. An international competition of proposals for saving of the Baikal region have been announced. The winner of the competition, D. Devis (USA), who had previously participated in the work on the programme for the restoration of the North American Great Lakes, proposed 12 programmes of regional land use and water protection, and provided 3 million dollars for its implementation (Shutov 1995). The main emphasis in these programmes was placed on the traditional methods of land use and economic activity in the region. Unfortunately there is no information available on the implementation of the programmes.

In 1991, Russian scientists applied to UNESCO with a request to include the Baikal region in the list of the Chapter of World Heritage. An international ecological inspection of the region showed that the lake already exhibited signs of degradation and could not be included in this list. The result of this inspection showed that all regulations issued and measures implemented in order to protect Lake Baikal and the Baikal region had not been effective and, apparently, would not lead to achievement of their goal. Baikal could not be considered as a regular water body, or treated as such. It required an individualistic approach. Baikal is a very fragile ecosystem, which has evolved to a strictly definite, uniquely balanced, stable condition, the consequences of any change from which might be catastrophic. Now, the threatened loss of this unique ecosystem and the world's greatest reserve of supreme-quality drinking water is all too real.

Several measures have been suggested for the reduction of the negative anthropogenic influences on Lake Baikal and the Baikal region (Galaziy 1990; Kozhova 1992). It is necessary to reduce sharply the input of pollutants in waste waters and gaseous dust discharges of all enterprises, including those located on the River Selenga. This requires improved systems of purification of industrial waste waters and gaseous and dust discharges. Baikal pulp and paper mill should be reorganised in order to become a wasteless enterprise and the Selenga pulp and cardboard mill should be switched to a closed cycle of water use and the deep underground discharge of wastes. Lead and zinc processing should be moved out of the Baikal catchment area altogether, and heat and electric power plants should be switched to electrical boiling equipment. The operations of the hydro-electric power plant should be adjusted to the regime of natural water level variations in Lake Baikal. The industrial use of water should be strictly controlled. Charges for water consumption and penalties for violation of environment-protecting regulations should be increased. Dumping of municipal waste waters should be prohibited.

It is necessary to introduce the strictest regulation of use of all natural resources. Turning the Baikal shoreline and water front into private ownership should be prohibited. It is urgent to address the problems of the lake at a unique, national park level. National nature reserves should be established on economically used territories.

It is necessary to change the system of agricultural land use, to prevent pollution of streams with fertilisers, pesticides and herbicides. More financial support to measures to protect against erosion is needed. Timber felling should be brought into balance with natural forest regeneration in the region. Uncontrolled tourism should be prohibited.

In order to find a solution to the ecological crisis in the Baikal region, it is essential to develop a long-term comprehensive regional programme and to provide it with adequate financial support. A network of scientific bases for the survey and evaluation of the status of the Lake Baikal ecosystem is necessary to improve further the present system of ecological control.

The principal solution to the problem of preserving Lake Baikal is the total protection of the lake and its catchment area from pollution via waste waters, even those which have been purified, and the deposition of industrial fallout from gaseous discharges (Galaziy 1996).

## 7.3 THE ARAL SEA

### 7.3.1 Introduction

Lake Aral, which is traditionally referred to as a sea, is one of the largest enclosed lakes in the world (Table 7.1). It is located in Kazakhstan, in the central part of the Turanian Plain, 45°00′N, 60°00′E. The Aral Sea was formed 15,000–18,000 years ago as a focus of surface and subsurface runoff in a continental catchment area (Williams 1995). It was once part of a larger water body which included both the Aral and the Caspian Seas (Murzaev 1991). Like other large water bodies, the Aral Sea is subject to alternating periods of transgression and regression. Natural variations of water level and salinity are governed by climatic cycles on scales of several centuries (Lamirev 1970; Shnitnikov 1985; Zolotokrylkin & Tokarenko 1991). Presently, the Aral Sea is in severe regression phase, which will continue well into the twenty-first century. As with most saline lakes, the Aral is highly sensitive to changes in climate.

There have been four transgression and three regression stages of dramatic variations of the water level and salinity during the Aral Sea's existence (Murzaev 1991). The fluctuations are expressed more sharply in the Aral than in other saline lakes because they are related to the channel processes of the river beds in the downstream parts of Amu-Darya and also to the fact that the Aral Sea is geologically relatively young. The amplitude of these natural variations in water level is 10 to 15 m and the salinity varies by 10 to 30‰.

### 7.3.2 General characteristics of the Aral Sea before 1960

Before 1960, the Aral was a unique brackish, warm-water, oligotrophic sea-lake with a distinctive fauna (Fig. 7.2). It was of great economic value as a fishery, for commercial navigation and as a site for hunting and recreation. The Aral Sea exerted a favourable climatic, hydrological and hydrogeological influence on the surrounding territory, extending 100 to 400 km from the lake shore (Kotlyakov 1991; Kuznetsov 1991; Williams &

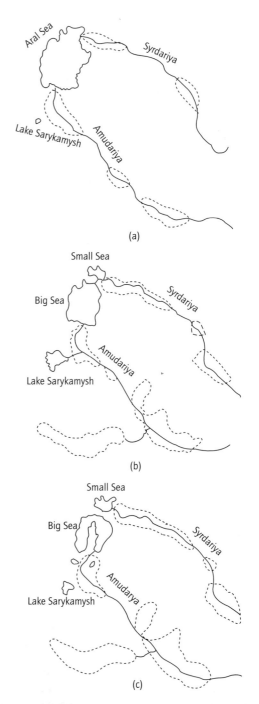

Fig. 7.2 Aral Sea (Aladin & Plotnikov 1995b): (a) before 1960; (b) at the present time; (c) in future. Irrigated areas are marked with a dotted line.

Aladin 1991). Before 1960, the hydrological and hydrochemical regimes of the Aral Sea were supposed to be quasi-stationary (Aladin & Plotnikov 1995a,b; Kotlyakov 1991; Kurnosov 1991; Kuznetsov 1991).

The major part of the inflow, 56 km$^3$ per year, was provided by the Syr-Darya and Amu-Darya rivers. Annual subsurface runoff brought in between 0.1 and 3.4 km$^3$ and direct atmospheric precipitation contributed about 10 km$^3$. Evaporation losses accounted for 66.1 km$^3$ per year. The lake water contained comparatively high sulphide and low chloride concentrations. The salinity of the lake water increased with the depth. Only a thin surface layer could be considered fresh. In summer, surface water temperatures reached 26–27°C, whereas the temperature of the near-bottom water scarcely exceeded 1 to 3°C. In winter, the Aral Sea would freeze over. During the open-water seasons, diurnal variations in the lake level of 20 to 25 cm were observed in calm weather, but strong, storm-driven surges of over 2 m have been observed during periods of strong winds. Because of the low input of biogenic compounds, the lake was not rich in phytoplankton. Its composition was close to that of the Caspian Sea phytoplankton (Rusakova 1995). Freshwater and brackish-water species constituted 54–63% of phytoplankton, representatives of euryhaline forms: *Acitinocyclus ehrenbergii*, *Exuviella cordata*, *Cyclotella caspia*, *Oocystis solitaria* and *Pseudoanaboena galeata* were the most abundant (Koroleva 1993; Rusakova 1995).

The fauna of the Aral Sea comprised 220 animal species. Freshwater and brackish-water species constituted 78% the fauna, Caspian species 17% and species of Mediterrenian or Atlantic origin 5% (Plotnikov *et al.* 1991). Zooplankton was represented by 21 Rotatoria, seven Cladocera and 22 Copepoda species. The most numerous (60%) were larvae of Mollusca (*Dreissena* and *Hypanis*) and Copepoda (*Arctodiaptomus salinus*). Average zooplankton biomass was 150 mg m$^{-3}$ (Plotnikov *et al.* 1991; Plotnikov 1993). Zoobenthos comprised 33 Mollusca, six Oligochaeta and five Crustacea species (Plotnikov *et al.* 1991; Plotnikov 1993; Filippov 1993). Zoobenthos biomass was 23 g m$^{-2}$, of which Mollusca and larval Chironomids constituted 63% and 33% respectively. *Dreissena polymorpha* and *Hypanis minima* (Mollusca), *Nais eligius* (Oligochaetae), *Cyprides torosa* (Ostracoda), *Dikerogammarus aralensis* (Crustacea), and larvae of the insect genera *Chironomus* and *Oecetis* flourished in the bottom sediments of Aral. The lake was inhabited by 20 species of commercially important fish. The Aralian and Caspian species constituted 45% of the fish fauna of the lake (Plotnikov *et al.* 1991). A basic part of the unique stock of commercial fish was represented by *Abramis brama orientalis* (Caspian bream), *Cyprinus carpio carpio* (common carp), *Rutilus rutilus* var. *aralensis* (Aral roach) and *Chalburnus chalcoides* var. *aralensis*. Some *Lucioperca luciopera* (zander), *Salmo trutta* var. *aralensis* (Aral brown trout), several species of *Silurus* (catfish) and *Esox* (pike) were fished from the lake as well. However, the lake was not rich in total stock of fish, owing to the lack of spawning areas and food for fry. The annual fish catch in the Aral did not exceed 4400 t. Many kinds of birds, including *Pelecanus* (pelecans), *Cygnus* (swans), various species of *Anatinae*, *Oxurini*, *Cairini* and *Dendrocygninae* (ducks), inhabited the littoral areas of the lake.

### 7.3.3 *The Aral crisis and its causes*

The Aral Sea began to dry out in 1960. Historically, the Aral and the Aral region have experienced previous ecological catastrophes. This modern catastrophe was made worse by the concurrence of the natural regression period with the onset of intensive and faulty human economic activity (Aladin & Plotnikov 1995b; Kotlyakov 1991; Kurnosov 1991).

Dramatic changes in the Aral ecosystem and its regional environment occurred as a result of acclimatisation works, during which scientific recommendations were ignored (Plotnikov *et al.* 1991) and as a result of irrigation and hydromelioration, which actually brought about a severe detriment to regional interests, mainly through placing a higher priority on irrigation and chemical promotion of agriculture than on the suitability of the crops to the location.

By 1927–1954 nine species of invertebrates (four species of *Paramysis*, two species of *Palaemon*,

*Calanipeda aquedulcis* (Crustacea), *Nereis deversicolor* (Polychaeta) and *Abra ovata* (Mollusca)) and 18 fish species were intentionally introduced into Lake Aral. Amongst the latter were *Acipenser stellatus* (starry sturgeon), *Clupea harengus* (Atlantic herring), *Ctenopharyngola idell*, *Hypophthalmus molitrix* (silver carp), *Mylopharyngodon piceus* (black carp) and *Pleuronectes flesus* (flounder). The introduction of six species of *Gobiidae* (gobies) and *Atherina mochon* (silverside) was accidental. Acclimatisation of such a number of fish led to a sharp increase in the feeding pressure over zooplankton and benthos. Although lake salinity did not change significantly, zooplankton biomass decreased by a factor of 3–3.5. Invasive species replaced the native Aralian taxa and forced them out of their spawning grounds.

Excessive expansion of areas used for growing of high water-demanding crops and a lack of crop rotation demanded unreasonable applications of mineral fertilisers, herbicides and insecticides. The combination of inappropriate uses to inappropriate areas, unprofessional and poor quality engineering, and uncontrolled use of irrigation water supplied free of any user charge all contributed to the despoliation of the environment and pollution exceeding all permissible limits by factors of 100 or more. The lack of any social programme aimed at improving the living conditions of the population led to a deep crisis for the Aral region.

Between 1960 and 1970, drying out and salinisation of the Aral Sea proceeded slowly. By 1971, water level had fallen by 1.7 m and the salinity had increased by 1.5‰ (Plotnikov *et al.* 1991). During this period, changes in the ecosystem were caused mainly by the acclimatisation works (Plotnikov *et al.* 1991). The introduced species quickly spread over the entire Aral Sea. By the autumn of 1970, *C. aquaedulcis* (Crustacea) accounted for 19–57% of the total numbers and 30–73% of the total biomass of zooplankton. Beginning in 1970, the rates of water level decline and salinisation of the lake noticeably increased. Between 1971 and 1976, the ecosystem of the Aral Sea underwent its first crisis period, related to salinisation (Plotnikov *et al.* 1991; Koroleva 1993). The water level had fallen by a further 5.3 m, the salinity increased by 6.5‰.

Changes were even sharper in the shallows compared with the open parts of the lake. The state of spawning grounds and nutritive base for fish fry sharply worsened. During the first crisis period freshwater and brackish-water species of a freshwater origin became extinct, whereas Caspian and marine, euryhaline species and halophyles survived. Salinisation began to adversely affect the fish eggs, juveniles and the process of fish reproduction. 1976–1985 were the years of a relative stabilisation of the lake ecosystem. The second crisis period, caused by further salinisation of the Aral, began in 1986 (Plotnikov *et al.* 1991). By this time the lake level had fallen by 8.2 m and the salinity had increased by 8–10‰ (Plotnikov *et al.* 1991; Koroleva 1993). At the shallows and in the bays water salinity increased by 15–20‰. Extinction of species of Caspian origin began. The stock of commercial fish became extinct.

### 7.3.4 General characteristics of the Aral Sea in the 1990s

By 1990, the total length of the irrigation system reached 150,000–200,000 km (Altunin *et al.* 1991; Kotlyakov 1991; Mirzaev & Rachinskiy 1991). Water consumption for irrigation was equal to 24.7 km$^3$ per year. Large-scale losses of water owing to infiltration and evaporation were occurring in this dense and branched hydrographic system. The largest in the system, the Karakum Canal, is a wide, artificial river, with a length of more than 1000 km. After construction of the Karakum Canal (begun during 1954) the irrigated area increased 2.5-fold and by 1990 reached 1,361,500 km$^2$ (Kurnosov 1991; Kotlyakov 1991; Aladin & Plotnikov 1995a,b). Annual withdrawal from the Amu-Darya river amounted to 12 km$^3$, of which 3–7 km$^3$ was lost directly to evaporation. Construction of the canal caused a rise in groundwater levels and an intensification of brine rising through soil capillaries. This led to formation of salt lakes and salt marshes, which became additional sites of useless water evaporation. Extensive widening of the canal led to the large-scale flooding and swamping of soils, which also intensified the process of their secondary salinisation. By 1961, river discharge into the lake had already almost halved to 29.9 km³ per year; in the period 1985–1990 residual river flow delivered under 5 km³ each year. The fall in water level of the Aral Sea appeared to be more rapid than had been anticipated.

### 7.3.5 Contemporary hydrological, hydrochemical and hydrobiological regimes of the Aral Sea in the 1990s

By 1988, the level of the lake had fallen by 15 m (Davydova et al. 1995). The volume of the lake had decreased by 60% and the area by 43% (Kotlyakova 1991; Kuznetsov 1991; Koroleva 1993). Water salinity of the lake increased two and a half to threefold and became as high as 24.9–25.6‰ (Plotnikov et al. 1991). In 1988, owing to a further decline in water level, the Aral Sea became separated into two almost isolated parts, the Big and the Small Seas, connected only by the narrow artificial Strait of Berg (Fig. 7.3). Although each basin continued to receive direct inflows from the now separated Amu-Darya and Syr-Darya rivers, there was a strong net flow through the Strait of Berg from the Small to the Big Sea. At the beginning of the 1990s the area of the Big Sea was approximately 33,500 km², with a volume of 310 km³ and an average salinity of c.39‰ (Plotnikov 1993). The corresponding figures for the Small Sea were: area c.3000 km², volume c.20 km³ and salinity ranging between 18 and 35‰ (Kotlyakov 1991; Plotnikov et al. 1991; Aladin & Plotnikov 1995a). The level of the Big Sea was lower than in the Small Sea, and the salinity was higher owing to a negative water balance of the Big Sea (Plotnikov 1993). Raising the water level in the Small Sea led to flooding of formerly dried areas of the sea.

The rate of salinisation of the Aral Sea increased with its drying. The lake has turned from a brackish to a saline, sea-like water body. The increase in salinity has been accompanied by a process of metamorphosis of the salt composition. There have been sharp increases in the concentrations of sodium, potassium, calcium, magnesium, sulphate and carbonate. Sulphate concentration is now several times higher than it is in the Caspian Sea or in the ocean (Orlova 1993). Precipitation of sulphates and carbonates occurs. Further increase in the salinity of the lake water will turn it to brine, and the lake may become a salt-depositing water body.

With the increasing salinity of water and the consequent increases to its density, changes to the patterns of circulation and convective mixing have occurred (Kotlyakov 1991). In turn, the hydrochemical regime has deteriorated, with decreased oxygen content, accumulating biogenic products and lowering of pH. The diminution of river inflows has altered the balance between biogenic loads, internal transformation and turnover.

In 1990 the lake salinity increased to 28–30‰, and even 40–43‰ in some areas (Rusakova 1995). The biological productivity of the Aral Sea collapsed. Phytoplankton became still poorer. The number of freshwater and brackish-water species decreased to 9.4% (Koroleva 1993; Rusakova 1995). Many Cyanophyta and Chlorophyta species became extinct. Phytoplankton was represented by 12 species of Bacillariophyta, five of Pyrrhophyta, two of Cyanophyta and two of Chlorophyta. Dinophyta and Bacillariophyta with wide halotolerance were abundant, including *Exuviella*

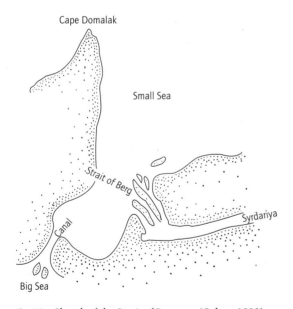

**Fig. 7.3** Sketch of the Strait of Berg area (Orlova 1993).

*cordata, Prorocentrum obtusum, Actinocyclus ehrenbergii*, and different species of *Chaetoceros, Nithzschia, Navicula* and *Cocconeis*. Their number varied from 8000 to 416,000 cells L$^{-1}$ depending on the hydrological and hydrochemical regime of local areas of the lake (Koroleva 1993).

Zooplankton was represented by one Rotatoria species (*Synchaeta*), three Copepods (*Calanipeda aquaedulcis, Halycyclops rotundipes, Halectinosoma abrau*), one Cladoceran (*Podonevadne camptonyx*), Molluscan larvae and one Polychaetae (*Nereis diversicolor*) (Plotnikov 1993). Zooplankton biomass varied from 39 mg m$^{-3}$ in the Big Sea to 642 mg m$^{-3}$ in the Small Sea and 11–15 mg m$^{-3}$ in partially isolated bays. Invasive *Calanipeda aquaedulcis* (Crustacea) and Molluscan larvae were abundant throughout the Aral. They constituted 85–95% of the total zooplankton numbers and 96–97% of its biomass (Plotnikov *et al.* 1991; Aladin & Plotnikov 1995a,b; Filippov 1995). Euryhaline species of Mediterranean and Atlantic origin and halophilic species were abundant in the zoobenthos. The latter comprised three species of Mollusca (*Abra ovata, Cerastoderma isthmicum, Caspiohydrobia* sp.), one species of Polychaeta (*Nereis deversicolor*) and three species of Crustacea (*Rhithropanopeus harrisii tridentatus, Palaemon elegans, Cyprideis torosa*). Zoobenthos biomass was 108 g m$^{-2}$ and 247 g m$^{-2}$ in the Big and the Small Seas respectively. Mollusca, *Abra ovata* and *Cerastoderma isthmicum* were abundant, constituting 85–95% of zoobenthos total numbers and 96–97% of its biomass. Freshwater fish and formerly numerous fish-eating birds became extinct. *Pungitius platygaster aralensis* (Black Sea flounder) and *Platichithus flesus luscus* (Aral (ninespine) stickleback) spread all over the Aral and were abundant amongst fish. In the 1990s the Big Sea, the Small Sea, the Strait of Berg, Butakov and Shevchenko Bays in fact became separate ecosystems (Plotnikov *et al.* 1991; Filippov 1993; Koroleva 1993; Orlova 1993, Plotnikov 1993; Rusakova 1995).

### 7.3.6  Consequences of the Aral crisis

At present, the Aral Sea may be recognised as a weakly mesotrophic, moderately polluted water body (Kotlyakov 1991; Kuznetsov 1991; Orlova 1993; Rusakova 1995). Its ecosystem appears to be severely damaged. More than 200 species of its fauna and flora have become extinct. Marine and euryhaline organisms of marine origin, as well as halophilic species originating from continental waters, maintain their existence in the lake (Plotnikov *et al.* 1991). The 15-m fall in water level of the Aral Sea led to the drying of 28,000 km$^2$ of former lake extent (Kotlyakov 1991; Kuznetsov 1991), losing a large former spawning area. Fishing was no longer practical. Navigation on the Aral Sea ceased. The biocoenoses characteristic of the delta area have been extinguished. Halophilic complexes with sparsely growing plants now populate the exposed lake bed. During the period 1965–1985, the areas overgrown by halophilic plants increased sixfold. At the same time, salinisation has affected 37,700 km$^2$ out of the 44,500 km$^2$ which were originally irrigated. Dried areas are now the source of dust storms which also export salt in the wind. Transformation of sandy relief towards desert has already commenced. The climate in the areas adjacent to the Aral Sea has become more continental. The groundwater surface (water table) is raised closer to the surface in the lake shore areas. An extremely unwelcome ecological and epidemiological situation has developed. There have been dramatic declines in human living conditions, nutrition, employment and housing opportunities, all of which have led to a sharp growth of morbidity and mortality, to become the highest in the country (Kotlyakov 1991). All this, together with the unemployment exported elsewhere by migration of the population out of the Aral region, has turned it to the status of a disaster area.

### 7.3.7  Measures to be taken to save the Aral Sea

Measures to save the Aral Sea need to be linked to a fundamental and holistic approach to the ecological, economic and social crisis in the Aral region as a whole. These measures must be applied simultaneously to the region in its entirety (Kotlyakov 1991; Kuznetsov 1991). The Aral Sea is located in the arid zone and it would be unrealistic to expect that it can be brought back to its former size. At present, it is essential to halt the decline in the levels of the lake itself and of local water tables.

In 1992, a dam was constructed to direct the flow of Syr-Darya into the Small Sea (Aladin & Plotnikov 1995b). This allowed its level to be raised by 1 m and to provide flow through the artificial canal from the Big Sea, thus filling several dry river beds, reversing the north-to-south flow and creating a gradient of salinity from the river delta toward the sea. As a result, the process of biodiversity restoration has begun: reeds have begun to grow in the fresher water and bird nesting areas have expanded. It is possible for freshwater fish to migrate back to the basin. However, some effects of these measures may also have been negative, as water levels in the Big Sea continue to fall, and salinity continues to increase.

In order to overcome the problems of the Aral region, to restore hydrological balance, geographical function and local climate, it is proposed to implement a general plan for a regional water supply which provides more effective purification, disinfection and desalinisation of water (Altunin et al. 1991; Bortnik et al. 1991; Kotlyakov 1991; Kuznetsov 1991; Kurnosov 1991; Mirzaev & Rachinskiy 1991). A concept of step-by-step stabilisation of the Aral Sea level was worked out, invoking a restoration of discharges of 17 to 50 km$^3$ of fresh water to the lake, each year for a period of 10 to 15 years. The sources of this water are to comprise: artificial glacial melt-water, retaining more of the natural flow of the Amu-Darya and Syr-Darya rivers, an artificial increase in winter atmospheric precipitation and a changed approach to water abstraction and consumption. Other steps to be taken include:

1 reduction in the amounts of mineral fertilisers and plant-protective chemicals applied;
2 banning the use of pesticides and defoliants;
3 curtailing saline drainage to the rivers;
4 prohibiting the discharge of drainage waters outside water-consumption areas;
5 stabilising exposed areas of the lake bottom in order to reduce the wind-induced export of salt;
6 reducing rapidly cotton and rice cultivation;
7 metering and charging for all water consumption;
8 reconstruction of the irrigation system;
9 increasing the proportion of abstracted subsurface water in the water supplied to consumers.

It is essential to continue close monitoring of both parts of the Aral Sea. It is necessary to create a legal basis for the solution of problems of improvement of environmental conditions, food products and medical care of the population of the Aral region. At present, all the resolutions issued continue to be violated. Even the dam was washed away 9 months after its construction. It was later re-erected, but not as high as it should be (12 to 14 m) in order to maintain the level in the Small Sea. None of the proposed additional inputs into the Aral Sea have yet been implemented. The Aral Sea keeps drying. The crisis in the Aral region continues to worsen.

## 7.4 LAKE BALKHASH

### 7.4.1 General characteristics of the lake

Lake Balkhash is located in the lowest part of the vast Balkhash–Alakul depression (Kazakhstan, central Asia), 46°00′N, 74°00′E. Its age is approximately 17,000 to 18,000 years. The lake basin was formed as a result of tectonic processes and was filled with surface- and groundwater flow (Davydova et al. 1995). The present lake is situated in the arid zone with little atmospheric precipitation (Shnitnikov 1976). Lake Balkhash is a unique natural formation, an oasis. A puzzling feature of the lake is the stability of the concentrations of some of its solutes.

Like other great lakes, Balkhash experiences successive periods of transgression and regression, when it either floods old terraces, dried-up nearshore lake basins and swampy floodplains, or it dries, shrinking until the western part of the lake virtually disappears (Shnitnikov 1976). During the past 100 years, the tendency has been towards a progressive drying of Lake Balkhash, as it has been for many other lakes of Central Asia and North America. With each succeeding regression, the water level in the lake has become lower. Within the general pattern, sharp variations in water level also follow alternations between drier and wetter phases of the local climate, which occur on a 12- to 13-year cycle (Shnitnikov 1975). Over the past 20 years, the average annual variations in the level of Lake Balkhash were 3.2 m. During the dry periods, the quality of the lake water becomes poorer, the flora and fauna of the lake experience change, with the vegetation and fauna in the dried margins of the shallows perishing accordingly.

Balkhash is a semi-freshwater, endorheic lake, fed by rivers emanating mainly from the surrounding mountainous areas. The hydrology and hydrochemistry of the lake depend mainly upon the surface runoff (Fig. 7.4). The river network to the north of the lake is as dense as 0.2 to 0.5 km km$^{-2}$, whereas to the south of the lake it varies from 0.6 to 3.0 km km$^{-2}$. Rivers flowing into the lake from the north are scanty and shallow. The northern shores of the lake are steep and rocky, with numerous inlets and bays (Shnitnikov 1976; Sevastianov et al. 1991). Southern shores are low, sandy, easily flooded, thickly overgrown with *Scirpus*, *Typha* and *Carex* and support numerous small saline lakes and salt marshes. The climate of the catchment is sharply continental with considerable daily and yearly temperature variation: down to −50°C in winter and up to +25°C in summer. Average annual air temperature is 6.7°C. The climate in the southern part of the lake is more humid than the northern part (Shnitnikov 1976). Average annual atmospheric precipitation is 120 mm in the northern area and 147 mm in the southern. The air humidity does not exceed 78%, and the minimum is 50%. Clear and cloudless conditions number 130 days a year.

### 7.4.2 Morphometric and hydrological characteristics of the lake

The Sary-Yesik peninsula located in the middle of Lake Balkhash separates western and eastern parts

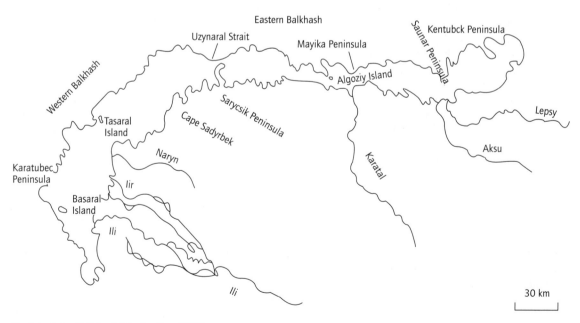

**Fig. 7.4** Lake Balkhash (Shnitnikov 1970).

of the lake (Fig. 7.4), which differ distinctly in their hydrological and hydrochemical characteristics (Shnitnikov 1976; Davydova et al. 1995). The western part of the lake is rather shallow; the eastern part is deeper, and the Uzynaral Strait, which connects them, is extremely narrow and shallow (see Table 7.2). Nevertheless, the strait provides a perpetual link as water flows slowly from west to east, playing an important role in maintaining the water quality and the salt regime of the lake, and balancing the ecological condition of either part and the lake as a whole.

Owing to peculiarities of the local climate, large area and relatively small depth of the lake, even small variations in the level of the lake cause substantial changes in its size, configuration and water quality. Thus, the area of the lake may be calculated only approximately. The data in Table 7.2 correspond to the lake when its surface stands at an altitude of 342 m above sea level.

The water in the west part of Balkhash is almost fresh. This may be explained by the influence of the River Ili, the input from which accounts for 73 to 80% of the total annual inflow (Shnitnikov 1970, 1976; Davydova et al. 1985; Sevastianov et al. 1991). The water of western Balkhash is turbid, yellow-grey. Its transparency does not exceed 1 m, and often falls as low as 0.2–0.4 m in summer. The bottom of western Balkhash is quite even, being covered with a thick layer of silts with sand as a considerable constituent. The rivers Karatal, Aksu and Lepsy, together with several smaller ones, flow directly into the eastern part of Balkhash (Fig. 7.4). The water in this part of the lake is brackish, with a blue to emerald-blue colour. Its transparency is up to 5.5 m, and even 10–12 m in some areas (Sevastianov et al. 1991). The bottom relief of eastern Balkhash is undulatory, and covered with a thick layer of light-grey lime silts or white dolomites.

River inflow provides the main water supply for the lake (Pozdnyakova & Shnitnikov 1981; Sevastianov et al. 1991). It accounts for 14.7 to 19.4 km$^3$ of fresh water annually. Annual subsurface input is 0.6 to 1.2 km$^3$. Direct precipitation on the lake contributes 2.1 to 3.3 km$^3$ per year and spring runoff from the shores brings between 0.5 and 1.8 km$^3$. Surface runoff accounts for 78 to 85% of the inputs to the western part of Balkhash. In eastern Balkhash, it accounts for 42 to 46% of the budgetary input. Flow from western Balkhash through the Uzynaral Strait brings in most of the balance (40 to 45%). The outputs from the western part are mostly evaporation (78 to 80%), flow into the eastern part of the lake making up only 19 to 21%. In eastern Balkhash, evaporation accounts for 97 to 99% of the hydrological input. Annual evaporation from the surface of Lake Balkhash is 17.1 to 17.6 km$^3$. Evaporation from the eastern part of the lake exceeds that from its western part by 8–11 mm.

The surface water temperature increases from 4.7°C in April to 23°C in July and falls to 3.3°C in November (Shnitnikov 1976; Sevastianov et al. 1991). Mean annual water temperature is 14.1°C. Owing to the relatively small depth and width of the lake and strong winds, the upper 5-m layer is generally well-mixed but there is generally a thermocline a little below that. It is particularly well-expressed during spring. From May till October the temperature gradient does not exceed 3.5°C. Full isothermy is observed in the lake during autumn.

Table 7.2 Morphometric features of subsections of Lake Balkhash

| Parts of the lake | Length (km) | Width (km) | Depth (m) | | | Area (1000 km$^2$) | Volume (km$^3$) |
|---|---|---|---|---|---|---|---|
| | | | Maximum | Average | Minimum | | |
| Western | 284–296 | 60 | 11 | 4.8 | – | 10.6 | 47.8 |
| Eastern | 309–318 | 9–48 | 27 | 9.0 | – | 7.5 | 57.6 |
| Uzynaral | – | 5–6 m | – | – | 2.8–3.3 | – | – |

From the end of November till the middle of March, the lake is frozen over.

### 7.4.3 Hydrochemistry and biology of the lake

The surface water layer is evenly oxygenated over the entire area of Lake Balkhash. Oxygen content in eastern Balkhash is 90% saturation and 100% in its western part, throughout the year (Shnitnikov 1976; Davydova et al. 1985, 1995; Sevastianov et al. 1991). In summer, the water at the westernmost area of eastern Balkhash is saturated with oxygen to 105–115%. In river mouths, oxygen content is lower than in the central part of the lake: 90–95%.

The chloride content of Lake Balkhash (14.4–19.5%) is considerably lower than that of the Black Sea or the Caspian Sea (Shnitnikov 1976; Sevastianov et al. 1991). Compared with the Ili River, water of the lake is characterised by a three to sevenfold lower content of calcium ($Ca^{2+}$ 0.4–8.7%), but a two to threefold higher content of alkaline metals (14.6–35.0%). The ionic composition of the lake water provides evidence of active processes of metamorphosis. This is reflected by a certain heterogeneity of mineral content and ionic composition between and within the main parts of the lake, in their annual and seasonal variations and in correspondence with the regularities of fluctuations of the water level in the lake.

Mineral content varies from 0.74 to 1.14 g L$^{-1}$ in western Balkhash and from 3.27 to 5.21 g L$^{-1}$ in eastern Balkhash (Shnitnikov 1970, 1976; Davydova et al. 1985, 1995). In the Uzynaral Strait, mineral content changes west to east from 1.4 to 3.3 g L$^{-1}$. Uneven water mineral content over the lake area may be explained by the riverine discharge to the southwestern part, as well as by the lack of outflow and the elongated shape of the lake bed (Sevastianov et al. 1991). Calcium accounts for 11.8% of the total cations in western Balkhash, whereas in eastern Balkhash it makes up no more than 1.2%. Precipitation of calcium carbonate ($Ca^{2+}$) in the form of dolomites occurs in western Balkhash. It is a rare phenomenon for the lakes located in the areas with the sharply continental climate (Shnitnikov 1976). In consequence, the total water mineralisation and the relative content of alkaline metals increase to 57.6–70.0%. This process, together with perpetual inflow of fresh water from western Balkhash through Uzynaral Strait govern a mechanism of self-regulation of mineral content of water in western Balkhash.

Nitrate content in the lake water varies between 0.01 and 0.7 mg L$^{-1}$ (Sevastianov et al. 1991). The highest concentrations in the lake are observed in Ili River delta and at the mouths of other large rivers. In permanently polluted areas (Berg and Buru-Baytal Bays), the concentration of ammonium ions is 0.6–3.4 mg N L$^{-1}$, and in the open part of the lake it is 0.05–0.6 mg N L$^{-1}$. Dissolved phosphorus concentrations vary between 0.002 and 0.4 mg L$^{-1}$. Its content increases from winter to summer. Silicon ion content varies from 2.4 to 4.6 mg L$^{-1}$. Its distribution over the water area of the lake is non-uniform: its lowest content is observed in eastern Balkhash. Iron content in lake water varies from 0.5 to 0.11 mg L$^{-1}$, with maxima in spring and autumn.

The biota of Balkhash are relatively poor, especially near the bottom (Shnitnikov 1976; Sevastianov et al. 1991; Shaporeno 1995). They comprise brackish–freshwater species (45%), typically freshwater species (38%) and typically brackish-water forms (17%). Phytoplankton are numerous and diverse. Bacillariophyta (211 species), Chlorophyta (42 species) and Cyanophyta (41 species) are abundant. The latter two are represented mainly by colonial forms. The average phytoplankton biomass is 2.1 g m$^{-3}$. Zooplankton comprises 50 species, including 28 rotatorians, 11 cladocerans and six copepods. Some 90% of the lake's zooplankton is made up by Crustacea, Cladocera being the most numerous. There is a west–east gradient of zooplankton composition which corresponds to the increase in salinity. Zooplankton of western Balkhash is more diverse and more abundant than in the eastern part of the lake (Sevastianov et al. 1991; Shaporenko 1995). Mean zooplankton biomass is 1.5 mg m$^{-3}$. The zoobenthos of Lake Balkhash is historically rather poor. It comprises not more than 20 species, and its biomass varies from 4.5 g m$^{-2}$ in western Balkhash to 2.1 g m$^{-2}$ in eastern Balkhash (Sevastianov et al. 1991). Larvae of Chironomidae and other insects and, to a lesser extent, Oligochaeta, Ostracoda and Mollusca are observed in benthos, as well as four artificially introduced species of Mysidacea, two species of Polychaeta and two species of Mollusca (*Monodacna*) (Sevastianov *et al.* 1991; Shaporenko 1995). Larvae of Chironomidae are the most abundant. Nowadays, Lake Balkhash is inhabited by 20 species of fish, six of them being endemic: *Schizothorax pseudaksaiensis* (snow trout), *Sch. argentatus* (Balkash marinka) *Perca schrenki* (Balkhash perch), *Nemachilus strauchi* (Stoliczka's roach), *N. labiatus* (thick-lipped roach) and *Phoxinus phoxinus*. Others species of fish have been introduced deliberately or by chance during the twentieth century. Amongst them are *C. carpio araliensis* (Aral common carp), *A. brama orientalis* (Caspian bream), *Aspius* (asp), *Ctizostedon lucioperca* (zanda), *Silurus glanis* (Wels catfish), *Leuciscus leuciscus baicalensis* (Siberian dace), *A. nudiventris* (fringe barbed sturgeon), and *Carassius auratus* (goldfish). The introduced species of fish constituted 84–90% of the annual commercial catch, whereas endemic species (*P. schrenki*, two species of *Schizothorax*) made up only 10–17%. The annual fish catch did not exceed 19,000 t, of which two-thirds were caught in western Balkhash. The acclimatisation of new species has severely affected the endemic species (Pavlov 1993): *Perca schrenki*, formerly of considerable commercial value, is almost extinct and rates an entry in the Red Book of Endangered Species.

Lake Balkhash was rich with birds, different species of *Aix* (ducks), *Pagohila* (gulls), *Sterna* and *Chlidonias* (terns), *Phalacrocoracidae* (cormorants), *Anser* (geese), *Gavidae* (loons), and *Ardeidae* (herons) being the most numerous. In 1934, *Ondatra* (muskrat) was acclimatised in the Ili River delta. The largest muskrat-breeding farm in the world, established in 1944, produces 750,000 skins per year.

### 7.4.4 Development of the Lake Balkhash region and economic use of the lake

Historically, the Lake Balkhash region has remained poorly developed. Aboriginal tribes practised cattle breeding and used irrigation to promote agriculture (Shaporenko 1995). However, intensive agricultural development began only after a Russian population settled in the region during the second half of the nineteenth century. By the beginning of the twentieth century the irrigated area was 440,000 ha, when the first fishing settlements were also being built on the lake. Construction of two railroads in the 1930s benefited greatly the economic development of the region. Now, navigation on the lake is well-established, although its shallowness and the variability and long-term decline in water level continue to present difficulties. Large volumes of water are abstracted from the lake by metallurgical, copper-smelting and other industrial enterprises, in addition to the supplies for municipal consumption and for the irrigation of rice, cotton and cattle pasture. Subsurface extraction supplements some of the demand. The quality of these waters is generally rather poor and the sources are generally insufficient and unreliable. In order to meet growing economic demands, a dam and hydroelectric power plant were erected on the River Ili in 1970, creating the Kapchagay reservoir. However, the project was not completed because of the inadequacy of the water resources: filling of the reservoir discontinued in 1982. In 1991, the volume of Kapchagay reservoir was only 36% of the capacity originally projected (Shaporenko 1995).

### 7.4.5 Environmental problems of Lake Balkhash

Increasing economic activity in the region had a strong impact on the ecosystem of Lake Balkhash. Construction of Kapchagay reservoir reduced the inflow volume from the River Ili by about 3.8 km$^3$ per year (Shaporenko 1995). The adverse effect on the hydrological balance of the lake was compounded by a change in the local climate towards drier conditions. An estimated total of 60.3 km$^3$ of river inflow and precipitation were lost from the hydrological balance over the period 1970–1985. Moreover, between 1960 and 1980, expansion of the areas receiving irrigation led to increased withdrawal of water, from around 3.3 to almost 5.0 km$^3$ per year. Water abstraction from affluent streams to supply industrial and municipal needs also increased during the 1970s, from 0.262 to 0.421 km³. This figure should be doubled if direct water consumption from the lake is accounted.

As a result of these simultaneous onslaughts on the hydrological budget, the surface of Lake Balkhash fell by 1.43 m and the shoreline receded by distances of between 2 and 8 km (Pozdnyakova & Shnitnikov 1981; Shaporenko 1995). Disruption of the natural flow of the River Ili also led to the drying out of the delta; by 1990, the area of delta lakes had been reduced by 40% (Shaporenko 1995).

Economic activity has also affected the salt input to the lake. Leaching from the bed of Kapchagay reservoir was one of the main sources of the increased salt content of the River Ili. Average annual total ion content of the river water over the period 1977–1981 was 373 mg L⁻¹. A considerable increase of salt content in eastern Balkhash affluents during recent decades is related to increases of drainage from irrigated rice-growing areas, exceeding 12 km³ per year. The salt content of these waters was 4 to 5 g L⁻¹. By the beginning of 1980, average salt content of the Karatal and Lepsy rivers had risen to 434 and 352 mg L⁻¹ respectively. Most of the increase was due to chloride, sulphate and sodium ions and, to a smaller extent, bicarbonates (Shaporenko 1995). The average salt content of Lake Balkhash increased in consequence: in western Balkhash, salinity rose by 0.5 to 0.8 g L⁻¹ over the similar period.

Disruption of water and salt regimes led to salinisation of soil and to desertification of exposed areas of the lake bottom and river deltas. Marshes which had been characteristic of these areas declined. The original vegetation now survives only as vestigial patches. Halophytes, weeds and poisonous plants have moved in. Animals have either migrated away or died. Decline in the meadows of the delta areas has led to a reduction in hay production and cattle breeding (Shaporenko 1995).

Dramatic changes in the environmental conditions have brought changes to the biota of the lake. Halophilic forms of phytoplankton have expanded all over the lake, and some Chlorophyta and Bacillariophyta have disappeared. The biomass of phytoplankton in the western part of the lake during the summer bloom has been reduced 1.6-fold. Changes in composition of phytoplankton caused a sharp decrease in the numbers of fine filter feeders, especially Rotatoria. On the other hand, coarser filter feeders have begun to increase in the zooplankton. During the 1970s, the average biomass of zooplankton reduced at least 2.5-fold. In the first half of 1980s the biomass increased with the rise of Cladocera and Copepoda, but in the mid-1980s the average biomass of zooplankton decreased 1.8-fold (Shaporenko 1995). Diminution in the availability of suitable spawning areas and the decline in benthic food organisms have adversely affected the growth and recruitment dynamics of fish. Catches of the commercially important species, such as *C. carpio araliensis*, have fallen by a factor of six to seven (Shaporenko 1995).

Economic activity has produced yet another adverse effect: pollution of the lake and its affluents. Agricultural runoff, industrial wastes and navigation are the main sources. Heavy metals and oil products are the principal pollutants in Lake Balkhash. Their concentrations in the lake water exceed the highest permissible limits in many instances. In 1988–1989, the average contents of copper and zinc in lake water were from 0.013 to 0.017 mg Cu L⁻¹ and from 0.015 to 0.016 mg Zn L⁻¹ and their maximum content was up to 0.07 mg L⁻¹. The average content of oil products was 0.15–0.16 mg L⁻¹, with a maximum of 1.22 mg L⁻¹. The lake is also polluted with pesticides, amongst which DDT appears to present the greatest problem: in 1989, its typical concentration in the lake was 0.009 fg L⁻¹ but maxima to 0.053 fg L⁻¹ have been noted. The bays appear to be much more polluted than the open part of the lake (Shaporenko 1995).

### 7.4.6 *Measures to improve the ecological situation*

Certain measures to correct the damage to Lake Balkhash have been proposed (Shnitnikov 1976; Shaporenko 1995). In 1980, the Institute of Lake Researches of the Academy of Sciences of USSR considered all existing projects for lake reconstruction and selected those considered to be most appropriate. These were as follows: (i) to construct a dam across Uzynaral Strait; (ii) to combine the construction of this dam with another, just to the east of the mouth of the Lepsy River; (iii) to construct two more dams — one to the east of the Karatal river mouth, almost in the middle of the eastern basin of Balkhash, and the other across Buru-Baytal Bay in the southernmost part of the western basin. All dams were designed to have locks. The projects would lead to stabilisation of local lake levels at altitudes around 341 m and salinity would be regulated not to exceed $2\,g\,kg^{-1}$. Other proposals include the transfer of River Ili flow by means of a pipeline to supply the industrial and municipal use on the northern shore of the lake. There have even been suggestions for the transfer of water from River Irtysh in Siberia. Unfortunately, none of these projects has been implemented nor is it likely that any will be implemented in the near future, owing to a lack of funding.

At the present time, it is more realistic to address the ecological situation of Lake Balkash and its region through improved management of the present resources. This could be approached through a 50% reduction in overall consumption, reconstructing the irrigation systems and switching production to crops less consumptive of water and optimising regulation of the water releases from the Kapchagay reservoir.

Some positive signs are already observable (Shaporenko 1995). The high inflows experienced in 1988 and 1989, together with the greater volumes available for release from Kapchagay, brought a sustained rise in lake level, which was kept stable at about 341.3 m above sea level during 1990–91. Unfortunately, the present structure of regional economy in Kazakhstan cannot support the necessary investment. There persists the threat of serious, perhaps catastrophic, degradation of Lake Balkhash.

## 7.5   REFERENCES

Agafonov, B.P. (1993) On the contemporary sediment accumulation in Baikal. *Vodnye Resursy*, **20** (2), 199–205.

Aladin, N.V. & Plotnikov, I.S. (1995a) On the problem of the possible conservation and rehabilitation of the Maloe Sea of the Aral Sea. In: Scarlato, O.A. (ed.), *Biologicheskie i prirodovedcheskie problemy Aral'skogo morya i Priaral'ya, Part 1. Proceedings of the Zoological Institute Russian Academy of Sciences St Petersburg*, **262**, 3–16.

Aladin, N.V. & Plotnikov, I.S. (1995b) Measurements of the level of Aral Sea: paleolimnological and archeological evidence. In: Scarlato, O.A. (ed.), *Biologicheskie i prirodovedcheskie problemy Aral'skogo morya i Priaral'ya, Part 1. Proceedings of the Zoological Institute Russian Academy of Sciences St Petersburg*, **262**, 17–46.

Altunin, V.S., Kupriyanova, E.I. & Tursunov, A.A. (1991) Internal reserves of water for stabilization of Aral Sea and restoration of the ecological balance in its catchment area. *Izvestiya Academy of Sciences SSSR, Series Geographic*, **4**, 118–24.

Anohin, Yu., Belova, N., Ostormogil'skiy, A. & Poslovin, A. (1984) Evolution and forecast of natural environment in the integrated background monitoring in the region of Lake Balkhash. In: Rovinskiyi, F.Ya. (ed.), *Problemy fonovogo monitoringa sostoyaniya prirodnoy sredi 2*. Gydrometioizdat, Leningrad, 176–85.

Bortnik, V.N., Kuksa, V.I. & Cycarin, A.G. (1991) Contemporary state and probable future of Aral Sea. *Izvestiya Academy of Sciences SSSR, Series Geographic*, **4**, 62–8.

Davydova, A.I., Pozdnyakova, G.V. & Shnitnikov, A.V. (1985) Peculiarities of hydrological and hydrochemical regimes of lake Balkhash. In: Alekin, O.A. & Smirnov, N.P. (eds), *Problemy issledovaniya krupnyh ozer SSSR*. Nauka, Leningrad, 239–43.

Davydova N.N., Martinson, G.G. & Sevastjanova, D.V. (eds) (1995) *The History of Lakes of Northen Asia*. Nauka, St Petersburg, 288 pp. (In Russian.)

Druyanov, V. (1984) All-Earth spring. *Nauka i Zhizn'*, **6**, 22–5. (In Russian.)

Feodorov, Yu.A., Grinenko, V.A. & Krauze, R.A. (1996) The state and prediction of the zone of influence of paper and pulp mill upon the water area of Lake Baikal. *Izvestiya Academy of Sciences SSSR, Series Geographic*, **1**, 106–15.

Filippov, A.A. (1993) Some structural peculiarities of macrozoobenthos of coastal zone of the north part of the Aral Sea. In: Skarlato, O.A. & Aladin, N.V. (eds), *Ekologicheskiy krizis na Aral'skom more. Proceedings of the Zoological Institute Russian Academy of Sciences St Petersburg*, **250**, 56–61.

Filippov, A.A. (1995) Macrobenthos of inshore zone of the Aral Sea North in modern polyhaline conditions:

quantity, biomass and spatial distribution. In: Scarlato, O.A. (ed.), *Biologicheskie i prirodovedcheskie problemy Aral'skogo morya i Priaral'ya, Part 1. Proceedings of the Zoological Institute Russian Academy of Sciences St Petersburg*, **262**, 103–67.

Galaziy, G.I. (1988) *Baikal in questions and answers.* Mysl', Moscow, 288 pp. (In Russian.)

Galaziy, G.I. (1990) A threat to Baikal ecosystem. In: Lemeshev, M.Ya. (ed.), *Ecologicheskaya al'ternativa.* Progress, Moscow, 312–49. (In Russian.)

Galaziy, G.I. (1996) Problems of ecological safety of water consumption and protection of water, Baikal being an example. *Gidrobiologicheskiy Zhurnal*, **32**(5), 3–8.

Izrael, Yu.A. & Anihina, Yu.A. (eds) (1991) *Monitoring of the Lake Baikal state.* Gidrometeoizdat, Lenigrad, 261 pp. (In Russian.)

Izrael, Yu.A. & Zenin, A.A. (1973) On the protection of Lake Baikal. *Meteorology and Hydrology* **1**, 15–19. (In Russian.)

Koroleva, N.N. (1993) Phytoplankton of the northern part of the Aral Sea in September 1991. In: Skarlato, O.A. & Aladin, N.V (eds), *Ekologicheskiy krizis na Aral'skom more. Proceedings of the Zoological Institute Russian Academy of Sciences St Petersburg*, **250**, 52–5.

Kozhov, M.M. (1963) *Lake Baikal and its Life.* W. Junk Publishers, Den Haag, 50 pp.

Kozhova, O.M. (1992) An estimate of the status of aquatic ecosystems: causes of crisis and the ways to overcome it (Baikal region). *Biologicheskie Nauki*, **11–12**, 5–8.

Kotlyakov, V.M. (1991) The Aral crisis—scientific and public meaning of the problem. *Izvestiya Academy of Sciences SSSR, Series Geographic*, **4**, 5–21.

Kurnosov, A.M. (1991) The verdict to the lake is not abolished yet. *Moskva*, **6**, 60–2. (In Russian.)

Kuz'min, M.I. (1994) Baikal bottom—a chronicle of the Earth. *Nauka v Rossii*, **6**, 53–7. (In Russian.)

Kuznetsov, N.T. (1991) Geographic and ecological aspects of hydrological functions of Aral Sea. *Izvestiya Academy of Sciences SSSR, Series Geographic*, **4**, 82–9.

Lamirev, V.I. (1970) The Aral Sea. In: Prohorov, A.M. (ed.), *Great Soviet Encyclopedia*, Vol. 2. Sovetskaya Entsiklopediya, Moscow, 465. (In Russian.)

Mirzaev, S.Sh. & Rachinskiy, A.A. (1991) Aral is our common responsibility. *Izvestiya Academy of Sciences SSSR, Series Geographic*, **4**, 113–8.

Murzaev, E.M. (1991) A short review of studies of Aral and Aral region. Responsibility. *Izvestiya Academy of Sciences SSSR, Series Geographic*, **4**, 22–35.

Orlova, M.I. (1993) Materials to the general estimate of the processes of production and destruction in the littoral zone of the northern part of Aral Sea. 1. Results of the field studies and experiments in 1992. In: Skarlato, O.A. & Aladin, N.V. (eds), *Ekologicheskiy krizis na Aral'skom more. Proceedings of the Zoological Institute Russian Academy of Sciences St Petersburg*, **250**, 21–38.

Pavlov, D.S. (1993) *Approaches to Protection of Rare and Endanged Species.* Russian Academy of Sciences, Pushchino, 25 pp. (In Russian.)

Plotnikov, I.S. (1993) The Aral Sea zooplankton in 1992. In: Skarlato, O.A. & Aladin, N.V. (eds), *Ekologicheskiy krizis na Aral'skom more. Proceedings of the Zoological Institute Russian Academy of Sciences St Petersburg*, **250**, 46–52.

Plotnikov, I.S, Aladin, N.V. & Filippov, A.A. (1991) The past and present of the Aral Sea fauna. *Zoological Journal*, **70**(4), 5–15.

Pozdnyakova, G.A. & Shnitnikov, A.V. (1981) On the subsurface (water) supply of Lake Balkhash. *Izvestia Vsesoyuznogo Geograficheskogo Obshchestva*, **110**(6), 473–78. (In Russian.)

Rusakova, O.M. (1995) Concise characteristic of the Aral Sea phytoplankton qualitative composition in spring and autumn 1992. In: Scarlato, O.A. (ed.), *Biologicheskie i prirodovedcheskie problemy Aral'skogo morya i Priaral'ya, Part 1. Proceedings of the Zoological Institute Russian Academy of Sciences St Petersburg*, **262**, 195–208.

Sevastianov, E.D., Mamedov, E.D. & Rumiantchev, V.A. (eds) (1991) *The History of Lakes Sevan, Issik-Kul', Balhash, Zaisan, Aral.* Nauka, St Petersburg, 304 pp. (In Russian.)

Shaporenko, S.I. (1995) Balkhash lake. In: Mandych, A.F. (ed.), *Enclosed Seas and Large Lakes of Eastern Europe and Middle Asia*, SPB Academic Publisher, Amsterdam, 155–97.

Shimarev, M.N. (1985) Possible natural changes of individual elements of the Baikal ecosystem by the end of this century. In: Alekin, O.A. & Smirnov, N.P. (eds), *Problemy issledovaniya krupnyh ozer SSSR.* Nauka, Leningrad, 180–4. (In Russian.)

Shnitnikov, A.V. (1970) Balkhash. In: Prohorov A.M. (ed.), *Great Soviet Encyclopedia*, Vol. 2. Sovetskaya Entsiklopediya, Moscow, 588. (In Russian.)

Shnitnikov, A.V. (1975) Variations of the climate and general humidity in 18–20 centuries and their future. *Izvestiya Vsesoyuznogo Geograficheskogo Obshchestva*, **104**(6), 473–84. (In Russian.)

Shnitnikov, A.V. (1976) Big lakes of the temperate zone

and some ways of their use. In: Shnitnikov, A.V. & Smirnov, N.P. (eds), *Ozera Sredinnogo Regiona*. Nauka, Leningrad, 5–134. (In Russian.)

Shnitnikov, A.V. (1985) Theoretical basis for multi-century variation of the general humidity and the state of lakes—contemporary possible future. In: Aleksin, O.A. & Smirnov, N.P. (eds), *Problemy issledovaniya krupnyh ozer SSSR*. Nauka, Leningrad, 5–23. (In Russian.)

Shutov, M. (1995) Clouds of the 'glorious sea'. *Priroda i Chelovek*, **2**, 10–12. (In Russian.)

Solomatina, E.K. (1995) Baikal at a closer look. *Nauka v Rossii*, **3**, 46–53. (In Russian.)

Suzyumov, E.M. & Tsiproruha, M.I. (1991) Discovering the mysteries of the ocean. *Znaniya*, **2**, 13–17. (In Russian.)

Tarasova, E.N. & Meshcheryakova, A.I. (1992) *Contemporary State of the Lake Baikal Hydrochemical Regime*. Nauka, Novosibirsk, 143 pp. (In Russian.)

Votintsev, K. K. (1992) On the characteristics of the self-purification potential of Lake Baikal. *Sibirskiy Biologicheskiy Zhurnal*, **4**, 36–40.

Zolotokrylkin, A.N. & Tokarenko, A.A. (1991) On the variations of the climate in Aral region for the recent 40 years. *Izvestiya Academy of Sciences SSSR, Series Geographic*, **4**, 69–76. (In Russian.)

Williams, W.D. (1995) The Aral sea: limnological prospects. In: Scarlato, O.A. (ed.), *Biologicheskie i prirodovedcheskie problemy Aral'skogo morya i Priaral'ya, Part 1. Proceedings of the Zoological Institute Russian Academy of Sciences St Petersburg*, **262**, 237–47.

Williams, W.D. & Aladin, N.V. (1991) The Aral Sea: recent limnological changes and their conservation significance. *Aquatic Conservation: Marine and Freshwater Ecosystems*, **1**, 3–23.

# 8 Lakes in Arid Environments

## W.D. WILLIAMS*

### 8.1 INTRODUCTION

Paradoxically, some of the world's largest lakes (and reservoirs), as well as many smaller bodies (but also including many with an area >100 km$^2$), occur in arid environments. They range from large, deep, permanent, freshwater lakes, to small, shallow, temporary, saline bodies. Many are of particular significance to local human populations, totalling some 10% of all people on Earth. However, despite their size, number and importance, comprehensive limnological knowledge of them lags far behind our knowledge of lakes and reservoirs elsewhere.

It is, of course, not difficult to see why our knowledge of the limnology of lakes in arid regions is in this state: most centres of limnological research are in temperate regions and most limnologists are interested in and study local lakes (Williams 1988). Nevertheless, some early investigators, notably Murray (1910), were keenly aware of the extent and general nature of lakes in arid environments, even if somewhat later ones (e.g. Welch 1935; Ruttner 1963) essentially chose to ignore them. Recent general texts (e.g. Burgis & Morris 1983; Margalef 1983) usually provide a more balanced treatment, and one (Serruya & Pollingher 1983), although less than comprehensive, pays particular attention to lakes situated in warm arid regions. Several reviews are also of interest in this connection, notably the early classic texts by Hutchinson (1937) on environments in the Lahonton basin, Cole (1968) on desert lakes, those edited by Robarts & Bothwell (1992) on aquatic ecosystems in semi-arid regions, and by Comm & Williams (1994) on dryland limnology. One kind of lake characteristic of arid environments, the salt lake, has attracted particular attention, and several texts discuss them (Kiener 1978; Nissenbaum 1980; Hammer 1986; Javor 1989); indeed, a relatively significant literature has accumulated.

Studies on a few lakes (both fresh and saline), e.g. Lake Kinneret (Serruya 1978), Lake Chad (Carmouze et al. 1983), the Gavish Sabkha (Friedman & Krumbein 1985), Mono Lake (Mono Basin Ecosystem Study Committee 1987), the Caspian (Kosarev & Yablonskaya 1994), several large lakes in central Asia (Mandych 1995), and the Aral Sea (Micklin & Williams 1996), have been reviewed. Other texts of indirect interest in the present context include those by Walton (1969), Neal (1975), Mabbutt (1977), Beaumont (1989), Thomas (1989), Agnew & Anderson (1992), Cooke et al. (1993), Abrahams & Parsons (1994), Murakami (1995), and Williams & Balling (1996); these discuss water resources, geomorphology, environmental management and related topics.

This chapter discusses four main subject areas. First, it deals with the nature and extent of arid environments. Of particular note here is that, depending on the definition of what constitutes an arid region, arid environments occupy between one-quarter and one-half of the world's total land area. Second, it provides a synopsis of major lakes and reservoirs in arid environments and discusses the nature of four important types of lentic waterbody there, namely, (i) permanent, freshwater lakes, (ii) reservoirs, (iii) temporary, freshwater lakes and (iv) saline lakes. Third, it discusses the

---

Sadly, Bill Williams also died before he was able to proof-read his chapter, which we have completed using his notes on our edited version, supplied by Bill not long before his death. We are proud to include this masterly chapter by a well-liked and well-respected scientist in The *Lakes Handbook* Volume 2.

uses and values of lakes and reservoirs in arid environments and threats to them from human impact. Finally, five case studies are presented which have been selected both to illustrate the range of lakes found in arid environments and to emphasise the significant limnological differences which exist between this category of lake and those distributed elsewhere. They are: Lake Chad in central Africa (a permanent, freshwater lake), the Aswan High Dam in Egypt and the Sudan (a reservoir), the Coongie Lakes in Australia (temporary to permanent freshwater lakes), the Aral Sea in central Asia (a permanent saline lake), and Lake Eyre in Australia (a temporary saline lake).

## 8.2 THE NATURE AND EXTENT OF ARID ENVIRONMENTS

Aridity is a concept linked to the lack of moisture, but this alone is too simple fully to illustrate what is intended by the use of the word arid. Pedological, geomorphological and biological factors, and a range of climatic features other than moisture, are those which more clearly identify arid environments. For present purposes, it is convenient to restrict definition largely to climatic criteria, as these, which are manifested and integrated by the hydrological budgets of lakes, are amongst the most significant determinants of lake occurrence and type.

In general climatic terms, an arid region is one where annual rainfall is low and evaporation high. Note that annual mean temperature may also be low ($<0°C$), as in the Antarctic, the world's most arid continent, or high ($>25°C$), as in some African deserts. More precise definitions of aridity range from relatively simple, early examples, which involved consideration of precipitation and temperature only (e.g. de Martonne & Aufrere 1928; Köppen 1931), to complex indices which also incorporate vegetation, climatic seasonality, transpiration and evaporation. Of the latter, those by Penman (1948) and Thornthwaite (1948) are best known. Thornthwaite's index was further developed by Meigs (1953) to produce a moisture index showing relationships between precipitation and

**Table 8.1** Criteria defining zones of bioclimatic aridity and indicative mean annual precipitation within them: $P$, mean annual precipitation (mm); $P_{et}$, mean potential evapotranspiration

|  | $P/P_{et}$* | $P/P_{et}$† | $P$‡ | $P$¶ |
|---|---|---|---|---|
| Hyper-arid | <0.03 | <0.05 | <25 | <80 |
| Arid | 0.03–0.20 | 0.05–0.20 | 25–200 | 80–350 |
| Semi-arid | 0.20–0.50 | 0.21–0.50 | 200–500 | 200–700 |
| Subhumid | 0.50–0.75 | 0.51–0.65 | [>500] | [>700] |

\* According to UNESCO (1977).
† According to UNEP (1992).
‡ According to Grove (1977).
¶ According to Beaumont (1989).

evapotranspiration, and on this basis he divided 'arid' regions of the world into three subregions: extremely arid, arid and semi-arid.

Further refinement led UNESCO (1977), and more recently UNEP (1992), to produce a map of the world's arid regions which involved four categories of bioclimatic aridity, based on the relationship between mean annual precipitation and potential evapotranspiration (calculated from Penman's index): hyper-arid, arid, semi-arid and subhumid. These categories, and the criteria on which they are based, are listed in Table 8.1, where also shown are two indicative **general guides** to mean annual precipitation in each category. A map of the distribution of these four categories was produced (UNESCO 1977), but the earlier and simpler map by Meigs (1953), showing the distribution of his hyper-arid, arid and semi-arid subregions, is the one most widely referred to, and for our purposes the most useful, and is reproduced in Fig. 8.1. Following Graf (1987), a convenient, shorthand way of referring to all three kinds of subregion is as 'drylands'.

Given the number of indices defining aridity, several values for estimating the global extent of drylands exist. Table 8.2 provides some for the most widely used indices. Even so, as can be seen, and notwithstanding the index used, total global dryland area is significant (from about one-quarter to almost a half of total land area). Drylands are not, of course, evenly spread geographically, and

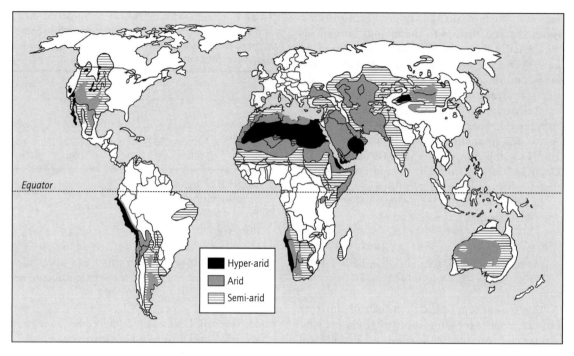

**Fig. 8.1** World distribution of dryland areas. (According to Meigs 1953.) (*UNESCO Arid Zone Program.*)

**Table 8.2** The global extent of drylands according to various criteria defining arid areas. Figures as percentage of total world land area

| Source | Total area |
|---|---|
| Köppen (1931) | 26.3 |
| Thornthwaite (1948) | 30.6 |
| Meigs (1953) | 36.3 |
| UNESCO (1977) | 32.8 |
| UNEP (1992) | 37.5* |

\* This figure does not include the dry subhumid zone (9.9%).

considerable intercontinental differences exist. Table 8.3, based on UNESCO estimates, illustrates this point. Broadly speaking, drylands are characteristic of western North America, eastern and southern South America, northern and southwestern Africa, central Asia, all non-coastal parts of Australia and all of Antarctica.

From a limnological viewpoint, an important feature of dryland climates, over and above the small absolute amount of annual rainfall, and the relationship between this factor and potential evapotranspiration, is their temporal variability or unpredictability. This feature, irrespective of its limnological importance, also distinguishes dryland climates from those of temperate arid tropical regions. Both seasonal and interannual (long-term or secular) variability is involved. Several measures have been proposed to describe this property, some of which are complex, and distinguish between key components such as constancy and seasonal regularity.

For present purposes it is sufficient only to note that, as a general rule, temporal climatic variability increases with aridity. The general relationship between aridity (as expressed by mean annual temperature, rainfall and net evaporation) and the likelihood ([un]predictability) of occurrence of waterbodies is shown in Fig. 8.2. This relationship is of considerable limnological significance, not least because temporal climatic variability is a sig-

Table 8.3  Geographical distribution of dryland areas. Based on UNESCO's (1977) criteria

| Continent | Hyper-arid | | Arid | | Semi-arid | | Total | |
|---|---|---|---|---|---|---|---|---|
| | $10^6$ km² | % | $10^6$ km² | % | $10^6$ km² | % | $10^6$ km² | % |
| Africa | 6.1 | 20.1 | 6.2 | 20.4 | 5.1 | 16.9 | 17.4 | 57.4 |
| America | 0.2 | 0.4 | 2.1 | 4.9 | 4.7 | 11.0 | 7.0 | 16.3 |
| Middle East | 1.1 | 18.3 | 3.0 | 49.7 | 1.0 | 16.0 | 5.1 | 84.0 |
| Asia | 0.4 | 1.0 | 4.0 | 10.5 | 5.3 | 13.9 | 9.7 | 25.4 |
| Australia | 0.0 | 0.0 | 3.8 | 49.0 | 1.5 | 20.0 | 5.3 | 69.0 |
| Europe | 0.0 | 0.0 | 0.0 | 0.1 | 0.2 | 2.3 | 0.2 | 2.4 |
| World | 7.8 | 5.7 | 19.1 | 14.1 | 17.9 | 13.2 | 62.5 | 33.0 |

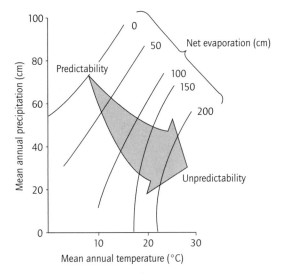

Fig. 8.2  The general relationship between aridity (expressed as mean annual temperature, rainfall and net evaporation) and the likelihood ([un]predictability) of occurrence of water-bodies. (After Langbein 1961.) *U.S. Geological Survey (Professional Paper* No 412).

nificant factor in determining the nature of biological communities in many dryland lakes. A fundamental dichotomy in this respect is between lakes in predictable seasonally variable climates, and those under unpredictable, **aseasonal** regimes.

An important feature associated with temporal climatic variability is spatial variability. The result is that, even over relatively small areas (<10 km²), considerable random differences in actual local climates may occur. For example, one region may receive significant rainfall, whereas another, nearby, and over the same period, may receive none. Such random differences are cumulative, and may compound local differences over time.

Overall, it cannot be emphasised too strongly that seasonal and secular patterns of climatic variability, and the consequences for hydrological patterns of flooding, filling and drying, are the major keys determining both physico-chemical and biological events in lakes of arid environments.

## 8.3  LAKES AND RESERVOIRS IN ARID ENVIRONMENTS

### 8.3.1  Number, size, limnic density and ratio

The largest lake in the world, the Caspian Sea (area 374,000 km²), as do several other lakes of area greater than 10,000 km² (Table 8.4), lies in an arid environment (see also Löffler, Volume 1, Chapter 2). Additionally, several extremely large reservoirs have been constructed in drylands, including the Aswan High Dam (volume 168,900 km³; area 6500 km²; one of the world's largest reservoirs), the Razza Dyke (volume 26,000 km³) in Iraq, and the Glen Canyon (33,000 km³) and Hoover (34,852 km³) reservoirs in the USA. How is it possible for such large bodies of water to exist in drylands? The paradox is easily explained: all are fed

**Table 8.4** Major morphometric features of large lakes (area >10,000 km$^2$) in arid environments. (Data drawn from Gleik 1995)

| Lake | Surface area (km$^2$) | Maximum depth (m) | Volume (km$^3$) | Drainage basin area (10$^6$ km$^2$) |
|---|---|---|---|---|
| Caspian Sea | 374,000 | 995–1025 | 78,200 | 3.625 |
| Aral Sea* | 64,000 | 68 | 1020 | 1.618 |
| Lake Chad | 7000–26,000 | 4–12 | 44 | 2.427 |
| Lake Balkhash | 17,000–19,000 | 26 | 112 | 0.176 |
| Lake Eyre | 0–40,000 | 0–20 | 0–23 | 1.122 |
| Lake Torrens | 0–30,000 | <5 | | |

\* Data for pre-1960, i.e. before recent anthropogenically induced decrease in size.

**Table 8.5** Lake distribution in drylands and temperate biomes: $L_D$, limnic density; $A_o$, lake area; $N$, lake number; $L_R$, limnic ratio. (Data derived and modified from Meybeck 1995)

| Lake area (km$^2$) | Dryland Biomes (total area = 65 × 10$^6$ km$^2$) | | | Temperate Biomes (Total area = 26 × 10$^6$ km$^2$) | | |
|---|---|---|---|---|---|---|
| | $L_D$ | $A_o$ | $N$ | $L_D$ | $A_o$ | $N$ |
| 0.01–0.1 | 6000 | 6500 | 250,000 | 10,000 | 7000 | 260,000 |
| 0.1–1 | 800 | 8500 | 33,000 | 2000 | 13,000 | 50,000 |
| 1–10 | 33–100* | 12,500 | 4770 | 200 | 13,000 | 5000 |
| 10–10$^2$ | 8–15* | 22,000 | 840 | 30 | 21,000 | 800 |
| 10$^2$–10$^3$ | 1.4–2.8* | 36,600 | 153 | 6 | 40,000 | 155 |
| 10$^3$–10$^4$ | 0.64–0.69* | 104,000 | 43 | 0.73 | 63,000 | 19 |
| 10$^4$–10$^5$ | 0.04–0.12* | 165,000 | 6 | 0.04 | 31,500 | 1 |
| >10$^5$ | – | 374,000 | 1 | 0.00 | 0.00 | 0 |
| Total lake area (10$^6$ km$^2$) | | 0.729†/0.164‡ | | | 0.188 | |
| $L_R$ (%) | | 1.121†/0.45‡ | | | 0.7 | |

\* Without access to all of Meybeck's data it is not possible to derive a common value for this parameter. The range given by Meybeck (1995) is shown.
† Values include data for the Caspian and Aral Seas.
‡ Values exclude data for the Caspian and Aral Seas.

by **allogenic** rivers (i.e. rivers which arise in well-watered areas and flow into drylands). Thus, the River Volga, which is the main source of water for the Caspian, rises in European Russia, and the Ainu- and Syr-Darya, which feed the Aral Sea, in humid regions of the Middle East and central Asia. The River Chari, which feeds Lake Chad, rises in humid west Africa, and the Aswan High Dam is fed, of course, by the River Nile, whose source lies in humid east Africa.

However, it is not only large lakes and reservoirs which occur in drylands; as Table 8.5 indicates (for lakes), lakes and reservoirs of all sizes are found there. This table is based on that given by Meybeck (1995), which in turn is largely compiled from a comprehensive analysis of lake censuses and literature, especially papers by Herdendorf (1982, 1984, 1990). In Table 8.5, Meybeck's data on lakes in (i) dry and arid biomes (dry temperate + arid + savanna, all with surface runoff

$<4\,L\,s^{-1}\,km^{-2}$) and (ii) desert biomes (surface runoff $>0.1\,L\,s^{-1}\,km^{-2}$) have been combined. Although the total global area of drylands thus considered is somewhat larger than the total area indicated in Table 8.3, these data nevertheless provide a useful basis for examination of the distribution of lakes in arid environments, with respect to number, size (area), and limnic density and ratio.

As the table indicates, there are indeed many thousands of lakes found in drylands, and although most are $<1\,km^2$ in area there are almost 6000 with areas $>1\,km^2$. These numbers are likely to be conservative, given the propensity in the limnological literature to underestimate the number and area of temporary lakes. Table 8.5 also gives data on limnic density and ratio (*sensu* Meybeck 1995), where limnic density is the number of lakes per million square kilometre of region considered, and limnic ratio the total lake area as a percentage of the total area of the region. It can be seen that limnic density in drylands ranges from 6000 (or one lake per $166\,km^2$) for small lakes (area $<0.1\,km^2$) to $\ll 1$ for large lakes (area $>10,000\,km^2$). The limnic ratio for lakes in drylands is 1.121 if the Caspian and Aral Seas are included in the calculations, but only about half this value if they are excluded.

How do these values compare with those for lakes in better-watered regions, such as temperate biomes? To enable comparison, relevant data (also from Meybeck 1995) are given in Table 8.5. These provide a number of surprises. Whereas there are rather more smaller to moderately sized lakes (area $<1000\,km^2$) in temperate regions than in drylands (c.316,000 to 289,000), there are more than twice as many larger lakes (area $>1000\,km^2$) in drylands (50:20). For limnic ratios, if all dryland lakes are considered (i.e. the Caspian and Aral Seas are included), then the value is greater for drylands than for temperate regions. Excluding the Caspian and Aral Seas produces a limnic ratio for drylands which is still almost two-thirds (64%) that of temperate regions.

Likewise, total lake area comparisons (Table 8.5) are also surprising. Including the Caspian and Aral Seas, the total global area of lakes located in drylands ($729,000\,km^2$) far exceeds that for temperate regions ($188,000\,km^2$). Even if they are excluded, the total area of lakes in drylands ($164,000\,km^2$) is not markedly less than that for lakes in temperate regions. Indeed, on a global basis, the number of lakes in drylands and in temperate regions, as a percentage of total world number, is about the same (6%). If lake areas are considered on a global basis, the percentage of global lake area for dryland lakes is 30% when the Caspian and Aral Seas are included, and 7% if not. For temperate lakes, the figure is only slightly more than this second value, namely 8%.

Similar comparisons could be produced for other biomes, e.g. those described by Meybeck (1995) as deglaciated and wet tropics. The main point of the present comparison is to underscore the central point that, whereas aridity is certainly a key climatic factor in drylands, its local limnological effects are masked to such an extent by hydrological and other features that lake number and size in these regions are not nearly as low as might be assumed from a priori principles. Indeed, they are not grossly dissimilar to values for better-watered areas.

### 8.3.2 Lake types

A large variety of standing (**lentic**) bodies of water are found in drylands, not all of which can be easily regarded as 'lakes'. These include permanent rock pools (e.g. pools on granitic outcrops in deserts), naturally impounded spring-fed waters (e.g. mound springs), extensive shallow swamps (e.g. the Okavango Delta, southwest Africa), surface exposures ('windows') of underlying aquifers, and small water-filled depressions on the floodplains of allogenic rivers. None of these is discussed here. Instead, treatment is largely confined to large, lentic water-bodies (area $>0.01\,km^2$). However, as in all parts of nature, it is recognised that boundaries between entities are not distinct, and that all gradations of size, depth and extent of water permanency and flow exist in dryland lentic waters. It is also recognised that the definition of a lake given here is as much open to argument as are most current definitions of 'wetland' (cf. Finlayson & Moser 1991).

Lakes themselves exhibit a variety of forms in

drylands, a variety greater than that characteristically found in any other biome, both tropical or temperate. Thus, whereas permanent, freshwater lakes (and reservoirs) are the characteristic lake type in more humid regions, in drylands, three other types also occur. In all, five categories may be distinguished:
1 permanent, freshwater lakes;
2 reservoirs (permanent, artificial, freshwater lakes);
3 temporary, freshwater lakes;
4 permanent, saline lakes;
5 temporary, saline lakes.

The reasons for this wider variety are essentially twofold. First, climatic aridity and its temporal variability explains the development of temporary waters. Second, the occurrence of **endorheic** drainage systems (hydrologically closed drainage networks), which terminate in a lake rather than discharging to the sea, lead to the concentration of salts and thus the development of saline lakes.

It is convenient to discuss the major types of lentic water-bodies found in drylands separately (although more convenient to consider permanent and saline lakes together). Before doing so, it is important to provide clear definitions of certain terms which have been widely used in the Anglophone literature to describe degrees of water permanency in dryland lakes. Considerable confusion has arisen in particular, over the use of the terms **temporary**, **intermittent**, **seasonal**, **periodic**, **episodic** and **ephemeral**. As Comm & Williams (1994) noted, in a discussion of these terms in relation to lotic waters in drylands, no agreed view of what these terms mean has emerged. They went on to provide a single set of definitions which can be easily modified to apply to dryland lakes, both fresh and saline (and indeed all dryland waterbodies). Accordingly, in order to reduce at least some confusion in this area, the following definitions are proposed and used in this chapter.

1 *Permanent lakes* are lentic systems which contain water at all seasons and over many years. If they dry up, they do so only infrequently and then only during severe droughts.
2 *Temporary lakes* are lentic systems which are frequently dry; when dry, no surface water remains, or water may persist in remnant pools. There are two types of temporary lake: *intermittent* and *episodic*.
(a) *Intermittent lakes* are lentic systems which contain water, or are dry, at more or less predictable times during an annual cycle. Intermittent lakes therefore can occur both in drylands and in humid regions.
(b) *Episodic lakes* are lentic systems which only contain water more or less unpredictably. They are therefore confined to arid and to hyper-arid regions.

As further noted by Comm & Williams (1994), the term 'ephemeral', which is often taken to be synonymous with episodic, and sometimes with intermittent, should not be used, as it is likely to confuse. Also to be discouraged is the use of 'temporary' as a synonym for intermittent. Both terms have been widely used by stream ecologists interested in lotic environments in drylands, several of whom have also attempted rationalisation of terms. Thus, Boulton (1988, 1989) distinguished 'temporary' or 'intermittent' rivers (those with seasonally regular discharges) from 'ephemeral' or 'episodic' rivers (which flowed only after irregular rains had fallen). Davies et al. (1994) simply distinguished 'temporary' and 'ephemeral'. Most proposals by stream ecologists are easily accommodated in the scheme of definitions offered here, however.

Also requiring definition, in order to avoid confusion, is the term 'saline'. Here, its usage follows the convention (see Williams & Sherwood 1994) that a saline or salt lake is one which contains a concentration of total salts greater than $3\,g\,L^{-1}$ (3‰).

### 8.3.2.1 Permanent, freshwater lakes

Dryland lakes originate in many ways: volcanic action, tectonic movements of the earth, river activities, solution and, frequently, aeolian action (**deflation**). Glacial action, so characteristic in the formation of lakes in temperate areas, has played no part. The nature of these origins is reflected in the morphometry of dryland lakes: most are shallow. In permanent lakes, however, depth is sufficient to enable retention of water between significant inflows. In freshwater lakes, these flows are greater than evaporation so that the lake is an open hydrological system and salts do not accumulate (see also Loeffler, Volume 1, Chapter 2; Winter, Volume 1, Chapter 3).

Shallowness leads to a number of physicochemical consequences. One of the most significant is that wind action becomes relatively more important, and water circulation patterns, nutrient dynamics and production processes reflect this trend. Wind action may lead to addition to the lake of large amounts of organic matter from the surrounding terrestrial environment, and this may serve as a source of food for the aquatic biota (a process referred to as **anemotrophy**). Wind action may also maintain large amounts of material suspended in the water column, and this too may serve as a source of food (**argillotrophy**). Both were first recognised as important limnological phenomena in studies of dryland lakes (Hutchinson 1937). The action of the wind in maintaining material suspended in the water column may also be important in at least two other ways: increased turbidity cuts down penetration of light (and thus decreases photosynthesis by phytoplankton), and enhances various interactions between sediments and the overlying water.

Shallowness also produces other effects. It increases the extent and rapidity of impact of many external physical phenomena on the lake, from which it is thus less well-buffered. This point is most readily illustrated by increased fluctuations in temperature; marked temperature fluctuations in the lake may occur over relatively short intervals. Shallowness, in combination with the temporal climatic variability characteristic of dryland regions (and manifest, in particular, via seasonal or secular changes in inflows, precipitation on the lake and evaporation), leads to much greater hydrological variability: significant temporal changes in lake area, depth and volume occur. **Astatic** water-levels are by no means confined to temporary lakes in drylands, and shallowness, of course, significantly influences thermal patterns.

With regard to thermal patterns in particular, the dimictic pattern so characteristic of temperate regions is virtually absent. Most permanent, freshwater lakes in drylands are monomictic, polymictic or warm **thereimictic** (*sensu* Bayly & Williams 1973). Even diurnal (**diel**) differences in horizontal and vertical circulation patterns may develop, if marked differences between day and night air temperatures occur and the lake is shallow enough: the lake may stratify by day and mix by night (or the reverse). As most solar energy is absorbed in the topmost layers of water in turbid lakes, thermal gradients, in any event, may often be abrupt.

An important point often forgotten when considering thermal patterns in lakes in warm regions is that the relationship between water density and temperature is different at higher temperatures, and density changes are more pronounced. This means that lakes in warm dryland regions are thermally more stable and possess shallower epilimnia than their morphometric analogues in cooler regions. Finally, with regard to thermal events, note that evaporative cooling affects heat budgets, with mean water temperature declining in many dryland lakes before seasonal heating has ceased to be effective (Neumann 1953), a phenomenon not found in temperate or tropical lakes.

The chemical properties of dryland lakes are also variable. Patterns and profiles of oxygen distribution follow those for temperature, although modified by photosynthesis, respiration and decomposition. Broadly speaking, the higher temperatures found in warm dryland lakes lead to reduced concentrations of dissolved oxygen in the water column, but enable increased rates of photosynthesis, respiration and decomposition—all of which affect dissolved oxygen concentrations. These may often fluctuate markedly on a diel basis, and from aerobic to anaerobic within a single day.

Total salt concentration generally remains low and below $3\,g\,L^{-1}$, but if seasonal restrictions in outflows (either surface or bottom) occur, evaporation may lead to minor increases in salt concentration. Of importance in this respect is the degree of association of the lake with underlying groundwaters; in many dryland lakes this is close. Subsurface waters are largely protected from the full effects of seasonal evaporation. Nutrient concentrations, unlike those of salt, are generally high, but both phosphorus and nitrogen may limit plant production.

Biological phenomena integrate, and are a product of, all of the above physico-chemical factors. Overall, rates of biological processes are greater, because of the usually higher temperatures. Primary productivity is generally high, reflecting both elevated temperatures and increased insolation and nutrient availability. Rates of sequential processes—secondary production, respiration, decomposition—are likewise generally high. However, physico-chemical variability, or the lack of stable environmental conditions, may act to limit certain biological processes. Astatic water-levels, for example, which in shallow lakes invariably lead to exposure of extensive littoral areas to desiccation (either seasonally, or at longer intervals), lead to destruction or restriction of macrophytes. High turbidity limits phytoplankton production. Marked fluctuations in oxygen concentrations exert stress on many faunal communities.

Notwithstanding the importance of such impacts on biological communities, so far as diversity, taxonomic structure and endemicity is concerned, these are no simpler in dryland lakes than in those of more humid regions. There are no features of this kind which are unique to, or distinctive of, permanent freshwater lakes in drylands. There are of course some taxonomic differences, but these are more a product of broadscale biogeographical trends than of regional ecological factors.

At the same time, distinct regional differences in the distribution of biota do occur. This is well illustrated by the fish fauna (cf. Sarvala *et al.*, Volume 1, Chapter 16). For example, the Salmonidae are (naturally) absent from lakes in the dryland regions of South America, Africa, Australia, and certain parts of Asia. This pattern of distribution is clearly the result of biogeographical events, not ecological factors, since certain species of salmonid are found in dryland lakes in North America, and other parts of Asia (as well as in humid areas throughout the northern hemisphere). The Cichlidae, so characteristic of African dryland lakes, are naturally absent from most of Asia and Australia. In Africa, however, the family occurs widely in humid temperate and tropical lakes, as well as dryland ones. Fish distributions, it may be added, have been greatly altered by artificial introductions (purposive or accidental) and many dryland lakes now contain fish which are naturally absent from the regions where the lakes occur. Amongst the more notable of such introductions are species of Cyprinidae (especially *Cyprinus carpio* L.) and Poeciliidae (especially *Gambusia affinis* Baird et Girard), many of which are extremely tolerant of high temperatures, elevated salinities, and other stressful physico-chemical conditions of dryland lakes, all of which may limit the occurrence of native fish species.

#### 8.3.2.2 Reservoirs

Reservoirs in drylands, as with permanent, freshwater lakes, span a wide range of size and depth, from small farm dams (area $\ll 0.01\,km^2$, maximum depth <10 m) to Lake Nubia (area >100,000 $km^2$, maximum depth >100 m). Small, shallow reservoirs behave essentially in the same ways as shallow lakes, but large, deep bodies do not; they display a number of differences. Many of these spring from the relative impact of the wind, which decreases as water-bodies become larger. Increasing depth, and thus water mass, leads to development of other physico-chemical differences.

Thus, in deep reservoirs, concentrations of suspended matter in the water column are less than those in shallow dryland lakes (and as turbidity falls, light penetration increases). The extent of interaction between bottom sediments and the overlying water is smaller. Buffering from external physical phenomena is greater; marked diurnal differences in thermal and oxygen conditions are absent, or less obvious. The predominant thermal pattern is warm monomictic (Fig. 8.3). Long periods of summer stagnation, it may be noted, may and frequently do lead to severe depletion of hypolimnetic oxygen concentrations (because high temperatures promote decomposition) and the production of hydrogen sulphide.

Although the larger mass of water per unit area suppresses some of the variability in physico-chemical features characteristic of lakes, opera-

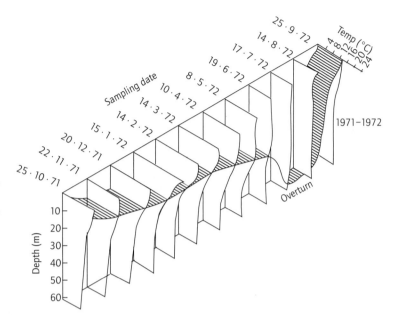

Fig. 8.3 Warm monomictic thermal pattern as displayed by Eildon Reservoir (Lake), Victoria, Australia, 1971–72. Hatched areas represent thermocline. (Modified from Powling 1980.) Australian National University Press, Canberra—for Williams, W.D. (ed) *An Ecological Basis for Water Resource Management.*

tional procedures for reservoirs are themselves a source of variability; for example, large releases of water for irrigation (leading to significant falls in water-level) may expose extensive littoral areas. Water storage and release patterns may produce water-level fluctuations quite different from natural patterns. It should not be forgotten, too, that operationally induced changes are added to those of natural origin: the larger mass of water does not suppress these altogether. Thus, in reservoirs in semi-arid areas of Australia, evaporation alone may lead to seasonal falls in water level of greater than 3 m, and reservoirs may need to be at least 4 m deep if they are to retain water all year (Timms 1980).

Again, physico-chemical features are reflected by biological phenomena which are distinct from those of lakes. Anemotrophy and argillotrophy are not important. Phytoplankton production occurs in a greater volume of water. Benthic communities are severely affected if hypolimnetic anoxis occurs. Rooted macrophytes are generally scarce (which in turn affects fish and invertebrate communities). Fluctuating shorelines frequently act to produce artificial floodplains.

Little can be said which is definitive with regard to more general biological attributes; almost any general statement could be repudiated by several contra-indications. Even so, in very broad terms, it appears that in dryland reservoirs, biological diversity, at least, is relatively depressed. As most are of recent construction, and dam closure is followed by long periods of physico-chemical and biological instability, diversity may ultimately increase once stable conditions develop. Alternatively, diversity may remain low either because the majority of native aquatic species are unable to adapt to the unnatural conditions which prevail in dryland reservoirs, or because there are usually few, if any, nearby deep freshwater lakes which can act as an appropriate source for colonisers.

### 8.3.2.3 Temporary, freshwater lakes

Temporary, freshwater lakes are amongst the most frequently encountered lentic dryland waterbodies. In certain areas, they represent the most obvious landscape feature: the **vleis** of the high veldt of South Africa, the numerous **pans** of north-western Texas and (formerly) the lakes of southeastern South Australia provide examples of such water-bodies and areas.

A number of factors—climatic, topographical, morphometric—determine hydrological patterns (particularly when and for how long water is present), so that examples range from small, shallow basins where water is present unpredictably (episodically) and not for long, to large, deeper basins where water is present on a seasonal, predictable basis (intermittently) and persists for several months. This variety of type has given rise to many names for local examples: **playas**, **pans**, temporary ponds, vernal pools and **vleis** are amongst the most common. No clear distinction exists between the localities described by these terms. Here, they are all simply regarded as temporary lakes.

The major physico-chemical features of temporary, freshwater lakes are, in large measure, essentially extensions of those which characterise permanent, freshwater lakes. In short, water circulation patterns, nutrient dynamics and sediment–water interactions (as well as several biological features such as productive processes and overall ecology) are even more variable and more closely linked to ambient environmental conditions in these aquatic habitats than they are in permanent bodies (Comm & Williams 1994).

Aeolian action is usually even more important in their genesis, and often continues to act as a geomorphological agent during the dry phase, when deflation of surface sediments occurs, especially when these are unvegetated. Deflation may lead to formation of marginal dunes or **lunettes**, to sorting of surface sediments, development of surface deposits, and so on (Thomas 1989).

Temporary freshwater lakes, being generally shallower than their permanent counterparts, are subject to even greater hydrological variability and wind action. Thus, turbidity is more responsive to wind action, sediment–water interactions are greater, environmental buffering is lower, only short-lived thermal stratification develops and oxygen concentrations usually lie at or near equilibrium (but may display wide diurnal changes). Many temporary lakes are closely associated with groundwaters, so that although temporal chemical changes may occur, they are more muted than hydrological and other phenomena. Only rarely, for example, do salinities increase beyond the $3\,g\,L^{-1}$ threshold, even when evaporation has left only a small volume of surface water.

Many of the physico-chemical features referred to above mean that temporary freshwater lakes in drylands are somewhat formidable habitats for most aquatic biota. Animals and plants which live in them must have adapted, *inter alia*, to at least occasional, and sometimes frequent and persistent occurrence of high temperatures, excessive insolation, environmental instability, the absence of water, habitat isolation and astatic water-levels (W.D. Williams 1985; D.D Williams 1987). However, 'nature abhors a vacuum', and many adaptations have evolved. These include behavioural, physiological, morphological and ecological traits. Thus, many animals avoid excessive stress by seeking protection in mud, beneath vegetation or in nearby refugia. Light protective pigments have evolved in some cases (both plant and animal), and many plants and animals have developed drought resistant stages (eggs, cysts, seeds, diapause). Lifecycle patterns and strategies have also evolved which take account of the temporal patterns of water availability. Finally, efficient dispersal mechanisms have been developed.

Despite the many stresses of life in temporary lakes, and the many adaptations needed in order to cope with them, it is abundantly clear that a large number of aquatic species find them suitable, and indeed thrive in them. Thus, overall, diversity is high (cf. King *et al.* 1996), and certainly much higher than suggested by earlier studies (e.g. Hartland-Rowe 1972), with representatives from most aquatic taxa present. Most notable, however, are notostracans, anostracans and conchostracans. Notably absent are those taxa which lack a drought resistant stage (e.g. amphipods, isopods, fish), although even these may be present if nearby permanent aquatic refugia are available from which colonisation can occur.

In addition to significant diversity, considerable regional **endemicity** and differences in the biota may be found. This is a result both of biogeographical and of ecological factors. The former operate both between and within continents. Thus, the composition of the fauna of temporary freshwater lakes in Australia is distinct from that of Africa. At the same time, the fauna within Australia exhibits marked regional differences. Even at local level, genetic differences between isolated populations may occur.

Some regional differences are the result of broad biogeographical events (e.g. fluctuations in past climates), but others are produced by more immediate, ecological factors. Amongst these, the timing and persistence of the aquatic phase is obviously one of the most important. In this regard, as indicated above, two major patterns exist—one in which the aquatic phase is seasonally predictable, and lasts for less than a year (**temporary**, **intermittent lakes**), and another in which it is unpredictable and, depending upon the volume of water initially present, lasts only a relatively short time, several months at most (**temporary**, **episodic lakes**). The full extent of the ecological differences between intermittent and episodic temporary lakes has yet to be determined, as are the ways in which the hydrological nature of the aquatic phase affects the ecology of temporary waters (as in: how ecologically important are the duration, amplitude of the flood pulse, timing, flushing, frequency, rates of rise and fall, and source of the water?; Boulton & Jenkins 1998).

#### 8.3.2.4 Saline lakes

Permanent and saline lakes represent the two remaining major types of lentic waterbody in drylands. As they share many features in common, they are treated together.

An important point to make in any discussion of salt lakes (i.e. lakes with salinities $>3\,g\,L^{-1}$) is to emphasise that they are a good deal more common than most general textbooks indicate (cf. Williams 1986, 1996a). Several estimates have been made of the fraction of inland waters which comprises saline waters, and all indicate that it is not insignificant. The most recent authoritative assessment (Shiklomanov 1995), based on the work of many former Soviet scientists, is that the total global volume of saline lakes is $85,000\,km^3$, or 0.006% of total biospheric water (the corresponding figures for freshwater lakes are $91,000\,km^3$ and 0.007%). At face value, there is not a great deal of difference between these two sets of figures, and on this basis some 48% of total lake water in the world resides in salt lakes.

However, it should be noted that the major proportion of inland salt water (91.5%) is located within the Caspian Sea, and if this is excluded from calculations then the percentage of global inland salt water declines to only 4% of total lake water. On the other hand, it may also be noted that a significant volume of the world's surface fresh water is found in just a few lakes: Baikal (volume $23 \times 10^9\,km^3$), Tanganyika ($18.9 \times 10^9\,km^3$) and the Great Lakes of North America ($22.1 \times 10^9\,km^3$). If water in these lakes is excluded, then the percentage of total lake water in saline lakes changes from 4 to 21%. These values are of particular interest in the present context, in that salt lakes are virtually confined to dryland areas. A few occur in humid areas but they are small, scattered and the result of special local circumstances (cf. Löffler, Volume 1, Chapter 2).

Given that there is a suitable depression or basin in which water can accumulate, two basic conditions are required in order for salt lakes to form. First, there must be a balance between hydrological inputs (precipitation on lake + inflows) and outflows (evaporation + seepage). Second, the basin must form part of a closed, endorheic drainage system. These two conditions are often found in arid and semi-arid regions, but not elsewhere. In drier regions (hyper-arid areas), drainage systems are generally **arheic** and outputs (mostly evaporation) far exceed inflows; such regions are too dry for lakes to form. In wetter (temperate and tropical) regions, drainage systems are mostly open and **exorheic**, and outflows exceed inputs so that salts do not accumulate: such regions are too wet for lakes to become saline.

As for the global extent of endorheically drained areas, this constitutes almost one-quarter of total land area (24.7%), a fraction not grossly different from that for total land area regarded as dryland; see Table 8.2). It would be surprising, of course, if it were. Table 8.6 indicates the geographical distribu-

Table 8.6 Regional distribution of endorheism. (Table based on data from Meybeck 1995)

| Region | Total endorheic area ($10^6$ km$^2$) |
|---|---|
| Eurasia | 14.1 |
| Africa | 22.2 |
| South America | 1.5 |
| North America | 0.9 |
| Australia | 4.2 |
| World | 33.2 |

tion of endorheic areas. Clearly there are marked differences between continents. Overall, approximately the same amount of water which falls on endorheic areas evaporates from them (in exorheic areas, more falls than evaporates, and in oceanic areas, more evaporates than falls). Note that, in endorheic areas, both fresh and salt lakes occur although salt lakes are more typical.

Salt lakes, like freshwater lakes in drylands, originate in many ways (but, again, glacial action is unimportant), and a wide range of size and shape attest to this observation. Thus, combined with the climatic variability characteristic of drylands, and the sensitivity of salt lakes to climatic events (as the amount of water present represents a rapid integration of changes in the balance between inputs and outputs), salt lakes are of many physical types: small to large, shallow to deep, temporary to permanent, intermittent to episodic. Not surprisingly, a number of terms have been used to describe particular types (e.g. **playas**, **salinas**, **salterns**, alkali lakes, closed basin lakes, terminal lakes) but distinctions are not clear, and 'saline lake' or 'salt lake' remains the least confusing general term. One word which does possess some utility, however, is **athalassic** (cf. Battarbee *et al.* 2001; O'Sullivan, Volume 1, Chapter 18). This term was coined by Bayly (1967) in order to distinguish inland saline lakes with no direct connection with the sea, from those with some connection to it (**thalassic**). A clear ecological distinction exists between athalassic and thalassic salt lakes.

Globally, there are several large, deep salt lakes (see above), but generally saline lakes are shallow, so that their physico-chemical features fluctuate widely on a diel, seasonal or secular basis. Wind action is important in such lakes. Consequently, turbidity is often high (Secchi disk transparency is often <2 cm), water temperatures are close to ambient air temperatures, the water column is isothermal and oxygen concentrations lie not far from equilibrium values (but note that these are lower than in fresh waters: Sherwood *et al.* 1992). Sensitivity to hydrological budgets can lead to marked and rapid changes in volume, depth and area, in response to seasonal or episodic climatic events. Lake Torrens, a huge and usually dry salt lake in Australia, filled to a mean depth of 0.5 m within weeks in 1989. The extent and nature of these changes depends upon the size and position of the lake involved; they are dampened in large, deep lakes and more prominent in small, shallow bodies.

Three major patterns may be distinguished. First, in shallow lakes in semi-arid regions, water depth, area and volume increase following predictable seasonal rains, and decrease during subsequent dry, evaporative periods. Well-defined wet and dry phases occur within one year. These are temporary and intermittent lakes. Second, in shallow lakes in more arid regions, water depth, area and volume increase following unpredictable, episodic rains, and subsequently decrease during a long evaporative period. There is no well-defined seasonal pattern, and water may persist for short periods or many months, the dry phase for many years. These are temporary, episodic lakes. Third, in large, deep lakes in both semi-arid and arid regions, water depth, area and volume increase following either seasonal or episodic rains, and decrease following subsequent evaporation. Such changes, however, are dampened by the size of the lake; they are relatively small. The volume of water present is sufficiently large for water to persist between rain events so that these lakes are permanent.

With regard to thermal patterns, shallow salt lakes do not stratify. Moderately shallow bodies are normally polymictic, and deep lakes are usually warm monomictic (Fig. 8.4). Salinity is an important factor determining thermal patterns, and

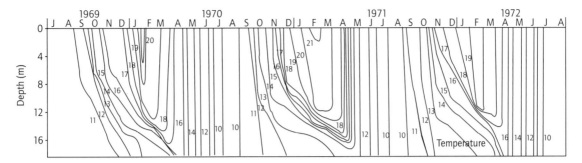

Fig. 8.4 Thermal pattern in Lake Gnotuk, a deep, permanent salt lake in Australia. (From Timms 1973.) Monash University, Melbourne (unpublished PhD Thesis *A comparative study of the limnology of three maars in western Victoria*).

if a salt lake is meromictic (that is, displays a salinity stratification, as occurs in some deep salt lakes), a reverse thermal pattern develops, in which the upper, less saline layer (the **mixolimnion**) is cooler than the lower, more saline layer (the **monimolimnion**). In such lakes, the **thermocline** and **chemocline** usually coincide.

Vertical differences in salinity only develop in deep salt lakes, where, however, they may persist for long periods. They may also not develop, however, and certainly do not occur in shallow lakes, except transiently. Horizontal spatial differences are not normal, either in deep or shallow salt lakes, but may occasionally occur (e.g. in Lake Balkhash where the two basins of the lake possess distinctly different salinities; see also Chapter 7, this volume).

Temporal differences in salinity are much more important and frequent. Generally, they follow hydrological fluctuations closely, as the mass of salt in a lake remains more or less (but not quite) constant. Thus, as the volume of the water decreases (and water-levels fall), salinity increases, and vice versa. Salinities encountered in salt lakes range from 3 to >300 g L$^{-1}$; maximum values depend upon the various ions involved. Actual temporal variation is usually less than this range, but sometimes by not much (e.g. in some intermittent salt lakes seasonal salinities may vary from c.50 to >300 g L$^{-1}$).

The major components of salinity in most salt lakes worldwide are sodium and chloride, but a variety of ionic compositions occur (Fig. 8.5), with some regional patterns (see also Löffler, Volume 1, Chapter 2). Thus, in almost all Australian salt lakes, salinity is due to the presence of sodium and chloride ions, whereas East African salt lakes are rich in sodium and carbonate/carbonate-chloride ions. Although temporal changes in the ionic composition of lakes may occur (as when, for example, the maximum solubilities of certain ion combinations are exceeded), this is not usual, and ionic composition is a remarkably stable attribute over large ranges of salinity. Values of pH vary from c.3.0, as in some acid saline lakes in Australia, to >10.0, as in some East African salt lakes. Little need be written about other ions. Nutrient concentrations span a range of values, and in certain cases (e.g. boron, uranium, selenium) minor elements may attain unusually high concentrations.

For the biota of freshwater lakes, salinities greater than about 20 g L$^{-1}$ are an insurmountable physiological barrier. This, together with hydrological instability, low oxygen concentrations and other physico-chemical features typical of many salt lakes, means that, except in salt lakes of low salinity, an aquatic biota has evolved (mostly from freshwater ancestors) which is restricted to, and distinctive of, salt lakes. Although this, in general terms, is less diverse than the freshwater biota, it is by no means impoverished (cf. Timms 1993, 1997). Indeed, most taxonomic groups found in fresh-

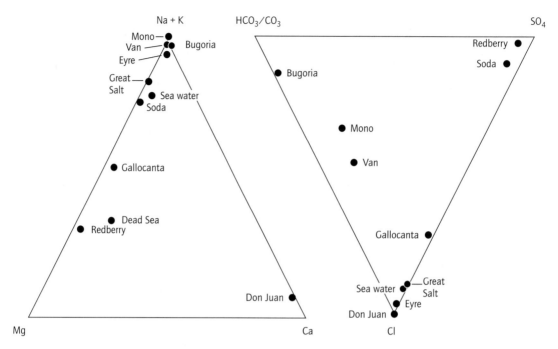

**Fig. 8.5** Ternary diagrams to illustrate the variety of ionic composition that can occur in salt lakes. (After various authors.)

water lakes are present in salt lakes, and all taxa found in salt lakes are represented in freshwater lakes. Notable taxa present in salt lakes are the Archaeobacteria, the Eubacteria (including the Cyanophyta), all major algal groups found in fresh waters (note especially *Dunaliella*) including charophytes, a small variety of angiosperm macrophytes (note especially *Ruppia*), rotifers (note especially *Brachionus plicatilis*), anostracans (note especially *Artemia* and *Parartemia*), cladocerans, copepods, ostracods, several insect groups (note especially *Ephydra*) and fish (note especially *Oreochromis alcalicus grahami* Boulenger). Other taxa are known, but are less important. Special mention should be made of flamingos (*Phoenicopterus* spp.) and the banded stilt (*Cladorhynchus leucocephalus* Viellot), for these birds are almost entirely restricted to salt lakes as breeding and feeding habitats. Mention also should be made of the biota (mostly fauna) of the dry beds of episodic salt lakes. This is well-defined and not simply an opportunistic input from surrounding terrestrial environments. A small list of spiders, beetles, ants and other insects have already been identified as endemic to Australian episodic salt lakes (see, e.g., Hudson & Adams 1996; Hudson 1996).

From this enumeration, it can be seen that the biota of salt lakes is a good deal more diverse than was thought to be the case some decades ago (cf. Macan 1963), and much more diverse than even some recent general literature would have us believe (e.g. Maltby 1991). Only a few of the taxa found in salt lakes are cosmopolitan at species or even generic level (a notable exception appears to be *Brachionus plicatilis* Müller, although, according to recent work, even this example now seems doubtful (R.J. Shiel, personal communication)); most exhibit various degrees of regional endemism. This is even the case for *Artemia*. Although this genus was formerly thought to comprise a single widespread species, *A. 'sauna'*, it actually contains several, and these exhibit distinct distributions.

Diversity is related in a general way to salinity; as salinity increases, diversity decreases. However, the relationship is not close, and a number of factors other than salinity determine biological composition within a particular salt lake, and the diversity of the total biota. These include ionic composition, pH, hydrological patterns, chance and biological interactions (Williams 1998).

No distinct salinity ranges can be distinguished, but some convenient classifications have been proposed. That by Hammer (1986) has a general use. He distinguished three ranges, **hyposaline** ($3–20\,g\,L^{-1}$), **mesosaline** ($20–50\,g\,L^{-1}$) and **hypersaline** ($>50\,g\,L^{-1}$). In the hyposaline range, most of the biota is essentially a **halotolerant** freshwater biota, and diversity is greatest. In the mesosaline range, most biota are restricted to salt lakes of moderate salinity, and diversity is less. In the hypersaline range, the biota are almost entirely restricted to highly saline lakes, and diversity is least. Many adaptations have evolved in order to cope with high salinities, including cellular halotolerance, effective osmoregulation and internal osmolytes (amino acids, glycerol, potassium salts, etc). All life-cycle stages must be adapted to ambient conditions (cf. Kokkinn & Williams 1988).

## 8.4 USES, VALUE AND IMPACTS

### 8.4.1 Uses and value

Both freshwater and salt lakes in drylands possess a large number of uses, some of which can be regarded as unique to that type of lake. The range of uses or value, in any event, is no less than that of lakes in more humid areas, and the extent of particular uses, given the nature of the background environment and—at least in semi-arid regions—burgeoning human populations, is frequently greater. The value of many dryland lakes, fresh and saline, has been recognised internationally by placing them on the list of Ramsar sites, that is, the global list of wetlands which the Ramsar Convention, an intergovernmental agreement, regards as of high conservation value (Frazier 1996). Over ninety countries are now party to this convention, and cover most land areas of the world.

Unfortunately, dryland lakes are significantly underrepresented on the list, an omission probably due to two factors. First, several dryland African and central Asian countries are not contracting parties to the Convention. Second, dryland lakes, especially salt lakes and temporary freshwater lakes, are perceived as less valuable than permanent freshwater lakes to waterfowl (the roots of the Convention lie in values relating to waterfowl usage). In this matter, it is hoped that all countries which are not yet signatories to the convention will be in the near future, and that for purposes of conservation the value of all dryland lakes will become better recognised.

For convenience, seven categories of value or use can be distinguished: economic, scientific, educational, aesthetic, cultural, recreational and ecological (cf. O'Sullivan, this Volume, Chapter 1). The first category aside, it is difficult, if not impossible, to arrive at monetary estimates of the 'worth' of each use or value (e.g. see Barbier et al. 1996), particularly those which represent indirect, 'existence' or 'heritage' value (e.g. Skinner & Zalewski 1995). What is quite certain, is that none of the uses, or the sources of value of dryland lakes, can be undervalued (Williams 1993a). The word inestimable could scarcely be more aptly used.

#### 8.4.1.1 Economic uses and value

The primary economic use of dryland freshwater lakes and reservoirs is as a source of fresh water for agricultural and other uses. By far the most important use, in terms of volume, is for irrigation and the amounts used, and areas irrigated are immense; the present global volumes of water used, and areas under irrigation, are, according to Postel (1995), respectively, about $2700\,km^3\,yr^{-1}$ and $2.4 \times 10\,km^2$. Most of the water used for irrigation comes from reservoirs. Although this is not an activity confined to dryland regions, it is much more important there, and indeed without it such regions could support far fewer people than they do. Other, often minor uses include domestic and industrial usage, and the generation of electricity.

Freshwater lakes, especially temporary bodies, are far less important in a relative sense as a source of water for irrigation. They are more important as sources of domestic and stock supplies, reeds and food (as fish and birds). A few temporary freshwater lakes provide additional food when their dry bed is cropped.

Water taken directly from salt lakes is of no value for irrigation, but the inflows of all of the world's largest, permanent salt lakes have been diverted in order to provide water for irrigation and other purposes. All of these of course are fresh water. Indirectly, therefore, salt lakes too are often an important source of water for human use in drylands. They also possess a number of other economic uses, some held in common with freshwater lakes (e.g. as sources of fish), others unique. Of the latter, mention should be made of their use as a source of minerals (especially halite (NaCl), but also many others, e.g. boron, lithium, uranium), food for the aquacultural industry (particularly *Artemia* cysts) and a variety of 'fine' chemicals (e.g. glycerol and β-carotene from *Dunaliella*).

### 8.4.1.2 Scientific value

Freshwater lakes in drylands, with some notable exceptions (e.g. Carmouze *et al.* 1983), have not attracted the interest of limnologists to the same extent as lakes in temperate or tropical areas. Undoubtedly this reflects in part their relative inaccessibility, but it also seems to reflect a perception that dryland lakes are less interesting scientifically: their biota is less diverse and more cosmopolitan than that of freshwater lakes elsewhere, and their limnological processes are little different from those of other lakes.

This perception should be firmly repudiated. There is now considerable evidence that the biota of freshwater lakes in drylands is much more diverse than previously thought (e.g. King *et al.* 1996), certainly more interesting (an explanation of cryptobiosis, the ability of some organisms of temporary fresh waters to survive extreme environmental extremes in a dehydrated state, remains to be provided), and displays considerable regional endemicity. Moreover, limnological processes in warm, shallow, temporary, turbid lakes, to take just one common kind of dryland lake, are not likely to resemble those of deep, permanent freshwater lakes, on which most of our limnological knowledge rests, except in the most fundamental respects. The dimictic thermal pattern, which is characteristic of temperate lakes, and which is so important in determining nutrient and other interactions in them, is scarcely known from dryland freshwater lakes.

The scientific value of salt lakes has been more clearly recognised, and, as a result, they have attracted attention from ecologists, physiologists, biochemists, geochemists, palaeolimnologists, hydrologists and other kinds of scientist. Ecological interest was aroused because salt lakes are more discrete, less taxonomically diverse and simpler, more tractable ecosystems for study than freshwater systems. For physiologists, interest has focused on the nature of adaptations to the stresses of life in salt lakes, particularly osmotic factors. The uniqueness of certain biochemical pathways in some biota of salt lakes has attracted the interest of biochemists. Geochemists are interested in the pathways and complexities of chemical reactions in the lake itself and with contiguous underground waters.

The sensitivity of salt lakes to climatic change, and thus their ecological sensitivity, provides the basis for most palaeolimnological interest (Last 2001); the sediments of salt lakes are seen, in some ways, as much more valuable sources of evidence of past events and climates than those of freshwater lakes. Hydrologists are interested in salt lakes, *inter alia*, because of the complex way in which budgets are determined both by surface (climate, hydrology) and subsurface (hydrogeological) events (cf. Winter, Volume 1, Chapter 3).

### 8.4.1.3 Educational value

Closely allied to scientific value and interest is the educational value of saline lakes. At a time when an environmentally educated citizenry is a prerequisite for any informed administration of natural resources, when environmental issues are becoming more and more important and when global climatic change is imminent, it is clearly sensible to use those parts of nature (lakes) where ecological value (and threats to it) can most simply, effectively and best be demonstrated and understood. It need hardly be added that it is in drylands where an informed citizenry is most needed, where water resources are in shortest supply and at most hazard, and where climatic change is likely to be soonest felt (given that dryland ecosystems are amongst the most fragile and sensitive).

### 8.4.1.4 Aesthetic value

Perhaps for deep, psychological reasons, lakes have long been regarded as elements of the natural landscape of particular aesthetic appeal. Whatever the reason, there is no doubt that the mere sight of many lakes evokes deep human emotions; there must be few who do not find lakes aesthetically satisfying to some degree.

In this respect, limnologists are rather more fortunate than many other scientists. In an arid environment, this satisfaction is heightened by contrast, at least so far as freshwater lakes are concerned. For salt lakes, it was less so in the past, when, at least through the eyes of many European explorers of arid environments, salt lakes were viewed with abhorrence. It is reputed that Warburton, one of the early European explorers of central Australia, on first sighting Lake Eyre, commented on its death-like stillness and vast expanse of unbroken sterility. Attitudes to natural beauty, however, have since undergone dramatic shifts. Now, many salt lakes, including Lake Eyre when dry (Serventy 1985), are regarded as objects of outstanding beauty. Special mention is made in this respect of Mono Lake (California), Lake Nakuru and other salt lakes of the Rift Valley of east Africa, and the Etosha Pan in Namibia. For several of these, their attractiveness is increased by associated wildlife, especially the flamingos in African salt lakes.

### 8.4.1.5 Cultural value

The aesthetic and cultural value of lakes in general are difficult to separate, because their aesthetic appeal has inspired numerous authors, artists and musicians whose works have often then become part of local human cultural heritage. Dryland lakes, however, have also inspired spiritual feelings, and are part of many religious writings. The special cultural significance of some aquatic plants and animals (e.g. flamingos) from dryland lakes also should not be overlooked. There are therefore many 'heritage' attributes of lakes which transcend aesthetics alone, as recognised in the criteria used by UNESCO to determine the value of 'world heritage areas' (including lakes).

Much of the cultural value of fresh and saline lakes to local peoples has been lost, or overwhelmed by historical accounts of events surrounding the 'discovery' of the lakes by European explorers. No mention, for example, is made by Carmouze (1983), in his account of the historical background to limnological study of Lake Chad, of the cultural value of the lake to local people, although it would be surprising indeed were there none. Likewise, there are few accounts of the cultural significance of lakes in arid areas of Australia to aboriginal Australians, although these were certainly important burial and ceremonial sites (cf. Mulvaney & Golson 1971). This kind of omission should not be allowed to obscure the fact that a firm cultural association undoubtedly existed between many dryland lakes and their neighbouring human populations.

In some well-known dryland lakes, cultural associations have been well-documented, most notably for those in central Asia (the Caspian and Aral Seas) and the Middle East (the Dead Sea and Lake Kinneret (the Sea of Galilee)). All these figure prominently in classical and religious literature, and all were associated with some of the earliest civilisations. Nissenbaum (1979) has given a comprehensive account of this subject with regard to the Dead Sea. More recent literature too, firmly entrenches them as part of our cultural heritage. There are few more evocative poems, for example, than that written by Matthew Arnold during the nineteenth century (*Sohrab and Rustum*), about the nature of the Aral Sea and its inflows. Writing of the Amu Darya (the River Oxus), he described its lower reaches thus:

*Oxus, forgetting the bright speed he had
In his high mountain cradle in Pamere,
A foil'd circuitous wanderer—till at last
The long'd-for dash of waves is heard, and wide
His luminous home of waters opens, bright
And tranquil, from whose floor the new-bathed stars
Emerge, and shine upon the Aral Sea.*

#### 8.4.1.6 Recreational uses

Many passive and active recreational activities take place on or near dryland lakes, both fresh and salt, although for the most part permanent rather than temporary. Passive recreation (nature appreciation, bird-watching, photography) involves both fresh and saline lakes, but more active recreational pursuits (fishing, swimming, sailing, rowing) are generally restricted to fresh or hyposaline lakes. Many lakes or their faunas, which are of significant cultural or aesthetic appeal, may be visited by large numbers of tourists, often from international destinations (and provide important additions to local economies). Even dry lake beds possess recreational value: in some countries they may be used for 'wind-surfing' or for speed trials. The use of lake muds for therapeutic purposes can also perhaps be referred to here. Bulgareanu (1996) gives an interesting and recent account of Romanian salt lakes used for this activity, and the sediments of the Dead Sea have long been used for similar purposes. Most dryland reservoirs were (originally) generally closed for recreational use. Many still are, but recent trends are to allow at least certain types of recreation on, or at least near, them.

#### 8.4.1.7 Ecological value

The ecological value of dryland lakes is the most difficult attribute of all to estimate, and yet perhaps the most important in the long term (see O'Sullivan, this volume, Chapter 1). Despite these difficulties, it is certain that this value is no less important than for lakes in more humid areas, and may well be more important. Like lakes in humid areas, dryland lakes possess significant biodiversity, represent an integral part of global hydrological and biogeochemical cycles, and possess high conservation values, despite being poorly represented in the Ramsar list (see above). It is often only when dryland lakes have been significantly degraded or destroyed that their ecological value becomes apparent. Nowhere is this more obvious than in the case of the Aral Sea; following degradation of this lake, widespread and unfavourable regional climatic and environmental changes ensued, there was a significant loss of biodiversity, and the lake no longer served as a staging post for central Asian migratory waterfowl (Williams & Aladin 1991; Löffler, Volume 1, Chapter 2).

### 8.4.2 Impacts

The nature and extent of human impact on the uses and values of dryland lakes are as diverse, pervasive and comprehensive as those on lakes in more humid areas. Given the fragility of many dryland environments, they are probably also more harmful and irreparable. In any event, almost every kind of human impact on lakes which occur in humid regions are also important for dryland lakes (lake acidification provides the most notable exception). In addition, several other kinds of impact either uniquely affect or exert heightened influence on dryland lakes.

Little point is served in documenting the nature and effects of impacts on dryland lakes which take place in common with those on lakes in humid areas, except to stress that the effects are often more severe given that the human demands upon a natural resource in shorter supply (e.g. water) are usually greater. This situation is exacerbated because management of these demands by local administrative bodies is often at a lower level than is adequate, more driven by immediate, local demands, and based more on the application of flawed ideas derived from the study of permanent lakes in temperate areas. Even greater management problems arise when individual lakes are shared by the territories of several countries (e.g. Lake Chad, the Aral Sea, the Dead Sea) and/or are fed by rivers rising in yet others (notably the Aral Sea).

Of the impacts unique to or particularly significant in dryland lakes, the following are the most important: salinisation, water diversion, catchment activities, pollution, activities directly affecting the biota, changes to the physical nature of lake beds, hydrological changes, and climatic and atmospheric changes.

#### 8.4.2.1 Salinisation

Many freshwater lakes located in semi-arid regions where irrigation, clearance of the natural vegetation and other land-use changes have occurred have consequently become much more saline. The process is generally referred to as **secondary salinisation**, and involves both permanent and temporary lakes. In the former case, only the halotolerant elements of the original biota survive, supplemented by elements from natural saline lakes in the vicinity. Basins formerly occupied intermittently by fresh water essentially become salt pans. Additionally, as part of salinity management schemes, some freshwater lakes or basins may receive mildly saline water drained from the surface or pumped from underground. These so-called **evaporation basins** or **ponds**, likewise, may ultimately develop high salinities.

Dryland salinisation is a worldwide problem of enormous proportions, involving not only dryland lakes but other parts of the natural environment (Ghassemi *et al.* 1996). Globally, over 10% of all irrigated areas are now subject to damage from salinisation (Postel 1995). In some particular dryland regions, salinisation is regarded as the single greatest threat to the natural character of freshwater lakes (e.g. in Western Australia; Halse 1998). Early signs of salinisation in permanent freshwater lakes include the death of riparian vegetation (including trees), and in temporary freshwater lakes the appearance of halotolerant terrestrial species.

#### 8.4.2.2 Water diversions

The greatest human impact on deep, permanent salt lakes has been the diversion of freshwater inflows for irrigation and other purposes. Because of the sensitivity of these lakes to any change in the balance between inputs and outflows, diversions have in almost all cases led to rapid and marked decreases in the volume of the lake, its water-level and its area. Often, such impacts compound local climatic trends towards increasing aridity.

Williams (1993b; 1996b) has drawn attention to the worldwide occurrence of this phenomenon, and documented several examples. Amongst these are Mono Lake, California (where water-levels have fallen 15 m since 1920), Pyramid Lake, Nevada (21 m since 1910), the Dead Sea (8 m since 1980), the Aral Sea, Kazakhstan (> 15 m since 1960), Qinghai Hu, China (10 m since 1908) and Lake Corangamite, Australia (3 m since the 1960s). Only in a relatively few lakes, notably the Great Salt Lake (Utah), the Caspian Sea (Russia/Iran) and Mar Chiquita in Argentina, have water-levels risen in the recent past. In almost all cases, this can be attributed to increased inflows following secular decrease in climatic aridity.

The immediate chemical effects of falling water-levels (decreased lake volume) are increased lake salinities. Thus, over the time periods cited above, salinity rose in Mono Lake from 48 to $90\,g\,L^{-1}$, in Pyramid Lake from 3.75 to $>5.5\,g\,L^{-1}$, in the Dead Sea from 200 to $340\,g\,L^{-1}$, in the Aral Sea from c.10 to $>30\,g\,L^{-1}$, in Qinghai Hu from 5.6 to $12\,g\,L^{-1}$ and in Lake Corangamite from 35 to c. $50\,g\,L^{-1}$. Maximum salinities given here are approximate for current values because the situation is dynamic, with ongoing impacts in some cases, mitigation of impacts by management in others and climatic moderation in yet others. Even so, it is obvious that salinities overall have undergone marked rises following water diversions.

Biological effects of falling water-levels and increased salinities depend largely upon original salinity. They are most profound when original salinities lie within the hyposaline or mesosaline ranges, and least profound when original salinities were hypersaline. Thus, the effects of salinity increase in Lake Corangamite, although less than $20\,g\,L^{-1}$, have been most significant on the biota: fish, amphipods, snails and *Ruppia* largely disappeared (Williams 1995), although above-average rains since 1992 have somewhat ameliorated the situation. In the Dead Sea, on the other hand, a much greater absolute increase in salinity (>100 g L$^{-1}$) has apparently had little effect on the fundamental nature of the biota and ecology of the lake.

Falling water-levels and rising salinities are by no means restricted to gross chemical and biological effects: a variety of physico-chemical and other environmental changes also follow. The point is best illustrated by reference to the Aral Sea. Here, following diversion of the Syr and Amu Darya, from the 1960s onwards the local climate became more severe, the fishery collapsed, salt and dust storms became more frequent as a result of the exposure of extensive areas of the sea bed, and groundwater levels fell. In turn, these changes affected the local human population so that morbidity rates became the highest in central Asia. Further information on this lake is given in the following section (Case Studies), in Butorina (this Volume, Chapter 7) and in Löffler (Volume 1, Chapter 2).

#### 8.4.2.3 Catchment activities

Although treated separately here, for purposes of emphasis, salinisation is largely a reflection of catchment activities. However, there are many catchment activities which affect dryland lakes in ways other than to increase their salinity. The most important of these are grazing, especially overgrazing (which in the case of Australian drylands has been exacerbated by the introduction of the rabbit), and excessive clearance of natural vegetation and its replacement by shallow-rooted growth, with effects other than secondary salinisation. Both activities usually lead to soil erosion and to changes in run-off patterns. Thus, sediment loads to dryland lakes and reservoirs may increase significantly, and in the case of reservoirs this increase may greatly shorten their useful life as water storage bodies. Changes in hydrological patterns usually involve larger but shorter inflows to lakes, in that the ability of land to retain water declines. Thus, many overgrazed and eroded drylands are characterised by severe erosion and the occurrence of floods, even after relatively small rain events.

#### 8.4.2.4 Pollution

Most kinds of pollutant discharged to temperate, freshwater lakes are also introduced into permanent dryland lakes, although there are differences, which reflect the generally decreased industrial activity in drylands. Domestic and agricultural wastes are most important, and contribute heavy organic loadings to many dryland lakes. Other pollutants include a variety of metals and other inorganic compounds, organochlorine residues and plant nutrients. The accumulation of selenium in evaporation ponds in western parts of North America is notable, and the problem it causes is difficult to resolve. In this regard, one important feature of salt lakes, often overlooked by water resource managers, should be restated and emphasised: salt lakes are parts of an essentially closed hydrological system, and so accumulate and biomagnify many pollutants in a way which does not occur in freshwater lakes (which are parts of open systems). There is little doubt that the effects of pollutants on dryland lake ecology are broadly in accord with what occurs in lakes elsewhere. However, high salinity may serve to modify some of these effects.

Although pollution of dryland lakes is an activity mostly confined to permanent lakes, temporary bodies have not escaped entirely. Many of these are used as dumping sites for domestic and agricultural wastes, and many mine dumps lie adjacent to temporary lakes even in the most remote areas.

#### 8.4.2.5 Activities directly affecting the biota

Reference has already been made to the purposive or accidental introduction of non-native fish to many dryland lakes, particularly *Cyprinus carpio* (common carp) and *Gambusia affinis* (western mosquito fish) whose tolerance of the kinds of conditions to be found in many permanent freshwater dryland lakes has facilitated their growth and dispersal there. There is much evidence to suggest that the effects of such introductions have been profound. In the case of *C. carpio*, primary effects appear to be on the physical nature of the aquatic environment (bottom sediments are disturbed, turbidity is increased, macrophytes are destroyed), but for G. *affinis* biological interactions (competitive exclusion, predation) seem to be more important.

Fish have also been deliberately introduced to many hyposaline lakes where, in some cases at least, populations can only be maintained by restocking. Often, viable recreational fisheries have nevertheless been produced. In the Aral Sea, whose salinity at the beginning of this century was about $10\,g\,L^{-1}$, wholesale introductions of fish and other marine biota (some accidentally) commenced from about 1927 onwards, and led to development of a moderately productive commercial fishery. Many marine fish and invertebrates (including barnacles) introduced to the Salton Sea, an artificial saline lake in California (salinity presently c.$45\,g\,L^{-1}$), have also thrived. On the other hand, none of the many marine species introduced earlier last century to Lake Corangamite, a natural permanent salt lake in Australia, ever survived.

For hypersaline lakes, introductions of *Artemia* spp. (brine shrimps) are more important. Most introductions of this crustacean have been by commercial salt producers, that is, those involved in the production of commercial quantities of salt from evaporated seawater in coastally located ponds. Introductions have largely been made on an *ad hoc* basis as part of salt pond management. They have paid scant attention to the possible effects on local brine shrimp populations. The potential dangers of these introductions have been noted, amongst others, by Persoone & Sorgeloos (1980) and Geddes & Williams (1987).

In permanent freshwater dryland lakes and reservoirs, introduced, non-native macrophyte species have often given rise to problems. Not the least of these is that their transpiration leads to increased and excessive loss of water from the lake or reservoir involved. Of other activities directly affecting the biota, brief mention must be made of the local exploitation of flamingo adults and eggs in certain areas of South America and the dangers posed by the silver gull (*Larus novaehollandiae* Stephens) to the continued viability of the banded stilt (*Cladorhynchus leucocephalus*) in Australia. The former is favoured by town rubbish dumps, and large populations have built up. This is important because the gull is a significant predator of the banded stilt during its breeding in episodic salt lakes (Robinson & Minton 1990).

### 8.4.2.6 Changes to the physical nature of lake beds

Extraction of minerals from the sediments of dry salt lakes frequently results in major physical disturbance. The effects remain indeterminate, but are likely to involve both physico-chemical and biological impacts. In some cases, unsightly causeways are constructed in order to facilitate mining operations either on or beneath the lake. They are costly to remove, and are usually left in place after mining has ceased. The direct effects of these too remain indeterminate. Levée banks constructed in order to help in production of salt from natural salt lakes also constitute a physical disturbance to lake beds.

A major physical disturbance to the dry beds of temporary freshwater lakes arises when these are cropped during the dry season. Their impacts upon the aquatic biota are unknown, but the widespread assumption that they are negligible remains to be rigorously tested.

### 8.4.2.7 Hydrological changes

Hydrological changes to lentic waters in drylands encompass a range of impacts which change the nature of, add to, or subtract from, the lentic waters there. A common change to the nature of freshwater lakes is brought about by raising the water-level by constructing low dams on outflows. In effect, the lakes then become shallow reservoirs. Both permanent and temporary lakes have been changed in this way. Once engineering structures have been put in place, the lakes are then managed in order to meet operational requirements for stored waters (usually for irrigation).

The limnological effects of such changes on lakes themselves have rarely been monitored.

Reservoirs behind high dams represent more dramatic and obvious additions to lentic waters in drylands. Many, perhaps most, are constructed in regions where deep, permanent, freshwater lakes are naturally absent. A good deal of information is now available concerning the impact of such reservoirs on downstream rivers in drylands (for a comprehensive introduction, see Davies *et al.* 1994). In brief, natural hydrological patterns downstream are changed (particularly by reducing the frequency of dominant discharges), temperatures are lowered, sediment transport regimes are altered, and profound changes to floodplain ecosystems adapted to unregulated and highly variable flows (and not the imposed, reverse pattern) are promoted. Additionally, reservoirs usually impede or halt fish migration, and increase the amount of water evaporated from the system as a whole.

In a few cases, unplanned additions to the lentic environment in drylands have been made. The most notable is the Salton Sea, a large body of saline water in California. This body was created when, in 1905, diversion of the Colorado River failed, and flooded the Salton Sea basin. Its present salinity is c.$45\,g\,L^{-1}$.

Alternatively, many freshwater dryland lakes, both permanent and temporary, have been lost by deliberate draining. Mostly, such lakes are 'reclaimed' for agricultural purposes by the construction of channels discharging into nearby streams and rivers. In certain regions, e.g. Mexico, destruction of dryland lakes has occurred following overabstraction of underground water (Alcocer & Escobar 1990).

### 8.4.2.8 Climatic and atmospheric changes

Current and likely future climatic and atmospheric changes will affect all lakes worldwide, but it seems likely that the greatest impacts will be on dryland lakes. Here, the greater sensitivity of both fresh and saline lakes to changes in rainfall, temperature and other climatic parameters (including variability) will mean that these lakes—more than others—are likely to respond rapidly to even small climatic changes.

The nature and extent of the projected changes remain uncertain, given the inexactitude of climatic numerical models (Watson *et al.* 1996). Even so, such models suggest that drylands will warm substantially over the next century (by +3 to +7°C), particularly in high latitudes of the northern hemisphere (Williams & Balling 1996). Estimates of dryland temperature increase, it may be noted, are generally greater than those predicted for the world as a whole. Predictions of changes in precipitation are particularly imprecise, but irrespective of change increased temperatures will raise evapotranspiration rates, thus increasing effective aridity.

Most authors agree that drought in drylands will increase in frequency, intensity, magnitude and duration. Most also agree that climatic change is likely to be rapid on a geological timescale. How the biota and ecosystems of dryland lakes will adapt remains unknown. It has been variously suggested that climate change might be too rapid for biological adaptation to occur naturally. Perhaps a more acceptable speculation is that only the most **euryoecious** taxa with efficient dispersal mechanisms will survive.

With regard to atmospheric changes, the most important effect is likely to be increased amounts of UV-B acting on biota, following decreased ozone concentrations in the upper atmosphere. Penetration of UV-B is considerably less in lakes than in the ocean, and although data are sparse, it appears that 10% UV-B (i.e. 10% irradiance of the value just below the surface) is usually <1 m in most fresh waters (Kirk 1994). Many factors determine UV-B penetration (e.g. dissolved organic matter, turbidity), and the interrelationship between this variable and ecosystems is complex (Williamson 1995; Rozema *et al.* 1997). However, as most dryland lakes are shallow, and thus offer little refuge to the biota below 1 m, the impact of increased amounts of UV-B on aquatic ecosystems in dryland lakes is likely to be greater than in deeper lakes elsewhere. The effects of salinity on UV-B penetration have yet to be determined.

## 8.5 CASE STUDIES

### 8.5.1 Lake Chad

Lake Chad (Fig. 8.6) is an extremely large, permanent, freshwater lake in the southern semi-arid area of the West African Sahara, where mean annual rainfall is c.300 mm, and mean annual evaporation >2000 mm. The lake lies within the territories of four African nations: Chad, Niger, Nigeria and Cameroon. It has been studied extensively by several French limnologists, whose results are summarised by Carmouze et al. (1983). General accounts of the lake, largely based on this work, but also on other, later data by Nigerian scientists, are to be found in the publications of Beadle (1981), Serruya & Pollingher (1983), Livingstone & Melack (1984) and Kira (1994). Problems associated with the lake are discussed by Dinar et al. (1995), and earlier in a comprehensive diagnostic study undertaken for the Lake Chad Basin Commission by Kindler et al. (1990).

Lake Chad arose by tectonic and aeolian action. Its total drainage catchment is about $2.4 \times 10^6$ km$^2$, with most water draining into the lake from the River Chari. The catchment is **endorheic**. Marked fluctuations in the morphometry of the lake have occurred (and continue to occur) in accord with large secular climatic changes. Changes in water-level of up to 3.5 m have taken place during the past 100 years alone (Fig. 8.7). Given that the maximum depth of the lake when full is less than 10 m, such fluctuations are significant.

Recorded secular fluctuations in lake area vary

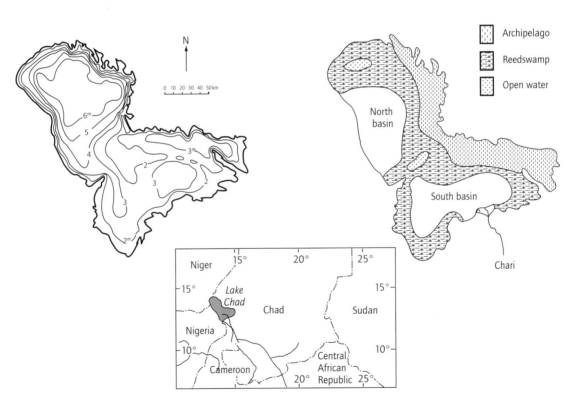

**Fig. 8.6** Lake Chad. Morphometry when full (left) and the three major ecological zones of the lake (right). The central diagram shows the position of the lake. (After various authors.)

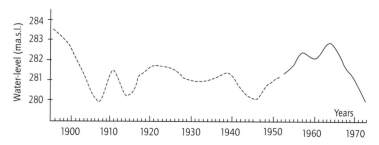

**Fig. 8.7** Fluctuations in the water-level of Lake Chad from 1895 to 1974: dashed line, estimates; full line, measurements. (Modified from Kira 1994.) International Lake Environment Committee, United Nations Environment Programme, Kusatsu, *Data Book of World Lake Environments—A Survey of the State of the World Lakes. 2. Africa and Europe.* Compact-size edition.

between c.10,000 and >25,000 km$^2$, but Lake Chad has been much larger in previous geological times. During dry periods, the northern basin dries out completely, and the southern basin becomes just a few metres deep. Annual precipitation is strongly seasonal with most rain falling in August. Seasonal changes in area, water volume and depth follow well-defined seasonal climatic changes.

The thermal pattern of Lake Chad is best described as **transiently polymictic**. Usually, the lake is isothermal overnight, but in the morning it stratifies weakly, a condition which persists during the rest of the day. Transparency is low, usually <50 cm, and nearly always <1 m, and declines even more as water-levels fall.

Lake Chad, despite being at the centre of an endorheic drainage basin, is quite fresh. Salinities in the northern basin are less than 1 g L$^{-1}$, and in the southern basin, less than 0.5 g L$^{-1}$. Anions are mainly carbonate, but significant amounts of Ca, Mg and Na are found amongst the cations. The constant low salinity of the lake has been explained as being due to loss of water and ions by seepage. Even so, considerable losses of water also occur by evaporation. Nutrient concentrations are high. However, oxygen concentrations in the water column are highly variable. They are often homogeneously distributed in the water column, at values between 80 and 100%, but stratification may occur during the warmer parts of the day with low concentrations (or anoxis) in the lower layers of water. Alternatively, concentrations may be low throughout the water column.

Three extensive ecological zones may be distinguished (Fig. 8.6): an archipelago in which the islands are the exposed crests of dunes; a thick reedswamp of *Cyperus papyrus*, *Typha australis* and *Phragmites*; and a zone of open water with submerged macrophytes in sheltered regions (mostly *Potamogeton*, *Vaiisneria* and *Ceratophyllum demersum*). The phytoplankton is highly productive, and comprises a diversity of blue-green algal, desmid and diatom taxa, in particular. Biodiversity is high for both plants and animals, although almost no endemic species occur. More than 140 species of fish have been recorded from the lake, and the fish provide a highly productive fishery. The lake is also visited by numerous species of birds, of which many are migrants whose paths include stop-overs at Lake Chad.

A principal use of the lake is for fishing, and as a source of irrigation water. Pressure on these uses has increased in the recent past following considerable human migration into the marginal areas of the lake. Threats to the lake ecosystem include diversion and blocking of the major inflows by dams, physical degradation of the catchment by overgrazing and the removal of vegetation. These are compounded by, and perhaps compound, increasing climatic aridity. At the time of writing, they were not being adequately addressed by the actions of the Lake Chad Basin Commission.

### 8.5.2 The Aswan High Dam

The Aswan High Dam (Fig. 8.8) is situated on the River Nile, some 200 km north of the Egypt–Sudan border, in a hyper-arid area (mean annual rainfall <5 mm, mean annual evaporation c.3000 mm). At this point, a rock-filled dam impounds a body of water which stretches southwards into the Sudan for some 500 km (Fig. 8.8).

In Egypt, the reservoir is known as Lake Nasser, in the Sudan, as Lake Nubia. It has been studied by Egyptian Government scientists and by Sudanese scientists, and many technical reports exist. Useful early summaries are given by Entz (1976) and Latif (1984). The most recent summary is that by Kira (1994), but this relies on much early data. There are many articles which focus on the negative environmental impacts of the reservoir.

Construction of the rock-filled dam began in 1959, and water began to be impounded in 1964. The reservoir is now some 500 km long, with a surface area of almost 6000 km$^2$, a volume of $162 \times 10^9$ km$^3$, a maximum depth of 110 m and a mean depth of 70 m. The normal annual range of fluctuation in water-level is 25 m. The area of the catchment is 2,850,000 km$^2$.

As for conditions in the impounded water, turbidity is relatively high and transparency is usually less than 0.3 m. The thermal pattern is warm monomictic; the reservoir stratifies in summer, and mixis occurs in winter. Oxygen varies seasonally, vertically and horizontally. In surface waters, con-

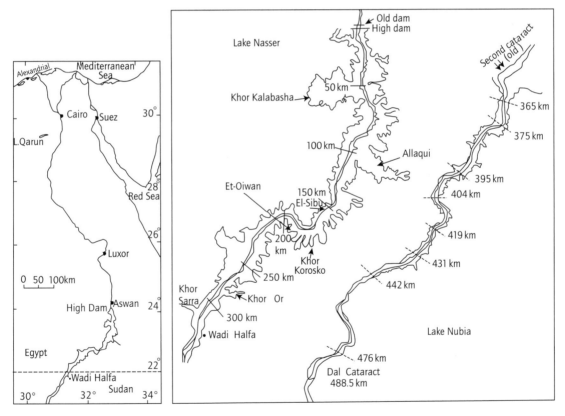

**Fig. 8.8** The Aswan High Dam. Left-hand map, geographical location; right-hand map, general outlines of Lake Nasser (Egypt) and Lake Nubia (Sudan). Kilometres indicate distance from dam wall. (After Latif 1984.) Elsevier, Amsterdam, for Taub, F.B. (ed) *Lakes and Reservoirs*. Ecosystems of the World No 23.

centrations often vary between 3 and 12 mg L$^{-1}$, but in deeper layers, they may be much less: 0–8 mg L$^{-1}$. The water column is well-oxygenated in winter, but when (in summer) the reservoir is thermally stratified, the hypolimnion is essentially anoxic. This is illustrated in Fig. 8.9, which shows the depth of the oxygenated and anoxic layers from a point near the dam wall back to the most southern reaches of the reservoir. Mean pH is generally >7.0.

Salinity is low at all times and generally <200 mg L$^{-1}$. The monovalent cations (Na + K) are generally less important than divalent species (Ca ± Mg). Recorded concentrations for orthophosphate have ranged from 0.001 to 2.2 mg L$^{-1}$. On the basis of the physico-chemical data available, the reservoir can be divided into three sections: a northern sector with essentially lacustrine characteristics; a middle sector where riverine conditions prevail during times of flood and lacustrine conditions at other times; and a southern sector where conditions are essentially riverine all year.

These three sectors are distinguished by their physico-chemical characteristics, and also their biological features. They have, however, undergone considerable changes with time, especially in the northern sector, as reservoir conditions have stabilised. Thus, the river crab *Potamonautus niloticus*, which was common during the 1970s, has now disappeared, whereas the shrimp *Caridina niloticus* remains widespread and common. Benthic fauna such as bivalves disappeared with the development of anoxic conditions. Macrophyte composition has also changed with time: formerly, the introduced species *Eichhornia crassipes*, a floating weed, was absent. The major macrophytes present are *Potamogeton spicatus* and three species of *Najas*. There is a variety of fish present. Commercial catches have increased with time, but have changed in composition. Cyprinids and cat-fishes are now less prominent.

The reservoir was constructed for four reasons: to provide dependable water for irrigation; to impound flood waters; to protect against flooding; and to generate cheap electricity. There is little doubt that it has achieved the ends for which it was constructed to a significant degree. Nevertheless, there is also much debate on the extent and significance of a number of negative environmental effects which were not fully considered before the dam was constructed.

Thus, although the total of irrigated land in Egypt has risen (by about 500,000 ha), this is essentially a modest increase, and the additional food produced is not sufficient to feed the increase in the size of the Egyptian population which has occurred since the dam was constructed. Waterlogging and salinisation are now said to affect some one-third of all irrigated land in Egypt, and crop productivity in some areas is said to have fallen by as much as 30%. The incidence of bilharzia, a significant human disease vectored by aquatic snails, has increased. Valuable nutrients and silt are no longer brought down naturally by the River Nile, and there has been a dramatic fall in the fishing productivity of the Nile delta. Finally, very large loads of silt are now being deposited in the reservoir (up to $200 \times 10^6$ t yr$^{-1}$).

### 8.5.3 *The Coongie Lakes*

The Coongie Lakes is a series of largely temporary freshwater lakes in the north-eastern arid part of

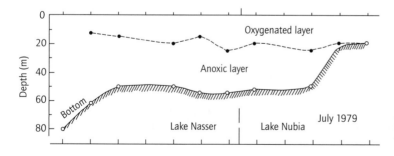

**Fig. 8.9** The Aswan High Dam. Horizontal and vertical extent of oxygenated and anoxic layers during thermal stratification (July 1979, summer). (After Latif 1984.) Elsevier, Amsterdam, for Taub, F.B. (ed) *Lakes and Reservoirs*. Ecosystems of the World No 23.

South Australia (Fig. 8.10), where the median annual rainfall is 100–150 mm, and the mean annual evaporation is >3600 mm. Rainfall variability of the region, both temporally and spatially, is amongst the greatest of any region of Australia. Until a few decades ago the lakes were almost unknown to limnologists; now, mainly owing to the work of Puckridge and others (Allan 1988; Puckridge & Drewien 1988; Reid & Gillen 1988; Roberts 1988; Reid & Puckridge 1990), they have become much better known. The importance of the lakes in a conservation sense has been recognised by their inclusion on the Ramsar list of wetlands of international importance. They are also listed on the Australian Register of the National Estate (Morelli & de Jong 1996).

The Coongie Lakes (Figure 8.11) are essentially of riverine origin, and form part of a mosaic of shallow lakes, channels, wetlands and floodplains, created by inundation of dunefields. The lakes themselves are a series of shallow basins which temporarily contain water derived on an intermittent basis from a northwestern branch of Cooper's Creek. Other lakes are present in the region, but these contain water less frequently. Thus, regionally, a series, from Lake Coongie (rarely dry) to Lake Marradibbadibba (mainly dry), is formed. When flooded, Lakes Coongie, Marroocoolcannie and Maroocutchanie fill first to a depth of c.1.5 m, then Lake Toontoowaranie to 1.5 m, then Lake Goyder to >1 m, and finally Lake Marradibbadibba.

Except during exceptional floods, all of these lakes are terminal. The occurrence of exceptional floods, however, is episodic as indicated in Fig. 8.12, which plots mean monthly discharge of Cooper's Creek, 150 km upstream of the Coongie Lakes. As can be seen, major floods during the time interval indicated are separated by more than 10 years. It is clear that seasonal and secular climatic changes are the main determinant of the frequency, duration and depth of water in this series of lakes. Whatever the case, maximum depths are less than 3–4 m, but usual depths are much less.

Thermal patterns are greatly dependent upon wind strength, and the intensity of solar radiation. In summer, stratification occurs during the day, or over longer periods; in winter, the lakes are largely isothermal. Despite the transient nature of thermal stratification, lake temperature profiles may exhibit marked departures from surface temperature (the latter sometimes as high as 40°C). Turbidity is high overall, and often extremely high, but greatly dependent upon wind speed. Oxygen concentration too is variable, but generally the whole of the water column is well-oxygenated, even during thermal stratification.

When the lakes are receiving no inflow, salinity and pH gradually increase (with gradients also apparent from the more frequently flooded, to the less frequently flooded lakes). Flooding diminishes these gradients. Even when near dry, however, the lakes remain fresh, with salinities <2 g L$^1$. Nutrient concentrations are relatively high but nitrogen : phosphorus ratios of 2.0 to 3.2 suggest that algal productivity is limited by the former element, not the latter.

Although the Coongie Lakes are not fully explored, faunal biodiversity is known to be high (though not as high, perhaps, as that of tropical lakes in northern Australia). The zooplankton community is rich in species, and there is an abundant insect fauna in which Chironomidae, Notonectidae and beetle families are prominent. *Macrobrachium australiense*, a large and widespread prawn, is regarded as the major detritus cycler. Also present is a diverse molluscan fauna, including both gastropods and bivalves. The invertebrate fauna as a whole is characterised by species which possess efficient dispersal abilities or life-cycle stages resistant to desiccation.

The fish fauna, too, is diverse and abundant, with only two introduced species recorded: *Gambusia holbrooki* (eastern mosquito fish), and *Carassius auratus* (goldfish). The most abundant native species are *Nematalosa erebi* (the bony bream) and *Retropinna semoni* (the smelt). Of other vertebrates, the Australian water-rat (*Hydromys chrysogaster*) is widespread but not common, the tortoise *Emydura macquari* is both widespread and common, and there are eight species of frogs. Some seventy species of waterbird have been recorded, fifteen of which are migrant waders from the northern hemisphere.

Apart from macrophytes, the aquatic flora is

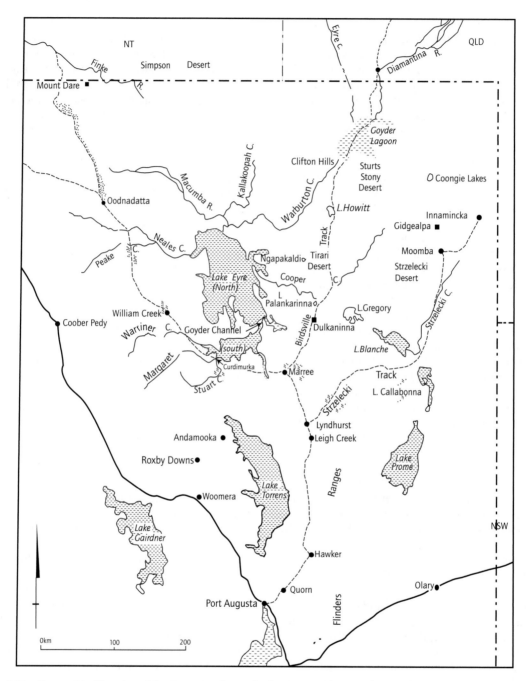

Fig. 8.10  Geographical location of the Coongie Lakes and Lake Eyre, South Australia. (Modified after several authors.)

*Lakes in Arid Environments* 229

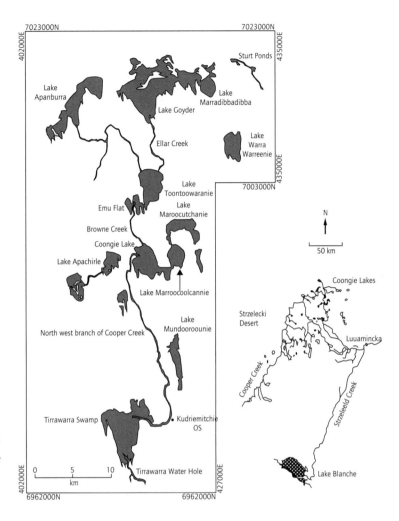

**Fig. 8.11** The major lakes of the Coongie Lakes series and, bottom right, the position of the lakes relative to Cooper and Strzelecki creeks. (Modified largely after Reid & Puckridge 1990.) Royal Society of South Australia, Adelaide, for Tyler, M.J., Twidale, C.R., Davies, M. & Wells, C.B. (eds) *Natural History of the North East Deserts*.

little known. Phytoplankton production, however, is recorded to be high, reflecting high nutrient concentrations. Because of the impermanence of the system, the long dry periods and the high turbidity, macrophytes are of limited diversity and abundance. There are two charophyte species, and five species of vascular plant, one of which (*Ludwigia peptoides*) has been introduced. The others are *Cyperus gymnocaulus*, *Myriophyllum verrucosum*, *Azolla fihiculoides* and *Lemna disperma*.

The lakes are of high conservation value as indicated by their inclusion on the Ramsar list. They are increasingly visited by tourists who, together with overgrazing, and the impact of feral horses and the rabbit, are potential threats to the ecological integrity of the system. Their relatively pristine nature at present gives them a particularly high scientific value as a system of temporary lakes in an arid landscape.

### 8.5.4 The Aral Sea

The Aral Sea (Fig. 8.13; see also Löffler, Volume 1, Chapter 2 and Butorina, this Volume, Chapter 7) is a large (area presently c.20,000 km$^2$), permanent, saline lake in arid central Asia. The lake receives a

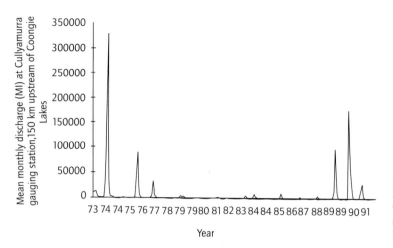

**Fig. 8.12** Mean monthly discharge in Cooper Creek 150 km upstream of Coongie Lakes. (Original graph supplied by J. Puckridge. Reproduced by author's permission.)

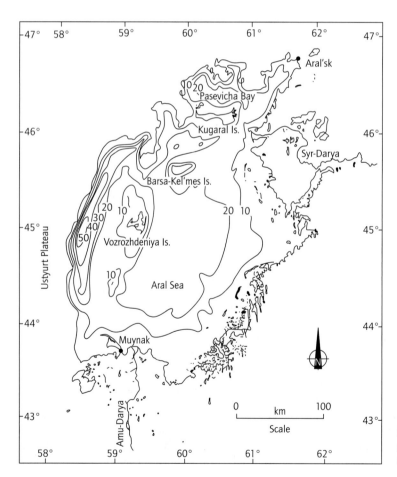

**Fig. 8.13** The Aral Sea. Bathymetry of lake prior to 1960. (Original map supplied by Dr N.V. Aladin.)

mean annual rainfall of <200 mm, and its annual evaporation is c.2000 mm. Its two major and virtually only inflows, the Syr Darya and Amu Darya, rise in humid mountainous regions elsewhere in central Asia. The entire catchment area is about $2 \times 10^6$ km$^2$, and includes not only all of Uzbekistan and Kazakhstan, but also parts of Tadzhikistan, Turkmenistan, Kirghizia, and northern Iran and Afghanistan.

The Aral Sea was studied relatively intensively before the 1960s, by limnologists from the former USSR. Following its marked decrease in size (from the 1960s onwards, after massive diversions of inflows for irrigation), limnological investigations were restricted, and only the persistence of a few Russian scientists, notably N.V. Aladin and his colleagues, has enabled documentation of the critical changes in the nature of the lake during the past 40 years. Since the break-up of the USSR, little ongoing comprehensive limnological work has been carried out. What has continued, despite the large sums of money expended by a number of international agencies in attempting to address the environmental changes caused by damage to the lake, has again mostly been undertaken by a few persistent Russian limnologists.

The attention of the western world was first drawn to the marked changes in the lake by the work of Micklin (1988) based on an analysis of satellite data, and publicised initially by Ellis (1990). Much of the earlier limnological work was summarised by Zenkevitch (1963). His account is essentially a description of the lake before it began to change. Changes which took place after the 1960s have been reviewed *inter alia* by Aladin & Kotov (1989), Williams & Aladin (1991) and Aladin *et al.* (1996). Micklin & Williams (1996) reported on a symposium on the lake, at which environmental issues were briefly summarised by Micklin (1996). Glazovsky (1990) has also addressed environmental problems, and later (1995) comprehensively reviewed these questions and our knowledge of the lake (see also Chapter 7, this volume).

Palaeolimnological views are that the modern lake is of relatively recent (post-Pleistocene) origin, and reached its pre-1960 size about 10,000 years BP, following discharge of the Amu Darya northwards into a depression enlarged by aeolian activity. The lake so formed was the terminus of two endorheic catchments drained by the Amu and Syr Darya. Lake bathymetry prior to 1960 is illustrated in Fig. 8.13.

Prior to the 1960s, the lake was the world's fourth largest, with an area of 68,000 km$^2$, a volume of 1090 km$^3$, and mean and maximum depths, respectively, of 16 and 69 m. All of these features changed after about 1960. Then, large diversions of water for irrigation began (mainly for cotton) from the Amu and Syr Darya. The effects on the lake were rapid, predictable and significant. Lake level and volume fell, large parts of the bed were exposed, and islands, of which there had been a great number, especially in the archipelago in the southeastern part of the lake, became peninsulas.

By 1990, the lake had separated into two, a smaller, northern body (the Small Aral Sea), and a larger, southern water mass (the Large Aral Sea). Water-levels of both had fallen by about 15 m, and the area and mean depth of the larger lake had halved (to 38,000 km$^2$ and 9.2 m) and its volume had decreased by two-thirds (to 310 km$^3$). The process of lake shrinkage is ongoing because diversions continue. Pre-1960, present (1998) and projected future (post-2000) sizes and drainage patterns are indicated in Fig. 8.14.

Thermal patterns, despite recent morphometric changes, have remained more or less constant. The deeper, western parts of the lake are isothermal during winter, and stratify in summer, with a weakly defined metalimnion between 10 and 20 m. The shallower (20 m), more central parts of the lake remain more or less isothermal, or without any well-defined stratification, all year. As the major part of both present lakes is less than 20 m deep, and becoming shallower, it is clear that areas of thermal stratification are decreasing in size.

Changes in morphometry have had some effect on surface thermal patterns; there has been an increase in mean annual surface temperature of 1 to 2°C, but the lake still freezes over in winter. Transparency of the lake both prior to and after 1960 was, and is, relatively high. In a series of observa-

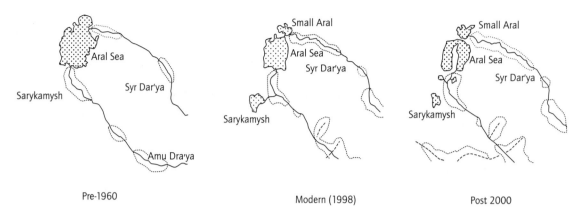

Fig. 8.14 The Aral Sea: past, present and future. The relative sizes and disposition of drainage lines are indicated. (Modified from Aladin et al. 1996.) Springer, Berlin, for Micklin, P.P. & Williams, W.D. (eds) *The Aral Sea Basin*. NATO ASJ series. Partnership sub-series 2, vol. 12.

tions carried out during July and August 1980, mean transparency was 9.5 m (Aladin & Kotov 1989).

Before 1960, the salinity of the Aral Sea was mostly homogeneous throughout, and stable at c.10 g L$^{-1}$. As lake volume decreased, salinity rose, at first slowly, then more rapidly. By 1970 it was c.11 g L$^{-1}$, by 1980, c.16 g L$^{-1}$ and by 1990, c.30 g L$^{-1}$. Predictions are (1998) that by 2000 it will be c.50 g L$^{-1}$, and will continue to rise. Sodium and chloride have been the most numerous ions at all times, although with significant amounts of Mg$^{2+}$ (27% equivalent cation sum), Ca$^{2+}$ (15%) and SO$_4^{2-}$ (39% of anions). Nutrient concentrations were naturally low, but have increased from inputs from the Syr and Amu Darya. Oxygen concentrations both pre-1960 and afterwards were relatively high. All parts of the water column are oxygenated.

The diversity, biomass and extent of endemicity of the biota of the Aral Sea was, and is, low. Prior to 1960, there were only some 20 fish species, 195 species of free-living invertebrates, 12 species of macrophytes and 82 species of other plants. Significant taxa in the zooplankton were *Podonevadne camptonyx*, *Evadne anonyx*, *Arctodiaptomus salinus*, *Mesocyclops Ieuckarti* and mollusc larvae. In the benthos, *Dreissenia* spp., *Hypanis minima*, oligochaetes, ostracods, amphipods and insects (especially chironomids) were important. Cyprinids made up most of the fish fauna, and diatoms the phytoplankton. The aquatic macrophytes present were mainly *Zostera nana*, *Polysiphonia violacea* and *Tolypella aralica*.

The composition of the biota has changed dramatically over the past four decades. Of course, salinity changes are implicated in this change, but also important have been many introductions of fish and other biota. Thus, only *Calanipeda aquaedulcis* and mollusc larvae became important in the zooplankton (with *Podonevadne camptonyx* important in the smaller Aral in 1990). In the zoobenthos, by the 1990s, only eight major taxa survived: *Cyprideis torosa*, *Palaemon elegans*, *Rhithropanopeus harrisii tridentatus*, *Abra ovata*, *Cerastoderma istmicum*, *Caspiohydrobia* sp., *Theodoxus* sp. and *Nereis diversicolor*. With regard to plants, the phytoplankton had become composed mostly of euryhaline forms of Bacillariophyta, the reedswamps had disappeared, and the only remaining submerged macrophyte was *Zostera nana*.

Broader environmental changes that were caused by changes to the lake have been referred to earlier. Here, all that need be added is that no amelioration of these changes can be seen for the immediate, or even the distant future. The Aral Sea and its environs are indeed an environmental crisis of global proportions.

### 8.5.5 Lake Eyre

Lake Eyre (Figs 8.10 and 8.15) is a very large (area c.9700 km$^2$), temporary, shallow, saline lake in central Australia. The lake basin lies in an arid region (median annual rainfall 100–150 mm, mean annual evaporation >3600 mm), with catchments extending into more humid regions, notably into Queensland (Warburton Creek). The lake remained virtually unexplored by scientists until the last century, but a series of recent fillings, better access, and modem support facilities (not least vehicular air-conditioning and better communications) have led to several explorations, although even these, for the most part, have merely been short expeditions.

As a result, although there is an extensive list of publications concerning the lake, comprehensive limnological explorations which address functional processes rather than descriptive matters remain outstanding. As an introduction to the literature of limnological interest, mention is made of the following, from which other accounts can be obtained: Johns (1963), Bayly (1976), Bye et al. (1978), Kotwicki (1986), Bonython & Fraser (1989), Williams (1990), and Williams & Kokkinn (1988). There are also several popular accounts of the natural history and general nature of the lake. Those by Dulhunty (1979) and Serventy (1985) are of particular interest.

Lake Eyre, as commonly referred to, actually consists of two lakes, a larger northern basin, Lake Eyre North (area 8430 km$^2$), and a smaller southern body, Lake Eyre South (area 1260 km$^2$). These are connected by the narrow Goyder Channel (Fig. 8.15). Both lakes arose from aeolian activity, following tectonic upwarping south of the present basin, during arid, post-Pleistocene times. During the Pleistocene, the region was occupied by the much larger, freshwater Lake Died. Both modern lakes are shallow, and occupy flat basins. When full, the maximum depth and mean depth (respectively) of Lake Eyre North are 5.7 m and 3.3 m, and

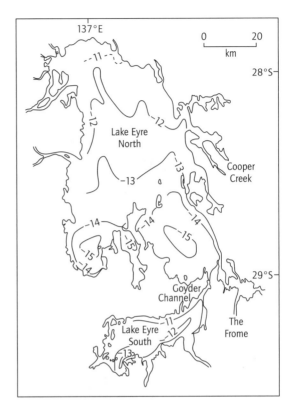

**Fig. 8.15** Lake Eyre. Bathymetry when full. (After Bye et al. 1978.) Royal Society of South Australia. *Transactions of the Royal Society of South Australia* 102.

those of Lake Eyre South 3.7 m and only 1.9 m (Fig. 8.15). Both lie below present sea-level (–15 m).

The lakes fill episodically, but the frequency of filling is only known with any certainty for this and the last century. During the periods 1949–50, 1974–78, 1984–85, 1985–86 and 1997+, they contained substantial amounts of water. Smaller bodies no doubt existed between these times. There is some evidence that the lakes were also full in 1890. Mostly, inflows arrive from Warburton Creek (Fig. 8.15), and flow first into Lake Eyre North. Occasionally, however (as in 1997), rains on the local catchments of Lake Eyre South fill this lake first.

Patterns of filling and drying are now evident from satellite images (Fig. 8.16), and from these it is clear that the lakes fill relatively rapidly and dry

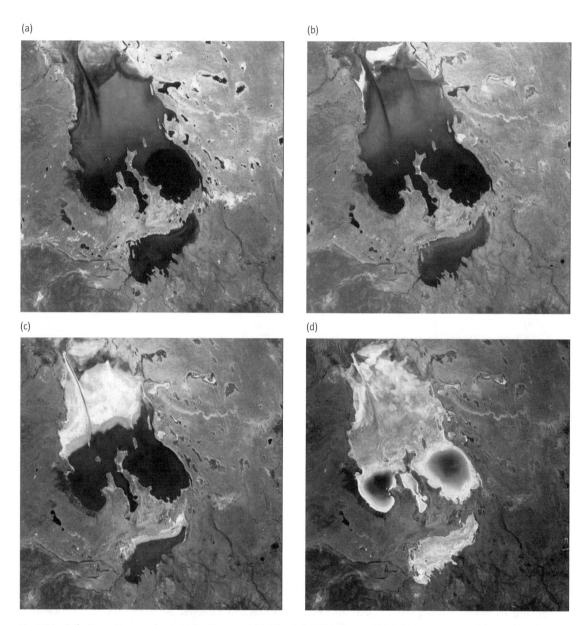

Fig. 8.16 Lake Eyre. Extent of water after January 1984 flood: (a) 22 February 1984; (b) 21 June 1984; (c) 12 November 1984; (d) 4 March 1985. (Landsat imagery provided by the Australian Centre for Remote Sensing, Australian Surveying and Land Information Group, Department of Administrative Services, Canberra. Reprinted with the permission of the Royal Society of South Australia.)

out progressively from the north over several months. Between inundations, the lakes are quite dry, with crusts either of saliniferous mud or halite deposits (up to 0.5 m thick). The salt crust dissolves entirely when the lakes are full.

Although, when they contain water, both lakes are always saline, salinities vary spatially and temporally. Thus, they are usually lowest close to regions of inflow, and increase progressively away from these sources. They may also be very high, at first, just above the salt crust, and lower in the overlying water column (leading to **meromixis**). When the lakes first fill, salinity rapidly increases, as the small volume of water present dissolves the salt crust, but then declines as the volume of water increases, and the salt mass is diluted. Figure 8.17 indicates temporal patterns of salinity in Lake Eyre South over the period 1984–85. This figure also illustrates the chemical character of the lake, which is remarkably stable, with sodium and chloride the most abundant ions at all times.

No data on thermal patterns exist, but there is little doubt that, given the huge expanse of exposed water surface, and the shallow nature of the lakes, thermal stratification would last for no more than a few days at most when the lakes are full. For a large part of the time, it is likely that the lake is isothermal. However, meromixis during the initial filling phase may lead to some thermal stratification, with higher temperatures just above the bottom salt crust. Presumably oxygen also exhibits little change in the water column, but absolute concentrations decline over time, as salinities rise (Sherwood et al. 1992). No data on nutrients are available.

Publications on the aquatic biota of the lake are fewer than those on its physical features, and are essentially limited to incomplete descriptions of the fauna and flora, and some speculations on the nature of the food chain. Only the observations of Williams & Kokkinn (1988) involved collections over several months. The phytoplankton includes several blue-green algal taxa (e.g. *Nodularia spumigena*), as well as several species of *Dunaliella*. No aquatic macrophytes occur, but a rich microbial community has been recorded, including photosynthetic sulphate-reducing bacteria, thiobacteria (which oxidise sulphur to sulphate), photosynthetic purple bacteria, *Halobacterium halobium*, and several species of protists. Microbial processes were briefly reviewed by Javor (1989).

Invertebrate studies suggest that there are differences between the two lakes, but that *Microcyclops* spp., *Moina baylyi* and *Diacypris* spp. are common to both. The same studies also suggest that the fauna is part of a widely distributed central Australian salt lake fauna together with a small

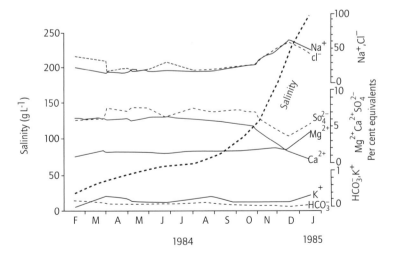

Fig. 8.17 Lake Eyre South. Salinity and ionic composition 1984–85. (After Williams & Kokkinn 1988.) Kluwer (for *Hydrobiologia* 158, 227–236).

number of cosmopolitan species (e.g. *Brachionus 'plicatilis'*). Notably, none of the taxa characteristic of intermittent salt lakes in southern Australia (e.g. *Calamoecia*, *Daphniopsis*, *Australocypris*, *Haloniscus* (an isopod), and *Coxiella* (a gastropod)) either occur, or are abundant. Williams & Kokkinn (1988) attributed this absence of southern forms to the episodic nature of the lake; the fauna, although resistant to desiccation and the other stresses of life in temporary salt lakes, cannot withstand prolonged desiccation.

With regard to fish, only two species (*N. erebi* (also present in the Coongie Lakes as indicated above) and *Craterocephalus eyresii* – chardhead) have been recorded in any abundance. Neither of these possesses any stage resistant to desiccation, so that they are constantly reintroduced to the lakes by flood-waters. Even so, very large biomasses have been recorded. These only occur when salinities are still relatively low; essentially, then, the fish fauna is one which comprises halotolerant freshwater forms which are easily dispersed.

More species of bird have been recorded, but few of these actually use the lake as a feeding resource. All possess extensive Australian ranges, and can be regarded as opportunistically associated with the lake. No other aquatic vertebrates whose occurrence is limited to the times when the lakes are flooded are present. Williams & Kokkinn (1988) speculated on the nature of the food chain (with fish present) at times when water salinity is still relatively low. They suggested that the top of the food chain was occupied by a few species of piscivorous birds Those fish present feed largely on invertebrates and organic detritus. Zooplankton graze on phytoplankton, but some ingest bottom detritus.

Finally, brief note is made of the occurrence of a small, taxonomically impoverished biota which is restricted to the lake, but which is active only during dry phases. Blue-green algae and algae occur just beneath the surface of the dry lake bed, together with spiders, beetles, ants and other insects. Most are predators. One vertebrate, the Lake Eyre dragon (*Amphibolurus maculosus*, a lizard), occurs.

## 8.6 ACKNOWLEDGEMENTS

A large number of colleagues have contributed, either directly or indirectly, to this chapter. I thank in particular for indirect help: Dr N.V. Aladin (Zoological Institute, St Petersburg, Russia), whose studies of the Aral Sea are a monument to his dedication to limnology; Professor Bryan Davies and Dr Jenny Day (University of Capetown), whose studies of waters in arid regions of southern Africa likewise are monuments to their dedication; and Associate Professor K.F Walker (University of Adelaide). Dr Aladin is also thanked for a bathymetric map of the Aral Sea. Mr J. Puckridge is thanked for the provision of information and diagrams. Mrs Kelly Fennell, University of Adelaide, prepared the originals of some of the figures; she too is thanked. Mr Ian McCullough of CEH was of invaluable assistance in locating some of the more obscure references in Professor William's manuscript (eds).

## 8.7 REFERENCES

Abraham, A.D. & Parsons, A.J. (eds) (1994) *Geomorphology of Desert Environments*. Chapman & Hall, London, 674 pp.

Agnew, C. & Anderson, E. (1992) *Water Resources in the Arid Realm*. Routledge, London, 329 pp.

Aladin, N.Y. & Kotov, S.V. (1989) The original state of the Aral Sea ecosystem and changes caused by man. *Trudy Zoologicheskogo Instituta Academiya Nauk SSSR (Leningrad)*, **199**, 4–25. (In Russian.)

Aladin, N.V., Plotnikov, I.S., Orlova, M.I., et al. (1996) Changes in the form and biota of the Aral Sea over time. In: Micklin, P.P. & Williams, W.D. (eds), *The Aral Sea Basin*. NATO ASI Series, Partnership Sub-series 2, Vol. 12, Springer-Verlag, Berlin, 33–55.

Alcocer, J. & Escobar, E. (1990) The drying up of the Mexican Plateau axalapazcos. *Salinet*, **4**, 34–6.

Allan, R. (1988) Meteorology and hydrology. In: Reid, J. & Gillen, J. (eds), *The Coongie Lakes Study*. Department of Environment and Planning, Adelaide, Australia, 31–50.

Barbier, E., Acreman, M. & Knower, D. (1997) *Economic Valuation of Wetlands. A Guide for Policy Makers and Planners*. International Union for the Conservation of Nature, Gland, 140 pp.

Battarbee, R.W., Jones, V.J., Flower, R.J., Cameron, N.G. et al. (2001) Diatoms. In: Smol, J.P., Birks, H.J.B. & Last, W.L. (eds) *Tracking Environmental Change using Lake Sediments. Volume 3, Terrestrial, Algal and Siliceous Indicators.* Kluwer, Dordrecht, 155–202.

Bayly, I.A.E. (1967) The general biological classification of aquatic environments with special reference to those of Australia. In: Weatherley, A.H. (ed.), *Australian Inland Waters: Eleven Studies.* Australian National University Press, Canberra, 78–104.

Bayly, I.A.E. (1976) The plankton of Lake Eyre. *Australian Journal of Marine and Freshwater Research,* **27,** 661–5.

Bayly, I.A.E. & Williams, W.D. (1973) *Inland Waters and their Ecology.* Longman, Camberwell, 316 pp.

Beadle, L.C. (1981) *The Inland Waters of Tropical Africa. An Introduction to Tropical Limnology* (2nd edition). Longman, London, 475 pp.

Beaumont, P. (1989) *Environmental Management and Development in Drylands.* Routledge, London, 505 pp.

Bonython, C.W. & Fraser, A.S. (eds) (1989) *The Great Filling of Lake Eyre in 1974.* Royal Geographical Society of Australasia (South Australian Branch), Adelaide, 119 pp.

Boulton, A.J. (1988) *Composition and dynamics of macroinvertebrate communities in two intermittent streams.* Unpublished PhD thesis, Monash University, Melbourne, 312 pp.

Boulton, A.J. (1989) Over-summering refuges of aquatic macroinvertebrates in two intermittent streams in Victoria. *Transactions of the Royal Society of South Australia,* **113,** 23–34.

Boulton, A.J. & Jenkins, K.M. (1998) Flood regimes and invertebrate communities in floodplain wetlands. In: Williams, W.D. (ed.), *Wetlands in a Dry Land: Understanding for Management.* Environment Australia, Canberra, 137–46.

Bulgareanu, V.A.C. (1996) Protection and management of anthroposaline lakes in Romania. *Lakes and Reservoirs: Research and Management,* **2,** 211–29.

Burgis, M.J. & Morris, P. (1987) *The Natural History of Lakes.* Cambridge University Press, Cambridge, 218 pp.

Bye, J.A.T., Dillon, P.J., Vandenberg, J.G. & Will, G.D. (1978) Bathymetry of Lake Eyre. *Transactions of the Royal Society of South Australia,* **102,** 85–9.

Carmouze, J.-P. (1983) Historical background. In: Carmouze, J-P., Durand, J.R. & Leveque, C. (eds), *Lake Chad. Ecology and Productivity of a Shallow Tropical Ecosystem.* Monographiae Biologicae, 53. Dr W. Junk, The Hague, 1–10.

Carmouze, J.-P., Durand, J.R. & Leveque, C. (eds) (1983) *Lake Chad. Ecology and Productivity of a Shallow Tropical Ecosystem.* Monographiae Biologicae 53. Dr W. Junk, The Hague, 575 pp.

Cole, G.A. (1968) Desert limnology. In: Brown, G.W. (ed.), *Desert Biology.* Academic Press, New York, 423–86.

Comin, F.A. & Williams, W.D. (1994) Parched continents: Our common future? In: Margalef, R. (ed.), *Limnology Now: a Paradigm of Planetary Problems.* Elsevier, Amsterdam, 473–527.

Cooke, R.U., Warren, A. & Goudie, A.S. (1993) *Desert Geomorphology.* University College London Press, London, 526 pp.

Davies, B.R., Thoms, M.C., Walker, K.F., O'Keefe, J.H. & Gore, J.A. (1994) Dryland rivers: their ecology, conservation and management. In: Calow, P. & Petts, G.E. (eds), *The Rivers Handbook,* Vol. 2, *Hydrological and Ecological Principles.* Blackwell Scientific Publications, Oxford, 484–511.

De Martonne, E., & Aufrere, L. (1928) L'extension des regions privees d'ecoulement vers l'ocean. *Publication de l'Union Géographique Internationale,* **3,** 197 pp.

Dinar, A., Seidl, P., Olem, H., Jorden, V., Duda, A. & Johnson, R. (1995) *Restoring and Protecting the World's Lakes and Reservoirs.* Technical Paper Number 289, World Bank, Washington, DC, 130 pp.

Dulhunty, R. (1979) *When the Dead Heart Beats Lake Eyre Lives.* Lowden, Kilmore, 259 pp.

Ellis, W.S. (1990) The Aral. A Soviet sea lies dying. *National Geographic,* **February,** 73–92.

Entz, B.A.G. (1976) Lake Nasser and Lake Nubia. In: Rzoska, J. (ed.), *The Nile. Biology of an Ancient River.* Dr W. Junk, The Hague, 271–98.

Finlayson, M. & Moser, M. (eds) (1991) *Wetlands.* Facts on File Ltd, Oxford, 224 pp.

Frazier, S. (1996) *An Overview of the World's Ramsar Sites.* Publication 39, Wetlands International, Wageningen, 58 pp.

Friedman, G.M. & Krumbein, W.E. (1985) *Hypersaline Ecosystems: the Gavish Sabkha.* Ecological Studies 53, Springer-Verlag, Berlin, 484 pp.

Geddes, M.C. & Williams, W.D. (1987) Comments on *Artemia* introductions and the need for conservation. In: Sorgeloos, P., Bengtson, D.A., Decleir, W. & Jaspers, E. (eds), *Artemia Research and its Applications.* Universa Press, Wetteren, 19–26.

Ghassemi, F., Jakeman, A.J. & Nix, H.A. (1996) *Global Salinization of Land and Water Resources: Human Causes, Extent and Management.* Centre for Resource and Environmental Studies, Canberra, 526 pp.

Glazovsky, N.F. (1990) *The Aral Crisis.* Academy of Science, Moscow, 134 pp. (In Russian)

Glazovsky, N.F. (1995) Aral Sea. In: Mandych, A.F. (ed.), *Enclosed Seas and Large Lakes of Eastern Europe and Middle Asia.* SPB Academic Publishing, Amsterdam, 119–54.

Gleik, P.H. (ed.) (1995) *Water in Crisis. A Guide to the World's Fresh Water Resources.* Oxford University Press, Oxford and New York, 473 pp.

Graf, W.L. (1987) *Fluvial Processes in Dryland Rivers.* Springer-Verlag, Berlin, 346 pp.

Grove, A.T. (1977) The geography of semi-arid lands. *Philosophical Transactions of the Royal Society of London, Series B*, **278**, 457–75.

Halse, S. (1998) Wetland R&D requirements in Western Australia. In: Williams, W.D. (ed.), *Wetlands in a Dry Land: Understanding for Management.* Environment Australia, Canberra, 59–66.

Hammer, U.T. (1986) *Saline Lake Ecosystems of the World.* Dr W. Junk, Dordrecht, 616 pp.

Hartland-Rowe, R. (1972) The limnology of temporary waters and the ecology of the Euphyllopoda. In: Clark, R.B. & Wootton, R.J. (eds), *Essays in Hydrobiology presented to Leslie Harvey.* University of Exeter, Exeter, 15–31.

Herdendorf, C.E. (1982) Large lakes of the world. *Journal of Great Lakes Research*, **8**, 379–412.

Herdendorf, C.E. (1984) *Inventory of the Morphometric and Limnologic Characteristics of the Large Lakes of the World.* Technical Bulletin No. 17, Ohio State University Sea Grant Program, Columbas, OH, 78 pp.

Herdendorf, C.E. (1990) Distribution of the world's large lakes. In: Tilzer, M.M. & Serruya, C. (eds), *Large Lakes, Ecological Structure and Function.* Springer-Verlag, Berlin, 3–38.

Hudson, P. (1996) Sympatric distribution of an Australian salt lake wolf spider and scorpion. *International Journal of Salt Lake Research*, **6**, 1–3.

Hudson, P. & Adams, M. (1996) Allozyme characterisation of salt lake spiders (*Lycosa*: Lycosidae: Araneae) of southern Australia: systematic and population genetic implications. *Australian Journal of Zoology*, **44**, 535–67.

Hutchinson, G.E. (1937) A contribution to the limnology of arid regions. *Transactions of the Connecticut Academy of Arts and Sciences*, **33**, 47–132.

Javor, B. (1989) *Hypersaline Environments. Microbiology and Biogeochemistry.* Springer-Verlag, Berlin, 328 pp.

Johns, R.K. (1963) Investigations of Lake Eyre. *Geological Survey of South Australia, Report of Investigations*, **24**, 1–24.

Kiener, A. (1978) *Ecologie, Physiologie et Economie des Eaux Saumâtres.* Masson, Paris, 220 pp.

Kindler, J., Warshall, P., Arnould, E.J., Hutchinson, C.F. & Varady, R. (1990) *The Lake Chad Convention Basin. A Diagnostic Study of Environmental Degradation.* Report to The Lake Chad Basin Commission (LCBC), Nairobi, 182 pp.

King, J.L., Simovich, M.A. & Brusca, R.C. (1996) Species richness, endemism and ecology of crustacean assemblages in northern California vernal pools. *Hydrobiologia*, **328**, 85–116.

Kira, T. (ed.) (1994) *Data Book of World Lake Environments — a Survey of the State of the World Lakes*, Vol. 2, *Africa and Europe* (compact-size edition). International Lake Environment Committee, United Nations Environment Programme, Kusatsu, 852 pp.

Kirk, J.T.O. (1994) Optics of UV-B radiation in natural waters. *Archiv für Hydrobiologie Beihefte (Ergebnisse der Limnologie)*, **43**, 1–16.

Kokkinn, M.J. & Williams, W.D. (1988) Adaptations to life in a hypersaline water-body: adaptations at the egg and early embryonic stage of *Tanytarsus barbitarsis* Freeman (Diptera, Chironomidae). *Aquatic Insects*, **10**, 205–14.

Köppen, W. (1931) *Die Klimate der Erde.* Teubner, Berlin, 388 pp.

Kosarev, A.N. & Yablonskaya, E.A. (1994) *The Caspian Sea.* SPB Academic Publishing, The Hague, 276 pp.

Kotwicki, V. (1986) *Floods of Lake Eyre.* Engineering and Water Supply Department, Adelaide, 99 pp.

Langbein, W.B. (1961) Salinity and hydrology of closed lakes. *U.S. Geological Survey Professional Paper*, **412**, 20 pp.

Last, W.M. (2001) Mineralogical analysis of lake sediments. In: Last, W.M. & Smol, J.P. *Tracking Environmental Change using Lake Sediments*, Vol. 2, *Physical and Geochemical Methods.* Kluwer, Dordrecht, 143–87.

Latif, A.F.A. (1984) Lake Nasser — the new man-made lake in Egypt (with reference to Lake Nubia). In: Taub, F.B. (ed.), *Lakes and Reservoirs. Ecosystems of the World No 23*, Elsevier, Amsterdam, 385–410.

Livingstone, D.A. & Melack, J.M. (1984) Some lakes of sub-Saharan Africa. In: Taub, F.B. (ed.), *Lakes and Reservoirs. Ecosystems of the World No 23*, Elsevier, Amsterdam, 467–97.

Mabbutt, J.A. (1977) *Desert Landforms.* Australian National University Press, Canberra, 340 pp.

Macan, T.T. (1963) *Freshwater Ecology.* Longmans, London, 343 pp.

Maltby, E. (1991) Wetlands and their values. In:

Finlayson, M. & Moser, M. (eds), *Wetlands*. Facts on File, Oxford, 8–26.

Mandych, A.F. (ed.) (1995) *Enclosed Seas and Large Lakes of Eastern Europe and Middle Asia.* SPB Academic Publishing, Amsterdam, 286 pp.

Margalef, R. (1983) *Limnologia.* Omega, Barcelona, 1010 pp.

Meigs, P. (1953) World distribution of arid and semi arid homoclimates. *UNESCO Arid Zone Program.*, **1**, 203–10.

Meybeck, M. (1995) Global distribution of lakes. In: Lerman, A., Imboden, D. & Gat, J. (eds), *Physics and Chemistry of Lakes.* Springer-Verlag, Berlin, 1–35.

Micklin, P.P. (1988) Desiccation of the Aral Sea: a water management disaster in the Soviet Union. *Science*, **241**, 1170–6.

Micklin, P.P. (1996) Introductory remarks. In: Micklin, P.P. & Williams, W.D. (eds), *The Aral Sea Basin.* NATO ASI Series, Partnership Sub-series No. 2, Vol. 12, Springer-Verlag, Berlin, 3–8.

Micklin, P.P. & Williams, W.D. (eds) (1996) *The Aral Sea Basin.* NATO ASI Series, Partnership Sub-series No. 2, Vol. 12, Springer, Berlin, 186 pp.

Mono Basin Ecosystem Study Committee (1987) *The Mono Basin Ecosystem. Effects of Changing Lake Level.* National Academy Press, Washington, DC, 272 pp.

Morelli, J. & de Jong, M.C. (1996) South Australia. In: Blackley, R., Usback, S. & Langford, K. (eds) (1996) *A Directory of Important Wetlands in Australia.* Australian Nature Conservation Agency, Canberra, 435–531.

Mulvaney, D.J. & Golson, J. (eds) (1971) *Aboriginal Man and Environment in Australia.* Australian National University Press, Canberra, 389 pp.

Murakami, M. (1995) *Managing Water for Peace in the Middle East: Alternative Strategies.* United Nations University Press, Tokyo, 309 pp.

Murray, J. (1910) The characteristics of lakes in general, and their distribution over the surface of the globe. In: Murray, J. & Pullar, L. (eds), *Bathymetrical Survey of the Scottish Fresh-water Lochs*, Vol. 1. Challenger Office, Edinburgh, 514–658.

Neal, J.T. (ed.) (1975) *Playas and Dried Lakes. Occurrence and Development.* Benchmark Papers in Geology 20, Dowden, Hutchinson & Ross, Stroudsburg, 411 pp.

Neumann, J. (1953) Energy balance and evaporation from the sweet-water lakes of the Jordan Valley. *Bulletin of the Research Council of Israel*, **2**, 337–57.

Nissenbaum, A. (1979) Life in the Dead Sea—fables, allegories, and scientific research. *BioScience*, **29**, 153–7.

Nissenbaum, A. (ed.) (1980) *Hypersaline Brines and Evaporitic Sediments.* Developments in Sedimentology No. 28, Elsevier, Amsterdam, 270 pp.

Penman, H. (1948) Natural evaporation from open water, bare soil and grass. *Proceedings of the Royal Society, Series A*, **193**, 120–45.

Persoone, G. & Sorgeloos, P (1980) General aspects of the ecology and biogeography of *Artemia.* In: Persoone, G., Sorgeloos, P., Roels, O. & Jaspers, E. (eds), *The Brine Shrimp Artemia*, Vol. 3. Universa Press, Wetteren, 3–24.

Postel, S. (1995) Water and agriculture. In: Gleik, P. (ed.), *Water in Crisis. A Guide to the World's Fresh Water Resources.* Oxford University Press, Oxford, 56–66.

Powling, I.J. (1980) Limnological features of some Victorian reservoirs. In: Williams, W.D. (ed.), *An Ecological Basis for Water Resource Management.* Australian National University Press, Canberra, 332–42.

Puckridge, J. & Drewien, M. (1988) The aquatic fauna. In: Reid, J. & Gillen, J. (eds), *The Coongie Lakes Study.* Department of Environment and Planning, Adelaide, 69–108.

Reid, J. & Gillen, J. (eds) (1988) *The Coongie Lakes Study.* Department of Environment and Planning, Adelaide, 220 pp.

Reid, J.R. & Puckridge, J.T. (1990) Coongie Lakes. In: Tyler, M.J., Twidale, C.R., Davies, M. & Wells, C.B. (eds), *Natural History of the North East Deserts.* Royal Society of South Australia, Adelaide, 119–31.

Robarts, R.D. & Bothwell, M.L. (1992) *Aquatic Ecosystems in Semi-arid Regions: Implications for Resource Management.* N.H.R.I. Symposium Series 7, Environment Canada, Saskatoon, 360 pp

Roberts, J. (1988) Aquatic biology of the Coongie Lakes. In: Reid, J. & Gillen, J. (eds), *The Coongie Lakes Study.* Department of Environment and Planning, Adelaide, 51–68.

Robinson, T. & Minton, C. (1990) The enigmatic banded stilt. *Birds International*, 72–85.

Rozema, J., Gieskes, W.W.C., van de Geijn, S.C., Nolan, C. & de Boois, H. (eds) (1997) *UV-B and the Biosphere.* Kluwer, Dordrecht, 320 pp.

Ruttner, F. (1963) *Fundamentals of Limnology*, (3rd edition). Translated by Frey, D.G. & Fry, F.E.J. University of Toronto Press, Toronto, 295 pp.

Serruya, C. (ed.) (1978) *Lake Kinneret.* Dr W. Junk, The Hague, 501 pp.

Serruya, C. & Pollingher, U. (1983) *Lakes of the Warm Belt.* Cambridge University Press, Cambridge, 569 pp.

Serventy, V. (1985) *The Desert Sea. The Miracle of Lake Eyre in Flood.* Macmillan, Melbourne, 174 pp.

Sherwood, J.E., Stagnitti, F., Kokkinn, M.J. & Williams, W.D. (1992) A standard table for predicting equilibrium dissolved oxygen concentrations in salt lakes dominated by sodium chloride. *International Journal of Salt Lake Research*, **1**, 1–6.

Shiklomanov, I.A. (1995) World fresh water resources. In: Gleik, P. (ed.), *Water in Crisis. A Guide to the World's Fresh Water Resources.* Oxford University Press, Oxford, 13–24.

Skinner, J. & Zalewski, S. (1995) *Fonctions et Valeurs des Zones Humides Mediterraneennes.* Tour du Valat, Arles, 78 pp.

Thomas, D.S.G. (ed.) (1989) *Arid Zone Geomorphology.* Belhaven, London, 250 pp.

Thornthwaite, C.W. (1948) An approach towards a rational classification of climate. *Geographical Review*, **38**, 55–94.

Timms, B.V. (1973) *A comparative study of the limnology of three maars in western Victoria.* Unpublished PhD thesis, Monash University, Melbourne, 228 pp.

Timms, B.V. (1980) Farm dams. In: Williams, W.D. (ed.), *An Ecological Basis for Water Resource Management.* Australian National University Press, Canberra, 345–59.

Timms, B.V. (1993) Saline lakes of the Paroo, inland New South Wales, Australia. *Hydrobiologia*, **267**, 269–89.

Timms, B.V. (1997) A comparison between saline and freshwater wetlands on Bloodwood Station, the Paroo, Australia, with special reference to their use by waterbirds. *International Journal of Salt Lake Research*, **5**, 1–27.

UNEP (1992) *World Atlas of Desertification.* Arnold, London, 69 pp.

UNESCO (1977) *Map of the World Distribution of Arid Climates.* MAB Technical Note 7, Paris, 1 p.

Walton, K. (1969) *The Arid Zones.* Hutchinson, London, 175 pp.

Watson, R.T., Zinyowera, M.C. & Moss, R.H. (eds) (1996). *Climate Change 1995: impacts, Adaptations, and Mitigation of Climate Change—Scientific–Technical Analyses.* Contribution of Working Group II to the Second Assessment Report of the Intergovernmental Panel on Climate Change. Cambridge University Press, New York, 878 pp.

Welch, P.S. (1935) *Limnology.* McGraw-Hill, New York, xiv + 471 pp.

Williams, D.D. (1987) *The Ecology of Temporary Waters.* Croom Helm, London, 205 pp.

Williams, M.A.J. & Balling, R.C., Jr (1996) *Interactions of Desertification and Climate.* Arnold, London, 270 pp.

Williams, W.D. (1985) Biotic adaptations in temporary lentic waters, with special reference to those in semiarid and arid regions. *Hydrobiologia*, **125**, 85–110.

Williams, W.D. (1986) Limnology, the study of inland waters: a comment on perceptions of studies of salt lakes, past and present. In: De Deckker, P. & Williams, W.D. (eds), *Limnology in Australia.* Dr W. Junk, Dordrecht, 471–84.

Williams, W.D. (1988) Limnological imbalances: an antipodean viewpoint. *Freshwater Biology*, **20**, 407–20.

Williams, W.D. (1990) Salt lakes: the limnology of Lake Eyre. In: Tyler, M.J., Twidale, C.R., Davies, M. & Wells, C.B. (eds), *The Natural History of the North East Deserts.* Royal Society of South Australia, Adelaide, 85–99.

Williams, W.D. (1993a) The conservation of salt lakes. *Hydrobiologia*, **267**, 291–306.

Williams, W.D. (1993b) The world-wide occurrence and limnological significance of falling water-levels in large, permanent saline lakes. *Verhandlungen der internationalen Vereinigung für Limnologie*, **25**, 980–3.

Williams, W.D. (1995) Lake Corangamite, Australia, a permanent saline lake: Conservation and management issues. *Lakes and Reservoirs: Research and Management*, **1**, 55–64.

Williams, W.D. (1996a) The largest, highest and lowest lakes of the world: saline lakes. *Verhandlungen der internationalen Vereinigung für Limnologie*, **26**, 61–79.

Williams, W.D. (1996b) What future for saline lakes? *Environment*, **38**, 12–20, 38–9.

Williams, W.D. (1998) Salinity as a determinant of the structure of biological communities in lakes. *Hydrobiologia*, **381**, 191–201.

Williams, W.D. & Aladin, N.V. (1991) The Aral Sea: recent limnological changes and their conservation significance. *Aquatic Conservation*, **1**, 3–23.

Williams, W.D. & Kokkinn, M.J. (1988) The biogeographical affinities of the fauna in episodically filled salt lake: a study of Lake Eyre South, Australia. *Hydrobiologia*, **158**, 227–36.

Williams, W.D. & Sherwood, J.E. (1994) Definition and measurement of salinity in salt lakes. *International Journal of Salt Lake Research*, **3**, 53–63.

Williamson, C.E. (1995) What role does UV-B radiation play in freshwater ecosystems? *Limnology and Oceanography*, **40**, 386–92.

Zenkevitch, L.A. (1963) *Biology of Seas of the USSR.* Allen and Unwin, London, 955 pp.

# 9 Floodplain Lakes and Reservoirs in Tropical and Subtropical South America: Limnology and Human Impacts

JOHN M. MELACK

## 9.1 INTRODUCTION

Lakes associated with rivers are numerous and conspicuous features of the landscape in tropical and subtropical South America. The two major types of such waters are floodplain lakes and reservoirs. Although these lakes share some limnological features, there are important differences among them. Both types are responsive to hydrological and chemical changes in rivers and both can experience large fluctuations in level. However, stage in reservoirs usually is managed for hydroelectric, flood control or water supply, whereas in floodplain lakes seasonality is regional and climate determines variations in water level. Many South American subtropical reservoirs lie in economically developed catchments and receive runoff affected by agricultural, industrial and urban activities. In contrast, most floodplain lakes and tropical reservoirs are only beginning to be influenced by developments in their catchments. However, the flooding of large areas of terrestrial forest by tropical reservoirs does create conditions fundamentally different from those in seasonally inundated forests fringing floodplain lakes.

Over 8000 floodplain lakes occur along the Amazon River and the lower courses of its major tributaries in Brazil (Melack 1984). Most are small (i.e. less than 1 km$^2$, Sippel et al. 1992), but some exceed 100 km in length. Many more lakes are found along the Orinoco River (Hamilton & Lewis 1990), the upper Paraguay River (Hamilton et al. 1996), the Parana River (Drago 1990) and along numerous smaller rivers. These lakes can be perturbed by changes to the river with which they are associated, by modifications to the uplands surrounding their local catchments, and by direct alterations to the lakes and fringing wetlands. Junk (1997) and Melack & Forsberg (2001) summarise much of the recent ecological and biogeochemical information on Amazon floodplain lakes.

Reservoirs currently inundate approximately 40,000 km$^2$ in tropical and subtropical South America (Tundisi 1990). Within the highly developed State of São Paulo (Brazil), hydroelectric and water supply reservoirs occupy about 15,000 km$^2$ (Tundisi & Matsumura-Tundisi 1990). In the Brazilian Amazon, over 12,000 km$^2$ are now flooded by hydroelectric dams, and proposed projects could cover tens of thousands of square kilometres more (Barrow 1988). Besides fundamentally changing the original fluvial ecosystems, the reservoirs themselves are affected by hydrological regulation, by discharges from their catchments and by decomposition of flooded forests within the water body. A large scientific literature has developed since the onset of construction of major dams throughout the tropics during the late 1950s (e.g. Lowe-McConnell 1966; Obeng 1969; Goldman 1976; Serruya & Pollingher 1983; Tundisi & Straskraba 1999), but much of the information on impoundments remains rather inaccessible in reports commissioned by utilities and their funding sources.

My objectives are to examine well studied examples of floodplain lakes in the central Amazon and of reservoirs in tropical and subtropical South America in order to illustrate perturbations associated with human activities. Impacts of human activities include eutrophication, deforestation of catchments, flooding of standing forests, introduction of toxic materials and of exotic species, increased sediment loading and altered hydrological conditions. Before considering individual lakes, general characteristics of lakes associated with rivers will be discussed.

Both floodplain lakes and reservoirs are subject to large changes in water level. As the water levels vary, the proportion of open water relative to other aquatic habitats can change and ecological conditions are altered (Junk 1997). At lowest water, floodplain lakes may be reduced to shallow, turbid pools which occasionally dry completely. During moderate to high water, the lakes expand in area, flood surrounding forests and support seasonal growth of aquatic macrophytes. Conversely, the terrestrial areas, which surround the aquatic environments and constitute a portion of the lakes' catchments, expand during low water and contract during high water. Junk et al. (1989) defined seasonally flooded land as the aquatic–terrestrial transition zone or ATTZ. Additionally, five types of aquatic interfaces, differing in horizontal and vertical gradients in hydrochemical and physical conditions, can occur in floodplain lakes and reservoirs: the downstream river–lake interface, the upland stream or river–lake interface, lake–lake connections, lake–sediment–porewater interface, and the lakeshore–groundwater interface. Diagrammatic representations of the ATTZ and some of the aquatic interfaces have appeared in Fisher et al. (1991) and Junk (1997).

Human influences on riverine lakes and their associated wetlands are numerous. Conversions of catchments for agriculture and urban development lead to major changes in hydrology, vegetation, soil characteristics and biogeochemical cycles. Downstream aquatic habitats are influenced both by on-site changes (e.g. impoundments and drainage) and by off-site modifications, which may lead to enhanced run-off, erosion, sedimentation, and inputs of organic and inorganic solutes and particulates. Floodplain lakes and reservoirs are important as a direct source of food (e.g. fish) and various other products (e.g. timber), as grazing lands during low water periods, and as sources of drinking and irrigation water. The single most important factor affecting these lakes is demographic pressure with related agricultural, industrial and urban land-use.

## 9.2 LIMNOLOGY OF SUBTROPICAL RESERVOIRS

### 9.2.1 *Hydroelectric and water supply reservoirs*

The limnology of the reservoirs in the State of São Paulo is well described thanks to the extensive and varied studies of J.G. Tundisi and his colleagues. A long-term investigation of Broa Reservoir has continued since its formation in 1971 (Tundisi & Matsumura-Tundisi 1995). Comparative studies of as many as 52 reservoirs were begun in the late 1970s (Arcifa et al. 1981a,b; Matsumura-Tundisi et al. 1981; Takino & Maier 1981; Tundisi 1981; Tundisi et al. 1991). More recently, reservoirs in the neighbouring states of Paraná (Agostinho & Gomes 1997) and Minas Gerais (Dabés et al. 1990; Rolla et al. 1990) have received attention from limnologists. Many of these reservoirs experience numerous impacts stemming from industrial, urban and agricultural activities. Studies by Tundisi and his associates as well as those by others in nearby states were designed to provide the scientific basis for improvement in management of the reservoirs and for detection and correction of deterioration in the aquatic ecosystems.

Tundisi (1981) summarised ranges of limnological conditions in 52 reservoirs sampled quarterly over 1 year. Transparency estimated as Secchi disk visibility ranged from 0.2 to 10 m in reservoirs with euphotic zones of 1 to 18 m. Much of this variation reflects differences in algal abundances and nutrient inputs. Chlorophyll *a* ranged from less than $1\,\mu g\,L^{-1}$ up to $250\,\mu g\,L^{-1}$ in polluted waters near the city of São Paulo. Reservoirs such as Pedreira (Billings), located in densely populated areas with considerable industry, contained relatively high solute concentrations (as high as $388\,\mu S\,cm^{-1}$ conductance) in comparison with reservoirs such as Itapeva, located at high altitude in areas with few people and vegetated catchments, where waters are dilute (as low as $6\,\mu S\,cm^{-1}$ conductance).

Sulphate and nutrient concentrations followed similar patterns.

Climatic conditions in the State of Saõ Paulo alternate between a hot, rainy season (October to March) and a cool, dry season (April to September), and strongly influence the thermal structure in the reservoirs (Matsumura-Tundisi et al. 1981). Although some are monomictic, others are polymictic (Arcifa et al. 1981b). Many possess low dissolved oxygen concentrations at depth when at least temporally stratified during the warm months. Rivers in the upper Paraná basin flow east to west, and the largest urban areas and pollution sources are in the east. Hence, the upstream reservoirs are often the most influenced by human activities, and water quality improves in downstream reservoirs. Barbosa et al. (1999) applied the cascading river continuum concept to the series of reservoirs along the Tietê River and illustrate the changes in water quality among the seven reservoirs.

The series of reservoirs in the middle Tietê River receive inputs from sugar cane processing plants and discharges from the city of São Paulo at their upstream end, hence the most polluted is the first in the cascade, Barra Bonita (Matsumura-Tundisi et al. 1981; Tundisi et al. 1991). Barra Bonita Reservoir was completed in 1963, is about 325 km$^2$ in area, with a maximum depth of about 25 m. Studies beginning during the late 1970s have demonstrated a rapid eutrophication of the reservoir and the importance of flushing rate and winds on water quality (Tundisi & Matsumura-Tundisi 1990). Thermal stratification is transient throughout the year, and the water column remains oxygenated. These conditions in combination with high concentrations of iron entering via tributaries tend to maintain low phosphate concentrations in the water, and the sediments possess elevated values (Esteves 1983). In contrast, nitrate concentrations are high, and enrichment experiments did not detect nitrogen limitation (Henry et al. 1985).

During periods of low flow, wastewaters from the city of São Paulo can constitute 25% of the total flow to Barra Bonita Reservoir (Tundisi & Matsumura-Tundisi 1990). Hence, a first step in improving its water quality is to treat the wastewater. Other options include increasing throughflow or selecting the depth of withdrawal at the dam.

### 9.2.2 Urban reservoirs

Reservoirs within the confines of cities often exhibit advanced states of eutrophication and other signs of pollution. Examples in Brazil include Paranoá, Brasília (Altafin et al. 1995), Pampulha, Belo Horizonte (Giani et al. 1988), Taquaral, Campinas (Matsumura-Tundisi et al. 1986) and the municipal reservoir of São José do Rio Preto (Bozelli et al. 1992). Of these, Paranoá Lake has received the most study and attempts to restore water quality.

Paranoá Lake was constructed in 1959 in order to beautify Brasília, provide recreation and generate hydroelectric power. Instead, discharge into the lake from sewage treatment plants caused eutrophication with blooms of *Microcystis aeruginosa* (Altafin et al. 1995). After numerous studies over the past two decades, a restoration programme is proceeding. Tertiary sewage treatment to remove phosphorus has reduced loading to the lake by about two-thirds, but has not led to significantly lower concentrations of total phosphorus. Based on experimental studies in mesocosms (Starling 1993), biomanipulation of planktivorous fish populations in order to reduce phytoplankton biomass is being implemented also. *Microcystis aeruginosa* appears less conspicuous and water quality is improving.

## 9.3 ENVIRONMENTAL IMPACTS ASSOCIATED WITH AMAZONIAN RESERVOIRS

Since the damming of the Suriname River to form Lake Brokopondo in 1964 (Heide 1982), several other large hydroelectric reservoirs have been built within tropical forested catchments of South America, including Curuá-Una, Samuel, Tucuruí and Balbina in Brazil (Junk et al. 1981; Fearnside 1989; Matsumura-Tundisi et al. 1991; Petrere & Ribeiro 1994) and Guri in Venezuela (Gonzalez et al. 1991). Many more are planned.

Pre- and post- impoundment studies of reservoirs within the Amazon have been sponsored by Electonorte, the electric utility responsible for operating the hydroelectric dams. Only a small portion of these studies have appeared in the open scientific literature.

The Tucuruí dam produced the first large hydroelectric reservoir in the Amazon. Located about 300 km south of Belém on the Tocantins River, the Tucuruí dam was closed in 1984 and created a 2800 km$^2$ reservoir with an average depth of 17 m, a maximum depth of 75 m, and an average residence time of 50 days (Petrere & Ribeiro 1994). Failure to clear the forest before flooding the area led to a loss of about 20 million m$^3$ of timber, and impeded fishing and transportation on the water (Barrow 1988). The standing trees offered protection to mats of floating plants which further obstructed the surface, provided habitat for disease-carrying mosquitoes and snails, and may have caused anoxis and augmented methane emission upon decay (see below). Decomposition of terrestrial organic matter inundated by the reservoir probably added nutrients to the water and decreased dissolved oxygen, but owing to rapid flushing of Tucuruí Reservoir these problems were probably short-lived. Soon after the reservoir filled, fishing within the lake was good, a common occurrence in new tropical reservoirs (Beadle 1981), but one not usually sustained as nutrient release from the submerged vegetation declines (Petrere & Ribeiro 1994).

Developments bordering the reservoir and those more distant within its catchment are likely to cause ecological and social problems. The water level of Tucuruí Reservoir is expected to rise and fall 12 to 14 m and create an alternately flooded and exposed area of about 1000 km$^2$ (Barrow 1988). This zone may create habitats for undesirable species and make shore access difficult. Fragmentation of the surrounding forest by roads and settlements, and changes in aquatic habitats are likely to increase the incidence of diseases such as malaria, yellow fever, schistosomiasis, leishmaniasis and onchoceriasis. Forest clearing in the upper and mid-Tocantins-Araguaia basin, and especially expanding mining and agriculture in the Serra do Carajás, have increased erosion, which will speed siltation of the reservoir.

Flows downstream of the dam have been altered with major implications for the floodplain ecosystems adapted to and dependent on the natural flood regime. Although detailed studies are not available for the lower and mid-Tocantins River, numerous interactions between the biota and the flood pulse documented elsewhere in the Amazon (Junk 1997) are fair warning that myriad impacts are likely. The regulated flows will reduce flood levels and maintain higher low flows with a net effect of decreasing the area of floodplain enriched with riverine sediments each year and available for agriculture and use by fish and other organisms (Barrow 1988). However, along the lower Tocantins, the rice-growing polders may be aided by reductions in unexpectedly high inundation (Barrow 1988).

Balbina Reservoir was formed behind the Balbina hydroelectric dam (closed in October 1987), built on the Uatumã River in order to supply power to Manaus (Fearnside 1989). Average depth of the reservoir, when full, is 7 to 8 m, and hydraulic residence time is about 14 months. Before closure of the dam, about 50 km$^2$ near the dam was cleared of vegetation; the remainder of an estimated flooded area of 2360 km$^2$ was left forested. The actual area flooded and the extent of possible aquatic habitats, i.e. open water, flooded forest or aquatic macrophytes, have been debated.

Melack & Wang (1998) used synthetic aperture radar (SAR) data obtained from a satellite-borne sensor to classify a SAR image of Balbina Reservoir into open water and rivers, aquatic macrophytes, upland forests and flooded forests. Adding these areas, they estimated the total area of the reservoir to be 2407 km$^2$. Furthermore, their analysis indicates a large area of emergent forest remaining in the reservoir (Fig. 9.1).

Decomposition of the submerged vegetation in tropical reservoirs is known to produce locally high concentrations of hydrogen sulphide, and is likely to lead to large emissions of methane, a potent greenhouse gas.

The repetitive, all weather coverage offered by satellite-borne SARs provides a valuable source of

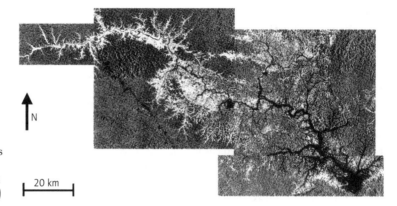

Fig. 9.1 Balbina Reservoir made from a mosaic of JERS-1 SAR images (October 1993). Light grey areas are flooded forests; black areas are open water. (From Melack & Wang 1998.)

information on the extent of inundation and the temporal changes of aquatic habitats in existing or newly formed reservoirs (Hess et al. 1995; Melack & Hess 1998; Alsdorf et al., 2001). Such information should aid both ecological and hydrological studies of tropical reservoirs, especially those in forested ecosystems. For example, Novo et al. (1995) have used SAR in order to evaluate coverage and distribution of macrophytes in Tucuruí Reservoir as part of an analysis of methane emissions.

Some estimates suggest that, per unit energy produced, fluxes to the atmosphere of greenhouse gases, such as carbon dioxide and methane, from reservoirs may be significant compared with greenhouse gas release during fossil fuel combustion (Rudd et al. 1993; Fearnside 2002). Methane emission from Amazonian reservoirs has been measured occasionally at several sites in the Samuel, Balbina and Curua-Una reservoirs and more extensively at Tucurui reservoir (Fearnside 1997; Rosa et al. 1997; Duchemin et al., 2000; Lima et al., 2000). It varies not only among reservoirs, but also within each reservoir, as a function of type and density of the drowned vegetation, macrophyte coverage, wind speed, temperature, oxygen saturation and water level. Fearnside (1997) proposed a method to compute the emissions from reservoirs using Tucuruí as an example. Although his calculations considered the initial stock and distribution of carbon, decay rates and pathways, and losses, he had few actual measurements to support his analyses. Fearnside (2002) has proposed a process-based model, which considers Tucuruí methane emissions by the spillways and turbines. Using empirical data from various sources, he concluded that the total methane released from water passing through the turbines in 1990 was two to eight times the total release from ebullition and diffusive emission across the surface of the reservoir, and may be of similar magnitude to the greenhouse warming potential of the gases emitted by the city of São Paulo.

## 9.4 NITROGEN AND PHOSPHORUS LIMITATION IN RESERVOIRS AND FLOODPLAIN LAKES

When the numerous data now available from temperate freshwater lakes are combined with information from lakes worldwide, phosphorus limitation is evidently not universal (Fisher et al. 1995). Regional differences in land use and geochemistry can lead to ample supply of phosphorus. Within individual lakes, seasonal variations in phosphorus and nitrogen supply are not uncommon. As lakes become eutrophic, there is a strong tendency for nitrogen limitation to become more important. Differences in nitrogen to phosphorus ratios (Smith 1979, 1982) and in trophic structure (Carpenter et al. 1985) among lakes cause substantial variability in the extent to which phosphorus controls algal biomass.

Although systematic evaluation of the role of nutrient limitation in tropical and subtropical South American reservoirs is not possible because too few of the wide variety of waters have been examined, information to guide management is improving. Arcifa et al. (1995) reviewed 13 studies which used experimental assays in four Brazilian reservoirs. Lobo Reservoir (São Paulo) has received the most attention; bioassays by Henry & Tundisi (1983), Henry et al. (1984) and others have detected phosphorus limitation at some times and nitrogen limitation at other times. In oligotrophic Jurumirim Reservoir and in eutrophic Barra Bonita Reservoir, phosphorus was found to be limiting (Henry 1990; Henry et al. 1985). In eutrophic Paranoá Lake, the main types of cyanobacteria declined after nitrogen and phosphorus additions singly or in combination, but chlorophytes exhibited a strong response to combined enrichment with nitrogen and phosphorus (Ibañez 1988).

The Pan American Center for Sanitary Engineering and Environmental Sciences (Lima, Peru) has conducted a regional assessment of eutrophication in warm-water reservoirs in South America and the Caribbean (Salas & Martino 1989). Eleven reservoirs in the State of São Paulo included in their study possessed total nitrogen to total phosphorus weight ratios in loading ranging from 6 to 34. Nine of these ratios lay above seven, a value indicative of phosphorus limitation. Further regression analyses demonstrated that concentrations of both total nitrogen and total phosphorus in South American reservoirs were correlated with chlorophyll.

The combination of experimental assays with regional surveys of limnological conditions and nutrient loading in Brazilian reservoirs indicates more evidence for phosphorus than for nitrogen limitation. However, conditions do occur when limitation by nitrogen or both nitrogen and phosphorus is detected. Further, the potential for augmented algal abundance in response to nutrient additions warns that these reservoirs are in various trophic states, and water quality will deteriorate if nutrient loading increases. Management of sources and dispersion of nitrogen and phosphorus is recommended.

In Amazon floodplain lakes, week to week and seasonal as well as lake to lake differences in the relative importance of nitrogen or phosphorus limitation of phytoplankton occur (Arcifa et al. 1995). In Lake Calado, Setaro & Melack (1984) fertilised natural populations in enclosures suspended in the lake and conducted physiological assays for nitrogen or phosphorus deficiency. When water levels were rising, they detected consistent positive responses to phosphorus addition with occasional enhancements after nitrogen enrichments (Fig. 9.2). Physiological assays indicated nitrogen and phosphorus deficiencies. During falling water, phosphorus alone, nitrogen and phosphorus, nitrogen alone or neither nitrogen nor phosphorus were implicated as limiting phytoplankton abundance. In Lake Jacarétinga, during low water, Zaret et al. (1981) found that chlorophyll concentrations increased after enrichment with nitrogen, increased further if nitrogen and phosphorus were added together, but did not increase with phosphorus only. In Lake Batata, a clear-water lake located along the Trombetas River, total nitrogen to total phosphorus molar ratios varied from 38 to 124 over the annual range of water levels (Huszar & Reynolds 1997); these ratios would indicate phosphorus limitation. Further studies by Farjalla et al. (2002) in Lake Batata and the nearby Trombetas River and Caranã stream tested for limitation of bacterial production by carbon, nitrogen or phosphorus. They found that phosphorus was the primary nutrient-limiting bacterioplankton production during high water, and that additions of glucose and phosphorus stimulated bacterial production, indicating low quality carbon in these clear waters.

Within the floodplains fringing the Amazon River and its sediment-laden tributaries, large expanses of floating macrophytes develop during rising water (Junk 1997), and the underwater stems and roots of these plants are colonised by periphyton. The epiphytic algae, whose biomass can exceed that of phytoplankton on a per unit area basis (Engle & Melack 1990), are responsible for high rates of nitrogen fixation and primary productivity (Doyle 1991), and are an important food for fishes

sediments, dissolved nutrients and epiphytic algal growth in a series of *in situ* experiments in Lake Calado. During rising and high water (May, June, July and August), chlorophyll per gram of roots was less than controls in experimental treatments bathed in river water. Addition of nitrogen and phosphorus without sediments increased chlorophyll above controls in August. Further, when epiphytic algae were treated with dilutions of river water (0%, 10%, 25% and 100%), only the 10% treatment produced significantly more algal chlorophyll. Hence, the influence of nutrient-rich, turbid water on epiphytic algal abundance will depend on differences in suspended sediment concentrations.

## 9.5 IMPACTS OF BAUXITE TAILINGS ON AN AMAZON FLOODPLAIN LAKE

Lake Batata, separated from the Trombetas River by a lateral leveé, received discharges of bauxite tailings totalling 18 million m³ per year from 1979 to 1989 (Esteves *et al.* 1990). The tailings now cover about 30% of the lake's bottom. These sediments contain a much higher proportion of very fine silts and clays with lower caloric content than do those not affected by tailings (Callisto & Esteves 1996a). Whereas Chironomidae and Chaoboridae larvae were most abundant in the natural sediments, the ephemeropteran *Campsurus notatus* thrived in areas affected by the bauxite tailings (Callisto & Esteves 1996b; Fonseca *et al.* 1998) and maintained productive populations (Leal & Esteves 2000). Phosphorus, carbon and nitrogen concentrations in the sediments were much lower in areas affected by tailings than in unaffected parts of the lake (Roland & Esteves 1993).

Concentrations of dissolved inorganic nitrogen differed between the region with tailings and that without. Ammonium varied from 565 µg L$^{-1}$ to 33 µg L$^{-1}$, most likely as a result of anoxia under the clay-rich tailings. Nitrate was higher above the tailings. In contrast, phosphorus fractions were similar at all the stations. Based on experiments conducted in microcosms, Leal *et al.* (2003) ob-

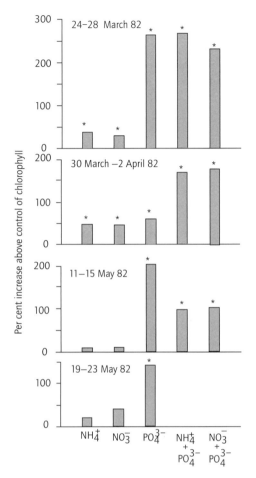

**Fig. 9.2** Changes in concentration of chlorophyll *a* as percentages above that of control in 4-L containers of Lake Calado water 4 days after enrichment. Asterisks denote changes significantly different from control by Student–Newman–Keuls test ($P < 0.05$). (From Setaro & Melack 1984.)

(Forsberg *et al.* 1993). Because attached algae retain their position close to the water surface, they may be better placed to exploit nutrients in turbid riverine floodwaters. Further, because particulate-bound nutrients are retained among the floating macrophytes, epiphytic algae possess access to these nutrients as well. Engle & Melack (1993) investigated the interactions between suspended

served that rates of ammonium released from sediment cores were at least fourfold lower in cores with tailings compared with natural sediments, and that bacterial production was lower in bauxite-affected sediments than in sediments not affected by bauxite.

During periods of low water, resuspension of the clays associated with the tailings impeded light penetration and probably reduced photosynthesis by phytoplankton. Chlorophyll concentrations were about ten times higher at stations not affected by tailings than at those with tailings. Phytoplankton collected mid-lake varied in composition and abundance over an annual cycle of filling and draining, but did not seem influenced by the altered sediments (Huszar & Reynolds 1997), although phytoplankton biomass was lower and species composition differed directly over the tailing-enriched sediments. Likewise, zooplankton abundances were much lower near the tailings.

Seasonally flooded forest was killed by the tailings. Hence, the fish that depended on the forest for food and shelter lost habitat. Moreover, high concentrations of suspended sediments in the region with tailings formed a further barrier to access by fish. Overall, the impact of discharge of bauxite mining residues on the ecology of floodplain environments is clearly negative.

## 9.6 IMPACTS OF DEFORESTATION ON AMAZON FLOODPLAIN LAKES

The deforested area in the Amazon basin increased throughout the 1980s (Skole & Tucker 1993), and continues to expand (Keller *et al.* 2001). Although accessible areas bordering navigable rivers and roads are particularly prone to deforestation, the effects of deforestation on the ecology of adjacent streams, rivers and lakes have received limited attention (Bruijnzeel 1991). Lake Calado, a moderate-sized lake located on the fringing floodplain of the central Amazon, was the site of a thorough examination of hydrology, limnology, and nutrient dynamics during the 1980s when its catchment was undergoing increasing development (Fig. 9.3;

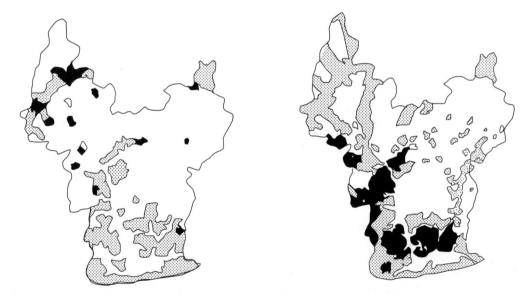

**Fig. 9.3** Deforestation in Calado basin. 1976 (left): dotted, clearing; black, recently burned clearing; white, mature forest or lake. 1987 (right): dotted, clearing; black, secondary vegetation; white, mature forest or lake. (From Melack *et al.* 1992.)

Melack & Fisher 1990; Melack 1996). Downstream lacustrine waters were influenced by increased inputs of water and solutes produced by deforestation. Nitrogen and phosphorus inputs from cut areas are of particular concern because of their role in aquatic eutrophication.

The hydrological characteristics of runoff in the Amazon determine the magnitude of the various hydrological pathways which transport nutrients and other solutes. Given the high frequency of storms in wet equatorial climates, it is important to understand how export of nutrients and solutes is influenced by storm size and frequency, and the volume of generated runoff. However, most hydrological studies in the Amazon basin have been carried out in areas of intact rain forest, and do not address the effects of deforestation (Nortcliff & Thornes 1981; Lesack 1993a).

Intrasystem processes and their effects on solute transport can be evaluated by measuring solution chemistry at different stages of a hydrological pathway. Studies of nitrogen transformations along hydrological pathways in Amazon catchments have been conducted by McClain et al. (1994) and Brandes et al. (1996). Only Williams et al. (1997) have addressed the effects of land conversion on solution chemistry for a suite of solutes at different stages of a hydrological pathway through an upland catchment. They determined that nitrogen to phosphorus ratios in stream water were lower in disturbed areas compared with those in intact forest. These studies indicated that there is a direct effect of land conversion on the ecology of receiving waters.

The small catchment studies conducted at Lake Calado provide some of the most extensive data on streamwater chemistry and catchment hydrology which exist for the Amazon basin. Lesack (1993a) measured all the water balance components for a first-order catchment draining a 23-ha stand of undisturbed upland rain forest adjacent to the Amazon floodplain. In a parallel study, Lesack (1993b) measured fluxes of solutes exported by streamflow and subsurface outflow. He provided the first evaluation for the Amazon basin of the relative roles of baseflow runoff versus storm flow runoff versus subsurface outflow in controlling the total export of nutrients and major ionic solutes from a catchment.

Williams & Melack (1997) and Williams et al. (1997) measured water and solute fluxes from the same catchment as Lesack after partial deforestation had occurred (Table 9.1). Large increases in solute mobilisation from the upper soil horizons to groundwater were observed from a 2-ha plot after slash-and-burning in a partially deforested catchment (Fig. 9.4). The first water collected from lysimeters in the cut and burned plot contained very high nitrate concentrations (about $1800\,\mu M$). In groundwaters nitrate increased by a factor of 5 to about $100\,\mu M$ after cutting and burning.

Streamwater solute concentrations increased and nutrient ratios were altered subsequent to deforestation. Rainfall was similar during the two periods of study, but runoff was higher in the catchment after disturbance. Although inputs of total nitrogen and total phosphorus via rain were slightly smaller in 1989–1990 than in 1984–1985, export via streamflow was greater in 1989–1990. In fact, after partial deforestation, more nitrogen and phosphorus were exported than added via rain. The nitrogen to phosphorus ratio of fluvial flux from the intact forest was 120:1; following deforesta-

Table 9.1 Comparison of precipitation, stream hydrology and chemistry at the Moto Brook watershed in Lake Calado during two periods. During the earlier period, the 23 ha drainage basin was undisturbed, whereas in the later period it was 60–80% cleared

| Parameter* | Units | 1984–1985 | 1989–1990 |
|---|---|---|---|
| Rain quantity | $cm\,yr^{-1}$ | 280 | 275 |
| Rain TN input | $kg\,N\,ha^{-1}\,yr^{-1}$ | 6.7 | 4.5 |
| Rain TP input | $kg\,P\,ha^{-1}\,yr^{-1}$ | 0.34 | 0.21 |
| TN:TP | Atomic ratio | 44:1 | 47:1 |
| Runoff | $cm\,yr^{-1}$ | 160 | 208 |
| Baseflow | $cm\,yr^{-1}$ | 151 | 196 |
| Storm flow | $cm\,yr^{-1}$ | 8 | 12 |
| Runoff TN export | $kg\,N\,ha^{-1}\,yr^{-1}$ | 4.3 | 9.1 |
| Runoff TP export | $kg\,P\,ha^{-1}\,yr^{-1}$ | 0.082 | 0.61 |
| TN:TP | Atomic ratio | 120:1 | 33:1 |

* TN, total nitrogen; TP, total phosphorus.

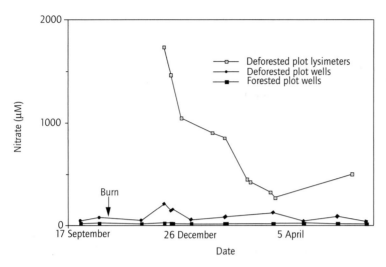

Fig. 9.4 Concentrations of nitrate in shallow throughflow (lysimeters) of experimental plot, and groundwater nitrate in mid-slope wells of experimental and control plots, September 1989 to June 1990; catchment of Lake Calado. (From Williams et al. 1997.)

tion, the ratio decreased to 33:1. Although the total nitrogen yield doubled after disturbance, the total phosphorus yield increased by a factor of seven, primarily owing to increased particulate phosphorus inputs.

The results from the Lake Calado catchment represent only one of the many types of land conversion occurring in the Amazon. Various logging practises, creation of livestock pastures and forest plantations, development of settlements and cities, and mining activities each cause specific hydrochemical and hydrological changes. For example, Forti et al. (2000) measured an increase in solute concentrations in a stream located in the northeastern Amazon as it passed from natural forest through an area disturbed by deforestation and manganese ore exploitation. Neil et al. (2001) compared concentrations of nitrogen and phosphorus in soil water and stream water draining forested catchments with others covered in pasture located in the southern Amazonian State of Rondônia (Brazil). Pasture streams contained lower concentrations of nitrate than forested streams. Low nitrogen to phosphorus ratios in pasture streams and periphyton bioassays indicated nitrogen limitation of algal growth in pasture streams. Biggs et al. (2004) examined regional patterns in nutrient concentrations in streams in the State of Rondônia as functions of soil properties, deforestation and urban populations. Although soil texture and nutrient status explained most of the variance in nitrate, total dissolved phosphorus and particulate phosphorus concentrations, the extent of deforestation and urban population density explained most of the variance in total dissolved nitrogen concentrations.

Bruijnzeel (1996) provided an excellent review of hydrological impacts of land cover transformations in the humid tropics, including examples from the Amazon. He offered empirical evidence and modelling results which demonstrated a variety of impacts depending on methods of land clearing and on subsequent land use. Further, he considered effects on dry season flows and on flooding. Most important, he outlined the need for research in order to permit quantitative understanding of hydrological responses. Such information is a prerequisite for developing predictive relations between land use changes and nutrient loading to tropical lakes.

Lesack & Melack (1996) have suggested that precipitation chemistry and nutrient balances of tropical forests could be affected by human disturbances such as large-scale deforestation in the Amazon basin (Fig. 9.5). Two scenarios were proposed, and further research is required in order to determine which, if either, is occurring. First, conversion of forest to pasture or cropland could lead

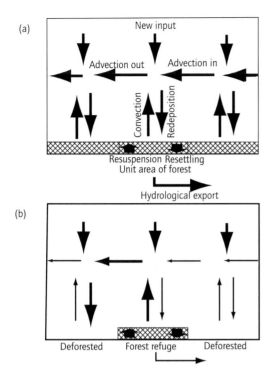

**Fig. 9.5** Schematic representation of the potential cycling of nutrients between a regional-scale rainforest ecosystem and the atmosphere (a), and how the cycling pathways are postulated to change if regional-scale tracts of forest are reduced to small islands within large deforested areas (b). Arrow widths reflect relative magnitudes of fluxes. (From Lesack & Melack 1996.)

to less material being entrained into the atmosphere by convection, which would then reduce the flux of recycled nutrients to the remaining islands of forest. The reduced amount of rainfall calculated by Shukla et al. (1990) after deforestation of the Amazon would exacerbate this scenario. Alternatively, conversion of forest to pasture or cropland could lead to an increase in dust in the atmosphere, with more material entrained into the atmosphere than from the former forest. Because repeated burning may be used in order to control secondary vegetation after conversion, emissions of some chemical species could be further enhanced relative to the uncut forest (Mackensen et al. 1996).

Hence, the flux of recycled nutrients to remnants of forest might increase as the surrounding landscape was converted from forest to pasture or cropland.

Estimates of yields of phosphorus and nitrogen per unit area ($kg\,ha^{-1}\,yr^{-1}$) for different ecosystems or land uses are valuable for calculating nutrient export from catchments of widely varying sizes and landscape complexity. Although unit-area yields of phosphorus and nitrogen are available for a variety of land uses and regions in North America and Europe (e.g. Dillon & Kirchner 1974; Reckhow et al. 1980; Prairie & Kalff 1986; Harper & Stewart 1987), these yields should not be applied casually to other climatic, hydrological or geological conditions. Few data on nutrient export are available with which to evaluate tropical catchments (Lewis 1986; Lesack 1993b) and most are from large rivers (Lewis et al. 1999).

A general problem with the application of unit-area yields to catchments of different sizes is the extent to which yields vary across differing hydrological paths and transit times with differing associated chemical and biological processing of nutrients. For example, some empirical models of phosphorus export have found a reduction in phosphorus delivery per unit area with increasing catchment size (Prairie & Kalff 1986). Moreover, the whole catchment usually does not contribute runoff, and the proportion which does varies with discharge. Furthermore, riparian vegetation can influence the amount of phosphorus which reaches a stream (Peterjohn & Correll 1984).

## 9.7 CONCLUSIONS

A large body of information describing reservoirs throughout tropical and subtropical South America, Africa and Asia exists. However, many of the reports are not readily available, some are considered proprietary and even the published literature is widely scattered. An integrated analysis of this information would be highly valuable to scientists and resource managers. Several key questions should guide such an analysis. What is the extent of climatic seasonality, variability and spatial heterogeneity influencing specific reservoirs and regions? What is the periodicity and vertical extent of water level fluctuations and how do these affect the ATTZ? How much clearing and burning of vegetation precedes flooding and what are the consequences for fertility of the reservoir? For what period and in what regions does anoxic water develop and how much production of reduced substances such as methane or hydrogen sulphide occurs? What factors are associated with development of massive mats of floating plants or algal blooms? How do fish yields vary as a function of time within the reservoir and downstream? Has the incidence of disease vectors changed? A superb example of such a comprehensive approach for tropical Africa is provided by Beadle (1981).

As new reservoirs are built and as the landscape surrounding existing reservoirs and floodplain lakes is modified by human developments, increasing demands for limnologically sound management are to be expected. The integrated analysis described above is the foundation for such guidance. Further site-specific advice will require applications of modern experimental assays for nutrient limitation and for responses to pollutants. The studies of impacts from bauxite at Lake Batata by Esteves and his students are a good example. Intensive studies spanning multiple years at representative sites should be complemented by comparative surveys. The intensive, long-term investigations at Broa Reservoir complemented by the extensive sampling of reservoirs throughout the State of Saō Paulo by Tundisi and his associates is an excellent example. Furthermore, monitoring efforts by government agencies should build on insights gained from research and should employ up-to-date data management systems. In order to ensure such enlightened monitoring will require considerable efforts by academic scientists to improve their communication with resource agencies. Fortunately, comprehensive, international programmes designed to encourage sustainable development of tropical ecosystems, such as the Amazon floodplain, are showing promise (e.g. Junk et al. 2000).

Remote sensing of limnological conditions in reservoirs and floodplain lakes and of land uses in their catchments is well suited to the large areas and considerable spatial heterogeneity typical of these systems. The combination of microwave estimates of inundation and vegetation structure with optical measures of water quality will be especially valuable (Melack & Hess 1998). Further, use of geographical information systems to organise, visualise and analyse remotely sensed (e.g. Alsdorf 2003) and ground observations will expedite management decisions.

Mathematical modelling is promising as a management tool, as well as offering heuristic value in scientific studies. Recently, physical models of stratification and mixing in combination with models of water quality have been applied to Amazonian reservoirs. Filho et al. (1990) attempted to help design operating criteria for the Cachoeira Porteria Reservoir on the Trombetas River with a model of how selection of the depth of withdrawal on the dam will influence the stratification as well as the degree of eutrophication and anoxia. Pereira (1995) developed a model of degradation of submerged vegetation which incorporated a one-dimensional model of thermal stratification in order to simulate distribution of dissolved oxygen and ammonium in Tucuruí Reservoir. Weber (1997) and Weber et al. (1996) have modelled seasonality of cations in an Amazonian floodplain lake, and provide a general scheme of nutrient fluxes in a floodplain. Although in need of further elaboration and validation, these modelling efforts are an encouraging and necessary step toward improved understanding and management of South American lakes and reservoirs.

## 9.8 ACKNOWLEDGEMENTS

I thank Patrick Osborne, Sally MacIntrye, Veronique LaCapra and Solange Filoso for careful review of an earlier draft. Opportunities to conduct research and attend meetings in Brazil were funded by grants from the U.S. National Science Foundation and NASA, including LBA.

## 9.9 REFERENCES

Alsdorf, D. (2003) Water storage of the central Amazon floodplain measured with GIS and remote sensing imagery. *Annals of the Association of American Geographers*, **93**, 53–66.

Alsdorf, D., Dunne, T., Hess, L., Melack, J.M. & Birkett, C. (2001) Water level changes in a large Amazon lake measured with spaceborne radar interferometry and altimetry. *Geophysical Research Letters*, **28**, 2671–4.

Altafin, I.G., Mattos, S.P., Cavalcanti, C.G.B. & Estuqui, V.R. (1995) Paranoá Lake—limnology and recovery program. In: Tundisi, J.G., Bicudo, C.E.M. & Matsumura-Tundisi, T. (eds), *Limnology in Brazil*. Brazilian Academy of Sciences, São Paulo, 325–49.

Agostinho, A.A. & Gomes, L.C. (1997) *Reservatório de Segredo—Bases Ecológicas para o Manejo*. EDUEM, Maringá, Brazil, 387 pp.

Arcifa, M.S., Carvalho, M.A.J., Gianesella-Galvão, S.M.F., Shimizu, G.Y., Froehlich, C.G. & Castro, R.M.C. (1981a) Limnology of ten reservoirs in southern Brazil. *Verhandlungen der Internationalen Vereinigung für theoretische und angewandte Limnologie*, **21**, 1048–53.

Arcifa, M.S., Froehlich, C.G. & Gianesella-Galvão, S.M.F. (1981b) Circulation patterns and their influence on physico-chemical and biological conditions in eight reservoirs in southern Brazil. *Verhandlungen der Internationalen Vereinigung für theoretische und angewandte Limnologie*, **21**, 1054–9.

Arcifa, M.S., Starling, F.L.R.M., Sipaúba-Tavares, L.H. & Lazzaro, X. (1995) Experimental limnology. In: Tundisi, J.G., Bicudo, C.E.M. & Matsumura-Tundisi, T. (eds), *Limnology in Brazil*. Brazilian Academy of Sciences, São Paulo, 257–81.

Barbosa, F.A.R, Padisak, J., Espindola, E.L.G., Borics, G.& Rocha, O. (1999) The cascading reservoir continuum concept (CRCC) and its application to the River Tietê-Basin, São Paulo State, Brazil. In: Tundisi, J.G. & Straskraba, M. (eds), *Theoretical Reservoir Ecology and its Applications*, Backhuys Publishers, Leiden, 425–37.

Barrow, C. (1988) The impact of hydroelectric development on the Amazonian environment: with particular reference to the Tucuruí Project. *Journal of Biogeography*, **15**, 67–78.

Beadle, L.C. (1981) *The Inland Waters of Tropical Africa*. Longman, London, 475 pp.

Biggs, T.W., Dunne, T. & Martinelli, L.A. (2004) Natural controls and human impacts on stream nutrient concentrations in a deforested region of the Brazilian Amazon basin. *Biogeochemistry*, **68**, 227–57.

Bozelli, R.L., Thomaz, S.M., Roland, F. & Esteves, F.A. (1992) Variações nictemerais e sazonais de alguns fatores limnológicos na represa municipal de São José do Rio Preto, São Paulo. *Acta Limnologica Brasiliensia*, **4**, 53–66.

Brandes, J.A., McClain, M.E. & Pimentel, T.P. (1996) $^{15}$N evidence for the origin and cycling of inorganic nitrogen in a small Amazonian catchment. *Biogeochemistry*, **34**, 45–56.

Bruijnzeel, L.A. (1991) Nutrient input–output budgets of tropical forest ecosystems: a review. *Journal of Tropical Ecology*, **7**, 1–24.

Bruijnzeel, L.A. (1996) Predicting the hydrological impacts of land cover transformation in the humid tropics: the need for integrated research. In: Gash, J.H.C., Nobre, C.A., Roberts, J.M. & Victoria, R.L. (eds), *Amazonian Deforestation and Climate*. Institute of Hydrology, Wallingford, 15–55.

Carpenter, S.R., Kitchell, J.F. & Hodgson, J.R. (1985) Cascading trophic interactions and lake productivity. *BioScience*, **35**, 634–9.

Callisto, M. & Esteves, F.A. (1996a) Composição granulométrica do sedimento de um lago Amazônico impactado por rejeito de bauxita em lago natural (Pará, Brasil). *Acta Limnologica Brasiliensia*, **8**, 115–26.

Callisto, M. & Esteves, F.A. (1996b) Macroinvertebrados bentônicos em dois lagos Amazônicos: Lago Batata (um ecossistema impactado por rejeito de bauxita) e Lago Mussurá (Brasil). *Acta Limnologica Brasiliensia*, **8**, 137–47.

Dabés, M.B.G.S., França, R.C., Gomes, M.C.S., Junqueira, M.V., Rolla, M.E. & Rosa, S.G. (1990) Caracterização limnológica da represa de Pontal, Itabira (MG). *Acta Limnologica Brasiliensia*, **3**, 173–99.

Dillon, P.J. & Kirchner, W.B. (1975) The effect of geology and land use on the export of phosphorus from watersheds. *Water Research*, **9**, 135–48.

Doyle, R.D. (1991) *The role of periphyton in organic matter production and nutrient cycling in Lake Calado, Brazil*. PhD thesis, University of Maryland, 269 pp.

Drago, E.C. (1990) Hydrological and geomorphological characteristics of the hydrosystem of the middle Parana River. *Acta Limnologica Brasiliensia*, **3**, 907–30.

Duchemin, E., Lucotte, M., Canuel, R., *et al.* (2000) Comparison of greenhouse gas emissions form an old tropical reservoir with those from other reservoirs worldwide. *Verhandlungen der Internationalen Vereinigung für theoretische und angewandte Limnologie*, **27**, 1391–5.

Engle, D.L. & Melack, J.M. (1990) Floating meadow epiphyton: biological and chemical features of epiphytic material in an Amazon floodplain lake. *Freshwater Biology*, **23**, 479–94.

Engle, D.L. & Melack. J.M. (1993) Consequences of riverine flooding for seston and periphyton of floating meadows in an Amazon floodplain lake. *Limnology and Oceanography*, **38**, 1500–20.

Esteves, F.A. (1983) Levels of phosphate, calcium, magnesium and organic matter in sediments of some Brazilian reservoirs and implications for the metabolism of the ecosystems. *Archiv für Hydrobiologie*, **96**, 129–38.

Esteves, F.A., Bozelli, R.L. & Roland, F. (1990) Lago Batata: Um laboratório de limnologia tropical. *Ciência Hoje*, **11**, 26–33.

Farjalla, V.F., Esteves, F.A., Bozelli, R.L. & Roland, F. (2002) Nutrient limitation of bacterial production in clear water Amazonian ecosystems. *Hydrobiologia*, **489**, 197–205.

Fearnside, P.M. (1989) Brazil's Balbina dam: Environment versus the legacy of the Pharaohs in Amazonia. *Environmental Management*, **13**, 401–23.

Fearnside, P.M (1997) Greenhouse-gas from Amazonian hydroelectric reservoirs: the example of Brazil's Tucuruí Dam as compared to fossil fuel alternatives. *Environmental Conservation*, **24**, 64–75.

Fearnside, P.M. (2002) Greenhouse gas emissions from a hydroelectric reservoir (Brazil's Tucurui dam) and the energy policy implications. *Water, Air and Soil Pollution*, **133**, 69–96.

Filho, M.C.A., de Jesus, J.A.O., Branski, J.M. & Hernandez, J.A.M. (1990) Mathematical modelling for reservoir water-quality management through hydralic structures: a case study. *Ecological Modelling*, **52**, 73–85.

Fisher, T.R., Lesack, L.F.W. & Smith, L.K. (1991) Input, recycling, and export of N and P on the Amazon floodplain at Lake Calado. In: Tiessen, H., Lopez-Hernandez, D. & Salcedo, I.H. (eds), *Phosphorus Cycles in Terrestrial and Aquatic Ecosystems*. Regional Workshop 3: South and Central America, Saskatchewan Institute of Pedology, University of Saskatchewan, Saskatoon, Canada, 34–53.

Fisher, T.R., Melack, J.M., Grobbelaar, J. & Howarth, R. (1995) Nutrient limitation of phytoplankton and eutrophication of inland, estuarine and marine waters. In: Tiessen, H. (ed.), *Phosphorus Cycling in Terrestrial and Freshwater Ecosystems*. Wiley, New York, 301–22.

Fonseca, J.J.L., Callisto, M.F.P. & Goncalves, J.F. (1998) Benthic macroinvertebrate community structure in an Amazonian lake impacted by bauxite tailings (Pará, Brazil). *Verhandlungen der Internationalen Vereinigung für theoretische und angewandte Limnologie*, **26**, 2053–5.

Forsberg, B.R., Araujo-Lima, C.A.R.M., Martinelli, L.A., Victoria, R.L. & Bonassi, J.A. (1993) Autotrophic carbon sources for fish of the central Amazon. *Ecology*, **74**, 643–52.

Forti, M.C., Boulet, R., Melfi, A.J. & Neal, C. (2000) Hydrogeochemistry of a small catchment in northeastern Amazonia: a comparison between natural with deforested parts of the catchment (Serra do Navio, Amapá, Brazil). *Water, Air and Soil Pollution*, **118**, 263–79.

Giani, A., Pinto Coelho, R.M., de Oliveira, S.J.M. & Pelli, A. (1988) Ciclo sazonal de parâmetros fisico-quimicos da água e distribuição horizontal de nitrogênio e fósforo no reservatório da Pampulha (Belo Horizonte, MG, Brasil). *Ciencia e Cultura*, **40**, 69–77.

Goldman, C.R. (1976) Ecological aspects of water impoundments in the tropics. *Revista de Biologia Tropical*, **24**, 87–112.

Gonzalez, E., Paolini, J. & Infante, A. (1991) Water chemistry, physical features and primary production of phytoplankton in a tropical blackwater reservoir (Embalse de Guri, Venezuela). *Verhandlungen der Internationalen Vereinigung für theoretische und angewandte Limnologie*, **24**, 1477–81.

Hamilton, S.K. & Lewis, W.M. Jr. (1990) Physical characteristics of the fringing floodplain of the Orinoco River, Venezuela. *Interciência*, **15**, 491–500.

Hamilton, S.K., Sippel, S.J. & Melack, J.M. (1996) Inundation patterns in the Pantanal wetland of South America determined from passive microwave remote sensing. *Archiv für Hydrobiologie*, **137**, 1–23.

Harper, D.M. & Stewart, W.D.P. (1987) The effects of land use upon water chemistry, particularly nutrient enrichment, in shallow lowland lakes: comparative studies of three lochs in Scotland. *Hydrobiologia*, **148**, 211–29.

Heide, J. van der. (1982) *Lake Brokopondo—filling phase limnology of a man-made lake in the humid tropics*. PhD thesis, Free University of Amsterdam, 427 pp.

Henry, R. (1990) Amônia ou fosfato como agente estimulador de crescimento do fitoplâncton na Represa de Jurumirim (Rio Paranapanema, SP)? *Revista Brasileira de Biologia*, **50**, 883–92.

Henry, R. & Tundisi, J.G. (1983) Responses of the phytoplankton community of a tropical reservoir (São Paulo, Brasil) to the enrichment with nitrate, phosphate and EDTA. *Internationale Revue der gesamten Hydrobiologie*, **68**, 853–62.

Henry, R., Tundisi, J.G. & Curi, P.R. (1984) Effects of phosphorus and nitrogen enrichment on the phytoplankton in a tropical reservoir (Lobo Reservoir, Brazil). *Hydrobiologia*, **118**, 177–85.

Henry, R., Hino, K. Gentil, J.G. & Tundisi, J.G. (1985) Primary production and effects of enrichment with nitrate and phosphate on phytoplankton in the Barra Bonita reservoir (State of S. Paulo, Brazil). *Internationale Revue der gesamten Hydrobiologie*, **70**, 561–73.

Hess, L.L., Melack, J.M., Filoso, S. & Wang, Y. (1995) Delineation of inundated area and vegetation along the Amazon floodplain with the SIR-C synthetic aperture radar. *Institute of Electrical and Electronics Engineering Transanctions of Geoscience and Remote Sensing*, **33**, 896–904.

Huszar, V.L.M. & Reynolds, C.S. (1997) Phytoplankton periodicity and sequences of dominance in an Amazonian floodplain lake (Lago Batata, Pará, Brasil): responses to gradual environmental change. *Hydrobiologia*, **346**, 169–81.

Ibañez, M.S.R. (1988) Response to artificial enrichment with ammonia and phosphate of phytoplankton from Lake Paranoá (Brasilia, DF). *Revista Brasileira de Biologia*, **48**, 453–7.

Junk, W.J. (1997) *The Central Amazon Floodplain*. Springer-Verlag, Berlin, 525 pp.

Junk, W.J., Robertson, B.A., Darwich, A.J. & Vieira, I. (1981) Investigações limnológicas e ictiológicas em Curuá-Una, a primeira represa hidreléctrica na Amazônica Central. *Acta Amazônica*, **11**, 689–716.

Junk, W.J., Bayley, P.B. & Sparks, R.E. (1989) The flood-pulse concept in river-floodplain systems. *Canadian Special Publication in Fisheries and Aquatic Science*, **106**, 110–27.

Junk, W.J., Ohly, J.J., Piedade, M.T.F. & Soares, M.G.M. (2000) *The Central Amazon Floodplain: Actual Use and Options for a Sustainable Management*. Backhuys, Leiden, 584 pp.

Keller, M., Victoria, R. & Nobre, C. (2001) Answers sought to big questions about the Amazon region. *Eos*, **82**, 405–6.

Leal, J.J.F. & Esteves, F.A. (2000) Life cycle and production of *Campsurus notatus* (Ephemeroptera, Polymitarcyidae) in an Amazonian lake impacted by bauxite tailings (Pará, Brazil). *Hydrobiologia*, **437**, 91–9.

Leal, J.J.F., Esteves, F.A., Farjalla, V.F. & Enrich-Prast, A. (2003) Effect of *Campsurus notatus* on $NH_4^+$, DOC fluxes, $O_2$ uptake and bacterioplankton production in experimental microcosms with sediment–water interface of an Amazonian lake impacted by bauxite tailings. *Internationale Revue der gesamten Hydrobiologie*, **88**, 167–78.

Lesack, L.F.W. (1993a) Export of nutrients and major ionic solutes from a rain forest catchment in the central Amazon basin. *Water Resources Research*, **29**, 743–58.

Lesack, L.F.W. (1993b) Water balance and hydrologic characteristics of a rain forest catchment in the central Amazon basin. *Water Resources Research*, **29**, 759–73.

Lesack, L.F.W. & Melack, J.M. (1996) Mass balance of major solutes in a rainforest catchment in the central Amazon: implications for nutrient budgets in tropical rainforests. *Biogeochemistry*, **32**, 115–42.

Lewis, W.M. (1986) Nitrogen and phosphorus runoff losses from nutrient-poor tropical moist forest. *Ecology*, **67**, 1275–82.

Lewis, W.M., Melack, J.M., McDowell, W.H., McClain, M.& Richey, J.E. (1999) Nitrogen yields from undisturbed watersheds in the Americas. *Biogeochemistry*, **46**, 149–62.

Lima, I.B., Novo, E.M., Ballester, M.V. & Ometto, J.P. (2000) Role of the macrophyte community in $CH_4$ production and emission in the tropical reservoir of Tucurui, Para State, Brazil. *Verhandlungen der Internationalen Vereinigung für theoretische und angewandte Limnologie*, **27**, 1437–40.

Lowe-McConnell, R.H. (1966) *Man-made Lakes*. Academic Press, New York, 218 pp.

Mackensen, J., Hölscher, D., Klinge, R. & Fölster, H. (1996) Nutrient transfer to the atmosphere by burning of debris in eastern Amazonia. *Forest Ecology and Management*, 86, 121–8.

Matsumura-Tundisi, T., Hino, K. & Claro, S.M. (1981) Limnological studies at 23 reservoirs in the southern part of Brazil. *Verhandlungen der Internationalen Vereinigung für theoretische und angewandte Limnologie*, **21**, 1040–7.

Matsumura-Tundisi, T., K. Hino & O. Rocha, (1986) Caracteristicas limnológicas da lagoa do Taquaral (Campinas, SP)—um ambiente hipereutrófico. *Ciência e Cultura*, **38**, 410–25.

Matsumura-Tundisi, T., Tundisi, J.G., Saggio, A., Oliveira Neto, A.L. & Espíndola, E.G. (1991) Limnology of Samuel Reservoir (Brazil, Rondônia) in a filling phase. *Verhandlungen der Internationalen Vereini-*

gung für theoretische und angewandte Limnologie, **24**, 1482–8.

McClain, M.E., Richey, J.E. & Pimentel, T.P. (1994) Groundwater nitrogen dynamics at the terrestrial–lotic interface of a small catchment in the central Amazon basin. *Biogeochemistry*, **27**, 113–27.

Melack, J.M. (1984) Amazon floodplain lakes: shape, fetch and stratification. *Verhandlungen der Internationalen Vereinigung für theoretische und angewandte Limnologie*, **22**, 1278–82.

Melack, J.M. (1996) Recent developments in tropical limnology. *Verhandlungen der Internationalen Vereinigung für theoretische und angewandte Limnologie*, **26**, 211–7.

Melack, J.M. & Fisher, T.R. (1990) Comparative limnology of tropical floodplain lakes with an emphasis on the central Amazon. *Acta Limnologica Brasiliensia*, **3**, 1–48.

Melack, J.M. & Forsberg, B.R. (2001) Biogeochemistry of Amazon floodplain lakes and associated wetlands. In: McClain, M.E., Victoria, R.L. & Richey, J.E. (eds), *The Biogeochemistry of the Amazon Basin and its Role in a Changing World*, Oxford University Press, Oxford, 235–76.

Melack, J.M. & Hess, L.L. (1998) Recent advances in remote sensing of wetlands. In: Ambasht, R.S. (ed.), *Modern Trends in Ecology and Environment*. Backhuys, Leiden, 155–70.

Melack, J.M. & Wang, Y. (1998) Delineation of flooded area and flooded vegetation in Balbina Reservoir (Amazonas, Brazil) with synthetic aperture radar. *Verhandlungen der Internationalen Vereinigung für theoretische und angewandte Limnologie*, **26**, 2374–7.

Melack, J.M., Sippel, S.J., Valeriano, D.M. & Fisher, T.R. (1992) Environmental conditions and change on the Amazon floodplain: an analysis with remotely sensed imagery. *The 24th International Symposium on Remote Sensing of the Environment*, Environmental Research Institute of Michigan, Ann Arbor, MI, 377–87.

Neil, C., Deegan, L.A., Thomas, S.M. & Cerri, C.C. (2001) Deforestation for pasture alters nitrogen and phosphorus in small Amazonian streams. *Ecological Applications*, **11**, 1817–28.

Nortcliff, S. & Thornes, J.B. (1981) Seasonal variations in the hydrology of a small forested catchment near Manaus, Amazonas, and the implications for its management. In: Lal, R. & Russel, E.W. (eds), *Agricultural Hydrology*. Wiley, Chichester, 37–57.

Novo, E.M.L.M., Lobo, F. & Calijuri, M.C. (1995) Remote sensing and geographical information system application to inland water studies. In: Tundisi, J.G., Bicudo, C.E.M. & Matsumura-Tundisi, T. (eds), *Limnology in Brazil*. Brazilian Academy of Sciences, São Paulo, 283–303.

Obeng, L.E. (1969) *Man-made Lakes: the Accra Symposium*. Ghana University Press, Accra, 389 pp.

Pereira, A. (1995) Mathematical modelling for Amazonian reservoirs. In: Tundisi, J.G., Bicudo, C.E.M. & Matsumura-Tundisi, T. (eds), *Limnology in Brazil*. Brazilian Academy of Sciences, São Paulo, 305–23.

Peterjohn, W.T. & Correll, D.L. (1984) Nutrient dynamics in an agricultural watershed: observation on role of a riparian forest. *Ecology*, **65**, 1466–75.

Petrere, M. Jr. & Ribeiro, M.C.L.B. (1994) The impact of a large tropical hydroelectric dam: the case of Tucuruí in the middle River Tocantins. *Acta Limnologica Brasiliensia*, **5**, 123–33.

Prairie, Y.T. & Kalff, J. (1986) Effect of catchment size on phosphorus export. *Water Resources Bulletin*, **22**, 465–70.

Reckhow, K.H., Beaulac, M.N. & Simpson, J.T. (1980) *Modeling Phosphorus Loading and Lake Response under Uncertainty: a Manual and Compilation of Export Coefficients*. 440/5-80-011, Environmental Protection Agency, Washington, DC.

Roland, F. & Esteves, F.A. (1993) Dynamics of phosphorus, carbon and nitrogen in Amazonian lake impacted by bauxite tailing (Batata Lake, Pará, Brasil). *Verhandlungen der Internationalen Vereinigung für theoretische und angewandte Limnologie*, **25**, 925–30.

Rolla, M.E., Rosa, S.G., Freitas, O.M.C., Gomes, M.C.S., Junqueira, M.V. & Souza, M.L.G. (1990) Composição físico-química e biológica do sedimento do reservatório do Volta Grande, Minas Gerais/São Paulo. *Acta Limnologica Brasiliensia*, **3**, 201–18.

Rosa, L.P., dos Santos, M.A., Tundisi, J.G. & Sikar, B.M. (1997) Measurements of greenhouse gas emissions in Samuel, Tucuruí and Balbina dams-Brazil. In: Rosa, L.P. & dos Santos, M.A. (eds), *Hydropower Plants and Greenhouse Gas Emissions*. COPPE, Cidade Universitaria, Rio de Janeiro, 41–55.

Rudd, J.W.M., Harris, R., Kelly, C.A. & Hecky, R.E. (1993) Are hydroelectric reservoirs significant sources of greenhouse gases? *Ambio*, **22**, 246–8.

Salas, H.J. & Martino, P. (1989) *Simplified Methodologies for the Evaluation of Eutrophication in Warm-water Tropical Lakes*. Pan American Center for Sanitary Engineering and Environmental Sciences, Lima, 51 pp.

Serruya, C. & Pollingher, U. (1983) *Lakes of the*

*Warm Belt*. Cambridge University Press, Cambridge, 569 pp.

Setaro, F.V. & Melack, J.M. (1984) Responses of phytoplankton to experimental nutrient enrichment in an Amazon lake. *Limnology and Oceanography*, **28**: 972–84.

Shukla, J., Nobre, C. & Sellers, P. (1990) Amazon deforestation and climate change. *Science*, **247**, 1322–5.

Sippel, S.J., Hamilton, S.K. & Melack, J.M. (1992) Inundation area and morphometry of lakes on the Amazon River floodplain, Brazil. *Archiv für Hydrobiologie*, **123**, 385–400.

Smith, V.H. (1979) Nutrient dependence of primary production in lakes. *Limnology and Oceanography*, **24**, 1051–64.

Smith, V.H. (1982) The nitrogen and phosphorus dependence of algal biomass in lakes: an empirical and theoretical analysis. *Limnology and Oceanography*, **27**, 1101–12.

Skole, D. & Tucker, C. (1993) Tropical deforestation and habitat fragmentation in the Amazon: Satellite data from 1978 to 1988. *Science*, **260**, 1905–10.

Starling, F.L.R.M. (1993) Control of eutrophication by silver carp (*Hypophthalmichthys molitrix*) in the tropical Paranoá Reservoir (Brasília, Brazil): a mesocosm experiment. *Hydrobiologia*, **257**, 143–52.

Takino, M. & Maier, M.H. (1981) Hydrology of reservoirs in the São Paulo State, Brazil. *Verhandlungen der Internationalen Vereinigung für theoretische und angewandte Limnologie*, **21**, 1060–5.

Tundisi, J.G. (1981) Typology of revervoirs in southern Brazil *Verhandlungen der Internationalen Vereinigung für theoretische und angewandte Limnologie*, **21**, 1031–9.

Tundisi, J.G. (1990) Perpectives for ecological modeling of tropical and subtropical reservoirs in South America. *Ecological Modelling*, **52**, 7–20.

Tundisi, J.G. & Matsumura-Tundisi, T. (1990) Limnology and eutrophication of Barra Bonita reservoir, S. Paulo, southern Brazil. *Archiv für Hydrobiologie Ergebnisse de Limnologie*, **33**, 661–76.

Tundisi, J.G. & Matsumura-Tundisi, T. (1995) The Lobo-Broa ecosystem research. In: Tundisi, J.G., Bicudo, C.E.M. & Matsumura-Tundisi, T. (eds), *Limnology in Brazil*. Brazilian Academy of Sciences, São Paulo, 219-3.

Tundisi, J.G. & Straskraba, M. (1999) *Theoretical Reservoir Ecology and its Applications*. Backhuys, Leiden, 585 pp.

Tundisi, J.G., Matsumura-Tundisi, T., Calijuri, M.C. & Novo, E.M.L. (1991) Comparative limnology of five reservoirs in the middle Tietê River, S. Paulo State. *Verhandlungen der Internationalen Vereinigung für theoretische und angewandte Limnologie*, **24**, 1489–96.

Weber, G.E. (1997) Modelling nutrient fluxes in floodplain lakes. In: Junk, W.J. (ed.), *The Central Amazon Floodplain*. Springer, Berlin, 109–17.

Weber, G.E., Furch, K. & Junk, W.J. (1996) A simple modeling approach towards hydrochemical seasonality of major cations in a central Amazonian floodplain lake. *Ecological Modelling*, **91**, 39–56.

Williams, M.R. & Melack, J.M. (1997) Solute export from forested and partially deforested catchments in the central Amazon. *Biogeochemistry*, **38**, 67–102.

Williams, M.R., Fisher, T.R. & Melack, J.M. (1997) Solute dynamics in soil water and groundwater in a central Amazon catchment undergoing deforestation. *Biogeochemistry*, **38**, 303–35.

Zaret, T.M., Devol, A.H. & dos Santos, A. (1981) Nutrient addition experiments in Lago Jacarétinga, central Amazon Basin, Brazil. *Verhandlungen der Internationalen Vereinigung für theoretische und angewandte Limnologie*, **21**, 721–4.

# Part III  Human Impact on Specific Lake Types

# 10 Eutrophication of Shallow Temperate Lakes

## G.L. PHILLIPS

### 10.1 INTRODUCTION

Shallow lakes are a ubiquitous freshwater habitat. Although they are much more numerous than deep lakes and arguably occupy a larger proportion of Earth's freshwater surface (Wetzel 1975; Mortensen et al. 1994), research has, until recently, concentrated on the limnology of deep lakes. In both types of lake, the general impact of eutrophication is similar and has been understood since the early descriptive work of Thienemann (1918), Naumann (1919) and Pearsall (1921). These authors laid the foundations of our current understanding that eutrophication is a process which enhances the production of both higher plants and algae in response to enrichment by the plant nutrients nitrogen and phosphorus. Much has been written subsequently (see, for instance, Rohlich 1969; Harper 1992; Sutcliffe & Jones 1992), although it was not until a conference in 1993 at Silkeborg, Denmark, that comprehensive attention was given to the limnology of shallow lakes (Mortensen et al. 1994).

The traditional approach to the control of eutrophication has been to reduce influent nutrient loads, particularly phosphorus, by diversion or tertiary treatment of effluents. A classic and successful example was the restoration of Lake Washington (Edmondson 1977; see also this Volume, Chapter 4). Numerous other attempts to restore lakes followed and, although many of these schemes were successful, several were not, or, following the reduction in nutrient inputs, took much longer to recover than had been predicted (Marsden 1989). Many were shallow lakes and it is now widely recognised that these lakes are more resilient to change than deep ones (Sas 1989; Scheffer et al. 1993). This is perhaps surprising as their smaller lake volume might be expected to reduce water retention times and thus produce a more rapid equilibration with the nutrient status of their inflowing streams.

Two factors are likely to be responsible. First, the return of nutrients lost through sedimentation via sediment release is faster, and its effects on the nutrient concentration in shallow lakes proportionately greater, relative to external load (Marsden 1989; Sas 1989). Second, shallow lakes appear to be more sensitive to trophic interactions such as the top-down control of the phytoplankton by grazing zooplankton (Jeppesen et al. 1997a). This chapter will focus on these issues and the implications of current understanding of shallow lake ecology for the management of these water bodies.

### 10.1.1 What do we mean by a shallow lake?

Osborne (this Volume, Chapter 11) reviews a number of definitions of shallow lakes and highlights the importance of functional issues such as mixing depth and light penetration. Sas (1989), in a quantitative analysis of 18 lakes, similarly separated lakes into two groups, based on the impact of temperature stratification during the growing season and on vertical homogeneity of the water column. If most of the epilimnion was regularly in contact with the sediment the lake was called 'shallow'; otherwise it was 'deep'. Similarly, the proportion of the lake volume in the photic zone will also be an important factor governing its ecology. The potential limitation of phytoplankton production by light availability is well understood (Talling 1957) and deep lakes are much more likely to experience a light-limited carrying capacity for phytoplankton than shallow lakes (Reynolds 1992) and, as a consequence, tend to support lower phytoplankton standing crops. However, recent studies on shallow lakes have also demonstrated the pivotal role of submerged macrophytes in lake ecology (Carpenter & Lodge 1986; Hootsmans & Vermaat 1991) and the proportion of the lake basin which these occupy is also an important factor. Finally, shallow lakes are more easily influenced by fluctuations in the physical environment caused by wind disturbance and temperature changes.

Thus, there is no simple definition of a shallow lake. A useful working concept might be a lake where wind-induced mixing was sufficient to prevent long-term stratification, enabling the bulk of the water volume to be in contact with the sediment surface; and where, in the absence of phytoplankton growth, water depth and transparency combine to allow macrophytes to occupy a substantial proportion of the lake area. In reality, this would probably apply to any lake with a mean depth of <3 m, although a larger, slightly deeper (<5 m) lake, with a substantial fetch and a low content of dissolved organic matter, might also behave in a similar way.

## 10.2 THE IMPORTANCE OF SEDIMENTS

As early as 1941, Mortimer recognised the importance of the release of phosphorus from deep, stratified, lakes (Mortimer 1941, 1942). Shallow lakes generally contain well-oxygenated water in direct contact with the sediment surface and, in these lakes, iron sorption was believed effectively to prevent phosphorus from being released into the overlying water. However, during the 1970s the failure of several eutrophication control attempts in shallow lakes such as Lake Trummen, Sweden (Bengtsson et al. 1975) and Barton Broad, UK (Osborne & Phillips 1978), demonstrated that phosphorus release could also occur in shallow lakes. Subsequently a large number of papers have been published on the subject and it would not be appropriate here to emulate a number of comprehensive reviews (Boström et al. 1988; Marsden 1989), although it is useful to summarise and identify how some of the suggested phosphorus release mechanisms may relate to the wider ecology of shallow lakes.

The mobilisation and release of phosphorus from aerobic sediments is dependent on the desorption of soluble phosphorus from inorganic complexes and the mineralisation of organic material. The rapid deposition of phytoplankton in shallow lakes provides a ready source of organic material and its decomposition close to the sediment surface provides a ready source of soluble phosphorus. Depending on the availability of chemical binding sites, mediated primarily through inorganic iron complexes and their sensitivity to both redox conditions and pH (Golterman et al. 1983), soluble phosphorus can build up in the interstitial water. The depth profile of this soluble phosphorus will vary depending on the depth at which the mineralisation is taking place. In shallow lakes, interstitial soluble phosphorus concentrations are often highest in the uppermost 10 cm of sediment (Boers & De Bles 1991; Phillips et al. 1994).

The ability of this phosphorus to enter the water column depends on a number of factors, including the concentration gradient across the sediment–water interface and the amount of physical disturbance, which can significantly increase the rates of release and dispersion (Boström et al. 1982). Where the sediment surface is aerobic, it is generally accepted that iron (III) complexes will counter the release of phosphorus but, in shallow lakes, wind stress may disturb the sediment surface sufficiently to provide an important mechanism by which phosphorus can be passed into the water column (Reynolds 1992). The surficial sediments in shallow lakes will consist of a semifluid layer composed of sediment particles and algal cells. Many of the latter will be benthic species, typically diatoms, but, in a shallow lake, the surface sediments will also contain recently sedimented phytoplankton, particularly cyanobacteria of the *Oscillatoria* group. In shallow lakes, these cells will be frequently resuspended and effectively returned to the plankton along with any phosphorus which they may have been able to accumulate when in the surficial sediments. Alternatively, resuspended sediment particles, rich in iron, may accumulate soluble phosphorus from an aerobic water column by adsorption to iron (III) hydroxy complexes (Lijklema 1980), effectively promoting the return of phosphorus to the sediment. The relative balance between these processes will depend on the concentrations of soluble phosphorus in the water column and the relative amounts of total iron and phosphorus in the sediments.

The importance of this ratio has been illustrated in a number of studies where the summer phosphorus concentrations in the water columns of shallow lakes have been negatively correlated with the sediment iron/phosphorus (Fe/P) ratio (sediment phosphorus increasing with decreasing ratios: Jensen et al. 1992; van der Molen & Boers 1994). Although the sediment Fe/P ratio is often a better predictor of internal phosphorus loading than the total phosphorus content of the sediment, not all sediment phosphorus is in the same potentially releasable form and this may weaken the general relationship (Boström 1984). Similarly, where sulphate reduction in the sediment is significant, the production of iron sulphides (De Groot 1991) can elevate phosphorus release rates, even when the total Fe/P ratio of the sediment is high (Phillips et al. 1994).

One of the characteristics of shallow hypertrophic lakes is the rapid change in the trophic structure that can take place. Dramatic collapses of phytoplankton have been reported for many lakes (Fott et al. 1980; Søndergaard et al. 1990a), often owing to zooplankton grazing (Jeppesen et al. 1990a; Phillips & Kerrison 1991). These can be associated with very rapid increases in the concentration of phosphorus in the water column (Osborne & Phillips 1978; Søndergaard et al. 1990a). The explanation for this is still not entirely clear but it is likely to be linked to the rapid increase in the availability of phosphorus from the decomposing phytoplankton, coupled with a decrease in the rate of particulate phosphorus sedimentation during the phytoplankton collapse. The additional organic load on the sediment may also change the redox potential of the sediment, promoting the release of iron-bound phosphorus. Temperature is likely to affect strongly the rates of decomposition and this may explain the large year-to-year variations in phosphorus release rates which are often reported (Søndergaard et al. 1990a).

It is difficult to obtain direct measurements of the rate of release of phosphorus from sediments and most work has involved the use of intact sediment cores incubated in laboratory conditions (Boers & van Hese 1988; Phillips & Jackson 1990). In these experiments, sediment disturbance by physical factors, such as wind or boat movements, cannot easily be simulated. However, even in laboratory conditions, release rates can be measured which are much greater than can be accounted for by diffusion. Although physical disturbance of the sediment does not occur, disturbance by burrowing benthic invertebrates, such as chironomid larvae, has been shown to be important (Holdren & Armstrong 1986; Tátrai 1988; Phillips & Jackson 1990; Andersen & Jensen 1991; Phillips et al. 1994;). Similarly, benthic fish can cause substantial disturbance to sediments and this form of bioturbation may also increase the rate of phosphorus release (Meijer et al. 1990; Tátrai et al. 1990).

The light climate at the sediment surface may also significantly affect exchange of phosphorus across the sediment–water interface. In shallow lakes, when phytoplankton populations are low, there is likely to be sufficient light for a significant benthic algal flora to develop. This can reduce nutrient release from sediments, both directly by phytobenthos uptake (Søndergaard et al. 1990b) and indirectly by changing the chemical environment in the sediment (Hannson 1989; Jansson 1989). With the onset of eutrophication, water transparency decreases in response to phytoplankton growth and insufficient light may reach the sediment surface to maintain benthic algae. As a result, phosphorus release may be increased further, so exacerbating the eutrophication problem.

The above discussion serves to illustrate the complex interactions which control the movement of phosphorus across the sediment–water interface. It is extremely difficult to predict the factors which will control the outcome in any particular lake but it is now clear that, in shallow lakes, there is likely to be a persistent return of phosphorus to the water column. This may continue to support the growth of phytoplankton for many years following a reduction in external nutrient loading.

Several attempts to overcome this have been made. The physical removal of phosphorus-rich sediments can produce substantial benefits. In Lake Trummen, a shallow lake in Sweden, the upper 0.5 m of sediment was removed by suction dredging, leading to an immediate reduction in phosphorus and algal biomass (Andersson et al. 1973; Björk 1988). However, sediment removal from the shallow lake Finjasjön, Sweden, did not reduce the internal phosphorus load (Eckerrot & Pettersson 1993) and smaller scale sediment removal exercises in the Norfolk Broads, UK, have met with mixed success (Pitt et al. 1997). Sediment removal is an expensive technique and although it will certainly reduce the amount of phosphorus present, and thus potentially available to support production, it cannot always be relied upon to reduce internal phosphorus loads. An alternative strategy is to attempt to increase the phosphorus retention capacity of the sediment by chemical dosing. Aluminium salts are attractive as they are not sensitive to redox changes and they have been used successfully in some lakes (Welch & Schrieve 1994). The potential toxicity of aluminium, however, raises some concerns and direct injection of iron salts into the sediment to increase the Fe/P ratio has been carried out in The Netherlands (Boers et al. 1994). Although this led to an immediate decrease in the total phosphorus concentration in the lake, the long-term success could not be assessed owing to unexpected short residence time of the test lake. Similar problems were experienced in a whole lake trial carried out in the Norfolk Broads. However, subsequent experiments in small field enclosures demonstrated that the majority of the added iron was rapidly lost to the overlying water and little evidence for enhanced phosphorus retention was found (Pitt et al. 1997).

## 10.3 IMPORTANCE OF MACROPHYTES

### 10.3.1 Changes with eutrophication

Typically, shallow clearwater lakes will be inhabited by submerged macrophytes. Although their distribution will be controlled by a variety of factors, such as wind/wave exposure (Weisner et al. 1997), sediment composition (Barko & Smart 1986) and grazing by water fowl (Søndergaard et al. 1996), invertebrates (Jacobsen & Sand-Jensen 1992) or fish (Van Donk & Otte 1996), it is light availability which has generally been considered to be the ultimate controlling factor (Spence 1982). The maximum depth of submerged plant colonisation will be approximately the same as the Secchi-disc depth (Chambers & Kalff 1985) and in very shallow, transparent oligotrophic or mesotrophic lakes (c. <3.0 m) the entire lake sediment surface is likely to be covered by macrophytic plants.

Eutrophication in shallow, macrophyte lakes is often characterised by a species change and an eventual decline in macrophyte growth (De Nie 1987). Typically, authors describe macrophyte communities in shallow, temperate lakes as often being mainly of Charophytes, but changing to a more diverse and productive community in richer lakes, before finally becoming devoid of almost all underwater vegetation (see also Volume 1, Chapter 11). For example, lake Veluwe, a shallow artificial, eutrophic lake in The Netherlands, contained a well-developed mixed macrophyte stand which was gradually replaced by monospecific stands of *Potamogeton pectinatus* L. during the 1980s (Hootsmans & Vermaat 1991). In the 1940s, the Loosdrecht lakes in The Netherlands contained an abundant, submerged characean community. By the 1960s, this had disappeared, being replaced by only a few submerged species, namely *Elodea*, *Myriophyllum*, *Ceratophyllum* and *Potamogeton* spp., and floating leaved species such as *Nuphar lutea*, *Nymphaea alba*, *Polygonum amphibium* and *Potamogeton natans*. By the 1980s, almost all of the submerged plants had disappeared (Best et al. 1984). Similar changes have taken place in the very shallow eutrophic lakes which form the

Norfolk Broads in the UK (Moss 1983). Numerous other examples can be found in the recent literature and, given the importance of these plants in providing habitat for a variety of fish and invertebrates, a clear understanding of the mechanisms which cause their disappearance is critical to the understanding of shallow-lake ecology.

### 10.3.2 Competition with algae

Until recently, the usual explanation for these changes was that, as nutrient enrichment took place, the growth of phytoplankton restricted light penetration, limiting the distribution of the submerged vegetation and ultimately causing its disappearance (Blindlow 1992). However, in very shallow lakes, macrophytes can escape this shading effect if the vegetation is able to reach to, or come close to, the water surface (Chambers 1987; Scheffer et al. 1992). For example, taxa such as *Myriophyllum* (Adams & McCracken 1974) and *Potamogeton pectinatus* (Van Wijk 1988; Hootsmans & Vermaat 1991) are able to grow rapidly to the water surface where their leaves form a canopy unaffected by water turbidity. The similar growth form has been seen in the very shallow Norfolk Broads (Phillips et al. 1978). It seems likely that the initial response to eutrophication in shallow lakes is an increase in macrophyte biomass and a shift from low-growing macrophytes, such as *Charophyta*, to upright species capable of maximising their photosynthetic activity in the upper water column (Moss 1983; Moss et al. 1996a).

The mechanisms which lead to the ultimate loss of these plants are less easy to identify. By combining evidence from palaeoecological studies, tank experiments and field observations, Phillips et al. (1978) were able to demonstrate that the smothering growth of epiphytic or filamentous algae may be an important factor. Although epiphytic algae will always form a component of the macrophytic flora, these authors suggested that, with increasing availability of nutrients in the water column, epiphytic and filamentous algae may obtain a competitive advantage for nutrients. As a result of their increased growth, the light available to the plants is significantly reduced.

### 10.3.3 Evidence for multiple stable states in shallow lakes and the importance of macrophytes

The mechanisms described above can explain how eutrophication in shallow lakes can lead to the eventual loss of submerged aquatic vegetation. It might be assumed that this change is a gradual process, with the proportion of the lake area occupied by macrophytes reducing as nutrient enrichment takes place and the combined impact of phytoplankton and epiphytic algae is exerted. In a deep lake, where macrophytes are restricted to the shallow littoral fringe, this may occur as macrophytes retreat into shallower water. However, as shallow lakes tend to lie in flatter basins, with much smaller proportions of their areas subject to significant gradients, there may be less opportunity for plants to adapt in this way. As a result, a decreasing light climate may lead to a relatively abrupt change in the proportion of the lake area covered by plants. In general, shallow lakes might be expected either to possess an extensive cover of macrophytes or to contain very few. This appears to be the case, as many recent studies have revealed that shallow lakes tend to be either rich in phytoplankton, with few submerged macrophytes, or contain clear water and abundant submerged vegetation (Jeppesen et al. 1990b; Moss 1990; Blindlow et al. 1993).

Surveys of large numbers of shallow lakes in Denmark have suggested that this change from macrophytes to phytoplankton takes place at phosphorus concentrations in the range 50–125 µg P L$^{-1}$. However, clearwater macrophyte lakes can still be found in which phosphorus concentrations are as high as 650 µg P L$^{-1}$ (Jeppesen et al. 1990b). It has been suggested that the apparent discontinuous response of shallow lakes to increasing nutrient concentrations may also occur during restoration, and that unless nutrients are reduced to concentrations well below 50–125 µg P L$^{-1}$ phytoplankton, rather than macrophytes, are likely to persist (Moss 1990, 1994).

The concept of two stable ecosystems with overlapping resource boundaries and a 'switch' mechanism operating to cause changes from one

state to another has been recognised for some time (May 1977) and these ideas have been explored in simple models relating macrophyte growth and turbidity in shallow lakes (Scheffer 1990). Scheffer's model assumes that:
1 algal growth and turbidity increase with enrichment;
2 the effect of vegetation on turbidity is negative;
3 vegetated area in a shallow lake declines with turbidity.

By simple graphical analysis, the model clearly demonstrates that, in contrast to deep lakes, the relationship between algal growth and nutrient supply in shallow lakes is sigmoidal, generating two alternative stable states. Thus, during a period of increasing eutrophication, macrophyte abundance is likely to persist despite rising nutrient concentrations, the presence of the vegetation tending to suppress the growth of algae and hence turbidity. In contrast, a eutrophic, algal-rich lake, with few macrophytes, will maintain this state even when nutrient reduction occurs.

The apparent existence of a 'switch mechanism', which operates at the ecosystem scale as shallow lakes respond to changing nutrient status, has led to considerable research centred around understanding the way in which submerged aquatic macrophytes influence both the physico-chemical environment and the biological structure of shallow lakes.

### 10.3.4 Influence of macrophytes on physico-chemical environment of shallow lakes

Macrophytes can affect their immediate environment in a number of ways. During the growing season, available phosphorus from both the water and the sediment will be taken up by the plants, thus potentially reducing the availability of phosphorus for algal growth (Boyd 1971). In calcareous marl lakes, elevated pH resulting from intensive photosynthesis may cause the co-precipitation of phosphorus with carbonates (Rørslett et al. 1985). Some whole-lake studies have found that following the growth of macrophytes, algal growth has been limited by nitrogen availability (van Donk et al. 1993;

Beklioglu & Moss 1996) and these authors have suggested that either nitrogen uptake by the plants or denitrification, which is known to be associated with the rhizosphere of aquatic plants (Risgaard-Petersen & Jensen 1997), is responsible.

Rooted macrophytes may also directly affect phosphorus retention in the sediment. Oxygen release from roots (Wium-Andersen & Andersen 1972; Carpenter et al. 1983) may lead to oxidation of iron and manganese, and subsequent adsorption of phosphate. However, the effectiveness of this, in what is often a highly reducing environment, is still uncertain (Wigand et al. 1997). Macrophyte beds may increase the rate of sedimentation, causing a build-up of organic matter (Carpenter 1981) and this, together with additional organic matter from decomposing macrophytes, may lead to increased mineralisation and the subsequent release of phosphorus to the water column (Moss et al. 1986). Alternatively, elevated pH may also enhance the release of phosphorus from the sediment through ligand exchange (Jensen & Andersen 1992). Thus, there is considerable uncertainty regarding the effect of macrophytes on the availability of nutrients, although it is clear that this can be significant. Macrophytes may also indirectly reduce water turbidity by preventing sediment disturbance (Vermaat et al. 1990; Blindlow et al. 1993) and it is generally considered that their presence is likely to reduce the potential for phytoplankton growth.

### 10.3.5 Direct influence of macrophytes on algal growth

Macrophytes are thought to be capable of producing allelopathic sustances which will suppress algal growth (Gopal & Goel 1993; Jasser 1995). In their theoretical model relating macrophyte loss to eutrophication, Phillips et al. (1978) suggested that, in clearwater conditions, allelopathy was an important factor which prevented the growth of epiphytes during the initial stages of eutrophication. Van Viersen et al. (1985) developed this model further, suggesting that, in addition to the impact on epiphytes, allelopathic chemicals produced by the plants also reduced phytoplankton growth. As macrophytes gradually disappear in response to periphyton shading, the negative influence of macrophytic allelochemicals on phytoplankton is simultaneously decreased and the loss of macrophytes contributes to phytoplankton increase. Current thinking places less emphasis on the gradual change in phytoplankton abundance. However, allelopathy has been suggested to be a possible factor in the return of Charophytes to the shallow lake Veluwe, where the mass emergence of *Chara* in the spring appears to determine the subsequent appearance of clearwater patches (Coops & Doef 1996).

Epiphytes are also subject to grazing by invertebrates such as snails. Field enclosure experiments have shown positive effects on macrophyte growth following reduction of fish predation on snails (Underwood 1991; Underwood *et al.* 1992). A similar conceptual model, which links changes in epiphyte shading, caused by invertebrate grazers, to regulation of grazers by their predators, has been proposed by Brönmark & Weisner (1992).

## 10.4 THE ROLE OF FISH

### 10.4.1 *The importance of fish–zooplankton interactions*

In recent years there has been a growing awareness of the importance of fish in structuring the zooplankton communities of lakes (Hrbáček *et al.* 1961; Brooks & Dodson 1965) and the cascading impacts, through grazing, on phytoplankton and nutrient status (Carpenter *et al.* 1985; Carpenter & Kitchell 1993). Enclosure experiments have demonstrated that the removal of planktivorous fish increases the numbers of large cladoceran grazers and reduces phytoplankton abundance (e.g. Phillips & Kerrison 1991). However, there has been considerable debate regarding the long-term stability of this 'top-down' control on the phytoplankton (De Melo *et al.* 1992; Carpenter & Kitchell 1992; Harris 1996; Sarnelle 1996). Much of this has been concerned with the degree to which control varies along a trophic gradient (McQueen *et al.* 1986). There is a growing consensus that the influence of the trophic cascade is likely to be greater in shallow lakes than deeper ones (Reynolds 1994; Jeppesen *et al.* 1997a). Moreover, these interactions are also likely to contribute to the switch mechanisms which are postulated to control the response of shallow lakes to changing nutrient status.

Typically, large cladocera, the main agents of significant grazing pressure on the phytoplankton, exhibit a bimodal distribution, with high biomass in the spring and autumn. Decline during summer may be due to a number of causes. One of the most important of these is predation by 0+ fish (Gliwicz & Pijankowska 1989) and, in many phytoplankton-rich lakes, these fish probably cut the grazing potential of the zooplankton, although factors such as edibility of phytoplankton, particularly cyanobacteria (e.g., Fulton & Paerl 1987), may also be important. In a recent review, Jeppesen *et al.* (1997a) point out that there is good evidence that fish biomass per unit volume at fixed total phosphorus concentration decreases with increasing mean depth and it therefore seems likely that predation on zooplankton will be proportionately greater in shallow lakes. In addition, the defence against predation that vertical migration offers to zooplankton in deep lakes (Lampert 1993) is weak in shallow lakes. Finally, Jeppesen *et al.* (1997a) point out that planktivorous fish species typical of shallow lakes, such as bream (*Abramis brama* L.), roach (*Rutilus rutilus* L.) and rudd (*Scardinius erythrophthalamus* L.), rely on benthic feeding. Hence, in shallow lakes, they are less sensitive to variations in zooplankton abundance and are able to maintain high numbers when zooplankton are not available. The overall impact is that fish can exert a higher predation pressure on the zooplankton in shallow lakes and, in the absence of aquatic macrophytes, this becomes an important reason for the maintenance of phytoplankton.

This apparent persistence of phytoplankton in shallow lakes has led to numerous experiments involving fish manipulations, designed to reduce this predation pressure and enhance the capacity of zooplankton to control phytoplankton (Gulati *et al.* 1990). Once free of predation pressure, large-bodied *Daphnia* quickly increase in numbers and rapidly reduce the phytoplankton standing crop. However, the zooplankton might be expected to then quickly become food-limited and, given the difference in population turnover times of zooplankton and phytoplankton, control of the phytoplankton might only last for short periods (Reynolds 1994). This does not appear to be the case in shallow lakes, where relatively long-term control of phytoplankton following the removal of planktivorous fish is reported by numerous authors (for a review, see Phillips & Moss 1994). One reason for this may be that, in shallow lakes, organic debris, benthic and epiphytic algae provide an alternative food supply, preventing the collapse of the zooplankton population when phytoplankton are exhausted, thus maintaining the potential grazing pressure. This is particularly important for larger species such as *Daphnia magna*, which are often reported close to the sediment (Gulati 1990) and are likely to be capable of exploiting a larger size range of food particles.

Although fish removal can reduce the growth of phytoplankton in shallow lakes over a growing season, it is clear that further interventions, in the form of subsequent fish removal, will be required to maintain this situation, unless other factors come into play which alter the balance between predator and prey. Many recent studies of experimentally manipulated lakes have demonstrated that one of the most important factors influencing the longer-term stability is the presence of macrophytes (Meijer *et al.* 1994).

### 10.4.2 How do macrophytes influence fish–zooplankton interactions?

There is good evidence that in shallow lakes the horizontal distribution of zooplankton varies from night to day. Timms & Moss (1984) demonstrated that pelagic zooplankton move into macrophyte beds during the daytime to avoid fish predation but move out into the open water at night to feed on the phytoplankton. Subsequently several other studies have confirmed that such horizontal migrations occur commonly and that the zooplankton species and size classes involved are related to the density of planktivorous fish in the open water (cf. review by Jeppesen *et al.* 1997b). In Lake Ring, Denmark, which is inhabited by large pike (*Esox lucius* L.) and perch (*Perca fluviatilis* L.), with low planktivorous fish density, the larger bodied *Daphnia magna* Straus exhibited a much greater difference in day/night densities in macrophyte beds than the smaller *Daphnia hyalina/galeata* Sars (Lauridsen & Buenk 1996). In the more fish-rich Lake Stigsholm, Denmark, even small species such as *Bosmina longirostris* and *Ceriodaphnia* spp. migrated (Jeppesen *et al.* 1997b).

Pelagic species of zooplankton tend to avoid macrophyte beds (Pennak 1973; Lauridsen & Lodge 1996) and it would appear that they only move into them to avoid predation. The success of this strategy depends on the ability of the zooplankton to avoid small planktivorous fish and, as a consequence, the density of the macrophyte bed is also important. Weed beds also offer refuges for planktivorous fish seeking cover from larger piscivorous species or birds (Carpenter & Lodge 1986). These prey fish tend to prefer sparse vegetation, where they can forage more effectively (Engel 1988). Thus, in less dense weed beds, the zooplankton is subject to greater predation than in the open water (Phillips *et al.* 1996; Stansfield *et al.* 1997). Large macrophyte beds are generally occupied by macrophyte-associated species such as *Sida crystallina*, *Eurycercus lamellatus* and *Simocephalus vetulus*, and there is also some evidence that small macrophyte beds are more important as daytime refuges for migrating cladocerans (Lauridsen *et al.* 1996). This is in accordance with the findings of Lauridsen & Buenk (1996), who found that *D. magna* and *D. hyalina/galeata* favour the edge zone between macrophytes and open water as a daytime refuge rather than the macrophyte bed itself.

Thus, macrophyte beds offer a means for grazing cladocerans to avoid predation from planktivorous fish, leaving them free to move into the open water at night and exert a grazing pressure on the zooplankton. In addition, some of the macrophyte-associated species collect seston by filtration when they are attached to plants (*Sida*) and others may be facultative filter feeders (*Chydorus*, *Euryc-*

*ercus*) (Jeppesen *et al.* 1997b). These non-pelagic species may contribute significantly to the removal of phytoplankton as water is swept into weedbeds. Those species which are scrapers on solid surfaces (*Eurycercus*) will also assist in reducing epiphytic algal growth and thus help to maintain the light climate for the vegetation.

Although it is clear that macrophytes can modify the open water community of shallow lakes substantially and, in so doing, create conditions which allow zooplankton grazing to assist in the process of maintaining low algal numbers, the absolute density of planktivorous fish is still of critical importance. Phillips *et al.* (1996) suggested that the piscivore/planktivore ratio was a critical factor in establishing stability in shallow, macrophyte lakes. One reason for this may be that, as the density of planktivorous fish increases, the refuge offered by macrophytes, even at high densities, may eventually disappear. Studies based on enclosure experiments suggest that this threshold may occur when the density of potentially planktivorous fish exceeds $2-5\,\text{m}^{-2}$. At present, there is insufficient evidence from field observations to confirm this (Jeppesen *et al.* 1997b). For example, Stansfield *et al.* (1997) suggested a loss of refuge for *Daphnia* spp. at a density of 0+ fish of only $0.25\,\text{m}^{-2}$.

Pike are clearly associated with macrophyte beds. Cook & Bergersen (1988) found that pike abundance varied in response to changes in macrophyte density and distribution. They may avoid very dense beds of macrophytes (Holland & Hutson 1984) and the larger individuals are generally found at the open-water–vegetation interface. The presence of such predators may reduce the activity of planktivorous fish (Bean & Wingfield 1995) or cause them to switch to other food sources. As an example, Persson (1993) found that roach switched from feeding on *Bosmina* to detritus/algae in the presence of piscivorous perch.

Stocking of piscivorous fish, strong catch restrictions on recreational fisheries (Benndorf *et al.* 1988) or leaving piscivorous species and removing planktivores (Annadotter *et al.* 1999) is a potential management strategy to reduce the number of planktivorous fish. Stocking with pike carried out in Lake Lyng, Denmark, successfully created a top-down control on the phytoplankton but this lasted for only one season as the autumn biomass of pike was not increased by the stocking (Berg *et al.* 1997; Søndergaard *et al.* 1997). Perhaps the carrying capacity for pike is largely determined by the amount of vegetation (Grimm & Backx 1990).

Thus, there is a complex series of interactions taking place in shallow, macrophyte lakes through which piscivorous fish, their planktivorous prey and the zooplankton position themselves to maximise resource availability and minimise the risk of predation. There is growing evidence that chemical cues are involved in this resource segregation (von Elert & Loose 1996), but the degree of their success and their impact on the trophic balance within the lake will depend on a large number of factors. Critical to this is probably the nutrient status of the lake. Insufficient quantitative data are yet available to give precise guidelines and many factors are likely to be involved. However, Grimm & Backx (1990) provide an example of the type of information which is required. They suggest that, in a lake containing an abundant submerged vegetation, $100\,\text{kg}\,\text{ha}^{-1}$ of pike can be maintained and that this stock is capable of consuming $300\,\text{kg}\,\text{ha}^{-1}$ of cyprinid fish. Such a planktivorous cyprinid population is associated with a total phosphorus concentration of about $400\,\mu\text{g}\,\text{P}\,\text{L}^{-1}$. They suggest that, beyond this level, cyprinid production will be too high to be consumed by the pike and, so, this nutrient concentration might represent the upper limit at which a shallow-lake community is stable. Clearly, other plant–nutrient relationships will also be important. For example, as explained earlier, submerged macrophytes living at very high nutrient concentrations tend to maximise their biomass in the form of floating canopies with long leafless stems, or only floating-leaved and emergent species survive. In these situations, the plant structure may provide only a relatively low refuge potential of low stem density (Winfield 1986), decreasing the stabilising effect of the macrophytes.

## 10.5 OVERVIEW AND MANAGEMENT IMPLICATIONS

### 10.5.1 The importance of 'forward switches'

Shallow lakes differ from deep lakes in many ways. Most occur in the lowlands and their catchments will usually be intensively farmed and even if there are no significant point sources of nutrients they are likely to be experiencing elevated nutrient loads. There is growing pressure to find ways of managing these water bodies to accommodate not only the need to maintain biodiversity and conservation value, but also to provide appropriate conditions for amenity and recreation. This management needs to focus on both the maintenance of currently pristine lakes as well as finding cost-effective ways of restoring those already impacted by eutrophication.

Much has been learned about the ecology of these systems in the past decade and perhaps the most important conclusion has been that the transition from 'pristine state' to degraded eutrophic water body is almost certainly not a steady change. This has been deduced from studies of whole lake systems where restoration attempts have been made (Meijer et al. 1994) and from detailed experimental work (see review by Jeppesen et al. 1997b) and the ideas have been placed in a theoretical context through the use of minimal models (Scheffer et al. 1993).

This resilience to change probably also existed during periods of increasing eutrophication. However, at the time that the majority of lakes were experiencing such change, insufficient knowledge and awareness of the process mean that few corroborative data are available, other than those from quantitative palaeolimnological studies (Jeppesen et al. 1996). However, attempts to demonstrate the onset of eutrophication by nutrient addition in a set of experimental ponds illustrated convincingly the stability of the system. Despite considerable nutrient additions it was only when aquatic plants were artificially removed from the ponds that the expected growth of phytoplankton occurred (Balls et al. 1989; Irvine et al. 1989).

Thus although the transition from an aquatic plant to algal lake clearly requires an increase in nutrient loading the change is most likely to take place if other changes, 'forward switches' (Moss et al. 1996b), occur. These switches may be the removal of plants by overzealous cutting or the impact of boat propellers from repeated boat movements. Fish stocking may upset the balance between piscivorous and planktivorous fish (Brönmark & Weisner 1992) and lead to a decline in zooplankton grazing. Alternatively organochlorine pesticides, now banned but widely used in the 1950s and 1960s, may have directly destroyed the zooplankton communities and thus enabled algal growth to escape grazing pressure (Stansfield et al. 1989). The important lesson for those charged with managing shallow lakes is to be aware of these switches and to recognise that preventing a lake from changing to the algal state is considerably easier than reversing the process. Thus, when a lake is experiencing an elevated nutrient load but is not exhibiting the normal symptoms of eutrophication, the operation of any of the switches described in this chapter will probably flip the lake into a new stable and probably less desirable state.

### 10.5.2 Reversing the process; can we use 'reverse switches'?

To stand any chance of rehabilitating a lake, it is essential to consider that, besides considering the reduction of nutrients, other forward switches may have destabilised the macrophyte state. These need to be identified and removed. Difficulties may be experienced in identifying the desired end-use of the lake. Although the restoration of clear water and abundant macrophytic vegetation may be the clear objectives for nature conservation, and acceptable for many amenity lakes, this may not be a satisfactory condition for many anglers and is unlikely to be welcomed by boat owners. Thus, not all aspirations can easily be met in a single lake and it is essential that this is clearly understood by all concerned prior to any rehabilitation work being undertaken.

Having established this, the second step in the process is to minimise nutrient loadings and concentrations. Fundamentally, this means a reduction in the input of nutrients from the catchment (see this Volume, Chapter 16) but internal sediment sources may also need to be considered. This is important, for although it is clear that two alternative stable states exist within a gradient of nutrient enrichment it is still very uncertain over what nutrient range these can occur. Currently the best evidence suggests that a macrophyte community may become progressively unstable once total phosphorus concentrations exceed 125–150 µg P L$^{-1}$.

Just as forward switches may have been required to overcome the effect of powerful mechanisms stabilising macrophytes, similar switches may be required to reverse the process. If nutrient condition can be very substantially reduced, to perhaps <30 µg P L$^{-1}$, these may operate naturally but, at the concentrations likely to occur in many lowland areas, will probably need operating by management actions. The most widely researched approach has been biomanipulation (see also this Volume, Chapter 17). Although the long-term effectiveness of the technique is still under debate, the selective removal of fish or addition of piscivorous species is already clearly capable of creating clearwater conditions. It is now well-established that such conditions are unlikely to be maintained as a stable system without continued intervention, unless there is a rapid re-establishment of submerged vegetation.

### 10.5.3 Re-establishment of aquatic macrophytes in the restoration process

In several lakes, biomanipulation led to the rapid regrowth of macrophytes (e.g. Lauridsen *et al.* 1994), although in others, macrophytes have been very slow to colonise (Moss *et al.* 1996a). Even in those lakes where macrophytes did return, they have often subsequently declined (Meijer *et al.* 1994; Van Donk & Otte 1996). This may stem from the failure of the fish/zooplankton community to form a new stable equilibrium, as a consequence, for example, of invading fish populations (Moss *et al.* 1996a) or direct herbivory by fish or by birds. Rudd are known to feed on macrophytes and, although this may stimulate macrophyte growth (Prejs 1984), enclosure experiments have suggested that they may counter macrophyte growth (Van Donk & Otte 1996). Other fish–macrophyte interactions may also be important: benthic fish species such as bream and carp (*Cyprinus carpio* L.) may interfere indirectly with macrophyte growth through sediment disturbance (Meijer *et al.* 1990) and tench (*Tinca tinca* (L)) may increase the impact of epiphytic algae by removing molluscs (Brönmark 1994).

The numbers of herbivorous water birds are known to be closely related to submerged vegetation abundance (Giles 1994; Hanson & Butler 1994) and, as macrophytes return to a lake, more birds may be attracted to the area. In Zwemlust, The Netherlands, following the removal of fish in 1987 and the subsequent growth of macrophytes, coot (*Fulica atra*) invaded the lake during the autumn and winter of 1989, to reach a maximum of c. 150 birds in a lake of 1.5 ha. Similar densities (6–18 ha$^{-1}$) have been counted in macrophyte lakes of the Norfolk Broads (Perrow *et al.* 1997). These birds are known to be high consumers of macrophytes (Cramp & Simmons 1980) and their grazing may be an additional important factor in the survival of the vegetation.

Waterfowl numbers are often highest in late summer or autumn as their numbers are supplemented by migrating birds. It has been argued that their impact on macrophyte standing crop occurs at a time when the plants are dying back anyway and that, therefore, there will be little impact on their long-term survival (Kiørboe 1980). Other studies have demonstrated that grazing can be important in the early summer (Lauridsen *et al.* 1993; Søndergaard *et al.* 1996), when reduction of the crop might still be important for the stability of the trophic cascade. In addition, it is possible that grazing during the late summer may be important for growth the following season (Lodge 1991), particularly for species which rely on overwintering vegetative material. For example, the production of overwintering tubers in *Potamogeton pectinatus* is related to its above-ground biomass (Van Dijk et al. 1992). Kiørboe (1980) reached his conclusions by combining estimates of plant growth and waterfowl consumption rates. This may underestimate the impact of grazing as, although birds may only ingest a small proportion of the plant, they may uproot whole stems and destroy them (Berglund et al. 1963). Other studies have used exclosure experiments using transplanted macrophytes (Lauridsen et al. 1993; Søndergaard et al. 1996), or enclosures around vegetation which was relatively sparse in the lake (Van Donk & Otte 1996). As a result, they may overemphasise the importance of grazing by providing birds with a limited food resource. A more detailed study of coot macrophyte populations has been carried out in the Norfolk Broads, UK (Perrow et al. 1997). These authors monitored macrophyte cover, coot populations, their diet and grazing pressure over a full year. During the breeding season, an important period of macrophyte colonisation, territorial behaviour limited bird numbers. In addition, despite the availability of macrophytes, their diet consisted mainly of invertebrates and filamentous algae, perhaps influenced by the food-quality requirements of their young chicks. As the birds dispersed in the summer, following breeding, Perrow et al. (1997) found a significant correlation between macrophyte cover and bird density. During this period, macrophytes were a major component of bird diets and consumption was 76-fold higher than in the spring. Despite this, grazing losses were shown to be less than the potential growth rate of the plants during both the spring and the summer. Thus, although macrophyte populations attract grazing waterbirds, bird grazing is unlikely to be a major factor limiting macrophyte growth in shallow lakes. However, in shallow lakes with vegetation which relies on overwintering vegetative material, it is conceivable that its removal would influence growth during the following season, particularly during the initial stages of plant colonisation.

### 10.5.4 *Restoration or rehabilitation*

Much of the work cited in this chapter has been carried out to support the restoration of currently degraded lake systems. However, to restore an ecosystem may actually be impossible, partly as we can never know in sufficient detail what the original was exactly like (Moss et al. 1996b). Moreover, once changed, it is very unlikely that all of the intricate links can be re-established, particularly when only a minority of them has been adequately described and understood by limnologists. What we must satisfy ourselves with is to rehabilitate the lake to a state which is more acceptable than that which obtains presently. This will not be easy but there is increasing pressure to improve our environment and the substantial amount of theoretical and practical work which has been carried out in the past decade has done much to increase our chances of success. More remains to be achieved, of course. Those lakes currently undergoing improvements need continued detailed monitoring if we are to test our hypotheses over the longer term. A better understanding of the spatial relationships of the different trophic levels and the factors driving them in shallow lakes may help to design future rehabilitation schemes. More quantitative information relating to interactions among fish, zooplankton, macrophytes and nutrients is required in order to provide the lake manager with appropriate targets. Particularly important are the parameters which might govern a stable community, such as macrophyte density and structure, and the biomass and composition of a stable fish community.

Osborne (this Volume, Chapter 11) concludes that there is still inadequate understanding of shallow tropical lakes to provide a basis for sound management. Fortunately, this is not the case in temperate shallow lakes. There is a continued interest among many limnologists in unravelling the intricate web of interactions which occur in shallow lakes and considerable progress has been made in making their findings available to lake managers (Moss et al. 1996b). With increasing knowledge come greater confidence and ability to tackle larger projects. These are likely to reap greater rewards and, although the risk of failure remains, we are now in one of the best positions to set about improving the quality of our shallow-lake environments.

## 10.6 REFERENCES

Adams, M.S., Titus, J. & McCracken, M.D. (1974) Depth distribution of photosynthetic activity in a *Myriophyllum spicatum* community in Lake Wingra. *Limnology and Oceanography*, **19**, 377–89.

Andersen, F.O. & Jensen, H.S. (1991) The influence of chironomids on decomposition of organic matter and nutrient exchange in a lake sediment. *Verhandlungen der Internationalen Vereinigung für theoretische und angewandte Limnologie*, **24**, 3051–5.

Andersson, G., Cronberg, G. & Gelin, C.I. (1973) Planktonic changes following the restoration of Lake Trummen, Sweden. *Ambio*, **59**, 9–15.

Annadotter, H, Cronberg, G. (1999) Multiple techniques for the restoration of Lake Finjasjön, a hypertrophic lake in southern Sweden—a management perspective. *Hydrobiologia*, **395/396**, 77–85.

Balls, H.R., Moss, B. & Irvine, K. (1989) The loss of submerged plants with eutrophication. 1. Experimental design, water chemistry, aquatic plant and phytoplankton biomass in experiments carried out in ponds in the Norfolk Broadland. *Freshwater Biology*, **22**, 71–87.

Barko, J.W. & Smart, R.M. (1986) Sediment related mechanisms of growth limitation in submersed macrophytes. *Ecology*, **67**, 1328–40.

Bean, C.W. & Winfield, I.J. (1995) Habitat use and activity patterns of roach (*Rutilus rutilus*), rudd (*Scardinius erythrophtalmus* (L.)), perch (*Perca fluviatilis*) and pike (*Esox lucius*) in the laboratory: the role of predation threat and structural complexity. *Ecology of Freshwater Fish*, **4**, 37–46.

Beklioglu, M. & Moss, B. (1996) Existence of a macrophyte-dominated clear water state over a very wide range of nutrient concentrations in a small shallow lake. *Hydrobiologia*, **337**, 93–106.

Bengtsson, L., Fleischer, G., Lindmark, G. & Ripl, W. (1975) Lake Trummen restoration project I. Water and sediment chemistry. *Verhandlungen der Internationalen Vereinigung für theoretische und angewandte Limnologie*, **19**, 1080–7.

Benndorf, J., Schultz, H., Benndorf, A., *et al.* (1988) Food-web manipulations by enhancement of piscivorous fish stocks: Long-term effects in the hypertrophic Bautzen Reservoir. *Limnologica*, **19**, 97–110.

Berg, S.E., Jeppesen, E. & Søndergaard, M. (1997) Pike (*Esox lucius* L.) stocking as a biomanipulation tool. 1. Effects on the fish population in Lake Lyng (Denmark). *Hydrobiologia*, **342/343**, 311–8.

Berglund, B.E., Curry-Lindahl, K., Luther, H., Olsson, V., Rohde, W. & Sellerberg, G. (1963) Ecological studies on the mute swan (*Cygnus olor*) in southern Sweden. *Acta Vertibratica*, **2**, 167–288.

Best, E.P., De Vries, D. & Reins, A. (1984) The macrophytes of the Loosdrecht Lakes: a story of their decline in the course of eutrophication. *Verhandlungen der Internationalen Vereinigung für theoretische und angewandte Limnologie*, **22**, 868–75.

Björk, S. (1988) Redevelopment of lake ecosystems. A case study approach. *Ambio*, **17**, 90–8.

Blindlow, I. (1992) Long- and short-term dynamics of submerged macrophytes in two shallow eutrophic lakes. *Freshwater Biology*, **28**, 15–27.

Blindlow, I., Andersson, G., Hargeby, A. & Johansson, S. (1993) Long-term pattern of alternative stable states in two shallow eutrophic lakes. *Freshwater Biology*, **30**, 159–67.

Boers, P. & De Bles, F. (1991) Ion concentrations in interstitial water as indicators for phosphorus release processes and reactions. *Water Research*, **25**, 591–8.

Boers, P. & van Hese, O. (1988) Phosphorus from the peaty sediments of the Loosdrecht Lakes (The Netherlands). *Water Research*, **22**, 355–63.

Boers, P., van der Does, J., Quaak, M. & van Vlugt, J. (1994) Phosphorus fixation with iron III chloride; a new method to control internal phosphorus loading in shallow lakes? *Archiv für Hydrobiologie*, **129**, 339–51.

Boström, B. (1984) Potential mobility of phosphorus in different types of lake sediments. *Internationale Revue der Gesamten Hydrobiologie*, **69**, 457–74.

Boström, B., Jansson, M.E. & Forsberg, C. (1982) Phosphorus release from lake sediments. *Ergebnisse der Limnologie*, **35**, 623–45.

Boström, B., Andersen, J.M., Fleischer, S. & Jansson, M. (1988) Exchange of phosphorus across the sediment–water interface. *Hydrobiologia*, **170**, 229–44.

Boyd, C.E. (1971) The limnological role of aquatic macrophytes and their relationship to reservoir management. *Special Publication of the American Fish Society*, **8**, 129–35.

Brönmark, C. (1994) Effects of tench and perch on interactions in a freshwater, benthic food chain. *Ecology*, **75**, 1818–28.

Brönmark, C. & Weisner, S.E.B. (1992) Indirect effects of fish community on submerged vegetation in shallow, eutrophic lakes: an alternative mechanism. *Hydrobiologia*, **243/244**, 293–301.

Brooks, J.L. & Dodson, S.I. (1965) Predation, body size and composition of the phytoplankton. *Science*, **150**, 28–35.

Carpenter, S.R. (1981) Submerged vegetation: an internal factor in lake ecosystem succession. *American Naturalist*, **118**, 372–83.

Carpenter, S.R. & Kitchell, J.F. (1992) Trophic cascade and biomanipulation: Interface of research and management—a reply to the comment by De Melo et al. *Limnology and Oceanography*, **37**, 208–13.

Carpenter, S.R. & Kitchell, J.F. (1993) *The trophic cascade in lakes*. Cambridge University Press, New York, 385 pp.

Carpenter, S.R. & Lodge, D.M. (1986) Effects of submersed macrophytes on ecosystem processes. *Aquatic Botany*, **26**, 341–70.

Carpenter, S.R., Elser, J.J. & Olson, K.M. (1983) Effects of roots of *Myriophyllum verticillatum* L. on sediment redox conditions. *Aquatic Botany*, **17**, 243–9.

Carpenter, S.R., Kitchell, J.F. & Hodgson, J.R. (1985) Cascading trophic interactions and lake productivity. *Bioscience*, **35**, 634–9.

Chambers, P. (1987) Light and nutrients in the control of aquatic plant community structure. II. *In situ* observations. *Journal of Ecology*, **75**, 621–8.

Chambers, P.A. & Kalff, J. (1985) Depth distribution and biomass of submersed aquatic macrophyte communities in relation to secchi depth. *Canadian Journal of Fisheries and Aquatic Sciences*, **42**, 701–9.

Cook, M.F. & Bergersen, E.P. (1988) Movements, habitat selection, and activity periods of northern pike in Eleven Mile Reservoir, Colarado. *Transactions of the American Fisheries Society*, **117**, 495–502.

Coops, H. & Doef, R.W. (1996) Submerged vegetation development in two shallow, eutrophic lakes. *Hydrobiologia*, **340**, 115–20.

Cramp, S. & Simmons, K.E.L. (1980) *Handbook of the Birds of Europe and the Middle East and North Africa. The Birds of the Western Paleartic*, Vol II, *Hawks to Bustards*. Oxford University Press, Oxford, 695 pp.

De Groot, C.J. (1991) The influence of FeS on the inorganic phosphate system in sediments. *Verhandlungen der Internationalen Vereinigung für theoretische und angewandte Limnologie*, **24**, 3029–35.

De Nie, H.W. (1987) *The Decrease in Aquatic Vegetation in Europe and its Consequences for Fish Populations*. EIFAC Occasional Paper 19, Food and Agriculture Organisation, Rome, 88 pp.

De Melo, R., France, R. & McQueen, D.J. (1992) Biomanipulation: hit or myth? *Limnology and Oceanography*, **37**, 192–207.

Eckerrot, A. & Pettersson, K. (1993) Pore water phosphorus and iron concentrations in a shallow, eutrophic lake—indications of bacterial regulation. *Hydrobiologia*, **253**, 165–77.

Edmondson, W.T. (1977) Recovery of Lake Washington from eutrophication. In: Cairns, J., Dickson, K.L. & Henricks, E.E. (eds), *Recovery and Restoration of Damaged Ecosystems*. University Press of Virginia, Charlottesville, 102–9.

Engel, S. (1988) The role and interactions of submerged macrophytes in a shallow Wisconsin Lake. *Journal of Freshwater Ecology*, **4**, 229–341.

Fott, J., Pechar, L. & Prazakov, M. (1980) Fish as a factor controlling water quality in ponds. In: Barica, J. & Mur, L.R. (eds) *Hypertrophic Ecosystems*. Developments in Hydrobiology, Vol. 2, Kluwer, Dordrecht, 255–61.

Fulton, R.S. & Paerl, H.W. (1987) Effects of colonial morphology on zooplankton utilization of algal resources during blue-green algal (*Microcystis aeruginosa*) blooms. *Limnology and Oceanography*, **32**, 634–44.

Giles, N. (1994) Tufted duck (*Aytha fuligula*) habitat use and brood survival increases after fish removal from gravel pit lakes. *Hydrobiologia*, **279/280**, 387–92.

Gliwicz, Z.M. & Pijanowska, J. (1989) The role of predation in zooplankton succession. In: Sommer, U. (ed.), *Plankton Ecology; Succession in Plankton Communities.*, Springer-Verlag, New York, 253–95.

Golterman, H.L., Sly, P.G. & Thomas, R.L. (1983) *Study of the Relationship between Water Quality and Sediment Transport*. UNESCO, Paris, 231 pp.

Gopal, B. & Goel, U. (1993) Competition and allelopathy in aquatic plant-communities. *Botanical Review*, **59**, 155–210.

Grimm, M.P. & Backx, J.J.G.M. (1990) The restoration of shallow eutrophic lakes, and the role of nothern pike, aquatic vegetation and nutrient concentration. *Hydrobiologia*, **200/201**, 557–66.

Gulati, R.D. (1990) Structural and grazing responses of zooplankton community to biomanipulation of some Dutch water bodies. *Hydrobiologia*, **200/201**, 99–118.

Gulati, R.D., Lammens, E.H.R.R, Meijer, M.-L. & van Donk, E. (eds) (1990) *Biomanipulation—Tool for Water Management*. Developments. Hydrobiology, Vol. 61, Kluwer, Dordrecht, 628 pp.

Hannson, L.A. (1989) The influence of a periphytic biolayer on phosphorous exchange between substrate and water. *Archiv für Hydrobiologie*, **115**, 21–6.

Hanson, M.A. & Butler, M.G. (1994) Responses of phytoplankton, turbidity and macrophytes to biomanipulation in a shallow prairie lake. *Canadian Journal of Fisheries and Aquatic Sciences*, **51**, 1180–8.

Harper, D. (1992) *Eutrophication of Freshwaters. Principles, Problems and Restoration.* Chapman & Hall, London, 327 pp.

Harris, G.P. (1996) A reply to Sarnelle (1996) and some further comments on Harris's (1994) opinions. *Freshwater Biology*, **35**, 343–7.

Holdren, G.C. & Armstrong, D.E. (1986) Interstitial iron concentration as an indicator of phosphorus release and mineral formation in lake sediments. In: Sly, P.G. (ed.), *Sediments and Water Interactions.* Springer-Verlag, New York, 133–47.

Holland, L.E. & Hutson, M.L. (1984) Relationship of young-of-the-year northern pike to aquatic vegetation types in backwaters of the upper Mississippi River. *North American Journal of Fisheries Management*, **4**, 514–22.

Hootsmans, M.J.M. & Vermaat, J.E. (1991) *Macrophytes, a key to understanding changes caused by eutrophication in shallow freshwater ecosystems.* PhD thesis, Agricultural University, Wageningen, 412 pp.

Hrbáček, J., Dvořaková, M., Kořínek, V. & Procházková, L. (1961) Demonstration of the effect of the fish stock on the species composition of zooplankton and the intensity of metabolism of the whole plankton assemblage. *Verhandlungen der Internationalen Vereinigung für theoretische und angewandte Limnologie*, **14**, 192–5.

Irvine, K., Moss, B. & Balls, H. (1989) The loss of submerged plants with eutrophication II. Relationships between fish and zooplankton in a set of experimental ponds, and conclusions. *Freshwater Biology*, **22**, 89–107.

Jacobsen, D. & Sand-Jensen, K. (1995) Variability of invertebrate herbivory on the submerged macrophyte *Potamogeton perfoliatus. Freshwater Biology*, **28**, 301–8.

Jansson, M. (1989) Role of benthic algae in transport of nitrogen from sediment to lake water in a shallow clearwater lake. *Archiv für Hydrobiologie*, **89**, 101–9.

Jasser, I. (1995) The influence of macrophytes on a phytoplankton community in experimental conditions. *Hydrobiologia*, **306**, 21–32.

Jensen, H.S. & Andersen, F.O. (1992) Importance of temperature, nitrate and pH for phosphate release from aerobic sediments in four shallow, eutrophic lakes. *Limnology and Oceanography*, **37**, 577–89.

Jensen, H.S, Kristensen, P., Jeppesen, E. & Skytthe, A. (1992) Iron:phosphorus ratio in surface sediments as an indicator of phosphorus release from aerobic sediments in shallow lakes. *Hydrobiologia*, **235/236**, 731–43.

Jeppesen, E., Søndergaard, M., Sortkjær, O., Mortensen, E. & Kristensen, P. (1990a) Interactions between phytoplankton, zooplankton and fish in a shallow, hypertrophic lake: a study of phytoplankton collapses in Lake Søbygård, Denmark. *Hydrobiologia*, **191**, 149–64.

Jeppesen, E., Jensen, J.-P., Kristensen, P., et al. (1990b) Fish manipulation as a lake restoration tool in shallow, eutrophic, temperate lakes 2: threshold levels, long-term stability and conclusions. *Hydrobiologia*, **200/201**, 219–27.

Jeppesen, E., Agerbo Madsen, E., Jensen, J.-P. & Anderson, N.J. (1996) Reconstructing the past density of planktivorous fish and trophic structure from sedimentary zooplankton fossils: a surface sediment calibration data set from 30 predominantly shallow lakes. *Freshwater Biology*, **36**, 115–27.

Jeppesen, E., Jensen, J.-P., Søndergaard, M., Lauridsen, T., Pedersen, L.J. & Jensen L. (1997a) Top-down control in freshwater lakes: the role of nutrient state, submerged macrophytes and water depth. *Hydrobiologia*, **342/343**, 151–64.

Jeppesen, E., Lauridsen, T.L., Kairesalo, K. & Perrow, M. (1997b) Impact of submerged macrophytes on fish-zooplankton interactions in lakes. In: Jeppesen, E., Søndergaard, Ma., Søndergaard, Mo. & Christoffersen, K. (eds), *The Structuring Role of Submerged Macrophytes in Lakes.* Ecological Studies Series, Vol. 131, Springer-Verlag, New York, 91–114.

Kiørboe, T. (1980) Distribution and production of macrophytes in Tipper Grund (Ringkøbing Fjord, Denmark) and the impact of waterfowl grazing. *Journal of Applied Ecology*, **17**, 675–87.

Lampert, W. (1993) Ultimate causes of diel vertical migration of zooplankton: new evidence for the predator-avoidance hypothesis. *Engebnisse der limnologie*, **39**, 79–88.

Lauridsen, T. & Buenk, I. (1996) Diel changes in the horizontal distribution of zooplankton in the littoral zone of two shallow eutrophic lakes. *Archiv für Hydrobiologie*, **137**, 161–76.

Lauridsen, T. L. & Lodge, D. (1996) Avoidance by *Daphnia magna* by fish and macrophytes: chemnical cues and predator mediated use of macrophyte habitat. *Limnology and Oceanography*, **41**, 794–8.

Lauridsen, T.L., Jeppesen, E. & Søndergaard, M. (1993) Colonization of submerged macrophytes in shallow fish manipulated Lake Vaeng: importance of sediment composition and waterfowl grazing. *Aquatic Botany*, **46**, 1–15.

Lauridsen, T.L., Jeppesen, E. & Søndergaard, M. (1994) Colonisation and succession of submerged macrophytes in shallow lake Vaeng during the first five years following fish manipalation. *Hydrobiologia*, **275/276**, 233–42.

Lauridsen, T.L., Pedersen, L.J., Jeppesen, E. & Søndergaard, M. (1996) The importance of macrophyte bed size for cladoceran composition and horizontal migration in a shallow lake. *Journal of Plankton Research*, **18**, 2283–94.

Lijklema, G.E. (1980) Interactions of orthophosphate with iron(III) and aluminium hydroxides. *Environmental Science and Technology*, **14**, 537–41.

Lodge, D.M. (1991) Herbivory on freshwater macrophytes. *Aquatic Botany*, **41**, 195–224.

Marsden, M.W. (1989) Lake restoration by reducing external phosphorus loading: the influence of sediment phosphorus release. *Freshwater Biology*, **21**, 139–62.

May, R.M. (1977) Thresholds and breakpoints in ecosystems with a multiplicity of stable states. *Nature*, **269**, 471–7.

McQueen, D.J., Post, M.R.S. & Mills, E.L. (1986) Trophic relationship in freshwater pelagic ecosystems. *Canadian Journal of Fisheries and Aquatic Sciences*, **43**, 1571–81.

Meijer, M.-L., de Haan, M.W., Breukelaar, A.W. & Buiteveld, H. (1990) Is reduction of the benthivorous fish an important cause of high transparency following biomanipulation in shallow lakes? *Hydrobiologia*, **200/201**, 303–15.

Meijer, M.-L., Jeppesen, E., van Donk, E., et al. (1994) Long term responses to fish stock reduction in small shallow lakes. Interpretation of five year results of four biomanipulation cases in the Netherlands and Denmark. *Hydrobiologia*, **275/276**, 457–66.

Mortensen, E., Jeppesen, E., Søndergaard, M. & Kamp-Nielsen, L. (eds) (1994) *Nutrient Dynamics and Biological Structure in Shallow Freshwater and Brackish Lakes.* Developments in Hydrobiology, Vol. 94, Kluwer, Dordrecht, 507 pp.

Mortimer, C.H. (1941) The exchange of dissolved substances between mud and water in lakes I. *Journal of Ecology*, **29**, 280–9.

Mortimer, C.H. (1942) The exchange of dissolved substances between mud and water in lakes II. *Journal of Ecology*, **30**, 147–201.

Moss, B. (1983) The Norfolk Broadland: experiments in the restoration of a complex wetland. *Biological Reviews*, **58**, 521–61.

Moss, B. (1990) Engineering & biological approaches to restoration from eutrophication of shallow lakes in which aquatic plant communities are important components. *Hydrobiologia*, **200/201**, 367–77.

Moss, B. (1994) Brackish and freshwater shallow lakes—different systems or variations on the same theme? *Hydrobiologia*, **275/276**, 1–14.

Moss, B., Balls, H., Irvine, K. & Stansfield, J. (1986) Restoration of two lowland lakes by isolation from nutrient-rich water sources with and without the removal of sediment. *Journal of Applied Ecology*, **23**, 391–414.

Moss, B., Stansfield, J.H., Irvine, K., Perrow, M.R. & Phillips, G.L. (1996a) Progressive restoration of a shallow lake—a twelve-year experiment in isolation, sediment removal and biomanipulation. *Journal of Applied Ecology*, **33**, 71–86.

Moss, B., Madgwick, J. & Phillips, G.L. (1996b) *A Guide to the Restoration of Nutrient-enriched Shallow Lakes.* Broads Authority/Environment Agency, Norwich, 180 pp.

Naumann, E. (1919) Några Synpunkter Anagående limnoplanktons ökologi med särskild hänsyn till fytoplankton. *Svensk Botanisk Tidskrift*, **13**, 129–63.

Osborne, P.J. & Phillips, G.L. (1978) Evidence for nutrient release from the sediments of two shallow and productive lakes. *Verhandlungen der Internationalen Vereinigung für theoretische und angewandte Limnologie*, **20**, 654–8.

Pearsall, W.H. (1921) The development of vegetation in English lakes, considered in relation to the general evolution of glacial lakes and rock basins. *Proceedings of the Royal Society of London, Series B*, **92**, 259–84.

Pennak, R.W. (1973) Some evidence for aquatic macrophytes as repellents for a limnetic species of *Daphnia*. *Internationale Revue der gesamten Hydrobiologie*, **58**, 569–76.

Persson, L. (1993) Predator-mediated competition in prey refuges: the importance of habitat dependent prey resources. *Oikos*, **68**, 12–22.

Perrow, M.R., Schutten, J.H, Howes, J.R., Holzer, T., Madgwick, F.J. & Jowitt, A.J.D. (1997) Interations between coot (*Fulica atra*) and submerged macrophytes: the role of birds in the restoration process. *Hydrobiologia*, **342/343**, 241–55.

Phillips, G.L. & Jackson, R. (1990) The control of eutrophication in very shallow lakes, the Norfolk Broads. *Verhandlungen der Internationalen Vereinigung für theoretische und angewandte Limnologie*, **24**, 573–5.

Phillips, G.L. & Kerrison, P. (1991) The restoration of the Norfolk Broads: the role of biomanipulation. *Memorie dell'Istituto italiano di Idrobiologia*, **48**, 75–97.

Phillips, G.L. & Moss, B. (1994) *Is Biomanipulation a Useful Technique in Lake Management?* R&D Note 276, National Rivers Authority, Bristol, 45 pp.

Phillips, G.L., Eminson, D. & Moss, B. (1978) A mechanism to account for macrophyte decline in progressively eutrophicated freshwaters. *Aquatic Botany*, **4**, 103–26.

Phillips, G.L., Jackson, R., Bennett, C. & Chilvers, A. (1994) The importance of sediment phosphorus release in the restoration of very shallow lakes (The Norfolk Broads, England) and implications for biomanipulation. *Hydrobiologia*, **275/276**, 445–6.

Phillips, G. L., Perrow, M.R. & Stansfield, J. (1996) Manipulating the fish-zooplankton ineraction in shallow lakes: a tool for restoration. In: Greenstreet, S.P.R. & Tasker, M. (eds), *Aquatic Predators and their Prey*. Fishing News Books, Blackwell Science, Oxford, 174–83.

Pitt, J., Kelly, A. & Phillips, G.L. (1997) Control of nutrient release from sediments. In: Madgwick, F.J. & Phillips, G.L. (eds), *Restoration of the Norfolk Broads—Final Report*. Broads Authority Research Series, Vol. 14a. Broads Authority and Environment Agency, Norwich, 82 pp.

Prejs, A. (1984) Herbivory by temperate freshwater fishes and its consequences. *Environmental Biology of Fishes*, **10**, 281–96.

Reynolds, C.S. (1992) Eutrophication and the management of planktonic algae: what Vollenweider couldn't tell us. In: Sutcliffe, D.W. & Jones, J.G. (eds), *Eutrophication: Research and Application to Water Supply*. Freshwater Biological Association, Ambleside, 4–29

Reynolds, C.S. (1994) The ecological basis for the successful biomanipulation of degraded aquatic communities. *Archiv für Hydrobiologie*, **130**, 1–33.

Risgaard-Petersen, N. & Jensen, K. (1997) Nitrification and denitrification in the rhizosphere of the aquatic macrophyte *Lobelia dortmanna* L. *Limnology and Oceanography*, **42**, 529–37.

Rohlich, G.A. (1969) *Eutrophication—Causes, Consequences, Correctives*. National Academy of Sciences, Washington, 661 pp.

Rørslett, B., Berge, D. & Johansen, S.W. (1985) Mass invasion of *Elodea canadensis* in a mesotrophic, South Norwegian lake—impact on water quality. *Verhandlungen der Internationalen Vereinigung für theoretische und angewandte Limnologie*, **22**, 2920–6.

Sarnelle, O. (1996) Predicting the outcome of trophic manipulation in lakes—a comment on Harris (1994). *Freshwater Biology*, **35**, 339–42.

Sas, H. (1989) *Lake Restoration by Reduction of Nutrient Loadings: Expectations, Experiences, Extrapolations*. Academia Verlag Richarz, Sankt Augustin, 497 pp.

Scheffer, M. (1990) Multiplicity of stable states in freshwater systems. *Hydrobiologia*, **200/201**, 475–86.

Scheffer, M., de Redelijkheid, M.R. & Noppert, F. (1992) Distribution and dynamics of submerged vegetation in a chain of shallow eutrophic lakes. *Aquatic Botany*, **42**, 199–216.

Scheffer, M., Hosper, S.H, Meijer, M.-L., Moss, B. & Jeppesen, E. (1993) Alternative equilibria in shallow lakes. *Trends in Ecology and Evolution*, **8**, 275–9.

Søndergaard, M., Jeppesen, E., Kristensen, P. & Sortkjaer, O. (1990a) Interactions between sediment and water in a shallow and hypertrophic lake: a study of phytoplankton collapses in Lake Søbygård, Denmark. *Hydrobiologia*, **191**, 139–48.

Søndergaard, M., Jeppesen, E., Mortensen, E., Dall, E., Kristensen, P. & Sortkjær, O. (1990b) Phytoplankton biomass reduction after planktivorous fish reduction in a shallow, eutrophic lake: a combined effect of reduced internal P-loading and increased zooplankton grazing. *Hydrobiologia*, **200/201**, 229–40.

Søndergaard, M., Bruun, L., Lauridsen, T., Jeppesen, E. & Madsen, T. V. (1996) The impact of waterfowl on submerged macrophytes: *in-situ* experiments in a shallow eutrophic lake. *Aquatic Botany*, **53**, 73–84.

Søndergaard, M., Jeppesen, E. & Berg, S. (1997) Pike (*Esox lucius*) stocking as a biomanipulation tool. 2. Effects on lower trophic levels in Lake Lyng, Denmark. *Hydrobiologia*, **342/343**, 319–25.

Spence, D.H.N. (1982) The zonation of plants in freshwater lakes. In: Macfadyen, A & Ford, E. (eds), *Advances in Ecological Research*, Vol. 12. Academic Press, New York, 37–125.

Stansfield, J.H., Moss, B. & Irvine, K. (1989) The loss of submerged plants with eutrophication III. Potential role of organochlorine pesticides: a palaeoecological study. *Freshwater Biology*, **22**, 109–32.

Stansfield, J.H., Perrow, M.R., Tench, L.D., Jowitt, A.J.D. & Taylor, A.A.L. (1997) Submerged macrophytes as refuges for grazing Cladocera against fish predation; observations on seasonal changes in relation to macrophyte cover and predation pressure. *Hydrobiologia*, **342/345**, 229–40.

Sutcliffe, D.W. & Jones, J.G. (1992) *Eutrophication: Research and Application to Water Supply*. Freshwater Biological Association, Ambleside, 217 pp.

Talling, J.F. (1957) The phytoplankton population as a compound photosynthetic system. *New Phytologist*, **56**, 29–50.

Tátrai, I. (1988) Experiments on nitrogen and phosphorus release by *Chironomus plumosus* from the sediments of Lake Balaton, Hungary. *Internationale Revue der gesamten Hydrobiologie*, **73**, 627–40.

Tátrai, I., Tóth, G., Ponyi, J.E, Zlinskzky, J. & Istvánovics, V. (1990) Bottom-up effects of bream (*Abramis brama* L.) in Lake Balaton. *Hydrobiologia*, **200/201**, 167–75.

Thienemann, A. (1918) Untersuchungen (ber dir Beziehungen zwischen der Sauerstoffgehalt des Wassers und der Zusammensetzung der Fauna in norddeutschen Seen. *Archiv für Hydrobiologie*, **12**, 1–24.

Timms, R.M. & Moss, B. (1984) Prevention of growth of potentially dense phytoplankton populations by zooplankton grazing in the presence of zooplanktivorous fish in a shallow wetland ecosystem. *Limnology and Oceanography*, **29**, 472–86.

Underwood, G.J.C. (1991) Growth enhancement of the macrophyte *Ceratophyllum demersum* in the presence of the snail *Planorbis planorbis*: the effect of grazing and chemical conditioning. *Freshwater Biology*, **26**, 325–34.

Underwood, G.J.C., Thomas, J.D, & Baker, J.H. (1992) An experimental investigation of interactions in snail–macrophyte–epiphyte systems. *Oecologia*, **91**, 587–95.

Van der Molen, D.T. & Boers, P.C.M. (1994) Influence of internal loading of phosphorus concentration in shallow lakes before and after reduction in the external loading. *Hydrobiologia*, **276**, 379–89.

Van Dijk, G., Breukelaar, A.W. & Gijlstra, R. (1992) Impact of light climate history on seasonal dynamics of a field population of *Potamogeton pectinatus* L. during a three-year period (1986–1988). *Aquatic Botany*, **43**, 17–41.

Van Donk, E. & Otte, A. (1996) Effects of grazing by fish and waterfowl on the biomass and species composition of submerged macrophytes. *Hydrobiologia*, **340**, 285–90.

Van Donk, E., Gulati, R.D., Iedema, A. & Meulemans, J.T. (1993) Macrophyte related shifts in the nitrogen and phosphorus contents of the different trophic levels in a biomanipulated shallow lake. *Hydrobiologia*, **251**, 19–26.

Van Viersen, W., Hootsmans, M.J.M. & Vermaat, J.E. (1985) Waterplanten: bondgenoten bij het waterkwaliteits-beheer? (The role of aquatic macrophytes in water quality management, in Dutch). $H_2O$, **18**, 122–6.

Van Wijk, R.J. (1988) Ecological studies on *Potamogeton pectinatus* L. I. General characteristics, biomass production and life cycles under field conditions. *Aquatic Botany*, **31**, 211–58.

Vermaat, J.E., Hootsmans, M.J.M. & van Dijk, G.M. (1990) Ecosystem development in different types of littoral enclosure. *Hydrobiologia*, **200/201**, 391–8.

Von Elert, E. & Loose, C.J. (1996) Predator-induced diel vertical migration in *Daphnia*: enrichment and preliminary chemical characterisation of a Kairomone exuded by fish. *Journal of Chemical Ecology*, **22**, 885–95.

Weisner, S.E.B., Strand, J.A. & Sandsten, H. (1997) Mechanisms regulating abundance of submerged vegetation in shallow eutrophic lakes. *Oecologia*, **109**, 592–99.

Welch, E.B. & Schrieve, G.D. (1994) Alum treatment effectiveness and longevity in shallow lakes. *Hydrobiologia*, **275/276**, 423–31.

Wetzel, R.G. (1975) *Limnology*. Saunders, Philadelphia, 743 pp.

Wigand, C., Stevenson, J.C. & Cornwell, J.C. (1997) Effects of different submersed macrophytes on sediment biogeochemistry. *Aquatic Botany*, **56**, 233–44.

Winfield, I.J. (1986) The influence of simulated aquatic macrophytes on the zooplankton consumption rate of juvenile roach, *Rutilus rutilus*, rudd, *Scardinius erythrophthalamus*, and perch, *Perca fluviatilis*. *Journal of Fish Biology*, **29**, 37–48.

Wium-Andersen, S. & Andersen, J.M. (1972) The influence of vegetation on the redox profile of the sediment of Grane Langsø, a Danish *Lobelia* lake. *Limnology and Oceanography*, **17**, 948–52.

# 11 Eutrophication of Shallow Tropical Lakes

PATRICK L. OSBORNE

## 11.1 TROPICAL LIMNOLOGY

Tropical limnology can trace its origins to the turn of the twentieth century with expeditions to Lake Nyasa (now Malawi) (Fülleborn 1900), Lake Titicaca (Neveu-Lemiare 1904) and Lake Tanganyika (Cunnington 1905). More detailed studies were only undertaken in the late 1920s with the Sunda expedition to Indonesia in 1929 (Ruttner 1931; Thienemann 1959; and see Göltenboth 1996) and the Cambridge expedition to Lakes Victoria, Albert and Kioga (Worthington 1929a, b, 1932; see also Worthington & Worthington 1933). A number of significant tropical limnology texts were published in the 1980s, including Beadle (1981), Payne (1986) and Lowe-McConnell (1987). Crisman & Streever (1996) pointed out that most early studies in tropical limnology focused on large, deep lakes of volcanic or tectonic origin with small, shallow lakes being largely ignored. Lewis (1987) noted that few tropical lakes had been studied comprehensively and that most work was descriptive rather than analytical. This gap between our knowledge and understanding of tropical versus temperate lakes remains despite a recent upsurge in interest in tropical limnology (see Dudgeon & Lam 1994; Gopal & Wetzel 1995, 1999; Schiemer & Boland 1996; Talling & Lemoalle 1998; Lewis 2000; Wetzel & Gopal 2001).

Lewis (1996) reviewed the effect of latitude on the distribution, origin and ecology of lakes. Only 10%, by number, of the world's lakes are located within the tropics. Whereas most temperate lakes have been formed by glaciation and permafrost (80%), tropical lakes are mostly of riverine origin (floodplain lakes) (40%) (see Melack, this Volume, Chapter 9) with the balance formed through other mechanisms (coastal, volcanic and aeolian lakes).

Lewis (2000) stressed that, because most lakes in the tropics are associated with rivers (either floodplain lakes or reservoirs), degradation of water quality in rivers will significantly and directly influence tropical lakes. There is a dearth of lakes within the subtropical latitudes which coincide with the low rainfall, high pressure zones of the world's arid regions.

## 11.2 STUDIES ON SHALLOW TROPICAL LAKES

### 11.2.1 Features of tropical, shallow lakes

Tropical lakes at low to moderate altitude are characterised by warm waters throughout the year and, with adequate nutrient supply, high rates of primary production. On cloudless days, intense solar radiation and high water temperatures promote phytoplankton photosynthesis and daily rates can be very high (Talling et al. 1973). Over equatorial regions, frequent cloudiness suppresses rates. Von Sperling (1997) suggested that tropical lakes are characterised by high rates of nutrient assimilation, nutrient cycling and decomposition but the rates of these processes have rarely been adequately measured in tropical lakes. Lewis (1996) identified four characteristic features of tropical lakes in comparison with temperate counterparts: (i) greater efficiency in producing phytoplankton biomass for a given nutrient supply; (ii) an inclination towards nitrogen, rather than phosphorus limitation; (iii) a lower efficiency in passing primary production to the highest trophic levels; and (iv) greater non-seasonal variation superimposed on a seasonal cycle. Lewis (1996) also noted a general similarity between tropical and temperate lakes in the composition of phytoplankton and zooplankton. Lewis (2000) also suggested that tropical lakes are more sensitive than temperate lakes to increases in nutrient supply and exhibit higher proportionate changes in water quality and biotic communities in response to eutrophication.

Shallow lakes have been defined in terms of water column mixing depth. If the maximum depth of the lake approximately equals the mixing depth, then the lake can be functionally regarded as shallow (Straškraba 1990). Mixing depth is a function of wind fetch and convective mixing. A very small lake (c.0.1 km long) may not be considered shallow unless it is less than 1.5 m deep whereas a larger lake (c.10 km long) could be regarded as shallow if its maximum depth was less than 15 m. However, MacIntyre & Melack (1995) demonstrated the importance of penetrative convective cooling on a daily to weekly basis in a tropical, floodplain lake. They measured water temperatures in Lake Calado, situated adjacent to the Solimões (Amazon) River near Manaus in Brazil, early in the morning over a 2-year period. When water depth was less than 3 m, the lake mixed to the bottom every night. Once lake depth exceeded 3 m, the lake began to develop a thermocline, with the depth of the thermocline at 0630 hours varying between 1 and 7 m. Although shallow, nocturnal mixing of the epilimnion occurred, dissolved substances accumulated below the thermocline until the next deep mixing event occurred. Consequently, this lake (and probably many like it), with an annual water depth fluctuation in excess of 10 m, and the change from diel mixing to seasonal stratification (and the effects this change imparts), confounds any definition of a shallow, tropical lake on the basis of mixing regime.

MacIntyre & Melack (1988) pointed out that the depth of vertical mixing not only depended on the magnitude of wind stress, heat flux at the surface and degree of water column stratification but also on whether these processes were acting in concert or not. With moderate wind speeds and positive heat flux into the lake, mixing may be damped, but a combination of the same wind speed and heat loss from the lake may result in mixing. These authors also indicated that few data are available with which to determine the frequency and extent of vertical mixing in shallow, tropical lakes.

Shallow lakes in general, and those situated on floodplains and in tropical regions with a monsoonal climate in particular, are strongly affected by changes in water level: a small decline in water level can expose a large area of lake bed. Hydrology strongly affects the ecology of the littoral zone and the rise and fall in water level generally promotes rates of nutrient cycling. Density-driven convective flows between nearshore and offshore waters possess considerable potential for transporting nutrients from the littoral zone to deeper, offshore waters (see Monismith et al. 1990).

Given the same areal loading rate and water residence time, nutrient concentrations will be higher in shallow lakes than deeper lakes. Therefore shallow lakes may be more responsive to an increase in limiting nutrient load. In shallow lakes the role of the sediment as a nutrient source is potentially greater than that in deeper lakes owing to the large ratio of sediment area to water volume and the greater chance of sediment disturbance by turbulent mixing. Conversely, the shallow water column reduces the chance of anoxic conditions developing, which would promote the release of nutrients. Processes such as sedimentation and resuspension, sorption, mineralisation in the water column and within the sediments, denitrification and cation exchange all undoubtedly play important roles in the nutrient dynamics in shallow, tropical lakes. Regrettably, the rates of these processes have not been adequately studied in such lakes.

### 11.2.2 The role of depth in shallow, tropical lakes

Despite small seasonal changes in air temperature, deep lakes in the tropics may nonetheless stratify and mix in a predictable way. Although the temperature differential between the surface and bottom waters may be small (1.5°C for equatorial lakes to 5°C for those nearer the Tropics of Cancer and Capricorn), the high temperature can generate sufficient density difference for stratification to be stable through a portion of the year. The length of the periods of significant mixing may be very short and in some lakes occurs only rarely (see Osborne & Totome 1992).

Talling (1992) and Dumont (1992) provide excellent reviews of the climatic, biogenic and geological determinants of environmental features in shallow, tropical lakes. Persistent thermal stratification is usually absent in shallow lakes because mixing by wind-induced turbulence and convection is usually strong enough to affect the entire water column. In contrast to deeper lakes, shallow, tropical lakes either stratify and mix diurnally (continuous polymixis), or follow an irregular sequence of stratification and mixing (discontinuous polymixis, oligomixis) depending on the sequence of climatic and hydrological conditions.

These alternating periods of water column mixing and stratification in shallow, tropical lakes, at the high temperatures prevailing, play a significant role in enhancing the rate of decomposition and nutrient regeneration. Decomposition may be impeded if oxygen is not replenished by water column mixing but in shallow lakes this will rarely occur and, therefore, the rate of nutrient cycling within shallow, tropical lakes should be more rapid than that in cooler, temperate lakes of similar depth. High rates of nutrient regeneration were recorded in equatorial, continuously polymictic, Lake George in Uganda (see Ganf & Viner 1973; Ganf 1974a, b; Viner & Smith 1973). Rapid rates of nutrient cycling within Lake George, coupled with little seasonal fluctuation in climate and plankton biomass ensured high rates of primary production throughout the year. These large phytoplankton populations and high rates of gross photosynthesis were sustained despite very low concentrations of dissolved inorganic nitrogen and phosphorus (Ganf 1972). Ganf & Blazka (1974) invoked high nutrient turnover rates, mediated by bacteria and zooplankton, to explain these high rates of phytoplankton production. Further work on nutrient turnover rates in shallow, tropical lakes is needed to determine whether the high rates recorded in Lake George are typical.

Although Lake George exhibited remarkable seasonal constancy, diurnal changes were shown to be significant (Ganf & Viner 1973). At dawn the lake was isothermal but by 1000 hours thermal stratification was established and became more intense as the day progressed. With the onset of thermal stratification, the algae began to sink and their distribution gradually changed, from uniform throughout the water column, to concentrated near the bottom of the lake. The water column also became stratified with respect to pH and oxygen. Vertical mixing at the end of the day occurred in response to decreased radiation input, transfer of heat to the cooler evening air and increased wind strength.

Lake Opi (mean depth 1–2 m) in Nigeria is situated $6°N$ and exhibited a diel cycle of daytime thermal stratification and complete or near-complete stratification breakdown at night (Hare & Carter 1984). Unlike Lake George, some seasonality was detected in this lake. During the dry season, strong prevailing winds (locally called the *harmattan*) ensured the complete circulation of the lake each night. In the wet, less windy season, when lake depth is greatest, complete vertical mixing did not take place and anoxia developed in near-bottom waters.

In productive lakes, respiration by decomposers may remove all the oxygen from the water overlying the sediment and sulphate may be reduced to sulphide. If the water column is subsequently mixed, fish and other aquatic animals may be asphyxiated. Such sporadic events have been recorded in Lake George, Uganda when storms occur after periods of calm weather, during which time nocturnal respiration had depleted oxygen in the lower part of the water column. The fish kills were usually restricted to one part of the lake and populations recovered rapidly.

Turbidity may often be high in shallow lakes owing to wind-induced turbulence disturbing bottom sediments. This is particularly true of lakes where sedimentary clays increase in abundance as resettling times are extended. In Lake Chilwa, Malawi the absence of aquatic plants in the open water areas was attributed to high light attenuation by suspended clay particles (Kalk et al. 1979).

### 11.2.3 Seasonality in tropical lakes

Although seasonality in shallow, equatorial Lake George may be muted and diurnal changes large, other shallow, tropical lakes exhibit more seasonal variability related to changes in water depth and flood events. Many such lakes exhibit marked seasonal and interannual fluctuations in their hydrological regime. Large-scale changes in lake area, water depth and volume result, with Lake Chilwa, Malawi providing an extreme example (Kalk et al. 1979). This shallow lake (mean depth 1–2 m) with an area of 700 km$^2$ dried up completely in 1968. Similar variations have been recorded in Lake Chad (Carmouze et al. 1983) and, in both these lakes, nutrient supply and water turbidity through sediment resuspension largely determined rates of algal photosynthesis. Gaudet & Muthuri (1981) stressed the importance of water level fluctuations in the regeneration of nutrients from sediments.

Seasonality in shallow tropical lakes is imposed more by variations in rainfall, windiness (speed and direction) and cloudiness than by seasonal variation in day length and temperature. Melack (1979) recognised three temporal patterns in the seasonal variation of phytoplankton in tropical lakes. He found that most tropical lakes exhibited pronounced seasonal fluctuations which usually corresponded with variation in rainfall, river discharges or vertical mixing. In a second type of lake, diel changes often exceeded month-to-month changes and the same phytoplankton assemblages persisted for many days. Two shallow African lakes (Elementeita and Nakuru) exhibited neither a regular seasonal cycle nor a near constant condition but one in which an abrupt change from one persistent algal assemblage to another persistent condition occurred. The actual cause of the switch in assemblages was not clear but probably related to an altered water chemistry.

### 11.2.4 The role of the littoral zone

Expansive areas of emergent wetland vegetation frequently surround shallow lakes in both temperate and tropical regions. The wetland–lake interface forms biologically-rich communities and the littoral zone may contribute the majority of productivity in these ecosystems (Wetzel 1999). Characteristic abundant plants of tropical wetlands include *Typha domingensis* Pers. in Lake Chilwa (Malawi); *Cyperus papyrus* L. in Lake Naivasha (Kenya) and *Phragmites karka* (Retz.) Trin. in Waigani Lake (Papua New Guinea). The study of shallow lakes, therefore, often centres on the relationship between the open water areas and the surrounding littoral vegetation (McLachlan 1975). The littoral vegetation constitutes a major source of organic matter which contributes significantly to the productivity as well as providing a diverse habitat for wetland animals. Inflowing waters are effectively filtered by the surrounding wetlands which act as a trap for allochthonous materials. An obvious difference between deep and shallow lakes is the absence of a profundal zone in the latter.

Howard-Williams & Gaudet (1985) indicated that papyrus wetlands possess a large capacity for nutrient uptake. This capacity for wetland plants to absorb nutrients has led to the use of constructed wetlands to purify wastewaters in both tropical and temperate regions (Kadlec 1994; Kansiime & Nalubega 1999). We require a lot more information on the nutrient-removal capacity and the mechanisms by which nutrients are stored or regenerated before the use of tropical wetlands for wastewater purification can be recommended. There are, as yet, too few detailed nutrient balance studies for tropical wetlands and little is known about the changes in growth characteristics of wetland plants which occur with higher nutrient loads. Rapid growth of plants under nutrient-rich conditions may, for example, reduce the amount of structural material laid down by the plant and this could make the plant and wetland more susceptible to physical damage. Certainly, significant changes in the vegetation of Waigani wetland in Papua New Guinea seem to be related to excessive nutrient enrichment from sewage effluent (Osborne & Polunin 1986).

In lakes with a seasonally variable water level, the littoral zone becomes colonised by organisms adapted to the alternating cycle of wetting and drying. With the exclusion of floodplain lakes, this feature is common in tropical lakes in monsoonal regions and less so in lakes found nearer the Equator, where periods without rain are short-lived. Water level in lakes on the floodplains of large rivers may be determined more by the rainfall pattern in the upper reaches of the river than by that of the lake environs. However, Melack (1996) found that local runoff contributed 57% of the water budget to Lake Calado and that the adjacent Solimões River provided only 21%. Further water budget studies of tropical lakes are required not only to determine the main sources of water but also to provide the basis for nutrient budget calculations.

### 11.2.5 Eutrophic lakes

A distinction needs to be made between lakes which are naturally eutrophic and those which have become, often through human activity within the catchment, nutrient-enriched and altered through the process of eutrophication. Naturally eutrophic lakes are of intrinsic interest; those which have become eutrophic through human activity attract additional attention because the often deleterious changes in water quality and productivity provoke a desire to reverse eutrophication and restore these lakes to their pre-enrichment condition.

Lakes can be arranged along a continuum of nutrient status from oligotrophic to eutrophic and a lake's position on this continuum can be altered by increasing or lowering the supply (loading) of key nutrients, most commonly phosphorus and nitrogen. Delineation of trophic state into oligotrophic, meostrophic, eutrophic and hypereutrophic categories is arbitrary but features of what constitutes a eutrophic lake can be illustrated by considering Lake George, one of the few well-studied, shallow, eutrophic lakes in the tropics.

Lake George in Uganda straddles the Equator where seasonal variations in day length and temperature are low. The lake is shallow (mean depth: 2.4 m) and covers an area of 250 km². Ganf (1974a) recorded dense phytoplankton populations (chlorophyll $a$ concentrations: 600 mg m$^{-2}$) throughout the year, mainly of cyanobacteria (*Microcystis* spp; *Anabaenopsis* spp). Concentrations of particulate organic carbon, nitrogen and phosphorus were high: 50 mg L$^{-1}$ C, 3 mg L$^{-1}$ N and 0.24 mg L$^{-1}$ P. Dissolved concentrations of inorganic nitrogen and phosphorus were very low. Phytoplankton photosynthesis was limited to the upper 1 m of the water column and rates declined rapidly below 15 cm depth (Ganf 1975). No seasonality was detected in photosynthetic production by phytoplankton. High rates of algal photosynthesis were coupled with high rates of respiration and therefore daily rates of net production were not particularly high.

Lake George maintained phytoplankton populations throughout the year similar to those found in nutrient-enriched, temperate lakes during the summer. However, high algal biomass in Lake George was maintained despite low nutrient income to the lake. This was achieved by rapid recycling of nutrients within the water column and small amounts of nutrients being permanently stored in the fluid mud.

Shallow lakes with their extensive littoral zone often attract large numbers of waterfowl and wading birds. In Africa, these water bodies are also home to hippopotamus (*Hippopotamus amphibius*). These birds and mammals can be responsible for significant transfer of nutrients from the surrounding terrestrial system to the lakes they roost on at night or reside in during the day. Approximately 5000 hippos lived around Lake George, Uganda, and Viner & Smith (1973) estimated that the hippo population brought into the lake approximately 2% of the nitrogen and 10% of the phosphorus measured leaving the lake via the outflow. Therefore significant quantities of nutrients, particularly phosphorus, were delivered to the lake by the hippos. Nutrients were removed from the lake by birds, which fed on fish and roosted, or at least spent much of the time, on land. These losses, unlike the gains from hippo activities, were regarded as insignificant in the budgets of these elements for the lake.

## 11.2.6 Eutrophication of shallow, tropical lakes

Rapidly expanding human populations, industrialisation and urbanisation are all leading to eutrophication becoming both more common and more significant as a water quality issue in tropical countries. Eutrophication of shallow lakes in Europe and North America has been well documented but much less information is available from shallow lakes in the tropics (see Crisman & Streever 1996). Reynolds (1992, p. 4), defined eutrophication as 'enrichment of biological systems by nutrient elements, notably nitrogen and phosphorus, and the enhanced production of algal and higher plant biomass that the added loads stimulate.' This definition, although suitable for deep, temperate lakes, needs to be broadened before it adequately describes the effects of nutrient enrichment of shallow, tropical lakes. Eutrophication in shallow lakes not only results in enhanced algal production but also is frequently accompanied by first an increase in aquatic plant biomass followed by the complete loss of submerged and floating-leaved plants. In tropical lakes, a once diverse flora may be replaced by phytoplankton and/or dense mats of surface floating weeds such as *Eichhornia crassipes* (C. Martius) Solms., *Salvinia molesta* Mitchell and *Pistia stratiotes* L., which may eventually cover the entire water body. Wetzel (2001), more broadly, associates eutrophication with 'increased productivity, simplification of biotic communities, and a reduction in the ability of the metabolism of the organisms to adapt to the imposed loading of nutrients.'

In this paper, eutrophication is **not** assumed to be a natural ageing process of lakes (see Rast & Thornton 1996). Although nutrient enrichment may enhance the rate of basin shallowing through increased autochthonous production, the timescales within which eutrophication (nutrient enrichment of lakes) and natural ageing (basin shallowing or lake succession) operate are usually very different. Basin shallowing and the successional processes which accompany it occur, inexorably, in almost all standing waters. Many of the symptoms of eutrophication can be removed by reducing nutrient inflows and although this may slow the process of lake succession it will not reverse its direction. Therefore these processes (eutrophication and lake succession) should be distinguished.

Limnologists studying temperate deep lakes have focused on the deterioration in recreational and drinking water quality resulting from eutrophication (Edmondson 1972, 1991). Those studying temperate shallow lakes have concerned themselves more with the replacement of diverse aquatic plant communities by phytoplankton (see Phillips, this Volume, Chapter 10). Although nutrient enrichment of tropical lakes frequently produces similar responses, the increase in fish production which accompanies this enrichment is both welcomed and often induced and the decline in water quality either ignored or tolerated.

Yusoff & Patimah (1994) compared two lakes in Malaysia: nutrient-rich, shallow (mean depth 0.3 m) Lake Aman and nutrient-poor and deeper (mean depth 3.4 m) Lake Titiwangsa. The phytoplankton of Lake Aman was composed mainly of euglenoids, and chlorophyll *a* concentrations ranged from 60 to 969 $\mu g L^{-1}$. In mesotrophic Lake Titiwangsa, phytoplankton communities were more diverse but rich in cyanobacteria and chlorophytes, with some dinoflagellates, chrysophytes and diatoms and few euglenophytes. Chlorophyll *a* concentrations were much lower in this lake and ranged between 0.5 and 6.3 $\mu g L^{-1}$. Yusoff & Patimah (1994) suggested that differences between these lakes were due to the higher nutrient loading (although this was not measured) received by Lake Aman and exacerbated by its shallowness. Although nutrient loading and depth may well be the major causes of difference in trophic status between these two lakes, the authors did not consider how differences in the fish faunas may play a role in determining phytoplankton species composition. Lake Aman contained only two species of fish: *Oreochromis mossambicus* (Peters) and *Channa striatus* Bloch. In contrast, in Lake Titiwangsa there were 11 species.

Loss of aquatic plant diversity and the invasion by water hyacinth (*Eichhornia crassipes*) are often the most noticeable changes in enriched, shallow, tropical lakes and the decline in water quality is often perceived as of less significance. Shallow lakes are less commonly used as a source of drinking water and are often more biologically productive even prior to enrichment and therefore it is not surprising that the perceived significance of eutrophication is different. Furthermore, shallow, tropical lakes, and particularly the more productive ones, provide varied resources to the humans who live around them: fish, shellfish, prawns and plant products (for thatching, weaving and papermaking). The organic-rich sediments of Rawa Pening Lake in central Java, Indonesia are mixed with ground mollusc shells in order to produce a compost for mushroom production (Göltenboth & Kristyanto 1987, 1994). Water hyacinth is harvested from this lake in order to provide a soil additive. Removal of this weed also maintains areas of open water in order to facilitate fishing and sediment removal. A thriving economy has thus been established around this eutrophic lake and Crisman & Streever (1996) stressed the importance of finding ways to turn management problems into locally-based commercial ventures. This has certainly been achieved on Rawa Pening Lake.

Shallow lakes generally possess shorter hydraulic residence times than deep lakes and this is particularly true of those lakes on the floodplains of large rivers. These lakes tend to be more influenced by the inflow of nutrients from the catchment area than by the regeneration of nutrients from the bottom sediments. The short water column precludes the development of an oxycline and the reduced conditions which generally promote nutrient release from sediments. Water column mixing in shallow, tropical lakes is usual but short-lived periods of stratification may occur. Rapid rates of decomposition, and hence oxygen consumption, may allow anoxis to develop during these short periods of stratification. For these reasons, Kilham & Kilham (1990) questioned the applicability of the 'old law of limnology' that shallow lakes are more productive than deep ones. Their assessment was based on the observation that *Stephanodiscus astraea* (Ehrenb.) characterised the fossil diatom assemblages in more than a dozen lakes when lake levels were maximal (c.9500 yr BP). *Stephanodiscus astraea* abundance also indicated low Si:P ratios and therefore eutrophic conditions. Kilham & Kilham (1990) concluded that tropical lakes appear to have experienced increasing epilimnetic phosphorus loading as lake levels increased. This contrasts with large, deep lakes in the temperate zone which are usually oligotrophic with high Si:P ratios.

Kilham & Kilham (1990) suggested that biological controls of element cycles characterise tropical lakes, whereas physical (light and temperature) controls are more important in temperate lakes. Shallow lakes, however, tend to be more influenced by catchment processes because of their shorter hydraulic residence times. Nutrient loss from the lake through the outflow is therefore likely to be greater. Many shallow, tropical lakes are surrounded by extensive wetlands, which further reduce nutrient inputs through uptake and storage in the plants or the peat that accumulates beneath the wetland. However, shallow lakes which receive enhanced nutrient loads are thought to be more susceptible to eutrophication than deep lakes (Jørgensen & Vollenweider 1989). In order to illustrate the conditions and changes which occur in shallow, tropical lakes undergoing eutrophication, three case studies are presented below.

## 11.3 CASE STUDIES

### 11.3.1 Waigani Lake, Papua New Guinea

Waigani Lake lies in the upper reaches of a small catchment which drains a large part of the urban area of Port Moresby, the capital city of Papua New Guinea. Water from the lake flows into an extensive wetland fed by the Laloki and Brown Rivers. The wetland is a mosaic of open water areas with floating-leaved and submerged plants, *Typha orientalis* Pers. and *Phragmites karka* marshes and *Melaleuca* swamp savanna. The vegetation within the Waigani basin has undergone extensive modification over the past 50 years (Osborne & Leach 1983; Osborne & Polunin 1986; Osborne & Totome 1994). During the 1940s the wetland was described as 'one of shallow lakes, choked with reeds, cat-tails, water lilies and other hydrophytic grasses' (Neill 1946, p. 17). An aerial photograph taken during the Second World War indeed shows emergent vegetation filling the area which subsequently became open water.

Aerial photographs taken in the 1960s and early 1970s show, at the start of the period, an almost solid cover of the lake with floating-leaved plants (*Nymphaea* spp and *Nymphoides indica* (L.) Kuntze). A gradual decline in the coverage of these plants is recorded by photographs taken in the 1970s and by 1978 the main lake basin was devoid of these plants and only a few patches remained in small lakes adjacent to it (Osborne & Leach 1983). The aquatic plants were replaced by dense phytoplankton populations (up to 400 mg chlorophyll $a$ m$^{-3}$) which persisted throughout the year and Secchi-disc depths varied between 0.11 and 0.34 m (Osborne 1991).

In the 1980s, the South American fern *Salvinia molesta* became established on some of the lakes within the wetland but, for some reason, never formed dense stands on Waigani Lake. A decline in the emergent, littoral vegetation was observed and at the end of the decade *Eichhornia crassipes* (water hyacinth) invaded the lake. By 1991, dense floating mats of this weed covered over half the lake. An aerial photograph taken in June 1997 shows a reduced expanse of water hyacinth on the main lake but over 70% of nearby Gerehu Lake is covered with this weed.

Osborne & Polunin (1986) were able to trace some of these changes in the aquatic vegetation through the analysis of sediment cores. The number of seeds, fruits and other plant remains in sediment layers paralleled the vegetation changes recorded in the aerial photography. The diatom species composition in the sediment cores switched from a diverse flora of mainly epiphytic species to one characterised by the planktonic *Cyclotella meneghiniana* Kütz. A marked increase in phosphorus deposition within the sediments was recorded in the sediment layers just below those in which this switch in the diatom flora occurred and an increase in photosynthetic degradation products was contemporaneous with the increase in the deposition of *Cyclotella* frustules. Significant limnological information can be obtained from the analysis of sediment cores and further work on sediment records in shallow, tropical lakes is encouraged.

Over 80% of the sewage effluent from the city of Port Moresby (population in 1980: 120,000) enters the Waigani wetland with disposal beginning in 1965. The decline in the populations of floating-leaved and submerged plants from Waigani Lake occurred soon after the advent of sewage disposal into the system. Osborne & Polunin (1986) concluded that nutrient enrichment stimulated phytoplankton growth and the resultant reduction in light penetration was the most likely cause of the decline in the distribution of aquatic plants.

Osborne *et al.* (1996) showed that whereas soluble reactive phosphorus concentrations in the surface waters of Waigani Lake were highest near the inflow of sewage effluent, the phosphorus concentrations in the sediments were higher near the centre of the lake. The distribution of iron and manganese in the sediments was also elevated near the centre of the lake. The sediments and layer of water immediately overlying them have been shown to be anoxic near the sewage outfall (R.G. Totome, personal communication, University of Papua New Guinea 1990). The high water temperatures (26–30°C) prevailing in this lake coupled with the high organic loadings on the sediments from the sewage and phytoplankton sedimentation are probable causes of this anoxia and significant phosphorus release from the sediments in the vicinity of the sewage outfall would be expected. The patterns of iron and manganese deposition in the surface sediments of Waigani Lake support this proposed redox-mediated mechanism for the redistribution of phosphorus. Towards the centre of the lake, phosphorus accumulation probably occurs through deposition of phosphorus-enriched phytoplankton (Osborne *et al.* 1996). The prevailing southeasterly trade winds blow down the length of the lake and therefore wind-induced turbulence is likely to be higher towards the centre of the lake, ensuring a more even distribution of oxygen throughout the water column in this region.

The open water areas in the Waigani wetland

support a productive fishery with two introduced species (*Oreochromis mossambicus* (Mozambique mouthbrooder) and *Cyprinus carpio* L (common carp)) making up most of the catch. Fishing effort (gill nets and hand-lines) was severely hampered by the mobile mats of water hyacinth. However, in 1998 three insect species were introduced as biological control agents of *Eichhornia crassipes* and these have reduced water hyacinth cover on Waigani Lake from around 70% of the lake's surface to about 20%. The remaining plants are smaller, less productive and produce fewer flowers, seeds and daughter plants than before control. The lake is again an important source of fish for the people in Port Moresby (Dr Mic Julien, CSIRO Entomology, Indooroopilly, Australia, personal communication 2001). The biological control of *Salvinia molesta* on Waigani Lake and elsewhere within Papua New Guinea has been spectacularly successful (Thomas & Room 1986).

Currently, there are no plans to reverse the effects of eutrophication on the Waigani system. Restoration would probably require reduction in both external and internal nutrient loads and re-establishment of submerged, floating-leaved and emergent plants. Reduction in the sewage effluent load on Waigani wetland is not being considered and, indeed, a proposal to expand the settling ponds is currently under investigation (R.G. Totome, University of Papua New Guinea, personal communication 1999).

Loss of aquatic plants has been recorded in other shallow, tropical lakes. A number of shallow lakes within the delta of the River Niger are eutrophic and the aquatic plant flora is depauperate. Dumont (1992) suggested that loss of aquatic plants may have occurred recently and that this may be the result of nutrient enrichment from the numerous cattle which graze adjacent to these lakes. This eutrophication stimulated the growth of cyanobacterial phytoplankton which may have limited the growth of submerged and floating-leaved plants through shading.

Marked changes in the aquatic flora of Lake Naivasha, Kenya have also been noted (Harper et al. 1990). Although eutrophication may have played a role in the demise of the water lilies and submerged plants in this lake, other factors, such as water level fluctuations and the introduction of coypu (*Myocaster coypus* (Molina)) and the Louisiana Red Crayfish (*Procambarus clarkii* (Girard)), may also have been important.

### 11.3.2 Laguna de Bay, Philippines

Laguna de Bay is a large (900 km$^2$) shallow lake on the outskirts of Manila in the Philippines. The lake has a mean water volume of $3.2 \times 10^9$ m$^3$ and a shoreline 220 km long. Over three million people live around the lake and use it for fish production, domestic water supply, irrigation, transport, power generation (cooling water), flood control and as a depository for domestic, industrial and agricultural wastes. The lake is drained by the Pasig River which is highly polluted with sewage and industrial effluents from Metro Manila, through which it flows to the sea. Prior to construction of a barrage, at the end of each dry season, when the level of Laguna de Bay fell below the high tide level, seawater flowed up the Pasig River. With this flow reversal, the polluted waters of the Pasig River entered Laguna de Bay. Twenty-one other rivers (excluding small streams) flow into the lake.

Human population growth within the Laguna de Bay catchment has been estimated at 5.2% (Pacardo 1994) and this rapid growth has placed significant environmental pressure on the lake and its catchment. The multipurpose use of the lake has increased food (mainly rice through irrigation) and fish production, generated more power and created employment within the basin. However, these gains have come at the expense of a marked decline in the water quality of the lake. The lake has been highly eutrophic for at least 50 years and algal blooms have occurred regularly during summer.

More recently, changes in the biology of the lake have become apparent. Rooted aquatic plants are now almost entirely absent from the lake, yet they were an important part of the ecosystem in the 1960s. Shading by phytoplankton is the probable cause of this plant loss but grazing by herbivorous fish and trawling for snails may also have played roles. During the 1960s, at least 23 species of fish were present in the lake even though several (*Cyprinus carpio, Oreochromis mossambicus*) were introduced. At that time, prior to the development of intensive aquaculture, the local fishery, with an annual catch of about 80 t, was largely dependent on just three species: *Therapon plumbeus* Kner (silver perch), *Glossogobius giurus* Hamilton (white goby) and *Arius manillensis* Valenciennes (catfish). Owing to overfishing, the open water fishery collapsed, with the catch in 1988 down to 15 t, leading to the economic demise of 13,000 fishermen and their families. In 1991, it was estimated that only 10 species of fish remained. Pacardo (1994) attributed this decline in the fish catch and species diversity to overfishing, pollution and the environmental effects of the barrage constructed on the Pasig River to prevent saline intrusions.

There are indications that turbidity in Laguna de Bay has increased over the past four to five decades. This increase can be attributed to greater phytoplankton biomass, loss of rooted plants, leading to less sediment stabilisation and greater sediment resuspension, and changes in the structure of the food web, leading to the promotion of smaller phytoplankton (Sly 1993).

Aquaculture started in the early 1970s and grew at an explosive rate and within a decade fish-pens and net-cages covered approximately 10% of the lake surface (Kock *et al.* 2000). The main cultured species were milkfish (*Chanos chanos* Forsskål), tilapia (*Oreochromis mossambicus*) and silver perch (*Therapon plumbeus*). However, for many reasons, both environmental and economic, fish-pen culture has not proven to be sustainable at the very high stocking rates initially used. Conflicts between fishermen and fish-pen operators have arisen and, to a certain degree, these have been resolved through enforcement of resource allocation programmes. However, since there is inadequate scientific information on the sustainable fish yield, the allocation of areas to aquaculture and capture fisheries has been somewhat arbitrary.

The hydrological regime of the lake has been altered through the construction of the Mangahan Floodway, which diverts floodwater from the Marikina River into Laguna de Bay. Furthermore, the Napindan Hydraulic Control Structure (NHCS) can prevent the back-flow into the lake of polluted and saline water in the Pasig River as well as control the water level within the lake. The NHCS was installed in 1982 so that the lake could become, by 2000, the primary source of drinking water for Metro Manila and also provide irrigation water for an additional 30,000 ha of farmland. The effects of this altered hydrological regime on the ecology of Laguna de Bay have not all been regarded as beneficial (Santos & Rabanal 1988).

Prior to the construction of these hydrological controls, the lake underwent an annual alternation between clear and turbid water phases (Santiago 1993). Turbidity tended to increase from October to March and decreased as the strong winds associated with the north-east monsoon declined. Turbidity was usually lowest following the onset of the wet season. These trends were interrupted by typhoons and periods of rapid algal growth, although turbidity was probably due more to suspended clay sediments and organic detritus than to phytoplankton (Sly 1993). During the early part of the wet season, when lake levels were still low, brackish waters entered the lake from flow reversal of the Pasig River. This back-flow usually occurred in June and July, after the reduction in water turbidity had begun. Saltwater intrusion may, through flocculation of clays, have facilitated the reduction in water turbidity. Phytoplankton production peaked during the clearwater phase and this has been attributed to increased light penetration coupled with greater nutrient availability.

Barril *et al.* (2001) showed that water quality continues to decline with concentrations of ammonia-nitrogen, nitrate-nitrogen, orthophosphate and chlorides, together with chemical oxygen demand, turbidity, conductivity and hardness, all exhibiting increasing trends and exceeding allowable water quality limits. They also showed a decline in dissolved oxygen. This deterioration in water quality occurred despite attempts to reduce inflows of organic pollutants. The Laguna Lake Development Authority implemented an Environmental User Fee System (EUFS) within the Laguna de Bay catchment in 1997. This scheme provides economic incentives to encourage lake users to reduce water pollution (polluter-pays principle) and also provides funds for developing remedial measures. The initial phase of the programme monitored the biochemical oxygen demand (BOD) of effluents and charged producers a disposal fee.

Santos & Rabanal (1988) argued that the fishery in Laguna de Bay declined as a result of the closure of the NHCS. They believed that the input of brackish water from the Pasig River was required to improve water clarity and stimulate phytoplankton production. These authors concluded that management authorities were faced with the stark choice between either providing potable water for Metro Manila or ensuring the continuation of a productive fishery. Following protests from interest groups (mainly people involved in the capture and culture fisheries), the NHCS has been kept open most of the time since 1986 (Santos-Borja 1993). Santos & Rabanal (1988) indicated that a return to higher fish production followed. Clearly, operation of the NHCS must optimise benefits to both agricultural and domestic water users as well as those involved in fish production. A more thorough understanding of the ecological significance of saltwater intrusions is required, and lake and catchment management practices which lead to sustainable resource use need to be developed.

The major environmental problems confronting the managers of Laguna de Bay are rapid siltation as a result of erosion from the deforested watershed, nutrient enrichment, industrial pollution and social conflicts between users of the lake and its catchment. These changes have culminated in poor water quality, lake shallowing, decreased productivity of the lake, loss of aquatic plants, declining harvests of fish and shrimps and more frequent fish kills from anoxia (Sly 1993). Management decisions need to be based on a holistic assessment of the lake and its catchment. Given the intense and diverse use of this lake and the rapid growth of the human population in the catchment, development of environmental management strategies which will allow sustainable use is an almost insurmountable prospect. The future of Laguna de Bay and its catchment is bleak indeed.

### 11.3.3 Okeechobee Lake, Florida, USA

Lake Okeechobee is a large ($1732\,km^2$), shallow (mean depth 2.7 m), nutrient-rich, subtropical lake in Florida, USA. The lake provides water, flood protection and recreation to a population of some 3.5 million people and, through its outflow to the Everglades, the lake attains both high conservation and economic value. The lake is surrounded by a levee and its hydrology is highly modified, with almost all of its surface water inflows and outflows controlled because it forms part of the South Florida Water Management District. The lake is eutrophic as a result of nitrogen and phosphorus inputs from the intensively farmed land which surrounds it. The catchment area ($22,500\,km^2$) is characterised by cattle ranching and dairy farming to the north and vegetable and sugarcane farming to the south of the lake (Flaig & Havens 1995).

The lake contains an extensive littoral zone with emergent and submerged vegetation which provides an important wildlife habitat. The lake is utilised by migratory waterfowl and supports a commercial and recreational fishery valued at over US$100 million (Havens *et al.* 1996a). Prior to human habitation of south Florida, the Okeechobee wetland was more extensive and contiguous with the Florida Everglades. Today, the lake is contained within a system of dykes and water-control structures, the Kissimmee River inflow has been channelised and the wetland is no longer integrated with the Everglades.

Agricultural development in the catchment of Lake Okeechobee has led to major impact on the water quality of the lake. The decline in water quality and the algal blooms which developed in the 1980s have been attributed to the high phosphorus loading on the lake ($0.3–0.4\,g\,P\,m^{-2}\,yr^{-1}$). Owing to its eutrophic condition and to its economic and conservation importance, Lake Okeechobee has been the subject of considerable ecological study (see Aumen 1995; Aumen & Gray 1995; Flaig & Havens 1995; Havens *et al.* 1995a, b, 1996b, c; James *et al.* 1995a, b).

Reductions in the external loads of both nitrogen and phosphorus have been achieved through the introduction of pollution control practices within the lake's catchment. These management practices include animal wastewater disposal on crop lands, fence construction to keep cattle away from tributaries, and use of constructed wetlands for nutrient removal. Areal export rates for phosphorus and nitrogen fell from 0.3–3.2 kg P ha$^{-1}$ yr$^{-1}$ and 2.7–26.0 kg N ha$^{-1}$ yr$^{-1}$ to 0.3–1.6 kg P ha$^{-1}$ yr$^{-1}$ and 2.1–3.0 kg N ha$^{-1}$ yr$^{-1}$ respectively (Flaig & Havens 1995). As a result, lake water total nitrogen concentrations have returned to early 1970s values and, although phosphorus concentrations have not declined, they appear to have stabilised.

Lake Okeechobee is polymictic, its sediments are flocculent and wave depths regularly exceed water column depth and water quality is significantly affected by wind, especially during the cooler months of the year. Three ecologically distinct regions of the lake have been recognised: (i) a large central region characterised by muddy bottom sediments and turbid water; (ii) a northern region with muddy sediments and nutrient-rich inflows; and (iii) a western region with firm sediments and frequent algal blooms. Phlips et al. (1995b) noted significant spatial differences in the relationships between phytoplankton biomass measured as chlorophyll a and light availability within Lake Okeechobee. In the northern and central regions of the lake, concentrations of non-algal suspended solids were high owing to the muddy sediments and polymictic conditions.

There has also been a massive build-up in the phosphorus content of the sediments in Lake Okeechobee over the past 150 years. The lake sediments are now estimated to contain over 28,000 t of phosphorus and the annual internal loads are equal to those from the catchment. The wind-driven resuspension of these phosphorus-rich sediments is probably the main mechanism driving this internal phosphorus loading. The vertical migration of buoyant, phosphorus-rich cyanobacteria from the sediment surface into the water column may also be an internal loading mechanism (James et al. 1995a).

Total nitrogen (TN) to total phosphorus (TP) ratios (by weight) were at or below 15 : 1 in 1996 and nitrogen limitation was the norm. These conditions favour both nitrogen fixation and abundance of cyanobacteria (Aldridge et al. 1995; Phlips & Ihnat 1995). In the late 1970s, TN : TP ratios exceeded 25 : 1, phosphorus limited phytoplankton production and nitrogen fixation rates were lower. The switch from phosphorus limitation to nitrogen limitation has probably come about through excessive phosphorus loads from external sources, coupled with enhanced internal loading and reduced nitrogen inputs from the catchment of nitrogen-rich agricultural drainage waters. Internal loading may have increased following a rise in mean water level, leading to more stable thermal stratification in summer and anoxic sediments and phosphorus release from organic- and phosphorus-enriched sediments.

James et al. (1995a) concluded that the reduction in the external phosphorus load has been accompanied by a change in the way phosphorus cycles within the lake. Increased internal phosphorus loading and/or decreased net phosphorus sedimentation from the water column may have compensated for the reduced external loads. The change in phosphorus sedimentation may be linked to the decline in calcium, a major component of precipitation of phosphorus to the sediments (James et al. 1995b). A problem encountered in developing predictive models for Lake Okeechobee has been habitat heterogeneity and wind-induced resuspension. The best predictive model for annual total phosphorus concentrations only explained 51% of the variation (James & Bierman 1994) and Phlips et al. (1995a) recommended that a spatially compartmentalised approach be taken in order to enhance the predictive capacity of future models.

Nutrient enrichment has also resulted in changes in the lake's biota. Cyanobacteria have replaced diatoms as the main phytoplankton group and the oligochaete *Limnodrilus hoffmeisteri* Claparède characterises the benthos. This worm thrives in microaerophilous waters. There have been concerns that large algal blooms might cause fish kills but, despite the changes in the plankton and benthos, there is no evidence of a declining fishery. Havens et al. (1996a) noted that a more gradual reduction in nitrogen loads from agricultural sources may have lessened the N:P ratio decline and may have avoided the nutrient conditions which favoured the development of the massive cyanobacterial blooms that occurred in the mid-1980s.

Initial efforts to restore Lake Okeechobee focused on reducing external nutrient loads. The Florida Department of Environmental Protection proposed setting total maximum daily loads (TMDL) for phosphorus which will limit in-lake total phosphorus concentrations to below $40 \, m L^{-1}$. However, as with shallow lakes elsewhere in the world (see Phillips, this Volume, Chapter 10), the response to nutrient load reduction was disappointing. The poor response was attributed to internal loading from phosphorus-rich sediments. James et al. (1995a) anticipated a slow response because input–output budgets suggested that the turnover time for phosphorus is greater than 3 years. Furthermore, internal phosphorus loading resulting from sediment resuspension is predicted to take more than 5 years to decline. Havens et al. (1996a) suggested that Lake Okeechobee may have entered a quasi-steady state following reduction in the external phosphorus load. They indicated that this state, with high fish yields (desirable) and frequent algal blooms (undesirable) may persist for decades until sedimentary reserves of phosphorus decline.

Efforts to restore Lake Okeechobee have centred on nutrient inflow reduction. The intense scientific study of Lake Okeechobee has followed a chemical rather than a biological bias. Work on shallow, eutrophic lakes in Europe has demonstrated the importance of shifts in community composition as not only an indicator of eutrophication but also, more importantly, an avenue for lake restoration techniques (see Scheffer et al. 1993; Phillips, this Volume, Chapter 10). However, Crisman et al. (1995) suggested that although fish predation may exert some control over zooplankton biomass, zooplankton are unlikely, even if freed from predation pressure, to reduce phytoplankton biomass significantly. Havens et al. (1996a) concluded that biomanipulation of fish stocks is unlikely to improve water quality in Lake Okeechobee because wind, rather than benthivorous fish, is the main mechanism of sediment resuspension. Aumen & Gray (1995) recommended the following research and management strategies for Lake Okeechobee: (i) consider ecological zones as the foundation for future research; (ii) maintain controls on phosphorus inflows to the lake; (iii) focus monitoring programmes on the western and southern regions of the lake as these are expected to exhibit early responses to management; (iv) lower water levels slightly during the wading bird nesting season; and (v) implement a water level management regime with the flexibility to allow more natural hydrological variability.

Havens & Schelske (2001) stressed the importance of considering biological processes when setting total maximum daily loads (TDML). They contended that the ability of shallow lakes to assimilate phosphorus is substantially reduced when surplus concentrations of phosphorus occur in the water column. Under these conditions, with nitrogen replacing phosphorus as the limiting nutrient, the phytoplankton becomes rich in nitrogen-fixing cyanobacteria, the benthos in oligochaetes and submerged plant growth restricted by poor light penetration. Oligochaetes, through bioturbation, promote internal nutrient loading which helps maintain dense phytoplankton populations. The dense phytoplankton populations preclude the re-establishment of submerged plants through light limitation. In a process of positive feedback, phytoplankton populations pervade, and the water column remains turbid. Havens & Schelske (2001) suggested that restoration of this lake would be facilitated by reduced phosphorus loading coupled with a series of changes in biological structure. These included a switch from cyanobacteria to diatoms, increased biomass of submerged plants, in the benthos from oligochaetes to vertically migrating insects with emerging adults, and a reduction in the population fish detritivores. With these biological changes, the lake might be able to support a higher TDML. Havens (2002) and Havens & Walker (2002) have described models to develop goals, respectively, for hydrological restoration and an in-lake total phosphorus concentration. These models are being used in the implementation of the Comprehensive

Everglades Restoration Programme (CERP). The CERP is a massive ecosystem engineering programme which aims to restore and protect the water resources of southern Florida, including Lake Okeechobee and the Everglades. The detailed studies undertaken on subtropical Lake Okeechobee should be used to guide research on the restoration of shallow, tropical lakes impaired by eutrophication.

## 11.4 CONCLUSIONS

Intensive, medium- to long-term (5–30 years) studies of shallow tropical lakes have not been carried out. Lakes George (Ganf 1974a, b; Ganf & Viner 1973) and Chilwa (Kalk et al. 1979) were studied intensively but only for around 5 years; no studies (to my knowledge) have followed a shallow (less than 3–4 m deep), tropical lake through the process of eutrophication to the development and application of a restoration plan. Although inadequate research personnel and resources may well explain this lack, the perception that eutrophication produces some benefits should not be discounted. In the tropical setting, eutrophication is not all bad and the balance between negative and positive aspects depends on the main use of the lake. Many tropical lakes are put to multipurpose use and some of these uses may benefit from nutrient enrichment (e.g., aquaculture) whereas for others the resource is impaired (e.g. provision of potable water). Achieving a balance is the challenge faced by aquatic resource managers.

Shallow lakes, because of their high surface-area to volume ratio, do not constitute a good reservoir for potable water. However, in the tropics many shallow lakes may be utilised as the primary source of water and therefore water quality is an issue. Nutrient enrichment not only enhances lake productivity but can also result in dangerous concentrations of dissolved substances. For example, high concentrations of nitrates can cause methaemoglobinaemia in infants and be fatal. Poor water purification or, as is often the case, no purification, can result in the spread of waterborne diseases and this is more likely to be a problem in nutrient-enriched water bodies. Water purification is more costly if the water contains high concentrations of suspended solids. Where this is the case a number of techniques centred on nutrient load control have been used to enhance water quality and reduce the magnitude of algal populations. Most shallow lakes in the tropics provide a storehouse of biological resources and productivity may be enhanced with moderate nutrient enrichment. Therefore, the most effective lake management strategy for multipurpose use often requires compromises in setting nutrient loading rates.

Eutrophication control strategies in developed countries have focused on technology-based measures such as nutrient removal from sewage effluents and catchment management techniques in order to reduce nutrient inflows from diffuse sources. Attempts to improve water quality in Lake Okeechobee provide a good example. Third world countries are generally unable to support the intense research programmes required to produce the information which underpins sound management strategies. Furthermore, even if such information were readily available, developing nations are rarely able to afford the nutrient control techniques, and even the infrastructure to direct effluents to treatment plants is often inadequate or non-existent.

Restoration of shallow, eutrophic lakes in temperate parts of the world simply through a reduction in nutrient loadings has not proved to be effective (see Phillips, this Volume, Chapter 10). This lack of success has been explained in terms of continued significant internal loading from nutrient-enriched sediments despite a reduction in external loads (van der Molen & Boers 1994). Internal loading is highest during the summer in many of these temperate lakes, the season when water residence times are usually at or near their maximum. In shallow, tropical lakes, restoration through external load reductions may be hastened if peak internal loadings coincide with the wet season and shorter water residence times.

Recovery of shallow, temperate lakes has also been hampered through the major change in the composition of trophic levels which often accompanies eutrophication of shallow lakes. Restoration cannot therefore always be effected through nutrient controls but may require assistance in the re-establishment of the pre-eutrophication fauna and flora, which may have disappeared entirely from the lake. Based on extensive studies of shallow, temperate lakes, Moss (1995) suggested that, over a wide range of nutrient loadings, shallow lakes tend to possess two main alternative states: clear water with aquatic plants and turbid water with phytoplankton. Different zooplankton and fish communities are associated with these two states and the details of each state vary from lake to lake. That these two states may exist over a wide range of nutrient loadings suggests that they are buffered by various mechanisms and that change from one state to the other is effected by a switch mechanism. A greater appreciation of the complexities involved in eutrophication of shallow lakes and the strategies required to reverse this process has led to the recent upsurge in interest in biomanipulation (see Kufel et al. 1997; see also Phillips, this Volume, Chapter 10).

Extrapolation of lake restoration techniques developed for temperate lakes to tropical ones needs to be undertaken with caution. Biomanipulation techniques require a sound understanding of food webs and a strong capacity to predict the impact of any top-down controls imposed. Regrettably, all too often this has not been carried out and the impact of the Nile perch (*Lates niloticus* L.) introduction to Lake Victoria provides a stark illustration of the consequences of food web manipulations, which, for whatever reason, can be serious and adverse (Kaufman 1992). However, early indications suggest that the introduction of silver carp (*Hypophthalmichthys molitrix* Valenciennes) may provide a technique to manage phytoplankton populations in shallow, tropical lakes (Starling 1993). Crisman & Beaver (1988) have shown that tilapiine species sequester water column nutrients in their slow-to-decompose mucilaginous faeces and effectively remove nutrients from the water column.

Ahlgren et al. (2000) explained differences in phytoplankton biomass and production in Lakes Cocibolca and Xolotlán in terms of differences in pelagic food web structure. Lake Cocibolca is the largest lake in Central America (7585 km$^2$) with a mean depth of 11 m. Despite a maximum depth of 43 m, the lake is mixed by strong winds for most of the year. Lake Xolotlán is also large (1000 km$^2$) and shallow enough to be fully mixed throughout the year. Both lakes are described as eutrophic (Erikson et al. 1998; Ahlgren et al. 2000). Erikson et al. (1998) concluded that phytoplankton populations in Lake Xolotlán were neither nutrient-limited nor significantly controlled by zooplankton grazing. They suggested that mixing depth and light limitation controlled algal production and biomass. Phytoplankton production in Lake Cocibolca was measured as 3050 mg C m$^{-2}$ day$^{-1}$ but a significantly higher rate was recorded for Lake Xolotlán: 6800 mg C m$^{-2}$ day$^{-1}$ (Ahlgren et al. 2000). Food chain dynamics may explain this difference. Greater zooplankton abundance in Lake Cocibolca may be due to greater predatory pressures on planktivorous fish. Lake Cocibolca contains many species of predatory fish, including some marine taxa such as the bullshark, tarpon and sawfish. Both lakes contain pelagic zooplanktivorous fishes (*Atherinella sardina* Meek (silverside), *Astyanax fasciatus* Cuiver (Mexican tetra) and *Dorosoma chavesi* Meek (Nicaraguan gizzard shad) but there is evidence that the populations of these fishes in Lake Xolotlán are large and not effectively controlled by piscivores. Ahlgren et al. (2000) suggested that energy transfer to higher trophic levels may be more efficient in Lake Cocibolca and that much of the energy fixed by photosynthesis in Lake Xolotlán is lost through phytoplankton and bacterioplankton respiration (Erikson 1998). However, Ahlgren et al. (2000) cautioned that food chain dynamics in these lakes are still poorly known and that increasing nutrient loadings on Lake Cocibolca is a threat to the use of the lake for fishing, water supply and tourism. They make no suggestion of food web manipulation as a means to restore either of these eutrophic lakes.

Another problem with biomanipulation technology transfer from temperate to tropical lakes lies in the comparative paucity of large *Daphnia* species in tropical lakes (Lewis 1979; Fernando 1980; Dumont 1994). Rotifera and herbivorous fishes are abundant in tropical lakes, whereas

Crustacea and non-herbivorous fishes are far more important in temperate lakes (Nilssen 1984; Fernando 1980, 1994). Zooplankters in tropical lakes are generally smaller and this is probably due to year-round predation by fish and a consequence of higher metabolic costs. Therefore, although large cladocerans may control algal biomass in temperate, eutrophic lakes, these organisms are less significant in tropical lakes. Phytoplankton-consuming fish are frequent in the tropics and therefore compete with zooplankton for food as well as consume them.

The phantom midge, Chaoborus, is of great importance in many tropical lakes. The growth of this midge larva over three orders of magnitude means that it feeds on the entire size spectrum of zooplankton in the lake, and, therefore, it can play a significant role in suppressing herbivore populations. However, dense populations of Chaoborus only seem to occur in lakes which develop an anoxic hypolimnion, as this layer provides Chaoborus with a refuge from fish predation. Chaoborus reaches only modest abundance when subject to predation throughout the water column. Saunders & Lewis (1988) showed that herbivore suppression by Chaoborus in Lake Valencia occurred primarily when the lake was stratified. Such intense herbivore suppression is unlikely to occur in shallow lakes because the development of a prolonged anoxic layer in the water column is less likely.

Pantropical introductions of Eichhornia crassipes (water hyacinth) and Salvinia molesta, both from South America, have infested many tropical lakes (Mitchell 1973; Thomas & Room 1986). The shallow and productive lakes in the Niger delta have been invaded by Eichhornia and this is hindering fishing on these lakes (Dumont 1992). Similarly many of the lowland lakes in Papua New Guinea were first covered with dense Salvinia mats and, following successful biological control, have since become covered with water hyacinth. Both these weeds disrupted the lives of the people living around these lakes, making it difficult to reach markets, health centres and schools, to catch fish and to harvest sago.

Dense growth of these weeds in nutrient-rich lakes makes biomass harvesting a possible mechanism of nutrient removal. Von Sperling (1997) recommended biomass harvesting rather than nutrient income reduction as the better technique for restoration of tropical eutrophic lakes. In Lake Rawa Pening in Central Java, water hyacinth, the main producer of organic matter entering the sediment, is not regarded as a weed but forms the basis of a thriving local economy (Crisman & Streever 1996). Use of these weeds to remove nutrients from sewage effluents needs further investigation and work also needs to be carried out on the use of water hyacinth and other aquatic plants in fibre production. Nutrient removal through biomass harvesting is labour intensive but with abundant labour is an option for eutrophication control. In southeast Asia, animal manure and human wastes are mixed with harvested water hyacinth to produce biogas. One hectare of water hyacinth can produce more than $70,000 m^3$ of biogas (Löffler 1990).

Scasso et al. (2001) described the restoration of subtropical Lake Rodó in Uruguay. Prior to the beginning of restoration efforts in 1996, dense phytoplankton populations were characterised, year-round, by Planktothrix agardhii (Gom.) Anagn. et Kom. Lake restoration included draining the lake and removing nutrient-enriched sediments and, following refilling the lake with groundwater, a nutrient removal system which involved circulating water through two pools from which surface-floating aquatic plants (Eichhornia crassipes and Spirodela intermedia W. Koch) were harvested at the end of the growing season. The phytoplankton community increased in diversity from 27 species to more than 150, a change which favoured grazing by mesozooplankton. Abundant populations of small omnivorous fish maintained a high predation pressure on zooplankton, reducing the number of large-bodied species. This led to increases in phytoplankton populations and a decrease in water quality. Scasso et al. (2001) recommended that a combination of nutrient removal by the mechanical harvest of aquatic plants and replacement of omnivorous fish with strict piscivores could lead to prolonged improvements in water clarity.

Limnology is still a science of temperate lakes with some examples from the tropics (see Lewis 1987). Information on tropical lakes is fragmentary both temporally and spatially. No single tropical lake has been studied intensively for more than a few years and those few where teams have worked for 3–5 years have not been studied comprehensively. Kalff (1991) indicated that the large temperate-lake literature cannot serve as a useful predictive model in the development of tropical limnology. He concluded that tropical limnology would advance most rapidly through regional studies leading to the development of models which can be used within that region and compared with those produced elsewhere.

Melack (1996) noted that recent advances in tropical limnology resulted from utilisation of stable isotopes to unravel food webs, isotopic assays of nutrient uptake and recycling rates coupled with nutrient enrichment experiments, improved analyses of vertical mixing, and remote sensing. Application of these techniques requires a strong cohort of well-trained scientists working in adequately equipped laboratories and field stations. In many developing countries these requirements are not met and significant international reallocation of resources will be required to address these impediments to sound limnological research in the tropics. Melack (1996) concluded that significant advances have been made over the past three decades and predicted a bright future for tropical limnology. However, with rapidly expanding human populations in many tropical countries, the management of water resources will continue to provide many challenges. Our level of understanding of tropical waters in general is still too inadequate to provide the basis for the development of sound conservation and management strategies. This point was made by Lewis in 1987 and almost two decades later it remains valid.

## 11.5 ACKNOWLEDGMENTS

I thank John Melack for his constructive comments on a draft of this paper.

## 11.6 REFERENCES

Ahlgren, I., Erikson, R., Moreno, L., Pacheco, L., Montenegro-Guillén, S. & Vammen, K. (2000) Pelagic food web interactions in Lake Cocibolca, Nicaragua. *Verhandlungen der Internationale Vereinigung für theoretische und angewandte Limnologie*, **27**, 1740–6.

Aldridge, F.J., Phlips, E.J. & Schelske, C.L. (1995) The use of nutrient enrichment bioassays to test for spatial and temporal distribution of limiting factors affecting phytoplankton dynamics in Lake Okeechobee, Florida. *Ergebnisse der Limnologie*, **45**, 177–90.

Aumen, N.G. (1995) The history of human impacts, lake management, and limnological research on Lake Okeechobee, Florida (USA). *Ergebnisse der Limnologie*, **45**, 1–16.

Aumen, N.G. & Gray, S. (1995) Research synthesis and management recommendations from a five-year, ecosystem-level study of Lake Okeechobee, Florida (USA). *Ergebnisse der Limnologie*, **45**, 343–56.

Barril, C.R., Tumlos, E.T. & Moraga, W.C. (2001) Seasonal variations in water quality of Laguna de Bay, Philippines: Trends and implications. *Philippine Agriculturist*, **84**(1), 19–25.

Beadle, L. (1981) *The Inland Waters of Tropical Africa*. Longman, London, 365 pp.

Carmouze, J.P., Durand, J.R. & Lévêque, C. (1983) *Lake Chad. Ecology and Productivity of a Shallow Tropical Ecosystem*. Dr. W. Junk, Den Haag.

Crisman, T.L. & Beaver, J.R. (1988) *Lake Apopka Trophic Structure Manipulation*. St. Johns River Water Management District, Palatka, FL.

Crisman, T.L. & Streever, W.J. (1996) The legacy and future of tropical limnology. In: Schiemer, F. & Boland, K.T. (eds), *Perspectives in Tropical Limnology*. SPB Academic Publishing, Amsterdam, 27–42.

Crisman, T.L., Phlips, E.J. & Beaver, J.R. (1995) Zooplankton seasonality and trophic state relationships in Lake Okeechobee, Florida. *Ergebnisse der Limnologie*, **45**, 213–32.

Cunnington, W.A. (1905) *Report on the Tanganyika Expedition 1904–05*. Cambridge University, Cambridge.

Dudgeon, D. & Lam, P.K.S. (1994) Inland waters of tropical Asia and Australia: Conservation and management. *Mitteilungen der internationale Vereinigung für theoretische und angewandte Limnologie*, **24**, 1–386.

Dumont, H.J. (1992) The regulation of plant and animal species and communities in African shallow lakes and wetlands. *Revue D'Hydrobiologie Tropicale*, **25**(4), 303–46.

Dumont, H.J. (1994) On the diversity of the Cladocera in the tropics. *Hydrobiologia*, **272**, 27–38.

Edmondson, W.T. (1972) Nutrients and phytoplankton in Lake Washington. In: Likens, G.E. (ed.), *Nutrients and Eutrophication: The Limiting-nutrient Controversy. Special Symposium, American Society of Limnology and Oceanography*, **1**, 172–93.

Edmondson, W.T. (1991) *The Uses of Ecology: Lake Washington and Beyond.* University of Washington Press, Seattle, 352 pp.

Erikson, R. (1998) Algal respiration and the regulation of phytoplankton biomass in a polymictic tropical lake (Lake Xolotlán, Nicaragua). *Hydrobiologia*, **382**, 17–25.

Erikson, R., Hooker, E., Mejia, M., Zelaya, A. & Vammen, K. (1998) Optimal conditions for primary production in a polymictic tropical lake (Lake Xolotlán, Nicaragua). *Hydrobiologia*, **382**, 1–16.

Fernando, C.H. (1980) The species and size composition of tropical freshwater zooplankton with special reference to the oriental region (South East Asia). *Internationale Revue der Gesamten Hydrobiologie*, **76**, 149–67.

Fernando, C.H. (1994) Zooplankton, fish and fisheries in tropical freshwaters. *Hydrobiologia*, **272**, 105–23.

Flaig, E.G. & Havens, K.E. (1995) Historical trends in the Lake Okeechobee ecosystem: I. Land use and nutrient loading. *Archiv für Hydrobiologie Supplementband*, **107**(1), 1–24.

Fülleborn, F. (1900) Über Untersuchungen im Nyassa-See und in den Seen im nördlichen Nyassa-land. *Verhandlungen der Gesellschaft für Erdkunde zu Berlin*, **28**, 332–8.

Ganf, G.G. (1972) The regulation of net primary production in Lake George, Uganda, East Africa. In: Kajak, A. & Hillbricht-Ilkowska, A. (eds), *Productivity Problems of Freshwaters*. PWN Polish Scientific Publications, Warsaw, 693–708.

Ganf, G.G. (1974a) Phytoplankton biomass and distribution in a shallow eutrophic lake (Lake George, Uganda). *Oecologia*, **16**, 9–29.

Ganf, G.G. (1974b) Diurnal mixing and vertical distribution of phytoplankton in a shallow equatorial lake (Lake George, Uganda). *Journal of Ecology*, **62**, 611–29.

Ganf, G.G. (1975) Photosynthetic production and irradiance-photosynthesis relationships of the phytoplankton from a shallow equatorial lake (Lake George, Uganda). *Oecologia*, **18**, 165–83.

Ganf, G.G. & Blazka, P. (1974) Oxygen uptake, ammonia and phosphate excretion by zooplankton of a shallow equatorial lake (Lake George, Uganda). *Limnology and Oceanography*, **19**, 313–25.

Ganf, G.G. & Viner, A.B. (1973) Ecological stability in a shallow equatorial lake (Lake George, Uganda). *Proceedings of the Royal Society of London Series B*, **184**, 321–46.

Gaudet, J. & Muthuri, F.M. (1981) Nutrient regeneration in shallow tropical lake water. *Verhandlungen der internationale Vereinigung für theoretische und angewandte Limnologie*, **21**, 725–9.

Göltenboth, F. (1996) Pioneering tropical limnology in South East Asia: The Sunda limnological expedition of 1928–1929 in retrospective. In: Schiemer, F. & Boland, K.T. (eds), *Perspectives in Tropical Limnology*. SPB Academic Publishing, Amsterdam, 1–18.

Göltenboth, F. & Kristyanto, A. (1987) Ecological studies on the Rawa Pening Lake (central Java): some biotic parameters of the semi-natural lake. *Tropical Ecology*, **28**, 101–6.

Göltenboth, F. & Kristyanto, A. (1994) Fisheries in the Rawa Pening Reservoir, Java, Indonesia. *Internationale Revue der Gesamten Hydrobiologie*, **79**, 113–29.

Gopal, B. & Wetzel, R.G. (1995) *Limnology in Developing Countries*, Vol. 1. International Association for Limnology and International Scientific Publications, New Delhi, 230.

Gopal, B. & Wetzel, R.G. (1999) *Limnology in Developing Countries*, Vol. 2. International Association for Limnology and International Scientific Publications, New Delhi, 330.

Hare, L. & Carter, J.C.H. (1984) Diel and seasonal physico-chemical fluctuations in a small natural West African lake. *Freshwater Biology*, **14**, 597–610.

Harper, D., Mavuti, K. & Muchiri, S.M. (1990) Ecology and management of Lake Naivasha, Kenya, in relation to climatic change, alien species' introductions, and agricultural development. *Environmental Conservation*, **17**, 328–36.

Havens, K.E. (2002) Development and application of hydrologic restoration goals for a large subtropical lake. *Lake and Reservoir Management*, **18**(4), 285–92.

Havens, K.E. & Schelske, C.L. (2001) The importance of considering biological processes when setting total maximum daily loads (TMDL) for phosphorus in shallow lakes and reservoirs. *Environmental Pollution*, **113**, 1–9.

Havens, K.E. & Walker, W.W. (2002) Development of a total phosphorus concentration goal in the TMDL process for Lake Okeechobee, Florida (USA). *Lake and Reservoir Management*, **18**(3), 227–38.

Havens, K.E., Bierman, V.J. Jr., Flaig, E.G., Hanlon, C., James, R.T., Jones, B.L. & Smith, V.H. (1995a) Historical trends in the Lake Okeechobee ecosystem: VI. Synthesis. *Archiv für Hydrobiologie Supplementband*, **107**(1), 101–11.

Havens, K.E., Hanlon, C. & James, R.T. (1995b) Historical trends in the Lake Okeechobee ecosystem: V. Algal blooms. *Archiv für Hydrobiologie Supplementband*, **107**(1), 89–100.

Havens, K.E., Aumen, N.G., James, R.T. & Smith, V.H. (1996a) Rapid ecological changes in a large subtropical lake undergoing cultural eutrophication. *Ambio*, **25**, 150–5.

Havens, K.E., Bull, L.A., Warren, G.L., Crisman, T.L., Phlips, E.J. & Smith, J.P. (1996b) Food web structure in a subtropical lake ecosystem. *Oikos*, **75**, 20–32.

Havens, K.E., East, T.L. & Beaver, J.R. (1996c) Experimental studies of zooplankton-phytoplankton-nutrient interactions in a large subtropical lake (Lake Okeechobee, Florida, U.S.A.). *Freshwater Biology*, **36**, 579–97.

Howard-Williams, C.H. & Gaudet, J.J. (1985) The structure and functioning of African swamps. In: Denny, P. (ed.), *The Ecology and Management of African Wetland Vegetation*. Dr. W. Junk, Dordrecht, 153–75.

James, R.T. & Bierman, V.J. (1994) A preliminary modeling analysis of water quality in Lake Okeechobee, Florida. Calibration results. *Water Research*, **29**, 2755–66.

James, R.T., Jones, B.L. & Smith, V.H. (1995a). Historical trends in the Lake Okeechobee ecosystem: II. Nutrient budgets. *Archiv für Hydrobiologie Supplementband*, **107**(1), 25–47.

James, R.T., Smith, V.H. & Jones, B.L. (1995b) Historical trends in the Lake Okeechobee ecosystem: III. Water quality. *Archiv für Hydrobiologie Supplementband*, **107**(1), 49–69.

Jørgensen, S.-E. & Vollenweider, R.A. (1989) *Guidelines of Lake Management*, Vol. 1, *Principles of Lake Management*. International Lake Environment Committee Federation, Otsu, 1174.

Kadlec, R.H. (1994) Wetlands for water polishing: free water surface wetlands. In: Mitsch, W.J. (ed.), *Global Wetlands: Old World and New*. Elsevier Science, Amsterdam, 335–49.

Kalff, J. (1991) The utility of latitude and other environmental factors as predictors of nutrients, biomass and production in lakes worldwide: Problems and alternatives. *Verhandlungen der internationale Vereinigung für theoretische und angewandte Limnologie*, **24**, 1235–9.

Kalk, M., Howard-Williams, C.H. & McLachlan, A.J. (1979) *Lake Chilwa. Studies of Change in a Tropical Ecosystem*. Monographiae Biologicae 35. Dr. W. Junk, Den Haag, 462.

Kansiime, F. & Nalubega, M. (1999) *Wastewater Treatment by a Natural Wetland: the Nakivubo Swamp, Uganda: Processes and Implications*. A.A. Balkema, Rotterdam, 300.

Kaufman, L. (1992) Catastrophic change in species-rich freshwater ecosystems. *BioScience*, **42**, 846–58.

Kilham, P. & Kilham, S.S. (1990) Endless summer: internal loading processes dominate nutrient cycling in tropical lakes. *Freshwater Biology*, **23**, 379–89.

Kock, M., Focken, U., Richter, H., Becker, K. & Santiago, C.B. (2000) Feeding ecology of silverperch, *Therapon plumbeus* Kner, and the impact of fish-pens in Laguna de Bay, Philippines. *Journal of Applied Ichthyology*, **16**(6), 240–6.

Kufel, L., Prejs, A. & Rybak J.I. (eds) (1997) Shallow Lakes '95. *Hydrobiologia* **342/343**.

Lewis, W.M. (1979) *Zooplankton Community Analysis*. Springer-Verlag, New York, 163.

Lewis, W.M. (1987) Tropical limnology. *Annual Review of Ecology and Systematics*, **18**, 159–84.

Lewis, W.M. (1996) Tropical lakes: how latitude makes a difference. In: Schiemer, F. & Boland, K.T. (eds), *Perspectives in Tropical Limnology*. SPB Academic Publishing, Amsterdam, 43–64.

Lewis, W.M. (2000) Basis for the protection and management of tropical lakes. *Lakes and Reservoirs: Research and Management*, **5**, 35–48.

Löffler, H. (1990) Human uses. In: Patten, B.C. (ed.), *Wetlands and Shallow Continental Water Bodies*, Vol. 1, *Natural and Human Relationships*. SPB Academic Publishing, Den Haag, 17–27.

Lowe-McConnell, R.H. (1987) *Ecological Studies in Tropical Fish Communities*. Cambridge University Press, Cambridge, 382.

MacIntyre, S. & Melack, J.M. (1988) Frequency and depth of vertical mixing in an Amazon floodplain lake (L. Calado, Brazil). *Verhandlungen der internationale Vereinigung für theoretische und angewandte Limnologie*, **23**, 80–5.

MacIntyre, S. & Melack, J.M. (1995) Vertical and horizontal transport in lakes: linking littoral, benthic, and pelagic habitats. *Journal of the North American Benthological Society*, **14**, 599–615.

McLachlan, A.J. (1975) The role of aquatic macrophytes in the recovery of the benthic fauna of a tropical lake after a dry phase. *Limnology and Oceanography*, **20**, 54–63.

Melack, J.M. (1979) Temporal variability of phytoplankton in tropical lakes. *Transactions of the American Fisheries Society*, **105**, 575–80.

Melack, J.M. (1996) Recent developments in tropical limnology. *Verhandlungen der internationale Vereinigung für theoretische und angewandte Limnologie*, **26**, 211–17.

Mitchell, D.S. (1973) Aquatic weeds in man-made lakes. In: Ackermann, W.C., White, G.F. & Worthington, E.B. (eds), *Man-made Lakes: their Problems and Environmental Effects. Geophysical Monograph*, **17**, 606–11.

Monismith, S.G., Imberger, J. & Morison, M.L. (1990) Convective motions in the sidearm of a small reservoir. *Limnology and Oceanography*, **35**, 1676–702.

Moss, B. (1995) The microwaterscape—a four-dimensional view of interactions among water chemistry, phytoplankton, periphyton, macrophytes, animals and ourselves. *Water Science and Technology*, **32**(4): 105–16.

Neill, W.T. (1946) Notes on *Crocodylus novaeguineae*. *Copeia*, **1**, 17–20.

Neveu-Lemiare, M. (1904) Le Titicaca et le Poopo. Contribution à l'étude des lacs des hauts plateaux boliviens. *La Géographie*, **9**(6), 409–30.

Nilssen, J.-P. (1984) Tropical lakes—functional ecology and future development: the need for a process-oriented approach. *Hydrobiologia*, **113**, 231–42.

Osborne, P.L. (1991) Seasonality in nutrients and phytoplankton production in two shallow lakes: Waigani Lake, Papua New Guinea, and Barton Broad, Norfolk, England. *Internationale Revue der Gesamten Hydrobiologie*, **76**, 105–20.

Osborne, P.L. & Leach, G.J. (1983) Changes in the distribution of aquatic plants in a tropical swamp. *Environmental Conservation*, **4**, 323–9.

Osborne, P.L. & Polunin, N.V.C. (1986) From swamp to lake: recent changes in the aquatic flora of a lowland Papuan swamp. *Journal of Ecology*, **74**, 197–210.

Osborne, P.L. & Totome, R.G. (1992) Influences of oligomixis on the water and sediment chemistry of Lake Kutubu, Papua New Guinea. *Archiv für Hydrobiologie*, **124**, 427–49.

Osborne, P.L. & Totome, R.G. (1994) Long-term impacts of sewage effluent disposal on a tropical wetland. *Water Science and Technology*, **29**, 111–17.

Osborne, P.L., Polunin, N.V.C. & Totome, R.G. (1996) Sediments as indicators of ecosystem function in four contrasting lakes in Papua New Guinea. In: Schiemer, F. & Boland, K.T. (eds), *Perspectives in Tropical Limnology*. SPB Academic Publishing, Amsterdam, 131–49.

Pacardo, E.P. (1994) Agroecosystem analysis of Laguna Lake in the Philippines. *Mitteilungen der internationale Vereinigung für theoretische und angewandte Limnologie*, **24**, 259–64.

Payne, A.I. (1986) *The Ecology of Tropical Lakes and Rivers*. Wiley, Chichester, 301.

Phlips, E.J. & Ihnat, J. (1995) Planktonic nitrogen fixation in a shallow subtropical lake (Lake Okeechobee, Florida, USA). *Ergebnisse der Limnologie*, **45**, 191–202.

Phlips, E.J., Aldridge, F.J. & Hanlon, C. (1995a) Potential limiting factors for phytoplankton biomass in a shallow subtropical lake (Lake Okeechobee, Florida, USA). *Archiv für Hydrobiologie*, **45**, 137–55.

Phlips, E.J., Aldridge, F.J., Schelske, C.L. & Crisman, T.L. (1995b) Relationships between light availability, chlorophyll $a$, and tripton in a large, shallow subtropical lake. *Limnology and Oceanography*, **40**, 416–21.

Rast, W. & Thornton, J.A. (1996) Trends in eutrophication research and control. *Hydrological Processes*, **10**, 295–313.

Reynolds, C.S. (1992) Eutrophication and the management of planktonic algae: What Vollenweider couldn't tell us. In: Sutcliffe, D.W. & Jones, J.G. (eds), *Eutrophication: Research and Application to Water Supply*. Freshwater Biological Association, Ambleside, 4–29.

Ruttner, F. (1931) Hydrographische und hydrochemische Beobachtungen auf Java, Sumatra, und Bali. *Archiv für Hydrobiologie Supplementband*, **8**, 197–454.

Santiago, A.E. (1993) Limnological behavior of Laguna de Bay: Review and evaluation of ecological status. In: Sly, P.G. (ed.), *Laguna Lake Basin, Philippines: Problems and Opportunities*. Environment and Resource Management Project, Halifax, Nova Scotia, Canada and University of the Philippines at Los Baños, Laguna, Philippines, 100–5.

Santos, M.S. Jr. & Rabanal, H.R. (1988) The Napindan hydraulic control structure and its effects on fish production in Laguna Lake, Philippines. *Journal of Aquaculture in the Tropics*, **3**, 47–62.

Santos-Borja, A.C. (1993) Water quality management of Laguna de Bay. In: Sly, P.G. (ed.), *Laguna Lake Basin, Philippines: Problems and opportunities*. Environment and Resource Management Project, Halifax, Nova Scotia, Canada and University of the Philippines at Los Baños, Laguna, Philippines, 143–67.

Saunders, J.F. & Lewis, W.M. (1988) Composition and seasonality of the zooplankton community of Lake Valencia, Venezuela. *Journal of Plankton Research*, **10**, 957–85.

Scasso, F., Mazzeo, N., Gorga, J., Kruk, C., Lacerot, G., Clemente, J., Fabián, D. & Bonilla, S. (2001) Limnological changes in a sub-tropical shallow hypertrophic lake during its restoration: two years of a whole-lake experiment. *Aquatic Conservation: Marine and Freshwater Ecosystems*, **11**, 31–44.

Scheffer, M., Hosper, S.H., Meijer, M.-L., Moss, B. & Jeppesen, E. (1993) Alternative equilibria in shallow lakes. *Trends in Ecology and Evolution*, **8**, 275–9.

Schiemer, F. & Boland, K.T. (eds) (1996) *Perspectives in Tropical Limnology*. SPB Academic Publishing, Amsterdam, 347.

Sly, P.G. (1993) Major environmental problems in Laguna Lake, Philippines: A summary and synthesis. In: Sly, P.G. (ed.), *Laguna Lake Basin, Philippines: Problems and Opportunities*. Environment and Resource Management Project, Halifax, Nova Scotia, Canada and University of the Philippines at Los Baños, Laguna, Philippines, 304–29.

Starling, F.L.M. (1993) Control of eutrophication by silver carp (*Hypophthalmichthys molitrix*) in the tropical Paranoá Reservoir (Brasília, Brazil): a mesocosm experiment. *Hydrobiologia*, **257**, 143–52.

Straškraba, M. (1990) Shallow lakes and reservoirs. In: Patten, B.C. (ed.), *Wetlands and Shallow Continental Water Bodies*, Vol. 1, *Natural and Human Relationships*, SPB Academic Publishing, den Haag, 425–44.

Talling, J.F. (1992) Environmental regulation in African shallow lakes and wetlands. *Revue D'Hydrobiologie Tropicale*, **25**, 87–144.

Talling, J.F. & Lemoalle, J. (1998) *Ecological Dynamics of Tropical Inland Waters*. Cambridge University Press, Cambridge, 441.

Talling, J.F., Wood, R.B., Prosser, M.V. & Baxter, R.M. (1973) The upper limit of photosynthetic productivity by phytoplankton: Evidence from Ethiopian soda lakes. *Freshwater Biology*, **3**, 57–76.

Thienemann, A. (1959). Erinnerungen und Tagebuchblätter eines Biologen. *Ein Leben im Dienste der Limnologie. IV. Die Deutsche Limnologische Sunda Expedition 1928–29*. E. Schweizbart'sche Verlagsbuchhandlung, Stuttgart, 129–261.

Thomas, P.A. & Room, P.M. (1986) Successful control of the floating weed *Salvinia molesta* in Papua New Guinea: A useful biological invasion neutralises a disastrous one. *Environmental Conservation*, **13**, 242–8.

Van der Molen, D.T. & Boers, P.C.M. (1994) Influence of internal loading on phosphorus concentration in shallow lakes before and after reduction of the external loading. *Hydrobiologia*, **275/276**, 379–89.

Viner, A.B. & Smith, I.R. (1973) Geographical, historical and physical aspects of Lake George. *Proceedings of the Royal Society of London, Series B*, **184**, 235–70.

Von Sperling, E. (1997) The process of biomass formation as the key point in the restoration of tropical eutrophic lakes. *Hydrobiologia*, **342/343**, 351–4.

Wetzel, R.G. (1999) Biodiversity and shifting energetic stability within freshwater ecosystems. *Archiv für Hydrobiologie – Advances in Limnology*, **54**, 19–32.

Wetzel, R.G. (2001) *Limnology: Lake and River Ecosystems*. Academic Press, San Diego, 1006.

Wetzel, R.G. & Gopal, B. (2001) Limnology in Developing Countries, Vol. 3. International Association for Limnology and International Scientific Publications, New Delhi, 189 pp.

Worthington, E.B. (1929a) *A Report on the Fishing Survey of Lakes Albert and Kioga*. Cambridge University Press, Cambridge, 136.

Worthington, E.B. (1929b) The life of Lake Albert and Lake Kioga. *Geography Journal*, **74**, 111–32.

Worthington, E.B. (1932) *A Report on the Fisheries of Uganda. Cambridge Expedition to the East African Lakes 1932–33*. Crown Agents for the Colonies, London, 88.

Worthington, E.B. & Worthington, S. (1933) *Inland Waters of Africa*. MacMillan, London, 259.

Yusoff, F.M. & Patimah, I. (1994) A comparative study of phytoplankton populations in two Malaysian lakes. *Mitteilungen der internationale Vereinigung für theoretische und angewandte Limnologie*, **24**, 251–7.

# 12 Reservoirs and other Artificial Water Bodies

## MILAN STRAŠKRABA*

### 12.1 INTRODUCTION

This chapter is devoted to the ecology of artificially constructed water bodies and to the application of this knowledge to sound management. The basic ecological processes in these artificial bodies of water are similar to those in natural lakes. However, there are substantial differences in the significance and in the consequences of these processes in reservoirs and other artificial waters. The main reasons for this are that reservoirs are constructed for particular purposes, possess specific morphometries, particular positions in the landscape and are subject to artificial hydraulic variation in their stored volumes. Focusing on differentiation, rather than neglecting it, invokes an understanding of the altered limnological functioning in reservoirs and constructed waters. A full knowledge of the features of each artificial water body is necessary for its effective management, if the functional outputs are to be achieved. To neglect behavioural differences or to treat problems as if they were identical to those of natural water bodies may lead, at best, to inadequate management and, more likely, mismanagement.

The categories of artificial water body adopted in this chapter are based mainly on the purposes for which they were constructed. Pollution or other substantial deleterious impacts are not used to distinguish among water bodies: although some cases are mentioned in passing, other chapters in this book cover these topics adequately (see especially, this Volume, Chapters 2–6). Here, the principal focus will be on larger reservoirs, because of their great extent, aggregate storage capacity and economic importance. Size, either in areal or volumetric terms, may also be used as a means of categorising reservoirs: the standards proposed by the International Commission on Large Dams (ICOLD 1984) are reproduced in Table 12.1. On this basis, Avakyan & Iakovleva (1998) were able to establish that there are currently 111 'large' reservoirs in the world, with volumes ($V$) of $>10\,km^3$ ($10^{10}\,m^3$), of which six contain between 100 and $200\,km^3$ ($>10^{11}\,m^3$): these include Bratskoye, Central Russia, Lake Nasser (Aswan) in Egypt and Lake Kariba (Zambia/Zimbabwe). 'Medium' reservoirs, containing $0.1$–$10\,km^3$, number around 2720 globally. The number of 'small reservoirs' holding between $10^6$ and $10^8\,m^3$ is estimated to be c.60,000 (Avakyan & Iakovleva 1998). 'Very small' artificial impoundments ($<1\,km^2$ or containing $<10^6\,m^3$) possibly number more than one million, if artificial farm ponds, fish cultivation tanks and industrial cooling or settling ponds are included, but these are not considered in this article.

Together, reservoirs exceeding $1\,km^2$ or storing $<1 \times 10^6\,m^3$ cover some $400{,}000\,km^2$ and retain some $6500\,km^3$ of water (Avakyan & Iakovleva 1998). This represents nearly 3% of the global store

---

* It was with great sadness that we learned that Milan Straškraba had died on 26 July 2000. Milan had agreed to contribute the chapter on reservoirs after one of his colleagues had been obliged to withdraw from his contract. We remain very grateful to Milan for his generosity. However, at the time of his death, we had received from him only a draft manuscript, from which most of the citations were omitted, and several of which proved very difficult to trace. In order to bring the draft to a publishable standard, we have edited the text and organisation as lightly as possible, so that Milan's thoughts and approaches have been preserved as far as possible. We claim to have done our best to fulfil what we believe to have been his bibliographic intentions. Nevertheless, we regret our inevitable shortcomings in this respect. We also apologise if our editorial efforts have unintentionally misrepresented his deductions and ideas in any way whatsoever.

of inland surface water but it is equivalent to c.22% of the volume of land run-off shed annually to the sea. As large and medium reservoirs contribute the largest share of this volume, it is evident that, limnologically, they also represent substantial bodies of enclosed fresh water. Interestingly, however, none is sufficiently large, on either area or volume criteria, to enter Beeton's (1984) 'Top 30' listings of the world's great lakes.

The diversity of ecological conditions found in artificial bodies of water corresponds with variations in their size which renders making general statements difficult. Any classification is more or less arbitrary: size is a somewhat fuzzy construct which allows many possible transitions and divergences to be accommodated. Consequential impacts on the operation of reservoirs need to be considered.

Table 12.1  The size categories of reservoirs (after ICOLD 1984) and the approximate numbers in each category, according to Avakyan & Iakovleva (1998)

| Category | Area (km$^2$) | Volume (m$^3$) | Number of reservoirs |
|---|---|---|---|
| Large | $10^4$–$10^6$ | $10^{10}$–$10^{12}$ | 111 |
| Medium | $10^2$–$10^4$ | $10^8$–$10^{10}$ | 2720 |
| Small | 1–$10^2$ | $10^6$–$10^8$ | c.60,000 |
| Very small | <1 | <$10^6$ | ? |

## 12.2 IMPACTS OF USE ON THE LIMNOLOGY OF RESERVOIRS

Reservoir uses may be divided according to the primary purpose of their construction and to any secondary uses to which they are later put, as subsequent demands for water have changed. Some newer reservoirs have been built in order to fulfil several objectives, although one is usually paramount. Some of these uses are distinguished in Table 12.2, which also summarises associated characters of morphometry and hydraulic exchange.

### 12.2.1 Ecological differences between reservoirs and natural lakes

Reservoirs are artificial lakes, but a number of characteristics differentiate them from natural lakes (Thornton et al. 1990; Straškraba et al. 1993). Some of these features, linked to the design and use of the reservoir, profoundly affect its limnological behaviour, and need to be taken into account in management. Usually, the ultimate purpose is to supply water—generally, reservoir catchments must be greater in area than the water body. Based on a sample of reservoirs in the USA, Thornton et al. (1990) found an average factor of 14 times. Besides the challenge of being able to fulfil an aspiration for water to be available at all times and in

Table 12.2  Features of reservoirs constructed for various primary purposes (after Straškraba et al. 1993)

| Primary use | Size | Depth | Retention time | Outflow | Water level |
|---|---|---|---|---|---|
| Flood protection and flow regulation | Small to medium | Shallow | Regionally dependent | Surface | Highly variable |
| Water storage | Small to medium | Various | Extremely variable | Usually below surface | Use dependent |
| Hydroelectricity | Medium to large | Deep | Variable | Near-bottom | Peak operation |
| Drinking-water supply | Small | Preferably deep | Short | Intermediate to deep | Natural variability |
| Fish cultivation | Small | Shallow | short | Surface | Natural or operational |
| Pump storage | Small to medium | Deep | Extremely variable | Near-bottom | Extremely Variable |
| Irrigation | Small | Shallow | Long | Surface | Highly variable |
| Navigation | Large | Deep | Short | Whole profile | Very low variability |
| Recreation | Small | Shallow | Long | Surface | Low variability |

sufficient quantities, designers and civil engineers face the problem of containing construction costs. Thus, there is a compromise to be struck such that schemes tend to be designed to provide a full supply only in those periods when low inflowing discharges are within the expected limits of the long-term fluctuations. During long droughts, significant net withdrawals of water storage from reservoirs are anticipated. In addition to major drawdown events, water levels may fluctuate routinely during the year, as a consequence of variable demands for power generation or water supply, as well as of periodic recharge following heavy or seasonal rainfall. Note that the impact of the same absolute variation in the stored volume of water is relatively greater on a shallow reservoir than on a deep one.

Another consequence of a relatively large watershed is a relatively short retention time ($RT = V/Q$, where $Q$ is the average hydraulic exchange per unit time). For instance, the $RT$ of half the large reservoirs in the USA is <110 days (Thornton *et al.* 1990). In the size category of reservoirs $V < 10^9 \, m^3$, long retention time reservoirs predominate.

Large watersheds generally contribute more pollutants to reservoirs and the consequences of higher watershed/waterbody area ratios are higher pollutant loads. Rivers are generally unable to digest all the materials washed out from the larger area. Even in relatively unpolluted areas, the suspended matter load of a reservoir is higher and sediment infill is sometimes a significant problem in reservoir management.

Explanations for fundamental differences in the limnological behaviour of lakes and reservoirs may be based on the behavioural distinctions between deep and shallow systems. Similarities among shallow reservoirs and shallow lakes are generally greater than those among deep reservoirs and deep lakes. Quantitative comparisons of deep lakes and reservoirs are available (Straškraba 1999) but, for shallow reservoirs, only qualitative explanations have been offered.

The hydrographical distinctions between shallow and deep reservoirs relate not to absolute depth but to the morphometric consequences of the reservoir morphometry and, in particular, the depth **relative** to the area of the reservoir. The reasons for this relate to the characteristics of vertical mixing, which are represented in Fig 12.1. Lakes and reservoirs defined as 'shallow' are, in almost all geographical conditions, polymictic and, actually, fully mixed for most of the time. Conversely, deep reservoirs are resistant to full mixing and are generally stratified for some of the time but the intensity and duration of stratification is highly dependent on geography and $RT$ (see later). Hydrologically, the maximum depths ($Z_{max}$) of shallow reservoirs lie within the observed or calculated mixing depth ($Z_{mix}$), determined by the shear stress of wind on the water surface. In deep reservoirs, $Z_{mix} \ll Z_{max}$. Various theoretical formulations for estimating the mixing depth of a waterbody are available (e.g. Gorham & Boyce 1989; Hanna 1990; Reynolds 1992a).

However, other variables influence the depth of mixing particularly in reservoirs: transparency (or rather the vertical extinction coefficient for radiation penetration) and $RT$ (Straškraba 1999). In more transparent waters, light penetrates significantly deeper than it does in coloured, turbid or phytoplankton-rich waters (Kirk 1994; Hocking & Straškraba 1994; Mazumder & Taylor 1994). If theoretical retention time is short (say <7–10 days), the high throughflow rates prevent stratification even in considerably deep reservoirs. Also, mixing depth seems to depend to a certain degree on $RT$, more so than on wind in some cases (Straškraba 1999).

Differences among lakes and reservoirs are affected also by latitude. In an inventive paper, Nilssen (1984) distinguished between tropical and temperate lakes on the basis of the role of fish. Many tropical species not only consume zooplankton but may compete with zooplankton for algal and sestonic food. Predation upon cladocerans, in particular, may decrease the ability of zooplankton to exploit planktonic algae. Some variables may exhibit continuous relationships to latitude, with intermediacy in the subtropics. Quirós (1998) found better fits to empirical regressions if the temperate and subtropical lakes are treated separately, although this finding was more closely

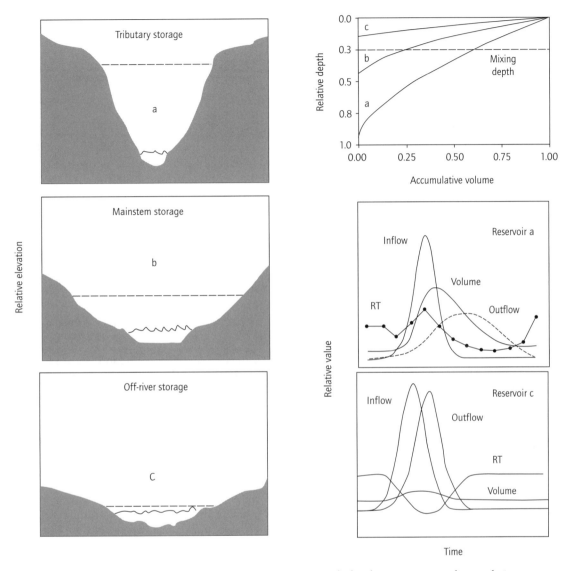

Fig. 12.1 Effects of depth and hydrology on reservoir environments. Idealised cross-sections indicate relative differences in bathymetry and storage in reservoirs constructed on (a) tributaries, (b) river mainstems and (c) shallow, off-river, lowland impoundments. Plots show differences in volume development in the three reservoir types with (top right) the relative extent of atmospherically forced mixing and (lower right panels) changes in storage volume, outflow and retention time (RT) against a similar relative input in type-a and type-c reservoirs. (Modified after Kennedy 1999.)

related to differences in relative abundance of visual planktivorous fish that the lakes supported rather than to climate differences (see also Quirós & Boveri 1999). The process of community maturation and ageing in neotropical reservoirs is also considered to respond to, rather than drive, the faster establishment of fish populations at low latitudes (Agostinho et al. 1999).

### 12.2.2 Deep reservoirs

Besides sharing these common distinctions, deep reservoirs are subject to some specific attributes. In temperate regions, water quality in deep reservoirs is influenced by:
- basin morphology characterised by marked deepening towards the dam
- longitudinal zonation
- mostly deep outlets
- temperature stratification and $RT$
- retention capacity for phosphorus and pollutants.

#### 12.2.2.1 Morphological zonation

The one-directional longitudinal gradient with maximum depth at the downstream end is typical of impoundments created by the damming of a river. The deepest part of a lake is usually more central. In lakes, the major flows are rotational, the main driving force being wind. The continuous inclination of the reservoir bottom from the inflow to the dam is often accompanied by continuous broadening, at least in the inflow part. The result is a predominantly unidirectional movement of the water mass from inflow to outflow. Flows diminish generally in the same direction. Similar directional sequences are seen in the changes of limnological variables along the reservoir length. The inflow zone receives high concentrations of particles, organic matter, nutrients and other biologically active chemical components. The biological activity is here lower than in more stagnant conditions down the reservoir, as the flow prevents full development of the pelagic community. In the transition zone, primary production is increased as a consequence of the slower flow, towards the capacity provided by nutrients, the underwater light availability and net of the predation pressure of zooplankton. Increased phytoplankton, chlorophyll *a* concentration, and decomposition of sedimenting matter result in lower oxygen concentrations near the bottom, perhaps even anoxia, low redox and enhanced concentrations of ammonia, manganese, iron and hydrogen sulphide. Tongues of water with differing characteristics may extend at intermediate levels into the lacustrine zone ('interflow').

#### 12.2.2.2 Outflow location

Whereas the main outflow from a lake is typically at the surface, many reservoirs allow release or abstraction from deep parts. For a selection of 26 reservoirs of the U.S. Army Corps of Engineers, Kennedy (1999) noted that 77% possessed outlets located in deeper strata, and only 23% an exclusive surface outlet. A deeper outlet is necessary especially where drawdown of water level during prolonged dry periods is anticipated and where power generation requires the high pressure which the head provides. Reservoirs with $RT$ of several days can possess surface outlets because there is no danger of prolonged drawdown.

#### 12.2.2.3 Dependence of the degree of stratification on $RT$

$RT$ critically affects the period and depth of stratification. If expressed by a fairly simple measure (the difference between surface and bottom temperature in the period of maximum stratification), stratification is asymptotic to increasing $RT$ (Fig.12.2). Models have succeeded in simulating the apparent critical value of $RT$, above which no marked greater stability of stratification is attained, to be about 1 year (300–400 days: Straškraba *et al.* 1993). Stratification may continue to vary interannually, in response to flow and to water-level conditions, but it requires retention times of longer than 1 year before the factors influencing stratification patterns coincide with those of lakes.

#### 12.2.2.4 Matter retention in deep reservoirs

Another characteristic feature of reservoirs is their high capacity to retain particulate matter. This is well documented for input sediments (Carvalho 1988). The percentage of the total phosphorus (TP) flowing into deep reservoirs which is shed through the outflow is broadly dependent on $RT$, inflow phosphorus concentration and areal phosphorus

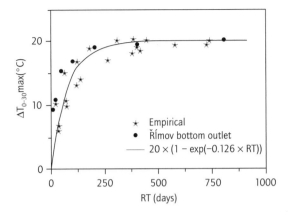

**Fig. 12.2** Differences in temperature at the water surface and at a depth of 30 m in large reservoirs as a function of retention time: comparison of empirical observations made in various reservoirs in the Czech Republic and southern Germany during summer (★) with predictions for the Římov Reservoir (●) using the model DYRESM (Straškraba et al., 1993b) and an earlier relationship modelled by Straškraba & Mauersberger (1988). (Redrawn, with permission, from fig. 5 of Straškraba (1999).

load (see Straškraba et al. 1995; Kennedy 1999). Retention of reactive phosphorus (orthophosphate) in deep reservoirs increases with $RT$ more steeply than retention of TP. A similarly intensive retention of nitrogen in deep reservoirs is not observed, although its elimination from the reservoir is influenced by the rate of its mineralisation. Organic matter is also retained (see Straškrabová et al. 1993). Few quantitative data exist on the retention of other elements, although it is supposed that most are retained to a considerable extent. The fates of these materials are influenced by internal transformations, involving primary production (Schulte & Lackey 1973), phytoplankton standing crop, turbidity and water colour (Javornický 1966; Townsend et al. 1996) and oxidative state (Straškraba 1973).

#### 12.2.2.5 Deep tropical reservoirs

The characteristics of temperate reservoirs may not be transposed uncritically to deep reservoirs in the tropical latitudes. The longitudinal differences seem to be similar but the stratification behaviour is often quite different, with a high incidence of polymixis. On the other hand, quite small temperature differences may be sufficient to induce stable water-column stagnation, often with sharp gradients of oxygen concentration and redox potential. When the density difference between the inflowing water and the reservoir surface is small, the structural impact is weak or, at least, weaker than in temperate reservoirs. Coupled with long retention times, ageing in tropical reservoirs is generally rapid (e.g. Balon & Coche 1974). A particular feature of many mesotrophic and eutrophic tropical reservoirs is the frequent overgrowth by floating macrophytes (*Eichhornia* sp., *Pistia stratiotes*, *Azolla* sp.).

### 12.2.3 Shallow reservoirs

Shallow reservoirs—those which are often or continuously fully mixed—behave differently from deep ones. As is true of shallow lakes, a relatively greater area of lake bottom is less than 5 m in depth and supposedly within the reach of significant wind-generated shear currents. The superficial bottom sediments are an integral part of the trophogenic zone and physical exchanges of materials and solutes are readily facilitated (enclosure experiments of Reynolds 1996). Phytoplankton and detritus are resuspended from the bottom during windy conditions and reduce the penetration of light for a period sufficient to impair phytoplankton reproduction, resulting in oscillations of biomass (Gons et al. 1992). Biologically available phosphorus may readily be recycled, although the precise mechanism is still not clear. Such recycling is the probable cause of their slow response of shallow lakes to external load reductions (Sas 1989) and for the evident behavioural differences between deep and shallow reservoirs. Phosphorus retention is weak in shallow reservoirs, whereas in deep reservoirs it is nitrogen retention that is poor. Unstratified lakes and reservoirs possess high capacity for nitrogen accumulation (Kelly et al. 1987; Howarth et al. 1996).

Another feature of shallow reservoirs is their

apparent ability to strike alternative states in relation to the external phosphorus load and to the biotic interactions of the fish which they support. In the Lake Burley Griffin reservoir, in Canberra, Australia, the switch from a macrophyte to a Cyanobacteria state occurred when the phosphorus loading increased but, after external loads were reduced, Cyanobacteria were replaced by a diversity of species and groups (Cullen 1991). The alternation is now fairly well understood on the basis of observations in English 'meres' in England (Scheffer et al. 1993; Søndergaard & Moss 1998) and of model studies by Scheffer (1998). The systematic decrease of productivity related phenomena (primary, secondary and fish production, as well as nutrient concentrations) with increasing depth, well known in lakes (e.g. Straškraba 1991), is presumed to be valid for reservoirs, although systematic observations do not exist. However, the very high productivity of some shallow reservoirs is unsurprising.

The longitudinal pattern characteristic for deep valley reservoirs is not seen among shallow examples, as the dispersion of solutes and suspensoids is influenced much more by wind. The effect is magnified by long retention. Horizontal distribution of both phytoplankton (in Rutland Water, England: Ferguson & Harper 1982) and zooplankton (Broa Reservoir, Brazil: Matsumura-Tundisi & Rocha 1983) is especially sensitive to fluctuations in the direction of wind. Interactions among nutrients and phytoplankton and phytoplankton and zooplankton are then reflected in the distribution of other variables. This is another reason why the effect of $RT$ on the limnology of shallow reservoirs is less systematic than it is in deep reservoirs. Yet another is the relative magnitude of water-level fluctuations: a given drawdown of the water level change in a shallow reservoir exposes a much larger area of the bottom of a shallow reservoir than a deep one.

The differences between deep (stratified) and shallow (mixed) reservoirs are summarised in Table 12.3 with respect to the effect of reservoir features. Each variable is represented with the assumption that other variables remain identical.

### 12.2.4 Reservoir ageing

Reservoirs are much younger than lakes and the process of their evolution is much more rapid, particularly owing to rapid filling, submerging of soil and terrestrial vegetation, and high load from the extensive watershed. The process of rapid water-quality changes during and after the reservoir is filled is known as 'ageing' (Straškraba & Tundisi 1999). Generally, impaired water quality is observed during the early life of a new reservoir. Recent research has focused particularly on productivity phenomena, the so-called 'trophic upsurge' (Ostrofsky 1978; see also Grimard & Jones 1982; Kimmel & Groeger 1986; Holz et al. 1997); the length of the ageing period depends on retention time, nutrient load, climate and character of the submerged soils and vegetation (Chang et al. 1996). Understanding the processes contributing

Table 12.3  The expression of limnological features in deep and shallow reservoirs

| Feature | Deep reservoirs | Shallow reservoirs |
| --- | --- | --- |
| Maximum depth | Codetermines whether deep or shallow | |
| Size | Codetermines whether deep or shallow | |
| | Influences mixing depth | Influences mixing frequency |
| Watershed area/reservoir area | Influences external loading | |
| Retention time | Load retention increases | Matter retention increases |
| Outlet depth | Stratification | Not applicable |
| Water level fluctuation | Washout of nearshore sediment | Resuspension and redeposition of sediment |

to the observed changes is necessary to the appreciation of the reactions of reservoir ecosystems to changes in the external organic and nutrient load and the internal adjustments of the developing biological communities. Both responses influence the development of given reservoirs and the extent of any perceived deterioration in water quality. The duration of the adverse conditions is of great importance for water quality management and reservoir use, particularly in the case of drinking water supply reservoirs.

The original hypothesis about reservoir ageing was based on the high rates of supply of organic matter deriving from the decaying submerged vegetation and of nutrients leaching from the soil. The most severe impacts are on the oxygen concentrations, particularly in the hypolimnion, and on water colour (Straškraba & Tundisi 1999). Both lead to more difficult and more expensive treatment for potability. Effects can be positive in that increased production during the first years may be manifest in enhanced fish production (Fernando & Holčik 1991; Kubečka 1993).

Severe oxygen deficits, leading to anoxia, are often consequent upon the decay of vegetation drowned at inundation. Investigations have demonstrated a quantitative dependence of the oxygen deficit on the amount of organic matter originating from the decay, especially with reference to the nature of the vegetation present (L.P. Rosa, quoted by Straškraba & Tundisi 1999). Organic matter leaching from soil cleared of vegetation, in particular if barren soil is exposed to contact with water, provides another source of organic matter. Experimental studies on this material have revealed the conditions leading to rapid leaching of organic matter and nutrients. Inundation of forest in the tropics can lead to such severe reducing conditions as to permit the generation of hydrogen sulphide and methane (Henry 1999). Corrosion of turbines is well known: in one case (Curuá Una, in the Brazilian Amazon) complete renewal was necessary in just 4 years (Tundisi 1984). Gas ebullition can produce technical consequences, particularly if it continues for several years: moreover, just high hypolimnetic concentrations of carbon dioxide can seriously affect dis-

**Table 12.4** Management classification of reservoirs

| Variable | Categories and subcategories |
| --- | --- |
| Geographical location | Temperate, subtropical or tropical |
| Position | Mainstem valley or off-river (bankside) |
| Size of dams | Very large, large, small or very small. |
| Hydrological depth | Deep (stratified) or shallow (mixed) |
| Retention time ($RT$) | Through flowing ($RT < 10$ days); flow sensitive ($10 < RT < 200$ days); flow insensitive ($RT > 200$ days) |
| Outlet | Surface or deep |

solution of concrete, even causing reservoir leakage (Strycker 1988).

### 12.2.5 Reservoir classification

It is apparent that several dimensions of variability characterise reservoirs and that any of them may be used as a basis for their classification. Here, I adopt a system based on management, especially those aspects affecting water quality (see Table 12.4). Reservoirs are sometimes distinguished by their position relative to the natural drainage (e.g. Søballe et al. 1992) — these being mainstem impoundments (created by building a dam on major rivers), tributary impoundments (constructed by damming smaller tributaries) and off-river impoundments, supplied by pumped flow from the river. However, there is a continuum from low order streams with tributary reservoirs to high order streams with mainstem reservoirs, so it is adequate to distinguish valley reservoirs from such off-river reservoirs as the Thames Valley reservoirs which supply drinking water to London (Steel & Duncan 1999) and the ancient impoundments of Sri Lanka (Parakrama Samudra: Newrkla & Duncan 1984). Chemical-based classifications also have been used, based upon geological considerations (Margalef 1975), although these may fail to distinguish reservoirs in the same geographical region (see, for instance, Margalef et al. 1976).

Management classification could be more comprehensive than shown. It may be valuable to distinguish reservoirs on the basis of hardness of their waters as this affects their trophic status. Mineral particle input (particularly of clays) is influential upon reservoir turbidity which, in turn, may reduce primary production through light limitation and through sorption of phosphorus. High natural humic loads may contribute to low pH, decreased primary production and increased difficulties with drinking water treatment. In respect to pollution and eutrophication, different management issues are raised by the support of an excessive plankton biomass. Whether productive capacity is limited by phosphorus, by nitrogen or by another constraint needs to be distinguished, although there may be cases where both nutrients limit intermittently or nearly simultaneously. Carbon limitation may occur in polluted tropical reservoirs. High concentrations of nitrates ($>11\,\mathrm{mg\,N\,L^{-1}}$) in drinking-water reservoirs exceed health limits on potability.

### 12.2.6 Multiple reservoir systems

Reservoirs are often linked hydrologically, usually as integral parts of the same drainage basins, and their operation is concerted to the fulfillment of some common goal. Chains of reservoirs along the same river course are referred to as **reservoir cascades**, and are exemplified by those of the Dnieper (Russia), Parapanema and Tietê (Brasil) and Negro (Argentina) rivers. Potentially, any effect of processes on an individual upstream reservoir is transferred to those downstream of it. Thus, we expect the water quality in a lower reservoir to be governed, in part, by the conditions in the upper one. The extent of this process may depend on depth and retention time, with the modifications wrought in deep, retentive reservoirs affecting downstream impoundments, with respect to the fluvial inflows, more profoundly than those of more flushed examples. In this way, eutrophication of the Tietê Cascade, driven by high nutrient and sediment loadings in the upper catchment (where the city of São Paulo is located), has so far been confined to the upper reservoirs (Barbosa et al. 1999). Distance between one reservoir and another is also very important: the effect of the upper on the lower reservoir is most pronounced where reservoirs are closely situated.

Groups of reservoirs located on separate tributaries or in separate drainage basins but whose releases or abstractions serve a common purpose are referred to as **reservoir multisystems**. For example, they may provide the storage needed to maintain a water supply in climates with markedly seasonal rainfall distribution. Their water quality is usually influenced by great variations in volume and in hydrological renewal. When subscribing reservoirs are located in geologically contrasted regions, blending supplies to a consistent chemical quality may become a challenging task.

The requirement to balance geographical inequalities in the demand for water and the supply of fluvial resources has led to the inception of another type of multisystem, based on between-catchment transfers of water flow. The approach is not new, for the construction of aqueducts was a feature of several ancient civilisations. The amounts of water transferred by these ancient systems were not large, compared with the scale of modern schemes. These affect not just water quality but entire regional water budgets. An unfortunate example is the case of the Aral Sea which, as a consequence of mismanagement during the communist era, was changed from a flourishing lake to a near dust bowl (Williams & Aladin 1991). In the semi-arid regions of southeast Australia, extensive water systems were implemented in the 1920s to transfer water from water-rich rivers near the sea coast to vast dry territories in New South Wales and South Australia. Irrigation of farmland and the raising of monocultural crops drawing superficial water has lifted deeper, solute-rich ground waters, leading to the salinisation of the soil and the abandonment of swathes of agricultural land. Moreover, water transfers may become major pathways for the transmission of waterborne diseases and contribute to complex chemical and biological changes.

Multiple reservoir systems also include schemes in which water is circulated among reservoirs by pumping. Such systems (such as the Dinorwic pumped storage, Wales) are used principally to generate electric power. Of course, the power consumption of pumping water to restore head cannot be less than the power drawn from the gravitation of its release. The purpose of the schemes is to meet peak demand for electric power, with the restorative pumping taking place at times of minimal demand. Two basic types of pumping scheme can be distinguished: in the first, water is lifted through pipework from a sump to an upper storage pond, pending release through the turbines; in the second, water is released through turbines in the dam to a downstream pool whence it is drawn back by the reverse running of the turbines. Apart from water temperature, limnological differences between the water removed from the store and that returned are normally quite small but changes in solute loads and in microbe content are possible.

## 12.3 THE ROLE OF RESERVOIRS IN FLUVIAL CATCHMENTS

### 12.3.1 Influence of the watershed on the reservoir

Reservoirs reflect the natural characteristics of the watershed as well as the impacts of anthropogenic activities. The natural influences on reservoirs are, first, geographical: the effects of latitude, altitude and continentality and their interactions with local geology and geomorphology (Straškraba et al. 1993a). In turn, these features influence profoundly the hydrological and hydrochemical properties of the rivers on which the reservoir is constructed and/or which contribute the impounded water. There are also considerable latitudinal differences in the management of reservoirs, particularly between the temperate region and tropics. This is important, as the knowledge acquired in the temperate regions cannot be applied uncritically to the tropics (see section 12.3.2). In considering rivers in the USA, Ward & Stanford (1983) recognised how the physical, hydrochemical and hydrobiological conditions are influenced by stream order: they deduced consequent order-dependent differences in the behaviour of stream impoundments (see also Margalef 1960).

Anthropogenic impacts are summarised in section 12.2.1 and are treated as separate topics else-where in this volume (see especially Chapters 2, 10, 11 and 13–18). Broadly, impacts upon the inflowing water will be similar for lakes and reservoirs but, as is also true for lakes, the same activities produce distinct effects on the immediate surroundings of a reservoir and on its fluvial inflows. From its source to its point of entry into standing water, the flowing river possesses a high capacity for the oxidation of organic inputs: any pollutant loads tend to be moderated during transport by a process of river self-purification. Organic pollutants contributed at margins of the reservoir are met by a limited self-purification capacity: particulate matter sediments from non-flowing water of impoundments and challenges a finite and weakly renewable oxidative potential. Therefore, it is desirable that reservoir shores are protected from pollution by buffer strips of meadow, scrub or woodland. Intact beds of littoral, semi-aquatic and aquatic vegetation beds fulfil a similar function provided that water-level fluctuations are modest or infrequent. Wetlands at the point of stream entry also contribute significant potential as sinks for organic and inorganic nutrient loads (see Istvánovics & Somlyódy 1998). Where recreational facilities are provided at reservoir shores, special care over hygienic arrangements is required.

Another respect in which tropical and temperate reservoirs differ is in the impact of increasing human populations and developing industrialisation. This leads to a deterioration in the quality of inflowing water and the introduction of novel pollutants, and deforestation, agricultural intensification and enhanced rates of soil erosion contribute to the introduced sediment loads.

### 12.3.2 Influence of the reservoir on the river

The effects of impoundment upon river biota are well-studied (see review of Agostinho et al. 1999). Here, the principal concern is with those impacts upon water quality which affect management. Water released from the reservoir into the river usually corresponds to a layer either overflowed at the dam or sucked from an outlet at depth (Fig. 12.3). The immediate source of this water is

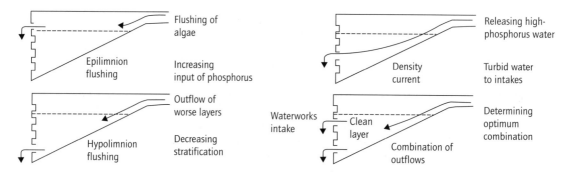

Fig. 12.3 How selective off-takes influence reservoir water quality. Near-surface off-take flushes the epilimnion of algal populations but allows phosphorus to accumulate in deep water; deep off-take decreases stratification but water may be poor in oxygen and may precipitate iron and manganese salts: creating density currents may be a more useful nutrient removal technique but may bring high loads of pollutant or turbid water to the waterworks treatment; the ideal will be to use the available options in the optimal combination. (Redrawn from an original figure in Straškraba & Tundisi 1999.)

affected by density stratification: the more intense it is, the less is the thickness of the layer of homogeneous density which is drawn off, but the greater must be its areal extent. The decisive variable for the height of the layer is the flow rate in the centre of the outflow layer, owing to the pressure of the overlying water given mainly by the depth and the size of the opening. In this way, the influence of density stratification of the reservoir is important, not just on the temperature of the outflowing water, but also its chemical and biological quality.

Moreover, this influence differs between shallow and deep reservoirs. The quality of outflow from a shallow, unstratified reservoir is likely to be similar to that of its inflow or, if not, to be poorer rather than improved. The concentration of bioavailable phosphate is not much altered, despite high primary production, because frequent resuspension of bottom sediment speeds the rate of phosphorus recycling. On the other hand, presently incomplete evaluations suggest that nitrogen may be significantly decreased. Because its production is vertically confined and there is relatively low light absorption in short water columns, the concentration of phytoplankton in shallow lakes is potentially much greater than that from unstratified deep reservoirs.

Strong density stratification in deeper reservoirs may lead to the formation of local depth maxima of phytoplankton but the compounding effects of stratification on long retention times produce other decisive impacts on the potential quality of the outflow. Surface water may benefit from precipitation of clays and other particles but they may be enriched by phytoplankton. Draw-off from the hypolimnion of a productive reservoir is liable to be deficient in oxygen, malodorous and liable to precipitate iron and manganese salts in the outflow. However, the characteristics of water in an outflow layer are not constant for the abstracted water is replaced, ultimately, from other layers. At the very least, there will be a direct gravitational substitution of water from the erstwhile more superficial layers. This will be compounded by associated turbulence. Potentially, the horizontal intrusion of a cold inflow to its density-determined depth (and called **interflow**) can short-cut the storage by finding its way, substantially intact, to the draw-off point.

Temperature conditions in the outflow reflect the geophysical site conditions, compounded by the hydraulic retentivity. If the outflow is located in the hypolimnion, the temperature of the outflowing water will be intermediate between those of the reservoir at its surface and at the draw-off depth; during the mixing period, the temperatures of each will be mutually similar. Oxygen conditions in the outflow depend on the relation between the production and respiration processes in the contributing water layers, subject to any equilibration which may take place in the turbines or other machinery. Surface release water can be supersaturated with oxygen as a result of intensive planktonic photosynthesis; at worst, deep draw-off is anoxic. Everywhere, organic pollution contributes to lowered oxygen concentrations.

## 12.4  WATER QUALITY IN RESERVOIRS

Reservoirs are exposed to many polluting influences, perhaps more so than are natural lakes. This is a consequence of their relatively larger catchments and may also reflect that, in some regions, reservoirs represent the main expanse of standing water. Reservoirs are often close to centres of population, are attractive to visitors and are intensively used for leisure purposes. There are few types of pollution specific to reservoirs or which are not observed among lakes. The extent to which they are realised in reservoirs and the consequences prejudicial to their optimal operation may be different: thus, the list of water-quality problems (below and Table 12.5) retains a relevance to the limnology of reservoirs. Moreover, the extent to which these problems are realised is highly variable, being governed by geography and by the intensity of the polluting influences. Climate (and the effects of aridity in particular) particularly influences the driving behaviour, whereas depth conditions some of the responses. Here, the consideration is subdivided among temperate and tropical reservoirs.

### 12.4.1  Temperate reservoirs

There are two major sources of non-industrial organic pollutants, namely human settlements and agriculture. In many developed countries, where the practice of secondary treatment to oxidise and disinfect urban sewage has been widely adopted, organic loadings are normally well contained. However, where intensive aquaculture is practised, normally requiring supplements of external foods, organic loads are raised. Mineralised effluents nevertheless continue to underpin ongoing and serious problems associated with eutrophication, where the *de novo* assembly of biogenic materials in receiving waters is stimulated by the inorganic nutrients liberated. Moreover, even where tertiary treatment is applied to remove a substantial part of the phosphate content of treated sewage, the contribution of diffuse, non-point sources from agricultural land is often sufficient to meet the aquatic uptake demands and the problems persist.

The responses of the biotic communities of reservoirs to water-level fluctuations also affect the perception of water quality. Reduction of the water level increases the relative depth in which net photosynthetic anabolism can occur and, thus, to a higher potential biomass of phytoplankton. It is a fact that as reservoirs are drawn down during summer droughts to meet higher water demands, the quality of the stored water may deteriorate simultaneously and require more extensive treatment. There may also be attendant changes in the phytoplankton, as diatoms are replaced by potentially toxic cyanobacteria (Reynolds 1999). More persistent drawdown effects increase the relative area amenable to macrophytic growth, and grass and scrub growth on exposed sediment and, in some cases, the grazing stock which it attracts can pose quality problems when the land is once again inundated.

A permanent threat to reservoirs is the accumulation of silt generated as a consequence of soil

Table 12.5  Water-quality problems specific to, or particularly important in, reservoirs

---
Decreased water level and volume in store
Turbidity problems arising from siltation
Health and waterborne diseases
Serious symptoms of eutrophication, especially in the tropics
Floating macrophytes, especially in the tropics
Deep-water anoxia
Abundance of toxic cyanobacteria
---

erosion and inappropriate land management. Rates of erosion (and, hence, siltation) are generally recognised to be greater in arid and semi-arid regions but exposed soils on freely draining slopes are a persistent source of entrained sands and silts. The rapid accumulation of inwashed sediment at the lake bottom, at rates measurable in centimetres per year, can shorten the life of small impoundments along the course of sediment-rich rivers to a few decades (Chalar & Tundisi 1999).

Clays and other fine particles washed from catchments cause water quality problems in some types of reservoir as a consequence of the fact that they do **not** sediment or, at least, they are kept in suspension for long periods and are readily resuspended thereafter. The fact that turbidity may restrict excessive phytoplankton growth is offset by its interference of zooplankton feeding (Arruda et al. 1983) and restriction of fish predation, and which thus contribute to a restricted biodiversity. Moreover, the costs of treating turbid water for potability are high, not least because of the problem of sludge disposal.

In-washed particles also carry latent risks of pollution by heavy metals. Whether loads are generated by mining or quarrying and the erosion of unconsolidated spoil, or result from industrial processing, there is an additional concern about the transport and sedimentary accumulation of toxic metals. Reservoir sediments act as traps; the problem is expressed when later changes in acidity or, especially, redox lead to mobilisation. In the case of the Fairmont Lakes, Minnesota, solution of copper accumulated over many years as a result of frequent use of copper sulphate as an algicide rendered water unpotable (Hanson & Stefan 1984).

Health and safety issues raised by microbiological quality challenge the supply of wholesome, potable water from reservoir storage (Szewzyk et al. 2000). Current difficulties are presented by such protistan parasites of the human digestive tracts as *Cryptosporidium* and *Giardia* whose spores are resistant to chlorine disinfection (Oda et al. 2000). Water quality is also threatened by the climatic changes predicted as a consequence of global warming. Altered fluxes of heating and precipitation will affect the duration and stability of thermal stratification and diminished deep water ventilation will lead to lower oxygen concentrations in reservoirs. Stefan et al. (1995) predicted average rises of about 3°C in the surface temperatures of lakes in the north-central USA against a doubling of atmospheric carbon dioxide, independent of lake morphometry; hypolimnetic temperatures for deep lakes were predicted to fall by 1°C. Indirect effects on water quality, emanating from vegetation and agricultural changes, were also predicted.

### 12.4.2  *Tropical reservoirs*

The warmer temperatures experienced in the tropics increase the intensity and onset of the same operational threats which occur in temperate reservoirs, and the longer duration of high water temperatures extends the period in which they may be confronted. Tropical reservoirs present additional difficulties of their own. They may suffer simultaneously the effects of organic pollution, eutrophication and siltation, and with exaggerated symptoms. High phytoplankton production at tropical temperatures is often mainly by cyanobacteria, constituting heavy and persistent blooms. A tendency for continental waters to be deficient in nitrogen favours the heterocystous atmospheric dinitrogen fixers. Among these, *Cylindrospermopsis*, a toxin-producing, nitrogen-fixing species, is expanding its range among eutrophic lakes and reservoirs in the tropical latitudes (Padisák 1997), where it contributes to poorer quality and increased treatment difficulties. Actual and threatened incidences of cyanobacterial toxins reaching the drinking water supply are also being documented (see Chorus & Bartram 1999, for review). An increasing need is for the routine application of simple toxicity monitoring.

One of the phenomena which are confined largely to the tropics is the development of floating vegetation, from the Lemnaceae and ferns (*Azolla, Riccia*) found in small, sheltered pockets, to the persistent, choking development of such free-floating plant species as *Eichhornia crassipes*, *Pistia stratiotes* or *Salvinia auriculata* which beset reservoirs as large as Hartbespoort, Kariba and Tucuruí, soon after their creation (Mitchell et al. 1990; Tundisi 1994). Under these conditions, boat passage is difficult, fishing impossible and floating mats are piled by wind against piers and other structures. Maximal biomass may exceed the 2.1 kg m$^{-2}$ of *E. crassipes* reported in the Pampulha Reservoir, Brasil, and others in the Paraná river catchment (Neiff 1986). Floating macrophytes generate sedimentary organic detritus and potential oxygen deficits, while shading prevents the compensatory generation of photosynthetic oxygen.

Health issues implicating tropical reservoirs include the spread of infective insect vectors of malaria, cholera and typhoid, especially where sewage treatment is inadequate. The incidence of schistosomiasis has increased in the wake of flooding of new tropical reservoirs.

## 12.5 APPROACHES TO RESERVOIR MANAGEMENT

The knowledge required to operate and manage reservoirs in a sustainable way—that is, the system is not spoiled during the lifetimes of the user community—already exists. The achievement of successfully applying that knowledge is still a future goal. Reservoir management is no different from managing other ecosystems in requiring the adoption of respect and understanding of the basic principles of their integrity. These are summarised in Table 12.6, in the context of reservoir management.

The application of these principles requires the substitution of current management approaches, focused on prevention and correction, by a positive outlook of integration and planned sustainability. Preventive management does not offend ecosystemic principles and is infinitely preferable to the expediency of corrective reactivity. Present strategies are not only fundamentally unsustainable, they are expensive and inefficient: we disperse organic and inorganic components of the system by dilution, then try to get them out of the water again, at high energetic cost, just to restore the quality of the original resource. This will continue so long as the responsibilities are divided and are tackled on a piecemeal basis. Ecosystem principles require the common management of the resource, to common objectives of clean production in all uses (domestic, industrial, agriculture) **for the good of the resource**. Savings of materials, costs and energy are beneficial to water quality and to a cleaner environment.

Reservoirs offer several opportunities for the application of systemic approaches to management and which, because they make use of features of design and engineering, including hydrodynamic manipulation, are not necessarily applicable or practical for natural water bodies.

### 12.5.1 *Management of the watershed*

Reservoir systems are open and sensitive to external inputs. Thus, management of the reservoir catchment to minimise pollution is the best way to guard good water: it is preferable, and cheaper, to focus on 'start-of-pipe' methods. Protecting vegetation and catchment soils from erosion is the surest way to restrict the passage of particles and plant nutrients to the water; the transfer of organic matter and biochemical oxidative demand (BOD) can be intercepted by constructed wetlands (Fisk 1989; Hammer & Knight 1994). Many of the design criteria are already known from experiences gained from the use of aquatic plant stands in the primary purification of sewage effluents (Magmedov et al. 1996).

### 12.5.2 *Management of the water column*

Many methods are available for managing water quality *in situ*; these are noted in Table 12.7. The first of these, the separation of a small impoundment at the reservoir inflow, adopts a technology devised by water scientists at the Technical University of Dresden (Uhlmann et al. 1971; Benndorf 1973). The small impoundment acts as a trap, intercepting and sedimenting coarse suspensoids before they can be dissipated throughout the main water body. Accumulation of material is correspondingly rapid but the small area is much more readily dredged or even replaced. Pre-reservoirs

**Table 12.6** Basic ecosystem properties and their relevance to reservoir management

| | |
|---|---|
| Ecosystems conserve energy and matter | Matter cannot disappear: it may only be converted to other forms or moved elsewhere. Saving of energy is important for water: while hydropower contributes electrical energy, creation of new reservoirs may contribute net deterioration of aquatic environments |
| | Energy available underwater is assimilated in additional plant and algal mass supported by eutrophication |
| | Self purification is the result of oxidation of organic matter or its sedimentary deposition. Sediments can represent dangerous accumulations of pollutants. Escape of pollutants back to the water brings catastrophic deterioration of water quality. Pollutants may escape as gases, causing bad smells |
| Ecosystems store information | Aquatic biodiversity is a source of system resistance to external variability. Many aquatic species of plant and animal are exploited by humans and new possibilities are open. The natural structure of watersheds, their flowing water and vegetation are the products of a healthy, functioning environment |
| Ecosystems are open systems | Inflows to reservoirs contribute most to the quality of the impounded water; hence, management of the watershed is a prime option for control of water quality. As the watershed is also open, prevention of adverse transfers is a preferable strategy to their correction. Materials are lost in the reservoir outputs |
| Ecosystems are dissipative | Plant and animal respiration is dissipative; the practice of epilimnetic mixing lowers the photosynthetic compensation and leads to lower average biomass. Decomposition and bacterial degradation of organic matter can lead to anoxia and associated water-quality difficulties |
| Ecosystems grow | Eradication of algal or plant biomass produces only a temporary effect, owing to regrowth (algae in days, plans in a few weeks). Prevention is preferred to repeated harvesting. Killing algae with copper sulphate is inefficient, especially in the tropics, as biomass is rapidly renewed. Moreover, the copper is conserved in the sediments (principle 1) where its toxicity endangers future water use |
| Ecosystems are constrained | Plant biomass does not grow indefinitely – there is a physiologically constrained upper limit. For algal populations, this means that a maximal capacity is reached. Decrease in nutrient loading may not achieve a reduction in biomass until a critical value, when the nutrient constraint is reimposed. Shallow waters and reservoirs which mix more deeply differ in supportive concentration capacity owing to differing constraints imposed by the differing light constraints imposed |
| Ecosystems are differentiated | Biological indication is locally dependent owing to geographical differentiation of species composition. Differentiation of function of the organisms within water bodies permits top-down controls to establish. Biomanipulative techniques must respect the composition of the local fish populations and their interactions. Water bodies are highly differentiated in terms of their microhabitats; features peculiar to certain reservoirs or common to most must be distinguished |
| Ecosystems are multiple-mediated feedback systems | Use of knowledge of regulatory mechanisms underpins many management methods: mixing invokes light constraints; critical loads impose nutrient constraints. The discovery of feedback controls on phytoplankton by zooplankton (Hrbáček 1961) underpins the development of modern techniques of biomanipulation, using controlled fish populations to keep the abundance and species composition of the zooplankton at a value maintaining the optimum possible control on the phytoplankton |
| Ecosystems possess homeostatic capability | The understanding of self-purification capacity of aquatic systems is used for achieving the maximum degree of pollutant loading without deterioration of their ability to continue to function. When capacity is exceeded, severe reactions may ensue |
| Ecosystems are adaptive and self-organising | Organismic adaptation leads to failure to respond to the continued use of chemical controls. Any intervention into dynamic systems leads to changes in their structure. Release of water from a certain depth in a reservoir will lead to altered water quality |

have been successfully introduced in order to cut phosphorus loads to South African reservoirs. Twinch & Grobler (1986) and Pütz (1995) found good agreement between theory and observation in the nutrient removal efficiency of 11 pre-reservoirs in Saxony. It is likely that the efficiency is greatest during seasonally low flows.

Further enhancement of phosphorus load interception is achieved by precipitation and filtration. The most sophisticated example is the elimina-

Table 12.7 Methods for in-lake management of water quality

| Measure | Location or means |
|---|---|
| Pre-reservoirs | At the inflow or submerged |
| Wahnbach reservoir* | Phosphorus elimination at inflow |
| Artificial mixing | Destratification, hypolimnetic aeration, epilimnetic mixing*, layer aeration, Speece cone, propellor mixing |
| Sediment removal | Dredging sediments |
| Sediment aeration | Sediment injection |
| Sediment capping | Covering sediment with sand, foil or any inert material |
| Phosphorus inactivation | Alum or iron precipitation |
| Biomanipulation | Zooplankton control of phytoplankton, by fish management or by water-level control* |
| Hydraulic regulation | Selective off-take withdrawal*, hypolimnion siphoning, curtains* |
| Use of algicide, biocide | Copper- or other poisoning of algae |
| Light reduction | Shade, covering, coloration, suspensions |
| Macrophyte control | Harvest, phytophagous fish, natural enemies |

* Methods specific to reservoirs.

tion plant at the Wahnbach Reservoir, Germany, which was constructed at a pre-reservoir formed where the main inflow river, the Sieg, enters the reservoir. The plant removes up to 96% of the total phosphorus and 92% of the dissolved phosphate. By keeping the phosphorus content in the main body of the reservoir very low, phytoplankton growth is comparable with that of an oligotrophic lake and the costs of subsequent treatment to supply drinking water to the cities of Bonn and Köln need to cover little more than microbial disinfection (Bernhardt & Schell 1979).

Understanding of reservoir hydrodynamics and the physics of thermal stratification assists the design and application of managed inflows, which may be used to control the routing of the incoming water, so as either to enhance or weaken thermal gradients. The use of deep draw-off points in concert with near-surface inputs also reduces the thickness of intermediate metalimnetic layers, and reduces the prospect of maintaining a deep, well-mixed and well-aerated epilimnion (Straškraba & Tundisi 1999). In general, processes which increase vertical mixing (but often avoiding complete destratification) are beneficial in containing algal growth (Steel & Duncan 1999).

Where multiple draw-offs are not available, it is possible to gain some control over flow and stratification through the use of plastic curtains (Asaeda et al. 1996). Such curtains deflect inflowing water in particular directions or prevent particular water masses from abstraction. They may be anchored at the reservoir bottom or suspended from the water surface.

The principle of multiple-mediated feedback control of phytoplankton, where the feeding activities of fish are regulated in order to encourage effective zooplankton herbivory (Hrbáček et al. 1961), is central to the development of biomanipulative management techniques (see Hosper et al., this Volume, Chapter 17). An interesting application specific to reservoirs has been the use of water-level regulation to interfere with the reproduction of planktivorous fish and so provide conditions providing maximum opportunity to zooplankton to graze down the phytoplankton (Zalewski et al. 1995).

### 12.5.3 Management of reservoir outflows

If it is not possible to control the watershed and it is impracticable to manage the water column of the whole reservoir, management might be concentrated on improving the water quality at the draw-off location, essentially by modification of the gas content. Some reported techniques are noted in Table 12.8.

### 12.5.4 Management of new reservoirs

The development of anoxia in newly filled reservoirs on good-quality rivers is not always anticipated. Usually the cause is the oxidation of flooded soils and vegetation, which can be mitigated by land clearance, with minimal soil disturbance, in advance of filling. This is not always practicable: it is often easier to tolerate the anoxia for a few years

Table 12.8  Techniques for management of reservoir outflows

| Technique | Reference |
|---|---|
| Selective withdrawal | Gaillard (1984), Pařízek (1984), Filho *et al.* (1990) |
| Aeration/oxygenation at outlet to hydropower plant | Cassidy (1989) |
| Spillwater re-aeration | Cassidy (1989) |
| 'Czech method' of oxygenation | Haindl (1973) |
| Epilimnetic pumps | Quintero & Garton (1973), Mobley & Harshbarger (1987) |

or, perhaps, to consider some means of hypolimnetic mixing. Anoxia carries the additional risk of metal mobilisation, which may prejudice the uses to which the water is put as well as to prejudice the edibility of its fish.

## 12.6 MONITORING AND WATER QUALITY EVALUATION

Monitoring of water quality is either driven by legislation or it is maintained by tradition but only rarely does it conform to a good, overall experimental design. Moreover, data are accumulated but detailed analysis, reasoned interpretation or application may be lacking (Ward *et al.* 1986). Ideally, monitoring and the collection of data should be treated as a systemic component of problem-solving and goal attainment. The determinands, methods and their precision, sampling locations and frequency, the scheduling and evaluation all need to be chosen with clear objectives. The merits of data evaluation over data accumulation are emphasised in this section.

### 12.6.1 Monitoring

For reservoirs, two basic types of monitoring approach may be separated: the determination of the state of the water resource and the determination of water-quality development within the reservoir. Investigations of the quality of raw resource, especially prior to the construction of reservoirs, reveal the influence of catchment processes on the quality of the stored water and the extent of desirable quality management. In the case of a proposed new reservoir, projections of the change in water quality consequent upon storage need to be predicted and, so far as possible, mitigated at the design stage. The ability to anticipate such changes may well draw on experiences elsewhere where the second approach to monitoring provides interpretable data about how water quality is modified during reservoir storage.

The design of monitoring programmes usually involves a necessary compromise in information gathering between that which is desirable and the costs in human time and in the capital and revenue charges for equipment. Good designs will focus on the most relevant analytical information and set the limits of the precision of the data recorded. Regular sampling frequencies make for easy statistical analysis but the timescales need to relate to the speed of the processes and the period required in order to react to deleterious changes which may be under way; some notion of site-specific risks of quality variability is needed. Thus, monitoring of some kinds of reservoir may require less intensive sampling during periods of low biological activity (such as those during temperate-zone winters) but the nutrient loading rates possible during flood periods may demand more frequent measurement of the inflows. Time-integrated automatic samplers offer the best overall precision, although data evaluation must take account of the variable integration periods.

Manual sampling, automatic data collection and satellite imaging each contribute relevant information to schemes of monitoring. Automatic monitoring systems provide the most consistent streams of certain kinds of data to managers, either through direct cable connections or through radio links. Systems are usually designed to yield adequately precise data but their interpretation is largely a matter of experience and judgement. Some monitoring systems do provide warning signals when predetermined values (e.g. of pH or oxygen content) are reached; the manager may also be advised of appropriate corrective action. At other times, data may be logged automatically and incorporated into retrospective plots (e.g. of temperature structure within the body of the reservoir, or the spatial distribution of phytoplankton chlorophyll). Appropriate reaction, such as the operation of pumps or helixors, is still largely dependent upon the knowledge of the manager.

Remotely sensed imagery is likely to be used only in larger reservoirs. There, subject to adequate calibration and verification (or 'ground truthing'), it may be applied to gauge the horizontal distribution of buoyant Cyanobacteria blooms, high mineral turbidity or organic suspensoids, the fate of fluvial plumes, macrophyte growth and even fish-kills.

### 12.6.2 Evaluation

Evaluation of water quality is usually based on each of the individual determinands achieving a predetermined level of acceptance. In turn, these standards may have been set against general criteria but the judgement about the scale of extraneous inputs and the internal concentrations are most likely conditioned by site-specific characteristics and by the proximal uses of the reservoir water. Properly, standards for drinking-water reservoirs are most critical, and require continuous conformity, but high water quality is also required in some industrial processes.

The following quality criteria are amongst those to be judged in reservoirs.

#### 12.6.2.1 Transparency

Each of the components influencing Secchi-disc transparency (colour, mineral turbidity, algal content) carries its own consequences on reservoir water quality. Colour, usually owing to organic (humic- or fulvic-acid) staining, is governed by criteria for the supply of drinking water, removal of which requires treatment with oxidants or bleaching. Clay and fine silt particles can interfere with filtration and increase costs significantly. However, the removal of algae is also a fundamental step in the treatment of potable water and high chlorophyll concentrations present obvious and expensive difficulties employing filtration or coagulation as a primary mechanism. Besides the problem of their potential toxicity, some kinds of Cyanobacteria are sometimes able to penetrate floc blankets and render the final product unacceptable. Even after algae have been removed, their erstwhile presence in abundance imparts surfactant products and off-flavours to the final product (Suffet et al. 1995). Where experience identifies a local difficulty, then monitoring must be able to provide warning and to trigger avoidance or corrective actions (Izaguirre & Devall 1995).

#### 12.6.2.2 Oxygen concentration

The natural oxygen concentration of reservoir water is the result of solution from the atmosphere, to around the equilibrium value for the given temperature and air pressure. Higher concentrations are generally the consequence of photosynthesis of submerged plants and algae. Lower concentrations are due to the respiration of organisms and consumption by chemical oxidation proceeding faster than replacement. Below particular species-specific thresholds, low oxygen concentrations are injurious to fish, are highly selective to benthic invertebrate communities and, at very low concentrations, conducive to mobility of reduced ions. Thus, the concentration of dissolved oxygen, either absolutely or relative to saturation at the ambient temperature and atmospheric pressure, is a commonly used indicator of water quality.

#### 12.6.2.3 Phosphorus concentration

The relevance of phosphorus concentration is its often key role as the capacity limitation on the autochthonously produced biomass. The strong relationship found between the average standing crop of phytoplankton chlorophyll and the mean in-lake concentrations of total phosphorus across a wide range of (mostly) temperate lakes and reservoirs (Sakamoto 1966) has been developed into one of the best-known general models (Vollenweider & Kerekes 1980) available to limnologists and to water-resource strategists. Its logarithmic formulation conceals what is widely recognised to be a sigmoid function (Straškraba & Gnauck 1985; Prairie et al. 1989): high concentrations of phosphorus may saturate the chlorophyll capacity determined by other factors, such as light; there are external concentrations (in the order of $10^{-8}$ molar) at which it becomes physically impossible to accumulate phosphate ions within cells of the producer organism (Falkner et al. 1989). The logarithmic format also disguises the high variability in the mean chlorophyll concentrations observed in lakes and reservoirs with high total phosphorus concentrations. There have been many attempts to explain this behaviour (zooplankton grazing and associated food-web effects being among the more common) but the relationship is descriptive rather than mechanistic. This makes it difficult and unwise to apply the model as a tool for managing an individual reservoir (Reynolds 1992b). It may be that the data collected in routine monitoring may reveal site-specific relationships between plankton biomass and fluctuations in phosphorus concentration but the main value is in the detection of altered supportive capacity. Even here there are interpretative difficulties because the routine determinands do not necessarily coincide with what is biologically available. Analyses of molybdate-reactive phosphorus in the water (variously referred to as 'orthophosphate phosphorus' or 'soluble reactive phosphorus') miss the phosphorus which is already intraorganismic. On the other hand, determinations of 'total phosphorus' include fractions which are sorbed on to clay and fine particles or flocs of metal oxides or which are chemically bound in organic and inorganic fractions. Neither are readily bioavailable. It is possible, by using tedious, serial fractionation, to approximate better the true bioavailability; however, these methods are scarcely practical adjuncts to a monitoring programme. In truth, the most useful data to come from monitoring are those which reveal the magnitude of the external load delivery (inflow concentrations × inflow discharges), the outflow resource loss and, by deduction, the phosphorus resource accumulated in the reservoir. Such budgetary data are invaluable to assessments and calculations of impacts of alternative management strategies at the design and evaluation stages.

#### 12.6.2.4 Nitrogen concentration

Inorganic nitrogen species (especially nitrate and ammonium) are routinely determined indicators of reservoir water quality, both as indications of polluting inputs and of acceptable outputs. Dissolution in rain water of nitrate concentrations exceeding $0.1–0.2\,mg\,N\,L^{-1}$ are indicative of atmospheric pollution. Leachates and ground waters from agricultural catchments may greatly enhance nitrate concentrations in drainage and in reservoir inflows (perhaps as high as $10–15\,mg\,N\,L^{-1}$), while substantial reduction to ammonium occurs at redox potentials of c.150mV. Poor sewage oxidation, industrial animal husbandry and some industrial processes may contribute to elevated ammonium concentrations in reservoir source waters. Verification of their net oxidation during storage is a further desirable output of monitoring.

#### 12.6.2.5 Organic matter

The array of organic constituents, natural and anthropogenic, autochthonous and allochthonous, dissolved and particulate, readily decomposed and refractory, are generally monitored by a bulk analogy. Biochemical oxidative demand (BOD), chemical oxidative demand (COD) and permanganate- or dichromate-reducing capacity have long been used. They are not sensitive to identical fractions and the resultant approximations are not necessarily well correlated. The standard 5-day BOD determination is perhaps the most useful in representing the readily decomposable sources of biomass carbon, much of which may be generated within the reservoir. According to Straškrabová et al. (1993), 1 mg chlorophyll a is stoichiometrically equivalent to a potential BOD of c.130mg of oxygen.

#### 12.6.2.6 Mineral composition, hardness and salinity

These are quality criteria in water supplied for drinking and for industrial processing of foodstuffs and beverages, where consistency of composition is often essential. Regeneration is especially important in regions experiencing marked seasonal differences in raw water flow and, especially, in semi-arid regions, where chloride and sulphate concentrations in the reservoir are liable to fluctuate substantially.

### 12.6.2.7  pH

Carbon-dioxide solution in non-buffered, air-equilibrated surface waters usually keeps them in a pH range 6.0 to 6.8. However, consumption of carbon dioxide during intense photosynthesis by phytoplankton, periphyton or submerged macrophytes can lead to rapid upward pH drift, perhaps to pH 9 or 10 (Talling 1976). In harder waters, the bicarbonate–carbon-dioxide system drives the ambient pH towards 8.3 and, so long as bicarbonate persists, buffers against the high pH attainable in soft waters. Acidity values commonly or persistently below pH 6.0 generally indicate dystrophic water sources (often emanating from bogs). Acidification of rain (through atmospheric dissolution of anthropogenically enhanced sulphate and nitrate concentrations), uncorrected during overland flow, also leads to depressed reservoir pH. Chemical reactions (with, for instance, pyrite) within the watershed can also generate low pH. The main concern about low pH for the reservoir manager may be the secondary consequences of metal lability (especially of aluminium and transition metals such as lead, zinc and copper). Corrective buffering is simple enough to devise but it is another reason for careful monitoring of the pH of water abstracted for treatment and distribution.

### 12.6.2.8  Heavy metals

Where there is a current risk of metal contamination or where the reservoir sediments already contain a burden of heavy metals accumulated as a consequence of anthropogenic activities, monitoring programmes must accommodate detection of toxic metals at limits in the nanomolar range.

### 12.6.2.9  Toxic organic compounds

Commonly encountered anthropogenic organic chemicals which are poisonous, carcinogenic or suspected of serious health consequences include petroleum products, heavy fractions of mineral oil, polychlorinated biphenyls, polycyclic aromatic hydrocarbons, phenols, various pesticide residues and a recently discovered group of endocrine disrupters. Many suppliers of drinking water are required not just to demonstrate that their products are wholesome but that they are capable of removing these compounds (for instance, by filtration through granular activated carbon) from potable water, whenever a need arises. This is an important driver for monitoring.

### 12.6.2.10  Colour

Because staining of water with dissolved humic or fulvic material is prevalent in some source waters but is often intolerable in the final product, colour must often be closely monitored. In between, interference with the penetration of the stored water by light may be welcome (fewer algae) or otherwise (poorer bleaching).

### 12.6.2.11  Bacteria

Psychrophilic and mesophilic bacteria respond to the presence of easily decomposable organic matter. Numbers fluctuate depending on trophic state, throughflow and vertical mixing. It is usually important in microbiological monitoring to distinguish coliforms and streptococcoids for, being primarily faecal in origin, their numbers are taken as good indicators of the extent of corruption by inadequately treated sewage.

### 12.6.2.12  Phytoplankton

Phytoplankton species composition in reservoirs responds to many seasonally variable hydrographic and hydrochemical drivers, most of which are geographically determined. For this reason, the only good systems for recognising and interpreting changes in composition and abundance are those which are locally devised: there are global outlines but no universal interpretative protocols. However, sound evaluation does depend everywhere on an adequate knowledge of the phytoplankton species likely to be present. This rarely requires a team of taxonomists-in-residence, however, but personnel who can distinguish the main groups of species likely to affect the treatment process or the quality of the final product are essential. Large populations of planktonic diatoms are notorious for clogging slow sand filters. Special care must be taken that mucilage-producing species are not allowed to dry out in treatment systems (where they will envelope everything in a polysaccharide matrix). Breakthrough of floc blankets by Cyanobacteria is particularly distressing to water suppliers, especially as many of the species are potentially toxic if consumed in sufficient quantities, although the effects of persistent exposure of consumers to sublethal doses of toxins are still debated (Codd 1995; but see Chorus & Bartram 1999).

Historically, the presence and abundance of phytoplankton in reservoirs have been monitored from microscope counts of cells or colonies filtered or sedimented from routinely collected water samples. Owing to the large disparity in cell size and in habit among the plankton, sequences of the raw numbers of cells per millilitre on a Sedgewick-Rafter make very poor records. It is possible to calculate (or recalculate) biomass as the sum of the products of each species by its cell volume (assuming a constant density), a process simplified by today's computer applications, some of which permit the sizes of individual organisms to be cumulated. However, most monitoring of phytoplankton quantity invokes the measurement of chlorophyll $a$ pigment as a surrogate of live biomass. In some cases, fluorescence is used as an index of the light harvesting capacity of the chlorophyll, and the spectral variance in the emitted light is sensitive to interphyletic differences among the main algae present.

### 12.6.2.13 Periphyton

Microscopic examination of periphytic growth on glass slides suspended at various depths in a reservoir can provide an index of water quality conditions. The habitat fidelity of those species of algae which grow on surfaces is at least as great as that of phytoplankton (Biggs et al. 1998).

### 12.6.2.14 Zooplankton

Water-quality evaluation by analysis of the species make-up of the zooplankton is not well-developed, although some attempts to rectify this have been made (Berzins & Bertilsson 1989). The presence and relative abundance of large *Daphnia* in reservoir plankton are usually indicative of effective phytoplankton grazing and/or minimal predation by fish. By implication, this is also taken by reservoir managers to be indicative of good water quality and minimal treatment problems.

### 12.6.3 *Water quality indices*

Owing to the large number of water quality criteria and to the wide extent of their variability, the idea of devising a single index of water quality has always seemed attractive. However, it is evident that the number of indices proposed is as large as the number of variables themselves (Thanh & Biswas 1990). Most of them also cover lakes or else they are fairly specific to particular geographical regions. The only system specific to reservoirs is that based on the components of the fish assemblage, developed by Dolman (1990).

None of these indices is generally accepted. The general lake water quality index (GLWQI), originally formulated by Malin (1984) for Finnish lakes, has attracted some following. The index is based upon variables selected through a principal components analysis: oxygen concentration; conductivity; pH; colour; manganese and total phosphorus concentration. The index is composited as a dimensionless number, each variable being represented by a graphically depicted value function (scaled from 0 to 1) based on a regression analysis. Then the value functions indicate how individual measurements compare relative to national or regional data, and the GLWQI is calculated as the volume-weighted mean of all the variables (and which is also located on the scale 0 to 1). No

**Table 12.9** Uses of models of reservoir water quality. (Modified after Straškraba & Tundisi 1999)

| | |
|---|---|
| In all situations | To estimate pollution sources in the watershed |
| Before reservoir construction | To provide reasonable estimates of water quality at alternative sites of construction, of alternative depth, dam height and outflow structures |
| | To estimate budgets of major water-quality components, in the finished reservoir and its outflows |
| | To predict conditions in reservoir and to compare consequences of alternative management options |
| For existing reservoirs | To predict consequences of altered environmental conditions in the watershed |
| | To provide estimates of water-quality yields from alternative options in long-term resource-use strategies |
| | To support short-term operational decisions for optimising quality of outputs |
| | To optimise sampling schedules for investigations and quality monitoring |

attempt has been made to separate classes of water body on the strength of the GLWQI score.

Another trend among general surveys is that they are applied within *ecoregions*. Ecoregions are defined as geographical units of relative homogeneity of topography, climate, soil, natural vegetation and land use. Within an ecoregion, all water bodies form a coherent group and are supposed to be similar to each other but for the influence of permanent human settlement. They are frequently less diverse than those of the entire nations (Hughes *et al.* 1992). Their value is the objective basis of the regional criteria for protection which are established (Biggs *et al.* 1990). Mostly, these are difficult to use in reservoir management, except where there is a high concentration of artificial water bodies. What tends to happen in many instances is that each reservoir belongs, often uniquely, in another ecoregion.

## 12.7 MATHEMATICAL MODELS AS TOOLS FOR RESERVOIR MANAGEMENT

Mathematical models represent a digestion of understanding into a concise formulaic form. The information included is generally restricted to a few key topics only; what is actually incorporated is the choice of model's author. There is, as yet, no general model for all purposes. Those which perform well are usually simple and directed to particular parts of whole systems and apply to one or a small number of sites only. Management of water quality in given reservoirs is highly subject to ecological behaviour but its controls are generally complex and difficult to predict, even by ecologists; often, the manager's experience and intuition provides the predictions and consequences of the regulatory actions available. The cost implications of risk may be large but the financial investment in operator experience is also seen as a costly overhead. So mathematical models, computer simulations and decision-supporting expert systems are seen as highly desirable tools for management.

A wide array of such models are available and are used extensively in the water industry (Alasaarela *et al.* 1993). The more successful of these relate to resource quantity. Quality models are more difficult to devise and validate but a number of these are in use (see Table 12.9). They fall into several categories.

### 12.7.1 Simple static calculation models

These comprise relatively simple algebraic equations and/or graphs. Usually, their derivation is statistical fitting based on large data sets. Thus, they are limited by the experienced behaviour represented within the data set and are not representative of events outside the period of observations (are more extreme events predicted or likely?). The applicability of these models increases with the time span of the data upon which they are based. Inputs based upon a small number of data points, seasons or locations must be considered unreliable. Moreover, such models rarely transport to

Table 12.10 Some simple models for estimating the responses and management consequences of variable features of reservoirs

| | |
|---|---|
| Nitrogen retention in shallow reservoirs | Kelly et al. (1987) |
| Pre-reservoir phosphorus reduction | Uhlmann et al. (1971), Benndorf et al. (1975) |
| Chlorophyll, as a function of phosphorus | Vollenweider & Kerekes (1980), Prairie et al. (1989) |
| Phosphorus retention in reservoirs | Straškraba et al. (1995) |
| Organic matter retention | Straškrabová (1976) |
| Hypolimnetic oxygen demand | Stauffer (1987) |
| Lake number – reservoir stratification | Imberger & Patterson (1990) |
| Temperature stratification | Straškraba & Gnauck (1985) |
| Dissolved oxygen and phosphorus in lakes | Chapra & Canale (1991) |
| Late summer oxygen profiles | Molot et al. (1992) |
| Hypolimnetic discharge | Horstman et al. (1983) |

Table 12.11 Complex dynamic ecological models permitting alternative scenario inputs to be simulated

| Model name | Main simulation | References |
|---|---|---|
| AQUAMOD 3 | Stratification | Straškraba & Gnauck (1985) |
| SALMO | Epilimnion and hypolimnion | Benndorf & Recknagel (1982) |
| SALMOSED | Phosphorus exchange with sediments | Recknagel et al. (1995) |
| COMPREHENSIVE | Kasumigaura lake model | Goda & Matsuoka (1986) |
| CE-QUAL-RIV1 | Reservoir eutrophication | Bedford et al. (1983) |
| WASP4 | Hydrodynamics, mass transport | Ambrose et al. (1988) |
| MINLAKE | Climate effects | Riley & Stefan (1988) |
| DYRESM | Reservoir hydrodynamics | Imberger & Patterson (1981) |
| DYRESM-WQ/DYRESM | For water quality | Hamilton & Schladow (1995) |
| Finnish 3-D model | Hydrodynamic–ecological coupling | Virtanen et al. (1986) |

other reservoirs, even within the same ecoregion, as many of the statistical components are peculiar to the site at which the data set was collected and accumulated, influenced by all the idiosyncrasies of its hydro-meteorological variability. Some examples are tabulated (see Table 12.10).

### 12.7.2 Complex dynamic prescriptive models

These provide analyses of time-dependent aspects of water quality. Calculations are not prescriptive but indicate impacts of input variables altered under scenario constructions. In this way, the consequences and refinements of altered management are tested experimentally within a wholly theoretical or simulated environment. Some examples are listed in Table 12.11.

### 12.7.3 Geographical information systems (GIS)

Storage and handling problems require the spatial resolution which GIS is able to provide. The basis of GIS is a series of computerised maps and procedures for accommodating and working with spatial data. Several applications of GIS to watershed management have been described (DePinto & Rodgers 1994; Wool et al. 1994).

Table 12.12  Models for use in water quality optimisation

| | |
|---|---|
| Optimal control by selective withdrawal | Fontane et al. (1981) |
| Optimising reservoir operation for downstream resources | Sale et al. (1982) |
| GIRL OLGA, minimisation of eutrophication abatement cost | Schindler & Straškraba (1982) |
| GFMOLP, a fuzzy multi-objective for planning watersheds | Chang et al. (1996) |

### 12.7.4  Management optimisation models

Management optimisation models (see Table 12.12) incorporate selection protocols for choosing the most suitable management option from among alternatives, according to a set of criteria appropriate to the problem. Such models permit the simultaneous analysis of several alternatives (multiparameter models) or several outcomes (multigoal models). Of course, the possibilities for optimisation are limited to those alternatives included in the model and the validity of outputs relies on the assumptions and constraints of the model formulations.

### 12.7.5  Expert systems

These use qualitative and quantitative formulations to guide the user towards relevant answers to complex water-quality questions. The most important advantage of expert systems is their ability to cover qualitative characteristics in setting and resolving complex decision rules.

### 12.7.6  Decision support systems (DSS)

These are a relatively new derivative from expert systems. They use computer software products to address complex sets of problems according to the multiple output formulations incorporated into their design. Graphics packages may be adjoined to assist with decision support. Generally, the user is led through the possible solutions on the basis of answers to questions prompted by the software. Two examples are noted in Table 12.12.

## 12.8  THE LIMNOLOGY OF RESERVOIRS—CONCLUSIONS

This chapter has attempted to highlight some of the features which distinguish reservoirs from natural lakes. Although both represent impoundments of natural gravitational hydraulic flow, the special design features and the hydrological control of reservoirs provide numerous distinguishing features.

The purpose for which the reservoir was constructed and the mode of its operation may affect, more or less profoundly, the hydraulics of the water flow and the composition of the solutes and suspensoids in the water itself. The same drivers underpin the practical requirements to monitor and manage the quality of the water in store.

Increasingly, the availability of models and expert systems is assisting the satisfaction of both requirements. As techniques improve and understanding is gained, it is inevitable that some of the new knowledge will flow in the opposite direction: the science of limnology and lacustrine systems will be enriched by lessons learned from experience gained peculiarly from the construction and operation of those special kinds of lakes called reservoirs.

## 12.9  REFERENCES

Agostinho, A.A., Miranda, L.E., Bini, L.M., et al. (1999) Patterns of colonization in neotropical reservoirs and prognoses on aging. In: Tundisi, J.G. & Straškraba, M. (eds), *Theoretical Reservoir Ecology and its Applications*. International Institute for Ecology, São Carlos, 227–65.

Alasaarela, E., Virtanen, M. & Koponen, J. (1993) The Bothnian Bay Project—past, present and future. *Aqua Fennica*, **23**, 117–24.

Ambrose, R.B., Wool, P.E.T.A., Connoly, J.P. & Schanz, R.W. (1988) *ASP 4, a Hydrodynamic and Water Quality Model: Model Theory, User's Manual and Programmer's Guide*. Environmental Research Laboratory, U.S. Environmental Protection Agency, Athens, GA, 48 pp.

Arruda, J.A., Marzolf, G.R. & Faulk, R.T. (1983) The role

of suspended sediments in the nutrition of zooplankton in turbid reservoirs. *Ecology*, **64**, 1225-35.

Asaeda, T., Priyantha, D.G.N., Saitoh, S. & Gotoh, K. (1996) A new technique for controlling algal blooms in the withdrawal zone of reservoirs using vertical curtains. *Ecological Engineering*, **7**, 95-104.

Avakyan, A.B. & Iakovleva, V.B. (1998) Status of global reservoirs: the position in the late twentieth century. *Lakes and Reservoirs, Research and Management*, **3**, 45-52.

Balon, E.K. & Coche, A.G. (eds) (1974) *Lake Kariba: a Man-made Tropical Ecosystem*. Dr W Junk, The Hague, 247 pp.

Barbosa, F.A.R., Padisák, J., Espíndola, E.L.G. *et al.* (1999) The cascading reservoir continuum concept (CRCC) and its application to the River Tieté basin, São Paulo State, Brazil. In: Tundisi, J.G. & Straškraba, M. (eds), *Theoretical Reservoir Ecology and its Applications*. International Institute for Ecology, São Carlos, 425-37.

Bedford, K.W., Sykes, R.M. & Libicki, C., 1983. Dynamic advective water quality model for rivers. *Journal of the Environmental Engineering Division, American Society of Civil Engineers*, **109**, 535-54.

Beeton, A.M. (1984) The world's great lakes. *Journal of Great Lakes Research*, **10**, 106-13.

Benndorf, J. (1973) Prognose des Stoffhaushaltes von Staugewässern mit Hilfe kontinuerliche und semikontinuerliche biologischer Modelle. *Internationale Revue des gesamten Hydrobiologie*, **58**, 1-18.

Benndorf, J. & Recknagel, F. (1982) Problems of application of the ecological model SALMO to lakes and reservoirs having various trophic status. *Ecological Modelling*, **17**, 129-45.

Benndorf, J., Zesch, M. & Weisner, E.M. (1975) Prognose der Phytoplankton-entwicklung in geplanten Talsperren durch Kombination von wachsturmkinerischen Modellvorstellungen und Analogiebetrachtungen zu bestehenden Talsperren. *Internationale Revue des gesamten Hydrobiologie*, **60**, 737-58.

Bernhardt, H. & Schell, H. (1979) The technical concept of phosphorus elimination at the Wahnbach estuary using floc-filtration (The Wahnbach system). *Zeitschrift fηr Wasser- und Abwasser Forschung*, **12**, 78-88.

Berzins, B. & Bertilsson, J. (1989) On limnic microcrustaceans and trophic degree. *Hydrobiologia*, **185**, 95-100.

Biggs, B.J.F., Duncan, M.J., Jowett, I.G., *et al.* (1990) Ecological characterization, classification and modeling New Zealand rivers; an introduction and synthesis. *New Zealand Journal of Marine and Freshwater Science*, **24**, 277-304.

Biggs, B.J.F., Stevenson, J.R. & Lowe, R.L. (1998) A habitat matrix conceptual model for stream periphyton. *Archiv für Hydrobiologie*, **143**, 21-56.

Carvalho, N. de O. (1988) Sediment yield in the Velhas River (Minas Gerais, Brazil). In: Bordas, M.P. & Walling, D.E. (eds), *Sediment Budgets*. International Association of Hydrological Sciences, Wallingford, 369-75.

Cassidy, R.A. (1989) Water temperature, dissolved oxygen and turbidity control in reservoir releases. In: Gore, J.A. & Petts, G.E. (eds), *Alternatives in River Regulation Management*. CRC Press, Boca Raton, FL, 27-62.

Chalar, G. & Tundisi, J.G. (1999) Main processes in the water column determined by wind and rainfall at Lobo (Broa) Reservoir. Implications for phosphorus cycling. In: Tundisi, J.G. & Straškraba, M. (eds), *Theoretical Reservoir Ecology and its Applications*. International Institute for Ecology, São Carlos, 53-65.

Chang, N.-B., Wen, C.G., Chen, Y.L. & Yong, Y.C. (1996) A grey, fuzzy multiobjective programming approach for the optimal planning of a reservoir watershed. Part A: theoretical development. *Water Research*, **30**, 2329-34.

Chapra, S.C. & Canale, R.P. (1991) Long term phenological model of phosphorus and oxygen for stratified lakes. *Water Research*, **25**, 707-15.

Chorus, I. & Bartram, J. (1999) *Toxic Cyanobacteria in Water*. E. & F.N. Spon, London, 416 pp.

Codd, G.A. (1995) Cyanobacterial toxins: occurrence, properties and biological significance. *Water Science and Technology*, **32**, 149-56.

Cullen, P. (1991) Responses to changing nutrient inputs over a twenty-year period in Lake Burley-Griffin. *Verhandlungen Internationale Vereinigung Limnologie*, **24**, 1471-76.

DePinto, J.V. & Rodgers, P.W. (1994) Development of GEO-WAMS: a modelling support system for integrating GIS with watershed analysis models. *Lake and Reservoir Management*, **9**, 68-9.

Dolman, W.B. (1990) Classification of Texas Reservoirs in relation to limnology and fish community associations. *Transactions of the American Fisheries Society*, **119**, 511-20.

Falkner, G., Falkner, R. & Schwab, A. (1989) Bioenergetic characterisation of transient state phosphate uptake by the Cyanobacterium, *Anacystis nidulans*. *Archives of Microbiology*, **152**, 353-61.

Ferguson, A.J.D. & Harper, D.M. (1982) Rutland Water

phytoplankton—development of an asset or a nuisance? *Hydrobiologia*, **88**, 117–33.

Fernando, C.H., Holčik, J. (1991) Fish in reservoirs. *Internationale Revue des gesamten Hydrobiologie*, **76**, 149–67.

Filho, M.C.A., de Jesus, J.A.O., Branski, J.M. & Hernandez, J.A.M. (1990) Mathematical modelling for reservoir water quality management through hydraulic structures: a case study. *Ecological Modelling*, **52**, 73–85.

Fisk, D.W. (1989) *Wetlands: Concerns and Successes*. American Water Resources Association, Bethesda, MD, 212 pp.

Fontane, D.F., Labadie, J.W. & Loftis, B. (1981) Optimal control of reservoir discharge quality through selective withdrawal. *Water Resources Research*, **17**, 1594–604.

Gaillard, J. (1984) Multilevel withdrawal and water quality. *Journal of the Environmental Engineering Division, American Society of Civil Engineers*, **110**, 123–5.

Goda, T. & Matsuoka, Y. (1986) Synthesis and analysis of a comprehensive lake model—with the evaluation of diversity of ecosystems. *Ecological Modelling*, **31**, 11–32.

Gons, H.J., Burger-Wiersma, T., Otten, J.H. & Rijkboer, M. (1992) Coupling of phytoplankton and detritus in a shallow eutrophic lake (Lake Loosdrecht, The Netherlands). *Hydrobiologia*, **233**, 51–9.

Gorham, E. & Boyce, F.M. (1989) Influence of lake surface area and depth on thermal stratification and the depth of the summer thermocline. *Journal of Great Lakes Research*, **15**, 233–45.

Grimard, Y. & Jones, H.G. (1982) Trophic upsurge in new reservoirs; a model for total phosphorus concentrations. *Canadian Journal of Fisheries and Aquatic Sciences*, **39**, 1473–83.

Haindl, K. (1973) Suitable location of bottom outlets of dams and oxidation outlets for the improvement of water quality in rivers. In: *Research and Practice in the Water Environment*, Vol. II, *Fifteenth Congress of the International Association for Hydraulic Research, Istanbul, 1973*. State Hydraulic Works of Turkey, Istanbul, 187–94.

Hamilton, D.P. & Schadlow, S.G. (1997) Prediction of water quality in lakes and reservoirs. Part I. Model description. *Ecological Modelling*, **96**, 91–110.

Hammer, D.A. & Knight, R.L. (1994) Designing constructed wetlands for nitrogen removal. *Water Science and Technology*, **29**, 15–27.

Hanna, M. (1990) Evaluation of models for predicting mixing depth. *Canadian Journal of Fisheries and Aquatic Sciences*, **47**, 940–7.

Hanson, M.J. & Stefan, H.G. (1984) Side effects of 58 years of copper sulphate treatment of the Fairmont Lakes, Minnesota. *Water Resources Bulletin*, **20**, 889–900.

Henry, R. (1999) Heat budgets, thermal structure and dissolved oxygen in Brazilian reservoirs. In: Tundisi, J.G. & Straškraba, M. (eds), *Theoretical Reservoir Ecology and its Applications*. International Institute for Ecology, São Carlos, 125–51.

Hocking, G. & Straškraba, M. (1994) An analysis of the effect of an upstream reservoir by means of a mathematical model of reservoir hydrodynamics. *Water Science and Technology*, **30**, 91–8.

Holz, J.C., Hoagland, K.D., Spawn, R.L., et al. (1997) Phytoplankton community response to reservoir aging, 1968–1992. *Hydrobiologia*, **346**, 183–92.

Horstman, H.K., Copp, R.S. & Browne, F.X. (1983) Use of predictive phosphorus model to evaluate hypolimnetic discharge scenarios of Lake Wallenpupack. In: *Lake and Reservoir Management*. Environmental Protection Agency, Washington, 165–70.

Howarth, R.W., Billen, G., Swaney, D., et al. (1996) Regional nitrogen budgets and riverine N and P fluxes for the drainages to the North Atlantic Ocean: natural and human influences. *Biogeochemistry*, **20**, 1–65.

Hrbáček, J., Dvořaková J., Kořínek, V. & Procháková, L. (1961) Demonstration of the effect of fish stock on the species composition of zooplankton and the intensity of the whole plankton association. *Verhandlungen International Vereinigung Limnologie*, **14**, 192–5.

Hughes, R.M., Whittier, T.R., Thiele, S.A. & Pollard J.E. (1992) Lake and stream indicators for U.S. EPA's environmental monitoring and assessment program. In: McKenzie, D. (ed.) *Ecological Indicators*. Elsevier, Amsterdam, 305–36.

ICOLD (1984) *World Register of Dams*. International Commission On Large Dams, Paris.

Imberger, J. & Patterson, J.C. (1981) A dynamic reservoir simulation model, DYRESM-5. In: Fischer, H.G. (ed.), *Transport Models for Inland and Coastal waters*. Academic Press, New York, 310–61.

Imberger, J. & Patterson, J.C. (1990) Physical limnology. *Advances in Applied Mechanics*, **27**, 303–475.

Istvánovics, V. & Somlyódy, L. (1998) The role of sediments in P retention of the Kis-Balaton reservoir. *Internationale Revue des gesamten Hydrobiologie*, **83**, 225–33.

Izaguirre, G. & Devall, J. (1995) Resource control and management of taste-and-odor problems. In: Suffet, I.H., Mallevialle, J. & Kawczynski, E. (eds), *Advances in Taste-and-odor Treatment and Control*. American Waterworks Association, Denver, 23–74.

Javornický, P. (1966) Light as the main factor limiting the development of diatoms in the Slapy Reservoir, 1958–1960. *Verhandlungen Internationale Vereinigung Limnologie*, **16**, 701–12.

Kelly, C.A., Rudd, J.W.M., Hesslein, R.H., et al. (1987) Prediction of biological acid neutralization in acid-sensitive lakes. *Biogeochemistry*, **3**, 85–90.

Kennedy, R.H. (1999) Reservoir design and operation: limnological implications and management opportunities. In: Tundisi, J.G. & Straškraba, M. (eds), *Theoretical Reservoir Ecology and its Applications*. International Institute for Ecology, São Carlos, 1–28.

Kimmel, B.L. & Groeger, A.W. (1986) Limnological and ecological change associated with reservoir ageing. In: Hall, G.E. & Van Den Avyle, M.J. (eds), *Reservoir Fisheries Management*. American Fisheries Society, Bethesda, 103–9.

Kirk, J.T.O. (1994) *Light and Photosynthesis in Aquatic Ecosystems* (2nd edition). Cambridge University Press, Cambridge, 528 pp.

Kubečka, J. (1993) Succession of fish communities in reservoirs of central and eastern Europe. In: Straškraba, M., Tundisi, J.G. & Duncan, A. (eds), *Comparative Reservoir Limnology and Water Quality Management*. Kluwer Academic Publishers, Dordrecht, 153–68.

Magmedov, V.G., Zacharenko, M.A., Yakovleva, L.I. & Ince, M.E. (1996) The use of constructed wetlands for the treatment and runoff and drainage waters: the UK and Ukraine experience. *Water Science and Technology*, **33**, 315–24.

Malin, V. (1984) A general lake water quality index. *Aqua Fennica*, **14**, 139–45.

Margalef, R. (1960) Ideas for synthetic approach to the ecology of running waters. *Internationale Revue des gesamten Hydrobiologie*, **45**, 133–53.

Margalef, R. (1975) Typology of reservoirs. *Verhandlungen Internationale Vereinigung Limnologie*, **19**, 1841–8.

Margalef, R., Planas, D., Armengol, J., et al. (1976) *Limnologia de los embalses españoles*. Dirección General des Obras Hidrólicas, Madrid, 421 pp.

Matsumura-Tundisi, T. & Rocha, O. (1983) Occurrence of copepods (Calanoida, Cyclopoida and Harpacticoida) from Broa Reservoir (São Carlos, São Paulo, Brasil). *Revista Brasileira de Biológia*, **13**, 1–17.

Mazumder, A. & Taylor, W.D. (1994) Thermal structure of lakes varying in size and water clarity. *Limnology and Oceanography*, **39**, 968–76.

Mitchell, D.S., Pieterse, A.H. & Murphy K. (1990) Aquatic weed problems and management in Africa. In: Pieterse, A.H. & Murphy, K. (eds), *Aquatic Weeds*. Oxford University Press, Oxford, 341–54.

Mobley, M.H. & Harshbarger, E.D. (1987) Epilimnetic pumps to improve reservoir releases. In: U.S. Army Corps of Engineers, *Proceedings of a Workshop on Reservoir Releases*. U.S Army Corps of Engineers, Vicksburg, 133–5.

Molot, L.A., Dillon, P.J., Clark, B.J. & Neary, B.P. (1992) Predicting end-of-summer oxygen profiles in stratified lakes. *Canadian Journal of Fisheries and aquatic Sciences*, **49**, 2363–72.

Neiff, J.J. (1986) Aquatic plants of the Paraná system. In: Davies, B.R. & Walker, K.F. (eds), *The Ecology of River Systems*. Dr W. Junk bv, Dordrecht, 557–71.

Newrkla, P. & Duncan, A. (1984) The biology and density of *Ehirava fluviatilis* (Clupeoid) in Parakrama Samudra (Sri Lanka). *Verhandlungen Internationale Vereinigung Limnologie*, **22**, 1572–78.

Nilssen, J.P. (1984) Tropical lakes: functional ecology and future development. The need for a process-orientated approach. *Hydrobiologia*, **113**, 231–42.

Oda, T., Sakagami, M. & Ito, H. (2000) Size-selective continuous-flow filtration method for the detection of *Cryptosporidium* and *Giardia*. *Water Research*, **34**, 4477–81.

Ostrofsky, M.L. (1978) Trophic changes in reservoirs; an hypothesis using phosphorus models. *Internationale Revue des gesamten Hydrobiologie*, **64**, 481–99.

Padisák J. (1997) *Cylindrospermopsis raciborskii* (Wolszynska) Seenayya et Suba Raju, an expanding, highly adaptive Cyanobacterium: worldwide distribution and review of its ecology. *Archiv für Hydrobiologie (Supplement)*, **107**, 563–93.

Pařlzek, J. (1984) Využití efektu čisté vrstvy. In: Straškraba, M., Brandl, Z. & Porcalová (eds), *Hydrobiologie a kvalita vody údolnich nadrží*. ČSTVS, České Budějovice, 72–83.

Prairie, Y.T., Duarte, C.M. & Kalff, J. (1989) Unifying nutrient-chlorophyll relationships in lakes. *Canadian Journal of Fisheries and Aquatic Sciences*, **46**, 1176–82.

Pütz, K. (1995) The importance of pre-reservoirs for the water-quality management of reservoirs. *Aqua*, **44**, 50–5.

Quintero, J.E. & Garton J.E. (1973) A low energy lake destratifies. *Transactions of the American Society of Agricultural Engineers*, **16**: 973–8.

Quirós, R. (1998) Fish effects on trophic relationships in the pelagic zone of lakes. *Hydrobiologia*, **361**, 101–11.

Quirós, R. & Boveri, M.B. (1999) Fish effects on reservoir trophic relationships. In: Tundisi, J.G. & Straškraba,

M. (eds), *Theoretical Reservoir Ecology and its Applications*. International Institute for Ecology, São Carlos, 529–46.

Recknagel, F., Hosomi, M., Fukushima, T. & Kong, D.-S. (1995) Short- and long-term control of external and internal phosphorus loads in lakes—a scenario analysis. *Water Research*, **29**, 1767–79.

Reynolds, C.S. (1992a) Dynamics, selection and composition of phytoplankton in relation to vertical structure in lakes. *Ergebnisse der Limnologie*, **35**, 13–31.

Reynolds, C.S. (1992b) Eutrophication and the management of planktonic algae: what Vollenweider couldn't tell us. In: Sutcliffe, D.W. & Jones, J.G. (eds), *Eutrophication: Research and Application to Water Supply*. Freshwater Biological Association, Ambleside, 4–29.

Reynolds, C.S. (1996) Phosphorus recycling in lakes: evidence from large limnetic enclosures for the importance of shallow sediments. *Freshwater Biology*, **35**, 623–45.

Reynolds, C.S. (1999) Phytoplankton assemblages in reservoirs. In: Tundisi, J.G. & Straškraba, M. (eds), *Theoretical Reservoir Ecology and its Applications*. International Institute for Ecology, São Carlos, 439–56.

Riley, M.J. & Stefan, H.G. (1988) Development of the Minnesota lake water quality management model, Minlake. *Lake and Reservoir Management*, **4**, 73–84.

Sakamoto, M. (1966) Primary production by the phytoplankton community in some Japanese lakes and its dependence upon lake depth. *Archiv für Hydrobiologie*, **62**, 1–28.

Sale, M.J., Brill, E.D. & Herricks, E.E. (1982) An approach to optimising reservoir operation for downstream aquatic resources. *Water Resources Research*, **18**, 705–12.

Sas, H. (ed, 1989) *Lake Restoration by Reduction of Nutrient Loading; Expectations, Experiences, Extrapolations*. Academia Verlag Richarz, Skt Augustin, 496 pp.

Scheffer, M. (1998) *Ecology of Shallow Lakes*. Chapman and Hall, Chichester, 357 pp.

Scheffer, M., Hosper, H.S., Meijer, M.-L., et al. (1993) Alternative equilibria in shallow lakes. *Trends in Ecology and Evolution*, **8**, 275–9.

Schindler, Z. & Straškraba, M. (1982) Optimální řízení eutrofizace údolních nádrží. *Vodohospodársky časopis SAV*, **30**, 536–48.

Schulte, T.L. & Lackey, R.T. (1973) Effect of rate of water discharge on phytoplankton in Claytor Lake, Virginia. *Proceedings of the Annual Conferences of the South Eastern Association of Game and Fisheries Commissioners*, **27**, 402–14.

Søballe, D.M., Kimmel, B.L., Kennedy, R.H. & Gaugash, R.F. (1992) Reservoirs. In: Hackney, C.T., Adams, S.M. & Martin, W.H. (eds), *Biodiversity of the Southern United States: Aquatic Communities*. Wiley & Sons, New York, 421–74.

Søndergaard, M. & Moss, B. (1998) Impact of submerged macrophytes on phytoplankton in shallow lakes. In: Jeppesen, E., Søndergaard, M., Søndergaard, M. & Christoffersen, K. (eds), *The Structuring Role of Submerged Macrophytes in Lakes*. Springer-Verlag, New York, 115–32.

Stauffer, R.E. (1987) Effects of oxygen transport on the areal hypolimnetic oxygen deficit. *Water Resources Research*, **23**, 1887–92.

Steel, J.A. & Duncan, A. (1999) Modelling the ecological aspects of bankside reservoirs and implications for management. *Hydrobiologia*, **395/396**, 133–47.

Stefan, H.G., Hondzo, M., Eaton, J.G. & McCormick, J.H. (1995) Predicted effects of global climate change on fishes in Minnesota lakes. In Beamish, R.J., (ed), *Climate Change and Northern Fish Populations*. National Research Council of Canada, Ottawa, 57–72.

Straškraba, M. (1973) Limnological basis for modelling reservoir ecosystems. In: Ackerman, C., White, E.G. & Worthington, E.B. (eds), *The Functioning of Freshwater Ecosystems*. Cambridge University Press, Cambridge, 517–35.

Straškraba, M. (1991) Geographical differences of lake and reservoir productivity. *Biologia acuatica*, **15**, 36–7.

Straškraba, M. (1999) Retention time as a key variable of reservoir limnology. In: Tundisi, J.G. & Straškraba, M. (eds) *Theoretical Reservoir Ecology and its Applications*. International Institute for Ecology, São Carlos, 385–410.

Straškraba M. & Gnauck, A. (1985) *Freshwater Ecosystems, Modelling and Simulation*. Elsevier, Amsterdam, 309 pp.

Straškraba, M. & Mauersberger, P. (1988) Some simulation models of water quality management of shallow lakes and reservoirs and a contribution to ecosystem theory. In: Mitsch, W.J., Straškraba, M. & Jørgensen, S.-E. (eds), *Wetland Modelling*. Elsevier, Amsterdam, 153–76.

Straškraba, M. & Tundisi, J.G. (1999) *Guidelines of Lake Management*, Vol.9, Reservoir Water Quality Management. International Lake Environment Committee Foundation, Kusatsu, 229 pp.

Straškraba, M., Tundisi, J.G. & Duncan, A. (eds) (1993a) *Comparative Reservoir Limnology and Water Quality Management*. Developments in Hydrobiology, Vol. 77, Kluwer Academic Publishers, Dordrecht, 291 pp.

Straškraba, M., Tundisi, J.G. & Duncan, A. (1993b). State-of-the-art of reservoir limnology and water quality management. In Straškraba, M., Tundisi, J.G. & Duncan, A. (eds), *Comparative Reservoir Limnology and Water Quality Management.* Developments in Hydrobiology, Vol. 77, Kluwer Academic Publishers, Dordrecht, 213–88.

Straškraba, M., Dostálková, M., Hejzlar, J. & Vynhálek, V. (1995) The effect of reservoirs on phosphorus concentration. *Internationale Revue des gesamten Hydrobiologie,* **80**, 403–13.

Straškrabová, V. (1976) Self-purification of impoundments. *Water Research,* **9**, 1171–7.

Straškrabová, V., Komárková, J. & Vynhálek, V. (1993) Degradation of organic substances in reservoirs. *Water Science and Technology,* **28**, 95–104.

Strycker, L. (1988) Decaying dam holds tide of trouble. *The Register-Guard, Eugene, Oregon,* 121, 1–8.

Suffet, I.H., Ho, J., Chou, D., et al. (1995) Resource control and management of taste-and-odor problems. In: Suffet, I.H., Mallevialle, J. & Kawczynski, E. (eds), *Advances in Taste-and-odor Treatment and Control.* American Waterworks Association, Denver, 1–21.

Szewzyk, U., Szewzyk, R. & Manz, W. (2000) Microbiological safety of drinking water. *Annual Reviews of Microbiology,* **54**, 81–127.

Talling, J.F. (1976) The depletion of carbon dioxide from lake water by phytoplankton. *Journal of Ecology,* **64**, 79–121.

Thanh, N.C. & Biswas, A. (1990) *Environmentally-Sound Water Management.* Oxford University Press, Oxford.

Thornton, K.W., Kimmel, B.L. & Payne, F.E. (1990) *Reservoir Limnology.* Wiley, Chichester.

Townsend, S.A., Luong-Van, T. & Boland, K.T. (1996) Retention time as a primary determinant of colour and light attenuation in two tropical Australian reservoirs. *Freshwater Biology,* **36**, 57–69.

Tundisi, J.G. (1984) Estratificacío hidráulica en reservatórios e suas consequéncias ecológicas. *Ciéncias e Cultura,* **36**, 1489–96.

Tundisi, J.G. (1994) Tropical South America; present and perspectives. In Margalef R. (ed.), *Limnology now; a paradigm of planetary problems.* Elsevier, Amsterdam, 352–424.

Twinch, A.J. & Grobler, D.C. (1986) Pre-impoundment as a eutrophication management option: simulation of the Hartbeespoort Dam. *Water South Africa,* **12**, 19–26.

Uhlmann, D., Benndorf, J. & Albert, W. (1971) Prognose des Stoffhaushaltes von tagewässern mit Hilfe kontinuierlicher oder semikontinuierlicher Modelle. I. Grundlagen. *Internationale Revue des gesamten Hydrobiologie,* **56**, 513–39.

Virtanen, M., Koponen, J., Dahlbo, K. & Sarkkula, J. (1986) Three-dimensional water-quality transport model compared with field observations. *Ecological Modelling,* **31**, 185–99.

Vollenweider, R.A. & Kerekes, J. (1980) The loading concept as basis for controlling eutrophication philosophy and preliminary results of the OECD programme on eutrophication. *Progress in Water Technology,* **12**, 5–38.

Ward, J.V. & Stanford, J.A. (1983) The serial discontinuity concept of lotic ecosystems. In: Fontaine, T.D. & Bartell, S.M. (eds), *Dynamics of Lotic Ecosystems.* Ann Arbor Science, Ann Arbor, MI, 29–42.

Ward, R.C., Loftis, J.C. & McBride, G.B. (1986) The 'data-rich but information-poor' syndrome in water quality monitoring. *Environmental Management,* **10**, 291–8.

Williams, W.D. & Aladin, N.V. (1991) The Aral Sea: recent limnological changes and their conservation significance. *Aquatic Conservation,* **1**, 3–23.

Wool, T.A., Martin J.L. & Schottman, R.W. (1994) The linked watershed/waterbody model (LWWM): a watershed management modeling system. *Lake and Reservoir Management,* **9**, 124–5.

Zalewski, M., Frankiewicz, P. & Nowak, M. (1995) Biomanipulation by ecotone management in a lowland reservoir. *Hydrobiologia,* **303**, 49–60.

# Part IV  Lake and Catchment Models

# 13 The Export Coefficient Approach to Prediction of Nutrient Loadings: Errors and Uncertainties in the British Experience

HELEN M. WILSON

## 13.1 INTRODUCTION

In order to determine the most effective means of reducing external phosphorus loadings on eutrophicated water bodies, it is necessary to identify the major sources of nutrient within their catchments, and the relative contribution of each source to the total load (Loehr et al. 1989). The most accurate method for estimating total loadings is through direct measurement at the mouth of inflowing rivers but this produces little information on the relative contributions of different sources. An alternative method, which is particularly useful for estimating loadings from non-point sources, is to use unit area loads or export coefficients. This approach assumes that, under average hydrological conditions, a given land use will export a reasonably constant nutrient load per unit of land area to the receiving waters. The area of each land-use category in a catchment is multiplied by the appropriate export coefficient and the results summed to give the total non-point load. Inputs from atmospheric deposition, and point sources, particularly effluent from sewage treatment, can be calculated in the same general way. The total load from all sources is taken to be the sum of these calculations.

Although the use of export coefficients to predict nutrient losses is an attractive, simple tool, requiring minimal catchment data input compared with more complex models, there are a number of limitations to the process which warrant examination. In Britain, the export coefficient approach has been developed since the late 1980s and it now provides the basis for lake classification in the implementation of national and European Union (EU) water quality legislation.

## 13.2 HISTORICAL PERSPECTIVE

The work of Vollenweider (1968) contributed an important step in the classification of lakes according to their trophic status and the recognition of phosphorus as the key element which controls eutrophication in most lakes. For a more detailed review of this work, see Rast & Thornton (this Volume, Chapter 14).

Although Vollenweider's earliest model was based on limnological evidence inferred from a small data set, it represented a first approximation of the relationship between phosphorus loading and trophic response (OECD 1982). It was an attempt to provide a basis for the development of a trophic status index with respect to nutrient loading which could be examined and developed. Vollenweider was able to demonstrate that if annual external nutrient inputs to lakes were expressed as loadings per unit area, lakes of different sizes could be compared. He recognised, however, that the relationship between nutrient loading and in-lake concentration is dependent on a number of hydrological and morphological factors, not least of which is lake depth. Vollenweider therefore plotted data on phosphorus loadings against mean depth, adding boundaries between oligotrophic and eutrophic lakes, and introducing the notion of 'dangerous' and 'permissible' loadings.

The work of Vollenweider was further developed through his involvement in the Cooperative Programme on Eutrophication of the Organisation for Economic Cooperation and Development (OECD) which was designed to quantify the relationship between nutrient load and trophic reaction. The Programme, begun in 1973, represented a unique and comprehensive international cooperative effort to generate an extensive and reliable limnological database for the development of sound management principles for the control of eutrophication. Data from the Programme were used to achieve new criteria for phosphorus loading. Improvements to the loading relationships were obtained through the inclusion of terms for hydraulic residence time and for rates of sedimentation (Vollenweider 1975), producing Vollenweider's third model (Vollenweider 1976).

Evaluating the data in 1978, Vollenweider and Kerekes were able to demonstrate a significant relationship between external phosphorus load and in-lake phosphorus concentration, as well as algal biomass (expressed as chlorophyll $a$ concentration) (Sas 1989). Expert opinion was then utilised to allocate a trophic category (oligotrophic, mesotrophic or eutrophic) to the lakes investigated. It was found that the trophic state of each lake lay within definitive ranges of phosphorus concentration and algal biomass, so that the relationship could be ascertained statistically. Contemporaneous and subsequent work, including that of Rast & Lee (1978), Janus & Vollenweider (1981) and Jones & Lee (1982), has confirmed that, with few exceptions, lakes generally conform to the relationships established by the Vollenweider–OECD model (Jones & Lee 1986).

The objective of the OECD Programme was to provide guidelines for eutrophication control and the restoration of water quality 'to a level of lower and more acceptable trophic conditions' (OECD 1982, p. 96). Verification of the idea that eutrophication was reversible led to the implementation of phosphorus control programmes in the majority of countries in Europe and North America. In qualitative terms, many of the lakes receiving reduced phosphorus input have responded in accordance with the predictions of the model by exhibiting improved water quality (Sas 1989). However, in quantitative terms, accordance has been verified for very few sites. In several cases, reductions in phosphorus inputs have led to delayed or only slight improvements in water quality.

The lack of response to phosphorus load reduction observed in some restoration programmes led to an evaluation of recovery for 18 eutrophicated lakes located in western Europe (Sas 1989). The Vollenweider–OECD model provided the reference for recovery evaluation, but despite its advantages it was claimed that there were some critical points in its application.

**1** The model is based upon the assumption that regression analyses from a large number of lakes in different steady-state conditions can be used to predict the effect of phosphorus load reduction on a particular lake. It was felt that this assumption was in need of validation.

**2** The ease with which the model can be applied without sufficient knowledge of factors peculiar to a particular lake may lead to the instigation of management programmes which may not achieve the desired objectives.

**3** Significant doubts about the practical application of the model have been raised because of the relatively wide confidence limits for predictions.

**4** The assumption of a steady state suggests that the model will not predict the transient phase after a restoration programme begins, in which a delay in response may occur.

**5** Changes in the composition of the algal community are not accounted for in the model.

With respect to the reasons for delayed or disappointing recovery following restoration programmes, the investigation concluded that the following mechanisms operate (Sas 1989):

**1** the flushing rate of a lake determines the degree to which the 'old stock' of phosphorus is removed from the system;

**2** in shallow lakes (those in which most of the epilimnion is regularly in contact with the sediment), net annual release of phosphorus from the sediment often occurs in the first years after restoration;

**3** in lakes receiving very high phosphorus loadings, algal biomass may not be limited by this element, so that an initial reduction in in-lake concentrations will produce no effect;

**4** when in-lake phosphorus concentrations are reduced to the point where this element becomes limiting, algal response is characterised either by overall biomass reduction or by migration to deeper water.

In relating these observations to the ability of the Vollenweider–OECD model to predict them, Sas concluded that the first mechanism, associated with flushing, was easily predicted. The second, that of 'sediment memory', was outside the scope of the model and therefore not predicted by it. With respect to the effects of phosphorus loading on algal biomass, the model appeared to predict response reasonably well.

A more recent critique of the Vollenweider–OECD model is presented by Johnes et al. (1994a). The authors emphasise the degree of variation exhibited among lakes and the continuous nature of the trophic spectrum. They decry the persistence of the notion that distinct lake types exist. It is true that spatial state schemes, such as that conceived by the OECD, are limited by the fact that they do not reflect changes over time and that naturally nutrient-rich lakes are placed in the same category as eutrophicated ones. In these respects, state-changed schemes, such as that devised by the authors, possess a distinct advantage.

The OECD results and, in particular, the diagnostic diagrams produced by Vollenweider do provide the impression that the boundary lines between 'dangerous' and 'permissible' loadings can be used as targets for nutrient control. The OECD report advises managers to 'establish a water quality objective (expected trophic response) taking into consideration the intended use of the water and the natural trophic conditions of the area' (OECD 1982, p. 98).

Following this advice, water quality objectives can be established merely by deciding upon the appropriate trophic category for a particular lake. Such an approach may be adequate in cases where the priority is to achieve water quality suitable for various human uses. In order to define realistic and appropriate objectives for ecological purposes, it may be necessary to obtain more specific information, such as data on pre-eutrophication phosphorus inputs. Direct methods of reconstructing changes in lake nutrient status may be essential, namely the use of palaeolimnological interpretation (e.g. Anderson et al. 1993; Bennion et al. 1996; see also O'Sullivan, Volume 1, Chapter 18, pp. 609–66).

In conclusion, the Vollenweider–OECD model possesses various advantages and limitations which can be summarised thus:

**1** as a steady state model, it cannot be used to assess past nutrient enrichment;

**2** the model is capable of predicting trophic and algal biomass responses to loading reductions reasonably well, except in cases where internal loadings become significant during the transient phase after restoration;

**3** in order to take full advantage of the model's predictive ability, knowledge of a lake's natural trophic condition is required;

**4** in the absence of an alternative tool, it remains the most readily available and widely applicable model of its kind.

Prior to refining the model to this level, Vollenweider had been aware of the usefulness for eutrophication management of some means of calculating nutrient loadings from different types of land use (Vollenweider 1968). The paucity of suitable scientific studies on nutrient export from land to water at that time led him to advocate the use of rough estimates as coefficients in order to assess relative contributions. Vollenweider's work in this area was expanded and refined, and led, eventually, to the adoption of what is now known as the 'export coefficient approach' to the prediction of nutrient loadings.

## 13.3 HOW THE EXPORT COEFFICIENT APPROACH WORKS

### 13.3.1 Selection of sites for application

The export coefficient approach (ECA) can be applied to the majority of lakes and reservoirs. However, there are certain catchment characteristics which may render sites unsuitable for this kind of analysis, or which present difficulties and uncertainties sufficient to undermine the reliability of the results. First, highly urbanised catchments, in which natural drainage patterns are often severely disrupted (House et al. 1993), present difficulties. For example, Sutton Park, West Midlands, incorporates the headwaters of two streams which have been dammed at intervals to form six main pools. The hydrology is further complicated by the fact that the catchment outside the park is entirely urban and subject to an artificial drainage system. Although export coefficients are available for urban areas, they should be applied with caution in such cases and preferably with knowledge of the sewerage system.

Second, export coefficients represent the transport of nutrients via surface and subsurface flow, rendering them unsuitable for application to lakes which are predominately groundwater-fed. Many examples of groundwater-fed lakes and ponds can be found in Britain, their hydrology influenced by the geology of the area. In some cases, such water bodies have been created by former extraction industries, such as at Upton Warren Pools in Worcestershire. Here, a small stream runs through the site, but is not connected via surface flow to any of the Pools. It must be assumed, therefore, that the water budget is almost entirely contributed by groundwater, the source of which may not be consistent with the surface catchment.

Third, the use of export coefficients is problematical where a site is one of a series of water bodies within the same catchment. The sediments of lakes upstream may retain a significant proportion of the nutrient loading (Canfield *et al.* 1989). Examples include streams dammed to create medieval fish ponds in the east Surrey area of Britain. Finally, although the ECA requires a relatively low level of data compared with more complex models, it is still necessary to acquire all the relevant information in order to generate realistic results. There are many reasons for data being unavailable. For sites which have received little scientific attention in the past, data on parameters which require longer term monitoring will be difficult to obtain. Conversely, for some well-studied sites, the necessary data may be unpublished and therefore unavailable.

### 13.3.2 Data collection

The use of the ECA for predicting total phosphorus (TP) loadings necessitates the collation of particular data for each lake. The extent of the catchment, its land use and livestock numbers, together with details of consented discharges, provide information on the sources of phosphorus. For diffuse sources, each type of land use and livestock is assigned an appropriate phosphorus export coefficient, based on literature values. Point sources, identified through information on consented discharges, are treated similarly, in that the phosphorus load normally associated with such effluent is used to calculate the amount from each discharge. The sum of all known sources within the catchment is taken to be the total potential total phosphorus loading on a particular lake. In order to predict the impact of such loadings on trophic status, data on mean depth and hydrological residence time are obtained or calculated for each lake. Water quality data, over as long a period as possible, are also necessary for the calibration and validation processes. These data can be acquired in Britain in a variety of ways, which will be outlined below.

#### 13.3.2.1 Catchment definition

For many sites the catchment boundary will have been defined and maps may be available in published literature or from statutory, commercial or voluntary organisations. For other sites the catchment can be determined from Ordnance Survey maps by inspection of contour lines and spot heights. The area must then be calculated, either by using a clinometer or by the cut and weigh method (Lind 1979).

#### 13.3.2.2 Morphometric data

Data for surface area, volume, mean depth and water residence time are required for each water body. For many sites these values are available from published literature. Alternatively, they can be calculated (volume equals surface area multiplied by mean depth), and the water residence time estimated, provided that sufficient information is available on river discharges.

#### 13.3.2.3 Land use

The extent of each type of land use within a catchment can be obtained in a number of ways. The method by which the most accurate and up-to-date information can be acquired is through a ground survey. However, this is time consuming and lack of co-operation from landowners may cause gaps in the data. Alternatively, land-use data can be obtained through satellite imagery co-ordinated with ground-truth information (Bunce et al. 1992; Fuller et al. 1994). A comparison between remotely sensed land-use data and those obtained through ground survey concluded that, for many applications, the former is perfectly adequate (Cherrill et al. 1995). There are, however, some drawbacks in using these data in conjunction with the ECA.

First, the land-use categories derived from remote sensing may not coincide with those for which export coefficients should be applied, or are available. Grassland, for instance, may simply be divided into improved and non-improved, rather than temporary, permanent and rough pasture. Conversely, the classes of land use generated by a remote sensing system may be so diverse that an adequate range of export coefficients may not be available and 'lumping' of land use types may be necessary. In an attempt to overcome this problem, researchers in Northern Ireland, utilising the CORINE (Co-ORdination of INformation on the Environment) land cover database, used regression analysis to derive export coefficients appropriate to the land-use classes provided (McGuckin et al. 1999). Second, some remote sensing data are only available based on the Ordnance Survey kilometre square grid. For smaller catchments this may not coincide adequately with the catchment, to the extent that averaging of the data may be necessary. Data based on a 25-m grid will therefore provide a more accurate set of results but can be more costly to obtain.

A further method of obtaining land-use data is through the County and Parish Summaries of Agricultural Returns taken by the Ministry of Agriculture, Fisheries and Food (MAFF, now incorporated into the Department for Environment, Food and Rural Affairs, DEFRA) and available for inspection at The National Archives (formerly the Public Record Office) at Kew, Surrey. The records consist of summaries of agricultural census returns, presented as totals for whole parishes. In order to safeguard the confidentiality of information for individual farms, where there are fewer than three holdings in a parish, the data are amalgamated at the county level (Clark et al. 1983). This may mean that data for important parishes within a catchment are not available. Furthermore, many farms do not lie wholly within a single parish. Such farms are normally included in the parish containing the greater part. It follows that the area of agricultural land shown in the parish summary is only approximately comparable with the geographical boundaries of the corresponding parish of the same name. This lack of comparability is generally greater for smaller parishes.

The main disadvantage of this method is that the data are collated on a parish basis. A comparison of Figs 13.1 and 13.2 illustrates the fact that although groups of kilometre squares often coincide roughly with the boundary of a reasonably sized catchment, the parishes bear little relation. It is therefore necessary to adjust the data for each parish to the percentage of agricultural land in the catchment. This adjustment assumes that land use and livestock numbers are homogeneously distributed throughout each parish (Mitchell 1990). The degree of error produced by this assumption will increase as the area of catchment within the parish decreases. Small catchments impinging on several large parishes will, of course, produce the least reliable results.

A further important limitation in the use of parish returns for data on catchment land use is that they provide information on **agricultural** uses only. The extent of settlements and roads is not provided. This lack of data means that potentially high nutrient exports from urban areas are not included, or that the area of urban land must be estimated.

A comparatively recent problem with the use of these returns is that data are available at the parish level only up to 1998 (G. Weber, personal communication). Data collected since then are aggregated into 'super-parishes', which consist of even larger areas, making it much more difficult to establish which farmland is within a catchment. The data can be disaggregated, because it is still collected on a parish basis, but it is yet to be seen whether

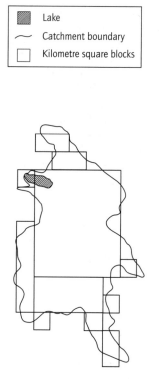

Fig. 13.1  The use of kilometre squares for land-use data for a typical catchment.

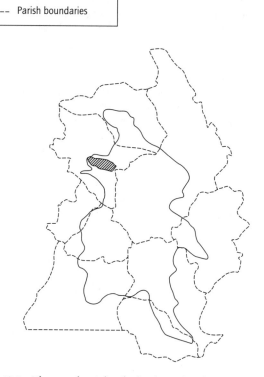

Fig. 13.2  The use of parishes for land-use data for a typical catchment.

DEFRA is willing to provide this service as routine. Agricultural parish returns may therefore remain useful only for historical studies.

#### 13.3.2.4 Livestock numbers

Data on livestock numbers within a catchment can again be obtained by farm surveys or through the County and Parish Summaries of Agricultural Returns. The accuracy of the latter data is subject to the same limitations as those for land use, in that parishes rarely coincide with catchments. Furthermore, an assumption implicit in reliance on data from the Agricultural Returns is that all animal waste produced within the parish is disposed of within the same area. In the case of small to medium size mixed farms this is probably a reasonable assumption. For other types of farming, however, the situation may be quite different. For instance, intensive poultry and pig units produce large quantities of waste which cannot be utilised locally (North & Bell 1990). Disposal of this waste to other farmers, or to manufacturers of garden fertilisers, or by dumping, may well involve transportation out of the parish. Added to this, animals reared in areas of intensive agriculture tend to be grouped in large units, which may or may not lie within the catchment. Inspection of the number

**Table 13.1** General phosphorus export coefficients (Reckhow & Simpson 1980)

| | Agriculture (kg ha$^{-1}$ yr$^{-1}$) | Forest (kg ha$^{-1}$ yr$^{-1}$) | Precipitation (kg ha$^{-1}$ yr$^{-1}$) | Urban | Input to septic tank (kg capita$^{-1}$ yr$^{-1}$) |
|---|---|---|---|---|---|
| High | 3 | 0.45 | 0.6 | 5 | 1.8 |
| Medium | 0.4–1.7 | 0.15–0.3 | 0.2–0.5 | 0.8–3 | 0.4–0.9 |
| Low | 0.1 | 0.02 | 0.15 | 0.5 | 0.3 |

and types of holdings in each parish may reveal the presence of such intensive livestock units, but determining their location can be problematic.

#### 13.3.2.5 Consented discharges

Details on consented discharges from sewage treatment works (STWs) and the population equivalents (p.e.) served by STWs or septic tanks can be obtained from the relevant Environment Agency office. The discharge consent will state whether a particular STW is subject to nutrient reduction requirements and, if so, the concentration range allowed in the effluent. For STWs not subject to such controls, the p.e. allows an estimate of phosphorus concentration in the discharge.

### 13.3.3 Deriving export coefficients

A large number of studies have been carried out in order to quantify nutrient losses from different land-use types. Several authors (Vollenweider 1968; Uttormark *et al.* 1974; Reckhow *et al.* 1980) have collated such information, converted the data into uniform units and tabulated them in a format convenient for the selection of relevant export coefficients. Uttormark *et al.* (1974) compiled a large amount of data from work conducted almost entirely in the USA. From this they produced an average phosphorus export coefficient for agricultural land of 0.3 kg ha$^{-1}$ yr$^{-1}$, within a 'typical' range of 0.1 to 1.0 kg ha$^{-1}$ yr$^{-1}$. Reckhow *et al.* (1980) also analysed the results of a large number of experimental studies, again mainly undertaken in the USA. A mean value for phosphorus export from agricultural land was calculated as 1.134 kg ha$^{-1}$ yr$^{-1}$, within a range of 0.08 to 3.25 kg ha$^{-1}$ yr$^{-1}$. These syntheses of results serve to illustrate the degree of variation possible, even where a large database is available.

As well as a compilation of export coefficients, Reckhow *et al.* (1980) provide a manual for the selection and application of export coefficients for a particular watershed. Reckhow & Simpson (1980) present a phosphorus model for lake quality management which uses generalised export coefficients to predict loading. The coefficients are based on data from Uttormark *et al.* (1974) and Reckhow *et al.* (1980) and are shown in Table 13.1.

The land-use categories defined by Reckhow & Simpson (1980) are very broad and do not differentiate between, for example, arable and grassland. Consequently, the range of values cited are extremely wide, because 'the broadness of the "source categories" leaves room for much variation, owing to basin geology, erosional patterns, and intensity and types of use' (Reckhow & Simpson 1980, p. 1442). From this range of values, the authors selected what were thought to be appropriate export coefficients for a particular example, namely Higgins Lake, Michigan, USA. In selecting these values, a wide range of factors were taken into consideration. The final figures for Higgins Lake catchment are reproduced in Table 13.2.

Reckhow & Simpson (1980) explained the procedure involved in using their phosphorus model as follows:

> 'The selection of appropriate phosphorus export coefficients is a difficult task ... In this example,

Table 13.2  Phosphorus export coefficients for Higgins Lake catchment (Reckhow & Simpson 1980)

|  | Agriculture (kg ha$^{-1}$ yr$^{-1}$) | Forest (kg ha$^{-1}$ yr$^{-1}$) | Precipitation (kg ha$^{-1}$ yr$^{-1}$) | Urban | Input to septic tank (kg capita$^{1}$ yr$^{-1}$) |
|---|---|---|---|---|---|
| High | 0.8 | 0.4 | 0.5 | 1.5 | 1.0 |
| Medium | 0.3 | 0.2 | 0.3 | 0.9 | 0.6 |
| Low | 0.1 | 0.02 | 0.15 | 0.5 | 0.3 |

*however, we have not explained our analyses or thought processes involved in the choice of export coefficients . . . In the interests of a concise presentation of this technique, it was not included. However, when this procedure is used . . . we strongly urge that the analyst present . . . all the information used in the selection of export coefficients' (Reckhow & Simpson 1980, p. 1443).*

The authors emphasise that the values selected by them 'describe conditions within the Higgins Lake watershed [and that] proper choice of export coefficients is a function of knowledge of . . . the watersheds of candidate export coefficients' (Reckhow & Simpson 1980, p. 1443). Reckhow et al. (1980) concede that, 'Despite the existence of pertinent literature, the selection of application export coefficients is still an unavoidably subjective task' (Reckhow et al. 1980, p. 21). They recommend consulting the original experimental work for details on the conditions under which literature values were generated. This example of the selection of export coefficients serves to underline the subjective nature of such processes. It follows that, without an explanation of the analysis involved, it is very difficult to assess the validity of values chosen.

The use of export coefficients to predict lake nutrient loadings has been widely used in North America and, to some extent, in continental Europe. In Britain, this approach has received limited attention, until relatively recently. Johnes & O'Sullivan (1989) and Johnes (1990) used export coefficients to quantify nutrient loadings on, respectively, Slapton Ley, Devon and the River Windrush, Oxfordshire. Johnes et al. (1994a, b; 1998a, b) developed the approach further and applied it to a wide range of lakes in England and Wales.

Phosphorus exports from catchments are commonly categorised as originating from either point or non-point sources. A further division can be made within the non-point category, in that phosphorus from organic sources (i.e. animal wastes) is distinct in origin from inorganic sources (i.e. artificial fertilisers). Vollenweider (1968) made this distinction; he calculated non-point organic and inorganic losses separately. However, the vast majority of subsequent work in the USA has employed export coefficients based on land use type only. This change in approach might have occurred for the following reasons:

**1** the need to reduce the amount of catchment data necessary to predict phosphorus loadings;
**2** the recognition that land use and management practices are often more important in determining phosphorus export than application rates of either organic or inorganic fertiliser (e.g. Lambert et al. 1985; Mostaghimi et al. 1992; Thomas et al. 1992; McIntyre 1993);
**3** the difficulty involved in measuring phosphorus export from organic and inorganic fertilisers separately, owing to the complex nature of soil phosphorus processes (White 1981; Sharpley & Smith 1990).

The main advantage of using a model which distinguishes between organic and inorganic sources is in the consequent ability to produce more specific recommendations for reducing loadings. For instance, if inorganic fertilisers are identified as the main source, then different management strategies would be recommended than if organic sources were predominant. The problem with this approach, however, is in obtaining appropriate, scientifically based data on rates of losses for organic and inorganic fertilisers. In over 40 research and review papers which contain export coefficients, very few specifically state the source of the phosphorus, that is, whether it is of organic or inorganic origin. The lack of relevant data has led to the use of assumed values, particularly for organic sources, in the work of Johnes & O'Sullivan (1989) and Johnes (1990).

The model used by these authors assumes a standard loss of 3% of phosphorus in animal wastes, applied as manure and directly voided. The references cited by Johnes & O'Sullivan (1989) for production and losses of phosphorus from livestock are Vollenweider (1968), Cooke (1976) and Gostick (1982). Johnes (1990) specifically cites Cooke (1976) as the source for the 3% figure, who gives no such detail. It seems likely that this 3% was derived from Vollenweider (1968) who envisaged a 'best-case' scenario of 1% loss of phosphorus from organic manures and a 'worst-case' one of 5% loss. The mean of these two figures is 3%. This hypothesis is strengthened by the use of 17% as the loss rate for nitrogen, which is roughly the mean of 10% and 25%, the 'best-case' and 'worst-case' figures presented by Vollenweider for this element. Phosphorus losses are, therefore, based on an average of values stated to be '... of the order of 1–5% at the most' (Vollenweider 1968, p. 106).

A further assumption is made in relation to animal wastes, in that 10% to 15% is allowed for losses during storage for both nitrogen and phosphorus. In the case of nitrogen it is very likely that losses of ammonia, dinitrogen and nitrous oxides will occur (Arden-Clarke 1988). However, the same loss rates are assumed for phosphorus but because, during its natural cycle, this element does not enter a gaseous phase (O'Neill 1985) such losses appear to be unlikely. Furthermore, Johnes & O'Sullivan (1989) cite Gostick (1982), who specifically discusses losses of **nitrogen** during storage, whereas Johnes (1990) refers to *The Nitrogen Cycle of the United Kingdom* (Royal Society 1983). Nevertheless, the authors adjust the standard 3% loss of phosphorus from organic manures for animals which spend some or all of the year indoors, in order to allow for supposed losses during storage. Such 'fine tuning' (based on an erroneous assumption) of a figure that is a rough estimate in the first place is at best questionable and at worst nonsensical, yet it appears to have been accepted, adopted and perpetuated in a number of publications (e.g. Johnes & Heathwaite 1997; Hanrahan *et al.* 2001; Hilton *et al.* 2002).

The lack of available data on phosphorus losses from organic sources proved to be a limitation in the necessary adjustment of the model during development. Johnes (1990) carried out a sensitivity analysis of the model by changing each export coefficient by an initial 10% and then by 25%, when the remaining parameters were held constant. The results of this analysis were then used to adjust sensitive parameters within the observed range of literature values to optimise the model, '... although the coefficients available for livestock did not provide the necessary range for adjustment' (Johnes 1990, p. 310). So, for phosphorus, the standard 3% loss rate for organic sources remained unamended, while some of the coefficients for inorganic sources were adjusted. The final model calibration for the River Windrush produced an error for phosphorus of only 0.15% but, bearing in mind the assumptions outlined above, it must be concluded that the coefficients for inorganic sources were adjusted in order to accommodate the lack of reliable data for organic sources.

As already discussed, there is a wealth of literature on export coefficients for different land-use types but very few of them specifically state whether the sources are organic or inorganic. Johnes (1990) selects a range of values for different land uses, citing Reckhow & Simpson (1980) as the reference. Contrary to the advice given in Reckhow & Simpson (1980) and quoted above, Johnes (1990) presents insufficient information to explain the processes involved in selecting these export coefficients. Table 13.3 details the phosphorus loss rates from inorganic sources used by Johnes. No value is selected for urban inputs, because, as noted above, data for this type of land use are not available from Agricultural Returns.

Direct comparison of the loss rates chosen by Johnes for organic and inorganic sources is hindered by the fact that for the former a percentage loss is given, whereas for the latter a rate of loss per hectare is used. However, the export coefficients for inorganic sources can be converted to a percentage loss, using data on fertiliser application rates given by Johnes (1990). The figures produced are shown in Table 13.4.

Comparisons can now be made between the assumed losses from inorganic and organic sources on different types of land use. Temporary and permanent grasslands receive applications of inorganic fertiliser as well as livestock waste directly voided onto the land. The fertiliser supply may be supplemented by additions of slurry and farmyard manure (FYM). According to Johnes (1990), losses on grasslands from inorganic sources are about 1.5%, compared with 3% from organic sources. There are no scientific data to support the idea that twice as much phosphorus is lost from livestock waste than from inorganic fertilisers on the same type of land. Indeed, a study in The Netherlands (Gerritse 1981) demonstrated that pig slurry consists of about 80% **inorganic** phosphorus, and that nearly all organic phosphorus compounds are strongly retained in the soil. In sheep faeces deposited onto New Zealand pastures, 56% to 82% of phosphorus is present in the **inorganic** fraction (Nguyen & Goh 1992). Such high proportions of inorganic phosphorus in organic wastes would suggest that similar loss values are applicable to both organic and inorganic sources of phosphorus on grassland. For other types of land use, losses of phosphorus from inorganic sources vary from 2.3% to as much as 9.2%. Yet it is assumed that a standard 3% of phosphorus from organic sources would be exported, regardless of land use.

It is evident from the above that when selecting export coefficients it is preferable to investigate the original sources of data. Table 13.5 provides the sources on which many of the nutrient export coefficients in current use are based.

Table 13.3 Phosphorus loss rates selected by Johnes (1990)

| Parameter | Optimised value (kg ha$^{-1}$ yr$^{-1}$) |
|---|---|
| Temporary grass loss | 0.3 |
| Permanent grass loss | 0.1 |
| Cereals loss | 0.8* |
|  | 0.65† |
| Root crops loss | 0.8 |
| Field vegetable loss | 0.8* |
|  | 0.65† |
| Oilseed rape loss | 0.8* |
|  | 0.65† |
| Woodland loss | 0.02 |
| Rough grazing and set-aside loss | 0.02 |
| Rainfall input loss | 0.2 |

\* First choice parameter for initial model calibration.
† Second choice parameter from final model calibration.

Table 13.4 Export coefficients for inorganic sources (from Johnes 1990) converted to percentage loss of fertiliser applied

| Land use | Phosphorus export (kg ha$^{-1}$ yr$^{-1}$) | Phosphorus applied (kg ha$^{-1}$ yr$^{-1}$) | Phosphorus export (%) |
|---|---|---|---|
| Temporary grassland | 0.3 | 20.3 | 1.5 |
| Permanent grassland | 0.1 | 6.8 | 1.5 |
| Cereals | 0.8 | 26.2 | 3.1 |
|  | 0.65 |  | 2.5 |
| Root crops | 0.8 | 8.7 | 9.2 |
| Field vegetables | 0.8 | 20.8 | 3.8 |
|  | 0.65 |  | 3.1 |
| Oil seed rape | 0.8 | 28.8 | 2.8 |
|  | 0.65 |  | 2.3 |

Table 13.5 Literature sources for nutrient export coefficients

| Type of source | References |
|---|---|
| Animal waste | Cooke 1976; Edwards & Daniel 1994; Gillespie 1989; McLeod & Hegg 1984; Nix 1992; Owens 1976; Richardson 1976; Uttormark *et al.* 1974; Vollenweider 1968 |
| Inorganic fertiliser | Cooke 1976; Delwiche & Haith 1983; Edwards & Daniel 1994; Jørgensen 1980; Loehr *et al.* 1989; McLeod & Hegg 1984; Reckhow *et al.* 1980 |
| Domestic sewage | Alexander & Stevens 1976; Devey & Harkness 1973; Smulders & Krings 1990; Wilson & Jones 1994 |
| Woodland | Clesceri *et al.* 1986; Cooke 1976; Delwiche & Haith 1983; Jørgensen 1980; Reckhow & Simpson 1980; Uttormark *et al.* 1974 |
| Settlements and roads | Owens 1976; Reckhow & Simpson 1980; Sheaffer & Wright 1982 |
| Precipitation | Allen *et al.* 1968; Loehr *et al.* 1989; Owens 1976; Rast & Lee 1983; Reckhow & Simpson 1980 |

When selecting export coefficients, a number of limitations need to be taken into account. Very few of the investigations included in Table 13.5 were carried out specifically in order to estimate losses to standing waters, the objective being to measure nutrient and often other material losses from distinct areas of land. Changes induced by subsequent transport to and within water courses are therefore not necessarily reflected in these results. In-stream processes can exert considerable influence over nutrient transport. However, using a mass-balance approach, House & Warwick (1998) found that phosphorus export coefficients for different land uses in Britain fall within the range of their measured values. They discovered, however, that storm events are very important in quantifying nitrate exports 'making the application of export coefficient models to some catchments difficult unless the coefficients are normalised with respect to rainfall intensity' (House & Warwick 1998, p. 151).

A further potential problem with the use of export coefficients is that they do not generally take into account the proximity of particular land uses to the water body. Some researchers do not consider this important, as the model 'accounts for the spatial variability of land use within a watershed since it operates on an individual land use patch and the resulting nutrient loads are aggregated to the watershed outlet' (Mattikalli & Richards 1996, p. 268). However, a study in the USA by Soprano *et al.* (1996) found that attenuation of nutrients transported over agricultural land can be significant and that this significance increases with catchment size. They state that neglect of such attenuation can lead to large errors and that it is necessary to recognise that 'critical areas' near surface water are important to total nutrient loading. The export coefficient model devised by these authors uses an additional parameter as a measure of attenuation.

Johnes (1996) also recognises the importance of attenuation of nutrients in some agricultural catchments and incorporates a 'distance decay function' into a model for the catchment of Slapton Ley, Devon. This is based on location within or outside the riparian area, defined as 50 m on either side of the surface drainage network. Most export coefficients for agricultural land uses inside the riparian area are significantly higher, as shown in Table 13.6.

These coefficients are designed to take account of 'the distance decay in nutrient export capacity inherent in a catchment underlain by impermeable bedrock, [which] is likely to lead to lower rates of nutrient loading on the surface drainage network' (Johnes 1996, p. 340) and in response to unacceptable initial calibration errors of 14.5% for nitrogen and 9.12% for phosphorus. The revised coefficients produced errors within 2% and 2.5% for nitrogen and phosphorus, respectively, of observed loads. In the same paper, the catchment of the River Windrush is modelled, producing acceptable results without the necessity of taking 'dis-

Table 13.6  Export coefficients* selected by Johnes (1996) for the catchment of Slapton Ley

| Nutrient source | Export coefficients for nitrogen: distance from stream | | Export coefficients for phosphorus: distance from stream | |
|---|---|---|---|---|
| | <50 m | >50 m | <50 m | >50 m |
| Permanent grass | 15% | 7.5% | 0.5 kg ha$^{-1}$ yr$^{-1}$ | 0.4 kg ha$^{-1}$ yr$^{-1}$ |
| Temporary grass | 15% | 7.5% | 0.5 kg ha$^{-1}$ yr$^{-1}$ | 0.4 kg ha$^{-1}$ yr$^{-1}$ |
| Cereals | 24% | 12% | 0.8 kg ha$^{-1}$ yr$^{-1}$ | 0.6 kg ha$^{-1}$ yr$^{-1}$ |
| Root crops | 50% | 25% | 0.9 kg ha$^{-1}$ yr$^{-1}$ | 0.7 kg ha$^{-1}$ yr$^{-1}$ |
| Field vegetables | 50% | 25% | 0.8 kg ha$^{-1}$ yr$^{-1}$ | 0.6 kg ha$^{-1}$ yr$^{-1}$ |
| Oil seed rape | 50% | 25% | 0.8 kg ha$^{-1}$ yr$^{-1}$ | 0.6 kg ha$^{-1}$ yr$^{-1}$ |
| Rough grazing (kg ha$^{-1}$ yr$^{-1}$) | 13 | 13 | 0.02 | 0.02 |
| Woodland (kg ha$^{-1}$ yr$^{-1}$) | 13 | 13 | 0.02 | 0.02 |
| Cattle (%) | 32.3 | 16.2 | 5.7 | 2.8 |
| Pigs (%) | 28.9 | 14.5 | 5.1 | 2.55 |
| Sheep (%) | 34 | 17 | 6 | 3 |
| Poultry (%) | 30.6 | 15.3 | 5.4 | 2.7 |
| Horses (%) | 32 | 16.2 | 5.7 | 2.85 |
| Humans (kg ha$^{-1}$ yr$^{-1}$) | 2.14 | 2.14 | 0.38 | 0.38 |
| Rainfall (%) | 56 | 56 | 56 | 56 |

* Values expressed as percentages are of the annual inputs to the system, which are determined from data on fertiliser use and estimates of total excreta produced.

tance decay' into account. In a similar paper, Johnes & Heathwaite (1997) applied the two-tiered export coefficients to the Slapton Ley catchment, but not to the Windrush catchment. Endreny & Wood (2003) argue that the two-tiered approach assumes that nutrient loading beyond the 50-m streamside area is always buffered, yet does not take into account the buffering capacity of riparian vegetation and wetlands within the 50-m zone. They state that topographic and land cover data gathered by remote sensing, together with an understanding of their controls on watershed nutrient runoff dynamics, provide the opportunity to weight export coefficients spatially. In this way, critical phosphorus loading areas can be identified across a catchment and more readily targeted. However, these authors, working in the USA, are able to draw on extensive research into riparian vegetative structure and buffering capacity (e.g. Delong & Brusven 1991), which is somewhat lacking in Britain.

The main limitations of the use of export coefficients can be summarised as:

**1** experimental work on nutrient export has been carried out mostly in the USA, where hydrological, climatic and agricultural conditions may be different from those of Britain;

**2** a wide range of figures are available for the same land use type, to the extent that a detailed knowledge of catchment characteristics and farming patterns is necessary before appropriate export coefficients can be selected;

**3** export coefficients are generally based on data which express losses from one land use type to another, or to surface streams, and do not necessarily take account of subsequent transport processes or of the proximity of particular land uses to the water body;

**4** the selection of export coefficients is necessarily a subjective exercise, in which choices need to be justified and assumptions made explicit.

### 13.3.4 Computing the data

The export coefficients selected, together with the data on land use, livestock numbers and lake morphology, are entered into a spreadsheet in order to calculate predicted nutrient loadings. Results from this part of the exercise are produced in terms of total load ($t\,yr^{-1}$) and areal load ($g\,m^{-2}\,yr^{-1}$). Predicted in-lake total phosphorus concentrations ($g\,m^{-3}\,yr^{-1}$) can be calculated from the areal loading using the following equation from Vollenweider (1976):

$$[\overline{P}]_\lambda = \left(\frac{L_p}{q_s(1+\sqrt{\tau_w})}\right) \quad (13.1)$$

where $L_p$ is areal loading ($g\,m^{-2}\,yr^{-1}$), $q_s$ is hydraulic load (mean depth divided by water residence time; $m\,yr^{-1}$), and $\tau_w$ is water residence time (yr). For consistency of units, the results of this calculation are converted to $\mu g\,L^{-1}\,yr^{-1}$ by multiplying by 1000.

In order to evaluate the effects of loading reduction strategies on lake trophic status, using the graph provided for this purpose in OECD (1982, p. 95), the results require conversion to mean inflow total phosphorus concentration. This is achieved by applying the OECD equation, which represents the generalised relationship between in-lake TP concentration and inflow total phosphorus concentration (OECD 1982):

$$[\overline{P}]_\lambda = 1.55\left(\frac{[\overline{P}]_i}{(1+\sqrt{\tau_w})}\right)^{0.82} \quad (13.2)$$

which can also be expressed as:

$$[\overline{P}]_i = 0.82\sqrt{\frac{[\overline{P}]_\lambda}{1.55}}(1+\sqrt{\tau_w}) \quad (13.3)$$

### 13.3.5 Evaluation of results

The method of model calibration used by Johnes (1990) and in subsequent work has been to compare the results with observed nutrient loads determined from monitoring data. If the predicted loads deviate by more than ±5% from the observed loads, then a sensitivity analysis is carried out. This involves varying the export coefficient for each source in turn by ±10%, while those for other sources are held constant, and observing the overall change in model predictions. This information is then used to revise the export coefficients exerting the greatest influence over the results (Johnes et al. 1994b, 1996). This process is continued until the model predicts within ±5% of observed loads. A further stage of validation is carried out by running the model over a longer period of time using historical data on land use and comparing the results with independent water quality data. The model is considered as valid when the overall degree of accuracy for the validation period does not exceed ±10% of observed loads.

The calibration and validation processes are explained in more detail in Johnes (1990), Johnes et al. (1994b, 1996) and Johnes (1996). Although calibration and validation of any model are necessary, the amount of data needed for these processes invalidates the claim that the model 'has relatively few data requirements, and can be easily calibrated' (Johnes 1996, p. 346). Johnes is aware that there are limitations to the approach, including the fact that 'export coefficients cannot be verified fully for the research site without considerable expenditure on field experimental work' (Johnes 1996, pp. 346–7). These factors effectively mean that the model cannot readily be used by anyone without extensive experience and considerable funding. Johnes (1996) also concedes 'that the importance of hydrological pathways in determining nutrient delivery to surface waters, and the variations in available transport mechanisms over the annual water cycle mean that the model cannot predict in real time.' (Johnes 1996, pp. 346–7). This is, however, not the purpose of the ECA, which is to predict changes in water quality on a yearly basis.

### 13.3.6 Using the results

Results obtained from modelling nutrient loadings on eutrophicated lakes are commonly used to evaluate the impact of proposed or potential reduction measures (e.g. O'Sullivan 1992; Wilson et al. 1993; Wilson 1995; Johnes 1996). However, the objective of a programme of such measures must first be decided upon. As discussed above, the objective may be **rehabilitation**, alleviating the problems associated with eutrophication and rendering the lake more suitable for human uses. In such cases, a pragmatic approach is called for, in which a cost–benefit analysis of potential options is carried out and the 'least cost—most benefit' measures adopted for implementation. If, however, the objective is **restoration** of the lake to its former, pre-eutrophication state, then a different approach is necessary.

Knowledge of the former trophic status of a lake can be gleaned from historical records. For instance, Loe Pool in Cornwall, presently suffering from the effects of advanced eutrophication, has been studied by botanists since the late nineteenth century. The early records (1874 to 1927) of floating and submerged macrophytes evoke an assemblage of plants mostly associated with mesotrophic conditions (Wilson & Dinsdale 1998). Palmer (1989) and Palmer et al. (1992) devised a system under which a wide range of aquatic species of plants are assigned to a trophic site type (dystrophic, oligotrophic, mesotrophic, eutrophic) and ascribed a trophic ranking score. The average value of all species found at a site provides an indication of trophic status. Although there are limitations to this approach, including the fact that relative abundance is not taken into account, it can be a useful tool when historical records of aquatic macrophytes are available. In the case of Loe Pool, palaeolimnological studies of sediment chemistry and diatoms confirm the change to a more eutrophic state during the 1930s (O'Sullivan 1992). However, in the absence of such research, historical records can prove invaluable.

Palaeolimnological studies, particularly of sediment chemistry and diatoms, have been used extensively to provide information on the pre-eutrophication state of lakes and the timing of the onset of enrichment. A brief description of the means by which this can be achieved will be provided here; for a fuller explanation of the methods and uses of palaeolimnology, see O'Sullivan (Volume 1, 609–66).

Lake sediment records provide site-specific insights into pre-eutrophication conditions. Changes in the historical phosphorus load can be deduced from sedimentary phosphorus profiles. However, in some cases, the geochemical phosphorus record may not relate directly to the phosphorus loading history, owing to post-depositional mobility (Anderson & Rippey 1994). Phosphorus fractionation techniques may help interpretation but different phosphorus fractions can also migrate upwards in the sediment. Diatoms are powerful indicators of trophic status and have been used successfully to infer historical changes. Until quite recently, however, such inference has entailed a qualitative interpretation of species change. Quantitative approaches have now been developed using multiple linear regression or weighted averaging (Bennion 1994). Diatom assemblages in surface sediments are correlated with, most commonly, total phosphorus concentrations to produce transfer functions, which are then used to infer past changes in total phosphorus from down-core changes in diatom assemblages (Anderson et al. 1993; Bennion et al. 1996). More recently, a chironomid—total phosphorus inference model has been developed (Brooks et al. 2001), adding a further tool with which to assess the sediment record. Results indicate that these methods offer a reliable means of determining pre-eutrophication phosphorus concentrations and of ascertaining objectives for lake restoration.

Once the objectives of restoration are decided upon, an assessment can be made of the measures necessary to achieve sufficient load reduction. The effects of such measures are calculated and incorporated into the model. The results can be assessed using the graph provided by OECD (1982, p. 95), or simply by comparing resultant values. However, obtaining literature values for reduction rates may pose problems, depending on the types of measures necessary to achieve the required trophic state. On the one hand, calculation of the impact on loadings of installing tertiary treatment at STWs is reasonably straightforward. Reduction rates are readily quantifiable and research in this area is relatively advanced, to the extent that appropriate values can be gleaned from the literature with confidence (e.g. Wilson & Jones 1994). On the other hand, calculations of the effect of riparian buffer zones designed to reduce diffuse losses from agricultural land require more investigation. Research in this area is, again, largely restricted to the USA and literature values vary considerably. Researchers report reduction rates for total phosphorus of as much as 93% (Dillaha *et al.* 1989), as well as much lower values of 47% (Edwards *et al.* 1983), although there does appear to be some correlation with width and slope. Knowledge in this area is increasing, although far more research needs to be carried out in this country (Muscutt *et al.* 1993). In the light of available data, approximate estimates can be made for the reduction effect of buffer zones. Other load reduction measures can present similar difficulties in terms of assessing their effect on loading, depending on the amount of reliable and definitive data in the literature.

## 13.4 CURRENT USES OF THE EXPORT COEFFICIENT APPROACH

Environmental legislation in the UK has been almost entirely driven by the necessity to comply with directives produced by the European Community (EC), now the European Union (EU). These directives have increasingly called for improvements in water quality. Until very recently, no single piece of legislation comprehensively addressed the issue of eutrophication. The development of European legislation and its impact on the problem of eutrophication in the UK are discussed in more depth in Wilson *et al.* (1996) and Wilson (1999).

The EC Directive on Urban Waste Water Treatment (UWWT) of 1991 addressed the problem of nitrogen and phosphorus inputs from larger sewage treatment works to 'sensitive areas' (i.e. surface waters liable to suffer from eutrophication). Implementation by the UK was inadequate, in that fewer sites were designated as 'sensitive' than was intended by the directive. The 1991 EC Directive on Nitrates proposed to deal with problems associated with this nutrient form. In the UK, it was interpreted narrowly as a measure to protect drinking water, even though the directive contains provisions for protection of freshwater lakes, estuaries and coastal waters liable to suffer from eutrophication. In 1994, the EC produced a draft Directive on the Ecological Quality of Surface Waters. The proposal was intended to supplement the Nitrate and UWWT Directives and to provide a framework for the control of pollution from point **and** diffuse sources, so as to maintain and improve the ecological quality of **all** surface waters. This draft has since been superseded by the EU Framework Directive on Water Resources 2000, which is designed to encompass the aims of the draft directive, and to broaden the scope to provide a comprehensive and coherent water policy framework. It was required to be transposed into UK law by December 2003 and the main environmental objectives are to be achieved by December 2015.

The Framework Directive requires the implementation of a monitoring and classification system of surface waters, followed by a qualitative and quantitative assessment of point and diffuse sources of pollution which may affect ecological quality. Operational targets must then be defined, and an integrated programme designed to achieve these targets should be implemented. The ultimate aim of the directive is for all Community waters to achieve good ecological quality, which is broadly defined as that which allows aquatic ecosystems to be self-sustaining. A list of parameters is provided for the determination of good ecological quality, which includes the provision that the diversity of populations of aquatic flora and fauna should resemble or reflect that of similar water bodies with insignificant anthropogenic disturbance. On this basis alone, eutrophicated inland waters will require significant remedial action to achieve good ecological quality. The Framework Directive will therefore provide the basis for eutrophication control in the UK.

In anticipation of increasingly stringent water quality standards from the EC, the UK introduced a system of statutory Water Quality Objectives (WQOs) under the Water Resources Act 1991. This system has yet to be fully devised and implemented, but it is likely to provide the basis for the integrated programme required by the Framework Directive. In parallel with the programme to develop the WQO system, the National Rivers Authority (NRA, now the Environment Agency) initiated a research project to develop a strategy for the classification of lakes, not only in terms of eutrophication, but also acidification. The ECA devised by Johnes & O'Sullivan (1989) and developed by Johnes (1990) forms the basis for a eutrophication classification scheme. The first phase of the research is reported in Johnes et al. (1994a, b) and the second in Johnes et al. (1998a, b).

In the process of developing a state-changed system of classification, Johnes and co-workers used historical data on land use, livestock numbers and population data to 'hindcast' nutrient loadings of a baseline state. The choice of baseline date (1931) was 'both practical and political' (Johnes et al. 1994a, p. 14) and represents the period prior to subsidy-controlled agriculture and major movement of the population from rural to urban areas. The export coefficients chosen for predictions of loadings on the River Windrush and Slapton Ley were applied to ten other sites, selected to reflect a wide range of environments. The results of this exercise were compared with actual data for the ten sites and the coefficients adjusted in order to reflect regional variations.

In order to reduce the amount of data necessary to run the model for a far greater number of sites, a regional approach was then adopted. Land-use regions in England and Wales were identified, using the Dudley Stamp Land Utilisation Survey of the 1930s. Average export coefficients were devised for these regions, of which there are typically five to ten per county. Results obtained using the coefficients based on land-use regions appear to correlate well with those from the catchment-specific approach. The model was then tested on a further 94 English and Welsh lakes and 'shown to give reasonable results' (Johnes et al. 1994a, p. VII), although available contemporary data were not always adequate for verification.

Comparisons were made between the values obtained for baseline and contemporary variables, and the percentage change calculated. Average percentage change for a total of eight variables, including lake total phosphorus and total nitrogen, was determined and placed in one of five categories, to provide an index of the degree of eutrophication over time. A table of provisional classification was then produced for the 94 lakes. Results of the research project were produced as reports to the NRA (i.e. Johnes et al. 1994a, b). Overviews of the proposed classification scheme were subsequently published (e.g. Johnes et al. 1996; Moss et al. 1996, 1997).

The provisional scheme was open to comment from a number of agencies with an interest in lake water quality and some concerns were raised. These included one from English Nature (the government's advisory body on nature conservation) that differing lake progressions can be masked by using only state-changed classification. For example, a lake categorised as being in Class 1, i.e. assessed as being subject to a small degree of change, can be representative of two different scenarios. It may be a productive lake which has undergone an increase in nutrient status of no real ecological consequence. Or it may be a relatively unproductive lake which has experienced the same degree of enrichment, but with more damaging repercussions for the ecosystem. This lack of distinction between sites in the same class can lead to a situation in which those at the lower end of the trophic spectrum, in danger of losing their high conservation interest, appear to require no remedial action.

Possibly as a result of this concern, the second phase of the project included the development of a spatial state classification scheme. Used in conjunction with the state-changed scheme, this enables the identification of nutrient-poor lakes which have undergone a small, but damaging degree of nutrient enrichment. However, the authors eschew the use of the OECD classifications of 'oligotrophic', 'mesotrophic', etc., as being too subjective and not allowing for extremely enriched lakes which are common in the UK. Instead, numbered categories (1–6) are adopted for the spatial state classification scheme, which also takes into account the fact that phosphorus may not be the limiting nutrient.

The second phase also developed further the classification of lakes by the state-changed scheme for eutrophication. As before, the scheme divides England and Wales into six natural land-use regions (Table 13.7) and assigns a set of export coefficients to each (Table 13.8). In contrast to previously published export coefficients, those used for the scheme are expressed entirely in terms of percentage loss of inputs. This is an interesting, but curious, development. Previously, export coefficients for nitrogen for the various types of land use were expressed as percentage loss of inputs and for phosphorus as $kg\,ha^{-1}\,yr^{-1}$, regardless of inputs. This would seem logical in terms of the behaviour of these elements in the environment, in that nitrogen is far more mobile than phosphorus; losses of nitrogen are more dependent on the intensity of inputs, whereas those for phosphorus are more strongly related to land use. The reasons for this change to a uniform percentage loss for all export coefficients are not explained. It should be noted that the rough estimate derived from Vollenweider (1968) of 3% for losses of phosphorus from animal waste still forms the basis for all export coefficients in these categories.

The state-changed classification scheme is again based on comparisons between the values obtained for baseline and for contemporary variables, and on the calculation of percentage change. Average percentage change in a number of variables is determined and allocated to one of six categories (Table 13.9). A trial classification was produced for 76 lakes in England and Wales, some of which are shown in Table 13.10.

A further development in the second phase of the project was that the 'hindcast' values obtained for nutrient concentrations using export coefficients were compared with those derived from determination of the diatom communities in dated sediment cores. The diatom transfer function approach (described above) was used for four sites, selected on the basis of the availability of long-term records of lake nutrient concentrations, and a three-way comparison was carried out. A further seven sites without historical water quality data were used for a two-way comparison under a project funded by English Nature. The three-way comparisons illustrated trends for both diatom models and export coefficient models that were broadly similar. The two-way comparison pro-

Table 13.7 Some characteristics of the land use regions used in the trial eutrophication classification scheme (Johnes et al. 1998b)

| Region | Location | Geology | Farming |
|---|---|---|---|
| 1 | Pennines, Lake District, Wales, southwest England, east Midlands | Impermeable igneous or metamorphic rock | Extensive livestock and upland regions |
| 2 | West England, west Wales, northwest Midlands, Lake District | Usually impermeable, metamorphic rock with flat topography | Lowland dairying regions |
| 3 | West Midlands, south England, east Midlands, northeast England | Permeable sedimentary rock with moderately sloping land | Mixed arable and dairying regions |
| 4 | West Midlands, southwest England | Impermeable metamorphic rock with steeply sloping land | Mixed arable and dairying regions |
| 5 | East England, east Midlands, southwest Lancashire, Home Counties | Permeable sedimentary rock often on glacial drift and very flat | Intensive arable regions |
| 6 | Urban land across England and Wales | Variable, normally impermeable surfaces | None |

Table 13.8  Export coefficients selected for regional type categories for England and Wales (Johnes et al. 1998a)

| Element | Source | Regional type | | | | | |
|---|---|---|---|---|---|---|---|
| | | 1 | 2 | 3 | 4 | 5 | 6 |
| Nitrogen | Permanent grass | 1.0 | 15.0 | 5.0 | 7.5 | 1.0 | 0 |
| | Temporary grass | 2.0 | 15.0 | 5.0 | 7.5 | 1.0 | 0 |
| | Cereals | 10.0 | 25.0 | 12.0 | 12.0 | 2.0 | 0 |
| | Other arable crops | 20.0 | 25.0 | 20.0 | 25.0 | 3.0 | 0 |
| | Rough grazing | 1.0 | 1.0 | 1.0 | 1.0 | 0.4 | 0 |
| | Woodland and orchard | 1.0 | 1.0 | 1.0 | 1.0 | 0.2 | 0 |
| | Bare fallow | 20.0 | 20.0 | 20.0 | 20.0 | 4.0 | 0 |
| | Cattle | 7.23 | 16.2 | 16.1 | 16.1 | 17.0 | 0 |
| | Pigs | 7.23 | 14.5 | 14.4 | 14.4 | 17.0 | 0 |
| | Sheep | 8.5 | 17.0 | 17.0 | 17.0 | 17.0 | 0 |
| | Poultry | 7.65 | 15.3 | 15.3 | 15.3 | 17.0 | 0 |
| | People | 54.5 | 54.5 | 54.5 | 54.5 | 54.5 | 0 |
| | Rainfall | 25.0 | 25.0 | 25.0 | 25.0 | 25.0 | 0 |
| Phosphorus | Permanent grass | 2.0 | 6.0 | 1.0 | 3.0 | 0.2 | 0 |
| | Temporary grass | 2.5 | 7.0 | 2.5 | 3.0 | 0.2 | 0 |
| | Cereals | 2.5 | 5.0 | 2.5 | 5.0 | 0.4 | 0 |
| | Other arable crops | 2.3 | 5.0 | 1.5 | 5.0 | 0.6 | 0 |
| | Rough grazing | 0 | 0 | 0 | 0 | 0 | 0 |
| | Woodland and orchard | 0 | 0 | 0 | 0 | 0 | 0 |
| | Bare fallow | 0 | 0 | 0 | 0 | 0 | 0 |
| | Cattle | 1.28 | 5.70 | 2.85 | 2.85 | 1.7 | 0 |
| | Pigs | 1.28 | 5.10 | 2.55 | 2.55 | 1.7 | 0 |
| | Sheep | 1.5 | 6.0 | 3.0 | 3.0 | 1.7 | 0 |
| | Poultry | 1.35 | 5.40 | 2.7 | 2.7 | 1.7 | 0 |
| | People | 32.8 | 68.0 | 68.0 | 32.8 | 68.0 | 0 |
| | Rainfall | 25.0 | 25.0 | 25.0 | 25.0 | 25.0 | 0 |

Table 13.9  Class boundaries for eutrophication in the state changed scheme (Johnes et al. 1998a)

| Class | Percentage change from baseline |
|---|---|
| 1 | <50 |
| 2 | 51–100 |
| 3 | 101–150 |
| 4 | 151–250 |
| 5 | 151–500 |
| 6 | >500 |

duced weaker correlations, and in the absence of observed data to validate the results it was concluded that the comparison was of limited value. An important aspect in a comparison of these two models relates to shallow productive lakes, where sedimentary phosphorus in the upper layers has been subject to redox changes and physical disturbance. Such processes induce phosphorus release from the sediment, increasing algal productivity. Neither model is expected to reflect this situation. The ECA model is not designed to include internal sources, whereas the diatom model is likely to overestimate total phosphorus concentrations

Table 13.10 Class allocation for the state changed scheme for a number of lakes (Johnes et al. 1998a)

| Site | Eutrophication class |
| --- | --- |
| Upton Broad | 1 |
| Llyn Brianne | 2 |
| Semerwater | 3 |
| Barton Broad | 3 |
| Hickling Broad | 4 |
| Derwent Water | 4 |
| Coniston Water | 5 |
| Llangorse Lake | 5 |
| Malham Tarn | 5 |
| Wastwater | 5 |
| Aqualate Mere | 6 |
| Esthwaite Water | 6 |
| Elterwater | 6 |
| Grasmere | 6 |
| Loe Pool | 6 |
| Rostherne Mere | 6 |
| Slapton Ley | 6 |
| Tabley Mere | 6 |

under such circumstances, owing to the high abundance of benthic taxa, many of which are poor indicators of lake trophic status. However, the diatom model is likely to predict total phosphorus concentrations more closely than the ECA, because diatoms reflect the internal loading, responding closely to within-lake total phosphorus.

The two approaches are seen by the authors as fulfilling different, but complementary, roles. The export coefficient model can be used to establish the origins and extent of changes in external load, and to assess the effect of load-reduction measures. The diatom model can be used to reflect the nature and timing of in-lake responses to changes in nutrient loading and to provide important ecological information. It is also less temporally limited and can produce information on lakes in pristine condition, prior to the advent of land-use data. The pragmatic approach adopted for the lake classification scheme, in using the 1930s as a baseline, is understandable, given the nature of the research contract, but such a temporal limitation does not permit the delivery of a true history of lake eutrophication in the UK. Palaeolimnological methods, limitations notwithstanding, will remain the means by which a more full and rounded record of lake history is obtained and thus provide objectives for **long-term** ecosystem stability.

The trial classification scheme is again open to comment from interested parties and will no doubt be refined further before adoption by the Environment Agency. It represents an important step forward in the management of lake water quality in this country and will be an invaluable tool in the implementation of the EU Framework Directive on Water Resources.

## 13.5 CONCLUSIONS

The export coefficient approach, as developed in England and Wales, has undergone a series of modifications and refinements. Nevertheless, it remains a relatively simple tool, requiring few data compared with more complex watershed models. There is no doubt that the model is sufficiently robust for the purposes to which it has been applied in Britain. However, the simpler the model, the less it represents the real world. This may be self-evident, but sometimes it seems we need reminding.

It is interesting that, after over a decade of calibration, validation and refinement of the ECA, the rough estimate proposed by Vollenweider (1968) for losses of phosphorus from animal waste of 1% to 5% (translated to 3%) is still used and appears to be valid. The question remains as to whether the persistence of this 'back of an envelope' value is an indication of the lack of scientific rigour in development of the ECA ('it works, so it must be right') or a lasting tribute to the brilliance and foresight of its originator, Richard Vollenweider.

## 13.6 ACKNOWLEDGEMENTS

This review is based partly on research funded by a SERC (CASE) award in collaboration with English

Nature, Peterborough. The author would like to thank Geraint Weber, Jocelyn Dela-Cruz and Helen Bennion for their assistance.

## 13.7 REFERENCES

Alexander, G.C. & Stevens R.J. (1976) Per capita phosphorus loading from domestic sewage. *Water Research*, **10**, 757–64.

Allen, S.E., Carlisle, A., White, E.J. & Evans, C.C. (1968) The plant nutrient content of rainwater. *Journal of Ecology*, **56**, 497–504.

Anderson, N.J. & Rippey, B. (1994) Monitoring lake recovery from point-source eutrophication: the use of diatom-inferred epilimnetic total phosphorus and sediment chemistry. *Freshwater Biology*, **32**, 625–39.

Anderson, N.J., Rippey, B. & Gibson, C.E. (1993) A comparison of sedimentary and diatom-inferred phosphorus profiles: implications for defining pre-disturbance nutrient conditions. *Hydrobiologia*, **253**, 357–66.

Arden-Clarke, C. (1988) *The Environmental Effects of Conventional and Organic/Biological Farming Systems*. Research Report RR-16. Political Ecology Research Group, Oxford, 109 pp.

Bennion, H. (1994) A diatom-phosphorus transfer function for shallow, eutrophic ponds in southeast England. *Hydrobiologia*, **275/276**, 391–410.

Bennion, H., Juggins, S. & Anderson, N.J. (1996) Predicting epilimnetic phosphorus concentrations using an improved diatom-based transfer function and its application to lake eutrophication management. *Environmental Science and Technology*, **30**, 2004–7.

Brooks, S.J., Bennion, H. & Birks, H.J.B. (2001) Tracing lake trophic history with a chironomid–total phosphorus inference model. *Freshwater Biology* **46**, 513–33.

Bunce, R.G.H., Howard, D.C., Hallam, C.J., et al. (1992) *Ecological Consequences of Land Use Change*. Final Report to the Department of Environment. Institute of Terrestrial Ecology, Grange-over-Sands, unpaginated.

Canfield, D.E., Jones, J.R., Ryding, S.-O. et al., (1989) Factors and processes affecting the degree of eutrophication. In: Ryding, S.-O. & Rast W. (eds), *The Control of Eutrophication of Lakes and Reservoirs*, UNESCO, Paris, 65–84.

Cherrill, A.J., McClean, C., Lane, A. et al., (1995) A comparison of land cover types in an ecological field survey in northern England and a remotely sensed land cover map of Great Britain. *Biological Conservation*, **71**, 313–23.

Clark, G., Knowles, D.J. & Phillips, H.L. (1983) The accuracy of the agricultural census. *Geography*, **68**, 115–20.

Clesceri, N.L., Curran, S.J. & Sedlak, R.I. (1986) Nutrient loads to Wisconsin lakes: Part I. Nitrogen and phosphorus export coefficients. *Water Resources Bulletin*, **22**, 983–90.

Cooke, G.W. (1976) A review of the effects of agriculture on the chemical composition and quality of surface and underground waters. In: Dermott W. & Gasser J.R. (eds), *Agriculture and Water Quality*. MAFF Technical Bulletin 32, pp 5–58. HMSO, London.

Delong, M.D. & Brusven, M.A. (1991) Classification and spatial mapping of riparian habitat with applications toward management of streams impacted by NPS pollution. *Environmental Management*, **15**, 565–71.

Delwiche, L.L.D. & Haith, D.A. (1983) Loading functions for predicting nutrient losses from complex watersheds. *Water Resources Bulletin*, **19**, 951–9.

Devey, D.G. & Harkness, N. (1973) The significance of man-made sources of phosphorus: Detergents and sewage. *Water Research*, **7**, 35–54.

Dillaha, T.A., Reneau, R.B., Mostaghini, S., et al. (1989) Vegetative filter strips for agricultural non-point source pollution control. *Transactions of the American Society of Agricultural Engineers*, **32**, 513–19.

Edwards, D.R. & Daniel, T.C. (1994) A comparison of runoff quality effects of organic and inorganic fertilizers applied to Fescue grass plots. *Water Resources Bulletin*, **30**, 35–41.

Edwards, W.M., Owens, L.K. & White, R.K. (1983) Managing runoff from a small paved beef feedlot. *Journal of Environmental Quality*, **12**, 281–6.

Endreny, T.A. & Wood, E.F. (2003) Watershed weighting of export coefficients to map critical phosphorus loading areas. *Journal of the American Water Resources Association*, **39**, 165–81.

Fuller, R.M., Groom, G.B. & Jones, A.R. (1994) The land cover map of Great Britain: automated classification of Landsat Thematic Mapper data. *Photogrammetric Engineering Remote Sensing*, **60**, 553–62.

Gerritse, R.G. (1981) Mobility of phosphorus from pig slurry in soils. In: Hucker, T.W.G. & Catroux, G. (eds), *Phosphorus in Sewage Sludge and Animal Waste Slurries*. Proceedings, EEC Seminar, Groningen, The Netherlands, 12–13 June 1980. D. Reidel, Dordrecht., 347–366.

Gillespie, J.R. (1989) *Modern Livestock and Poultry Production* (3rd edition). Delmar, New York, 957 pp.

Gostick, K.G. (1982) ADAS recommendations to farmers on manure disposal and recycling. *Philosophical Transactions of the Royal Society of London Series B*, **296**, 321–39.

Hanrahan, G., Gledhill, M., House, W.A., et al. (2001) Phosphorus loading in the Frome catchment, UK: seasonal refinement of the export coefficient modeling approach. *Journal of Environmental Quality*, **30**, 1738–46.

Hilton, J., Buckland, P. & Irons, G.P. (2002) An assessment of a simple method for estimating contributions of point and diffuse source phosphorus to in-river phosphorus loads. *Hydrobiologia*, **472**, 77–83.

House, M.A., Ellis, J.B., Herricks, E.E., et al. (1993) Urban drainage—impacts on receiving water quality. *Water Science and Technology*, **27**, 117–58.

House, W.A. & Warwick, M.S. (1998) A mass-balance approach to quantifying the importance of in-stream processes during nutrient transport in a large river catchment. *The Science of the Total Environment*, **210/211**, 139–52.

Janus, L.L. & Vollenweider, R.A. (1981) *The OECD Cooperative Programme on Eutrophication—Canadian Contribution*. Scientific Series No. 131 and 131-S, Canada Centre for Inland Waters, Burlington, Canada, 24 pp.

Johnes, P.J. (1990) *An investigation of the effects of land use upon water quality in the Windrush catchment—an export coefficient approach*. Unpublished DPhil thesis, University of Oxford, Oxford.

Johnes, P.J. (1996) Evaluation and management of the impact of land use change on the nitrogen and phosphorus load delivered to surface waters: the export coefficient approach. *Journal of Hydrology*, **183**, 323–49.

Johnes, P.J. & Heathwaite, A.L. (1997) Modelling the impact of land use change on water quality in agricultural catchments. *Hydrological Processes*, **11**, 269–86.

Johnes, P.J. & O'Sullivan, P.E. (1989) The natural history of Slapton Ley Nature Reserve XVIII. Nitrogen and phosphorus losses from the catchment—an export coefficient approach. *Field Studies*, **7**, 285–309.

Johnes, P.J., Moss, B. & Phillips, G. (1994a) *Lakes—Classification and Monitoring: a Strategy for the Classification of Lakes*. R&D Note 253, National Rivers Authority, Bristol, 84 pp.

Johnes, P.J., Moss, B. & Phillips, G. (1994b) *Lakes—Classification and Monitoring: a Strategy for the Classification of Lakes*. R&D Project Record 286/6/A, National Rivers Authority, Bristol, 139 pp.

Johnes, P.J., Moss, B. & Phillips, G. (1996) The determination of water quality by land use, livestock numbers and population data-testing of a model for use in conservation and water quality management. *Freshwater Biology*, **36**, 451–73.

Johnes, P.J., Curtis, C., Moss, B., et al. (1998a) *Trial Classification of Lake Water Quality in England and Wales: a Proposed Approach*. R&D Technical Report E53, Environment Agency, Bristol, 81 pp.

Johnes, P.J., Bennion, H., Curtis, C., et al. (1998b) *Trial Classification of Lake Water Quality in England and Wales: a Proposed Approach*. R&D Project Record E2-i731/5, Environment Agency, Bristol, 240 pp.

Jones, R.A. & Lee, G.F. (1982) Recent advances in assessing the impact of phosphorus loads on eutrophication-related water quality. *Journal of Water Research*, **16**, 503–15.

Jones, R.A. & Lee, G.F. (1986) Eutrophication modeling for water quality management: an update of the Vollenweider–OECD model. *Water Quality Bulletin*, **11**, 67–74.

Jørgensen, S.-E. (1980) *Lake Management*. Pergamon, Oxford, 167 pp.

Lambert, M.G., Devantier, B.P., Nes, P., et al. (1985) Losses of nitrogen, phosphorus, and sediment in runoff from hill country under different fertiliser and grazing management regimes. *New Zealand Journal of Agricultural Research*, **28**, 371–9.

Lind, O.T. (1979) *Handbook of Common Methods in Limnology*. C.V. Mosby, St Louis, 199 pp.

Loehr, R.C., Ryding, S.-O. & Sonzogni, W.C. (1989) Estimating the nutrient load to a waterbody. In: Ryding, S.-O. & Rast, W. (eds), *The Control of Eutrophication of Lakes and Reservoirs*. UNESCO, Paris, 115–46.

Mattikalli, N.M. & Richards, K.S. (1996) Estimation of surface water quality changes in response to land use change: application of the export coefficient model using remote sensing and Geographical Information System. *Journal of Environmental Management*, **48**, 263–82.

McGuckin, S.O., Jordan, C. & Smith, R.V. (1999) Deriving phosphorus export coefficients for CORINE land cover types. *Water Science and Technology*, **39**, 47–53.

McIntyre, S.C. (1993) Reservoir sedimentation rates linked to long-term changes in agricultural land use. *Water Resources Bulletin*, **29**, 487–95.

McLeod, R.V. & Hegg, R.O. (1984) Pasture runoff water quality from application of inorganic and organic nitrogen sources. *Journal of Environmental Quality*, **13**, 122–6.

Mitchell, D.J. (1990) The use of vegetation and land use parameters in modelling catchment sediment yields. In: Thornes, J.B. (ed.), *Vegetation and Erosion*. Wiley, Chichester, 289–316.

Moss, B., Johnes, P.J. & Phillips, G.L. (1996) The monitoring of ecological quality and the classification of standing waters in temperate regions: a review and proposal based on a worked scheme for British waters. *Biological Reviews*, **71**, 301–39.

Moss, B., Johnes, P. & Phillips, G.L. (1997) New approaches to monitoring and classifying standing waters. In: Boon, P.J. & Howell, D.L. (eds), *Freshwater Quality: Defining the Indefinable?* The Stationary Office, Edinburgh, 118–33.

Mostaghimi, S., Younos, T.M. & Tim, U.S. (1992) Effects of sludge and chemical fertilizer application on runoff water quality. *Water Resources Bulletin*, **28**, 545–52.

Muscutt, A.D., Harris, G.L., Bailey, S.W., et al. (1993) Buffer zones to improve water quality: a review of their potential use in UK agriculture. *Agriculture Ecosystems and Environment*, **45**, 59–77.

Nguyen, M.L. & Goh, K.M. (1992) Nutrient cycling and losses based on a mass-balance model in grazed pastures receiving long-term superphosphate applications in New Zealand. *Journal of Agricultural Science*, **119**, 89–106.

Nix, J. (1992) *Farm Management Handbook*. Wye College, Ashford, 284 pp.

North, M.O. & Bell, D.D. (1990) *Commercial Chicken Production Manual* (3rd edition). Van Nostrand Reinhold, New York, 913 pp.

O'Neill, P. (1985) *Environmental Chemistry*. George Allen & Unwin, London, 232 pp.

OECD (1982) *Eutrophication of Waters: Monitoring, Assessment and Control*. Organisation for Economic Co-operation and Development, Paris, 154 pp.

O'Sullivan, P.E. (1992) The eutrophication of shallow coastal lakes in southwest England—understanding and recommendations for restoration, based on palaeolimnology, historical records, and the modelling of changing phosphorus loads. *Hydrobiologia*, **243/244**, 421–34.

Owens, M. (1976) Nutrient balances in rivers. In: Dermott, W. & Gasser, J.R. (eds), *Agriculture and Water Quality*. MAFF Technical Bulletin 32, HMSO, London, 257–75.

Palmer, M. (1989) *A Botanical Classification of Standing Waters in Great Britain and a Method for the Use of Macrophyte Flora in Assessing Changes in Water Quality*. Nature Conservancy Council, Peterborough, 20 pp.

Palmer, M., Bell, S.L. & Butterfield, I. (1992) A botanical classification of standing waters in Britain: applications for conservation and monitoring. *Aquatic Conservation: Marine and Freshwater Ecosystems*, **2**, 125–44.

Rast, W. & Lee, G.F. (1978) *Summary Analysis of the North American (US Portion) OECD Eutrophication Project: Nutrient Loading—Lake Response Relationships and Trophic State Indices*. EPA-600/3-78-008, Environmental Protection Agency, Corvallis, 455 pp.

Rast, W. & Lee, G.F. (1983) Nutrient loading estimates for lakes. *Journal of Environmental Engineering*, **109**, 502–17.

Reckhow, K.H. & Simpson, J.T. (1980) A procedure using modeling and error analysis for the prediction of lake phosphorus concentrations from land use information. *Canadian Journal of Fishery and Aquatic Science*, **37**, 1439–48.

Reckhow, K.H., Beaulac, M.N. & Simpson, J.T. (1980) *Modeling Phosphorus Loading in Lake Response Under Uncertainty: a Manual and Compilation of Export Coefficients*. EPA-440/5-80-011. US Environmental Protection Agency, Washington, 226 pp.

Richardson, S.J. (1976) Animal manures as potential pollutants. In: Dermott, W. & Gasser, J.R. (eds), *Agriculture and Water Quality*. MAFF Technical Bulletin 32, HMSO, London, 405–17.

Royal Society (1983) *The Nitrogen Cycle of the United Kingdom: Report of a Royal Society Study Group*. Royal Society, London, 264 pp.

Sas, H. (1989) *Lake Restoration by Reduction of Nutrient Loading: Expectations, Experiences, Extrapolations*. Academia Verlag Richarz, Skt Augustin, 497 pp.

Sharpley, A.N. & Smith, S.J. (1990) Phosphorus transport in agricultural runoff: the role of erosion. In: Boardman, J., Foster, I.D.L. & Dearing, J.A. (eds), *Soil Erosion on Agricultural Land*. Wiley, Chichester, 351–66.

Sheaffer, J.R. & Wright, K.R. (1982) *Urban Storm Drainage Management*. Dekker, New York, 271 pp.

Smulders, E. & Krings, P. (1990) Detergents for the 1990s. *Chemistry and Industry*, **6**, 160–3.

Soprano, P.A., Hubler, S.L., Carpenter, S.R., et al. (1996) Phosphorus loads to surface waters: a simple model to account for spatial pattern of land use. *Ecological Applications*, **6**, 865–78.

Thomas, M.L., Lal, R., Logan, T., et al. (1992) Land use and management effects on nonpoint loading from Miamian soil. *Soil Science Society of America Journal*, **56**, 1871–5.

Uttormark, P.D., Chapin, J.D. & Green, K.M. (1974) *Estimating Nutrient Loadings of Lakes from Non-Point Sources*. EPA-660/3-74-020. Environmental Protection Agency, Washington, 112 pp.

Vollenweider, R.A. (1968) *Scientific Fundamentals of Stream and Lake Eutrophication, with Particular Reference to Nitrogen and Phosphorus*. OECD Technical Report DAS/DST/88, Organization for Economic Co-operation and Development, Paris, 159 pp.

Vollenweider, R.A. (1975) Input–output models with special reference to the phosphorus loading concept. *Schweizerisches Zeitschrift für Hydrologie*, **37**, 58–84.

Vollenweider, R.A. (1976) Advances in defining critical loads for phosphorus in lake eutrophication. *Memorie dell'Istituto Italiano di Idrobiologica*, **33**, 53–83.

White, R.E. (1981) Pathways of phosphorus in soils. In: Hucker, T.W.G. & Catroux, G. (eds) *Phosphorus in Sewage Sludge and Animal Waste Slurries*. Proceedings, EEC Seminar, Groningen, The Netherlands, 12–13 June 1980. D. Reidel, Dordrecht, 21–41.

Wilson, B. & Jones, B. (1994) *The Phosphate Report. A Life Cycle Study to Evaluate the Environmental Impact of Phosphates and Zeolite A-PCA as Alternative Builders in UK Laundry Formulations*. Landbank, London, 91 pp.

Wilson, H.M. (1995) *An evaluation of alternative management strategies for shallow eutrophicated lakes and reservoirs*. Unpublished PhD thesis, University of Plymouth, Plymouth, 326 pp.

Wilson, H. (1999) Legislative challenges for lake eutrophication control in Europe. *Hydrobiologia*, **395/396**, 389–401.

Wilson, H. & Dinsdale, J. (1998) *Loe Pool Catchment Management Project*. Final Report, Environment Agency, Bodmin, 116 pp.

Wilson, H.M., Gibson, M.T. & O'Sullivan, P.E. (1993) Analysis of current policies and alternative strategies for the reduction of nutrient loads on eutrophicated lakes: the example of Slapton Ley, Devon. *Aquatic Conservation: Marine and Freshwater Ecosystems*, **3**, 239–51.

Wilson, H.M., Payne, S., O'Sullivan, P.E. & Gibson, M.T. (1996) Policy and legislation relevant to the conservation of freshwater SSSIs subject to eutrophication. *Journal of the Chartered Institution of Water and Environmental Management*, **10**, 348–54.

# 14 The Phosphorus Loading Concept and the OECD Eutrophication Programme: Origin, Application and Capabilities

## WALTER RAST AND JEFFREY A. THORNTON

### 14.1 INTRODUCTION

Because it represents a natural phenomenon with negative impacts that may be greatly exaggerated by human activities, lake eutrophication causes degradation of water quality on a global scale. In temperate-zone countries, the word 'eutrophication' often invokes an image of a weed-infested, scum-encrusted lake which looks, smells and feels unwholesome. On the other hand, fish farmers in some developing countries work hard to enhance the biological productivity of lakes that characterises eutrophication, for the purpose of producing a high yield of fish or prawns for food or economic gain, and may well be puzzled by the thought that eutrophication should be considered deleterious.

Eutrophication is frequently said to describe a natural ageing process of lakes, in which external (allochthonous) sources of nutrients, organic matter and sediment from the land accumulate in the basin of the lake, gradually filling it in, decreasing its depth and enhancing its biological production. Over time, the lake will take on a marsh-like character. Over a longer period of time, many lakes (although not all; see Wetzel 1983) will change into terrestrial systems (Fig. 14.1). In the absence of human influences, this process typically takes place within geological time. The process, however, can be greatly accelerated by human activities, which may shorten it to mere decades. This effect is termed 'cultural eutrophication' in order to distinguish it from the natural process, and it is characterised by sharply increased external loads of nutrients, especially phosphorus and nitrogen.

These extra resources can exert profound effects on water quality, biological communities and the trophic status of a lake. Major symptoms of cultural eutrophication include excessive growths of algae and aquatic plants, degraded water quality, and enhanced sedimentary fluxes of decomposing algae that can induce deoxygenation of the bottom water and increased fish biomass at the expense of decreased quality in the fish harvest (Table 14.1).

The objective of this chapter is to discuss the development and subsequent application of the simplified phosphorus-load–lake-response models produced and refined within the context of the international Eutrophication Programme conducted by the Organization for Economic Cooperation and Development (OECD 1982) during the late-1970s and early-1980s. The ultimate goal of the OECD Eutrophication Programme was to provide simplified, practical methodologies for the environmentally sound and cost-effective management of lake and reservoir eutrophication. These models were derived from the most systematic study ever undertaken to measure, analyse and quantify the eutrophication process, and subsequently they have been applied in efforts to improve lake and reservoir management elsewhere, as discussed in this chapter.

### 14.2 THE EUTROPHICATION PROCESS

The Organization for Economic Cooperation and Development study (OECD 1982) characterised cultural eutrophication as the nutrient

# Phosphorus Loading Concept and the OECD Eutrophication Programme

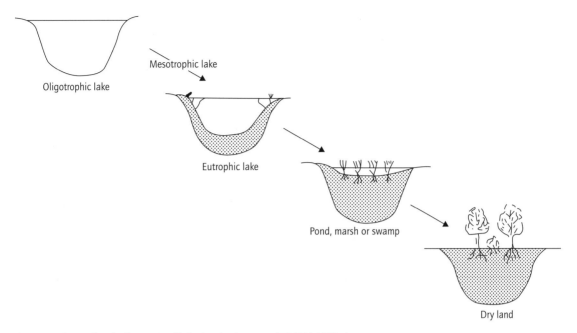

**Fig. 14.1** Generalised schematic of lake 'ageing' process (US EPA 1973a).

**Table 14.1** Water quality and trophic criteria and their responses* to increased eutrophication. (Modified from Ryding & Rast 1989)

| Physical | Chemical | Biological† |
|---|---|---|
| Transparency (D) (e.g., Secchi-disc depth) | Nutrient concentration (I) (e.g., spring maximum) | Algal bloom frequency (I) Algal species diversity (D) |
| | Chlorophyll concentration (I) | Phytoplankton biomass (I) |
| Suspended solids (I) | Electrical conductance (I) | Littoral vegetation (I)‡ |
| | Dissolved solids concentration (I) | Zooplankton (I) |
| | Hypolimnetic oxygen deficit (I) | Fish (I)§ |
| | Epilimnetic oxygen supersaturation (I) | Bottom fauna (I)¶ |
| | | Bottom fauna diversity (D) |
| | | Primary production (I) |

\* (I) signifies that the parameter generally increases with the degree of eutrophication; (D) signifies that it generally decreases with the degree of eutrophication.
† The biological criteria undergo important qualitative (e.g., species) and quantitative (e.g., biomass) changes with increasing eutrophication.
‡ Aquatic plants in shallow, nearshore areas may decrease in the presence of a high phytoplankton density.
§ Fish may be decreased in numbers and species in the hypolimnion beyond a certain level of eutrophication, as a result of hypolimnetic oxygen depletion.
¶ Bottom fauna may be decreased in numbers and species in high concentrations of hydrogen sulphide ($H_2S$), methane ($CH_4$) or carbon dioxide ($CO_2$), or low concentrations of oxygen ($O_2$) in the hypolimnion.

enrichment of waters which stimulates an array of symptomatic changes, including increased algae and macrophyte production, deterioration of water quality and other symptomatic changes, which are considered undesirable and which interfere with beneficial human water uses. This characterisation highlights the concept of the impairment or loss of beneficial human uses of freshwater resources as a principal defining feature of eutrophication, particularly for developed countries (see Gregor & Rast 1981; Ryding & Rast 1989).

Some lakes are naturally eutrophic, in that they receive sufficient nutrient supplies from natural sources to produce abundant growths of algae (phytoplankton) and aquatic plants (macrophytes). Human activities which accelerate the transport of nutrients and other materials from the drainage basin to a lake or reservoir, however, can greatly accelerate this process. The resulting additional growths of algae and macrophytes can cause significant negative changes in a lake, including a dramatic deterioration in its water quality, which also hinders its utility for many beneficial human purposes (e.g. domestic and industrial water supply, irrigation, recreation). Eutrophication ranks as one of the most pervasive causes of water quality deterioration on a global scale.

The symptoms of eutrophication are most pronounced in lakes and reservoirs, and in some wide, slowly flowing river systems. The standing water environment of these water systems provides the opportunity for excessive algal and aquatic plant growths to accumulate and become visible symptoms of eutrophication. Lakes and reservoirs are sinks for many inflowing substances from their drainage basin, including nutrients (Rast 2003). Lakes and reservoirs are both effective integrators and sensitive barometers of human activities within their respective drainage basins and, sometimes, from activities originating outside them (e.g., the long-range transport of airborne pollutants). Although algal and macrophyte growths also can occur in the flowing water environment of rivers, the algae continue to be moved downstream with the flowing water, rather than accumulating in the standing water environment of lakes to become visible signs of eutrophication.

### 14.2.1 Eutrophication and nutrient loads

In the same manner as terrestrial plants, the production of algae and aquatic plants stems from the photosynthesis reaction, involving the same components (e.g. nutrients, sunlight energy, appropriate temperature). The role of phosphorus and nitrogen as essential aquatic plant nutrients in this process and as causative factors for eutrophication has been well-documented over time (e.g. Sawyer 1947, 1952, 1966; Mackenthun et al. 1964; Fruh et al., 1966; American Water Works Association 1966, 1967; Stewart & Rohlich 1967; Vollenweider 1968, 1975; National Academy of Sciences 1969; Bartsch 1970, 1972; Edmondson 1970; Lee 1971, 1973; Schindler 1971; Brezonik 1972, 1973; Likens 1972; Maloney et al. 1972; Martin & Goff 1972; Powers et al. 1972; US EPA 1973a, 1974; Bachmann & Jones 1974; Porcella et al. 1974; Schindler & Fee 1974; Uttormark et al. 1974; Vallentyne 1974; Vollenweider & Dillon 1974; Rast & Lee 1978; Rast & Thornton 1997). The effects of human-induced (anthropogenic) nutrient loads in accelerating the eutrophication process also have been studied in detail (e.g. Sawyer 1952, 1966; Curry & Wilson 1955; Maloney 1966; Stewart & Rohlich 1967; Vollenweider 1968; Bartsch 1970, 1972; Lee 1971; Schindler 1971; Beeton & Edmondson 1972; Edmondson, 1972; Schindler & Fee 1974; Vallentyne 1974; Schindler et al. 1976; US EPA 1976; Rast & Lee 1978; Ryding & Rast 1989; Rast & Thornton 1997). It is virtually impossible to control significantly the other factors which fuel the growth of algae and macrophytes in the photosynthesis reaction (e.g. sunlight energy, temperature). From the perspective of eutrophication management, therefore, it is generally accepted that the major eutrophication control measures should be directed to reducing the external nutrient loads (mainly phosphorus and nitrogen) to a lake or reservoir from their major sources in the drainage basin.

As a general observation, the regional chemical characteristics of natural waters generally reflect a close relationship to the geological characteristics of their drainage basins. The phosphorus and nitrogen chemical species of concern within the context of eutrophication management are those which are soluble in water. The phosphorus species of concern usually are the inorganic phosphates, including orthophosphate $(PO_4)^{3-}$ and the condensed phosphate forms. Orthophosphate is especially important as a control target because it is readily available for uptake by algae and aquatic plants. In contrast, bacterial action is required in order to release dissolved orthophosphate from dissolved and particulate organic phosphorus forms. Particulate inorganic phosphates also must be dissolved in order to become available for algal uptake. Bacteria can hydrolyse condensed phosphates to dissolved orthophosphate (Clesceri & Lee 1965a,b; Shannon & Lee 1966; Porcella et al. 1974).

The atmosphere is a major source of nitrogen in the biosphere, originating from the fixation of atmospheric nitrogen in a molecular form. The soil contains stable organic nitrogen or complexed compounds, resulting mainly from bacterial nitrification processes in the upper soil layers and from direct aeolian supply and atmospheric inputs. The Earth's crust itself contains virtually no nitrogen compounds, except from organogenic layers in sediment rocks. The nitrogen species of concern within the context of eutrophication comprise the inorganic ammonium $(NH_4)^+$, nitrate $(NO_3)^-$ and nitrite $(NO_2)^-$ ions (American Water Works Association 1966; Vollenweider 1968; Lee 1971).

### 14.2.2 The limiting-nutrient concept

The basis of most eutrophication control programmes is the limiting-nutrient concept. The term 'limiting nutrient' refers to the process whereby the maximum biomass of algae and/or macrophytes produced in a waterbody is proportional to the quantity of the nutrient present in the waterbody in the least quantity, relative to their needs. A range of elements or compounds previously has been suggested as the limiting nutrient for eutrophication, including iron, molybdenum, nitrate, sulphate, vitamins and other organic growth factors, carbon and silicon (Goldman 1960a,b, 1964; Menzel & Ryther 1961; Goldman & Wetzel 1963; Lange 1967; Kuentzel 1969; Provasoli 1969; Kerr et al. 1970; Schelske &

Stoermer 1972). However, most limitations attributable to these elements and compounds appear to be either site-specific or temporal in nature. It is generally accepted that phosphorus and nitrogen are most frequently found to be the limiting nutrient in lakes and reservoirs. Further, not only are the absolute quantities of phosphorus and nitrogen important, but their relative quantities are also a key factor in determining which of these two elements will limit the maximum algal biomass in lakes and reservoirs.

The limiting-nutrient concept is based on the observation that a nutrient will be consumed or assimilated by an organism (e.g., plant) in proportion to the organism's need for the nutrient. As early as 1840, Justus Liebig observed that the growth of a crop was not generally limited by the nutrients needed in large quantities, which were often abundant in the environment. Rather, the growth was limited by the nutrients needed in the least quantities, which were often scarce. This observation forms the basis for Liebig's 'Law of the Minimum', one of the oldest laws of plant nutrition. The law highlights the fact that an organism's growth is limited by the substance or foodstuff available to it in the minimal quantity, relative to its need for that substance for growth and reproduction. Although this notion, in principle, can be applied to other growth-facilitating factors (e.g., sunlight energy, temperate), these latter factors are typically outside of human control.

### 14.2.3 Carbon, nitrogen and phosphorus as limiting nutrients

The nutrients needed in relatively large quantities by aquatic plants include carbon, hydrogen, oxygen, sulphur, potassium, calcium, magnesium, nitrogen and phosphorus. A range of trace elements also is required (Table 14.2).

There was considerable controversy in the late 1960s over the possible role of carbon as a limiting nutrient (Lange 1967; Kuentzel 1969; Kerr et al. 1970). However, this hypothesis was subsequently refuted by a number of researchers. Goldman et al. (1972) suggested that the conclusions regarding carbon limitation were based mainly on laboratory

Table 14.2 Summary of aquatic plant micronutrient requirements (Fruh 1967)

| Process trace | Element required |
| --- | --- |
| Photosynthesis | Manganese, iron, chloride, zinc, vanadium |
| Nitrogen fixation | Iron, boron, molybdenum, cobalt |
| Other functions | Manganese, boron, cobalt, copper, silicon |

data with samples containing surplus phosphorus and limited carbon dioxide. As a result, carbon limited the algal growth almost from the beginning of the experiments. A similar situation is often seen in wastewater stabilisation ponds, which typically exhibit carbon limitation because of the presence of excessive quantities of phosphorus and nitrogen, relative to carbon. Maloney et al. (1972), using laboratory assays, and Powers et al. (1972), using field experiments, also demonstrated that carbon addition to the water samples produced no effect on algal growth rates. Further, Shapiro (1973) demonstrated that low carbon dioxide concentrations in water do not necessarily limit algal biomass, but rather shifted the main algal types from green to blue-green without significantly affecting the overall algal biomass. James & Lee (1974) demonstrated similar results in examining the possibility that inorganic carbon limitation could occur in low-alkalinity waters.

Schindler (1977) provided the most convincing evidence against the notion of carbon limitation of algal growth, based on whole-lake experiments in the Canadian Experimental Lakes Area. In his experimentally fertilised lakes, atmospheric carbon dioxide provided sufficient carbon to support algal biomass proportional to the phosphorus concentrations in the lakes over a wide range of values. The initial bottle bioassay experiments supporting carbon limitation were carried out in small, closed or semi-closed containers, in which water turbulence and interaction with the overlying atmosphere were restricted. The proportion of alkalinity supplied by hydroxyl ions in the water also affected the rate at which carbon was supplied to natural waters. Based on these results, he concluded that lakes appear to possess biological mechanisms which eventually corrected algal deficiencies for carbon, and that these mechanisms also can occur with algal nitrogen deficiencies. Thus, phosphorus limitation, or phosphorus proportionality, was commonly maintained in his study lakes, even where the $C:N:P$ ratios appeared to favour nitrogen or carbon limitation. Such work was the basis for subsequent resolutions by the assemblies of both the International Limnological Congress and the International Ecology Congress recommending widespread phosphorus control as the primary control measure for combating cultural eutrophication.

### 14.2.4 Eutrophication and lake trophic status

As natural or artificially constructed water-filled depressions on the Earth's surface, lakes and reservoirs are transient features of the global landscape. They are analogous to giant sedimentation basins, most of which are destined to become filled with soil, nutrients and organic materials over time. Thus, they also act as giant reactor vessels ('stirring pots') for the physical, chemical and biological phenomena of eutrophication.

Weber (1907, as cited in Hutchinson 1969; Rodhe 1969; Brezonik 1969) was amongst the first to introduce the terms 'eutrophic' and 'oligotrophic' into science, using them to describe the general nutrient conditions of bog soils in Germany. Weber's scheme ran from eutrophic to oligotrophic, as a submerged bog was built up into a raised bog. The former was characterised as well nourished, and the latter as poorly nourished.

Naumann (1919, as cited in Hutchinson 1969; Rodhe 1969) subsequently introduced these terms into limnological science, using the expression 'eutrophic formation' to describe the phytoplankton assemblage in nutrient-rich waters. He later refined his definition (Naumann, 1931, as cited in Stewart & Rohlich 1967) as 'an increase of the nutritional standards (of a body of water), especially with respect to nitrogen and phosphorus.' Although originally referring to water quality, the terms eutrophic and oligotrophic have subsequently come to refer to general lake types as well (Brezonik 1969).

## 14.3 ALGAL NUTRIENT STOICHIOMETRY

Vallentyne (1974) highlighted the significance of the 15 to 20 elements needed by aquatic plants for growth, by calculating a demand:supply ratio for the elements. Consistent with Liebig's Law of the Minimum, he observed that aquatic plants exert a certain demand for nutrients proportional to the quantities of the nutrients in their cells. When one or more of the nutrients was present in short supply, relative to the others, the maximum biomass of the aquatic plants would be limited by the supply rates of these nutrients. Thus, the demand:supply ratio can reveal the nutrient most likely to limit lake productivity—the higher the ratio of a given nutrient, the more likely that it will limit algae and/or macrophyte growths. Vallentyne calculated the demand:supply ratios by determining the average chemical composition of aquatic plant communities, and dividing them by the mean chemical composition of the river waters of the world. The key role of phosphorus and nitrogen, particularly during the summer growing season, was clearly illustrated by this technique (Table 14.3).

Table 14.3 Demand:supply ratios for major aquatic plant nutrients (Vallentyne 1974)

| Element | Demand:supply ratio | |
|---|---|---|
| | Late winter* | Mid-summer† |
| Phosphorus | 80,000 | Up to 800,000 |
| Nitrogen | 30,000 | Up to 300,000 |
| Carbon | 5000 | Up to 6000 |
| Iron, silicon | Variable, but generally low | |
| All other elements | Less than 1000 | |

\* Prior to spring bloom.
† At algal maximum growth period.

It is a long-recognised principle in ecology that interactions between organisms and their environment are reciprocal (e.g. Redfield 1958; Odum 1971). The environment determines the conditions under which an organism can live. In turn, organisms can respond to changes in their physical environment by altering their metabolism or growth requirements, thereby also influencing their physical environment. For example, phytoplankton can directly change the nutrient concentrations in a waterbody by withdrawing them for growth and reproduction, usually accompanied by reciprocal changes in algal biomass.

The synthesis phase for algae and macrophytes includes phosphorus and nitrogen withdrawal from a waterbody during the photosynthesis reaction. The proportions removed can vary widely under certain nutrient extremes (Ketchum & Redfield 1949; Fitzgerald 1969; Fuhs et al. 1972; Maloney et al. 1972). Phosphorus and nitrogen are generally withdrawn from the water in proportions required for phytoplankton growth. The regeneration phase for the nutrients occurs when the elements are returned to the water as decomposition products and excretions of phytoplankton, the higher trophic level organisms which feed on them, and the microorganisms which decompose their organic debris (Redfield et al. 1963).

The proportions in which nutrients in natural waters enter into this cyclical process of synthesis and regeneration are generally determined by the elementary composition of the phytoplankton biomass. The traditional view (Redfield 1958; Redfield et al. 1963; Vollenweider 1968; Ketchum 1969) is that phytoplankton need a relatively fixed atomic ratio of carbon to nitrogen to phosphorus of 106 to 16 to 1 (i.e. 106 C : 16 N : 1 P), based on the simple stoichiometry of the photosynthesis–respiration reaction as it is believed to occur in nature. A general, simplified representation of this reaction (Stumm & Morgan 1981) is as follows:

$$106\ CO_2 + 16\ NO_3 + HPO_4 + 122\ H_2O + 18\ H + \text{trace elements} + \text{energy} \underset{\text{respiration}}{\overset{\text{photosynthesis}}{\rightleftharpoons}} \{C_{106}H_{263}O_{110}N_{16}P_1\} + 138\ O_2 \quad (14.1)$$

algal protoplasm

Table 14.4 Carbon:nitrogen:phosphorus atomic ratio in plankton (Redfield et al. 1963)

|  | Carbon | Nitrogen | Phosphorus |
|---|---|---|---|
| Zooplankton | 103 | 16.5 | 1 |
| Phytoplankton | 108 | 15.5 | 1 |
| Average value | 106 | 16 | 1 |

As part of the basic process of primary productivity in natural waters, inorganic carbon, nitrogen and phosphorus become components of organic algal protoplasm, under the appropriate conditions of light and temperature.

The 106 C:16 N:1 P atomic ratio of aquatic plants was based on the relative quantities of these elements in the organic matter of plankton samples in seawater (Redfield et al. 1963). The C:N:P ratio values represented an average of the carbon, nitrogen and phosphorus content present in phytoplankton and zooplankton (Table 14.4).

Dismissing carbon as a limiting nutrient, a number of primarily laboratory-based studies indicated that the average N:P atomic ratio can change significantly as a function of the condition of the aquatic environment (e.g. Ketchum & Redfield 1949; Harris & Riley 1956; Fitzgerald 1969; Fuhs et al. 1972; Hutchinson 1973; Shapiro 1973; Keenan & Auer 1974; Jensen & Sicko-Goad 1976). Under normal conditions in natural waters, however, neither phosphorus nor nitrogen is usually present in relative excess.

Rast & Lee (1978) simplified the analysis of the phosphorus and nitrogen content of lakes and reservoirs, utilising the atomic weights of these elements in order to convert them from atomic ratios to concentrations which can be measured in a waterbody. Thus, the above-noted 16 N:1 P atomic ratio corresponds to a concentration ratio of 7.2 N:1 P. Rast & Lee (1978) and Ryding & Rast (1989) also pointed out that the limiting nutrient can best be determined by measuring the concentrations of the biologically available forms of phosphorus (soluble orthophosphate, expressed as P) and nitrogen (ammonia + nitrate + nitrite nitrogen, expressed as N) in a waterbody during the period of maximum algal biomass. Measurement at that time would identify the nutrients 'left over' after they have been extracted to the maximum extent in the algal bloom event. They also pointed out that both the relative and absolute quantities of phosphorus and nitrogen are important in identifying the limiting nutrient in a waterbody. As a rule-of-thumb, they suggested that if the orthophosphate concentration in a waterbody decreases below approximately $5\mu g L^{-1}$, phosphorus was likely to be the limiting nutrient. The corresponding number for the inorganic nitrogen concentration was approximately $20\mu g L^{-1}$, below which nitrogen was likely to be the limiting nutrient. If both nutrients are below these values, then both may be at algal growth-limiting concentrations.

There also are other factors which can affect or control the effects of the nutrients on a waterbody's trophic state, and lakes of similar nutrient concentrations can vary widely in their productivity and water quality (Brezonik 1969). These factors, mainly physical and chemical in nature, affect lake productivity mainly by altering the distribution, availability and utilisation of nutrients by aquatic plants. Rawson (1939) provided an example of the complex linkage of physical, chemical and biological factors affecting lake productivity (Fig. 14.2). An interesting observation is that consideration of these linkages is as timely today as it was when Rawson developed his model. In fact, following the intensive lake and reservoir eutrophication study undertaken by the U.S. Environmental Protection Agency in its National Eutrophication Survey (US EPA 1974), and the subsequent OECD Eutrophication Programme (OECD 1982) on selected lakes and reservoirs around the world, little additional systematic study has since been carried out on the topic of eutrophication.

Based on these and other observations, phosphorus and/or nitrogen are prominent in nearly every lake study relating nutrients to productivity (e.g. Sawyer 1947, 1966; Fruh et al. 1966; Vollenweider 1968; Lee 1971; Schindler et al. 1971; Vallentyne 1974; Schindler & Fee 1974;

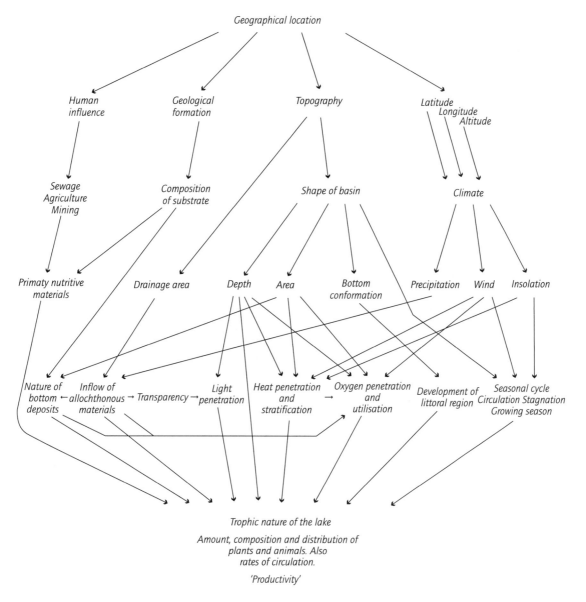

**Fig. 14.2** Interrelated factors affecting eutrophication and trophic status of lakes and reservoirs. (From Rawson 1939.)

Vollenweider & Dillon 1974; Schindler 1977; Lee et al. 1978, 1980; Rast & Lee 1978; Vollenweider et al. 1980; OECD 1982; Ryding & Rast 1989). Phosphorus is usually singled out for primary attention in eutrophication control programmes, partly because it is relatively easily removed from municipal and other wastewaters by standard and relatively inexpensive water treatment methods. In fact Golterman (1975) suggested that, for a given lake or reservoir, it is not important whether or not phosphorus is the nutrient which controls or limits the maximum algal biomass—it is the only essential element which can easily be made to limit algal growth. As discussed further below, this 'controllability' factor is of major importance in developing effective eutrophication management programmes. Hitherto, particular attention has been given to the so-called 'critical' nutrient concentrations as a basis for predicting eutrophic versus oligotrophic conditions in a lake or reservoir. Studying the effects of agricultural and urban runoff on lake fertility in Wisconsin, Sawyer (1947) found that when the inorganic nitrogen (i.e. ammonia, nitrate and nitrite combined, expressed as N) and inorganic phosphorus (i.e. soluble orthophosphate, expressed as P) concentrations exceeded 0.3 mg N $L^{-1}$ and 0.01 mg P $L^{-1}$, respectively, at spring overturn, there was a high likelihood the waterbody would experience algal bloom problems during the following summer growing season. Sawyer did not suggest that algal bloom problems would not occur at lower nutrient concentrations, but rather that their probability during the following summer season was high when these critical nutrient concentrations were exceeded.

## 14.4 RELATIONSHIP BETWEEN NUTRIENT LOADS, CONCENTRATIONS AND TROPHIC STATUS

The delineation of critical nutrient concentrations is based on their causing nuisance growths of algae and aquatic plants which produce aesthetically unpleasing waters, or for which boating, fishing, swimming and other recreational uses can become impaired. In identifying these criteria, the water quality impairments characterising the recreational functions of a lake or reservoir are not usually the same as those faced by a water treatment plant operator or commercial fisherman. Thus, the criteria for designation of 'nuisance' conditions can vary, depending on the specific beneficial water use(s) being impacted.

The in-lake nutrient concentrations control the maximum algal biomass and aquatic plant growth and, therefore, the eutrophication process. However, many factors can affect the relationship between the external phosphorus load and the resultant in-lake phosphorus concentration in a lake or reservoir, including the morphology, geology and hydrology of the lake or reservoir drainage basin, the extent of the littoral zone and shoreline, the degree and duration of thermal stratification, internal nutrient loadings, the concentrations and chemical forms of the nutrient influx, etc. (Vollenweider & Dillon 1974; Vollenweider 1975).

From a eutrophication management perspective, however, control of the external nutrient load (especially phosphorus) is the factor meriting the primary attention for the vast majority of lakes. Although previous researchers (e.g., Sawyer 1947; Sakamoto 1966) identified qualitative relationships between in-lake nutrient levels and algal growth, Vollenweider (1968) was amongst the first to develop a quantitative expression of the nutrient loading concept. He initially derived a quantitative relationship between the external phosphorus load to a waterbody and its trophic response or degree of fertility, using its areal phosphorus load and mean depth as the primary eutrophication-defining parameters (Fig. 14.3). Lakes and reservoirs of similar trophic states tended to group together in the same general area of the diagram. Based on a waterbody's mean depth as a trophic classification factor, Vollenweider empirically determined the boundary phosphorus loading ('excessive') above which a lake's nutrient load exceeded its ability to assimilate it without producing nuisance algal growths (i.e., eutrophic zone), and a boundary ('permissible') below which a lake's phosphorus load was small relative to its assimilative capacity (i.e., oligotrophic zone). His model also contained an intermediate mesotrophic transition zone between oligotrophic and eutrophic conditions in a lake or reservoir. Assuming that algal nitrogen requirements were related to phosphorus requirements in the ratio of 15:1 by weight, he also derived a similar relationship for the external nitrogen load.

The approximation for the 'permissible' phosphorus loading (Vollenweider 1968) was empirically determined to be:

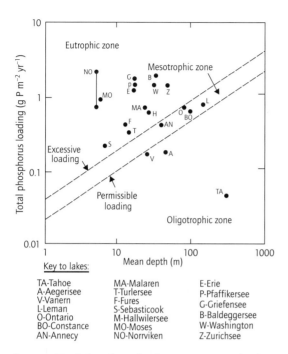

Fig. 14.3 Total phosphorus loading versus mean depth model. (From Vollenweider 1968.)

$$L_c(P) = 25\, z^{0.6} \quad (14.2)$$

where $L_c(P)$ is the areal permissible phosphorus loading (mg P m$^{-2}$ yr$^{-1}$), and $z$ is mean depth (m). The excessive phosphorus loading was empirically determined to be approximately twice the permissible loading:

$$L_c(P) = 50\, z^{0.6} \quad (14.3)$$

The slope of the boundary lines illustrated the greater dilution capacity of deeper lakes and reservoirs, which increases their ability to assimilate more nutrients than shallow lakes without increasing their degree of fertility (i.e. trophic status). The relative degree of eutrophy or oligotrophy of a water body was proportional to its vertical displacement above or below the 'permissible' loading line, consistent with earlier work by Sakamoto (1966), who reported a general tendency for shallower lakes to be more productive than deeper lakes, based on studies on Japanese lakes exhibiting a wide range of depths.

This simple model was a significant advance at the time of its introduction, being amongst the first attempts to provide a quantitative expression of the relationship between the external phosphorus load and eutrophication response of a lake or reservoir. For most of the lakes for which sufficient phosphorus loading data were available, the trophic state predicted by the loading diagram agreed with the trophic state indicated by the standard (and arbitrary) indicators available at the time (phosphorus and chlorophyll concentrations, primary productivity, Secchi-disc depth, hypolimnetic oxygen depletion, etc.). Notwithstanding, Vollenweider (1968) also suggested that other parameters should be considered in delineating a waterbody's trophic status, including the extent of the shoreline and littoral zone, degree of nutrient mixing in the water column, internal loadings from the sediments, and water flushing rate (Vollenweider & Dillon 1974).

Dillon (1975) was amongst the first to report cases in which Vollenweider's initial model did not provide accurate delineations of lake trophic status. Although some of his study lakes possessed external phosphorus loads placing them in the eutrophic zone of the initial diagrams, they also contained low chlorophyll $a$ concentrations, large Secchi-disc depths and no significant hypolimnetic oxygen depletion, indicating they were less fertile waterbodies than predicted by the diagrams. This discrepancy was attributed to the fact that the ratios of the drainage basin areas to the study lake surface areas were very large, and that the lakes possessed small mean depths. This gave them rapid flushing rates and a faster cycling of water (and the phytoplankton contained in it) through the lake systems. Thus, the waterbodies could assimilate a larger phosphorus load, with no adverse eutrophication responses, than slower-flushing waterbodies because, although the phytoplankton in the lake interacted with the in-lake phosphorus, they also were being flushed from the waterbody before they could accumulate to nuisance levels. Thus, rapidly flushed and slowly flushed waterbodies can exhibit different trophic responses for the same phosphorus load, resulting in the observed discrepancies between the actual and predicted trophic states of Dillon's waterbodies.

In order to account for the effects of fast versus slow flushing rates on his nutrient-load–lake-response relationships, Vollenweider (1975, 1976) modified his initial phosphorus diagram to incorporate the water flushing rate, as manifested in the hydraulic residence time. This modification allowed the effects of the hydraulic load to be incorporated into evaluating the trophic response of a waterbody to its external phosphorus load.

His refinement was a phosphorus diagram relating a waterbody's areal total phosphorus load ($\text{mg P m}^{-2} \text{yr}^{-1}$) and the ratio of the mean depth (m) to the hydraulic residence time (yr). This ratio was represented as $z/\tau_\omega$. With this new relationship, a lake's critical phosphorus load is directly proportional to its volume (expressed as mean depth, in metres) and inversely proportional to the hydraulic residence time (expressed in years). The direct proportionality of the critical phosphorus load to the mean depth reflected the dilution of the phosphorus load by the lake volume, and the inverse hydraulic residence time reflected the residence time of the phosphorus in the lake (Fig. 14.4). This revised model also delineated permissible and excessive phosphorus loading boundaries based on Sawyer's (1947) critical nutrient concentrations. The permissible phosphorus loading boundary was based on the relationship:

$$L_c(P) = 100 \left( z/\tau_\omega \right)^{0.5} \quad (14.4)$$

where $L_c(P)$ is the areal permissible total phosphorus load ($\text{mg P m}^{-2} \text{yr}^{-1}$), $z$ is mean depth (m) and $\tau_\omega$ is the hydraulic residence time (= waterbody volume ($\text{m}^3$)/annual inflow volume ($\text{m}^3 \text{yr}^{-1}$)). The excessive phosphorus loading line also was assumed to be equal to twice the permissible loading (Vollenweider 1975, 1976; Dillon 1974, 1975). This model was subsequently used to determine critical phosphorus loadings for lakes in the USA (US EPA 1973b, 1976).

Elaborating on earlier work by Biffi (1963) and Piontelli & Tonolli (1964), Vollenweider also developed an input–output mass balance equation for total phosphorus, including the external loads to the lake and the losses from the lake through outflow and sedimentation. It represented an 'accountability model' focusing on the balance of phosphorus between its sources and sinks in lake systems. This model modification included terms for the mean depth, hydraulic residence time and a sedimentation parameter. Based on this conceptualisation, Vollenweider (1975) expressed the phosphorus loading dynamics of a waterbody as:

$$\frac{d[P]}{dt} = P \text{ supply} - P \text{ lost to sedimentation} - P \text{ lost to outflow}$$

$$= \frac{v_j [P]_j}{V} - \sigma_P [P] - \rho_\omega [P] \quad (14.5)$$

where $[P]$ is lake phosphorus concentration ($\text{M L}^{-3}$), $v_j$ is the flow rate of the $j$th tributary ($\text{L}^3 \text{T}^{-1}$), $[P]_j$ is the phosphorus concentration in the $j$th tributary ($\text{M L}^{-3}$), $V$ is lake volume ($\text{L}^3$), $\rho_\omega$ is the flushing rate ($\text{T}^{-1}$) (= annual inflow volume ($\text{L}^3 \text{T}^{-1}$)/lake volume ($\text{L}^3$)) and $\sigma_P$ = phosphorus sedimentation coefficient ($\text{T}^{-1}$). This revised model assumed a well-mixed lake, constant lake volume, outflow phosphorus concentration equal to in-

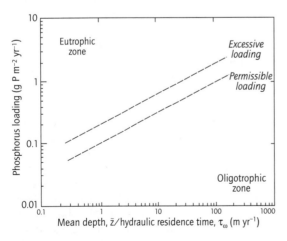

Fig. 14.4 Initial phosphorus load versus mean-depth/hydraulic-residence-time model. (From Vollenweider 1975.)

lake phosphorus concentration, equivalent inflow and outflow rates, no net phosphorus load from the sediments and a phosphorus sedimentation proportional to in-lake phosphorus concentration (Dillon 1974; Vollenweider 1975).

The revised model contained both a time-dependent and steady-state solution, with the latter (i.e., $t \to \infty$) being:

$$[P]_\infty = \ell(p)/(\rho_\omega + \sigma_P) \quad (14.6)$$

where $[P]_\infty$ is the steady-state total phosphorus concentration $(ML^{-3})$ and $\ell(p)$ is the volumetric total phosphorus load $(ML^{-3}T^{-1})$. As a further refinement,

$$\ell(p) = L(P)/z \quad (14.7)$$

where $L(P)$ is the surface area phosphorus loading $(ML^{-2}T^{-1})$ and $z$ is mean depth (L). Thus:

$$[P]_\infty = \ell(p)/(\rho_\omega + \sigma_P) = [L(P)/z]/(\rho_\omega + \sigma_P)$$
$$= L(P)/[z(\rho_\omega + \sigma_P)] \quad (14.8)$$

Rearranging, eqn. 14.8 becomes:

$$L(P) = [P]_\infty z(\rho_\omega + \sigma_P) \quad (14.9)$$

Sawyer's (1947) critical spring overturn concentration of $10\,\mu g\,L^{-1}$ was used as the steady-state total phosphorus concentration, $[P]_\infty$, in eqn. 14.9. The hydraulic flushing rate, $\rho_\omega$, is equal to the reciprocal of the hydraulic residence time $(= 1/\tau_\omega)$. Although the phosphorus sedimentation rate coefficient, $\sigma_P$, cannot easily be measured directly, Vollenweider (1975) suggested it could be approximated by the quantity $10/z$.

Rearranging, eqn. 14.9 becomes:

$$L_c(P) = [P]_c^{sp} z([1/\tau_\omega] + \sigma_P)$$
$$= 10([z/\tau_\omega] + [z(10/z)])$$
$$= 10([z/\tau_\omega] + 10)$$
$$= 100 + 10(z/\tau_\omega) \quad (14.10)$$

where $L_c(P)$ is the areal critical total phosphorus load (mg P m$^{-2}$ yr$^{-1}$), $[P]_c^{sp}$ is the critical spring over-

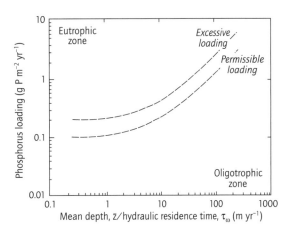

Fig. 14.5 Modified phosphorus load versus mean-depth/hydraulic-residence-time model. (From Vollenweider 1975, 1976.)

turn phosphorus concentration (mg m$^{-3}$), $\tau_\omega$ is the hydraulic residence time (yr) (= waterbody volume (m$^3$)/annual inflow volume (m$^{-3}$ yr$^{-1}$)) and $z$ = mean depth (m) (= volume (m$^3$)/surface area (m$^2$)). The revised phosphorus diagram based on eqn. 14.10 is presented in Fig. 14.5. The phosphorus loading which would be likely to produce nuisance algal blooms in a waterbody during the growing season was again considered to be approximately twice the permissible loading (Vollenweider 1975, 1976). The equation for the excessive loading line in Fig. 14.5 was:

$$L(P) = 200 + 20(z/\tau_\omega) \quad (14.11)$$

where $L(P)$ is the areal excessive total phosphorus load (mg P m$^{-2}$ yr$^{-1}$). This modified phosphorus diagram indicates that, below a certain combination of mean depth and flushing time, a waterbody's phosphorus loading tolerance becomes constant. This condition would hold even though, based on mean depth alone, waterbodies may appear to possess a higher phosphorus assimilation capacity. With this modified diagram, the phosphorus load boundary lines flatten at $z/\tau_\omega$ values of less than about 2. At $z/\tau_\omega$ values greater than approximately

80, the tolerable phosphorus loading capacity becomes proportional to $z/\tau_\omega$.

Another refinement of this basic approach was an alternative expression for the sedimentation rate coefficient (Vollenweider 1976). An equation relating the phosphorus residence time and water residence time was used to derive a sedimentation rate coefficient:

$$\sigma_p = \sqrt{(z/q_s)}/\tau_\omega = \sqrt{(z/[z/\tau_\omega])}/\tau_\omega) = \sqrt{\tau_\omega}/\tau_\omega \quad (14.12)$$

where $q_s$ is the hydraulic loading $(\text{m yr}^{-1}) = z/\tau_\omega$. The hydraulic flushing rate $(\rho_\omega)$ is the reciprocal of the hydraulic residence time $(\tau_\omega)$.

Recalling that $\rho_\omega = 1/\tau_\omega$, $\sigma_p = \sqrt{\tau_\omega}/\tau_\omega$, and $\tau_\omega = z/q_s$, the steady-state expression of eqn. 14.12 can be rearranged, as follows:

$$[P]_\infty = [L(P)/q_s](1/[1+\sqrt{z/q_s}]) \quad (14.13)$$

This relationship equates the predicted steady-state in-lake phosphorus concentration to an equivalent expression relating the phosphorus load as modified by the hydraulic load (Fig. 14.6).

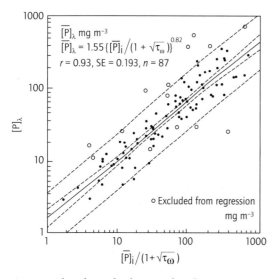

Fig. 14.6 Phosphorus load versus phosphorus concentration model (OECD 1982; see reference for definition of terms).

The term $[L(P)/q_s]$ represents the average inflow phosphorus concentration, $[P]_i$ (Vollenweider 1975; Larsen & Mercier 1976). The complete phosphorus loading expression on the right-hand side of eqn. 14.13 was termed the 'flushing corrected average annual phosphorus inflow concentration' (OECD 1982), being expressed as $[P]_i/(1+\sqrt{\tau_\omega})$ in subsequent phosphorus diagrams.

## 14.5 OTHER PHOSPHORUS-LOAD–LAKE-RESPONSE RELATIONSHIPS

The original phosphorus diagram (Vollenweider 1968) identified a positive relationship between a lake's external phosphorus load and its in-lake phosphorus concentration. From the perspective of eutrophication management, however, neither the phosphorus load nor in-lake concentration has a direct impact on the beneficial water uses for a lake or reservoir. Rather, it is the lake's water quality and trophic responses to its phosphorus inputs which are of concern, including the chlorophyll concentration (a measure of algal biomass), Secchi-disc depth (a measure of water transparency) and hypolimnetic oxygen depletion (a measure of algal decomposition).

### 14.5.1 Phosphorus load and chlorophyll concentration

Previous researchers (Sawyer 1947; Sakamoto 1966; Dillon 1974; Dillon & Rigler 1974; Schindler et al. 1976; Oglesby & Schaffner 1978; Schindler 1978; Canfield & Bachman 1981; Thornton & Rast 1987) demonstrated a positive relationship between the mean phosphorus concentration in single waterbodies and groups of waterbodies at spring overturn and the mean chlorophyll a concentration during the following summer growing season. Recalling that the term on the left-hand side of eqn. 14.13 represents the predicted steady-state in-lake phosphorus concentration (as a function of a lake's phosphorus load, mean depth and hydraulic residence time), a relationship also should exist between the phosphorus loading char-

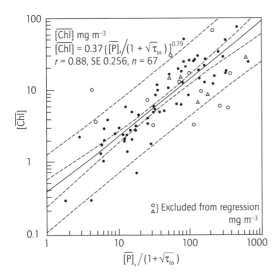

Fig. 14.7 Phosphorus load versus mean chlorophyll model (OECD 1982; see reference for definition of terms).

acteristics of a lake, expressed as the equivalence on the right-hand side of eqn. 14.13.

Accordingly, Vollenweider (1976) provided an empirical relation between a waterbody's flushing corrected average annual phosphorus inflow concentration, $[P]_i/(1 + \sqrt{\tau_\omega})$, and its mean summer epilimnetic chlorophyll $a$ concentration (Fig. 14.7). Because the phosphorus loading expression is equivalent to the predicted steady-state, in-lake phosphorus concentration, Fig. 14.7 relates the summer chlorophyll $a$ concentration to the total phosphorus concentration, in the same manner as previous researchers. Because of the equivalence in eqn. 14.13, Fig. 14.7 also relates the chlorophyll $a$ concentration in a waterbody to both its phosphorus concentration and load.

Another factor to consider in eutrophication management efforts is that a lake or reservoir will not respond immediately to a reduction in its external phosphorus load with an accompanying reduction in its chlorophyll concentration (or other trophic state parameter). Rather, there will be a lag period during which the in-lake phosphorus concentrations and the resultant chlorophyll concentrations are adjusting to the new in-lake phosphorus equilibrium. When a new equilibrium is reached, the phosphorus–chlorophyll loading diagram (Fig. 14.7) can be used validly to predict the expected chlorophyll biomass in the waterbody. Sonzogni et al. (1976) presented a simplified approach to calculate the expected lag period for the new phosphorus equilibrium state. Their model illustrates that a lake subjected to a reduction in its phosphorus load will typically reach 95% of its new steady-state, in-lake phosphorus concentration in a period equivalent to three times the waterbody's phosphorus residence time (assuming no new phosphorus additions during this period).

### 14.5.2 Phosphorus load and Secchi-disc depth

Water clarity or transparency was previously proposed as an indicator of algal biomass in lakes and reservoirs (e.g., Edmondson 1972; Carlson 1977). The use of this parameter is based largely on the public's perception of good versus bad water quality (Barker 1971; David 1971; Thornton & McMillan 1989; Thornton et al. 1989; Thornton 1993). Waterbodies containing transparent water are usually considered to be of better quality, and more useful for beneficial human uses, than waterbodies containing turbid water. The degree of water transparency or clarity, in fact, has probably become the most frequently cited general indicator of water quality. Exceptions to this general rule in a scientific sense are waterbodies with high colour content. Public perceptions vary, but generally support this exception.

The basis for this model is that increased algal growth in a lake is typically accompanied by decreased water clarity or increased water turbidity, owing to increased concentrations of algal cells and other particles in the water. Rast & Lee (1978) developed an inverse relationship between water transparency (expressed as Secchi-disc depth) and algal biomass (expressed as chlorophyll concentration) for a number of USA lakes and reservoirs. Although some of the decreased water transparency was doubtless due to light scattering by non-algal particles in the water column, there was a definite negative hyperbolic relationship between the Secchi-disc depth and chlorophyll concentration. The slope of the curve was steepest at lower algal biomass levels (reflected in the chlorophyll concentrations), indicating that biomass changes are more easily detected in clear (oligotrophic) waters than in eutrophic waters. Edmondson (1972) noted that, above a chlorophyll concentration of about $20\,\mu g\,L^{-1}$, a further large increase in the chlorophyll concentration did not produce a proportionally large decrease in Secchi-disc depth in Lake Washington. This suggests that, above a certain degree of eutrophication, Secchi-disc depth loses its value as an indicator of changes in algal biomass (other than a low Secchi-disc depth typically indicating a relatively eutrophic condition). This is especially true in lakes in the semi-arid zone, where concomitant increases in salinity generally lead to increased transparencies even under eutrophic or hypertrophic conditions (Thornton & Rast 1993).

An advantage of the Secchi-disc depth as an algal biomass indicator for a lake or reservoir is that it is an easily measured parameter, involving a minimum of time and cost. Its general meaning also is easily understood by the general public, and it is a parameter which can be evaluated over time in correlation with a lake's general trophic condition. Hence, a number of volunteer lake monitoring programmes worldwide make use of this indicator (Carlson 1977).

Based on the previously derived relationship between the external phosphorus load and in-lake phosphorus concentration (eqn. 14.13), and the relationship between the in-lake phosphorus and chlorophyll concentration (Fig. 14.7), an empirical model was developed between the external phosphorus load to a waterbody and its predicted water transparency (expressed as Secchi-disc depth). This model is illustrated in Fig. 14.8, using data collected by Rast & Lee (1978) on USA lakes and reservoirs. Because of the non-linear relationship between Secchi-disc depth and chlorophyll, it is not surprising that a non-linear relationship also exists between the phosphorus load expression in eqn. 14 13 and the Secchi-disc depth. This model assumes that the primary factor controlling water clarity is phytoplankton, that a lake does not contain large quantities of inorganic turbidity or colour, and that the phosphorus load to the lake is approximately in an equilibrium state.

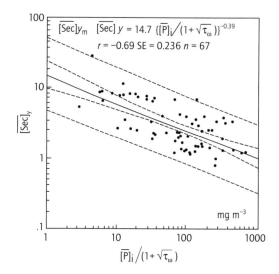

Fig. 14.8 Phosphorus load versus mean Secchi-disc depth model (OECD 1982; see reference for definition of terms).

### 14.5.3 Phosphorus load and areal hypolimnetic oxygen depletion

Hypolimnetic oxygen depletion is of concern because of its implications for the development of anoxic conditions in the bottom waters of a lake or reservoir, especially eutrophic ones. Consequences of anoxic conditions include destruction of the deep, coldwater fish habitat, and development of chemically reducing conditions in the hypolimnion, which can facilitate the remobilisation of nutrients and heavy metals back into the water column (Rast & Lee 1978).

Hutchinson (1938) was amongst the first to use areal hypolimnetic oxygen depletion (AHOD) as an indicator of lake trophic status. Based on their previously derived 106 C : 16 N : 1 P ratio in phytoplankton, Redfield et al. (1963) determined that approximately 276 atoms of oxygen were required for the oxidation (decomposition) of organic

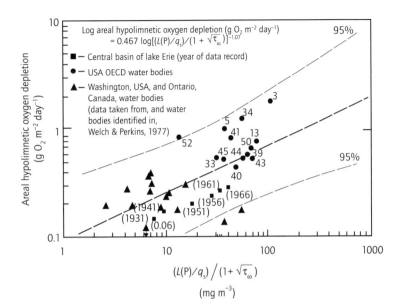

Fig. 14.9 Phosphorus load versus hypolimnetic oxygen depletion model. The numbers against data points refer to individual lakes, which are identified in the original reference. (Taken from Rast & Lee 1978; see eqn. 14.13 for definition of terms.)

matter in seawater, corresponding to a −276 O : 106 C : 16 N : 1 P stoichiometric atomic ratio in plankton. They also reported that the phosphorus and oxygen concentrations were stoichiometrically equivalent to a −110.3 O : 1 P mass ratio, meaning that approximately 110 mg $O_2$ $L^{-1}$ were required for the decomposition of each milligram of organic phosphorus per litre (i.e., algal phosphorus). With this ratio, one can relate the oxygen requirements for the decomposition of a given concentration of algae (expressed as total phosphorus equivalence) to the hypolimnetic oxygen concentration. With knowledge of the total mass of oxygen in the hypolimnion, and the algal and phosphorus concentrations and algal settling rates, one can make a rough estimate of the hypolimnetic oxygen depletion rate.

As with previous models, however, the major goal was to relate the controllable element (i.e., external phosphorus load) and the response of a lake to this variable (i.e., hypolimnetic oxygen depletion rate). Gilbertson et al. (1972) previously reported a linear correlation between municipal wastewater treatment plant phosphorus loads and hypolimnetic oxygen depletion rates in the central basin of Lake Erie.

Lasenby (1975) subsequently derived a relationship between AHOD and Secchi-disc depth, assuming that the quantity of seston sinking into the hypolimnion was proportional to the quantity in the epilimnion. Because the linear development of his model suggested that oxygen consumption in the hypolimnion of his study lakes was not sensitive to brief changes in productivity, relatively few measurements should give a relatively good estimate of oxygen depletion rates. Rast & Lee (1978) utilised Lasenby's relationship, as well as data of Welch & Perkins (1977), to relate the external phosphorus load (expressed in the phosphorus loading term in eqn. 14.13) to the areal hypolimnetic oxygen depletion. Using the Secchi-disc depth as the common variable, they developed the model illustrated in Fig. 14.9. Whether or not the hypolimnetic oxygen will be depleted can be determined by calculating the total hypolimnetic oxygen depletion expected for the thermal stratification period (expressed as g $O_2$ depleted per $m^2$ of hypolimnion area per day). This calculated value can be compared with the total volume of oxygen in the hypolimnion during the period of thermal stratification (calculated as the oxygen concentration in the hypolimnion multiplied by the hypolimnetic volume). This approach also can be used to estimate the number of days required to deplete the oxygen in the hypolimnion after thermal stratification has occurred.

### 14.5.4 *Phosphorus load and fish yield*

Lee & Jones (1991) provided another application of this basic approach, with a model that estimated the fish yield which can be sustained for a given phosphorus load, as well as changes in fish yields resulting from increased or reduced phosphorus loads. A general effect of eutrophication is increased fish stocks, consistent with increased primary production (Oglesby 1977). Hanson & Leggett (1982) examined the relationship between fish yield and various physical and chemical characteristics, including mean epilimnetic total phosphorus concentration. It is again most useful from the perspective of eutrophication management, however, to relate lake-response parameters (e.g., fish yield) to the external phosphorus load causing them. Thus, based on a statistical adaptation of the flushing corrected average annual phosphorus inflow concentration (OECD 1982), Lee & Jones (1991) combined the phosphorus-loading equivalent (eqn. 14.13) and the data of Oglesby, Hanson & Leggett produce a predictive model relating a waterbody's external phosphorus load and wet-weight fish yield (g wet wt m$^{-2}$ yr$^{-2}$). However, the predicted increased fish yield with increasing phosphorus load is not without limits. Beyond a certain degree of eutrophication, the quality of the fishery decreases, owing to such factors as changes in the quality of planktonic food organisms (increased chlorophyll concentrations), shifts in the nature or composition of the fishery from game fish to bottom-feeding species (decreased water transparency) and likelihood of fish kills (increased hypolimnetic oxygen depletion). Results include the production of stunted pan fish populations and/or elimination of coldwater fish in thermally stratified waterbodies exhibiting hypolimnetic oxygen depletion. A detailed discussion of their model and its assumptions and application is provided by Lee & Jones (1991).

## 14.6 THE OECD EUTROPHICATION STUDY

Against this background of research findings, and with the realisation that eutrophication produces economic- and development-related impacts, the Organization for Economic Cooperation and Development (OECD) embarked upon a major synthesis of the knowledge of eutrophication, seeking to identify ways and means to manage this serious global problem. The OECD, an independent, international organisation headquartered in Paris, focuses on the economic growth of its member nations, which comprise the world's more highly developed countries. Among its activities is the need to address the reality that economic productivity frequently can lead to significant environmental problems and that environmental problems can limit economic productivity.

The OECD noted that varying degrees of eutrophication symptoms have occurred in lakes, reservoirs and flowing waters in most of the world's developed countries for many years. Following a comprehensive review of lake and reservoir eutrophication, and the role of phosphorus and nitrogen as causative factors (Vollenweider 1968), the OECD formed the Water Management Sector Group (WMSG) in 1971. Among its activities, the WMSG produced the report, 'Summary Report of the Agreed Monitoring Projects on Eutrophication of Waters' (OECD 1973), which outlined the working plan for an international co-operative Eutrophication Programme involving an agreed group of more than 200 lakes and reservoirs around the world. The study goals included: (i) promotion of an agreed common system of lake response parameters, and analytical sampling methods, to allow comparison of eutrophication data between waterbodies; (ii) application of the common measuring system to selected categories of waterbodies, to obtain a better understanding of the causes of eutrophication and the influence of nutrient loads on trophic status; and (iii) promotion of a systematic exchange of information and experience on eutrophication and eutrophication control (OECD 1973, 1975).

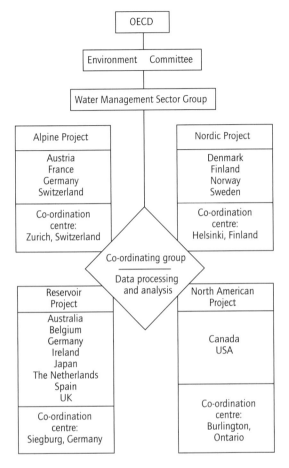

Fig. 14.10 Organisation outline of OECD Eutrophication Programme (Rast & Lee 1978).

Eighteen OECD member nations agreed to participate in the Eutrophication Programme. Because geographical, ecological, geological and morphometric factors are of major importance in the eutrophication process (see Fig. 14.2), the WMSG utilised three regionally based projects and one functionally based project. The regional organisation and participating countries are illustrated in Fig. 14.10, with the co-ordination centres serving as information exchange vehicles between the four projects. The overall assessment and co-ordination of the four projects was the responsibility of a group of nationally nominated delegates from the participating countries. The Eutrophication Programme would run for 4 years, with an overall analysis of the results to be prepared in 1977. The results from the four projects would be used to synthesise an optimal eutrophication control strategy. The specific study objectives were to:

1 develop detailed nutrient budgets (phosphorus and nitrogen) for a selected group of lakes and reservoirs;

2 assess the physical, chemical and biological characteristics of the selected waterbodies;

3 relate the trophic state of the waterbodies to their nutrient budgets and their limnological and environmental characteristics;

4 synthesise an optimal strategy, based on data from all four projects, for controlling eutrophication.

The four regional projects (Fig. 14.10) were characterised as follows (Rast & Lee 1978):

• Nordic Project. Reasonably comparable conditions exist in this project, including the cool climate zone of the Baltic and North Sea areas. Many lakes were formed from the retreat of the great Quaternary glaciers. The region contains comparable ecological conditions and equivalent levels of economic development and pollution, as well as close political, cultural and scientific links.

• Alpine Project. A source of water for many European countries, the Alpine waters are of great social and economic significance, representing a great natural amenity and source of considerable tourism. Their ecology is characterised by an abundant variety of species vulnerable to human interventions. The Alpine regions represent similar hydrological conditions owing to comparable geography, geology and ecology. The Alpine zones share certain river basins and commissions.

• Reservoir and Shallow Lakes Project. This project focused on constructed lakes and comparable waterbodies (i.e., shallow lakes, lagoons and estuarine waters). All the waterbodies were relatively shallow, with great economic and social values (i.e. used as water supply reserves, water sports, fishing, navigation, etc.). Control of water quality by manipulation of hydrological and other factors was more feasible for these waterbodies than for larger waterbodies.

• North American Project. In contrast to the other projects, this one was not restricted to specific types of lakes. Rather, these waterbodies differed significantly in their limnological characteristics, spanning the trophic spectrum from oligotrophic to hyper-eutrophic. Some of the study waterbodies also were included in the National Eutrophication Survey conducted by the U.S. Environmental Protection Agency (US EPA 1973a, 1974).

### 14.6.1 OECD phosphorus load and lake response models

No single model was found to be satisfactory for assessing eutrophication for the range of lakes and reservoirs in the four projects. Thus, the models derived within the OECD (1982) Eutrophication Programme represent simplified quantitative methodologies to describe the average statistical behaviour of a large population of lakes relative to the factors analysed, rather than a description of the behaviour of an individual lake.

The data from the OECD waterbodies were used to refine and supplement the previously described phosphorus-load–lake-response models. Utilising the complete OECD data set, as well as those from the four individual regional projects, the analyses included the relationship between the phosphorus loading term in eqn. 14.13 and the mean in-lake phosphorus concentration, mean and maximum chlorophyll concentration, mean Secchi-disc depth and hypolimnetic oxygen depletion. The models provided an easily usable means of estimating lake and reservoir response applicable to a wide range of waterbodies based upon a number of key eutrophication indicators. Of particular importance was the utility of these models in regard to alternative eutrophication management programmes based on reducing a waterbody's external phosphorus load. The resulting OECD eutrophication models and relevant statistical information are summarised in Table 14.5.

### 14.6.2 OECD trophic boundary systems

A further advance achieved with the OECD data was the development of boundary values between different trophic conditions for the study lakes. Both a fixed and an open trophic boundary system were elaborated. The fixed boundary system was based on the best judgement of the researchers regarding appropriate transition values between different trophic conditions for several water quality parameters (e.g. phosphorus and chlorophyll concentrations, water transparency). The resulting fixed boundary system is illustrated in Table 14.6. A distinct advantage of the fixed boundary system is that it is easily understood and applied by lake managers. A disadvantage is that lakes are dynamic ecosystems, and nature does not typically exhibit the distinct boundaries suggested by the numbers in Table 14.6. In fact they often exhibit conditions for some water-quality parameters suggestive of one trophic condition, and conditions for other parameters suggestive of a different trophic condition. In this regard, Rast & Thornton (1997) pointed out that, because there is no universal acceptance of the scientific or technical basis for applying the terms oligotrophic, mesotrophic and eutrophic, these terms remain somewhat subjective in their application. The same chlorophyll concentration in Wisconsin lakes and Texas reservoirs, for example, can be interpreted in completely different ways. The primary difference is the interpretation of the degree to which beneficial water uses are being impaired.

An open boundary trophic classification also was developed, in order to address this reality. With this system, the data for specific water-quality parameters were grouped, and the group mean values and standard deviations for each parameter were calculated. After removing values which were more than, or less than, two standard deviations (where applicable) in the first calculation, the geometric means of the OECD data were calculated on the basis of log-10 transformations (OECD 1982), as summarised in Table 14.7.

The open trophic boundary system incorporates a probabilistic aspect in its classification scheme, taking the uncertainty regarding individual water quality parameters into account in delineating a lake's trophic status. Based on the values in Table 14.7, which represent the collective opinion of a large group of limnologists on how trophic

**Table 14.5** Selected OECD (1982) models of phosphorus-load–lake-response relationships. (From Rast & Thornton 1997)

|  | OECD project | Derived relationship | n | r |
|---|---|---|---|---|
| Annual mean total phosphorus concentration (g L$^{-1}$) | Combined OECD study* | $1.55P^{0.82}$ | 87 | 0.93 |
|  | Shallow lakes and reservoirs† | $1.02P^{0.88}$ | 24 | 0.95 |
|  | Alpine lakes‡ | $1.58P^{0.83}$ | 18 | 0.93 |
|  | Nordic lakes§ | $1.12P^{0.92}$ | 14 | 0.86 |
|  | USA lakes¶ | $1.95P^{0.79}$ | 31 | 0.95 |
| Annual mean chlorophyll concentration (g L$^{-1}$) | Combined OECD study | $0.37P^{0.79}$ | 67 | 0.88 |
|  | Shallow lakes and reservoirs | $0.54P^{0.72}$ | 22 | 0.87 |
|  | Alpine lakes | $0.47P^{0.78}$ | 12 | 0.94 |
|  | Nordic lakes | $0.13P^{1.03}$ | 13 | 0.82 |
|  | USA lakes | $0.39P^{0.79}$ | 20 | 0.89 |
| Annual maximum chlorophyll concentration (g L$^{-1}$) | Combined OECD study | $0.74P^{0.89}$ | 45 | 0.89 |
|  | Shallow lakes and reservoirs | $0.77P^{0.86}$ | 21 | 0.88 |
|  | Alpine lakes | $0.83P^{0.92}$ | 11 | 0.96 |
|  | Nordic lakes | $0.47P^{01.00}$ | 13 | 0.77 |
|  | USA lakes | – | – | – |
| Annual mean Secchi-disc depth transparency (m) | Combined OECD study | $14.7P^{-0.39}$ | 67 | −0.69 |
|  | Shallow lakes and reservoirs | $8.5P^{-0.26}$ | 26 | −0.55 |
|  | Alpine lakes | $15.3P^{-0.30}$ | 18 | −0.74 |
|  | Nordic lakes | – | – | – |
|  | USA lakes | $20.3P^{-0.52}$ | 22 | −0.82 |
| Areal hypolimnetic oxygen depletion (g O$_2$ m$^{-2}$ day$^{-1}$) | Combined OECD study | $\approx 0.1P^{0.55}$ | – | – |
|  | Shallow lakes and reservoirs | – | – | – |
|  | Alpine lakes | – | – | – |
|  | Nordic lakes | $0.085P^{0.47}$ | – | – |
|  | USA lakes | $0.115P^{0.67}$ | – | – |

\* OECD (1982). † Clasen & Bernhardt (1980). ‡ Fricker (1980). § Ryding (1980). ¶ Rast & Lee (1978).
$P = \{[L(P)/q_s]/(1+\sqrt{\tau_\omega})\}$ = 'flushing corrected average annual phosphorus inflow concentration' (OECD 1982); see eqn. 14 13 for definition of terms.
$n$ = number of data points; $r$ = correlation coefficient; –, data not available.

**Table 14.6** OECD Fixed Boundary Trophic Classification System (OECD 1982)

| Trophic category | Mean phosphorus concentration (µg L$^{-1}$) | Mean chlorophyll concentration (µg L$^{-1}$) | Maximum chlorophyll concentration (µg L$^{-1}$) | Mean annual Secchi-disc depth (m) | Minimum annual Secchi-disc depth (m) |
|---|---|---|---|---|---|
| Ultra-oligotrophic | ≤4.0 | ≤1.0 | ≤2.5 | ≥12.0 | ≥6.0 |
| Oligotrophic | ≤10.0 | ≤2.5 | ≤8.0 | ≥6.0 | ≥3.0 |
| Mesotrophic | 10–35 | 2.5–8 | 8–25 | 6–3 | 3–1.5 |
| Eutrophic | 35–100 | 8–25 | 25–75 | 3–1.5 | 1.5–0.7 |
| Hypereutrophic | ≥100 | ≥25 | ≥75 | ≤1.5 | ≤0.7 |

Table 14.7  General OECD trophic classification values. (Modified from OECD 1982)

| | | Annual mean values | | | |
|---|---|---|---|---|---|
| | | Oligotrophic | Mesotrophic | Eutrophic | Hypereutrophic |
| Total phosphorus concentration (µg L$^{-1}$) | Mean | 8.0 | 26.7 | 84.4 | – |
| | Range | 3.0–17.7 | 10.9–95.6 | 16–386 | 750–1200 |
| | N | 21 | 19 | 71 | 2 |
| Total nitrogen concentration (µg L$^{-1}$) | Mean | 661 | 753 | 1875 | – |
| | Range | 307–1630 | 361–1387 | 393–6100 | – |
| | N | 11 | 8 | 37 | – |
| Chlorophyll $a$ concentration (µg L$^{-1}$) | Mean | 1.7 | 4.7 | 14.3 | – |
| | Range | 0.3–4.5 | 3–11 | 3–78 | 100–150 |
| | N | 22 | 16 | 70 | 2 |
| Maximum chlorophyll $a$ concentration (µg L$^{-1}$) | Mean | 4.2 | 16.1 | 42.6 | – |
| | Range | 1.3–10.6 | 4.9–49.5 | 9.2–275 | – |
| | N | 16 | 12 | 46 | – |
| Secchi-disc depth (m) | Mean | 9.9 | 4.2 | 2.45 | – |
| | Range | 5.4–28.3 | 1.5–8.1 | 0.8–7.0 | 0.4–0.5 |
| | N | 13 | 20 | 70 | 2 |

N = number of lakes comprising annual mean values.

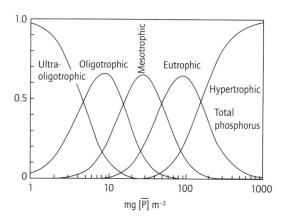

Fig. 14.11  Probabilistic trophic state classification scheme, based on mean phosphorus concentration [P] (OECD 1982; see reference for definition of terms).

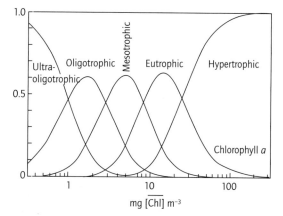

Fig. 14.12  Probabilistic trophic state classification scheme, based on mean chlorophyll concentration [Chl] (OECD 1982; see reference for definition of terms).

terminology should be applied in practice, the OECD (1982) developed a graphic form of presentation for predicting the probable trophic status of a lake based on these water quality parameters (Figs 14.11–14.14). These trophic probability diagrams illustrate that, based on a specific value of a given water quality parameter, there is a certain probability of lakes being in more than one trophic condition. In Fig. 14.12, for example, for a lake with a mean chlorophyll concentration of 10 µg L$^{-1}$, there

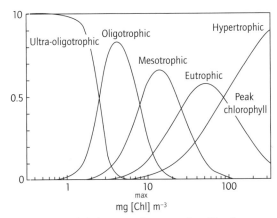

**Fig. 14.13** Probabilistic trophic state classification scheme, based on maximum chlorophyll concentration [Chl]$^{max}$ (OECD 1982; see reference for definition of terms).

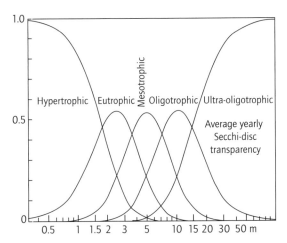

**Fig. 14.14** Probabilistic trophic state classification scheme, based on annual average Secchi-disc depth transparency (OECD 1982; see reference for definition of terms).

is about a 55% probability of being eutrophic, based on the experiences with the OECD waterbodies. There is also approximately a 30% probability of being mesotrophic and about a 10% probability of being hypertrophic. There is also a minute probability (about 3%) of being oligotrophic. In the practical application of this scheme, therefore, a waterbody is classified on the basis of all these water quality parameters. Its most likely trophic state would be that with the highest probability when all the parameters are considered collectively. For this example, the lake would most likely be considered eutrophic or meso-eutrophic.

A similar approach was developed in a comparable study of warm-water lakes in Latin America (Salas & Martino 1991), the results of which are more likely relevant to many lakes and reservoirs in developing countries than those developed in the OECD Eutrophication Programme. Figure 14.15 compares the predictive capability of the two trophic classification schemes, based on the mean chlorophyll concentrations. As can be seen, the distinction between trophic states is more sensitive in the warm-water classification scheme.

## 14.7 PHOSPHORUS-LOAD–LAKE-RESPONSE MODELS AND EUTROPHICATION MANAGEMENT

Practical management tools are essential for addressing the problems of lake and reservoir eutrophication. As stated in the introduction to this chapter, cultural eutrophication ranks amongst the more ubiquitous causes of water quality degradation on the global scale, and interferes with a wide range of beneficial water uses. A particular need, therefore, is for simplified models which can provide quantitative information on the potential effectiveness of alternative eutrophication control measures based on reducing the external phosphorus load.

In contrast to the traditional process-oriented studies prevailing prior to the late 1960s, which were typically complex and data-intensive, Vollenweider (1968) developed simplified mass-balance calculations and statistical regressions. His original work, as well as its elaboration within the context of the OECD Eutrophication Programme, demonstrated that eutrophication could be reversed for most lakes by reducing their exter-

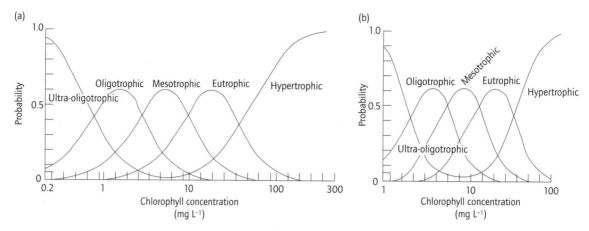

Fig. 14.15 Comparison of probabilistic trophic state classification scheme for (a) OECD (1982) lakes and (b) warm-water lakes (Salas & Martino 1991), based on mean chlorophyll concentration (Rast & Thornton 1997).

nal phosphorus loads. His simplified approach provided both predictive power and practical applicability in designing lake and reservoir eutrophication management programmes (Håkanson & Peters 1995). As seen with the results of the OECD (1982) Eutrophication Programme, the models of Vollenweider and others subsequently based upon them (e.g. Dillon & Rigler 1974; Larsen & Mercier 1976; Rast & Lee 1978; Clasen & Bernhardt 1980; Fricker 1980; Lee et al. 1980; Ryding 1980; Janus & Vollenweider 1981) have demonstrated the practical application of this approach within the context of eutrophication management. Håkanson & Peters (1995) also discuss this aspect of predictive limnology in more detail.

In a separate study, Sas (1989) specifically examined the phosphorus-load–lake-response models elaborated in the OECD Eutrophication Programme. In evaluating the relationship between the external phosphorus load and in-lake phosphorus concentration, and those between the phosphorus concentration and other eutrophication response parameters (e.g. chlorophyll, Secchi-disc depth), he concluded that eutrophic lakes recover 'reluctantly', but steadily, after their phosphorus load is reduced. The 'reluctance' stems from the time necessary to flush the phosphorus out of a lake, as well as that from the remobilisation of phosphorus from the lake bottom sediment, after its external phosphorus load is reduced. Upon reaching a new phosphorus equilibrium in his case studies, the positive responses of the lakes were relatively rapid. His ultimate conclusion was that the OECD models, within their predictive capabilities, were reliable predictors of lake responses to changes in their external phosphorus loads.

Other researchers provided additional evidence of the predictive capability of the OECD (1982) eutrophication models. Rast et al. (1983) used the USA OECD data base (Rast & Lee 1978) in order to compare the measured and model-predicted responses of a number of temperate lakes and reservoirs which had undergone external phosphorus load reductions. They found that the predicted mean chlorophyll concentrations for most of their waterbodies (Fig. 14.16) were within a factor of about ±1.5 of the measured values. A similar good agreement was found for the predicted and measured Secchi-disc transparency for the USA OECD waterbodies. The analyses of Thornton & Walmsley (1982), Walmsley & Thornton (1984), Ryding & Rast (1989) and Thornton & Rast (1993) demonstrated that these approaches were largely appropriate for temperate, tropical and subarctic waterbodies. Although the models may require some calibration or adjustment with respect to trophic boundary conditions when used in non-temperate settings, Rast & Thornton (1997) pro-

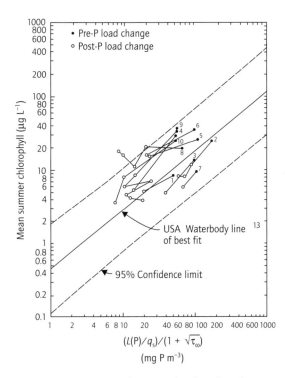

Fig. 14.16 Comparison of measured and predicted changes in chlorophyll concentrations in USA OECD waterbodies, following reduction of external phosphorus loads (Rast *et al.* 1983). The numbers against data points refer to individual lakes, which are identified in the original reference.

vided ample evidence that these differences are generally more a matter of degree than of substance. The previously cited work of Salas & Martino (1991) in the Regional Programme for the Development of Simplified Methodologies for the Evaluation of Eutrophication in Warm-Water Tropical Lakes, similar to that carried out for the primarily temperate-zone lakes and reservoirs within the OECD Eutrophication Programme (1982), offers further evidence that this type of nutrient-load–lake-response model is a useful, accurate, quantitative management tool for assessing eutrophication-related water quality, and the probable responses of lakes and reservoirs to alternative control programmes based on reducing their external phosphorus loads.

In a broadly based survey of lakes and reservoirs around the world, Jones & Lee (1986) also provided evidence that the phosphorus-load–lake-response models exemplified by the OECD study were useful predictive tools for eutrophication management. They focused on waterbodies with characteristics different from those in the USA OECD study, including lakes which seemingly should not fit the derived load–response relationships. Their expanded data base included hyper-eutrophic reservoirs in Colorado (USA), a water supply reservoir in Kansas (USA) which received large quantities of non-point source phosphorus, Lake Mjøsa (Norway) with a mean depth greater than 150 m, 20 reservoirs in Spain, 21 reservoirs in South Africa, Salto Grande Reservoir on the Uruguay/Argentina border, Lacar Lake on the Chile/Argentina border, and ultra-oligotrophic Lake Vanda in Antarctica (Jones & Lee 1986). Thus, their data base comprised small and large, shallow and deep, warm-climate and cold-climate, hypereutrophic and oligotrophic, bowl-shaped and run-of-the-river waterbodies, as well as two estuarine systems. The important point to be emphasised is that, in virtually all cases, the lakes and reservoirs in their analysis generally followed the phosphorus-load–lake-response relationships (Fig. 14.17) predicted from the original Vollenweider (1976) approach and its elaboration in the OECD Eutrophication Programme (1982). Jones & Lee (1986) also provided guidance for applying this modelling approach for lakes and reservoirs with unusual characteristics, or those exhibiting conditions outside those of their original data base. Using a group of Canadian lakes, Janus & Vollenweider (1981) evaluated the predictive capability of the load–response models utilised in the OECD Eutrophication Programme. As an independent evaluation of the models, conducted after the completion of the OECD Eutrophication Programme, these researchers came to essentially the same conclusions regarding the value and utility of the OECD (1982) phosphorus-load–lake-response models for eutrophication management.

### 14.7.1 Operational use of OECD phosphorus-load–lake-response models

Since the 1980s, the OECD suite of eutrophication

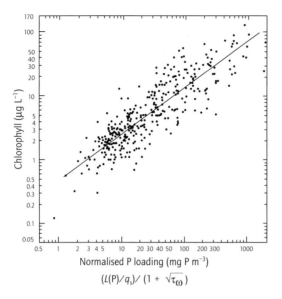

Fig. 14.17 Updated phosphorus load versus mean chlorophyll diagram (Jones & Lee 1986).

Rast et al. (1979) used the phosphorus-load–lake-response models developed under the USA OECD study in an independent evaluation of the predicted response of the Great Lakes to changes in their external phosphorus loads, as well as to determine the phosphorus load reductions necessary to achieve specific target loads and/or desired in-lake water quality conditions. Their study results illustrated reasonable agreement between the predicted water quality attainable under the target phosphorus loads, and the stated water quality objectives, in the Great Lakes Water Quality Agreement (International Joint Commission 1978). They also pointed out, however, that their models suggested a lower phosphorus load of approximately 7000 t yr$^{-1}$ would be necessary in order to achieve the stated water quality objectives for Lake Erie, compared with the 11,000 t yr$^{-1}$ target load in the Agreement. They also concluded that even this lower phosphorus load target for Lake Erie was likely to be unattainable, given that it would require a substantial reduction in the phosphorus load from non-point sources in the Lake Erie drainage basin.

In another application of the USA OECD data set, Rast & Lee (1983) made use of an equivalence described by Vollenweider (1975) in order to determine the accuracy of phosphorus loading estimates to lakes and reservoirs. Vollenweider previously noted that the ratio of a waterbody's mean in-lake phosphorus concentration to its inflow phosphorus concentration, $[(P)_l/(P)_i]$, was theoretically equivalent to the value of the hydraulic residence time expression, $(1/1+ \sqrt{\tau_\omega})$ (see eqns. 14.10–14.13 for definition of terms). The inflow phosphorus concentration is a function of the phosphorus load (i.e. $(P)_i = L(P)/q_s$). Thus, any major deviations between these two terms would make the reported phosphorus loading data suspect. Vollenweider (1975) used this relationship to trace loading errors in the phosphorus load estimates for Lake Constance and the Lunzer See. Rast & Lee (1983) developed a graphic form of this equation (Fig. 14.18), using it to evaluate the USA OECD waterbodies. Their evaluation showed that 65% of the predicted phosphorus load estimates were within a factor of about ±2 of the reported

response models has been deployed in numerous situations. In an early application, the models were used, together with more complex eutrophication response models, to evaluate nutrient management strategies in the Laurentian Great Lakes. In this application, a phosphorus-load–lake-response model derived by Vollenweider (1968, 1975) was used, along with other models, to develop the phosphorus target loads for the Laurentian Great Lakes of North America under the Great Lakes Water Quality Agreement (IJC 1978). The USA and Canadian governments convened an expert model group to determine the target loads necessary in order to achieve the desired water quality conditions in the Great Lakes Basin Ecosystem. Using five different models, ranging from complex, dynamic, multidimensional models to the simplified load–response models, the expert group found that all the model predictions tended to converge to the same values. This convergence was particularly dramatic for Lake Erie, the most eutrophic of the Great Lakes, with the prediction that, in order to eliminate the anoxic zone in its central basin, its external phosphorus load should not exceed 11,000 t yr$^{-1}$.

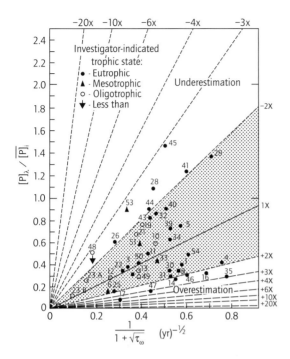

Fig. 14.18 Use of Vollenweider (1975) relationship to evaluate accuracy of phosphorus load estimates (Rast & Lee 1983; see reference for definition of terms). The numbers against data points refer to individual lakes, which are identified in the original reference.

phosphorus loads, which was very close to the theoretical 67% statistical value of one standard deviation for their data set. They also pointed out that a major disagreement between the predicted and reported phosphorus loads did not automatically signify a significant error in the phosphorus load estimate. Rather, the possibility of errors in the values of the other parameters in the model, or in the way that a particular waterbody utilising phosphorus in the production of planktonic algae may be different from similar waterbodies located in other parts of the USA and the world, should be considered. Nevertheless, considering the multitude of methods used by the USA OECD investigators in determining their annual phosphorus load estimates, Rast & Lee (1983) concluded that there was remarkably good agreement between the phosphorus-load estimates made with the phosphorus-load–lake-response models and those reported by the investigators.

In similar applications to inland lakes, Panuska & Kreider (2002) utilised the OECD relationships in a spreadsheet format, known as the Wisconsin Lake Model Spreadsheet (WILMS), to assist lake managers in the midwestern USA to evaluate the efficacy of lake management actions in a wide range of river-run and internally drained lakes. This model has been applied variously in determining the distribution of non-point sources of phosphorus to the lakes, with the goal of allocating lake management measures to control the main sources of external phosphorus loads, and to evaluating a range of remedial measures aimed at achieving phosphorus-load reductions. The Southeastern Wisconsin Regional Planning Commission routinely utilises these models in order to prioritise management interventions in lake drainage basins in the Milwaukee metropolitan area, just north of Chicago.

Based on this proven utility, Thornton & Harding (1998) applied a previous version of the WILMS model to a data set compiled for a range of African lakes being studied under the auspices of the Food and Agriculture Organization (FAO) Committee for Inland Fisheries in Africa (CIFA). Application of this modelling approach showed promise as a means of forecasting potential eutrophication problems, and diagnosing existing problems targeted for remediation. This exercise again demonstrated the validity and utility of the OECD suite of phosphorus-load–lake-response models in addressing the global problem of eutrophication.

## 14.8 CONCLUSIONS

The simplified phosphorus-load–lake-response models based on the initial work of Vollenweider (1968, 1975; 1976), as well as its subsequent elaboration by others (e.g. Dillon & Rigler 1974; Larsen & Mercier 1976; Rast & Lee 1978; Clasen & Bernhardt 1980; Fricker 1980; Lee et al. 1980; Ryding 1980; OECD 1982), represent extremely useful and easily understood lake and reservoir eutrophication management tools, based both on their predictive power and their ease of application. Their predictive capacity has been demonstrated in a number of situations, including direct comparisons between measured and predicted values (Thornton & Walmsley 1982; Walmsley & Thornton 1984; Rast *et al.* 1983). They can be used to determine the external phosphorus load corresponding to a range of water-quality conditions in a lake or reservoir. Alternatively, they can be used to calculate the magnitude of the phosphorus-load reduction necessary to achieve desired in-lake water quality conditions. They also can be used to evaluate the impacts of future land use or phosphorus loading conditions, and allow remedial measures to be established prior to the occurrence of eutrophication-related water quality problems in lakes and reservoirs. The OECD (1982) Eutrophication Programme, in which the initial models were elaborated and tested, remains the most systematic study to date to identify and quantify phosphorus-loading–lake-response relationships for assessing and predicting the responses of lakes and reservoirs to their external nutrient loads. Further, sufficient experience with their use since the OECD Eutrophication Programme in a wide spectrum of lakes and reservoirs of differing size, area, depth, climatic and trophic condition now exists to allow their modification for atypical conditions (e.g. Thornton & Walmsley 1982; Walmsley & Thornton 1984; Jones & Lee 1986; Ryding & Rast 1989; Rast & Thornton 1997). Indeed, what is probably most remarkable is the utility and predictive capability of these models across such a wide range of lakes and reservoirs exhibiting a spectrum of eutrophication characteristics. It is recommended that these models continue to be used to identify, develop, evaluate and refine eutrophication management strategies for waterbodies.

## 14.9 REFERENCES

American Water Works Association (1966) Nutrient-associated problems in water quality and treatment. Task Group 2610-P Report. *Journal of the American Water Works Association*, **58**, 1337–55.

American Water Works Association (1967) Sources of nitrogen and phosphorus in water supplies. Task Group 2610-P Report. *Journal of the American Water Works Association*, **59**, 344–66.

Bachmann, R.W. & Jones, J.R. (1974) Phosphorus inputs and algal blooms in lakes. *Iowa State Journal of Research*, **49**, 155–60.

Barker, M.L. (1971) Beach pollution in the Toronto region. In: Sewell, W.R.D. & Burton, I. (eds), *Perceptions and Attitudes in Resources Management*. Information Canada, Ottawa, 37–47.

Bartsch, A.F. (1970) Accelerated eutrophication of lakes in the United States: ecological response to human activities. *Environmental Pollution*, **1**, 133–40.

Bartsch, A.F. (1972) *The Role of Phosphorus in Eutrophication*. Report EPA-R3-72-001, U.S. Environmental Protection Agency, National Environmental Research Center, Corvallis, OR, 45 pp.

Beeton, A.M. & Edmondson, W.T. (1972) The eutrophication problem. *Fisheries Research Board Canada*, **29**, 673–82.

Biffi, F. (1963) Determinazone del fattore tempo come cratteristica del potere di antodepurazione del Lago d'Orta in relazione ad im inquinamento constante. *Attidell'Istituto Veneto di Scienzi, Lettere ed Arti*, **121**, 131–36.

Brezonik, P.L. (1969) Eutrophication: the process and its modeling potential. In: *Federal Water Quality Administration, Workshop Proceedings*, St Petersburg, Florida, 19–21 November, 68–110.

Brezonik, P.L. (1972) Nitrogen: sources and transformations in natural waters. In: Allen, H.E. & Kramer, J.P. (eds), *Nutrients in Natural Waters*. Wiley, New York, 1–50.

Brezonik, P.L. (1973) Nitrogen sources and cycling in natural waters. Report EPA-660/3-73-002, U.S. Environmental Protection Agency, Washington, DC, 167 pp.

Canfield, D.E. & Bachman, R.W. (1981) Prediction of total phosphorus concentrations, chlorophyll *a*, and Secchi depths in natural and artificial lakes. *Canadian Journal of Fisheries and Aquatic Science*, **38**, 414–23.

Carlson, R.E. (1977) A trophic state index for lakes. *Limnology and Oceanography*, **22**, 361–9.

Clasen, J. & Bernhardt, H. (1980) *OECD Eutrophication Programme. Shallow Lakes and Reservoir Project*. Final Report, Water Research Centre, Medmenham Laboratory, Medmenham, Marlow, Bucks SL7 2HD, UK, 289 pp.

Clesceri, N.L. & Lee, G.F. (1965a) Hydrolysis of condensed phosphates I: non-sterile environment. *Air and Water Pollution*, **9**, 723–42.

Clesceri, N.L. & Lee, G.F. (1965b) Hydrolysis of condensed phosphates II: sterile environment. *Air and Water Pollution*, **9**, 743–751.

Curry, J.J. & Wilson, S.L. (1955) Effect of sewage-borne phosphorus on algae. *Sewage and Industrial Wastes*, **27**, 1262–9.

David, E.L. (1971) Public perceptions of water quality. *Water Resources Research*, **7**, 453–57.

Dillon, P.J. (1974) A critical review of Vollenweider's nutrient budget model and other related models. *Water Resources Bulletin*, **10**, 969–89.

Dillon, P.J. (1975) The phosphorus budget of Cameron Lake, Ontario: the importance of flushing rate to the degree of eutrophy in lakes. *Limnology and Oceanography*, **19**, 28–39.

Dillon, P.J. & Rigler, F.H. (1974) A test of a simple nutrient budget model predicting the phosphorus concentration in lake water. *Journal of the Fisheries Research Board Canada*, **31**, 1771–8.

Edmondson, W.T. (1970) Phosphorus, nitrogen and algae in Lake Washington after diversion of sewage. *Science*, **169**, 690–1.

Edmondson, W.T. (1972) Nutrients and phytoplankton in Lake Washington. In: Likens, G.E. (ed.), *Nutrients and Eutrophication: the Limiting Nutrient Controversy*. Limnology and Oceanography, Special Symposium, Vol. I–93.

Fitzgerald, G.P. (1969) Field and laboratory evaluation of bioassays for nitrogen and phosphorus with algae and aquatic weeds. *Limnology and Oceanography*, **14**, 206–12.

Fricker, H. (1980) *OECD Eutrophication Programme. Regional Project. Alpine Lakes*. Swiss Federal Board for Environmental Protection (Bundesamt für Umweltschutz), Bern, 234 pp.

Fruh, E.G. (1967) The overall picture of eutrophication. *Journal of the Water Pollution Control Federation*, **39**, 1449–63.

Fruh, E.G., Stewart, K.M., Lee, G.F. & Rohlich, G.A. (1966) Measurements of eutrophication and trends. *Journal of the Water Pollution Control Federation*, **38**, 1237–58.

Fuhs, G.W., Demmerle, S.D., Canelli, E. & Chen, M. (1972) Characteristics of phosphorus-limited plankton algae (with reflections on the limiting nutrient concept). In: Likens, G.E. (ed.), *Nutrients and Eutrophication: the Limiting Nutrient Controversy*. Limnology and Oceanography, Special Symposium, Vol. I–33.

Gilbertson, M., Dobson, H.H. & Lee, T.R. (1972) Phosphorus and hypolimnial dissolved oxygen in Lake Erie. In: Burns, N.M. & Ross, C. (eds), *PROJECT HYPO—an intensive Study of the Lake Erie Central Basin Hypolimnion and Related Surface Water Phenomena*. Paper No. 6, Canada Centre for Inland Waters, Burlington, Ontario, 141–5.

Goldman, C.R. (1960a) Primary productivity and limiting factors in three lakes of the Alaskan peninsula. *Ecological Monographs*, **30**, 207–30.

Goldman, C.R. (1960b) Molybdenum as a factor limiting primary productivity in Castle Lake, California. *Science*, **132**, 1016–17.

Goldman, C.R. (1964) Primary productivity and micronutrient limiting factors in some North American and New Zealand lakes. *Verhandlungen Internationale Vereinigung Limnologie*, **15**, 365.

Goldman, C.R. & Wetzel, R.G. (1963) A study of the primary productivity of Clear Lake, Lake County, California. *Ecology*, **44**, 283–94.

Goldman, J.C., Porcella, D.B., Middlebrooks, E.J. & Toerien, D.F. (1972) The effects of carbon on algal growth—its relationship to eutrophication. *Water Research*, **6**, 637–79.

Golterman, H.L. (1975) *Physiological Limnology. An Approach to the Physiology of Lake Ecosystems*. Elsevier, New York, 366–402.

Gregor, D.J. & Rast, W. (1981) Benefits and problems of eutrophication control. In: *Restoration of Lakes and Inland Waters*. Report EPA-440/5/81-010, U.S. Environmental Protection Agency, Washington, DC, 166–71.

Håkanson, L. & Peters, R.H. (1995) *Predictive Limnology. Methods for Predictive Modelling*. SPB Academic Publishing, Amsterdam, 401–3.

Hanson, J.M. & Leggett, W.C. (1982) Empirical prediction of fish biomass and yield. *Canadian Journal of Fisheries and Aquatic Science*, **39**, 257–63.

Harris, E. & Riley, G.A. (1956) Oceanography of Long Island Sound, 1952–1954. VIII. Chemical composition of the plankton. *Bulletin of the Bingham Oceanographic Collection*, **15**, 315–23.

Hutchinson, G.E. (1938) On the relation between the oxygen deficit and the productivity and typology of lakes. *Internationale Revue der gesamte Hydrobiologie*, **35**, 336–55.

Hutchinson, G.E. (1969) Eutrophication, past and present. In: National Academy of Sciences, *Eutrophication: Causes, Consequences, Correctives*, Symposium Proceedings, Washington, DC, 197–209.

Hutchinson, G.E. (1973) Eutrophication. *American Scientist*, **61**, 269–79.

International Joint Commission (1978) *Great Lakes Water Quality Agreement of 1978. Agreement with Annexes and Terms of Reference*. Great Lakes

Regional Office, International Joint Commission, Windsor, Ontario.

James, D.W. & Lee, G.F. (1974) A model of inorganic carbon limitation in natural waters. *Water, Soil and Air Pollution*, **3**, 315–20.

Janus, L.L. & Vollenweider, R.A. (1981) *Summary Report: the OECD Eutrophication Programme on Eutrophication—Canadian Contribution*. Scientific Series No. 131, Canada Centre for Inland Waters, Burlington, Ontario.

Jensen, T.E. & Sicko-Goad, L. (1976) *Aspects of Phosphate Utilization by Blue-green Algae*. Report EPA-600/3-76-103, U.S. Environmental Protection Agency, Washington, DC, 121 pp.

Jones, R.A. & Lee, G.F. (1986) Eutrophication modeling for water quality management: an update of the Vollenweider-OECD model. *Water Quality Bulletin* (World Health Organization), **11**, 67–74.

Keenan, J.D. & Auer, M.T. (1974) The influence of phosphorus luxury uptake on algal bioassays. *Journal of the Water Pollution Control Federation*, **46**, 532–42.

Kerr, P.C., Paris, D.F. & Brockway, D.L. (1970) *The Interrelation of Carbon and Phosphorus in Regulating Heterotrophic and Autotrophic Populations in Aquatic Ecosystems*. Water Pollution Control Research Series No. 16050 FGS, U.S. Department of Interior, Federal Water Quality Administration, U.S. Government Printing Office, Washington, DC, 53 pp.

Ketchum, B.H. (1969) Eutrophication of estuaries. In: National Academy of Sciences, *Eutrophication: Causes, Consequences, Correctives*, Symposium Proceedings, Washington, DC, 197–209.

Ketchum, B.H. & Redfield, A.C. (1949) Some physical and chemical characteristics of algae grown in mass cultures. *Journal of Cellular and Comparative Physiology*, **33**, 291–300.

Kuentzel, L.E. (1969) Bacteria, carbon dioxide and algal blooms. *Journal of the Water Pollution Control Federation*, **41**, 1737–47.

Lange, W. (1967) Effects of carbohydrates on the symbiotic growth of plankton blue-green algae with bacteria. *Nature*, **215**, 1277–8.

Larsen, D.P. & Mercier, H.T. (1976) Phosphorus retention capacity of lakes. *Journal of the Fisheries Research Board Canada*, **33**, 1742–50.

Lasenby, D.C. (1975) Development of oxygen deficits in 14 southern Ontario lakes. *Limnology and Oceanography*, **20**, 993–9.

Lee, G.F. (1971) Eutrophication. In: *Encyclopedia of Chemical Technology* (2nd edition). Wiley, New York, 315–38.

Lee, G.F. (1973) Role of phosphorus in eutrophication and diffuse source control. *Water Research*, **7**, 111–28.

Lee, G.F. & Jones, R.A. (1991) Effects of eutrophication on fisheries. *Reviews in Aquatic Science*, **5**, 287–305.

Lee, G.F., Rast, W. & Jones, R.A. (1978) Eutrophication of waterbodies: insights for an age-old problem. *Environmental Science and Technology*, **12**, 900–8.

Lee, G.F., Jones, R.A. & Rast, W. (1980) Availability of phosphorus to phytoplankton and its implications for phosphorus management strategies. In: Loehr, R.C., Martin, C. & Rast, W. (eds), *Phosphorus Management Strategies for Lakes*. Interscience, Ann Arbor, MI, 259–308.

Likens, G.E. (ed.) (1972) *Nutrients and Eutrophication: the Limiting Nutrient Controversy*. Limnology and Oceanography, Special Symposium, Vol. I 328 pp.

Mackenthun, K.M., Ingram, W.M. & Porges, R. (1964) *Limnological Aspects of Recreational Lakes*. Publication No. 1167, Public Health Service, Washington, DC, 176 pp.

Maloney, T.E. (1966) Detergent phosphorus effects on algae. *Journal of the Water Pollution Control Federation*, **38**, 38–45.

Maloney, T.E., Miller, W.E. & Shiroyana, T. (1972) Algal responses to nutrient addition in natural waters. I. Laboratory assays. In: Likens, G.E. (ed.), *Nutrients and Eutrophication: the Limiting Nutrient Controversy*, Limnology and Oceanography, Special Symposium, Vol. I 134–40.

Martin, C.M. & Goff, D.R. (1972) The role of nitrogen in the aquatic environment. Contribution No. 2, Limnology Department, Academy of Natural Sciences of Philadelphia (USA), 46 pp.

Menzel, D.W. & Ryther, J.H. (1961) Nutrients limiting the production of phytoplankton in the Sargasso Sea, with special reference to iron. *Deep Sea Research*, **7**, 276–81.

National Academy of Sciences (1969) *Eutrophication: Causes, Consequences, Correctives*. Symposium Proceedings, Washington, DC, 661 pp.

Naumann, E. (1919) Nagra synpunkter planktons ökologi. Med. särskild hänsyn till fytoplankton. *Svensk Botanisk Tidskrift*, **13**, 129–58.

Naumann, E. (1931) *Limnologische terminologie*. Urban und Schwarzenberg, Berlin-Wien.

Odum, E.P. (1971) *Fundamentals of Ecology*. W.G. Saunders Co., Philadelphia, 106–39.

OECD (1973) *Summary Report of the Water Management Sector Group on Agreed Monitoring Project on Eutrophication of Waters*. Environment Directorate,

Organization for Economic Cooperation and Development, Paris, 19 pp.

OECD (1975) *OECD Co-operative Programme for the Monitoring of Eutrophication in Inland Waters*. Progress Report, Environment Directorate, Organization for Economic Cooperation and Development, Paris, 35 pp.

OECD (1982) *Eutrophication of Waters. Monitoring, Assessment and Control*. Organization for Economic Cooperation and Development, Paris, 154 pp.

Oglesby, R.T. (1977) Relationships of fish yield to lake phytoplankton standing crop, production, and morphoedaphic factors. *Journal of the Fisheries Research Board Canada*, **34**, 2271–82.

Oglesby, R.T. & Schaffner, W.R. (1978) Phosphorus loadings to lakes and some of their responses. Part 2. Regression models of summer phytoplankton standing crops, winter total P, and transparency of New York lakes with phosphorus loadings. *Limnology and Oceanography*, **23**, 135–45.

Panuska, J.C. & Kreider, J.C. (2002) *Wisconsin Lake Modeling Suite Program Documentation and Users Manual: Version 3.3 for Windows*. Publication No. PUBL-WR-363-94, Department of Natural Resources, Madison, State of Wisconsin, 32 pp.

Piontelli, R. & Tonolli, V. (1964) Il tempo di residenza delle aqua lacustri in relazione ai fenomeni di arricchimento in sostanze immesse, con particolare riguardo al Lago Maggiore. *Memorie dell'Istituto Italiano di Idrobiologia*, **17**, 247–66.

Porcella, D.B., Bishop, A.B., Anderson, J.C., et al. (1974) *Comprehensive Management of Phosphorus Water Pollution*. Report EPA-600/5-74-010, U.S. Environmental Protection Agency, Washington, DC, 411 pp.

Powers, C.F., Schults, D.W., Malueg, K.W., Brice, R.M. & Schuldt, M.D. (1972) Algal responses to nutrient additions in natural waters. II. Field experiments. In: Likens, G.E. (ed.). 1972a. *Nutrients and Eutrophication: the Limiting Nutrient Controversy, Limnology and Oceanography, Special Symposium*, Vol. I 141–56.

Provasoli, L. (1969) Algal nutrition and eutrophication. In: National Academy of Sciences, *Eutrophication: Causes, Consequences, Correctives*, Symposium Proceedings, Washington, DC, 574–93.

Rast, W. (ed.). (2003) *World Lake Vision: a Call to Action*. International Lake Environment Committee, United Nations Environment Programme, and Shiga Prefecture, Otsu, Japan, 36 pp.

Rast, W. & Lee, G.F. (1978) *Summary Analysis of the North American (U.S. Portion) OECD Eutrophication Project: Nutrient Loading—Lake Response Relationships and Trophic State Indices*. Report EPA-600/3-78-008, U.S. Environmental Protection Agency, Environmental Research Laboratory, Corvallis, OR, 455 pp.

Rast, W. & Lee, G.F. (1983) Nutrient loading estimates for lakes. *American Society of Civil Engineers, Journal of the Environmental Engineering Division*, **109**, 502–17.

Rast, W. & Thornton, J.A. (1997) Trends in eutrophication research and control. In: Peters, N.E., Bricker, O.P. & Kennedy, M.M. (eds), *Water Quality Trends and Geochemical Mass Balance*, Wiley, New York, 171–90.

Rast, W., Lee, G.F. & Jones, R. A. (1979) Use of the OECD eutrophication modeling approach for assessing Great Lakes' water quality. In: Lee, G.F. & Jones, R. A. (eds), *Water Quality Characteristics of the US Waters of Lake Ontario During the IGYFL and Modeling Contaminant Load—Water Quality Response Relationships in the Nearshore Waters of the Great Lakes*. Technical Report, National Oceanic and Atmospheric Administration, Ann Arbor, MI, 184–233.

Rast, W., Jones, R.A. & Lee, G.F. (1983) Predictive capability of U.S. OECD phosphorus loading—eutrophication response models. *Journal of the Water Pollution Control Federation*, **55**, 990–1003.

Rawson, D.S. (1939) Some physical and chemical factors in the metabolism of lakes. In: *Problems of Lake Biology*. Publication No. 10, American Association for Advancement of Science, Washington, DC, 9–26.

Redfield, A.C. (1958) The biological control of chemical factors in the environment. *American Scientist*, **46**, 205–21.

Redfield, A.C., Ketchum, B.J. & Richard, F.A. (1963) The influence of organisms on the composition of seawater. In: Hill, M.M. (ed.), *The Sea*, Vol. 2. Wiley-Interscience, New York, 26–77.

Rodhe, W. (1969) Crystallization of eutrophication concepts in Northern Europe. In: National Academy of Sciences, *Eutrophication: Causes, Consequences, Correctives*, Symposium Proceedings, Washington, DC, 50–64.

Ryding, S.-O. (1980) *Monitoring of Inland Waters: OECD Eutrophication Programme—The Nordic Project*. Publication 1980:2, Nordic Co-operative Programme for Applied Research (NORDFORSK), Helsinki, 207 pp.

Ryding, S.-O. & Rast, W. (1989) *The Control of Eutrophication of Lakes and Reservoirs*. Man and the Biosphere Programme, Vol. 1, United Nations Educational, Scientific and Cultural Organization (UNESCO), Paris, 314 pp.

Sakamoto, M. (1966) Primary production by phytoplankton community in some Japanese lakes and its dependence on mean depth. *Archiv für Hydrobiologie*, **62**, 1–28.

Salas, H.J. & Martino (1991) A simplified phosphorus trophic state model for warm-water tropical lakes. *Water Research*, **25**, 341–50.

Sas, H. (1989) *Lake Restoration by Reduction of Nutrient Loading. Expectations, Experiences, Extrapolations*. Academia Verlag, Richarz, 497 pp.

Sawyer, C.N. (1947) Fertilization of lakes by agricultural and urban drainage. *Journal of the New England Water Works Association*, **61**, 109–27.

Sawyer, C.N. (1952) Some new aspects of phosphates in relation to lake fertilization. *Sewage and Industrial Wastes*, **24**, 768–76.

Sawyer, C.N. (1966) Basic concepts of eutrophication. *Journal of the Water Pollution Control Federation*, **38**, 737–44.

Schelske, C.L. & Stoermer, E.F. (1972) Phosphorus, silica and eutrophication of Lake Michigan. In: Likens, G.E. (ed.), *Nutrients and Eutrophication: the Limiting Nutrient Controversy, Limnolology and Oceanography, Special Symposium*, Vol. I 157–71.

Schindler, D.W. (1971) Carbon, nitrogen and phosphorus and the eutrophication of freshwater lakes. *Journal of Phycology*, **7**, 321–9.

Schindler, D.W. (1977) Evaluation of phosphorus limitation in lakes. *Science*, **195**, 260–2.

Schindler, D.W. (1978) Predictive eutrophication models. *Limnology and Oceanography*, **23**, 1080–1.

Schindler, D.W. & Fee, E.J. (1974) Experimental Lakes Area: whole-lake experiments in eutrophication. *Journal of the Fisheries Research Board Canada*, **31**, 937–53.

Schindler, D.W., Armstrong, F.A.J., Holmgren, S.K. & Brunskill, G.J. (1971) Eutrophication of Lake 227, Experimental Lakes Area, northwestern Ontario, by addition of phosphate and nitrate. *Journal of the Fisheries Research Board Canada*, **28**, 1763–82.

Schindler, D.W., Fee, E.J. & Ruszczynski, T. (1976) *Phosphorus Inputs and its Consequences for Phytoplankton Standing Crop and Production in the Experimental Lakes Area and in Similar Lakes*. Technical Report, Freshwater Institute, Environment Canada, Winnipeg, Manitoba, 15 pp.

Shannon, J.E. & Lee, G.F. (1966) Hydrolysis of condensed phosphates in natural waters. *Air and Water Pollution*, **10**, 755–6.

Shapiro, J. (1973) Blue-green algae: Why they become dominant. *Science*, **179**, 382–4.

Sonzogni W.C., Uttormark, P.D. & Lee, G.F. (1976) The phosphorus residence time model. *Water Research*, **10**, 429–35.

Stewart, K.M. & Rohlich, G.A. (1967) *Eutrophication—a Review*. Publication No. 34, State Water Quality Control Board, State of California, Sacramento, 188 pp.

Stumm, W. & Morgan, J.J. (1981) *Aquatic Chemistry. An Introduction Emphasizing Chemical Equilibrium in Natural Waters* (2nd edition). Wiley-Interscience, New York, 562.

Thornton, J.A. (1993) Perceptions of public waters: Water quality and water use in Wisconsin. In: van Valey, T, Crull, S. & Walker, L. (eds), *The Small City and Regional Community*, Vol. 10. Stevens Point Foundation Press, Stevens Point, WI, 469–78.

Thornton, J.A. & Harding, W.R. (1998) *Phosphorus Modeling of African Lakes*. Report to the Committee for Inland Fisheries of Africa: 5th Session of Working Party on Pollution and Fisheries. FAO Fisheries Report 587, Accra, Ghana.

Thornton, J.A. & McMillan, P.H. (1989) Reconciling public opinion and water quality criteria in South Africa. *Water SA*, **15**, 221–6.

Thornton, J.A. & Rast, W. (1987) Application of eutrophication modelling technique to manmade lakes in semi-arid southern Africa. In: Nix, S.J. & Black, P.E. (eds) *Monitoring, Modeling and Mediating Water Quality*. American Water Works Association, Bethesda, MD, 547–58.

Thornton, J.A. & Rast, W. (1993) A test of hypotheses related to the comparative limnology and assessment of eutrophication in semi-arid man-made lakes. In: Straskraba, M.L., Tundisi, J. & Duncan, A. (eds), *Comparative Reservoir and Water Quality Management*. Kluwer, The Hague, 219 pp.

Thornton, J.A. & Walmsley, R.D. (1982) Applicability of phosphorus budget models to southern African man-made lakes. *Hydrobiology*, **89**, 237–45.

Thornton, J.A., McMillan, P.H. & Romanovsky, P. (1989) Perceptions of water pollution in South Africa: case studies from two waterbodies (Hartbeespoort Dam and Zandvlei). *South African Journal of Psychology*, **19**, 199–204.

US EPA (1973a) *Measures for the Restoration and Enhancement of Quality of Freshwater Lakes*. Technical Report, U.S. Environmental Protection Agency, Washington, DC, 238 pp.

US EPA (1973b) *Proposed Criteria for Water Quality*, Vols. I & II. U.S. Environmental Protection Agency, Washington, DC, 425 pp. (Vol. I), 164 pp. (Vol. II).

US EPA (1974) *The Relationship of Nitrogen and*

Phosphorus to the Trophic State of Northeast and Northcentral Lakes and Reservoirs. National Eutrophication Survey Working Paper No. 23, Pacific Northwest Environmental Research Laboratory, Corvallis, OR, 28 pp.

US EPA (1976) *Quality Criteria for Water*. Report EPA-440/9-76-023, U.S. Environmental Protection Agency, Washington, DC, 501 pp.

Uttormark, P.D., Chapin, J.D. & Green, K.M. (1974) *Estimating Nutrient Loading of Lakes from Non-point Sources*. Report EPA-660/3-74-020, U.S. Environmental Protection Agency, Washington, DC, 112 pp.

Vallentyne, J.R. (1974) *The Algal Bowl. Lakes and Man*. Miscellaneous Special Publication No. 22, Fisheries and Marine Service, Department of the Environment, Environment Canada, 185 pp.

Vollenweider, R.A. (1968) *Scientific Fundamentals of the Eutrophication of Lakes and Flowing Waters, with Particular Reference to Nitrogen and Phosphorus as Factors in Eutrophication*. Technical Report DAS/CSI/68.27, Organization for Economic Cooperation and Development, Paris, 159 pp.

Vollenweider, R.A. (1975) Input–output models, with special reference to the phosphorus loading concept in limnology. *Schweizerische Zeitschrift für Hydrologie*, **37**, 53–84.

Vollenweider, R.A. (1976) Advances in defining critical loading levels for phosphorus in lake eutrophication. *Memorie dell'Istituto Italiano di Idrobiologia*, **33**, 53–83.

Vollenweider, R.A. & Dillon, P.J. (1974) *The Application of the Phosphorus Loading Concept to Eutrophication Research*. NRCC No. 13690, NRC Associate Committee on Scientific Criteria for Environmental Quality, National Research Council, Ottawa, Canada, 42 pp.

Vollenweider, R.A., Rast, W. & Kerekes, J. (1980) The phosphorus loading concept and Great Lakes eutrophication. In: Loehr, R.C., Martin, C.S. & Rast, W. (eds), *Phosphorus Management Strategies for Lakes*. Ann Arbor Science, Ann Arbor, MI, 207–34.

Walmsley, R.D. & Thornton, J.A. (1984) Evaluation of OECD-type phosphorus eutrophication models for predicting the trophic status of southern Africa manmade lakes. *South African Journal of Science*, **80**, 257–9.

Weber, C.A. (1907) Aufbau und Vegetation der Moore Norddeutschlands. *Beiblatt zu den Botanischen Jahrbuchern*, **90**, 19–34.

Welch, E.B. & Perkins, M.R. (1977) Oxygen deficit rate as a trophic state index. Technical Report, Dept. of Civil Engineering, University of Washington, Seattle, WA, 12 pp.

Wetzel, R.G. (1983) *Limnology* (2nd edition). Saunders College Publishing, Philadelphia, PA, p. 737.

# 15 Models of Lakes and Reservoirs

SVEN-ERIK JØRGENSEN

## 15.1 INTRODUCTION

Models are increasingly used in environmental management and in research, mainly because they are the only means by which it is possible to relate quantitative impact upon an ecosystem, and the consequences for the state of that system. As models of aquatic ecosystems in particular have been developed during the past few decades, it is not surprising that they have found a wide application in lake and reservoir management.

The idea behind the use of ecological management models is demonstrated in Fig. 15.1. The impact of urbanisation and technological development on nature is increasing. Energy and pollutants are released into ecosystems, where they may cause more rapid growth of algae or bacteria, damage to species, and/or change the entire ecological structure. Ecosystems are extremely complex, and so to predict the environmental effects of such emissions is an overwhelming task. It is here that models come into play. With sound ecological knowledge, it is possible to isolate/highlight those features of the ecosystem affected by the pollution problem under consideration. Results can be used to select the remedy best suited to the solution of specific environmental problems, including legislation, or reducing or eliminating the perturbation.

Figure 15.1 represents the idea behind the introduction of ecological modelling as a management tool in about 1970. Modern environmental management is more complex and must apply environmental technology in combination with environmental legislation. Technology is particularly important when trying to solve the problems of non-point or diffuse pollution, mainly originating from agriculture. Before about 1980, the importance of such pollution was hardly acknowledged. Global environmental problems also play a more important role today than 25 years ago. Abating global warming and the depletion of the ozone layer are widely discussed topics and several international conferences at governmental level have taken the first steps toward the use of international standards to solve these crucial problems. Figure 15.2 attempts to illustrate the more complex picture of environmental management today.

Ecological models focus only on those objects of interest to the problem under consideration. To include too many 'irrelevant' details would disturb the main objectives of the model. In that the model is selected according to desired goals, many different ecological models of the same ecosystem may be developed.

The field of ecological and environmental modelling has changed rapidly during the past three decades, owing essentially to three factors:
1 the development of computer technology, which has enabled us to model very complex mathematical systems;
2 an increased understanding of the social context of pollution problems, including the point that complete elimination of pollution ('zero discharge') is not feasible, but that a proper pollution control with the limited economical resources available requires serious considerations of the influence of pollution impacts on ecosystems;
3 our knowledge of environmental and ecological problems has increased significantly.

We have particularly gained more understanding of the quantitative relations and the relations between ecological properties and environmental factors in ecosystems.

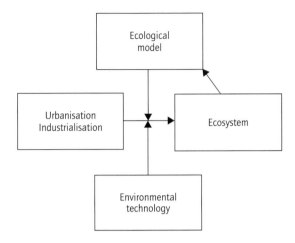

Fig. 15.1 Relationships between environmental science, ecology, ecological modelling and environmental management and technology.

## 15.2 MODELS AS A MANAGEMENT TOOL

Management problems can often be formulated as follows: if certain forcing functions (management actions) are changed (**varied**), what will be the influence on the ecosystem? The model is used to address this question, by attempting to predict which factors will change in the system when forcing functions are varied with space and time.

The term **control functions** is used to denote those forcing functions, such as consumption of fossil fuel, regulation of water level in a river by a dam, discharge of pollutants or fishery policy, which can be controlled. Control models therefore differ from other similar models by incorporating the following two elements:

1 a quantitative description of control processes;

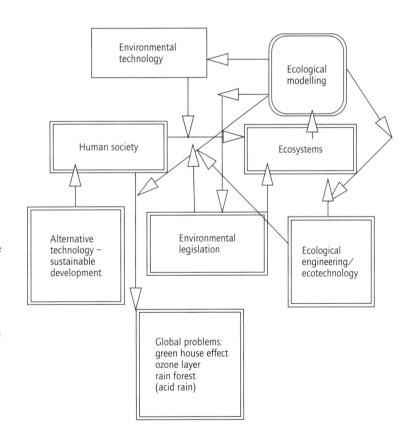

Fig. 15.2 The idea behind the use of environmental models in environmental management. The environmental management of today is very complex and must apply environmental technology, alternative technology and ecological engineering or ecotechnology. In addition, global environmental problems play an increasing role. Environmental models are used to select environmental technology, environmental legislation and ecological engineering.

2 a formalisation of objectives and evaluation of achievements.

The difference between control models and other environmental management models can best be illustrated by an example. A simple eutrophication model can still be used as a management model. If we determine its model response to various nutrient input changes, we obtain corresponding scenarios as model output. The model is used as an environmental management tool, but it is not a control model, in which more specific goals (e.g. a given lake transparency within a certain time) are formulated in order to achieve this level of control. We must introduce a variable input of nutrients into the model and find the relation between these.

Managers find, directly from this relationship, which inputs they must include in order to achieve their goal. They can then select which model they prefer, using ecological–economic criteria. Some additional equations or concepts need to be introduced in order to construct a control model. In many cases this is quite feasible, but adds to the complexity. In cases where control functions can be varied continuously, the advantages of control models are often sufficient to justify the additional complexity. If only a few constraints are available, it hardly pays to construct a control model.

In the case of eutrophication, only a few methods are available with which to reduce nutrient inputs. The management problem is which to select. Such questions may be answered simply by comparison of corresponding scenarios. If objectives are multiple, the goals formulated might not all be achieved simultaneously. Some might even be contradictory. Several methods in operational research, linear transformation, use of control indices, use of metrics in goal function space, Pareto methods, etc., are available to solve such multigoal problems. Nevertheless, final selection of a control function may ultimately be determined by subjective criteria, such as aesthetics, which cannot be formulated. The final decision is to a considerable extent political.

A further step in complexity involves construction of ecological–economic models. As we gain experience, more and more of these will be developed. It is often feasible to find a relation between a control function and the economy, but it is in most cases quite difficult to assess a relationship between the economy and the ecosystem state. What, for instance, are the **economic** advantages of increased transparency? Ecological–economic models are useful in some cases, but should be used with much precaution. The relationship between economy and environmental conditions should be critically evaluated before the results are applied.

This discussion could give the impression that environmental models are always more complex than scientific models. This is not the case. The objectives of environmental management models are often more clearly formulated than those of scientific models, which might render it more easy to select the level of complexity. Knowledge as to the required predictive value of environmental management models might also enable modellers to reduce their complexity, and the **scientific** use of models implies that modellers rigorously question any possibilities of complexity reductions.

Data collection is the most costly part of model construction. For many lake models it has been found that this activity amounts to 80–90% of total modelling costs. Complex models require far more data than simple ones. Selection of complexity of environmental management models therefore should be related closely to the environmental problem to be solved. It is therefore not surprising that the most complex environmental management models have been developed for large ecosystems, where economic involvement is great.

The predictive capability of environmental models can always be improved in a specific case by expansion of the data collection programme, and by increased complexity, provided, of course, that modellers are sufficiently skilled to know in which direction to develop the programme use. The relation between the economics of a project and the accuracy of a model is somewhat as shown in Fig. 15.3. The discrepancy between the model and reality diminishes for each extra monetary unit invested in the project. It is also clear from the shape of the curve that error will never be completely eliminated in all model predictions; there

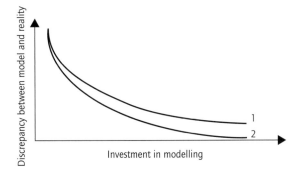

**Fig. 15.3** The more modellers invest in a model and in data collection, the closer they will come to realistic predictions, but they will always gain less accuracy for the next monetary unit invested and will never be able to provide completely accurate predictions. Curve 1 represents a less experienced modeller, whereas curve 2 gives the same relationship for a more experienced operator.

is a standard deviation. This point is not surprising to scientists, or economists, but it is often not realised by decision makers, to whom modellers present their results.

Validation, if carried out properly, can be used to determine the standard deviation of the model. Results used in environmental decisions should always be accompanied by an indication of the standard deviation on the prognosis, and it is important to clarify the meaning of this statistic to decision makers. Modellers should even give recommendations as to how to use results and standard deviations in their proper context. Decision makers have often used standard deviations wrongly, for example as numbers which indicate the extent to which costs could be reduced without any effect on environmental quality.

The onus is therefore upon modellers to explain to decision makers the consequences of various choices. Standard deviations of prognosis for an environmental management model cannot, however, always be translated into probabilities, because we do not know the probability distribution. It is possible, however, to use the standard deviation qualitatively or semiquantitatively, and to translate by use of words the meaning of the results. Civil engineers have succeeded in convincing their decision makers. Why should the environmental modeller not be able to do the same?

It is often useful to address environmental problems in the first instance by use of simple models. These require very few data and can give the modeller and the decision maker some preliminary results. If the project is ended at this stage, a simple model is still better than none at all, because it will at least give a preliminary assessment of the problem.

Simple models are useful starting points for construction of more complex models. In many cases construction is carried out as an iterative process. Figure 15.4 shows how step-wise development of a complex model might take place. The first step is, as mentioned above, the conceptual model, which is used to conduct a survey of the processes and state variables in the system. The next stage is development of a simple model, which is calibrated and validated. This is used to establish a data collection programme for a more comprehensive study, the outcome of which is close to the final version of the model. Often, however, as also shown in Fig. 15.4, the third stage will reveal some weaknesses, which the fourth version attempts to eliminate.

This seems at first glance to be a very cumbersome procedure, but as data collection is the most expensive part of modelling, fewer resources will be required to construct a preliminary model to use for optimisation of the data collection programme. It might be added that Fig. 15.4 can be considered a formalisation of the iterative procedure which many modellers are forced to use anyhow, and that planning of these steps at an initial phase of the project is always advantageous.

A simple mass balance is also recommended for use in biogeochemical models. The balance will highlight possibilities for reduction or increase in lake concentration, which is a crucial factor in environmental management.

Point sources (Table 15.1) are usually more easy to control than artificial non-point sources, which again usually can be controlled more easily than natural inputs. We may also distinguish between local, regional and global sources. Using the mass

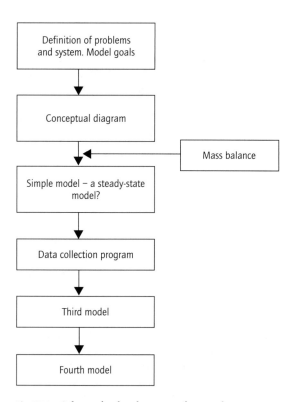

**Fig. 15.4** Scheme for development of a complex management model.

**Table 15.1** Examples of sources

| Source | Examples |
|---|---|
| Point sources | Wastewater (N, P, BOD), $SO_2$ from fossil fuel and discharge of toxic substances from industries |
| Artificial, non-point sources | Agricultural use of fertilisers, deposition of lead from vehicles and contaminants in rain water |
| Non-point natural sources | Run-off from natural forests and deposition on land of salt originated from the sea |

balance approach, it is possible to decide which sources to concentrate on first. If, for instance, a regional non-point source is greatest, it might be pointless to reduce small, local, point sources first (unless, of course, this might exert some political influence on the regional decisions) or unless they are especially ecologically dangerous.

Modellers and decision makers should understand each other. Ideally, decision makers should follow the process of model construction from the outset, in order to be acquainted with the strength and shortcomings of the model. It is also important that modellers and decision makers formulate the objectives of the model, and interpret the results together. Holling (1978) has demonstrated how such teamwork develops and runs, phase by phase.

Communication between decision makers and modellers can be facilitated in many ways. If models are built as menu systems (e.g. Mejer 1983), it may be possible in a few hours to teach decision makers how to use them. This will, of course, increase their understanding of the model and its results. If an interactive approach is applied (e.g. Fedra 1984), it is possible for decision makers to visualise a wide range of possible decisions. The effect of this approach is increased by the use of various graphic methods to give the best possible decision as to what happens in the system using various management strategies.

Model results should never be used alone. Rather, they should be considered a useful tool in the decision-making process. This implies that modelling results should be presented clearly and illustratively, and considered an important component in the discussion of which decision to select. 'Side effects', i.e., the interpretation of the prognosis, implications of the prognosis accuracy, etc., are elements which might be considered in such a discussion.

A wide range of environmental problems has been modelled during the past 15–20 years, and all the models have provided important assistance to the decision makers. With the rapid growth in the use of environmental models, the situation will only improve in the nearest future. Obviously, we have not achieved the same level of experience for all environmental problems.

The use of models in environmental management is growing. They are widely used in several European nations, in North America and in Japan, but more and more countries are applying models produced by environmental agencies. Through the journal *Ecological Modelling*, and the ISEM (International Society for Ecological Modelling), it is possible to follow progress in the field. This infrastructure facilitates communication, and accelerates exchange of experience and thereby growth of the entire field of ecological modelling. It is seldom possible to transfer a model from one case study to another, and it even may be difficult to transfer it from one computer to another, unless it is exactly the same type, but it is often a great help to obtain the experience gained by somebody else in modelling a similar situation at some other place in the world. Jørgensen *et al.* (1995) contains information on more than 400 models focusing on a wide range of environmental problems.

## 15.3 METHODOLOGY

It is clear from the above that selection of model complexity is a matter of balance. On the one hand, it is necessary to include the state variables and the processes essential to the problem in focus. On the other, it is important not to make the model more complex than the data set can bear. Current knowledge of processes and state variables, together with the data set, will determine model complexity. If knowledge is poor, the model will lack detail and possess relatively high uncertainty. If our knowledge of the problem we wish to model is profound, we can construct a more detailed model, with a relatively lower uncertainty. Many researchers claim that models cannot be developed before we attain a certain level of knowledge, and that it is unwise to attempt to model 'data poor' contexts. However, models can always assist the researcher by synthesis of the present knowledge, and by visualisation of the system. Researchers must, of course, always present the shortcomings and the uncertainties of models, and not try to pretend that they are complete pictures of reality. Models will often be fruitful instruments to test, but only if their incompleteness is fully acknowledged.

Models have always been applied in science. The difference is that today, with computers, we are able to develop very complex models. However, the temptation is to construct models which are too complex. It is easy to add more equations and state variables to the program, but much harder to obtain the data needed for calibration and validation. Even if knowledge is very detailed, we will never be able to develop models capable of accounting for the complete input–output behaviour of real ecosystems (Zeigler 1976). Such models are referred to by Zeigler as 'the base model'. These are very complex, and require such a great number of computational resources that they are almost impossible to use. The base model of a problem in ecology will never be fully known, because of the complexity of the system and the impossibility to observe all states. Up to a point, a model may be made more realistic by adding ever more connections. Additions of new variables after that point do not contribute further to improved simulations; on the contrary, addition of more implies more uncertainty, because of the lack of information about the flows which the parameters quantify.

Given a certain amount of data, the addition of a new state beyond a certain complexity does not add to our ability to model the ecosystem, but only adds to unaccountable uncertainty. Costanza & Sklar (1985) examined 88 different models. They were able to show that the more theoretical discussion behind Fig. 15.5 is actually valid in practice. The relationship between knowledge gained through a model and its complexity is shown for two levels of data quality and quantity. The question under discussion can be formulated with relation to this figure. How can we select the complexity and the structure of the model in order to assure the optimum knowledge gained or the best answer to the question posed?

Selection of the appropriate complexity is of great importance in environmental and ecological models (Jørgensen 1994). Selection will always demand that the application of methods is combined with knowledge of the system being modelled. The methods must work hand in hand with an intelligent answer to the question: **Which components and processes are most important for the problem in focus?** The conclusion is therefore:

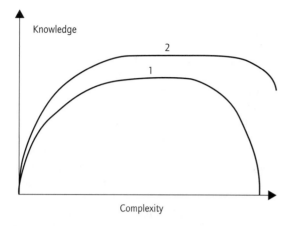

Fig. 15.5 Knowledge plotted versus model complexity measured, for instance, by the number of state variables. The former increases up to a certain level, but increased complexity beyond this point will not add to the knowledge gained. At a certain level the knowledge might even decline owing to uncertainty caused by too many unknowns. Curve (2) corresponds to a data set which is more comprehensive or of better quality than (1). Therefore the knowledge gained and the optimum complexity achieved is higher for data set (2) than for (1). (Reproduced from Jørgensen 1994.)

**know your system and your problem before you select your model, including its complexity**. Although we can never know everything needed to make a complete model (i.e. inclusion of all details) we can produce workable models which expand our knowledge of ecosystems, particularly of their properties as systems. Ulanowicz (1979) points out that the biological world is a sloppy place. Very precise predictive models will inevitably be wrong. It would be more fruitful to build a model that shows the general trends, and takes into account the probabilistic nature of the environment.

All in all models should be considered as tools with which to take an overview of complex systems and to acquire a view of properties at the system level (Jørgensen 1986). Even a few interactive state variables make it impossible to see how the system reacts to perturbations or other changes.

Only two possibilities exist by which to circumvent this dilemma: either to limit the number of state variables, or to describe the system by the use of holistic methods and models, preferably by use of higher level scientific laws. **The trade off for the modeller is between knowing much about little or little about much**.

More complex models require more data and knowledge about the lake or reservoir in focus. They imply higher costs, but they may be justified by the urgency of the eutrophication problem. It is better, however, to proceed step-wise toward complex models, and to learn from simpler examples before complex ones are constructed. Such development could proceed along the following lines.

**1** Determining mass balances and lake morphology is the first natural step. A general assessment of the sources of inputs, the area of the lake, its depth and retention time will highlight the main problems of the lake or reservoir under investigation.

**2** Set up a simple model, for instance a Vollenweider plot (Vollenweider 1975) for a eutrophication problem. This helps evaluate possible alternative solutions to the problem. On a simple basis, which nutrient is limiting? What are its main sources? What will be the impact of treating or diverting wastewater? With such plots, it may be feasible to define initial hypotheses, and even to make coarse prognoses as to solutions. They will probably also indicate which processes further work should concentrate on—whether point sources or non-point sources are more important and therefore which of the two it is necessary to determine with greater accuracy. They may also indicate the importance of storage, or whether retention time is an important factor. It is of course important to compare these initial hypotheses and prognoses with results, obtained by use of more elaborate models at a later stage. If prognoses are not confirmed, it should be possible to explain discrepancies.

**3** Set up a conceptual diagram. The procedure for development of models implies that goals are defined and the model confined in time and space. The next step is to develop a conceptual diagram, containing all important state variables and processes. Selection of major components is made on the basis of the mass balance (i.e., processes are selected because of their relative contribution). For a eutrophication model, this means that nutrient balances or at least the limiting nutrient must be included. It is also necessary to consider the characteristic features of the lake or reservoir. Which are the main biological species? What are its hydrological properties? It is generally important to consider model complexity at this stage, although this should also be reconsidered when the model has been calibrated and validated.

**4** Assess the data requirement. This step is also part of normal procedure. However, economic considerations should be added at this stage. The scale on which a given problem operates (knowing which economic implications should be addressed) should be chosen. It is also possible to assess the optimum amount of funds which may be invested in a modelling project, including data collection, if we know the economic parameters of the project (e.g. wastewater treatment costs, the economic value of the lake or reservoir as a recreational area (though this is often an open question)). **Experience has shown that 2–10% of investment in wastewater treatment may be used for development of models to select the appropriate management strategy.** From this coarse rule it is then possible to assess whether the model selected under points 3–4 can be realised from an economic point of view, whether it is possible to expand the project, or whether it is necessary to scale it down.

When the model has been developed, calibrated and validated, we know its standard deviation. We can therefore test whether it is sufficiently accurate for choosing between two or more management strategies. Modelling is an iterative process; therefore it cannot be ruled out that a model developed in a particular context cannot be used, and that a more complex model is needed.

It is important to confront managers with this problem during the initial phase of model development. Managers should therefore be prepared to accept that projects should be enlarged, or cannot be continued, if model accuracy is insufficient. An experienced modeller will in most cases be able to estimate the required complexity at an early stage, but unforeseen difficulties may lead to inaccuracy, even for experienced modellers. Generally, a continuing dialogue between the modeller and the manager is recommended. It is absolutely necessary that managers are aware of the strengths and shortcomings of modelling, including the expected accuracy of the focal model. Furthermore, modellers should attempt to demonstrate to managers the simulation results obtained from different management strategies. This will initiate the necessary discussion between the manager and the modeller on all the possibilities that the model can offer, including alternative changes in the management strategy.

**5** Follow strictly the generally recommended modelling procedure. The procedure (for details, see Jørgensen 1994) should be followed very carefully, as already pointed out. All steps should be included. Verification is particularly important. Modellers should give themselves sufficient time to test the model, and whether its parameters are consistent with the literature. As will be discussed in the last section of this chapter, 'ecological constraints' should be introduced. The ability of ecosystems to incorporate other species better fitted to the prevailing conditions may require the use of a structural dynamic model (see section 15.4). The use of a model in order to optimise data collection is often recommendable, particularly when this is expensive. Kettunen (1988) has presented a means to assess the value level of data, which of course means that the first complex model becomes provisional, and used only to find the data collection optimum. When additional data are available, the final model can be developed, using all the data available for model development.

Can models developed in one case be used without change in other case studies? Experience shows that rather simple models can be used more generally than more complex examples. Simple models contain a description of the processes characteristic of all lakes and reservoirs (e.g. nutrient uptake by phytoplankton, mineralisation of detritus). More complex models inevitably include more site-specific process descriptions which may not be applicable to all lake and reservoir case studies. Therefore, more complex models will in most instances need to be modified from case to case. This point is illustrated in Table 15.2, where the experience gained in the general use of a eu-

Table 15.2  Survey of eutrophication studies based upon the application of a modified model of Glumsø, Denmark

| Ecosystem | Modification | Level* |
|---|---|---|
| Glumsø, version A | Basic version | 6 |
| Glumsø, version B | Non-exchangeable nitrogen | 6 |
| Ringkøbing Firth | Boxes, nitrogen fixation | 5 |
| Lake Victoria | Boxes, thermocline, other food chain | 4 |
| Lake Kyoga | Other food chain | 4 |
| Lake M. Sese Seko | Boxes, thermocline, other food chain | 4 |
| Lake Fure | Boxes, nitrogen fixation, thermocline | 3 |
| Lake Esrøm | Boxes, Si-cycle, thermocline | 4 |
| Lake Gyrstinge | Level fluctuations, sediment exposed to air | 4–5 |
| Lake Lyngby | Basic version | 6 |
| Lake Bergunda | Nitrogen fixation | 2 |
| Broia Reservoir | Macrophytes, two boxes | |
| Lake Great Kattinge | Resuspension | 5 |
| Lake Svogerslev | Resuspension | 5 |
| Lake Bue | Resuspension | 5 |
| Lake Kornerup | Resuspension | 5 |
| Lake Balaton | Adsorption to suspended matter | 2 |
| Roskilde Fjord | Complex hydrodynamics | 4 |
| Stadsgraven, Copenhagen | Four to six interconnected basins | 5 (level 6: 93) |
| Internal lakes of Copenhagen | Five to six interconnected basins | 5 |

* Level 1, conceptual diagram selected; level 2, verification carried out; level 3, calibration using intensive measurements; level 4, calibration based on a longer period (at least 1 yr); level 5, validation; level 6, prognosis validation.

trophication model with 17–20 state variables is shown. As can be seen, it was necessary to modify the model for almost every new case study. The core, however, was used in all cases. To summarise: the simpler the model, the more generally it can be applied. The more complex a model, the more modellers should be prepared to modify it from case to case.

The problems of interest for lake management (Table 15.3) are (i) eutrophication, (ii) ecotoxicological effects, (iii) lake acidification, (iv) fishery management, (v) oxygen concentration of lake water, (vi) bacteriological quality of lake water, (vii) hydrodynamic problems related to discharge of water, or its use for cooling or production of drinking water, and (viii) management of wetlands surrounding the lake. Most of these have been modelled intensely, or very intensely. As outlined

Table 15.3  Notional scale of availability of lake models for different environmental problems

| Problem | Modelling effort* |
|---|---|
| Eutrophication | 5 |
| Heavy metal pollution | 4 |
| Pesticide pollution | 4 |
| Acid rain | 4–5 |
| Fishery | 3 |
| Oxygen balance | 4–5 |
| Wetlands | 4 |
| Bacteriological water quality | 3–4 |
| Hydrodynamic problems | 5 |

* 5, very intense modelling effort; 4, intense effort; 3, some effort; 2, a few models fairly well studied; 1, one detailed study or a few not sufficiently well calibrated and validated models; 0, no modelling effort.

above, selection of appropriate complexity, characteristics of the ecosystems, and quality and quantity of available data are crucial. It is therefore beneficial that lake ecosystems have been modelled intensely, and a wide experience of different complexities is available.

## 15.4 EUTROPHICATION MODELS

Eutrophication models are represented by a particularly wide spectrum of complexity. Table 15.4 indicates the characteristic features of each model, the number of state variables used, the range of nutrients studied, the number of segments and layers in the model, whether constant stoichiometric or independent nutrient cycles were applied, whether the model has been calibrated and validated, and the number of case studies to which it has been applied. Particularly for more complex models, it is assumed that some modifications from case to case will be made, as site-specific properties will be reflected in the model (see also the discussion above on the generality of models). Many of the processes incorporated in the models (for instance, mineralisation of detritus) are described as

Table 15.4 Selected eutrophication models

| Originator/ name | Number of state variables per layer or segment | Nutrients | Segments | Dimension (D) or Layer (L) | CS or NC* | C and/or V† | Number of case studies |
|---|---|---|---|---|---|---|---|
| Vollenweider | 1 | P(N) | 1 | 1L | CS | C+V | Many |
| Imboden | 2 | P | 1 | 2L, 1D | CS | C+V | 3 |
| O'Melia | 2 | P | 1 | 1D | CS | C | 1 |
| Girl | 3 | P | 1 | 1–2D | CS | C+V | Many |
| Larsen | 3 | P | 1 | 1L | CS | C | 1 |
| Lorenzen | 2 | P | 1 | 1L | CS | C+V | 1 |
| Thomann 1 | 8 | P,N,C | 1 | 2L | CS | C+V | 1 |
| Thomann 2 | 10 | P,N,C | 1 | 2L | CS | C | 1 |
| Thomann 3 | 15 | P,N,C | 67 | 2L | CS | – | 1 |
| Chen and Orlob | 15 | P,N,C | Several | 2L | CS | C | Minimum of 2 |
| Patten | 33 | P,N,C | 1 | 1L | CS | C | 1 |
| Di Toro | 7 | P,N | 7 | 1L | CS | C+V | 1 |
| Biermann | 14 | P,N,Si | 1 | 1L | NC | C | 1 |
| Canale | 25 | P,N,Si | 1 | 2L | CS | C | 1 |
| Jørgensen | 17 | P,N,C | 1 | 1–2L | NC | C+V | 21 |
| Cleaner | 40 | P,N,C,Si | Several | Several L | CS | C | Many |
| Lavsoe | 7 | P,N | 3 | 1–2L | NC | C+V | 25 |
| Aster/Melodia | 10 | P,N,Si | 1 | 2L | CS | C+V | 1 |
| Baikal | >16 | P,N | 10 | 3L | CS | C+V | 1 |
| Chemsee | >14 | P,N,C,S | 1 | Profile | CS | C+V | Many |
| 3DWFGAS | ≥70 | P,N,C,S | Many | Many | CS | C+V | 135 |
| Reward | 16 | P,N | 1 | Profile | CS | C+V | 5 |
| Mahamah | 8 | C,P,N | 1 | 1–2L | CS | C+V | 2 |
| Lake ecosystem | 11 | C,P,N | 1 | 1 | CS | C+V | 1 |
| Minlake | 9 | P,N | 1 | 1 | CS | C+V | >10 |
| Salmo | 17 | P,N | 1 | 2L | CS | C+V | 16 |

\* CS, constant stoichiometric; NC, independent nutrient cycle.
† C, calibrated; V, validated.

first-order reactions. Most use Michaelis–Menten equations in order to account for the relationship between growth and resources (nutrients or food).

Three possible core **submodels**, which represent typical considerations in the selection of the complexity for eutrophication models (Jørgensen 1976), and which should always be considered for inclusion, will be presented here and their applicability discussed. However, characteristics of individual case studies, as discussed above, should determine whether these submodels should actually be included or not.

### 15.4.1 Independent nutrient cycles

Application of independent nutrient cycles increases the complexity of the model significantly. It requires that nitrogen, phosphorus, carbon and perhaps also silica are included as state variables in each trophic level. In most models of this type, however, only one state variable is considered for zooplankton and fish. Application of independent nutrient cycles also implies that the growth of phytoplankton is described as a two-step process:
1 uptake of nutrients in accordance with Monod's kinetics;
2 growth determined by the internal concentrations of nutrients.

This complication obviously requires that data are of sufficient quality and quantity. Di Toro (1982) has shown that the application of independent nutrient cycles is particularly important in the case of shallow, very productive lakes, whilst it can be omitted for deep less productive or unproductive bodies.

### 15.4.2 Sediment–water submodel

It is important to describe quantitatively the processes determining mass flows of nutrients from sediment to water, particularly when this process is significant compared with the other focal mass flows. As the amount of nutrients stored in the sediment is most significant for shallow lakes, more detailed description, as presented below, should always be included in models of such lakes.

The sediment–water submodel attempts to answer the following question: To what extent (and when) will compounds accumulated in the sediment be redissolved in the lake water? Exchange processes of phosphorus and nitrogen between mud and water have been extensively studied, as these are important for eutrophication of lakes. Chen & Orlob (1975) ignored exchange of nutrients between mud and water and, as pointed out by Jørgensen (1976), this will inevitably give a false prognosis. Ahlgren (1973) applied a constant flow of nutrients between sediment and water. Dahl-Madsen & Strange-Nielsen (1974) used a simple first-order kinetic to describe the exchange rate.

A more comprehensive submodel (Fig. 15.6) for the exchange of phosphorus has been developed by Jørgensen (1976). Settled material ($S$) is divided into $S_{detritus}$ and $S_{net}$, the first being the amount of the element mineralised by microbiological activity in the water body, and the latter the amount of that material actually transported to the sediment. The $S_{net}$ material can also be divided into two flows:

$$S_{net} = S_{net,s} + S_{net,e} \qquad (15.1)$$

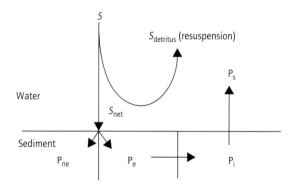

**Fig. 15.6** Settlement, $S$, divided into $S_{detritus}$ and $S_{net}$; $P_{ne}$, non-exchangeable phosphorus in unstabilised sediment; $P_e$, exchangeable phosphorus in unstabilised sediment; $P_i$, phosphorus in interstitial water; $P_s$, dissolved phosphorus in water.

where $S_{net,s}$ is flow to the stable non-exchangeable sediment and $S_{net,e}$ is mass flow to the exchangeable unstable sediment.

Correspondingly, non-exchangeable and exchangeable phosphorus concentrations, $P_{ne}$ and $P_e$, both based on the total dry matter in the sediment, can also be separated via analysis of phosphorus influx to sediment cores. The application of $^{210}$Pb dating, for example, is a fast and reliable method (Appleby 2001).

Exchangeable phosphorus is then converted to detritus. A first-order reaction gives a reasonably good description of this process. Finally, interstitial phosphorus, $P_i$, is transported by diffusion from the pore water to the lake water. This process, which has been studied by Kamp-Nielsen (1974), can be described by means of the following empirical equation (valid at 7°C):

$$\text{release of P} = 1.21(P_i - P_s) \\ -1.7 \left( \text{mg P m}^{-2} \, 24 \, \text{h}^{-1} \right) \quad (15.2)$$

where $P_s$ is the dissolved phosphorus in the lake water.

This submodel of water–sediment exchange was validated in three case studies (Jørgensen 1976), by examining sediment cores in the laboratory. Kamp-Nielsen (1975) has added an adsorption term to these equations. Figure 15.7 shows a sediment profile from these studies. The interpretation of the profile to be used in the model to distinguish between exchangeable and non-exchangeable phosphorus is illustrated on the figure. A similar submodel for nitrogen release has been set up by Jacobsen & Jørgensen (1975). Release from sediment is expressed as a function of nitrogen concentration in the sediment and temperature, considering both aerobic and anaerobic conditions.

### 15.4.3 Grazing of phytoplankton by zooplankton

Grazing of phytoplankton by zooplankton $(Z)$ and predation of zooplankton by fish $(F)$ are both expressed by a modified Monod expression, which

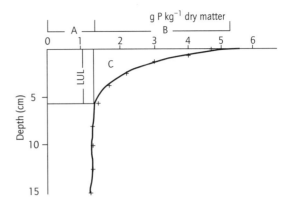

Fig. 15.7 Analysis of a core from Lake Esrøm. Value of mg P g$^{-1}$ dry matter is plotted against depth. The area C represents exchangeable phosphorus; f = (B·A − 1); LUL is the unstabilised layer.

considers a threshold concentration, $KT$, below which grazing or predation does not occur. For grazing the expression is as follows:

$$\mu Z = \mu Z_{max} \times (\text{Phyt} - KT)/(\text{Phyt} + KM) \quad (15.3)$$

where KM is the Michaelis-Menten constant. These expressions are according to Steele (1974).

Carrying capacity of zooplankton must often be introduced in order better to simulate changes in zooplankton and phytoplankton. Carrying capacities are often observed in ecosystems, but their necessity in this case may be due to oversimplified simulation of the grazing process. Phytoplankton might not be grazed by all zooplankton species present, and some might use detritus as a food source. The zooplankton growth rate, $m_Z$, is computed in accordance with these modifications as

$$m_Z = m_{Z\,max} \times \text{FPH} \times \text{FT2} \times \text{F2CK} \quad (15.4)$$

where FPH corresponds to the expression in eqn. (15.3), FT2 is a temperature regulation expression and F2CK accounts for carrying capacity:

$$\text{F2CK} = (\text{CK} - \text{ZOO})/\text{CK} \quad (15.5)$$

where CK is the carrying capacity.

If there are insufficient data to include in the above processes, it is recommended that they be obtained by means of an intensive measuring period. This should provide high quality data from the period during which the eutrophication processes are most dynamic.

Several sets of measurements should be recorded per week during the spring and summer bloom period. This then can be used to improve parameter estimation, which is often a focal problem in development of ecological models. Experience suggests the following advantages.

**1** Different optional expressions of simultaneous limiting factors have been tested. Only two expressions give an acceptable maximum growth rate for phytoplankton and an acceptably low standard deviation. These are (i) multiplication of the limiting factors, and (ii) averaging the limiting factors.

**2** The expression used for the influence of temperature on phytoplankton growth gave unacceptable parameters with too high a standard deviation. After an intensive measuring programme a better expression, $\exp[A\,(T - T_{opt})/(T_{max} - T_{opt})]$, has been determined.

**3** It has been possible to improve parameter estimation, which gives, for some, more realistic values. Whether this would produce improved validation when observations from a period with drastic changes in nutrient loading cannot be stated.

**4** Other expressions applied for process descriptions have been confirmed.

It is important to validate models against independent set of measurements. Table 15.5 gives results of a typical validation of a model with a medium to high complexity, developed on the basis of a good data set. Figure 15.8 shows the conceptual diagram for the nitrogen cycle in this model. In Table 15.5, $R$ is the standard deviation for the mean value of the state variables, and $A$ is the standard deviation for the maximum value: $Y$, $R$ and $A$ give the errors in relative terms. By multiplication by 100, the errors are obtained as percentages. The standard deviation, $Y$, for all measured state variables is 16. As the standard deviation for a comparison of $n$ sets of model values and measured values is $\sqrt{n}$ times smaller, and $n$ is in the order

**Table 15.5** Numerical validation of a medium to complex model

| Validation criterion | State variable | Value |
|---|---|---|
| Y | All | 0.16 |
| R | Ptotal (P4) | 0.18 |
| R | Psoluble (PS) | 0.16 |
| R | Ntotal (N4) | 0.02 |
| R | Nsoluble (NS) | 0.14 |
| R | Phytoplankton (CA) | 0.08 |
| R | Zooplankton (Z) | 0.20 |
| R | Production | 0.03 |
| A | Ptotal (P4) | 0.12 |
| A | Psoluble (PS) | 0.15 |
| A | Ntotal (N4) | 0.07 |
| A | Nsoluble (NS) | 0.03 |
| A | Phytoplankton (CA) | 0.15 |
| A | Zooplankton (Z) | 0.00 |
| A | Production | 0.08 |

of 200, the overall average modelling of the lake is given with a standard deviation of about 1%, which is very acceptable: $Y$ is, for instance, generally five to ten times larger for hydrological models (WMO 1975).

The relative error, $R$, is 3% for production, 8% for phytoplankton and 2% for nitrogen (again very acceptable); but that for total phosphorus is 18% and for zooplankton 20%, which can be considered too high. The relative error of maximum values, $A$, varies from 0% to 15%, which is acceptable. The ability to predict maximum production and maximum phytoplankton concentration is of special interest for eutrophication modellers. Here the relative errors are 8% and 15%, respectively, which are fully acceptable.

### 15.4.4 Example: development of eutrophication in Glumsø

A prognosis for development of eutrophication in Glumsø (cf. Table 15.5) was achieved by calculating simultaneous removal efficiencies for phosphorus, nitrogen or phosphorus and nitrogen. Validation of this prognosis is presented here in order to illustrate the reliability of well-developed

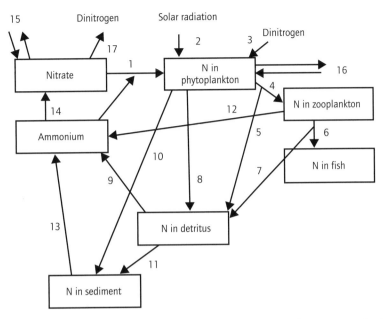

**Fig. 15.8** The conceptual diagram of the model applied as the basis for the case studies mentioned in Table 15.4, used for the validation presented in Table 15.5 and furthermore applied for the prognosis in Table 15.6 and the prognosis validation in Table 15.7. The model is based on an independent cycling of phosphorus, carbon and nitrogen, the grazing equations given above, and the sediment–water exchange processes illustrated in Fig. 15.7. The conceptual diagram shows only the nitrogen cycle. Similar cycles are used for phosphorus and carbon, although only phosphate and carbonate are considered. In the case of phosphorus, pore-water concentration is included in addition to the state variables of the nitrogen cycle (see Fig. 15.7).

eutrophication models. The characteristic features of the model are those mentioned above, i.e.,
- two-step description of phytoplankton growth, determined by internal concentrations of nutrients and application of independent nutrient cycles;
- the sediment–water exchange processes;
- grazing by zooplankton.

It was found that removal of nitrogen inputs to the lake would register little or no effect, whereas reduction of phosphorus loading would produce substantial lowering of phytoplankton concentration.

The results of two case studies are summarised in Table 15.6. In case A, treated wastewater concentration of total phosphorus was $0.4\,\text{mg}\,\text{L}^{-1}$, corresponding to about 92% removal efficiency. This could be achieved by proper chemical precipitation. In case B, the treated wastewater concentration is $0.1\,\text{mg}\,\text{P}\,\text{L}^{-1}$, corresponding to about 98% removal efficiency. This would require chemical precipitation in combination with, for example, ion exchange.

**Table 15.6** Predictions by means of a model in two cases for phosphorus concentration of treated wastewater (A: $0.4\,\text{mg}\,\text{P}\,\text{L}^{-1}$; B: $0.1\,\text{mg}\,\text{P}\,\text{L}^{-1}$)

|  | Third year | | Ninth year | |
|---|---|---|---|---|
|  | Case A | Case B | Case A | Case B |
| $\text{g cm}^{-2}\,\text{yr}^{-1}$ | 650 | 500* | 500 | 320* |
| Minimum transparency (cm) | 50 | 60 | 60 | 75 |

* An error of 3% on this value could be expected if the validation results hold (see $R$ for production in Table 15.5).

Table 15.7  Comparison of prognosis data and measured data

| Variable | | Prognosis (Case A, 92% P reduction) | Measurement approximately (88% reduction) |
|---|---|---|---|
| Minimum transparency (cm) | First year | 20 | 20 |
| | Second year | 30 | 25 |
| | Third year | 45 | 50 |
| g C 24 h m$^{-2}$ maximum | First year | 9.5 ± 0.8 | 5.5 ± 0.5 |
| | Second year: | | |
| | spring | 6.0 ± 0.5 | 11 ± 1.1 |
| | summer | 4.5 ± 0.4 | 3.5 ± 0.4 |
| | autumn | 2.0 ± 0.2 | 1.5 ± 0.2 |
| | Third year (spring) | 5.0 ± 0.4 | |
| Chlorophyll in spring maximum (mg m$^{-3}$) | First year | 750 ± 112 | 800 ± 80 |
| | Second year | 520 ± 78 | 550 ± 55 |
| | Third year | 320 ± 48 | 380 ± 38 |

As seen, lake-water quality improves significantly in accordance with the prognosis. Case B (98% removal) must be preferred. During the third year, case B would produce a reduction in biomass production from 1100 to 500 g C m$^{-2}$ yr$^{-1}$. Transparency would increase from 20 to 60 cm. The ninth year production would be 320 g C m$^{-2}$ yr$^{-1}$, which corresponds to that of a 'mesotrophic' lake. This is an acceptable improvement for a shallow lake situated in an agricultural area. The measure involving 98% phosphorus removal should therefore be recommended to the environmental authorities. Further improvements after 9 years should not be expected.

Diversion of wastewater was also considered but possessed the following disadvantages:
1 diversion is slightly more expensive than case B, taking into consideration interests, depreciation and running costs;
2 by diversion, phosphorus is **not** removed but only transported downstream to the Susan River, where its effects have not been considered;
3 the sludge produced at the biological treatment plant will be less valuable as a soil conditioner, because its phosphorus concentration will be lower than that produced when phosphorus removal at the treatment works is implemented;
4 fresh water is not retained in the lake, from where, after storage for some time, it could be reclaimed, if needed — water supply is not a problem in this area at present, but it is foreseen that it might be in 20–40 years time.

In spite of these scientific **and** economic arguments, the community, in this case, chose to convey wastewater to the Susan River, owing to a preference for traditional methods. The pipeline was constructed in 1980, and began operation in April 1981. This process has enabled a validation of the presented prognosis.

Glumsø is ideal for such studies, owing to its limited depth and size, but also because a reduced nutrient input to the lake 'could be foreseen'. The limited retention time (about 6 months) makes it realistic to obtain a validation of a prognosis within a relatively short time interval (a few years). On 1 April 1981, input of wastewater directly to the lake was discontinued. As the capacity of the local sewerage system is still too small, a minor input of mixed rainwater and wastewater is, however, from time to time received through an upstream tributary. Phosphorus loading is therefore not reduced by 98% but rather only by 88% (determined by a phosphorus balance). Case A (92% removal) may therefore be a more valid comparison.

During the third year after reduction in loading (1984), a pronounced effect in the lake was observed. Table 15.7 compares some of the most important prognoses, and also includes observations obtained during the first two months of that year. Errors are indicated as ± for gC 24 h m$^{-2}$, and chlorophyll maximum mg m$^{-3}$.

For the prognosis values, the results from Table 15.7 for production (8%) and phytoplankton concentration (15%) are used in order to determine standard deviations. For the measured values, a standard deviation of 10% is estimated.

Previously, *Scenedesmus* had been abundant in the lake, but once inputs of wastewater began this genus was replaced by diatoms, which are able to metabolise at lower optimum temperatures (and therefore to bloom earlier in the spring than *Scenedesmus*). This event explains the discrepancy between the prognosis and the outcome. Predictions might improve if it was possible to account for shifts in species composition. Results published by Jørgensen (1981, 1992a,b) and Jørgensen & Mejer (1981) indicate that this might be achieved by introduction of terms for maximum growth rate of phytoplankton (see section 15.6).

Since growth of diatoms depends greatly on uptake of silica, it is also necessary to introduce a silica cycle similar to that of nitrogen in Fig. 15.8. Production and chlorophyll are predicted accurately, except for during the spring production during the second year (Table 15.7). Prediction of minimum transparency, with a difference of 5 cm or less, is acceptable (Table 15.7). Predicted trends in nutrient concentrations accord with measured values.

This case study shows that it is possible to make reliable prognoses by the use of eutrophication models, provided that data of sufficient quality and quantity are available. On the other hand an attempt to develop eutrophication models should always be made, even when available data are not sufficient to use complex types. Results can then be used to compare different management strategies, with an indication of the uncertainty resulting from the validation.

If uncertainties are not acceptable, the model can be used to assess the weak points, and where amendments should be introduced to improve the model results. Models should be considered as tools able to synthesise knowledge of lake ecosystems. By following the recommendations given in this section it should be possible to obtain the best possible model under the given circumstances.

## 15.5 ECOTOXICOLOGICAL MODELS

Ecotoxicological models have emerged during the past 20 years as a result of increasing interest in environmental management of toxic substance pollution. Such models attempt to assess the fate and/or effect of toxic substances in ecosystems. Toxic substance models are usually biogeochemical, in that they attempt to describe mass flows of toxic substances, although some population dynamics models include the influence of toxic substances on the birth rate, growth rate and/or the mortality, and therefore should be considered here (Jørgensen *et al.* 1991, 2000).

Toxic substance models share certain characteristic properties.

**1** The need for parameters to cover all possible toxic substance models is great, and general estimation methods are therefore used widely. Methods based upon the chemical structure of given chemical compounds (the so-called QSAR (quantitative structure analysis research) and SAR methods) have been developed.

**2** The safety margin should be high, when, for instance, expressed as the ratio between the actual concentration and the concentration that gives undesired effects.

**3** An effect component which relates output concentration to its effect should be included. It is easy to include an effect component in the model, but it is often a problem to find a well-examined relationship to base it on.

**4** There is a tension between the need for simple models and the limited knowledge of processes, parameters and sublethal, antagonistic and synergistic effects.

Because of the character of ecotoxicological models, it is recommended that a few questions are clarified before entering the modelling procedure:

**1** obtain the best possible knowledge about the processes in which the toxic substance under consideration participates;

Table 15.8  Examples of ecotoxicological lake models

| Toxic substance model class | Model characteristics | Reference |
|---|---|---|
| Cadmium | Food chain similar to a eutrophication model | Thomann et al. 1974 |
| Mercury | Six state variables: water, sediment, suspended matter, invertebrates, plant and fish | Miller 1979 |
| Vinyl chloride | Chemical processes in water | Gillett et al. 1974 |
| Methyl parathion | Chemical processes in water and benzothiophene-microbial degradation, adsorption, 2–4 trophic levels | Lassiter 1978 |
| Methyl mercury | A single trophic level: food intake, excretion metabolism growth | Fagerstrøm & Aasell 1973 |
| Heavy metals | Concentration factor, excretion, bioaccumulation | Aoyama et al. 1978 |
| Pesticides in fish: DDT and methoxychlor | Ingestion, concentration factor, adsorption on body, defaecation, excretion, chemical decomposition, natural mortality | Leung 1978 |
| Lead | Hydrodynamics, precipitation, toxic effects of free ionic lead on algae, invertebrates and fish | Lam & Simons 1976 |
| Radionuclides | Hydrodynamics, decay, uptake and release by various aquatic surfaces | Gromiec & Gloyna 1973 |
| Polycyclic aromatic hydrocarbons | Transport, degradation, bioaccumulation | Bartell et al. 1984 |
| Cadmium, PCB | Hydraulic overflow rate (settling), sediment interactions, steady state food chain submodel | Thomann 1984a |
| Hydrophobic organic | Gas exchange, sorption/desorption, hydrolysis compounds, photolysis, hydrodynamics | Schwarzenbach & Imboden 1984 |
| Mirex | Water-sediment exchange processes, adsorption, volatilisation, bioaccumulation | Halfon 1984 |
| Toxins (aromatic hydrocarbons, Cd) | Hydrodynamics, deposition, resuspension, volatilisation, photooxidation, decomposition, adsorption, complex formation (humic acid) | Harris et al. 1984 |
| Persistent organic chemicals | Fate, exposure and human uptake | Paterson & Mackay 1989 |
| Mirex and Lindane | Fate in Lake Ontario | Halfon 1984 |
| pH, calcium and aluminium | Survival of fish populations | Breck et al. 1988 |
| Pesticides and surfactants | Fate in rice fields | Jørgensen et al. 1997 |
| Toxicants | Migration of dissolved toxicants | Monte 1998 |
| Growth promoters | Fate, agriculture | Jørgensen et al. 1998 |
| Toxicity | Effect on eutrophication | Legovic 1997 |
| Pesticides | Mineralisation | Fomsgaard 1997 |
| Mecoprop (3) | Mineralisation in soil | Fomsgaard & Kristensen 1999 |

2 attempt to obtain parameters for these processes in the environment from the literature;
3 estimate all parameters using various estimation methods;
4 compare the results from 2 and 3 and attempt to explain discrepancies, if present;
5 use sensitivity analysis widely in order to estimate which processes and state variables it would be most feasible and relevant to include in the model.

Table 15.8 gives an overview of a number of ecotoxicological models applied to lake and reservoir management.

Ecotoxicological models differ from ecological models in general by:
1 being, most often, more simple;

2 requiring more parameters;
3 using parameter estimation methods more widely;
4 including an effect component.

They may be divided into six classes. The decision on which to apply will be based upon the ecotoxicological problem that the investigation aims to solve. The present classification is based upon actual problems modelled. Definitions of the model classes are given below and the most appropriate context in which to use each type is indicated.

### 15.5.1 Food chain or food web dynamic models

This class of models considers the flow of toxic substances through the food chain or food web. Such models will be relatively complex, and contain many state variables and parameters. They will typically be used when a great number of organisms are affected by the presence of a toxic substance, or the entire structure of the ecosystem is threatened. Because of their complexity, they have not been used widely. They are similar to the more complex eutrophication models which consider the flow of nutrients through the food chain, or even through the food web. In some instances, they are even constructed as submodels of a eutrophication model (see, e.g., Thomann *et al.* 1974). Figure 15.9 shows a conceptual diagram of an ecotoxicological foodchain model for lead, incorporating flow of the element from atmospheric fall-out and wastewater to an aquatic ecosystem, where it is concentrated through the food chain—the so-called 'bioaccumulation'. Simplification is hardly possible for this model, because the aim is to describe and quantify bioaccumulation through the food chain.

### 15.5.2 Static models of mass flows of toxic substances

If seasonal changes are minor, or of little importance, a static model of mass flows will often be sufficient to indicate the expected results of reduced or enlarged input of toxic substances. This type of model is based upon a mass balance. It will

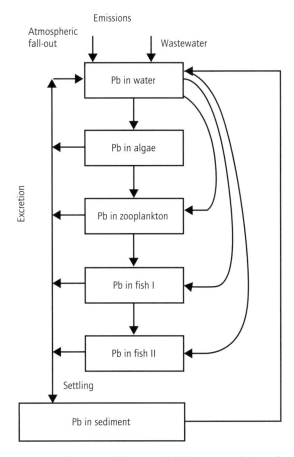

**Fig. 15.9** Conceptual diagram of the bioaccumulation of lead through a food chain in an aquatic ecosystem.

often, but not necessarily, contain more trophic levels, and the modeller is often concerned with the flow of the toxic substance through the food chain. If there are some seasonal changes, this type, which in most cases is simpler than the type described above, can still be used (e.g., if the modeller is concerned with the worst case and not the changes).

### 15.5.3 A dynamic model of a toxic substance in a trophic level

Often, it is only at one trophic level that toxic substance concentration is of concern. This includes

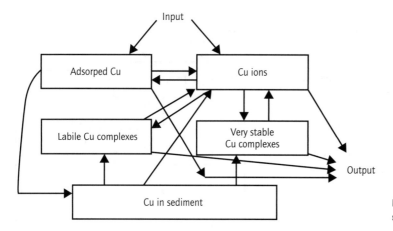

Fig. 15.10 Conceptual diagram of a simple copper model.

the zero trophic level, which is the medium (i.e. $u < u_{lake\ water}$). Figure 15.10 gives an example of copper contamination in a lake ecosystem. The main concern is concentration in the water, which may become toxic for phytoplankton. Zooplankton and fish are much less sensitive to copper contamination, so warning signs are first seen when the concentration is harmful to phytoplankton. However, only the ionic form of the element is toxic, so that it is necessary to model partition of the element between the ionic, complex bound, and adsorbed forms. Exchange between water and sediment is also included, where relatively large amounts of heavy metals are able to accumulate. Re-release of these from the sediment may be significant under certain circumstances—for instance under low pH.

Figure 15.11 gives another example. Here the main concern is DDT (dichlorodiphenyltrichloroethane ) concentration in fish, especially where these may be so high, according to WHO (World Health Organization) standards, as to be unfit for human consumption. The model therefore can be simplified by including not the entire food chain but only the fish. Some physical–chemical reactions in the lake water are, however, of importance, and they are incorporated in the conceptual diagram (Fig. 15.11).

As seen from these examples, simplifications including which component is most sensitive to toxic matter, and which processes are most impor-

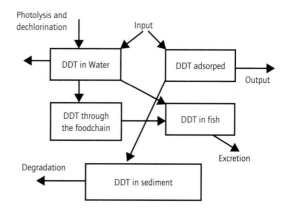

Fig. 15.11 Conceptual diagram of a simple dichlorodiphenyltrichloroethane (DDT) model.

tant for concentration changes are often feasible when the problem is well defined.

### 15.5.4 Ecotoxicological models in population dynamics

The concentration of toxic substances in the environment or in organisms influences natality and mortality. If the relationships between toxic substance concentration and population dynamics are included in the model, it becomes an ecotoxicological model of population dynamics.

Population dynamic models may include two or more trophic levels. Ecotoxicological population dynamic models will include influence of toxic substance concentration on natality, mortality and interactions between populations. In other words, an ecotoxicological model of population dynamics is a general model of population dynamics with the inclusion of the relation between toxic substance concentrations and some of the model parameters.

### 15.5.5 Ecotoxicological models with effect components

Although the models just discussed already include relations between concentrations of toxic substances and their effects, these are limited to population dynamic parameters. Models of more comprehensive relations between toxic substance concentrations and effects may include not only lethal and/or sublethal effects, but also those on biochemical reactions or on the enzyme system. In many problems, it may be necessary to go into more detail in order to answer the following relevant questions.

**1** Does the toxic substance accumulate in the organism?
**2** What will be the long-term concentration in the organism when uptake rate, excretion rate and biochemical decomposition rate are considered?
**3** What is the chronic effect of this concentration?
**4** Does the toxic substance accumulate in one or more organs?
**5** What is the transfer between various parts of the organism?
**6** Eventually, will decomposition products cause additional effects?

A detailed answer to all these questions may require a model of processes within the organism, and a translation of concentrations in various parts of the organism into effects. This implies, of course, that the intake (= uptake by the organism × efficiency of uptake) is known. Intake may either be from water or air, which also may be expressed by the use of concentration factors, which are the ratios between the concentration in the organism and in the air or water.

However, if all of the above processes are taken into consideration even for just a few organisms, the model will easily become too complex, contain too many parameters to calibrate, and require more detailed knowledge than it is possible to provide. Therefore, most models in this class will not consider too many details of partition of the toxic substances in organisms and their corresponding effects, but rather will be limited to simple accumulation in the organisms. In most cases accumulation and its effects are rather easy to model.

### 15.5.6 Fate models with or without a risk assessment component

These models focus on the fate of the toxic substances—where in the ecosystem will they be found? In which concentration? They attempt to answer the following questions using the concept of **fugacity**. In which of the six compartments, corresponding to the spheres, can we expect the greatest effect on the environment? Which concentration will be expected in each compartment? What are the implications?

Fugacity models are to a large extent based on physical–chemical parameters and their use to estimate other variables. They are easy to use, but give only rough estimations, which, however, in many cases are completely satisfactory. Mackay (1991) used this approach to model the distribution of PCBs (polychlorinated biphenyls) between air, suspended matter and water in the North American Great Lakes. He predicted that significant amounts $(2\,\text{ng}\,\text{L}^{-1})$ would be found in the water, which is approximately according to observations.

Uncertainties relating to information on which models are based are crucial in risk assessment. These may be classified into one or more of the following categories.

**1** Direct knowledge and statistical evidence of the important components (state variables, processes and interrelations of the variables) of the model are available.
**2** Reliable knowledge and statistical evidence on the important submodels are available, but the aggregations of the submodels are less certain.

**3** No reliable knowledge of the model components is available, but data on the same processes from a similar system exist and it is estimated that these may be applied directly or with minor modifications.
**4** Some, but insufficient, knowledge is available from other systems. Attempts are made to use these data without the necessary transferability, and to eliminate gaps in knowledge by use of additional experimental data as far as it is possible within the limited resources available for the project.
**5** The model is to a large extent based on the subjective judgement of experts.

Uncertainty is of great importance and may be taken into consideration, either qualitatively or quantitatively. Another problem is of course where to take the uncertainty into account. Should the economy or the environment benefit from the uncertainty? Unfortunately, up to now, most decision makers have used the uncertainty to the benefit of the economy. This is of course completely unacceptable. The same decision makers would never, for a moment, consider for a civil engineering project whether uncertainty should be used for the benefit of the economy or the strength of a bridge.

Until 15–20 years ago, researchers had developed very little understanding of the processes by which people actually perceive the exposures and effects of toxic chemicals. These processes are just as important for the assessment as exposures and effects. These may be summarised as follows.
**1** Characteristics of risk:
  (a) Voluntary or involuntary?
  (b) Are the levels known to the exposed people or to science?
  (c) Is it novel, or old and familiar?
  (d) Is it common or dreaded (for instance does it involve cancer)?
  (e) Does it involve death?
  (f) Are mishaps controllable?
  (g) Are future generations threatened?
  (h) Global, regional or local?
  (i) Function of time? How (whether for instance increasing or decreasing)?
  (j) Can it easily be reduced?

**2** Characteristics of effects:
  (a) Are they immediate or delayed?
  (b) Do they affect many or a few people?
  (c) Are they global, regional or local?
  (d) Do they involve death?
  (e) Are effects of mishaps controllable?
  (f) Are they observable immediately?
  (g) Are they a function of time?

Several risk management systems are available, but no attempt will be made here to evaluate them. However, some recommendations should be given for the development of risk management systems. Investigators should:
**1** Consider as many of the characteristics listed above as possible and include the human perceptions of these in the model.
**2** They should not focus too narrowly on certain types of risks. This may lead to suboptimal solutions. Attempt to approach the problem as broadmindedly as possible.
**3** Choose strategies which are pluralistic and adaptive.
**4** Recognise that cost–benefit analysis is an important element of the risk management model, but it is far from being the only one. Uncertainty in evaluation of benefit and cost should not be forgotten. The variant of this analysis applicable to environmental risk management may be formulated as

net social benefit = (social benefits of the
  project – 'environmental' cost of the project)

**5** Use multi-attribute utility functions, but remember that people in general find it difficult to think about more than two or three, at the most four, attributes to each outcome.

### 15.5.7 Summary

Presentation of the six classes of ecotoxicological model clearly illustrates these advantages **and** limitations. The simplifications used in classes described in sections 15.5.2, 15.5.3 and 15.5.6 (at least without risk assessment components) often offer great advantages. They are sufficiently accurate to give a reliable overview of concentrations of toxic substances in the environment, owing to their incorporation of great safety factors. Application of estimation methods renders it feasible to construct such models, even when our knowledge of their parameters is limited. Such methods are obviously highly uncertain, but a large safety factor renders uncertainty acceptable. On the other hand, our knowledge of the effects of toxic substances **is** very limited—particularly at organism and organ level. It must not be expected, therefore, that models with effect components give more than a first approximation of current knowledge.

## 15.6 MODELS OF LAKE ACIDIFICATION

Acidification of lakes and reservoirs is caused by acid rain originating from emissions of sulphur and nitrogen oxides. Management of acidification therefore requires a chain of models, linking the energy policy with emissions to the atmosphere, then to the effect of acid rain on soil chemistry in catchments, and then on to pH changes in lakes (Fig. 15.12). Further development of soil models is still needed, as the soil model must be considered the weakest model in the chain.

Models of soil processes and of chemical composition, including concentration of various ions and drainage water pH, are complex. It is not intended to present such models here in detail—but rather the difficulties and the basic ideas behind them and to give some understanding of the considerations behind very complex regional models. Further details can be found in the references given in this section, and in Henrikson *et al.* (this

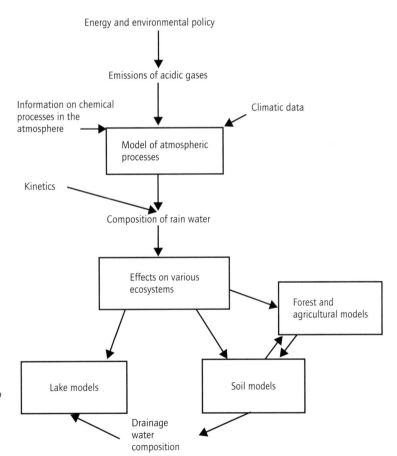

**Fig. 15.12** A series of models must be used to relate energy, and environmental protection policy, to effects on soil, plants, water chemistry and aquatic ecosystems, which serve as feedback to political issues.

Volume, Chapter 18). In this account, only the two last submodels of the chain (i.e. soil and lake models) are considered. The former determines the composition of the drainage water flowing into lakes and reservoirs. Those interested in the entire chain can refer to Alcamo et al. (1990) and Jørgensen (1994).

Several models have been developed which translate emission of sulphur and of nitrogen compounds into changes in soil chemistry and soil-water pH. Kauppi et al. (1986) used knowledge about buffer capacity and velocity to relate emissions to alter the pH. A critical pH of 4.2 is usually applied to interpret the results. From data on deposition, it is possible to estimate acid load in equivalents of numbers of hydrogen ions $m^{-2} yr^{-1}$. In the European model RAINS (Alcamo et al. 1990), acid load is computed from deposition after accounting for forest filtering and atmospheric deposition of cations.

The resolution of RAINS is such that each grid element includes one to seven soil types, with a mean of 2.2. A grid using $1°$ longitude $\times 0.5°$ latitude is applied, and the fraction of each soil type within each grid element is computerised to an accuracy of 5%. The basic computations of the model follow six steps.

**1** The weathering rate of silicate minerals is subtracted from the annual acid load.

**2** The result of step **1** is then subtracted from the soil buffering capacity to account for depletion of the acid neutralising capacity of the soil. Base saturation and pH are computed using these comparisons.

**3** In calcareous soils containing free carbonate, base saturation is not depleted and the soil pH is assumed to stay above 6.2. In non-calcareous soil, as long as silicate weathering keeps pace with the acid load, no decline of pH or base saturation is assumed to occur.

**4** If acid load exceeds the buffering capacity of the silicates, base saturation begins to decline. Cation exchange capacity is depleted at a rate equal to the difference between the acid load and the silicate buffer. pH is computed to decrease according to decline of base saturation.

**5** When base cations are almost depleted, an equilibrium is computed between solid phase aluminium and soil $H^+$.

**6** A recovery of soils is computed in terms of base saturation and pH, when silicate buffer rate exceeds the rate of acid deposition.

The aim of the model is to keep track of the development of soil pH and buffer capacity. In order to simulate flows within the catchment, the soil is segmented into two layers: the uppermost 0.5 m, and the deeper parts. Run-off is divided into quick flow and base flow. The first mainly represents contact with the upper mineral and humus layers, whereas base flow is assumed to come from the saturated soil zone. In order to compute stream ion concentration, the same six step approach is used as in the soil acidification model (see above). Complete mixing, and chemical equilibrium to be reached according to computed saturation, are assumed. The acid load in the soil participates in two processes, cation exchange and release of inorganic aluminium species, the net effect being to provide a buffer for hydrogen ions in the soil solution.

Arp (1983) developed an even more complex model, incorporating $n$ soil layers (in practice between 10 and 50), which includes the carbon, nitrogen and sulphur cycles, and the many soil reactions caused by acid rain. He finds close accordance between model results and actual measurements for several case studies.

Henriksen (1980) developed a model based upon the assumption that, in acidified lakes, titration of the hydrogen carbonate (bicarbonate) buffer takes place. Hydrogen carbonate ions are replaced by sulphate, which mobilises aluminium ions with a corresponding biological effect. Buffer capacity is estimated on the basis of concentrations of non-marine calcium and magnesium ions, and acid load from that of non-marine sulphate. Henriksen distinguishes three categories of water body: hydrogen carbonate lakes, transition lakes and acid lakes with pH and low buffering capacity (and consequently low concentrations of non-marine calcium and magnesium ions). His diagram (Fig. 15.13) can be used to predict pH of rainwater from in-lake concentration of sulphate, and non-marine calcium and magnesium ions. Conversely pH of rain water and lake concentrations of

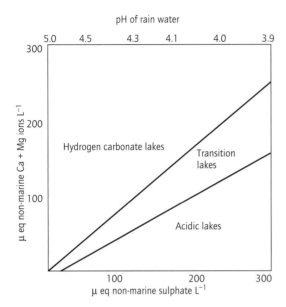

Fig. 15.13  Diagram relating pH of rainwater, concentrations of non-marine calcium and magnesium ions, and concentrations of non-marine sulphate ions in lake water.

calcium and magnesium ions can predict lake sulphate concentration, and hence the type of lake.

The simple empirical approach presented in Fig. 15.13 is used without including terms for stream-water chemistry. Weathering processes can, however, be incorporated into the empirical approach, although the results are difficult to apply in a general context because they are based on regression analysis of local measurements.

RAINS (Reuss 1983) is a more complex model which does include inputs of ions from the catchment. A non-linear relationship between soil base saturation, β, and soil pH is used

$$-\log[H^+] = 4.0 + 1.6 \times \beta^{3/4} \quad (15.6)$$

where soil base saturation is given by β = buffer capacity (BC)/total cation exchange capacity (CEC). If cation exchange plays no role in buffering inputs to the soil solution, i.e. if BC is zero, it is assumed that equilibrium with gibbsite (Al[OH]$_3$) controls soil buffering. If not, then aluminium will be dissolved or precipitated until gibbsite equilibrium is reached:

$$[Al^{3+}] = Kg \times [H^+]^3 \quad (15.7)$$

where the gibbsite equilibrium constant Kg = $10^{2.5}$.

Over the long term, the flux of acids and bases into and out of the soil changes its chemical state. In non-calcareous soils, weathering rates of base cations largely control the long-term response of the catchment. So long as input of bases from weathering ($w_r$) and from base deposition ($d_b$) is larger than the acid load, there will be no change in soil pH.

If, however, acid load exceeds the base cation input, the capacity of the cation exchange buffer system is depleted by the rate:

$$d_{BC}/d_t = d_b + w_r - d_f \quad (15.8)$$

where $d_f$ is the acid load equal to the total deposition of acid, and $d_t$ is corrected for the throughfall deposition on forested land and for the fraction of forested land in the relevant area.

Lake-water pH and alkalinity are found from the convective flows $Q_a$ and $Q_b$ from the two soil reservoirs, and the direct input from the atmosphere to the lake:

$$F_H = Q_a \times [H^+] + Q_b + [H^+] + d_f \times A_L \quad (15.9)$$

where $F_H$ is the flux of hydrogen ions to the lake and $A_L$ is the area of the lake.

Fluxes of hydrogen carbonates originating from the terrestrial catchment ($F_{HC}$) will contribute to the alkalinity of the lake:

$$F_{HC} = d_b + w_r - d_f \quad (15.10)$$

The contribution from the lake's own internal alkalinity ($F_L$) must be added to $F_{HC}$ in order to find the total flux of hydrogen carbonates: $F_L$ is found with the following equation

$$F_L = d_t \times k_s/(Q_t/A_L + k_s) \quad (15.11)$$

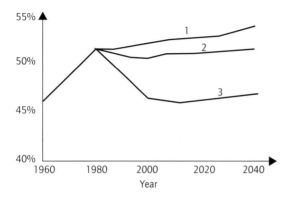

**Fig. 15.14** Time trend of percentage of lakes in Swedish Götland with pH < 5.3. Curve 1 corresponds to a scenario based on no control, 2 on current reduction plans, and 3 to best available technology (Alcamo et al., 1990).

where $k_s$ is the sulphate retention rate coefficient, and $Q_t$ is the total flow of water to the lake.

The fluxes of ions mix with the lake water, and cause a change in the hydrogen carbonate and hydrogen ion concentrations, until equilibrium is reached:

$$[HCO^{3-}] = K_1 \times H \times pCO_2 / [H^+] \quad (15.12)$$

where $K_1$ is the first dissociation constant for carbonic acid, $H$ is Henry's law constant for carbon dioxide and $pCO_2$ is the partial pressure of carbon dioxide in the lake water. Figure 15.14 gives the results of an application of this approach to acidification in Swedish Götland (the region between Göteborg and Stockholm, Sweden). It can be seen that the effect on lake-water chemistry of applying control of emissions is unambiguous.

Most biological models of lake acidification focus on fish. Brown & Salder (1981) have developed an empirical model, which relates fish population to pH of the water, on the basis of data for 719 lakes. They found that for the lakes in southern Norway, a 50% reduction of sulphate emission would give an average increase in pH of 0.2, which would only improve the fish populations in 9% of the lakes. The model has, however, been criticised for underestimating the relationship between reduction of sulphate emissions and pH. Muniz & Seip (1982) distinguish between lakes of different conductivity. Chen et al. (1982) have developed a very comprehensive model considering the effect on all levels in the food chain and the total effect on the ecosystem.

## 15.7 OXYGEN MODELS

A wide spectrum of BOD–DO (biochemical oxygen demand–dissolved oxygen) models have been developed for streams, based on a description of all the oxygen depleting or generating processes (i.e. decomposition of organic matter in water and sediment, nitrification, respiration and chemical conversion of oxygen), and those generating or transferring oxygen to the water (i.e. primary production and re-aeration). The processes are described for two types of aquatic ecosystems by similar equations. The main difference for lakes and reservoirs is that wind may be significant for the rate of re-aeration, whereas in streams it is the rate of the flow. Banks (1975) and Banks & Herrera (1977) have studied the effect of wind on the re-aeration coefficient, and suggest the use of the following relationship:

$$K_a = 0.782\sqrt{W} - 0.317W + 0.0372W2 \quad (15.13)$$

where $W$ is the wind speed (m s$^{-1}$) 10 m above the water surface.

Oxygen production by photosynthesis can be estimated from the following equation:

$$p(\text{in mg O}_2\, L^{-1} \times \text{day}) = a \times \mu \times P(1.066)^{T-20} \times G \quad (15.14)$$

where $a$ is the ratio of mg O$_2$ chl $a^{-1}$ (range 0.1–0.3, average 0.18), $\mu$ is the growth rate of phytoplankton (found for instance from equations usually applied in eutrophication models, omitting the function of temperature, as $(1.066)^{T-20}$ considers the influence of the temperature, $T$), $P$ is the concentration of phytoplankton expressed as μg chl $a$ L$^{-1}$ and $G$ is the light attenuation factor. It can be estimated from:

Table 15.9 Selected values of sediment oxygen demand $(g\,O\,m^{-2}\,day^{-1})$

| Bottom type | Range | Average |
|---|---|---|
| Filamentous bacteria (10 g dry wt m$^{-2}$) | 5–10 | 7 |
| Municipal sewage sludge – outfall vicinity | 2–10 | 4 |
| Municipal sewage sludge – downstream of outfall | 1–2 | 1.5 |
| Estuarine mud | 1–2 | 1.5 |
| Sandy bottom | 0.2–1.0 | 0.5 |
| Mineral soils | 0.05–0.1 | 0.07 |

$$G = 2.718 \times f \times LA/e \times H \quad (15.15)$$

where $f$ is the photoperiod (duration of daylight in hours/24), $LA = (\exp(-\beta) - \exp(-\delta))$, $\beta = I_a \exp(-e \times z)/I$, $\delta = I_a/I$, $e$ is the extinction coefficient (m$^{-1}$), $I_a$ is the average solar radiation during the day (ly day$^{-1}$), $I$ is the light at which phytoplankton grows at maximum rate (ly day$^{-1}$), $H$ is depth, and $e$ may be estimated as $1.8/z_s$ (Beeton 1958), where $z_s$ is the Secchi-disc depth.

Phytoplankton respiration, $R$ (mg L$^{-1}$ day$^{-1}$), is given approximately by

$$R = 0.1 \times a \times P \times (1.08)^{T-20} \quad (15.16)$$

Finally, sediment oxygen demand, s, can be found as s = SOD/H, where SOD is the g O m$^{-2}$ day$^{-1}$ at the water temperature: SOD either can be measured or estimated from values given in Table 15.9 (Thomann 1984b), where a temperature of 20°C is assumed.

It is generally assumed that sediment oxygen-uptake rate is independent of oxygen concentration in the overlying water. Edwards & Owens (1965) indicate that it varies with oxygen concentration, and Fillos & Molof (1972) proposed the following relationship for the dependence of uptake on oxygen concentration:

$$SOD_a = SOD \times (C/(0.7 + C)) \quad (15.17)$$

where $SOD_a$ is actual sediment oxygen demand at oxygen concentration $C$ (mg L$^{-1}$).

If we consider a lake to be completely mixed, the basic equations for the BOD–DO system are:

$$V d_L/d_t = W - Q \times L - V \times K_1 \times L \quad (15.18)$$

$$V d_C/d_t = Q \times C_{in} - Q \times C + K_a A(C_s - C) - V \times K_1 \times L \pm W_O \quad (15.19)$$

where $W$ is the BOD-loading per unit of time, $Q$ is the flow rate through the lake, $V$ is the volume, $A$ the area, $C_s$ is the oxygen concentration at saturation, and $W_O$ equals all other sources and sinks of oxygen, including photosynthesis, respiration and sediment oxygen demand. Oxidation of ammonia is not considered in this case, but an equation similar to (15.18) is valid for nitrogen.

For a steady-state situation (i.e., $d_L/d_t = d_C/d_t = 0$), eqns. (15.18) and (15.19) yield:

$$L = W/(Q + K_1 \times V) \quad (15.20)$$

$$C = Q \times C_{in}/(Q + K_a A) + C_s \times K_a A/(Q + K_a A) - V \times K_1 \times L/(Q + K_a A) \pm W_O/(Q + K_a A) \quad (15.21)$$

For deeper lakes and reservoirs it is necessary to simulate the oxygen profile, which is achieved using a hydrodynamic model, and dividing the lake or reservoir into several layers. As it is outside the scope of this chapter to present complex hydrodynamic models, the interested reader is referred to Thomann & Mueller (1987) and to section 15.8, which deals with development of thermocline models.

Formation of a thermocline separates the lake into two layers (or more) with different chemical composition. Differences in concentrations between layers can be equalised only by molecular diffusion, and not by advection or conduction. As chemical processes and biological activity also differ between layers, major differences in chemical composition will inevitably occur. In the epilimnion, major photosynthetic activity will take place. Re-aeration is therefore only possible with this layer, and there will therefore be high oxygen concentrations. In the hypolimnion, major decomposition of detritus (for instance dead algae settling from the epilimnion, organic matter in the sediment) occurs. Anoxic conditions will therefore easily emerge. Description of the formation of a thermocline can be translated into a description of the oxygen profile.

## 15.8 MODELS OF STRATIFICATION

Stratification divides the lake into two or more layers of different chemical composition. Oxygen may be absent from the hypolimnion, but there may also be elevated nutrient concentrations owing to a low photosynthetic activity and release of nitrogen and phosphorus from the sediment. It is therefore of utmost importance to model the appearance of stratification in lakes and reservoirs in order to account for these differences in chemistry and biological activity.

Stratification in salt lakes (see Williams, this Volume, Chapter 8) may be caused by a **halocline**, but is most often a result of formation of a thermocline, which requires simulation of the heat balance cycle. A few older, but well tested approaches to assess the formation of the thermocline in deeper lakes and reservoirs is presented in Jørgensen (1994). Several thermocline models are available and can be found in the literature, but they are all more or less based on the same concepts: to assess the heat balance and account for the wind mixing.

Details of thermocline models will not be presented in this context, as they are based upon a description of the lake hydrodynamics which would require comprehensive presentation of basic hydrodynamic processes and their related equations (see Imboden, Volume 1, Chapter 6). Those interested in the details of thermocline models may also refer to Orlob (1983) and Jørgensen (1994).

The present or absence of a thermocline plays an important role in lake management, which emphasises the need for a thermocline model to be integrated with one or more of the other models presented in this chapter. Release of nutrients and heavy metals from sediment is governed by the redox potential at the sediment surface, which is also highly dependent on the presence or absence of a thermocline. Selection of lake restoration methods is also dependent on the presence of a thermocline. Drainage of hypolimnion water is based on an assumption that difference in concentration between epilimnion and hypolimnion can be predicted only if a model of the differences due to formation of the thermocline can be developed. The effect of hypolimnion aeration is similarly dependent on a proper model of the redox condition in the hypolimnion.

The need for a thermocline model is therefore obvious, although an alternative method would be to measure the temperature profile which is usually quite similar from year to year, and to then use the measurements as a forcing function in biogeochemical models of the same lake.

## 15.9 MODELLING FISH PRODUCTION IN LAKES

Augmented exploitation of freshwater fish resources by sport and commercial fishery, and the general deterioration of water quality, have stimulated concern about depletion of fish stocks. This has inevitably intensified the work on developing models which take into account the effect of harvesting and water quality on the fish population. Models with a wide spectrum of complexity attempt to provide a management tool to assess an optimum fishery strategy (see also Sarvala et al., Volume 1, Chapter 16).

For lakes with important commercial fisheries, landing records over decades are often available, and may open the possibilities for a statistical type of model. However, such models generally do not take a number of important factors (such as interaction among species, water quality and changes in the concentrations of fish food) into account. A statistical model would build upon the assumption that past and present properties of the environment, and the population, will be maintained. Statistical models will not be presented here, mainly because they do not consider the influence of water quality, which is of great importance for management of lakes and reservoirs. It should, however, be mentioned that ILEC (1992), which represents software for a simple eutrophication model with four state variables, also computes fish populations by a simple regression equation related to primary production.

Two fish production models, representing different complexity, are presented below. The more simple approach assumes that the entire fish population is homogeneous, and does not consider population dynamics and the related age-structure, which is essential for fishing policy. More complex approaches consider the influence of water quality on dynamics and age-structure of the fish population. Ways in which information from a concrete case study could be applied to set up a fish production model for a lake in this context are also discussed.

### 15.9.1 Fish models without and with age structure

Jørgensen (1994) developed a model which assumes that the fish population is homogeneous, and does not consist of different age classes. This assumption is obviously not realistic, but simple fish models may in some cases lead to workable policy. For unexploited fish populations the logistic growth equation is often valid:

$$d_N/d_t = r \times N(1 - N/K) \quad (15.22)$$

where $N$ is the biomass concentration or the number of fish, $r$ is the growth rate (1 day$^{-1}$) and $K$ is the carrying capacity of the lake. The fish population being exploited may then be modelled by the following equation:

$$d_N/d_t = r \times N(1 - N/K) - H(N, E) \quad (15.23)$$

where $H(N, E)$ is the harvesting function, which is dependent on the population size, $N$, and the fishery effort, $E$, and which expresses the total catch per unit of time. Jørgensen (1994) considers the harvest in the lake with the simplification that there are $B$ identical boats. The fraction of the population which each boat can catch, with given density of fish per unit of time $p(B)$, is then a function of the number of boats, $B$, as in:

$$p(B) = kC'/(1 + BC') \quad (15.24)$$

Here, total fishing power of the entire fleet is $Bp(B)$, $C'$ is a measure of the degree of competition between boats (the number of boats required to halve the fishing power per boat is $C'^{-1}$), and $k$ is a constant.

If we consider catch per boat $c$ in time interval $\Delta t$, then catch can be considered as being proportional to:
1 the time spent on fishing, $\Delta t$;
2 the fishing power per unit boat, $p(B)$, expressed as fraction of the population, $N$;
3 the fish density.

We then have:

$$c = \Delta t p(B) N \quad (15.25)$$

The total catch per unit of time is $B \times c/\Delta t = f \times H(N, E)$, where $f$ is the fraction of time the boats are fishing on average. We therefore obtain:

$$H(N, E) = Bc/\Delta t = f \times B \times k \times C' \\ \times N/((1 + B) \times C') \quad (15.26)$$

These simple equations have been used to set up fish-population management strategies, but the model will still in most cases be inadequate, as it does not account for the role of water quality and the age structure of the fish population. It can be used with advantage as a first approximation to give a suitable basis for development of more complex fishery models.

Age structure is best modelled by the use of a matrix model (see, e.g., Jørgensen 1994). The following expression may be used:

$$\begin{vmatrix} f_0, f_1, f_2, \ldots f_{n-1}, f_n \\ g_0, 0, 0, \ldots 0, 0 \\ 0, g_1, 0, \ldots 0, 0 \\ \cdots \\ \cdots \\ \cdots \\ 0, 0, 0 \ldots g_{n-1}, 0 \end{vmatrix} * \begin{vmatrix} n_{t,0} \\ n_{t,1} \\ n_{t,2} \\ \cdots \\ \cdots \\ \cdots \\ n_{t,n} \end{vmatrix} = \begin{vmatrix} n_{t+1,0} \\ n_{t+1,1} \\ n_{t+1,2} \\ \cdots \\ \cdots \\ \cdots \\ n_{t+1,n} \end{vmatrix} \quad (15.27)$$

where $f$ indicates the fecundity for the age classes $0,1,1,3,4,\ldots n$, $n$ is the number of fish, $n_{t,n}$ is the number of fish at time $t$ in the $n$th age class, and $g$ indicates the fraction of fish transferred from one age class to the next. All values of $g$ are $<1$. The following equation can be set up for determination of $g$:

$$g_n + p_n + m_n + e_n = 1 \quad (15.28)$$

where $p$ indicates the fraction removed by predation, $m$ the fraction exposed to natural death, $e$ the mortality due to fishing, and $n$ the age class. The term $p$ can be found by a general Michaelis–Menten expression, and $m$ can be considered a constant, dependent on water quality (for instance the oxygen concentration, the temperature, the ammonia concentration and pH). The term $e$ can be found from fishery effort, for instance by the use of the fishery statistics.

In addition to the number of fish, we need to know their biomass (i.e. the weight of the average fish in each age class from 0 to $>n$). Fish growth can be modelled by use of von Bertalanaffy's equation:

$$dW/dt = UF - kW^r \quad (15.29)$$

where $W$ is the weight of one fish, $k$ and $r$ are constants, and $UF$ is the utilised feed of the fish. The value of $r$ usually lies between 2/3 and 1, but many literature values point towards the range 0.75–0.8. The term $k$ is dependent on the ambient temperature, is known for various fish species, and can be found in the literature (for instance Jørgensen et al. 1991). Growth is calculated for each age class for the period between two promotions to the next age class.

The value of $UF$ is found from the following expression:

$$UF = F(1 - NDF)(1 - ALC) \quad (15.30)$$

where $F$ is the intaken food, $NDF$ the non-digested fraction and $ALC$ the fraction of food energy used for assimilation processes; $NDF$ is dependent on the available feed relative to the appetite, $A$, and $ALC$ on temperature. Appetite is found as $HW^{2/3}$,

**Table 15.10** Relations of AFC/A versus F/A and NDF

| AFC/A | F/A | NDF |
|---|---|---|
| 0.1 | 0.1 | 0.020 |
| 0.2 | 0.2 | 0.0225 |
| 0.3 | 0.3 | 0.025 |
| 0.4 | 0.4 | 0.0275 |
| 0.5 | 0.5 | 0.03 |
| 0.6 | 0.6 | 0.05 |
| 0.7 | 0.7 | 0.08 |
| 0.8 | 0.8 | 0.11 |
| 0.9 | 0.9 | 0.14 |
| 1.0 | 1.0 | 0.17 |
| 1.1 | 1.07 | 0.20 |
| 1.2 | 1.12 | 0.23 |
| 1.3 | 1.17 | 0.26 |
| 1.4 | 1.21 | 0.29 |
| $\geq 1.5$ | 1.25 | 0.32 |

where $H$ also is temperature dependent. An expression relating $F$ and $NDF$ when the available feed ($AFC$) concentration is known relative to $A$ is shown in Table 15.10. The data in this table are based upon fish farming experiments.

The principle behind this modelling approach is shown in Fig. 15.15. The feed is used for growth, for assimilation, for respiration or is not digested. As shown in Table 15.10, the fish eat more when there is more food available, but more is also lost as non-digested food. If $AFC$ falls below 0.5, mortality will increase rapidly. The model can allow for this relation by adjustment of the natural mortality $m$ in accordance with the $AFC/A$ ratio, for instance by the use of a table function.

A case study of the fish population of Lake Victoria using this approach can be found in Jørgensen (1994). The food web corresponding to this model is shown in Fig. 15.16.

## 15.10 WETLAND MODELS

Nitrogen and phosphorus balances have shown that agriculture and other non-point sources contribute significantly to overall pollution, and par-

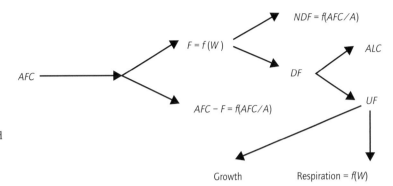

Fig. 15.15 The utilisation of the available fish food for feeding, F, which is divided into digested and non-digested food. The digested food is used for growth, to cover the energy cost of assimilation processes and for respiration.

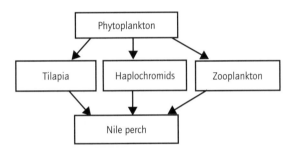

Fig. 15.16 The main food web in Lake Victoria.

ticularly to eutrophication. This finding implies that environmental technology alone is not sufficient, but must be supplemented with other methods in order to cope with the problems of diffuse (in North America 'non-point') sources. These methods are covered by the term **ecological engineering**, or **ecotechnology**.

Mitsch & Jørgensen (1989) give an overview of the methods used to abate eutrophication of lakes. They compare their efficiency for a particular lake by the use of a eutrophication model. The result of their case study (others have given similar results) is that the application of constructed wetlands to reduction of non-point inputs is a very effective method, at least where nitrogen plays a role in eutrophication. Whereas most other methods do not reduce phytoplankton concentration, or primary production, by more than 10%, construction of wetlands for removal of nitrogen, and partially of phosphorus, from the inflowing main stream was able to lower the latter by about 33%, giving an increase in lake transparency from 0.6 to 1.0 m.

Mitsch (1976, 1983) has given a more comprehensive review of wetland models than it is possible to include here. He distinguishes between **energy/nutrient**, **hydrological**, **spatial ecosystem**, **tree growth**, **process**, **causal** and **regional energy** models. Mitsch *et al.* (1988) have reviewed several types of wetland models.

Nitrogen balances for agricultural regions reveal that nitrogen from non-point sources plays a major role in eutrophication. Solutions therefore cannot be found without addressing the problems associated with pollution from non-point sources.

A wide spectrum of methods of ecological engineering (touched on in Chapters 6 and 8, this Volume) have been implemented so far in an attempt to solve the problem, but as indicated above their effects may be different, depending on the mass balance and general properties of the lake. In this context, there is obviously a need for a general model which is able to make predictions regarding nutrient removal capacity of given types and capacities of wetlands. As ecological models only possess limited generality (see Jørgensen 1994), it is necessary to distinguish between general relations and more site-specific parameters and forcing functions. Thus, it may not be possible to achieve complete generality for wetland models.

Below is presented a relatively general wetland model able to predict nitrogen removal. It could also be expanded relatively easily in order to cover heavy metal and phosphorus removal by adsorption, and removal of phosphorus and nitrogen by harvest. In the former case, a simple adsorption isotherm is added, and in the latter, a harvest function which removes nitrogen and phosphorus at day $M$, corresponding to the content of these two nutrients in the plants and the plant biomass removed. According to Mitsch's classification (see above), this is a causal process model.

The model is based upon previous approaches by Dørge (1991). It differs from the previous models by being more simple, which is necessary in order to make it more general. Furthermore, the model is dynamic in respect of its hydrological as well as its biological components, whereas Dørge's model is steady state for the latter. Dynamic models are more difficult to calibrate, but calibration will often reveal bias relations more clearly. This feature of dynamic models has been used to make a site-specific calibration, as will be demonstrated below. The results of model application for two case studies are presented. A procedure for more general application in environmental management is proposed.

The conceptual diagram of the model and its equations are presented in Fig. 15.17 and Table 15.11. The software STELLA is applied. The climatic forcing functions are: precipitation, evaporation, temperature and solar radiation. The last is given as a cosine function (see Dørge 1991) and the three former as tables. The same functions are applied to both case studies. The site-specific forcing functions are nitrate and ammonium concentrations in in-flowing water and the flow rate.

The model considers conversion of nitrogen in 1 m$^2$ of wetland. The results will therefore predict the amount of nitrogen removed, accumulated and/or released per unit area. Two hydrological state variables are applied, one representing the surface layer, where nitrification takes place, and the second the active zone, where there is pronounced denitrification and accumulation. The depth of this layer is not very important, because in the great majority of cases the limiting factor is hydraulic conductivity. Amounts of organic matter and volume occupied by denitrifying microorganisms are not limiting.

The nitrogen state variables are nitrate and ammonium in the surface layer, and nitrate, ammonium, detritus-nitrogen, plant-nitrogen and adsorbed nitrogen in the active zone. Cycling of nitrogen takes place in the active layer. Ammonium and nitrate are taken up by plants. Plant-nitrogen forms detritus-nitrogen by decay and, after mineralisation, ammonium. Nitrification and denitrification are described by Michaelis–Menten equations, and the uptake of nitrate and ammonium by the plants, which is proportional to light, is formulated by first-order kinetics. There are no differences between the uptake rates for ammonium and nitrate. Uptake is therefore proportional to concentration of inorganic nitrogen = ammonium + nitrate. Mineralisation also follows first-order kinetics.

Decay is dependent on uptake, and also on a mortality function which can be formulated as a table according to seasonal variations. All biological rates are dependent on temperature, with more pronounced dependence for nitrification and denitrification. The following site-specific parameters are used: hydrological conductivity, nitrification capacity, denitrification capacity, detritus-nitrogen pool (the initial value of this state variable), and initial and maximum value of plant-nitrogen.

The following parameters are calibrated: uptake rates for nitrate and ammonium, and mineralisation rate. These are adjusted to give observed trends in detritus-nitrogen and the maximum value of plant-nitrogen.

The model has been applied in several case studies, two of which are shown. Site-specific parameters, which are the basis for model application, are shown in Table 15.12. Uptake rates for nitrate and ammonium, and mineralisation rate, are found by calibration. These two parameters are presented in Table 15.13. Calibration of the two case studies was easy to perform and gave reasonable values, as seen in Table 15.14.

Fig. 15.17 STELLA diagram of the model of nitrogen removal by wetlands.

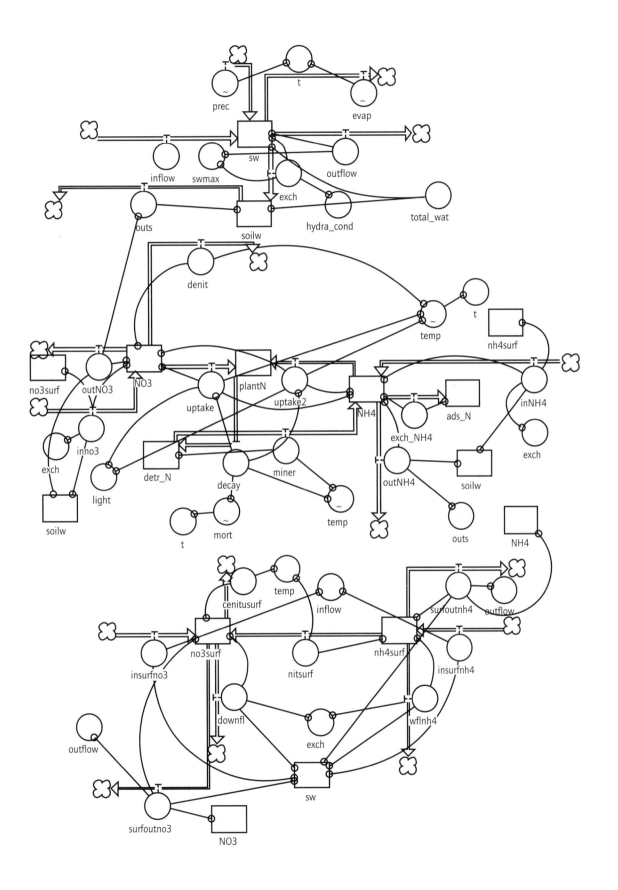

**Table 15.11** Equations for wetland model

ads_N = ads_N + $d_t$ × (exch_NH4)
INIT(ads_N) = 200/9
detr_N = detr_N + $d_t$ × (decay − miner)
INIT(detr_N) = 1200
NH4 = NH4 + $d_t$ × (−uptake2 + miner − exch_NH4 − outNH4 + inNH4)
INIT(NH4) = 1.0
nh4surf = nh4surf + $d_t$ × (−nitsurf + insurfnh4 − wflnh4 − surfoutnh4)
INIT(nh4surf) = 0.1
NO3 = NO3 + $d_t$ × (−uptake1 − outNO3 − denit + inno3)
INIT(NO3) = 10
no3surf = no3surf + $d_t$ × (insurfno3 + nitsurf − downfl − denitsurf − surfoutno3)
INIT(no3surf) = 5
plantN = plantN + $d_t$ × (uptake1 + uptake2 − decay)
INIT(plantN) = 20
soilw = soilw + $d_t$ × (exch − outs)
INIT(soilw) = 2.0
sw = sw + $d_t$ × (inflow − outflow + prec − evap − exch)
INIT(sw) = 0.015
decay = (1.04^(temp-20)) × mort × (uptake1 + uptake2)
denit = (1.12^(temp-20)) × 8 × NO3/(12 + NO3)
denitsurf = (1.12^(temp-20)) × 8 × no3surf/(12 + no3surf)
downfl = exch × no3surf/sw*
exch = IF sw > swmax THEN hydra_cond ELSE sw × hydra_cond/swmax
exch_NH4 = IF ads_N < 200 × NH4/(8 + NH4) THEN NH4/(8 + NH4) ELSE 0

hydra_cond = 0.09
inflow = 0.035
inNH4 = (exch × nh4surf + 0.01 × (nh4surf − NH4))/soilw
inno3 = (exch × no3surf + 0.01 × (no3surf − NO3))/soilw
insurfnh4 = inflow × 0.2/sw
insurfno3 = inflow × 5/sw
light = 1.91 − 1.68 × COS(6.1 × (TIME-355)/365)
miner = 0.0001 × detr_N × 1.07^(temp-20)
nitsurf = 8 × (1.12^(temp-20)) × nh4surf/(8 + nh4surf)
outflow = IF sw > swmax THEN 1.0 × (sw − swmax) ELSE 0
outNH4 = outs × NH4/soilw
outNO3 = outs × NO3/soilw
outs = IF soilw > 2.45 THEN 0.1 ELSE 0
surfoutnh4 = (nh4surf × outflow + 0.01 × (nh4surf − NH4))/sw
surfoutno3 = (outflow × no3surf + 0.01 × (no3surf-NO3))/sw
swmax = 0.05
t = TIME
total_wat = soilw + sw
uptake1 = IF NO3 > 0.05 THEN light × 0.15 × (1.05^(temp-20)) × NO3/(NO3 + NH4) ELSE 0
uptake2 = IF NH4 > 0.05 THEN light × 0.15 × (1.05^(temp-20)) × NH4/(NO3 + NH4) ELSE 0
wflnh4 = exch × nh4surf/sw
evap = graph(t)
mort = graph(t)
prec = graph(t)
temp = graph(t)

\* sw = surface water.

**Table 15.12** Wetland properties (based on 1 m²)

| Parameter | Rabis wet meadow | Glumsø reed-swamp |
|---|---|---|
| Hydrological conditions (m 24 h⁻¹) | 0.009 | 0.009 |
| Production (N yr⁻¹) | 7.0 | 40.0 |
| Detritus (N g) | 800 | 1200 |
| Maximum nitrification (g N 24 h⁻¹) | 11 | 7 |
| Maximum denitrification (g N 24 h⁻¹) | 22 | 72 |

The most interesting results are nitrate concentration in the outflowing water (shown in Figs 15.18 & 15.19) and the nitrogen balance, given in Table 15.14. Accordance with previous results is acceptable, particularly in the light of the uncertainty, which should be accepted in environmental planning.

The aim is to construct a model with a general applicability. Given some pertinent information about the wetland, what will be the capability of the wetland to remove nitrogen? The environmental planner will thus be able to decide how much wetland is needed in order to achieve certain goals for the removal of nitrogen from non-point sources.

A tentative procedure for wider application is summarised in Fig. 15.20. The methods to be applied if the wetland does not exist, but is planned for construction, are similar. Climatic forcing functions used are regionally based, but the properties of a non-existent wetland cannot, of course, yet be found.

Hydrological conductivity can still be estimated from soil characteristics and by comparison

## Models of Lakes and Reservoirs

**Table 15.13** Calibrated parameters

| Parameter | Rabis wet meadow | Glumsø reed-swamp |
|---|---|---|
| Uptake rate (L 24 h$^{-1}$) | 0.025 | 0.125 |
| Mineralisation rate (L 24 h$^{-1}$) | 0.00005 | 0.00025 |

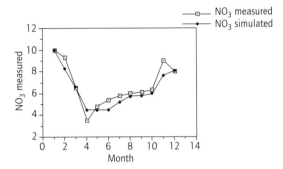

**Fig. 15.18** Comparison of measured and simulated values of nitrate (mg L$^{-1}$) for Rabis wet meadow.

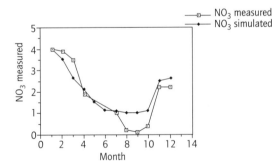

**Fig. 15.19** Comparison of nitrate measured and simulated for Glumsø reed-swamp.

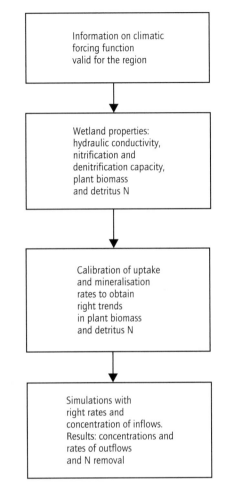

**Fig. 15.20** A procedure applicable for development of a wetland model for a specific site, from the general model presented in the text.

**Table 15.14** Nitrogen balance values found by the simulations (based on 1 m$^2$). Numbers in parentheses are previously found values

| Nitrogen flow (g N yr$^{-1}$) | Rabis wet meadow | Wet Glumsø reed-swamp |
|---|---|---|
| Loading (L) | 55 | 64 |
| Removed by denitrification (1) | 24 (20) | 89 (92) |
| Released (2) | 0 (0) | 37 (40) |
| Accumulate (3) | 3 (5) | 7 (5) |
| Per cent (1) + (3) − (2)/L | 49 (45) | 92 (89) |

with wetlands of similar vegetation and soil types; as may initial and maximum values of plant-nitrogen and trends in detritus-nitrogen. The thickness of the surface layer is estimated from wetlands with similar vegetation and from the slope of the landscape where the wetland is planned.

## 15.11 MODELLING BIOMANIPULATION

The success of biomanipulation (see Hosper et al., this Volume, Chapter 17) varies considerably from case to case. Modelling therefore seems the obvious tool to use in selection of biomanipulation techniques, as the primary task of management models is always to predict the results of any alterations to the ecosystem. However, biomanipulation implies that the structure of the ecosystem (of the lake) is changed, which is much more difficult to capture in a model than simple changes in forcing functions, for instance inputs of nutrients. Structurally dynamic models are, on the other hand, emerging (see, e.g., Jørgensen 1992a,b), and models which sometimes can be used to explain ecosystem behaviour are increasingly used as an experimental tool (Jørgensen 1990, 1992b), for instance in association with the Chaos Theory and Catastrophe Theory.

Model developments based upon case studies with structural changes are still very rare. We know, however, that ecosystems are able to adapt to changed forcing functions and to shift to species better fitted to emerging conditions. When radical changes are imposed, leading to pronounced changes in ecosystem structure, models with dynamic structure seem to be of increasing importance in environmental management and are particularly interesting to apply. This type of model is therefore discussed in this section and in relation to the use of biomanipulation.

Biomanipulation is based on imposed changes in ecosystem structure. In this context, it is of interest to explain when and why top-down control is working (i.e. when the imposed structural changes can be predicted to work). This is possible by the use of catastrophe theory in lake eutrophication models. A short presentation of catastrophe theory is therefore included, in order to give the basis for the understanding of biomanipulation models.

The short-term results of biomanipulation have been encouraging, but it is unclear whether manipulated systems inevitably return to initial nutrient and turbidity conditions. Some observations (see Hosper 1989; van Donk et al. 1989) seem to indicate that if low nutrient concentration is combined with a relatively high biomass of predatory fish, a stable steady state will be attained. High nutrient concentration and high predator biomass will lead to an unstable clear-water state. On the other hand, turbid conditions may prevail even at medium nutrient concentrations, provided that predatory fish concentration is low. With introduction of more predators, however, conditions may improve significantly, even at medium nutrient concentrations.

Willemsen (1980) distinguishes two possible conditions.

**1** A 'bream state', characterised by turbid water and high nutrient concentration. Submerged vegetation is largely absent. Large amounts of bream (*Abramis abramis* L.) are found, whereas pike (*Esox lucius* L.) hardly occur.

**2** A 'pike state' characterised by clear water and low nutrient concentrations. Pike are abundant, whereas significantly fewer bream are found.

Willemsen's work shows that the pike/bream ratio is strongly correlated with water transparency, and that separation between the two states is relatively distinct (see Fig. 15.21).

Scheffer (1990) has used a mathematical model to describe shifts between these states, which may be catastrophic in character. The conceptual diagram of this model is presented in Fig. 15.22. Its components are connected by four relationships, which are incorporated in the model as simple Monod and Hill functions (Fig. 15.23). The general features of the model are reflected by the plot of the isoclines (Fig. 15.24). The low position of the pike isocline at high bream densities is explained by the decrease of vegetation which accompanies this state.

The rise of the bream isocline at low bream den-

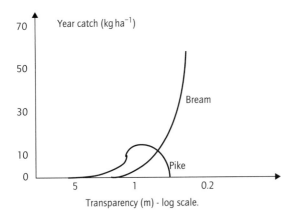

**Fig. 15.21** Annual catches (kg ha$^{-1}$) of pike (*Esox lucius* L.) and bream (*Abramis abramis* L.) plotted versus transparency (m).

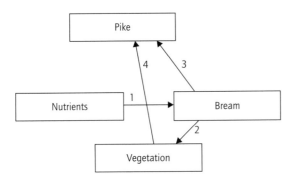

**Fig. 15.22** Conceptual model applied to describe shifts from 'bream state' to 'pike state' and vice versa.

sity is a result of the functional response used in the model. The isoclines intersect at three points, at which net growth of both pike and bream is equal to zero. The three points therefore represent equilibrium. However, although points 1 and 3 are stable equilibria, point 2 represents an unstable condition. The slightest perturbation will therefore cause the system to shift away from this point. The two stable points appear to correspond to Willemsen's pike (1) and bream (3) states described above.

The effect of eutrophication on the system may be visualised by plotting isoclines for different values of nutrient loading. The pike isocline is not affected by nutrient concentration, but the bream isocline changes, as shown in Fig. 15.25. The greater the nutrient concentration, the higher the position of the isocline. The behaviour of an oligotrophic system also can be derived easily from Fig. 15.25, with the lower bream isocline applying to that state. The only possible steady state under these conditions is low pike equilibrium.

With increasing nutrient concentration the equilibrium will then shift slowly upwards. Pike density will increase, whereas bream density does not change much. This response continues until the bream isocline reaches a position where the intersection point disappears. At this point the pike population will collapse.

**Turbid bream equilibrium** is then attained. If we change nutrients from high to low concentration, however, the result will be different. The system will remain within bream equilibrium to a very low nutrient concentration, although bream numbers will decrease slightly. Only at very low nutrient concentration does the intersection point representing bream equilibrium disappear, and the system return to a pike state.

Clearly this behaviour is analogous to other examples described by the application of catastrophe theory to biological systems (Jørgensen 1992a, 1994). Figure 15.26 shows the catastrophe fold where bream isocline is plotted versus nutrient concentration, assuming that pike numbers are in steady state. The isocline consists of the stable parts 1 and 3 of Fig. 15.24, and the unstable part 2 to the 'jump' between the two stable conditions.

Discontinuous response to increase and decrease of nutrient concentrations implies that reduction will not cause a significant decline in productivity, or a significant increase in water transparency, before rather low values are attained. However, it may be possible to 'push' equilibrium from point 3 to point 1 by addition of predatory fish. Thus two different concentrations of planktivorous fish may coexist at the same nutrient concentration, which explains the hysteresis reaction shown in Fig. 15.26.

The general modelling experience is that a given set of forcing functions produces a certain set

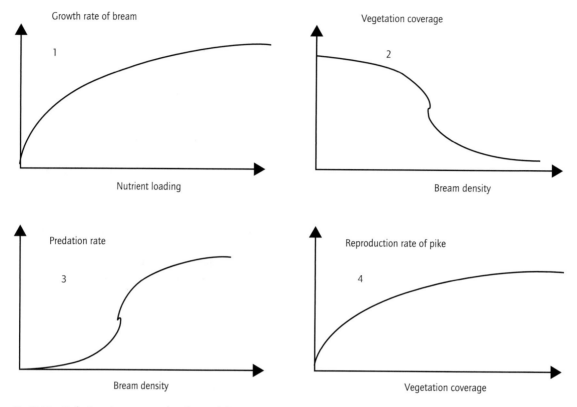

**Fig. 15.23** Relations incorporated in the model in Fig. 15.22.

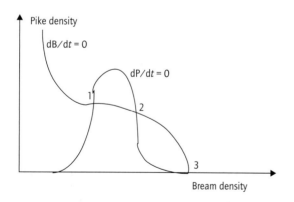

**Fig. 15.24** Zero isoclines of the bream (dB/dt = 0) and pike (dP/dt = 0).

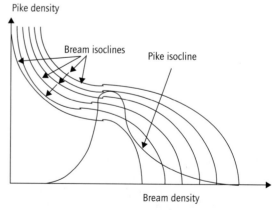

**Fig. 15.25** The position of isoclines and stable equilibria at different nutrient concentrations. The highest position of the bream isocline corresponds to the highest nutrient concentration.

*Models of Lakes and Reservoirs* 423

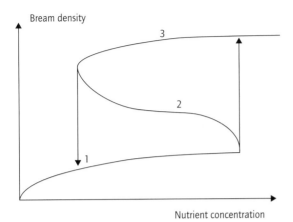

**Fig. 15.26** A catastrophe fold is shown by means of projection of the intersection line of isocline planes of bream and pike on the nutrient–bream plane.

of the state variables, including concentration of planktivorous fish, but when the set of equations which describes the ecosystem is formulated to cause catastrophic behaviour (in the mathematical sense!) we observe the reactions described above. In order to explain them further, models have been used as an experimental tool, in the sense that the model description is in accordance with lake models known to work well. Simulations with forcing functions for which there are no data available to control the model outputs are carried out.

A model used for such experiments is shown in Fig. 15.27. Only phosphorus is considered as a nutrient in this case, but it is of course feasible to consider both nitrogen and phosphorus. The model encompasses the entire food chain. The problem of modelling in the case of different inputs of nutrients is, however, that phytoplankton, zooplankton and planktivorous and carnivorous fish are all able to adjust their growth rate within certain

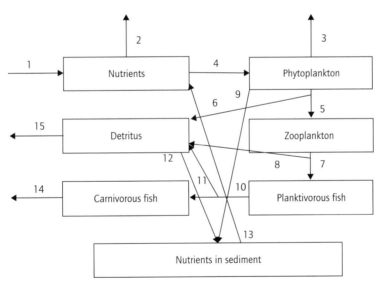

**Fig. 15.27** The model used as experimental tool in this section is shown. The state variables are nutrients (only phosphorus is considered in this case), phytoplankton, zooplankton, planktivorous fish, carnivorous fish, nutrients in sediment and detritus. The processes are: 1, inflow of nutrients; 2, outflow of nutrients; 3, outflow of phytoplankton; 4, uptake of nutrients by phytoplankton for growth; 5, grazing; 6, loss of detritus by grazing; 7, predation on zooplankton; 8, loss of detritus by this predation; 9, settling of phytoplankton; 10, predation by carnivorous fish; 11, loss of detritus by this predation; 12, settling of detritus; 13, release of nutrients from the sediment; 14, catch of fish; 15, outflow of detritus.

ranges. It is therefore necessary to test a series of simulations with different combinations of growth rates in order to find which gives the highest probability of survival for all four classes of species.

The results of these simulations can be summarised in the following points.

**1** Between total phosphorus concentration 70 μg to 180 μg L$^{-1}$, two values for numbers of planktivorous fish give a stable situation, with high probability of survival for all four classes of species. The lower value gives the highest concentration of zooplankton and of carnivorous fish, whereas phytoplankton concentration is low. Values for total phosphorus concentration are (of course) dependent on the model-description and must not be taken as fixed values for all lakes. They indicate only that, within an interval corresponding approximately to mesotrophic conditions, there are two, more or less stable configurations for lake ecosystem structure.

**2** The stable configuration with the lowest population of planktivorous fish corresponds to the smallest growth rate of zooplankton and phytoplankton This normally implies that it is species which are larger in size which make up the population (Peters 1983; Winfield, Volume 1, Chapter 15; Sarvala et al., Volume 1, Chapter 16).

**3** Above 180 μg, or below 70 μg total P L$^{-1}$, only one stable condition in each case is achieved. These correspond respectively to low numbers of planktivorous fish, or of phytoplankton. Zooplankton are relatively abundant below 70 μg and relatively scarce beyond 180 μg total P L$^{-1}$.

**4** When numbers of planktivorous fish are low (at either below 70 μg total P L$^{-1}$ or between 70 μg and 180 μg total P L$^{-1}$), phytoplankton are clearly controlled by relatively high zooplankton numbers (i.e. grazing pressure). Beyond 70 μg total P L$^{-1}$, phytoplankton are controlled by nutrient concentrations (see Pahl-Wostl, Volume 1, Chapter 17).

This exercise should be considered only qualitatively, but the results indicate that biomanipulation seems to be successful only over an intermediate range of nutrient loading, which is in accordance with many other results. The range is most probably dependent on conditions in specific lakes, which unfortunately makes it problematic to use modelling results more than qualitatively. The example uses many combinations of growth rates to find the best combination from the point of view of survival. Adjustment of the parameters corresponds to adaptation within some ranges.

In cases where significant changes in nutrient concentrations will take place, or where biomanipulation is considered for other reasons, it is necessary to develop models which can account for structural dynamic changes. Such models are, however, in their infancy and only limited experience is therefore available.

## 15.12 STRUCTURALLY DYNAMIC MODELS

Our present models are built on generally rigid structures and a fixed set of parameters. No change or replacement of components is usually possible. We need, however, to introduce parameters (properties) which can vary according to changing general conditions for the state variables (components). The idea is currently to test whether a change of the most crucial parameters produces a higher so-called goal function of the system and, if that is the case, to use that set of parameters (Jørgensen 2002).

Models which can account for change in species composition as well as the ability of the species (i.e. the biological components of our models) to change their properties (i.e. to adapt to the prevailing conditions imposed on them) are, as indicated in the introduction to this section, called **structural dynamic models**. They also may be called the **next** or **fifth generation** of ecological models, in order to underline that they are radically different from previous modelling approaches and can do more, namely describe changes in species composition.

It could be argued that the ability of ecosystems to replace current species with others which are better adapted can be addressed by construction of models which incorporate all possible species for the entire period studied. There are, however, two essential disadvantages. First, such models become very complex, as they will contain many state variables for each trophic level. Thus many more need to be calibrated and validated, which will introduce a high uncertainty and render application of the model very case specific (Nielsen 1992a,b). In addition, the model will still be rigid and not replicate the ecosystem property of parameters continuously changing even without changing species composition.

Several goal functions have been proposed, but only very few models have been developed which account for change in species composition or for the ability of the species to change their properties within some limits. Straškraba (1979) used maximisation of biomass as the governing principle (the goal function). The model adjusts one or more selected parameters in order to achieve maximum biomass at every instance. A routine is included which computes biomass for all possible combinations of parameters within a given realistic range. The combination which gives the maximum biomass is selected for the next time-step and so on. Biomass, however, can hardly be used in models with more trophic levels. To add biomass of fish and phytoplankton together will lead to biased results.

The thermodynamic variable **exergy** has been used most widely as a goal function in ecological models. One of the few lake case-studies available will be presented and discussed below. There are two pronounced advantages of exergy as the goal function compared with entropy and maximum power (Odum 1983). These are (i) that it is defined far from thermodynamic equilibrium and (ii) is related to state variables which are easily determined or measured. As exergy is not a common thermodynamic function, however, we need first to describe it.

The term exergy (i) expresses energy with a built-in measure of quality, (ii) quantifies natural resources (Eriksson et al. 1976) and (iii) can be considered as fuel for any system which converts energy and matter in a metabolic process (Schrödinger 1944). Ecosystems therefore convert exergy, and **exergy flow** through the system is necessary in order to keep the system functioning. Exergy (as the amount of work the system can perform when it is brought into equilibrium with its environment) measures the distance from the 'inorganic soup' in energy terms, as will be further explained below. Exergy is therefore dependent on both the environment and the system and not entirely on the system. It is therefore not a state variable, as for instance are free energy and entropy.

If we assume a reference environment which represents the system (ecosystem) at thermodynamic equilibrium, the configuration illustrated in Fig. 15.28 is valid. All components are inorganic, and at the highest possible oxidation state, and homogeneously distributed in the system. Thus there is as much free energy as possible utilised in order to perform work (no gradients). As the chemical energy embodied in the organic components and the biological structure contributes by far the most to the exergy content of the system, there seems to be no reason to assume a (minor) temperature and pressure difference between it and the reference environment. Under these circumstances we can calculate the exergy content of the system as provided entirely by chemical energy: $\Sigma_c(\mu_c - \mu_{ceq})N_i$, where $\mu$ is chemical potential and $N$ is number of molecules. We can also determine its exergy compared with that of a similar system at the same temperature and pressure but in the form of an inorganic soup without any life, biological structure, information or organic molecules. As $(\mu_c - \mu_{ceq})$ can be found from the definition of the chemical potential (replacing activities by concentrations), we obtain the following expressions for the exergy:

$$Ex = RT \sum_{i=0}^{i=n} C_i \ln C_i / C_{i,\mathrm{eq}} \quad (15.31)$$

As can be seen, exergy measures the difference between free energy (given the same temperature and pressure) of an ecosystem and that of the surrounding environment. If the system is in equilibrium with its surroundings, exergy is zero: $n = 0$ accounts for inorganic compounds and $n = 1$ corresponds to detritus.

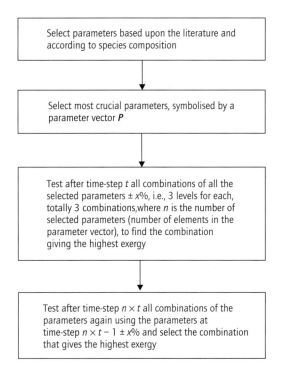

Fig. 15.28 The procedure used for the development of structural dynamic models.

As the only way to shift systems away from equilibrium is to perform work on them, and because the available work in a system is a measure of the ability to do so, we need to distinguish between the system and its environment (or thermodynamic equilibrium alias the inorganic soup). Therefore it is reasonable to use the available work (i.e. the exergy) as a measure of the distance from thermodynamic equilibrium.

Survival implies maintenance of biomass and growth means its increase. Exergy is needed in order to construct biomass, which therefore possesses exergy, which is transferable to support other processes. Survival and growth therefore can be measured by use of the thermodynamic concept exergy, which may be understood as **the free energy relative to the environment**; see eqn.(15.31).

Darwin's theory of natural selection (1859) therefore may be reformulated in thermodynamic terms and expanded to system level, as follows: **The prevailing conditions of an ecosystem steadily change. The system will continuously select those species which contribute most to the maintenance or even to the growth of the exergy of that system.**

Notice that the thermodynamic translation of Darwin's theory requires **populations** to possess properties of reproduction, inheritance and variation. Selection of species which contribute most to the exergy of the system under the prevailing conditions requires that there are sufficient individuals with different properties for selection to take place. Reproduction and variation must be high, and once a change has taken place it must be conveyed to the next generation via better adaptation. Notice also that change in exergy is not necessarily ≥0, but depends on the resources of the ecosystem. The above proposition claims, however, that an ecosystem attempts to reach the highest possible exergy level under the given circumstances with the genetic pool available (Jørgensen & Mejer 1977, 1979).

It is not possible to measure exergy directly, but it is possible to compute it if the composition of the ecosystem by eqn. (15.31) is known: $C_i$ represents the $i$th component expressed in a suitable unit (e.g. for phytoplankton in a lake, $C_1$ milligrams nutrient in the phytoplankton per litre of lake water) and $C_{i,eq}$ is the concentration of the $i$th component at thermodynamic equilibrium. For detritus, exergy can be found on the basis of equilibrium constants which give the ratio between the concentration of detritus in the ecosystem and that at thermodynamic equilibrium. The exergy content of detritus is approximately $18\,kJ\,g^{-1}$, which can be compared with the exergy (chemical energy) content of mineral oil of about $42\,kJ\,g^{-1}$.

For more complex compounds, such as unicellular organisms, $C_{i,eq}$ will be smaller. The probability of forming one *coli* bacterium or a simple phytoplankton cell can be estimated from the number of genes, which again determine the sequence of amino acids. It has been found by this method (Jørgensen 2002) that the exergy content of phytoplankton is 3.4 times greater than that of detritus, owing to their information content. The

Table 15.15  Approximate number of non-repetitive genes. (Sources: Cavalier-Smith 1985; Li & Grauer 1991; Lewin 1994)

| Organisms | Number of information genes | Conversion factor* |
|---|---|---|
| Detritus | 0 | 1 |
| Minimal cell (Morowitz 1992) | 470 | 2.7 |
| Bacteria | 600 | 3.0 |
| Algae | 850 | 3.9 |
| Yeast | 2000 | 6.4 |
| Fungus | 3000 | 10.2 |
| Sponges | 9000 | 30 |
| Moulds | 9500 | 32 |
| Plants, trees | 10,000–30,000 | 30–87 |
| Worms | 0.500 | 35 |
| Insects | 0.000–15,000 | 30–46 |
| Jellyfish | 0.000 | 30 |
| Zooplankton | 0.000–15,000 | 30–46 |
| Fish | 0.000–120,000 | 300–370 |
| Birds | 20,000 | 390 |
| Amphibians | 120,000 | 370 |
| Reptiles | 130,000 | 400 |
| Mammals | 140,000 | 430 |
| Human | 250,000 | 740 |

* Based on number of information genes and the exergy content of the organic matter in the various organisms, compared with the exergy contained in detritus: 1 g detritus contains about 18 kJ exergy (= energy which can do work).

probability of forming multicell organisms at thermodynamic equilibrium is of course even lower, in that additional exergy is required to make up the more complex structure and to provide the information embedded in its structure. The exergy content of zooplankton is therefore provisionally estimated to be around 35 times that of detritus, and of fish to be more than 300 times. Such conversion factors, for several classes of organisms, are listed in Table 15.15.

Proposals for the concentrations of various biological components at thermodynamic equilibrium do not, of course, lead to any exact value of exergy or even of its exact relative change. They will, however, account for relative changes caused by variations in the properties (parameters) of **organisms**. The inorganic constituents of an ecosystem do not create similar computational difficulties, but the thermodynamic equilibrium concentrations will be the total concentrations of the various elements, corresponding to the fact that all chemical compounds in the primeval soup were in inorganic form, which we often use as the reference state for our exergy calculations.

So far, this section has presented the theoretical background for the application and development of structurally dynamic models. These are important tools in environmental management, as they account for current changes of species composition, and the properties of the organisms in the focal ecosystem. The idea of the new generation of models presented here is to find a new set of parameters (limited for practical reasons to the most crucial, i.e., the most sensitive) better suited to the prevailing conditions of the ecosystem, defined in the Darwinian sense by the ability of the species to survive and grow. This may, as indicated above, be measured by the use of exergy (Jørgensen 1982, 1986, 1990, 1994; Jørgensen & Mejer 1977, 1979; Mejer & Jørgensen 1979). Figure 15.28 shows the proposed modelling procedure which has been applied in the cases presented below.

Thirteen case studies of biogeochemical modelling have used exergy calculations continuously to vary parameters. One of these (Søbygaard Lake) will be described here as an illustration of what can be achieved by this approach. The results (Jeppesen et al. 1989) are particularly fitted to test its applicability to structural dynamic models.

Søbygaard is a shallow lake (depth 1 m) with a short retention time (15–20 days). Nutrient loading, namely for phosphorus, from 30 g to 5 g P m$^{-2}$ yr$^{-1}$ was significantly lowered in 1982. The decreased load did not, however, produce reduced nutrient and chlorophyll concentrations in the period 1982–1985, owing to internal loading caused by storage of nutrients in the sediment (Jeppesen et al. 1989; Søndergård 1989).

Radical changes were then observed during the period 1985–1988. Recruitment of planktivorous fish was significantly reduced during the interval

1984–1988, owing to very high pH. Zooplankton increased, and phytoplankton decreased in concentration and the summer average of chlorophyll $a$ was reduced from $700\,\mu g\,L^{-1}$ in 1985 to $150\,\mu g\,L^{-1}$ in 1988. The phytoplankton population even collapsed during shorter periods, owing to extremely high zooplankton concentrations.

Simultaneously, phytoplankton species increased in size. Growth rates declined, and higher settling rates were observed (Kristensen & Jensen 1987). The case study shows, in other words, pronounced ecosystem structural changes, caused by biomanipulation-like events. Primary production was, however, not higher in 1985 than in 1988, owing to pronounced self-shading by smaller algae. It was therefore very important to include a self-shading effect in the model. Simultaneously more sloppy feeding (see Gliwicz, Volume 1, Chapter 14) of zooplankton was observed, as a shift from *Bosmina* to *Daphnia* took place.

The model contains six state variables which are all forms of nitrogen: fish, zooplankton, phytoplankton, detritus-nitrogen, soluble-nitrogen and sedimentary-nitrogen. The equations are given in Table 15.16. As nitrogen is the nutrient controlling eutrophication in this particular case, it may be sufficient to include only this element.

The aim of the study is to describe by use of a structural dynamic model the continuous changes in the most essential parameters using the procedure shown in Fig. 15.28. The data from 1984 to 1985 were used to calibrate the model. The two parameters which it was intended to change, for the period 1985 to 1988, received the following values:

| | |
|---|---|
| maximum growth rate of phytoplankton: | 2.2 day$^{-1}$ |
| settling rate of phytoplankton: | 0.15 day$^{-1}$ |

The state variable fish-nitrogen was kept constant at 6.0 during the calibration period, but during the period 1985–88 an increased fish mortality was introduced in order to reflect the increased pH. Fish stock was thereby reduced to $0.6\,mg\,N\,L^{-1}$ — notice the equation 'mort = 0.08 if fish >6 (may be changed to 0.6) else almost 0'. A time-step of $t = 5$ days and $x\% = 10\%$ was applied (Fig. 15.28). This means that nine runs were needed for each time-step in order to select the parameter combination which gives the highest exergy. The results are shown in Fig. 15.29. Changes in parameters from 1985 to 1988 (summer) are summarised in Table 15.17. It may be concluded that the proposed pro-

**Table 15.16** Model equations for Søbygaard Lake

fish = fish + $d_t$ × (–mort + predation)
INIT(fish) = 6
na = na + $d_t$ × (uptake – graz – outa – mortfa – settl – setnon)
INIT(na) = 2
nd = nd + $d_t$ × (–decom – outd + zoomo + mortfa)
INIT(nd) = 0.30
ns = ns + $d_t$ × (inflow – uptake + decom – outs + diff)
INIT(ns) = 2
nsed = nsed + $d_t$ × (settl – diff)
INIT(nsed) = 55
nz = nz + $d_t$ × (graz – zoomo – predation)
INIT(nz) = 0.07
decom = nd × (0.3)
diff = (0.015) × nsed
exergy = total_n × (Structura-exergy)
graz = (0.55) × na × nz/(0.4 + na)
inflow = 6.8 × qv
**mort = IF fish > 6 THEN 0.08 × fish ELSE 0.0001 × fish**
mortfa = (0.625) × na × nz/(0.4 + na)
outa = na × qv
outd = qv × nd
outs = qv × ns
pmax = uptake × 7/9
predation = nz × fish × 0.08/(1 + nz)
qv = 0.05
setnon = na × 0.15 × (0.12)
settl = (0.15) × 0.88 × na
Structural exergy = (nd + nsed/total_n) × (LOGN(nd + nsed/total_n) + 59) + (ns/total_n) × (LOGN(ns/total_n) – LOGN(total_n)) + (na/total_n) × (LOGN(na/total_n) + 60) + (nz/total_n) × (LOGN(nz/total_n) + 62) + (fish/total_n) × (LOGN(fish/total_n) + 64)
total_n = nd + ns + na + nz + fish + nsed
uptake = (2.0 – 2.0 × (na/9)) × ns × na/(0.4 + ns)
zoomo = 0.1 × nz

Table 15.17 Parameter combinations giving the highest exergy

| | Maximum growth rate (day$^{-1}$) | Settling rate (m day$^{-1}$) |
|---|---|---|
| 1985 | 2.0 | 0.15 |
| 1988 | 1.2 | 0.45 |

Table 15.18 Exergy and stability by different combinations of parameters and conditions

| | Parameter 1985 | Conditions 1988 |
|---|---|---|
| 1985 | 75.0 stable | 39.8 (average) violent fluctuations (chaos) |
| 1988 | 38.7 stable | 61.4 (average) only minor fluctuations |

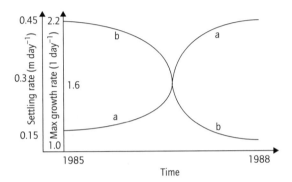

Fig. 15.29 Continuously changed parameters obtained from application of a structural dynamic modelling approach to Søbygaard Lake: a, the settling rate of phytoplankton; b, the maximum growth rate of phytoplankton.

cedure (Fig. 15.28) can simulate approximately the observed change in ecosystem structure.

Maximum growth rate of phytoplankton is reduced by 50% from 2.2 day$^{-1}$ to 1.1 day$^{-1}$, approximately according to increase in size. It was observed that average size was increased from a few hundred µm$^3$ to 500–1000 µm$^3$, a factor of 2–3 (Jeppesen et al. 1989). This would correspond to a specific growth reduction by a factor $f = 2^{2/3} - 3^{2/3}$ (Jørgensen 1994).

Thus:

grow rate in 1988 = growth rate in 1985/$f$

(15.32)

where $f$ is between 1.58 and 2.08, whereas 2.0 is found in Table 15.17 by use of the structural dynamic modelling approach. Kristensen & Jensen (1987) observed that settling was 0.2 m day$^{-1}$ (range 0.02–0.4) during 1985, but 0.6 m day$^{-1}$ (range 0.1–1.0) in 1988. Using the structural dynamic modelling approach, the increase was found to be 0.15 m to 0.45 m day$^{-1}$, a slightly lower set of values, but the same (3) phytoplankton concentration as chlorophyll a was simultaneously reduced from 600 µg to 200 µg L$^{-1}$, approximately in accord with observations.

It may be concluded that, in this instance, structurally dynamic modelling gave an acceptable result. Validation of the model, and the procedure in relation to structural changes, was positive. The approach, is, of course, never better than the model applied, and the model presented here may be criticised for being too simple, and not accounting for changes in zooplankton.

For further elucidation of the importance of introducing parameter shifts, an attempt was made to run data for 1985 with parameter combinations for 1988 and vice versa. These results (Table 15.18) show that it is of great importance to apply the appropriate parameter set to given conditions. If those for 1985 are used for 1988, significantly less exergy is obtained, and the model behaves chaotically. Parameters for 1988 used on 1985 conditions give significantly less exergy.

Experience presented in this section shows that models can be applied to explain why biomanipulation may work under some circumstances, and not others. Catastrophe theory is able to explain the appearance of hysteresis in the relation between nutrient level and eutrophication. Qualitatively, results can be used to explain that hysteresis exists over an intermediate range of nutrient loadings, so that biomanipulation has worked properly over this range, but not above or below.

The structurally dynamic approach has been used recently to calibrate eutrophication models. It is known that different phytoplankton and zooplankton species are abundant during different periods of the year. Therefore, a calibration based upon one parameter set for the entire year will not capture the succession which we know takes place over the year. By using exergy optimisation to capture the succession, i.e. the parameter giving the best survival for phytoplankton and zooplankton over the year, it has been possible to improve the calibration results (see Jørgensen *et al.* 2002).

Models have, in other words, been an appropriate tool in our effort to understand the results of structural change. In addition to the use of a goal function, it is also possible to base the structural changes on knowledge, for instance under which conditions specific classes of phytoplankton are dominant. This knowledge can be used to select the correct combination of parameters, and is very well illustrated in Reynolds (1996; see also Reynolds *et al.*, 2001). That the combined application of expert knowledge and the use of exergy as a goal function will offer the best solution to the problem of making models more accordant with the properties of real ecosystems cannot be ruled out. Such combinations would draw upon the widest possible knowledge at this stage.

## 15.13 EXPLANATION OF SUCCESS AND FAILURE OF BIOMANIPULATION BY STRUCTURALLY DYNAMIC MODELS

Eutrophication and remediation of lacustrine environments does not proceed according to a linear relationship between nutrient load and vegetative biomass, but rather displays a sigmoid trend, with delay, as shown in Fig. 15.30. This **hysteresis reaction** is completely consistent with observations (Hosper 1989; Van Donk *et al.* 1989), and can be explained by changes in ecosystem structure (de Bernardi 1989; Hosper 1989; Sas 1989; de Bernardi

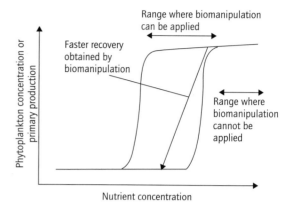

Fig. 15.30 Hysteresis relationship between nutrient concentration and eutrophication, measured by the phytoplankton concentration. The possible effect of biomanipulation is shown. An effect of biomanipulation can hardly be expected above a certain concentration of nutrients, as indicated on the diagram. Biomanipulation can only give the expected results in the range where two different structures are possible.

& Giussani 1995). Lake ecosystems exhibit a marked buffering capacity to increasing nutrient concentration, which can be explained by an increased removal rate of phytoplankton by grazing and settling. Under these circumstances, zooplankton and fish are maintained at relatively high abundance.

At a certain level of eutrophication, however, it is not possible for zooplankton to increase their grazing rate further. Phytoplankton concentration will then increase very rapidly, even with only a slight further increase in nutrient concentrations. When nutrient inputs are decreased under these conditions, a similar buffering capacity to variation is observed. Ecosystem structure has now changed, to high concentrations of phytoplankton and planktivorous fish, which causes a resistance and delay to a change. The second and fourth trophic levels regain their importance.

The existence of two possible ecosystem states over a certain range of nutrient concentrations may explain why biomanipulation has not always been successful. According to observations referred to in the literature, success is associated with total phosphorus concentration below 50 µg L$^{-1}$ (Lammens 1988), or at least below 100–200 µg L$^{-1}$ (Jeppesen et al. 1990). Disappointing results are often related to phosphorus concentration above approximately 120 µg L$^{-1}$ (Benndorf 1987, 1990), beyond which it is difficult to control standing stocks of planktivorous fish (Shapiro 1990; Koschel et al. 1993).

Scheffer (1990) has used a mathematical model based on catastrophe theory in order to describe these shifts in ecosystem structure. This model does not, however, consider shifts in species composition, which is of particular importance for biomanipulation. According to several authors (de Bernardi & Giussani 1995; Giussani & Galanti 1995), when we increase nutrient concentrations, zooplankton populations undergo structural change (e.g., replacement of calanoid copepods by small cladocera and rotifers). Hence, a test of structurally dynamic models could be used in order to develop better understanding of the relationship between concentrations of nutrients and the vegetative biomass, and to explain the possible results of any given biomanipulation.

Jørgensen & de Bernardi (1998) have developed a structurally dynamic model whose aim is to understand precisely these changes in ecosystem structure and species composition. The model contains six state variables, namely dissolved inorganic phosphorus (DIP), phytoplankton (phyt), zooplankton (zoopl), planktivorous fish (fish 1), predatory fish (fish 2) and detritus (detritus). The forcing functions are (i) input of phosphorus (in P), and (ii) throughflow of water ($Q$), determining water retention time, and also outflow of detritus and phytoplankton.

Simulations have been carried out for phosphorus concentrations in inflowing water of 0.02, 0.04, 0.08, 0.12, 0.16, 0.20, 0.30, 0.40, 0.60 and 0.80 mg L$^{-1}$. For each of these cases, the model was run for any combination of phosphorus uptake rate of 0.06, 0.05, 0.04, 0.03, 0.02, 0.01 per 24 hour period ($(24 h)^{-1}$), and a grazing rate of 0.125, 0.15, 0.2, 0.3, 0.4, 0.5, 0.6, 0.8 and 1.0 $(24 h)^{-1}$. When these two parameters were varied, simultaneous changes of phytoplankton and zooplankton mortalities were made, according to allometric principles (Peters 1983). Those parameters which are changed in order to account for dynamic changes in structure are therefore phytoplankton growth rate (uptake rate of phosphorus) and mortality, and zooplankton growth rate and mortality.

Settling rate of phytoplankton was made proportional to (length)$^2$. Half of the additional sedimentation (which takes place when the size of phytoplankton increases according to a decrease in uptake rate) was allocated to detritus, in order to account for resuspension or faster release from the sediment. Sensitivity analysis has revealed that exergy is most sensitive to changes in these five selected parameters, which also represent those which change significantly by size. Six of the nine levels respectively selected represent approximately the range in size for phytoplankton and zooplankton.

For each phosphorus concentration, 54 simulations were carried out in order to account for all combinations of the two key parameters. Simulations over 3 years (1100 days) were applied in order to ensure that either steady state, limit cycles, or chaotic behaviour would be attained.

This structurally dynamic modelling approach presumes that the combination with the highest exergy should be selected as representing process rates in the ecosystem. If exergy oscillates even during the last 200 days of the simulation, the average value for the last 200 days was used in order to decide on which parameter combination would give the highest exergy. Those combinations of the two parameters (uptake rate of phosphorus for phytoplankton, grazing rate of zooplankton) giving the highest exergy at different phosphorus inputs are observed and plotted for zooplankton in Fig. 15.31. The uptake rate of phosphorus for phytoplankton is gradually decreasing when the phosphorus concentration increases. As can be seen, the zooplankton grazing rate changes radically (from 0.4 $(24 h)^{-1}$ to 1.0 $(24 h)^{-1}$, i.e. from larger species to smaller species) at a phosphorus concentration of 0.12 mg L$^{-1}$, which is according to the expectations.

Figure 15.32 shows the exergy (named on the diagram 'information'), with an uptake rate according to the results, and respective grazing rates of

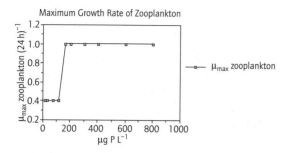

**Fig. 15.31** The maximum growth rate of zooplankton obtained by the structurally dynamic modelling approach is plotted versus phosphorus concentration.

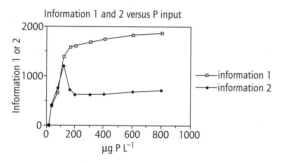

**Fig. 15.32** The exergy is plotted versus the phosphorus concentration. Information 1 corresponds to maximum zooplankton growth rate of 1 $(24\,h)^{-1}$, and information 2 to maximum zooplankton growth rate of 0.4 $(24\,h)^{-1}$. The other parameters are the same for the two plots, including the maximum phytoplankton growth rate as a function of the phosphorus concentration taken from Fig. 15.4.

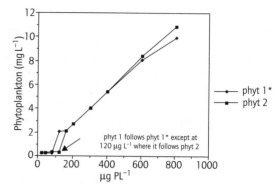

**Fig. 15.33** Phytoplankton concentration as a function of phosphorus concentration for parameters corresponding to 'information 1' and 'information 2' (see Fig. 15.32). The plot named 'phyt 1*' coincides with 'phyt 1', except for phosphorus concentration of $0.12\,mg\,L^{-1}$, where the model exhibits limit cycles. At this concentration, information 1* represents greater phytoplankton concentration, whereas information 1 indicates the lower phytoplankton concentration. Notice that the structural dynamic approach can explain the hysteresis reactions.

1.0 $(24\,h)^{-1}$ (information 1) and 0.4 $(24\,h)^{-1}$ (information 2). Below phosphorus concentration $0.12\,mg\,L^{-1}$, information 2 is slightly higher, whereas above this concentration, information 1 is significantly higher. Phytoplankton concentration increases with rising phosphorus input for both parameter sets (see Fig. 15.33), and planktivorous fish exhibit significantly increased numbers (by a grazing rate of 1.0 $(24\,h)^{-1}$) when phosphorus concentration is $\geq 0.12\,mg\,L^{-1}$ (valid for the high exergy level). The difference is minor below this concentration. Concentration of carnivorous fish is higher for case 2, corresponding to a grazing rate of 0.4 $(24\,h)^{-1}$ for phosphorus concentrations below $0.12\,mg\,L^{-1}$. Above this value, differences are minor, but at phosphorus concentration of $0.12\,mg\,L^{-1}$, numbers are significantly higher for a grazing rate of 1.0 $(24\,h)^{-1}$, particularly for the lower exergy level. Here also, zooplankton numbers are greatest.

If it is presumed that exergy indices can be used as a goal function in ecological modelling, the results seem to be able to explain why we observe a shift in grazing rate of zooplankton at phosphorus concentrations in the range of $0.1-0.15\,mg\,L^{-1}$. Above this concentration, the ecosystem selects the smaller species of zooplankton because it means a higher level of the exergy index, which can be translated to a higher rate of survival and growth. It is interesting that this shift in grazing rate only gives slightly increased zooplankton, but the exergy index level becomes significantly higher. This may be translated as survival and growth for the entire ecosystem. Simultaneously, a shift from a zooplankton, predatory fish system to one made up of phytoplankton and particularly planktivorous fish takes place.

It is interesting that the exergy indices, and the four biological components of the model for phosphorus concentrations at or below $0.12\,\mathrm{mg\,L^{-1}}$, are only slightly different for the two parameter combinations. This finding can explain why biomanipulation is often more successful in this concentration range. Above $0.12\,\mathrm{mg\,L^{-1}}$, differences are much more pronounced, and the exergy index is clearly higher for a grazing rate of 1.0 $(24\,\mathrm{h})^{-1}$. It should therefore be expected that, after biomanipulation, ecosystems may easily revert to abundance of planktivorous fish and phytoplankton. These observations are consistent with the general experience of success and failure of biomanipulation (see above).

An interpretation of the results points toward a key shift at $0.12\,\mathrm{mg\,P\,L^{-1}}$, where a grazing rate of $1.0\,(24\,\mathrm{h})^{-1}$ yields **limit cycles**. It indicates an instability, and a probable easy shift to a grazing rate of $0.4\,(24\,\mathrm{h})^{-1}$, although the exergy level is on average highest for the greater grazing rate. A preference for a grazing rate of $1.0\,(24\,\mathrm{h})^{-1}$ at this phosphorus concentration therefore should be expected, but lower or higher zooplankton biomass is dependent on the initial conditions.

If concentrations of zooplankton and carnivorous fish are low, and high for fish 1 (planktivorous fish) and phytoplankton (i.e., the system is changing from higher phosphorus concentrations), the simulation gives (with high probability) a low concentration of zooplankton and fish 2. When the reverse occurs (i.e. the system is changing from high concentrations of zooplankton and fish 2), it gives (again with high probability) high concentration of zooplankton and fish 2, which correspond to an exergy index level slightly lower than obtained by a grazing rate of $0.4\,(24\,\mathrm{h})^{-1}$. This grazing rate will therefore prevail. As populations of zooplankton and particularly of carnivorous fish and (in the other direction) of fish 1 take time to recover, these observations also explain the presence of hysteresis reactions.

The model is considered to be generally applicable, and has been used to discuss the general relationship between nutrient concentration and vegetative biomass, and the general experiences of application of biomanipulation. When it is used in specific cases, it may, however, be necessary to include a greater number of local details, and to change some of the process descriptions in order to account for site-specific properties, a procedure which accords with general modelling strategy (section 15.3). It could be modified in order to include two state variables to cover zooplankton, one for larger and one for smaller species. Both zooplankton state variables should of course possess a current change of the grazing rate according to the maximum value of the goal function.

The model probably also could be improved by introduction of size preference for the grazing and the two predation processes, which is in accordance with numerous observations. In spite of these shortcomings, it has been possible to give accurate qualitative descriptions of the reaction to changed nutrient concentration and biomanipulation, and even to indicate approximately the key phosphorus concentration at which structural changes may occur. This may be due to the increased robustness of the structural dynamic modelling approach.

A further hysteresis behaviour obtained by the use of structurally dynamic models of lakes has been published recently (Zhang et al. 2003a,b). This study focuses on the well-known switch between submerged vegetation and phytoplankton in shallow lakes. The model results indicate that between about 100 and $250\,\mathrm{\mu g\,P\,L^{-1}}$ both types of ecosystem structure may exist—i.e., they exhibit hysteresis in this range. This result is completely in accordance with observations from many shallow lakes (Scheffer et al. 2001).

Ecosystems are very different from physical systems, mainly due to their enormous adaptability. It is therefore crucial to develop models which are able to account for this property if we wish to obtain reliable results. The use of exergy as goal functions to cover the concept of fitness seems to offer an excellent prospect of developing a new generation of models, which will be able to consider the adaptability of ecosystems and to describe shifts in species composition. The latter advantage is probably the most important, because a description of the major species in an ecosystem is often more essential than to assess the concentration of the focal state variables.

## 15.14 REFERENCES

Ahlgren, I. (1973) *Limnologiske studier av Sjöen Norrviken. Avlastnings effekter.* Scripta Limnologica Upsalienca No. 333.

Alcamo, J., Shaw, R. & Hordjik, L. (eds) (1990) *The RAINS Model of Acidification.* Kluwer, Dordrecht, 280 pp.

Aoyama, I., Yos. Inoue & Yor. Inoue. (1978) Simulation analysis of the concentration process of trace heavy metals by aquatic organisms from the viewpoint of nutrition ecology. *Water Research*, **12**, 837–42.

Appleby, P.G. (2001) Chronostratigraphic techniques in recent sediments. In: Last, W.M. & Smol, J.P. (eds), *Tracking Environmental Change using Lake Sediments*, Vol. 1, *Basin Analysis, Coring, and Chronological Techniques.* Kluwer, Dordrecht, 171–203.

Arp, P.A. (1983) Modelling the effects of acid precipitation on soil leachates: a simple approach. *Ecological Modelling*, **19**, 105–18.

Banks, R.B. (1975) Some features of wind action on shallow lakes. *American Society of Civil Engineers, Journal of the Environmental Engineering Division*, **101**, 813–27.

Banks, R.B. & Herrera, F.F. (1977) Effect of sun and rain on surface reaeration. *American Society of Civil Engineers, Journal of the Environmental Engineering Division*, **103**, 489–503.

Bartell, S.M., Gardner, R.H. & O'Neill, R.V. (1984) The fates of aromatics model. *Ecological Modelling*, **22**, 109–23.

Beeton, A.M. (1958) Relationship between Secchi disk readings and light penetration in Lake Huron. *American Fisheries Society Transactions*, **87**, 73–9.

Benndorf, J. (1987) Food-web manipulation without nutrient control: A useful strategy in lake restoration? *Schweizisches Zeitschrift für Hydrologie*, **49**, 237–48.

Benndorf, J. (1990) Conditions for effective biomanipulation. Conclusions derived from whole-lake experiments in Europe. *Hydrobiologia*, **200/201**, 187–203.

Breck, J.E., DeAngelis, D.L., Van Winkle, W. & Christensen, S.W. (1988) Potential importance of spatial and temporal heterogeneity in pH, Al and Ca in allowing survival of a fish population: a model demonstration. *Ecological Modelling*, **41**, 1–16.

Brown, D.S. & Salder, K. (1981) The chemistry and fishery status of acid lakes in Norway and their relationships to European sulfur emission. *Journal of Applied Ecology*, **18**, 434–41.

Cavalier-Smith, T. (1985) *The Evolution of Genome Size.* Wiley, Chichester, 438 pp.

Chen, C.W. & Orlob, G.T. (1975) Ecological simulations of aquatic environments. In: Patten, B.C. (ed.), *System Analysis and Simulation in Ecology*, Vol. 3. Academic Press, New York, 476–588.

Chen, C.W., Dean, J.D., Gherini, S.A. & Goldstein, R.A. (1982) Acid rain model: hydrological module. *American Society of Civil Engineers, Journal of the Environmental Engineering Division*, **108**, 455–72.

Costanza, R. & Sklar, F.H. (1985) Articulation, accuracy and effectiveness of mathematical models: a review of freshwater wetland applications. *Ecological Modelling*, **27**, 45–69.

Dahl-Madsen, K.I. & Strange-Nielsen, K. (1974) Eutrophication models for ponds. *Vand*, **5**, 24–31.

Darwin, C. (1859) *The Origin of Species.* John Murray, London, 494 pp.

De Bernardi, R. (1989) Biomanipulation of aquatic food chains to improve water quality in eutrophic lakes. In: Ravera, O. (ed.), *Ecological Assessment of Environmental Degradation, Pollution and Recovery.* Elsevier, Amsterdam, 195–215.

De Bernardi, R. & Giussani, G. (1995) Biomanipulation: Bases for a Top-down Control. In: de Bernardi, R. & Giussani, G (eds), *Guidelines of Lake Management*, Vol. 7, *Biomanipulation in Lakes and Reservoirs*. ILEC and UNEP, Otsu, Shiga, Japan, 1–14.

Di Toro, D.M. (1980) Applicability of cellular equilibrium and Monos theory to phytoplankton growth kinetics. *Ecological Modelling*, **8**, 201–18.

Dørge, J. (1991) *Model for nitrogen cycling in freshwater wetlands.* MSc thesis, University of Copenhagen, 228 pp.

Edwards, R.W. & Owens, M. (1965) The oxygen balance of streams. In: Goodman, G.T. (ed.), *Ecology and the Industrial Society.* Fifth Symposium of the British Ecological Society. Blackwell, Oxford, 149–72.

Erikson, B., Erikson K.E. & Wall, G. (1976) *Basic Thermodynamics of Energy Conversions and Energy Use.* Institute of Theoretical Physics, Göteborg.

Fagerstrøm, T. & Aasell, B. (1973) Methyl mercury accumulation in an aquatic food chain. A model, and implications for research planning. *Ambio*, **2**, 164–71.

Fedra, K. (1984) Interactive water quality simulation in a regional framework: a management oriented approach to lake and watershed modelling. *Ecological Modelling*, **21**, 209–20.

Fillos, J. & Molof, A.H. (1972) Effect of benthal deposits on oxygen and nutrient economy of flowing waters. *Journal of the Water Pollution Control Federation*, **44**, 644–62.

Fomsgaard, I. (1997) Modelling the mineralisation kinetics for low concentrations of pesticides in surface and subsurface soil. *Ecological Modelling*, **102**, 175–208.

Fomsgaard, I. & Kristensen, K. (1999) Influence of microbial activity, organic carbon content, soil texture and soil depth on mineralisation rates of low concentrations of $^{14}C$ mecoprop—development of a predictive model. *Ecological Modelling*, **122**, 45–68.

Gillett, J.W., *et al.* (1974) *A Conceptual Model for the Movement of Pesticides through the Environment*. Report EPA 600/3-74-024, National Environmental Research Center, U.S. Environmental Protection Agency, Corvallis, OR, 79 pp.

Giussani, G. & Galanti, G. (1995) Case study: Lake Candia (Northern Italy). In: De Bernardi, R. & Giussani, G. *Guidelines of Lake Management*, Vol. 7, *Biomanipulation in Lakes and Reservoirs*. ILEC and UNEP, Otsu, Shiga, Japan, 135–46.

Gromiec, M.J. & Gloyna, E.F. (1973) *Radioactivity Transport in Water*. Final Report No. 22, to U.S. Atomic Energy Commission, Washington, DC, Contract AT(11–1)-490, 212 pp.

Halfon, E. (1984) Error analysis and simulation of *Mirex* behavior in Lake Ontario. *Ecological Modelling*, **22**, 213–53.

Harris, J.R.W., Bale, A.J., Bayne, B.L., *et al.* (1984) A preliminary model of the dispersal and biological effect of toxins in the Tamar estuary, England. *Ecological Modelling*, **22**, 253–85.

Henriksen, A. (1980) Acidification of freshwaters—a large titration. In: Drabløs, D. & Tollan, A. (eds), *Ecological Impacts of Acid Precipitation*. SNFS-Project, Norwegion Science Foundation, Oslo, 68–74.

Holling, C.S. (1978) *Adaptive Environmental Assessment and Management*. Wiley, New York, 388 pp.

Hosper, H.S. (1989) Biomanipulation, new perspective for restoring shallow, eutrophic lakes in The Netherlands. *Hydrobiological Bulletin*, **73**, 11–18.

ILEC (1992) *Pamolare Software. Eutrophication Modelling* (1st edition): 2nd edition, third version, released 2003 by International Lake Environment Committee, Otsu and International Environmental Technological Center.

Jacobsen, O.S. & Jørgensen, S.E. (1975) A submodel for nitrogen release from sediments. *Ecological Modelling*, **1**, 147–51.

Jeppesen, E., Mortensen, E., Sortkjær, O., *et al.* (1989) *Restaurering af søer ved indgreb i fiskebestanden. Status for igangværende undersøgelser. Del 2: Unsdersøgelser i Frederiksborg slotsø, Væng sø og Søbygård sø*. Danmarks Miljøundersøgelser, Silkeborg, 114 pp.

Jeppesen, E.J., Søndergaard, M., Sortkjær, O., Mortensen, E. & Kristensen, P. (1990) Fish manipulation as a lake restoration tool in shallow, eutrophic temperate lakes. Cross-analysis of three Danish Case Studies. *Hydrobiologia*, **200/201**, 205–18.

Jørgensen, S.E. (1976) A eutrophication model for a lake. *Ecological Modelling*, **2**, 147–65.

Jørgensen, S.E. (1981) Application of exergy in ecological models. In: Dubois, D. (ed.), *Progress in Ecological Modelling*. Cebedoc, Liège, 39–47.

Jørgensen, S.E. (1982) A holistic approach to ecological modelling by application of thermodynamics. In: Mitsch, W., Ragade, R.K., Bosserman, R.W. & Dillon, J.A., Jr. (eds), *Systems and Energy*. Ann Arbor Science Publications, Ann Arbor, MI, 61–72.

Jørgensen, S.E. (1986) Structural dynamic model. *Ecological Modelling*, **31**, 1–9.

Jørgensen, S.E. (1990) Ecosystem theory, ecological buffer capacity, uncertainty and complexity. *Ecological Modelling*, **52**, 125–33.

Jørgensen, S.E. (1992a) Parameters, ecological constraints and exergy. *Ecological Modelling*, **62**, 163–70.

Jørgensen, S.E. (1992b) Development of models able to account for changes in species composition. *Ecological Modelling*, **62**, 195–208.

Jørgensen, S.E. (1994) *Fundamentals of Ecological Modelling* (2nd edition). *Developments in Environmental Modelling*, Vol. 19. Elsevier, Amsterdam, 628 pp.

Jørgensen, S.E. (2002) *Integration of Ecosystem Theories: a Pattern* (3rd edition, revised). Kluwer, Dordrecht, 428 pp.

Jørgensen, S.E. & de Bernardi, R. (1998) The use of structural dynamic models to explain successes and failures of biomanipulation. *Hydrobiologia*, **359**, 1–12.

Jørgensen, S.E. & Mejer, J.F. (1977) Ecological buffer capacity. *Ecological Modelling*, **3**, 39–61.

Jørgensen, S.E. & Mejer, H.F. (1979) A holistic approach to ecological modelling. *Ecological Modelling*, **7**, 169–89.

Jørgensen, S.E., Nors Nielsen, S. & Jørgensen, L.A. (1991) *Handbook of Ecological Parameters and Ecotoxicology*. Elsevier, Amsterdam, 1263 pp.

Jørgensen, S.E., Halling-Sørensen, B. & Nielsen, S.N. (1995) *Handbook of Environmental and Ecological*

*Modeling.* CRC Lewis Publishers, Boca Raton, FL, 672 pp.

Jørgensen, S.E., Marques, J.C. & Anastatcio, P.M. (1997) Modelling the fate of surfactants and pesticides in a rice field. *Ecological Modelling*, **104**, 205–14.

Jørgensen, S.E., Lützhøft, H. & Halling Sørensen, B. (1998) Development of a model for environmental risk assessment of growth promoters. *Ecological Modelling*, **107**, 63–72.

Jørgensen, S.E., Nors Nielsen, S. & Jørgensen, L.A. (2000). *ECOTOX* (CD). Elsevier, Amsterdam.

Jørgensen, S.E., Ray, L., Berec, L. & Straškraba, M. (2002) Improved calibration of a eutrophication model by use of the size variation due to succession. *Ecological Modelling*, **153**, 269–78.

Kamp-Nielsen, L. (1974) Mudwater change of phosphate and exchange rate. *Archiv für Hydrobiologie*, **2**, 218–37.

Kamp-Nielsen, L. (1975) A kinetic approach to the aerobic sediment–water exchange of phosphorus in Lake Esrøm. *Ecological Modelling*, **1**, 153–60.

Kauppi, P., Kämäri, J., Posch, M., Kauppi, L. & Matzner, E. (1986) Acidification of forest soils: model development and application for analysing impacts of acidic deposition in Europe. *Ecological Modelling*, **33**, 231–54.

Kettunen, J. (1993) Design of limnological observations for detecting processes in lakes and reservoirs. In: Strakraba, M., Tundisi, J.G. & Duncan, A. (eds), *Comparative Reservoir Limnology and Water Quality Management*. Kluwer, Dordrecht, 136–46.

Koschel, R., Kasprzak Krienitz, L. & Ronneberger, D. (1993) Long term effects of reduced nutrient loading and food-web manipulation on plankton in a stratified Baltic hard water lake. *Verhandlungen Internationale Vereinigung Limnologie* **25**, 647–51.

Kristensen, P. & Jensen, P. (1987) *Sedimentation og resuspension*. Master thesis, Aarhus University.

Lam, D.C.L. & Simons, T.J. (1976) Computer model for toxicant spills in Lake Ontario. In: Nriago, J.O. (ed.), *Metals Transfer and Ecological Mass Balances. Environmental Biochemistry*, Vol. 2. Ann Arbor Science, Ann Arbor, MI, 537–49.

Lammens, E.H.R.R. (1988) Trophic interactions in the hypertrophic Lake Tjeukemeer: top-down and bottom-up effects in relation to hydrology, predation and bioturbation, during the period 1974–1988. *Limnologica* (Berlin), **19**, 81–5.

Lassiter, R.R. (1978) *Principles and Constraints for Predicting Exposure to Environmental Pollutants.* Report EPA 118-127519, U.S. Environmental Protection Agency, Corvallis, OR, 186 pp.

Legovic, T. (1997) Toxicity may affect predictability of eutrophication models in coastal sea. *Ecological Modelling*, **99**, 1–6.

Leung, D.K. (1978) *Modelling the Bioaccumulation of Pesticides in Fish.* Report 5, Center for Ecological Modelling, Polytechnic Institute, Troy, NY, 156 pp.

Lewin, B. 1994. *Genes V.* Oxford University Press, Oxford, 754 pp.

Li, W.-H. and Grauer, D. (1991) *Fundamentals of Molecular Evolution.* Sinauer, Sunderland, MA, 696 pp.

Mackay, D. (1991) *Multimedia Environmental Models.* Lewis Publishers, Boca Raton, 257 pp.

Mejer, H.F. (1983) A Menu Driven Lake Model. *International Society for Ecological Modelling Journal* **5**, 45–50.

Mejer, H. & Jørgensen, S.-E. (1979) Energy and ecological buffer capacity. In: Jørgensen S.-E. (ed.), *State-of-the-art of Ecological Modelling, Environmental Sciences and Applications.* Proceedings of the Seventh Conference, International Society for Ecological Modelling, København, 829–46.

Miller, D.R. (1979) Models for total transportation. In: Butler, G.C. (ed.), *Principles of Ecotoxicology. Scope*, Vol. 12. Wiley. New York, 71–90.

Mitsch, W.J. (1976) Ecosystem modeling of water hyacinth management in Lake Alice, Florida. *Ecological Modelling*, **2**, 69–89.

Mitsch, W.J. (1983) Ecological models for management of freshwater wetlands. In: Jørgensen, S.E. & Mitsch, W.J. (eds), *Application of Ecological Modelling in Environmental Management*, Part B. Elsevier, Amsterdam, 283–310.

Mitsch, W.J. & Jørgensen, S.E. (1989) *Ecological Engineering. An Introduction to Ecotechnology.* Wiley, New York, 472 pp.

Mitsch, W.J., Straškraba, M. & Jørgensen, S.E. (eds) (1988) *Wetland Modelling. Developments in Environmental Modelling*, Vol. 12. Elsevier. Amsterdam, 228 pp.

Monte, L. (1998) Predicting the migration of dissolved toxic substances from catchments by a collective model. *Ecological Modelling*, **110**, 269–80.

Muniz, I.P. & Seip, H.M. (1982) *Possible Effects of Reduced Norwegian Sulphur Emissions on the Fish Population in Lakes in Southern Norway.* SI-Report 8103. 13–2, Scientific Information, Oslo, 198 pp.

Nielsen, S.N. (1992a) *Application of maximum exergy in structural dynamic models.* PhD thesis, National Environmental Research Institute, Denmark, 51 pp.

Nielsen, S.N. (1992b) Strategies for structural-dynamical modelling. *Ecological Modelling*, **63**, 91–102.

Odum, H.T. (1983) *System Ecology.* Wiley, New York, 510 pp.

Paterson, S. & Mackay, D. (1989) A model illustrating the environmental fate, exposure and human uptake of persistent organic chemicals. *Ecological Modelling,* **47**, 85–114.

Peters, R.H. (1983) *The Ecological Implications of Body Size.* Cambridge University Press, Cambridge, 329 pp.

Reuss, J.O. (1983) Implication on the Ca–Al exchange system for the effect of acid precipitation on soils. *Journal of Environmental Quality,* **12**, 15–38.

Reynolds C.S. (1996) Plant life in the pelagic. *Verhandlungen der internationale Vereinigung für theoretische und angewandte Limnologie* **26**, 97–113.

Reynolds, C.S., Irish, A.E. & Elliott, J.A. (2001). The ecological basis for simulating phytoplankton responses to environmental change (PROTECH). *Ecological Modelling* **140**, 271–91.

Sas, H. (Coordinator) (1989) *Lake Restoration by Reduction of Nutrient Loading. Expectations, Experiences, Extrapolations.* Akademie Verlag Richarz, St Augustin, 497 pp.

Scheffer, M. (1990) *Simple Models as Useful Tools for Ecologists.* Elsevier, Amsterdam, 192 pp.

Scheffer, M., Carpenter, S., Foley, J.A., Folke, C. & Walker, B. (2001) The Response of Shallow Lakes to increased nutrient loading. *Nature,* **413**, 591–6.

Schrödinger, E. (1944) *What is Life?* Cambridge University Press, Cambridge, 212 pp.

Schwarzenbach, R.P. & Imboden, D.M. (1984) Modelling concepts for hydrophobic pollutants in lakes. *Ecological Modelling,* **22**, 171–213.

Shapiro, J. (1990) Biomanipulation. The next phase—making it stable. *Hydrobiologia,* **200/210**, 13–27.

Steele, J.H. (1974) *The Structure of the Marine Ecosystems.* Blackwell, Oxford, 128 pp.

Straškraba, M. (1979) Natural contral mechanisms in models of aquatic ecosystems. *Ecological Modelling,* **6**, 305–22.

Søndergård, M. (1989) Phosphorus release from a hypertrophic lake sediment; experiments with intact sediment cores in a continuous flow system. *Archiv für Hydrobiologie,* **116**, 45–59.

Thomann, R.V. (1984a) Physico-chemical and ecological modelling the fate of toxic substances in natural water systems. *Ecological Modelling,* **22**, 145–70.

Thomann, R.V. (1984b) *Systems Analysis and Water Quality Management.* McGraw Hill, New York, 286 pp.

Thomann, R.V. & Mueller, J.A. (1987) *Principles of Surface Water Quality Modeling.* Harper Collins, New York, 644 pp.

Thomann, R.V., et al. (1974) A food chain model of cadmium in western Lake Erie. *Water Research,* **8**, 841–51.

Ulanowicz, R.E. (1979) Prediction chaos and ecological perspective. In: Halfon, E.A. (ed.), *Theoretical Systems Ecology.* Academic Press, New York, 107–17.

Van Donk, E., Gulati, R.D. & Grimm, M.P. (1989) Food web manipulation in lake Zwemlust: positive and negative effects during the first two years. *Hydrobiological Bulletin,* **23**, 19–35.

Vollenweider, R.A. (1975) Input–output models with special reference to the phosphorus loading concept in limnology. *Schweizeriches Zeitschrift für Hydrologie,* **37**, 53–83.

Willemsen, J. (1980) Fishery aspects of eutrophication. *Hydrobiological Bulletin,* **14**, 12–21.

WMO (1975) *Intercomparison of Conceptual Models used in Operational Hydrological Forecasting.* World Meteorological Organization, Geneva, 160 pp.

Zeigler, B.P. (1976) *Theory of Modelling and Simulation.* Wiley, New York, 344 pp.

Zhang, J., Jørgensen, S.E., Tan, C.O. & Beklioglu, M. (2003a) A structurally dynamic model—Lake Mogan, Turkey as a case study. *Ecological Modelling,* **164**, 103–20.

Zhang, J., Jørgensen, S.E., Tan C.O. & Beklioglu, M. (2003b) Hysteresis in vegetation shift—Lake Mogan prognoses. *Ecological Modelling,* **164**, 227–38.

# 16 The Assessment, Management and Reversal of Eutrophication

## HELMUT KLAPPER

### 16.1 ASSESSMENT OF TROPHIC STATE AND ITS RELEVANT FACTORS

Natural eutrophication of lakes is, according to one theory, a consequence of their ageing and it may require hundreds to thousands of years for a deep, oligotrophic lake to change to a shallow, eutrophic one or to be finally extinguished by accumulated siltation. In contrast, eutrophication caused by human activities (**anthropogenic eutrophication**) has afflicted many water bodies within a remarkably short period of time. Algal blooms, oxygen deficiency, fish-kills, impairment of water use have been the drivers of public pressure for the eutrophication to be controlled and abated. Such management of water quality needs to be firmly based on limnological fundamentals and an understanding of component processes underpinning the target of restorative measures. The trophic state which may be realistically achievable depends upon lake-specific factors, including hydrographic factors, morphometry, hydraulic exchanges, stratification, the nature and intensity of land use in the drainage basin and, of course, the extent to which nutrient inputs and nutrient concentration can be regulated.

There have been various approaches to defining the present condition and setting targets for the trophic states achievable through rehabilitation. Discrimination among the various trophic levels has, for a long time, provided the basis for describing the condition of a water body. Even though this still provides the central core of many systems of lake classification, it is nevertheless inadequate for diagnosing the suitability of the water for a variety of uses. In countries where the exploitation of water resources and their catchment areas is intense, the pollution load and pollutant capacity (carrying capacity) are especially important measures of the state of a lake and its amenability to change. Those factors which do not affect the trophic status directly but which are, nevertheless, important from the standpoint of the utility of the water body in question (such as its salt content and hardness, the sensitivity of its pH to acid incursion in precipitation or, in the case of basins created by opencast mining, from geogenic, ground-water sources, the content of humic substances in brown water lakes, health-oriented factors and the presence of nitrates and other hazardous solutes) are often included in assessments of water quality.

In general, the assessment of trophic condition should fulfil the following requirements:
- the body of water should be considered as part of a larger system, that is, the entire catchment area;
- the interactions between the configuration of the lake basin and properties of the water should be taken into account;
- the water quality must be viewed as an amalgam of physical properties, chemical composition and biological components;
- when selecting the times of sampling and field investigations, the temporal and spatial dynamics of numerous factors affecting the perception of quality must be borne in mind, such as the seasonal stratification in the vertical plane, as well as its horizontal variability, if a false representation of the water body is to be avoided;
- criteria for subdividing information (classification categories) should be sensitive to relevant changes imposed by pollution (eutrophication) and invoked by rehabilitation (oligotrophication);
- the classification should contain indicators of the suitability of the body of water for its various

uses and, if possible, an indication of their direct impacts on its condition and pollutant capacity;

- decision-makers responsible for water quality management should be provided with intelligible information and advice for applying critical assessments.

In this chapter, a relatively simple classification method is outlined. It has been developed to satisfy the needs of water-quality management and of use regulation of natural waters (Standard TGL 27885/01 1982; Ryding & Rast 1989; Klapper 1991). The approach is founded on a similar scheme, applied by Kudelska et al. (1981) in Poland. The class boundaries for chemical and biological criteria are almost identical in the two methods. Both approaches were devised and tested against lakes of the Baltic basin but they have been applied successfully to water bodies located outside the temperate zone, including Lake McIlwaine, Zimbabwe (by Thornton & Nduku 1982; Uhlmann 1985) and the Funil Reservoir in Brasil (see Klapper 1998).

Evaluation of the state of a body of water is undertaken with respect to three main considerations (categories):
1 hydrographic and topographical criteria;
2 trophic criteria;
3 salt content and other special or hygienically relevant criteria.

Within each of these three main categories, three separate groups of characteristics may be distinguished. For assignment to one or other of the classes, the information for a given lake must be processed against the criteria of each of these subdivided groups. User applications may be assessed within the same framework.

## 16.2 DETERMINATION OF HYDROGRAPHIC AND TOPOGRAPHIC CRITERIA

The first main category is subdivided into groups:
I configuration of the body of water;
II hydrographic relations between the catchment area and the body of water;
III anthropogenic pollution.

Groups I and II are indicative of the body of water in its natural state, and the criteria in group III represent the artificial effects responsible for the deterioration from the natural to the poorer, impacted state. The hydrographic and topographical classification yields a ranking corresponding to the trophic state of the non-impacted lake water. Contrasting this expectation for the natural condition with the current trophic state reveals the extent of the pollution and the scope for rehabilitation (see Table 16.1).

The configuration of the water body includes the pattern of thermal-stratification behaviour. In water bodies in which the trophogenic and tropholytic processes are mutually separated during the stagnation periods there is little circulation of material, and hence declining productivity during the stratified phases (Patalas 1968; Richardson 1975).

In temperate climates, seasonal abundance of nutrient-limited phytoplankton growth typically exhibits two peaks. The first follows the spring overturn. Towards the end of the summer stratification, when a deepening mixed layer entrains deeper, **hypolimnetic** water, with the nutrients accumulated there, the restored fertility supports a second algal peak in the autumn. This pattern gives low phytoplankton biomass in the warmest part of the year, which is especially convenient in lakes used for bathing! Independently of nutrient loadings, grazing of phytoplankton by zooplankton to very low biomass similarly leads to high transparency. However, such 'Daphnia-Klarwasserstadien' (Uhlmann 1958), or clear-water stages, need simultaneously low fish predation on Daphnia.

By common agreement, the quality of water made turbid by the mass development of plankton algae is considered to be poor, whereas that of a water body with low plankton-supporting capacity would be considered to be high. Early limnological investigations of lake typology revealed that, generally, quality is proportionately higher in deeper and more stably stratified water bodies. Conversely, shallower polymictic lakes tend to be more prone to massive plankton growth, plant-generated discoloration and algal blooms.

**Table 16.1** Classification of static bodies of water according to their hydrography (lake configuration), hydrographic interrelation with the catchment area and use-related pollutant loading

| Group | Criterion | Quality classification | | | | | |
|---|---|---|---|---|---|---|---|
| | | 1 | 2 | 3a | 3b | 4 | 5 |
| I – Configuration | Stratification | Stratified – holomictic or meromictic* | | | | No stable stratification – polymictic | |
| | Average depth $h$ (m) | >15 | >10 | <10 | 2–10 | 1–2 | <1 |
| | Maximum depth $h_{max}$ (m) | >30 | >20 | <20 | Not considered | | |
| | Volume ratio = $\frac{\text{hypolimnion}}{\text{epilimnion}}$ (mean of stagnation periods) | >1.5 | >1.0 | <1.0 | Not considered | | |
| | Mean retention time (yr) $t = \frac{\text{Total volume}}{\text{Annual inflow}}$ | >10 | >1 | >0.2 | >0.1 | <0.1 | <0.1 |
| II – Catchment area | Volume ratio (km² 10⁶ m⁻³) $Vq = \frac{\text{Catchment area}}{\text{Total volume}}$† | <3 | <5 | <10 | <10 | <10 | Not considered |
| | Area ratio $Fq = \frac{\text{Catchment area}}{\text{Area of body of water}}$ | <30 | <60 | <300 | <300 | <300 | Not considered |
| | Forested area as percentage of total land area | >80 | >50 | >20 | >20 | >10 | <10 |
| III – Pollution loading‡ | $B = \frac{\text{Population equiv}}{\text{Total water volume}}$ ($PE\ 10^6\ m^{-3}$) | <50 | <500 | <2500 | <2500 | <5000 | >5000 |
| | $P$ = import (orthophosphate) (g P m⁻² × yr⁻¹) | The upper limits are a function of the average depth and the mean retention time of the body of water and are derived from Fig. 16.1. For seston-rich inflows it is preferable to use 50% of total phosphorus in place of the orthophosphate-phosphorus | | | | | |
| | Nitrogen import (combined inorganic N)§ (g N m⁻² × yr⁻¹) | 5 | 10 | 15 | 15 | Not considered | |

\* For meromictic water bodies, classification parameters $h_{max}$ and $h$ and the values obtained from them are derived after subtracting the non-circulating deep-water region of the monimolimnion.
† For reservoirs and storage ponds the values are based on the standard filling depth.
‡ Pollution index $B_1$ based on direct uses (see Table 16.2). Pollution index $B_2$ from uses in the catchment area (see Table 16.3). For the class ranking the $PE$ (population equivalent) for direct usage is multiplied by 10 and added to the loading from the catchment, namely $B = 10B_1 + B_2$.
§ This is taken into account only where nitrogen is the growth-limiting nutrient.

Polymixis—where wind-induced mixing of the full water column occurs frequently—is a variable characteristic which relates to depth, exposure and wind-fetch across the lake. Clear-water stages are exceptional occurrences in shallow lakes, unless they are caused by grazing pressures from zooplankton, or macrophyte abundance over phytoplankton (see also this Volume, Chapter 17). Thus, the average and greatest depths of a water body convey important information about its likely characteristics and classification. Hence these quantities are specified in many of the systems for classifying and modelling lake ecosystems. Among stratified lakes the extent of the hypolimnion relative to that of the epilimnion is a reliable indicator of the natural oxygen dynamics in the summer hypolimnion and, thus, of the redissolution of nutrients and their possible consequences for productivity.

The retention time is also a valuable measure relating to the hydraulic loading and the supply of nutrients from the catchment area (Kerekes 1975; Margalef 1975; Hillbricht-Ilkowska & Zdanowski 1978).

Although not forming part of the present classification, further configuration factors which influence water quality and fishery yields may be noted:
1 the percentage of the lake bottom which is in direct contact with the epilimnion;
2 the bank configuration, that is, the divergence of the lake surface from a circular shape;
3 the slope of the immediate margins of the lake as a measure of the direct inputs from surface erosion;
4 the steepness of the natural banks as a measure of the extent of the littoral zone.

Group II properties address the sources of polluting substances in the water body, that is, the catchment area. It is considered in relation to the lake volume (or volume ratio, $Vq$) and its surface area (area quotient, $Fq$). It should be noted that some authors include the area of the lake in the catchment area used to calculate volume ratio (Schindler 1971). Although vegetation cover is of considerable importance for nutrient runoff, only the afforested portion is included in the evaluation, because of its protective, nutrient-retentive effect. Erosion and nutrient losses to drainage water increase in the order: permanent grassland, fodder and cereals, root crops.

In those countries where both land and water resources are used intensively, point-source phosphorus inputs assume a special significance (Müller 1978). They are included in the third subdivision of site properties (Group III). Two indices for classifying anthropogenic pollution loads are indicated in Table 16.1. The first requires a statistical survey of uses polluting a body of water ($B_1$) and the various water uses in the catchment area ($B_2$). The population equivalent ($PE$) values are expressed relative to lake volume. The nominated $PE$ values are exemplified and summarised in Tables 16.2 & 16.3. So long as suitable data are available, the areal loading of the body of water and the ratio of average depth to the mean retention time can be interpolated into the expanded, orthophosphate-based Vollenweider scheme to give a direct indication of the expected phosphorus availability and ascription to the appropriate quality class (see Fig. 16.1).

With regard to the boundaries distinguishing the quality classes 1, 2, 3a, 3b, 4 and 5, it may be remarked that they were calibrated in a temperate climate. In a world context, the hydrographic factors peculiar to the environment must also take in factors such as geographical latitude and height above sea level. In polar and mountainous regions, the available nutrient supply will be less efficiently utilised for biological production than it might be in (say) the tropics. Nevertheless, successful application of the scheme to tropical Lake McIlwaine (Uhlmann 1985) and also to the subtropical Hartbeespoort Dam in South Africa (Thornton 1987) encourages confidence in its wider application.

## 16.3 DETERMINATION OF TROPHIC CRITERIA

This second main category comprises factor groups relating to criteria of biological activity:
I   oxygen conditions;
II  nutrient relationships;
III biological production (see also Table 16.4).

For determination of the extent of diel variation of oxygen saturation in the epilimnion, daytime and night-time measurements at the height of the main vegetative period are needed. The criterion is decisive, especially in the higher trophic ranges, for fisheries-related uses. In stratified bodies of water, the oxygen content of the hypolimnion provides information about the depletion processes taking place during the stagnation period and is particularly aimed at distinguishing among the lower trophic levels.

**Table 16.2** Pollutant loading indices $B_1$ for direct consumption or usage of the body of water

| Class | Nature of pollutant source | Population equivalent* values relative to: | | |
|---|---|---|---|---|
| | | $BOD_5$ | Total nitrogen | Total phosphorus |
| 1 | Duck-keeping with free access to the water, per 100 ducks per day | 14 | 8–12 | 16 |
| 2 | Duck-keeping on slatted floors with swimming channels, per 100 ducks per day | 11 | 7 | 10 |
| 3 | Goose-rearing with free access to water, per 100 geese per day | 42 | 24–36 | 48 |
| 4a | Trout-rearing raceways, per tonne actual fish stocks per day | 85 | 110 | 110 |
| 4b | Intensive rearing (mesh cages, raceways) Average loading per day for an annual output of 1 t fish flesh | 30 | 30 | 30 |
| 5 | Recreational use; pollution due to primary body contact, 100 bathers per day in mid-season | 2 | 2 | 2 |

\* 1 $PE \approx$ 54 g $BOD_5$, 13 g N, 2 g P; for phosphorus-limited water bodies the phosphorus-dependent values and for nitrogen-limited waters the nitrogen-dependent values should be used. For waters which are not nutrient-limited the $BOD_5$ values are used.
About 50% of nutrients are bound to particulates, so that a reduction of around 50% in the pollution load can be achieved by sedimentation facilities.

**Table 16.3** Pollutant loading indices $B_2$ for estimation of the effects of catchment uses on the quality of the body of water for various levels of sewerage provision

| Nature of pollutant source | Population equivalent values relative to: | | |
|---|---|---|---|
| | $BOD_5$ | Total nitrogen | Total phosphorus |
| For 1 inhabitant discharging untreated sewage | 1.0 | 1.0 | 1.0 |
| For 1 person, from small sewage plants | 0.7 | 0.8 | 0.8 |
| For 1 person, from mechanical treatment plants | 0.7 | 0.8 | 0.7 |
| For 1 person, from biological treatment plants | 0.2 | 0.4 | 0.4 |
| For 1 person, from sewage applied to land or from tertiary treatment with phosphorus removal | 0.1 | 0.3 | 0.1 |
| For 1 livestock equivalent* via diffuse runoff following agricultural use (indoor feeding) | 1 | 5 | 1 |
| For 1 livestock equivalent via diffuse runoff following agricultural use (liquid manuring) | 2 | 12 | 5 |
| For 1 livestock equivalent with animals grazed on permanent pasture | 0.5 | 3 | 0.6 |
| Wastewaters from industrial and crop production systems to be calculated similarly according to the method of treatment used | | | |
| Area inputs from arable crop production (without the effects of organic manure from stock) per hectare of cultivated surface | – | 9 | 0.5 |

\* Livestock equivalent = 1 cow or horse, 6 pigs, 14 sheep or 150 head of poultry.
6 months grazing, 6 months stall feeding, no direct access to the water body, little slope.
The annual nitrogen runoff is dependent on the nature of the crop and the rate of fertiliser application; grassland about 10, clover 30, cereals 40, root crops 70, vegetables 80 kg N ha$^{-1}$.

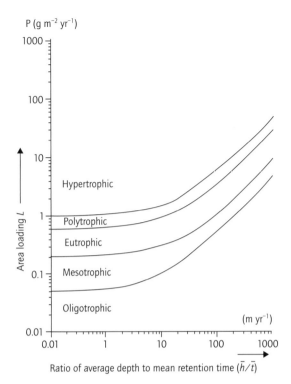

Fig. 16.1 Graphical estimation of the expected trophic status of a static body of water on the basis of the area loading with o-$PO_4$-P. (From Standard TGL 27885/01.)

The nutrient relationships are prerequisite to anticipating biological productivity in the water body. Pronounced seasonal fluctuations may make classification difficult. In the temperate zones, at least, relevant nutrient concentrations may be obtained during the cold season, when consumptive phytoplankton growth is limited by light. Moreover, the 'starting conditions' obtaining immediately prior to the onset of algal growth provide a relatively reliable indication of the productive capacity. As is the case in hard-water lakes, rich in calcium, part of the phosphorus may be co-precipitated out in the course of biogenic lime removal, with the result that the quantity entering the biomass falls short of the expected value. In the standard system, the classification by phosphate content is carried out differently for lakes of high and low carbonate hardness. At the other extreme, nutrient redissolution processes and the special biological conditions discount the value of nutrient concentrations measured during the summer stagnation.

Since the introduction of the radiocarbon method, the production conditions have been best suited to provide a biological basis for the trophic categories (Vollenweider 1968). The user, however, is more concerned with the biological product—the plankton biomass—than the assimilation capacity of the body of water. Both the $^{14}C$ method for measuring primary productivity and the counting of plankton are very tedious. Gravimetric determination of the seston is, however, unreliable on account of the allochthonous fraction, especially in shallow or fast-flowing waters. In practice, chlorophyll determinations and the very simple measurement of Secchi-disc depth have assumed the most useful role. In particular, frequent measurement of the Secchi-disc depth, for example by anglers, provides useful information for classifying lakes.

## 16.4 DETERMINATION OF SALT CONTENT AND OTHER SPECIAL OR HEALTH-RELATED CRITERIA

This third main category contains some very important characteristics from the standpoint of water use:
I  salt content;
II special criteria;
III health-related criteria (as listed in Table 16.5).
Owing to the heterogeneity of the characteristics summarised here, it is useful to distinguish in Table 16.5 those criteria of decisive importance in classification by reference to their chemical symbols.

Salinity (Group I) is of special significance for industrial water usage (e.g., boiler feedwater) and for irrigation. The principal cations are calcium, magnesium and sodium, and the corresponding anions are chloride, sulphate and bicarbonate (expressed as carbonate hardness).

The group of special criteria encompasses chiefly 'abnormal' water body characteristics,

Table 16.4 Classification according to trophic criteria

| Group | Criteria | Quality classification | | | | | |
|---|---|---|---|---|---|---|---|
| | | 1 | 2 | 3a | 3b | 4 | 5 |
| I – oxygen conditions | Range of variation of $O_2$-saturation value at the surface (summer, still air, day–night) (%) | 90–120 | 80–150 | | 60–200 | 20–300 | 0–500 |
| | $O_2$ content in the hypolimnion* (end of summer stagnation) | >6 | >1 | Anaerobic | Not considered | Not considered | Not considered |
| II – nutrient conditions† | At the start of the spring turnover (average value for $h_{mix}$)‡ orthophosphate-P (mg L$^{-1}$) | <0.005 | <0.01 | | <0.03 | <0.05 | >0.05 |
| | total phosphate-P§ (mg L$^{-1}$) | <0.005 | <0.015 | | <0.2 | <1.2 | >1.2 |
| | | <0.015 | <0.025 | | <0.04 | <0.06 | >0.06 |
| | | <0.015 | <0.045 | | <0.3 | <1.5 | >1.5 |
| | combined inorganic N (mg L$^{-1}$) | <0.3 | <0.5 | | <1.0 | <1.5 | >1.5 |
| | During summer stagnation (average values for the epilimnion)¶ orthophosphate-P (mg L$^{-1}$) | 0–0.002 | 0–0.005 | | 0–0.1 | 0.1 | >0.5 |
| | total phosphate-P (mg L$^{-1}$) | <0.015 | <0.04 | | 0.04–0.3 | >0.3 | >0.5 |
| | inorganic N (mg L$^{-1}$) | <0.01 | <0.03 | | <0.1 | >0.1 | >0.5 |
| III – productivity conditions | Primary productivity of phytoplankton annual primary productivity (g C m$^{-2}$ yr$^{-1}$) | <120 | 120–250 | | 250–400 | 400–500 | >500 |
| | ratio of primary productivity in the plane of maximum productivity to the total productivity (%) (mean value April–September) | <15 | 15–30 | | 30–75 | 75–90 | >90 |
| | Phytoplankton biomass (in epilimnion) phytoplankton volumes (cm$^3$ m$^{-3}$) | <1.5 | <5 | 5–10 | 10–20 | 20–30 | >30 |
| | chlorophyll $a^{665}$ content (mg m$^{-3}$) (mean value April–September) | <3 | <10 | 10–20 | 20–40 | 40–60 | >60 |
| | Transparency Secchi-disc depth (m) | >6 | >4 | >1 | >1 | >0.5 | <0.5 |
| | vertical extinction coefficient at 400–700 nm wavelength (m$^{-1}$) (mean April–September) | <0.5 | <0.6 | <1.3 | <1.3 | >1.3 | >2.5 |
| | Zooplankton biomass (in epilimnion) Zooplankton (dry wt) (g m$^{-3}$) (mean April–September) | <0.1 | <0.3 | | <0.8 | 0.8 | 0–>0.8 |
| Trophic level | | Oligotrophic | Mesotrophic | Eutrophic stratified | Eutrophic Non-stratified | Polytrophic | Hypertrophic |

\* Dissolved $O_2$ readings, 5 m above lake bottom.
† For lakes without throughflow of water or retention periods of >1 year (for those bodies with more pronounced flow and impounding reservoirs only the area loading (P g m$^{-2}$ . a) is used.
‡ $h_{mix}$ = total water column affected by circulation (excluding monimolimnion).
§ For bodies of water of low carbonate hardness (many reservoirs in the region of igneous rocks) the low values apply, and for bodies of water with a higher carbonate hardness (many lakes in regions of sedimentary rocks) the high values of the quoted pairs are used.
¶ Only taken into consideration where nitrogen is clearly the growth limiting factor.

Table 16.5  Classification according to salt content, special and health-related criteria

| Group | Criteria | Quality classification | | | | | |
|---|---|---|---|---|---|---|---|
| | | 1 | 2 | 3a | 3b | 4 | 5 |
| I – salt content | Calcium $Ca^{2+}$ (mg $L^{-1}$) | <60 | <100 | <150 | | <250 | >250 |
| | Magnesium $Mg^{2+}$ (mg $L^{-1}$) | <25 | <50 | <100 | | <150 | >150 |
| | Sodium $Na^{+}$ (mg $L^{-1}$) | <30 | <70 | <150 | | <300 | >300 |
| | Chloride $Cl^{-}$ (mg $L^{-1}$) | <50 | <100 | <250 | | <500 | >500 |
| | Sulphate $SO_4^{2-}$ (mg $L^{-1}$) | <100 | <150 | <350 | | <500 | >500 |
| | Total hardness as CaO (mg $L^{-1}$) | <100 | <150 | <300 | | <500 | >500 |
| | Carbonate hardness as CaO (mg $L^{-1}$) | <70 | <120 | <250 | | | |
| | Total salt content (mg $L^{-1}$) | <350 | <750 | <1500 | | <2500 | >2500 |
| II – special criteria | Iron, total Fe (mg $L^{-1}$) | <0.1 | <0.5 | <1.0 | | 3.0 | 3.0 |
| | Manganese Mn* (mg $L^{-1}$) | <0.02 | <0.1 | <0.2 | | 0.5 | 0.5 |
| | pH values | | | | | | |
| | in epilimnion of neutral lakes | 6.5–8 | 7–8.5 | 7–9 | 7–9.5 | 6.5–10 | 6–11 |
| | in acidic mineral workings | >6 | 5–6 | 4–5 | 4–5 | 3–4 | <3 |
| | Ammonium in epilimnion $NH_4^{+}$ (mg $L^{-1}$) | n.d | n.d | <0.1 | | <1.0 | >1.0 |
| | Hydrogen sulphide in epilimnion $S^{2-}$ (mg $L^{-1}$) | n.d | n.d | <0.005 | | n.d | >0.01 |
| | $H_2S$ in hypolimnion $S^{2-}$ (mg $L^{-1}$) | n.d | n.d | >0.01 | | – | – |
| | Brown substances as COD-Mn (mg $L^{-1}$ $O_2$) (humic acids) | <3 | <5 | <10 | | <20 | >20 |
| | Humic standard | Oligohumous | Oligohumous | Oligohumous | | Mesohumous | Polyhumous |
| III – health-related criteria | Nitrate max concentration $NO_3^{-}$ (mg $L^{-1}$) | <15 | <30 | <40 | | >40 | |
| | Mean annual concentration (mg $L^{-1}$) | <10 | <20 | <30 | | >30 | |
| | Fluoride (mg $L^{-1}$) | <1.0 | <1.0 | <1.2 | | <5.0 | >5.0 |
| | Phenols, volatile in steam (mg $L^{-1}$) | n.d | n.d | <0.005 | | <0.5 | >0.5 |
| | Anionic detergents (mg $L^{-1}$) | n.d | <0.1 | <0.2 | | <2.0 | >2.0 |
| | Dissolved metals | | | | | | |
| | copper (Cu) (mg $L^{-1}$) | n.d | n.d | 0.05 | | <1.0 | >1.0 |
| | chromium ($Cr^{3+}$) (mg $L^{-1}$) | n.d | <0.1 | <0.2 | | <1.0 | >1.0 |
| | chromium ($Cr^{6+}$) (mg $L^{-1}$) | n.d | <0.01 | <0.02 | | <0.1 | >0.1 |
| | lead (Pb) (mg $L^{-1}$) | n.d | <0.03 | <0.05 | | <0.5 | >0.5 |
| | arsenic (As) (mg $L^{-1}$) | n.d | <0.01 | <0.05 | | <0.2 | >0.2 |
| | zinc (Zn) (mg $L^{-1}$) | n.d | n.d. | <0.01 | | <0.1 | >0.1 |
| | cadmium (Cd) (mg $L^{-1}$) | n.d | n.d. | <0.001 | | <0.01 | >0.01 |
| | cobalt (Co) (mg $L^{-1}$) | n.d | <0.01 | <0.1 | | <1.0 | >1.0 |
| | nickel (Ni) (mg $L^{-1}$) | n.d | <0.05 | <0.2 | | <1.0 | >1.0 |
| | mercury (Hg) (mg $L^{-1}$) | n.d | n.d | <0.001 | | <0.01 | >0.01 |
| | Polycyclic aromatic hydrocarbons (PAH) (mg $L^{-1}$) | n.d | n.d | <0.0001 | | <0.001 | >0.001 |
| | Organophosphorus insecticides (Pol) (mg $L^{-1}$) | n.d | n.d | <0.001 | | <0.01 | >0.01 |
| | Bacteriological criteria | | Not detectabe in: | | | Colony-forming units (cfu $mL^{-1}$) | |
| | coliforms | 10 mL | 1 mL | <100 | | <1000 | >1000 |
| | enterococci | 100 mL | 10 mL | <10 | | <100 | >100 |

* For stratified lakes, the mean value for the hypolimnion, otherwise mean value for entire water column.

such as those encountered in acidic and heavily ferruginous artificial lakes in mining districts, as well as the presence of brown substances, which would, at one time, have led to the designation of a 'dystrophic' state. In the main, the category is concerned with criteria which disrupt trophic interactions between nutrient supply and biological productivity which are familiar in neutral, clearwater lakes.

The group of health-related criteria was assembled chiefly with reference to drinking-water and bathing-water applications. It conforms to the appropriate standards for these uses. At this point it should be remarked that, for fishery applications, the maximum permissible concentration of toxins is, in some cases, lower than that tolerated in drinking water.

## 16.5 ASSIGNMENT OF PERMISSIBLE USES

The assignment of standing bodies of water to particular uses stems from the overall valuation of the three groups of properties applicable to the respective categories (see Table 16.6). For reservoirs and lakes, management objectives should be stipulated and, in the case of multiple uses, priorities for the various uses should be specified. For a dedicated evaluation for a particular use, the relevant stipulations should be mandatory. Where more demanding requirements must be compounded with those corresponding to the particular use category, then an improvement in the overall water quality, using appropriate management approaches, should be demanded. In certain special cases, the use category may be determined by a single criterion, for example, for nitrate in connection with drinking water supply.

## 16.6 PERFORMING THE CLASSIFICATION

### 16.6.1 *Obtaining data*

The hydrographic–configurational relationships coupled with the mapping of depth contours and the use-related pollutant loads are required at an initial stage. In order to obtain the data necessary for classification according to the relevant trophic, salinity and health-related criteria, measurements should be made in accord with the following specification.

- For drinking-water sources and reservoirs, at least ten times a year, or monthly from April to October inclusive.
- For all other lakes and storage reservoirs at least six times a year, including four occasions between April and October.
- Where particular health-related criteria are not detectable using standard methods, then, for subsequent periods, one measurement only following the disappearance of ice cover is sufficient.
- One sampling should be carried out directly after disappearance of ice cover, if possible at the commencement of the spring turnover, in order to ascertain the initial situation at the commencement of the growth season.
- Other samplings must be carried out in the form of depth profiles.
- Sample collection should be performed at the deepest point and out of the range of influence of direct discharges.
- Bodies of water with hydrographically distinct basins should be investigated separately.

### 16.6.2 *Determination of the class rating*

In the groups of characteristics indicated in sections 16.2, 16.3 and 16.4, the particular class ratings are ascribed to each criterion, as set out in Tables 16.1, 16.4 & 16.5. Then, for each main category, the arithmetic mean of the group cores is solved as a group average. The average value for each of the groups I to III is then rounded to the nearest whole number, in order to provide a mean score for the particular main category. The process of averaging in three stages facilitates the goal of decision-support for the implementation of measures to improve water quality, at the same time ensuring that the subgroups appropriate to each class are given appropriate weight.

The overall classification of water quality for a standing water body is the arithmetic mean of the

Table 16.6  Allocation of possible uses

| Use | Quality classification | | | | | |
|---|---|---|---|---|---|---|
| | 1 | 2 | 3a | 3b | 4 | 5 |
| Drinking water* | Simple treatment | Normal treatment | Extended treatment, sometimes complex methods | Complex methods of treatment, may be difficult | Not suitable | Not suitable |
| Process water | Simple treatment | Normal treatment | As above | As above | Only for limited uses | Not suitable |
| Cooling water | Acceptable | Acceptable | Acceptable | Usable | Limited use | Limited use or unsuitable |
| Irrigation water: | | | | | | |
| organic pollution | Acceptable | Acceptable | Acceptable | Acceptable | Usable | Limited use |
| inorganic pollution | Acceptable | Acceptable | Usable | Usable | Limited use | Unsuitable |
| hygienic pollution | Acceptable | Acceptable | Usable | Usable | Limited use | Unsuitable |
| Recreation | Acceptable | Acceptable | Acceptable | Usable | Limited use | Unsuitable |
| Bathing | Not permitted when used for drinking water, otherwise acceptable | | | | Undesirable | Undesirable |
| Fishing | Under natural feeding conditions | | When used for bathing or drinking, under natural feeding conditions | | Intensive rearing, limited ($NH_4^+$, pH) | Intensive rearing, feeding risky ($O_2$, $NH_4^+$, pH) |
| Waterfowl stocking | Not permissible | Not permissible | Not permissible | Not permissible when used for drinking water; limited use coupled with bathing | Permissible | |
| Navigation and boating with internal combustion engines | Not permissible | | | | Permissible | |

* Examples for drinking water treatment using raw waters drawn from lakes and reservoirs of increasing trophic level: simple treatment – rapid filtration, chlorination; normal treatment – Al sulphate coagulation, floc filtration, possibly with powdered carbon, lime dosing, chlorination; extended treatment – microscreening, Al sulphate coagulation, floc filtration, filtration through activated carbon (or slow sand filtration), lime dosing, chlorination; complex treatment – denitrification, Al sulphate coagulation with addition of flocculant aid agents, filtration through activated carbon, ozonisation.

non-rounded ratings for each main category, then rounded to the nearest whole number. (The mean value may be quoted as well.) The overall classification constitutes the suitability for particular uses. Where the actual rating for a particular health-related criterion is smaller than the overall classification, then it is substituted for the overall score for all uses in which the health-related criterion is critical and, so, sets the quality rating for the site in all aspects of use in which it is relevant.

A worked example is shown in Box 16.1 and an example of classification is shown in Fig. 16.2.

## 16.7 OTHER EUTROPHICATION INDICES AND ASSESSMENT SCHEMES

The system of classification outlined in sections 16.1 to 16.6 is entirely oriented towards the practical demands of water-quality management and use allocations of lakes. The decisive characteristics appropriate to the description of water quality for these purposes are represented by six-point scales (Table 16.1: 1, 2, 3a, 3b, 4 and 5) for each part of the four-component index (section 16.6.2). Previous classifications have relied chiefly on a purely trophic dimension, which perhaps also included weightings for hydrography, salinity or organic pollution, or they were wholly biological in nature. The manipulation of such variables to yield a single index often leads to a high degree of information loss, added to which very widely differing sets of individual parameters may lead to the same numerical value. Other systems, heavily biased in the direction of phosphorus, also tend to break down at the higher trophic levels, because these are more often nitrogen- or even carbon-limited (Uhlmann & Hrbáček 1976).

Purely configurational indices were used by Sylvester et al. (1974) and Kerekes (1977). Zafar (1959), Patalas (1960), Lueschow et al. (1970), McColl (1972) and Sheldon (1972) all used statistical analyses or point recordings of varying degrees of comprehensiveness. In the US EPA Working Paper No. 24 (USEPA 1974), a trophic index number (TIN) was used, derived from the total phosphorus, orthophosphate-phosphorus, inorganic nitrogen, Secchi-disc depth, minimum dissolved oxygen concentration in the hypolimnion and chlorophyll a concentrations. Shannon & Bresonik (1972) constructed a trophic state index from seven criteria, namely: Secchi-disc depth, conductivity, organic nitrogen, total phosphorus, orthophosphate-phosphorus, primary productivity, chlorophyll a, as well as the so-called cationic ratio ([Ca + Mg]: [Na + K]). In addition, a trophic state index was developed by Carlson (1977), which was obtained in the form of a numerical value lying between 0 and 100 by selective use of the Secchi-disc depth, the total phosphorus or the chlorophyll content.

Uttormark & Wall (1975) proposed to assess lakes on the basis of use-limiting factors, such as hypolimnetic oxygen-consumption, Secchi-disc depth, fish-kills and growth of macrophytes or algae. The lake condition index was given a value ranging from 0 (no use-limiting factors present) to 23 (all restrictions obtained).

Heinonen (1980) distinguished eight categories for the biomass of phytoplankton, assigning them to six trophic classes for entry on a lake inventory (see Table 16.7). This classification has also been ecologically calibrated, by analogy to the steps described in section 16.6. The primary aim is a high performance capability in the body of water for beneficial uses. It is also a useful starting point for establishing strategies for water-quality restoration.

## 16.8 EUTROPHICATION UNDER DIFFERING CLIMATIC AND CHEMICAL BACKGROUNDS

On the planetary scale, environmental conditions vary widely and to an extent which impinges on the normal processes of eutrophication, and we should recognise the following factors.
1 Very few problems with eutrophication occur under habitually cold climates of polar or high montane regions. A long period of ice- and snow-cover hinders the physical and biogenic oxygen intake and may be the reason for oxygen depletion with its well-known consequences. The ice-free

## Box 16.1 Worked example

LAKE: Name

| Criteria | Group | Description | Average value | Water quality classification | Rounded value |
|---|---|---|---|---|---|
| Hydrographic and topographic | I | Configuration of lake | 2.6 | 2.4 | 2.0 |
| | II | Catchment hydrography | 1.9 | | |
| | III | Pollutant loading | 2.7 | | |
| Trophic | I | Oxygenation | 2.0 | 3.3 | 3 |
| | II | Nutrient regime | 4.2 | | |
| | III | Productivity | 3.7 | | |
| Salinity, special features, health | I | Salt content | 1.2 | 2.7 | 3 |
| | II | Special features | 3.0 | | |
| | III | Health-related criteria | 4.0 | | |
| Overall classification | | | | 2.8 | 3 (2.8) |

| Most adverse health-related criterion | Proposed use | Class rating | Classification for proposed use |
|---|---|---|---|
| Nitrate | Drinking water | 4 | 4 |

In order to illustrate the application of ratings for each lake, with respect to each of the three main categories, as well as the overall use category, four-part flags are affixed appropriately to the map. From left to right, the flag parts are coloured in order to give immediate information on the hydrography, trophic status, special features and use classification. The colour designation is kept consistent for all illustrative and mapping purposes.

| Class | Colour | Representation in black and white |
|---|---|---|
| 1 | Blue | (dotted) |
| 2 | Green | (vertical lines) |
| 3a | Yellow | (diagonal lines) |
| 3b | Brown | (cross-hatched diagonal) |
| 4 | Red | (crosshatch) |
| 5 | Black | (solid black) |

450 H. KLAPPER

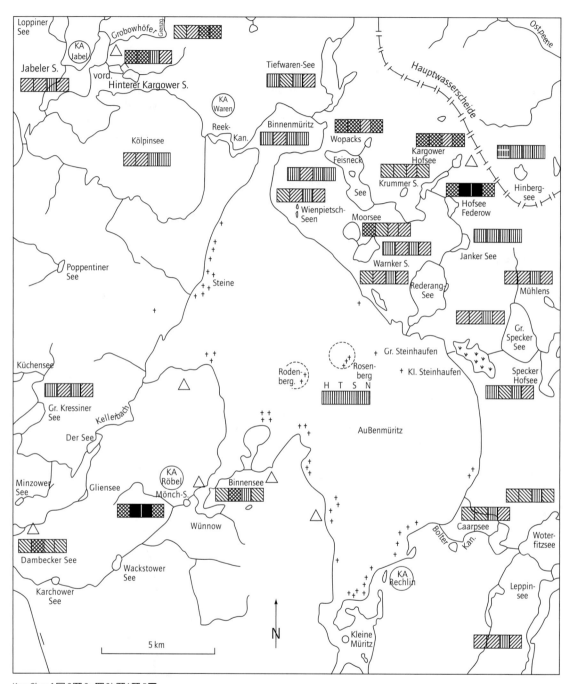

Key: Class 1 ▭ 2 ▨ 3a ▥ 3b ▧ 4 ▩ 5 ■
KA, main sewage treatment works; △ pumping station for water irrigation.

**Fig. 16.2** Classification of inland bodies of water taking the Müritz district in Germany as an example. The four-part bars have the following significance (from left to right): H, hydrography; T, trophic state; S, salt content; N, use or overall classification. (Modified from Klapper 1991.)

Table 16.7 Results of a lake survey in Finland, 1963 and 1965 (Heinonen 1980)

| Group | Phytoplankton biomass (mg L$^{-1}$ wet wt) | Classification | Number of samples | Per cent |
|---|---|---|---|---|
| I | 0–0.20 | Ultra-oligotrophic | 212 | 25.7 |
| II | 0.21–0.50 | Oligotrophic | 322 | 39.0 |
| III | 0.51–1.00 | Weakly mesotrophic | 129 | 15.6 |
| IV | 1.01–1.50 | Strongly mesotrophic | 49 | 5.9 |
| V | 1.51–2.50 | | 40 | 4.9 |
| VI | 2.51–5.00 | Eutrophic | 35 | 4.2 |
| VII | 5.01–10.00 | | 25 | 3.0 |
| VIII | 10.00 | Hypereutrophic | 14 | 1.7 |
| Total | | | 826 | 100.0 |

period may last as little as a few weeks (this Volume, Chapter 6, gives some documented examples). However, the long days of polar summers allow diatoms, desmids and some zooplankton to develop during the short vegetation period, even though persistent low temperatures impede growth rates.

2 In large tropical reservoirs, soil erosion and eutrophication present serious operational problems. Prior to damming, the rivers had been extremely turbid and the primary productivity was strongly light-limited, especially during rainy seasons. The reservoirs formed in their valleys function incidentally as settling basins. They may remain turbid at the main and tributary inlet points but, elsewhere, the plankton in clear water is released from the constraint of light deficiency. The upper parts of a reservoir may be highly eutrophic but fertility decreases with further distance down the reservoir and with increasing depth. Thus, these reservoirs typically exhibit a longitudinal gradient, with best quality near the dam (see also this Volume, Chapter 12). Because the particulate load of the inflows increases the specific mass of inflowing water, some of this 'short-circuits' the upper convection and traverses the reservoir without mixing. Under these conditions, adequate assessments of the trophic state of large reservoirs are difficult without time-consuming campaigns of sampling and investigation of spatial variation (Klapper 1998).

3 In arid regions of the globe or in the vicinity of geological salt layers, the salinisation governs the ecosystem behaviour. In terms of area, nearly half the world's lakes belong to this category (Hammer 1986; see also this Volume, Chapter 8). The main anions may be chloride, sulphate and carbonate; sodium, potassium, calcium and magnesium are the most abundant cations. Concentrations may vary over orders of magnitude between rainy and dry seasons, at times increasing to saturation and salt crystallisation. Selection is stringent and bioproduction may be performed by only few but abundant species and with high intensity. Food chains are simple with few links. The shallow salt lakes of the tropics are highly productive because of rapid and continuous nutrient recycling and turnover by huge populations of birds. Deeper salt lakes, such as the Dead Sea, are meromictic and possess anaerobic monimolimnia. Mixing events are sometimes catastrophic, because of large quantities of hydrogen sulphide, accumulated over years of stable stratification.

4 Acidification of lakes by acid rain often leads to lowered bioproductivity. Acidification is often related to the solution of aluminium from the catchment and its precipitation with phosphorus in receiving lakes. Few fish can survive pH values <5.5. Between 4.5 and 5.5, pH is buffered by aluminium. Remediation is possible by lime application, which has been used in over 1000 instances (Olem 1991).

In geogenic acidic lakes, formed in basins created by opencast mining, acidity may be even more extreme. The water is rich in sulphate and dissolved iron hydroxide. pH may be stabilised between 2 and 3.5 by the iron buffer. Because of high base-binding capacity, neutralisation with lime is impracticably expensive. A more sustainable approach to neutralisation of the acidity is provided by microbially mediated sulphate reduction (the reversal of sulphate-releasing pyrite oxidation). At these extremes, primary production is carbon limited. Besides limitation by inorganic carbon, low assimilation rates also restrict the supply of organic substrate to sulphate-reducing bacteria. The precondition of oxygen-free environments in the deep water and the sediments, required by these obligate anaerobes, cannot be satisfied until oxidant (oxygen and nitrate) has been exhausted, and then only if the organic carbon pool persists. Clearly, encouragement of sulphate consumption in these circumstances can be encouraged by applying organic carbon (in the form of sewage) or by promoting its generation (by phosphorus addition). So, in these special circumstances, the goal of moving towards pH neutrality, a controlled eutrophication is the treatment to be recommended (Klapper & Schultze 1997).

## 16.9 STRATEGIES AND ECOTECHNOLOGICAL MEASURES TO CONTROL EUTROPHICATION IN LAKES AND RESERVOIRS

The planning of rehabilitation proceeds in a fashion analogous to the practice of medicine. Firstly there is the investigation and diagnosis. The result may be expressed in the form of a water quality report, following the classification system in sections 16.2–16.6, or on the basis of limnological expertise. Possible methods of ecosystem treatment should be considered for effectiveness, in the sequence suggested below. Here too, as for medical science, there are analogous well-tried rules: 'prevention is better than cure', for it is better to fight the causes than merely palliate the damage which has already occurred.

The preventative measures indicated in section 16.9.1 are to be considered only as rehabilitation measures in the wider sense. They make use of the legal options for protecting the water body from harmful effects. It is often only through statutory regulations that the use of a lake and its catchment can be so regulated on a planned, long-term basis that more costly clean-up measures are avoided. In this way, 'prophylaxis' is the first priority, followed by purposeful 'dieting'. Dieting provides for the maintenance of a healthy state, in as much as the causes of pollution are dealt with and the nutrients are diverted before they enter the water body (see section 16.9.2).

Therapeutic measures become essential if elimination at the point of origin is insufficient or ineffective, such as in instances where the pollution originates substantially from diffuse sources. Accordingly, internal, within-lake mechanisms are invoked or are otherwise adjusted by technological measures in a way which reverses the decline in water quality (section 16.9.3). 'Curative treatments' must be used when injury has already occurred and repair is necessary. Thus, removal of accumulated limnogenic sediments may reverse the effects of successional processes by the equivalent of several thousand years (see section 16.9.4).

In the following overview only keywords and brief statements are given. For details, please consult the appropriate chapters of the handbook.

### 16.9.1 Prophylaxis: legal regulations to avoid or counteract eutrophication

**1** Legal approval procedures need to accept that standing waters are far more sensitive to pollution than running waters. Wherever possible, therefore, a policy of zero discharges to lakes should be adopted.
**2** Supplementary conditions permitting rehabilitation may need to be superimposed on existing water abstraction rights, respecting multiple use where necessary; examples might include the fitting of clean-water pumphouses with long intake pipes for abstraction of lake water at depth from deep lake water for irrigation.

**3** Undertaking rehabilitation measures in order to compensate polluting uses such as farming fish feeding in enclosures, via sediment removal, or applying artificial aeration.

**4** Protective zones should be established in order to serve as buffer ecosystems, interposed between intensively utilised land in the catchment area and surface drainage channels, thus defending the less biologically productive state of the receiving water.

**5** Statutory controls are needed in order to govern the priorities of use in specially threatened waters, establishing rules with respect to public activities and prohibiting sewage discharge, powered boating activities and access to ecologically sensitive marginal zones and controlling public boathouses and moorings.

**6** Nature conservation to protect the oligotrophic state of water quality, which has become a rarity in many regions. The relevant biocenoses can survive only if the quality is maintained. Similarly conditions need to be applied to internationally revered sites, such as Lake Baikal, Lake Tahoe and Lake Ochrid, for which concern is shared, through international institutions (UNESCO, World Heritage Sites), by everyone.

**7** Regional development plans may and should take into account the protection of lakes and reservoirs. Water quality is a central consideration in the drainage basins of drinking-water reservoirs and in landscapes where water-oriented recreation is a main economic activity.

**8** Standards for the control of water quality and protection of the quality demanded by particular users may guide and drive the implementation and compliance with water-quality standards. They are developed to differing extents in different countries. A few are united in adopting international (EU) standards for bathing waters, fisheries and raw water for drinking purposes.

### 16.9.2 *Dietary measures: limitation of the input of harmful substances*

**1** Advanced treatment of sewage eliminates not only the settleable matter (primary) and the organic substances (secondary) but also plant nutrients (tertiary treatment). The target may be achieved by the pre-treatment of wastewater for irrigation (thus gaining the recycling of the water, organic carbon and natural fertilisers). Today, various chemical means are available for removing phosphorus from wastewater, as well as techniques for removing nitrates by reduction to nitrogen, with safety and often high efficiency. According to the polluter-pays-principle the producer needs to carry the expense.

**2** Diversion of sewage away from the catchment area is the surest way of lake protection and, where the local situation warrants it, is the preferred solution. Surplus costs for diversion pipes may be offset by dispensing with the tertiary treatment which would otherwise be necessary. The latest concepts of group solutions for sewage treatment offer ways to keep sewage away from the lake or reservoir by building treatment plants downstream.

**3** Land cultivation with reduced nutrient runoff has been scientifically investigated and generalised in standards for farming in drinking-water protection zones. Diffuse nutrient runoff depends on the main products and crop rotation sequence (area-specific nutrient export increases in the order grassland, fodder cropping, cereal, potato and beet production). The amount, frequency and delivery technology of organic and inorganic fertiliser application affect export: irrigation as well as improved drainage accelerate hydraulic transport of nutrients to lakes. Soil husbandry is crucial, as ploughed soils without vegetation cover are especially vulnerable to erosion and phosphorus loss. Relief and slopes relative to watersheds compound these considerations.

The sanitary engineer needs to produce drinking water in the same territory where the losses from arable land are polluting the ground and surface waters. Minimisation of nutrient losses from farmland is first of all a task for the farmer, who needs to earn from high yields but is required to remove as much of the applied nutrients as is possible with the harvest. Thus:

**4** Treatment of inflow is the preferred solution if the nutrient load originates from diffuse sources. Phosphorus elimination plants (PEP) are comparable to water treatment works. The process of phosphorus precipitation and floc separation may be effected by a variety of technologies. An efficiency of over 90% elimination is achievable but there is a need for reserve capacity at times of high throughflow when the greatest loads of phosphorus are carried. The first PEP was installed at the Wahnbach Reservoir near Bonn, Germany, where a pre-reservoir was constructed upstream of the plant, in order to provide the extra holding capacity at time of flood (Bernhardt & Clasen 1985).

Pre-reservoirs may remove 50–60% of the nutrients following their incorporation into biomass and subsequent sedimentation. The detention time must allow plankton–algae to grow and to compensate the flushing losses (<2 days during high throughflow). The inflow needs to be distributed over the whole cross-section of the reservoir and the outflow to be discharged on the surface over the crest of the dam (Standard TGL 27885/02 1983).

Algae barriers and macrophytic biofilters are offered as relatively cheap, ecotechnological aids but, save in certain cases, they have not been shown to be efficient. In lake chains, the near-surface algae and, especially, cyanobacterial (blue-green algal) 'blue-green water blooms' may be restrained from passing to the next lake with the help of floating algae shields or booms. They may be made from plastic, like oil barriers but with hanging plastic sheets, or constructed as floating reed grown on coco-fleece.

The plankton also may be screened out by macrophytic biofilters. On shallow parts of the outlet-region *Typha*, *Phragmites*, etc. are planted. Their stems carry dense growths of filamentous algae which support microbial biofilms. It must be admitted that it is not simple to force water mass to pass through macrophyte stands if there is an alternative route.

### 16.9.3 *Therapy: in-lake measures to regulate the production of matter*

**1** Biomanipulation (biomelioration) involves biological measures to influence the material balances in water bodies, by controlling animals or plants. Biomanipulation in the strictest sense is understood to be the combating of phytoplankton abundance by enhanced zooplankton grazing. The most effective means of securing this outcome is through the top-down control exerted by introduced predatory fishes. These decrease the biomass of small, planktivorous prey fishes so that zooplankton may increase under the influence of reduced consumption, and augment the grazing pressure on the phytoplankton.

**2** The introduction of plant-feeding fish may be used for certain types of weed control (grass carp, *Ctenopharyngodon idella*, feeds on water plants) or for direct consumption of primary plankton production (silver carp, *Hypophthalmichthys molithrix*). Fish yield may also be increased through the introduction of more attractive food animals for fish, such as amphipod crustaceans.

**3** Biomelioration using macrophyte stands makes use of the switch between phytoplankton and higher water plants in shallow lakes. The water contains less plankton particulate matter if the nutrients are used and retained in macrophyte tissue. Implementation may depend upon lowering the water level in the early spring in order to allow light to penetrate to the bottom and preparing shallow 'bioplateaus', which are then planted with the desired macrophyte species (Oksijuk & Stolberg 1986; Klapper 1991).

**4** The biological elimination of nutrients by incorporation and deposition of algae with the nutrients in the permanent sediment is the main elimination mechanism in pre-basins, oxidation ponds and 'polishing' ponds. Biogenic decalcification contributes to phosphorus precipitation in hardwater lakes. Nitrate elimination was shown to be feasible under anaerobic conditions in the hypolimnion of drinking-water reservoirs. An organic substrate must be added for heterotrophic microbial nitrate respiration. After consumption of the dissolved oxygen, nitrate is reduced first to nitrite and then to nitrogen $N_2$ (Fichtner 1983, Klapper 1991). Nitrate elimination with the help of nitrophilic plants (nitrophyte method) is possible but, in temperate climates, treatment is limited to the growth period. There is a general requirement, too, to devise suitable means of harvesting the wetland surfaces not accessible to machines (Niemann & Wegener 1976).

**5** Chemical methods for controlling matter balance are based on the low solubility of some phosphorus compounds. Phosphorus precipitation is carried out by the application to the water of such flocculants as aluminium sulphate and iron chloride, supported by polyacrylamide or powdered bentonite or clay to create nuclei for floc formation. Different means for spreading the floccing agents are available. Orthophosphate is easier and more usefully eliminated than is the total phosphorus (Klapper 1991).

**6** Long-term effectiveness of sediment sealing is more reliable when aluminium salts are used in preference to iron salts, owing to the tendency of iron phosphate to dissolve under reducing conditions. Given the acidity of the flocculants, it is generally more practical to use this treatment in hard, calcareous waters.

**7** Chemical conditioning of the uppermost sediment layers aims at increasing redox, in order to oxidise organic substances and sulphides to decrease the phosphorus redissolution.

**8** Especially in shallow lakes of higher trophic levels, the internal fertilisation may need to be curbed. Addition of Fe-III salts, sometimes in combination with lime in order to avoid $H_2S$ being released, or nitrate as salt granules, has proved successful (Ripl 1978).

**9** Calcite flushing makes use of the natural lake lime, often found accumulated in thick layers in the littoral zone. Calcite is excavated by suction dredgers and spread by pipes and sprinkler jets, in order to cover those areas of the lakes with soft, phosphorus-rich sediments. The capping by calcite intensifies the natural biogenic decalcification and decreases the phosphorus release from the sediments (Klapper 1991, 1992).

**10** Hydromechanical methods utilise concentration differences in time and space in order to control the matter balance in lakes. For instance, deep-water diversion provides a way to decrease nutrient content in stratified lakes. Nutrient-rich and oxygen-deficient hypolimnic waters are diverted in order to take the place of natural surface outflow. The same strategy is used with selective take-off from reservoirs, in order to export as much nutrient as possible via the deep draw-off. This method is most effective with diversion at the end of the summer stagnation period when the concentration differences between the surface and bottom waters are at their greatest. Various methods have been developed and applied, using plastic or steel pipes, exploiting gradient, under pressure or through siphons, with or without pumping (Klapper 1991).

**11** Dilution with water of better quality is not often used. However, in some urban bathing lakes, water sometimes must be exchanged. Water is pumped out to a sewer so that ground water seeps in in its place, or space is created for refilling with mains water. Consideration may be given to the scale of annual variations in rivers with which lakes are interconnected and the control of hydraulic exchanges in order to minimise the nutrient load to the lake. Diversion of flowing water, in order to by-pass the lake at relevant times, is another way to reduce loads to the lake.

**12** Increasing the depth improves a lot of quality characteristics of a lake. In the past 100 years, lake management was mostly aimed at winning arable land by lowering water tables. Only for hydropower generation was damming considered but improved water quality was often an incidental benefit. Because of agricultural overproduction in the European Union, options for water-quality improvement by increased water levels and higher water tables may be planned realistically. Deepening by dredging, which may give similar benefits of increased depth, is considered in the next section.

### 16.9.4 Curative treatments: rehabilitation methods for counteracting harmful lake properties

This section reviews methods for rehabilitating properties of individual lakes by modifying their internal components. They include:

**1** Dredging is one of the most expensive options available for rehabilitating lakes. However, it may be the only one available for those bodies of water which have already suffered serious encroachment from sediment accumulation and siltation. Re-use of the excavated materials, as natural fertiliser or soil-conditioner, may help to offset the cost of the lake restoration. Where lakes are situated in cities lacking nearby suitable sites for disposal of dredgings, or where the sediment is affected by toxic compounds, separation of usable sand and the mechanical drying of the polluted sediments may be necessary, perhaps even involving filter pressing and sediment backwashing. This kind of 'high-tech' dredging costs (up to ten times) more than the value realised in its use as a soil conditioner.

2 Limnological objectives of sediment dredging may include an improvement of hydrography with greater area, volume, average or maximum depth, the occurrence of stratification with a stable hypolimnion, and more favourable area or volume quotients. With an improved hydrographic class (see Table 16.1), the deeper lake better disperses nutrients, gains the improved phosphorus binding capacity which stems from the exposure of older, unsaturated sediments, and possesses a more ample oxygen reserve (and, hence, better potential for self-purification). Sometimes the desired results may be obtained through partial excavation, say, of only the uppermost (polluted) sediment layer, coupled with the digging of sedimentation traps and/or oxidising the sediment surface by chemical or hydromechanical means. A great variety of equipment is available for the excavation, transport, deposition, treatment and re-use of lake sediments. All technological steps within the restoration process need to be carefully planned, co-ordinated and optimised.

3 Aeration is applied in order to stabilise or recover a disturbed oxygen regime. So long as the mechanisms driving the oxygen deficiency persist, aeration needs to be maintained. The oxygen demand to be satisfied stems from the saturation deficit set by the degradable organics and reduced chemical compounds present, approximately as follows (simplified):

$$\text{oxygen demand}(\text{mg L}^{-1}\text{O}_2) = \text{oxygen saturation deficit}(\text{mg L}^{-1}) + \text{BOD}_5(\text{mg L}^{-1}) + (\text{mg L}^{-1}\text{NH}_4) \times 2.8 + (\text{mg L}^{-1}\text{H}_2\text{S}) \times 1.9 + (\text{mg L}^{-1}\text{Fe II}) \times 0.29 + (\text{mg L}^{-1}\text{Mn II}) \times 0.29 + (\text{mg L}^{-1}\text{CH}_4) \times 4.0$$

Surface aeration is best developed in aerated sewage lagoons. In polytrophic, shallow lakes and intensively used fish-rearing ponds the diurnal oxygen variations are sometimes high, so that artificial aeration is necessary, at least during the night hours, or when there is more continuous oxygen depletion, to concentrations risking fish-kills. The equipment may be floating centrifugal or brush aerators, with diffusers for generating fine or medium bubbles of compressed air. Self-priming venturi-aerators, hydropneumatic pumps, fountains, etc., are also used.

4 Destratification or artificial circulation is a commonly applied technique in order to overcome harmful effects arising from the temperature and chemical stratification of a body of water. Where the artificial breakdown of an already established stratification system is concerned, all oxygen-consuming compounds must be considered (see above). This applies especially to meromictic lakes, where the products of decomposition of several years will have accumulated in the monimolimnion and may be so reducing that admixture to the epilimnion might lead to complete exhaustion of the oxygen, with attendant fish mortalities. Artificially enhanced circulation of the water may inhibit phytoplankton, owing to its light limitation in the deeper water column. Mixing depth imposes the inhibiting effect when it extends beyond the compensation level, below which no net photosynthetic production is possible. This level, which may be regarded as the critical depth for destratification, must be determined in each particular case. Because destratification may, at the same time, increase the availability and distribution of nutrients, it is important that the stimulus towards added productivity is more than compensated by the poorer insolation. A useful biological side-effect of artificial circulation is its suppression of blue-green algae in favour of diatoms and green algae, to the benefit of fishery, drinking-water supply and recreational uses of the water body (Teerink & Martin 1969; Shapiro 1979).

Suitable equipment is available, working with air bubbles, or air-lifting of oxygen-deficient waters to the surface, or by pumping oversaturated surface water to the lake bottom; circulation may be enhanced by the use of jetted inflows at depth. For more details, see Klapper (1991) or Cooke et al. (1993).

5 Hypolimnetic aeration offers certain advantages. The volume of water to be treated is relatively small and the low temperature of the hypolimnion can be maintained even during the warm season. Linked with this is a greater solubility of oxygen and a reduced uptake of oxygen at the sediment–water interface. Hypolimnetic aeration increases the tenability of habitat for cold-water fish and their food animals. Draw-off from depth for drinking-water supply may also be more attractive, as aeration decreases the content of iron and manganese in the hypolimnetic raw-water. The nutrients which have accumulated in the hypolimnion are not displaced into the euphotic layers, as often happens with full destratification.

Various types of hypolimnetic aerators are available. Near-bottom water is aerated during the air-lifting. Gases are exchanged at the surface and the aerated water returns to the hypolimnion, where it spreads horizontally (for details, see Klapper 1991; Cooke et al. 1993). Supplying pure (liquid) oxygen or other oxidants, such as sodium nitrate, is about twice as expensive as mechanical aeration, but is particularly appropriate for use in water bodies where there is a risk of hydrogen sulphide emission. In order to oxidise 1 t of hydrogen sulphide to sulphate, 1.8 t of oxygen or 2.55 t of sodium nitrate is required to be added. If there is a risk of accident, the polluter can be required to introduce the necessary amounts of nitrate into the discharge immediately.

For the purpose of oxygen injection, the industry supplies complete systems comprising vacuum-insulated storage vessels for keeping the liquid oxygen at −270°C, together with evaporators and heat exchangers. The oxygen is mixed with water under pressure in order to give a solution of 200 mg $L^{-1}$ $O_2$, which may cater for all respiratory consumption in the hypolimnion, even at quite low rates of flow (Fast et al. 1975).

6 Eradication of undesirable organisms from a lake as a target of rehabilitation is never more than the successful treatment of the symptoms. Description of organisms as 'harmful' is invariably based on an implicit assumption that bodies of water are intended only to benefit humans. It is relatively recent that ecosystem protection and species diversity came to be regarded as objectives of lake management. A species may become harmful if it develops to such an overriding extent that numerous other members of the biocenosis are squeezed out and the materials budget is changed adversely. Suppression of unwelcome organisms in temperate latitudes, however, does not assume anything like the same importance and economic significance as they do in tropical countries, where macrophytes clog up the irrigation canals, render navigable channels impassable and make large reservoirs impossible to fish. Parasitic trematodes such as *Schistosoma haematobium*, the causative organism of bilharzia, may be suppressed at any point of the life cycle by interruption of the chain of transmission from one host to another. More promising as a control is the regulation of the molluscs acting as intermediate hosts, for which the most effective option is to suppress the growth of the macrophytes on which they live.

Mechanical methods for macrophyte control, i.e. cutting and harvesting, are in use worldwide. The plants rooting in the sediments cannot be curbed by decreasing the concentrations of nutrients in lake water. Removal of plant material and its incorporated nutrient content does, however, diminish the quantity available in the whole system. It may prolong the life time of shallow lakes overgrown by aquatic weed. Rapid re-colonisation is a risk unless roots and rhizomes are harvested.

7 Chemical combating of phytoplankton with copper sulphate was and is commonly applied in drinking water reservoirs but its use cannot be advocated. Repeated applications over a long period lead to selection of copper-tolerant species and thus failing benefit. After application, killed algae remain in the water body and are degraded with consequences to the oxygen budget and for the taste and odour of the water. Because of the toxicity to fish, the desirability of even small applications must remain questionable (Klapper 1998). Similarly, herbicide application to floating weed (*Eichhornia*, *Pistia*, *Salvinia*, etc.) carries draw-

backs as dead plant material persists in the water body where it is in conflict with preferred methods of biological control (Bethune 1996). Biological control using phytophagous fish has been mentioned earlier: at least 12 species of insect have been bred for the control of such nuisance species as alligator weed, water lettuce, *Hydrilla* and water hyacinth (Cooke *et al.* 1993). In Namibia, the leaf-mining beetle *Cyrtobagous singularis* was spread in order to reduce the Kariba weed, *Salvinia molesta* (Bethune 1996). Large snails (*Pomacea* sp.) feeding on the eggs of *Biomphalaria glabrata*, the intermediate hosts of bilharzia, have helped to suppress this dangerous helminthosis in Brazil (E. Paulini, Universidade Federal de Minas Gerais, personal communication, 1995).

## 16.10 MAINTENANCE, SUCCESS AND ECONOMY OF WATER-QUALITY MANAGEMENT

The success of approaches to restore and improve the quality of lakes depends on the thoroughness of the diagnosis and evaluation prior to initiating the treatment. The user-oriented limnological investigation methodology, including a classification of the activities of water-quality management, as outlined in this chapter, is intended to assist the implementation of a beneficial treatment.

The assessment of the restoration strategy needs to take into consideration the trophic class in relation to the hydrographic class, whether the lake is shallow or deep, whether it is governed by nutrient imports or by internal mechanisms, whether algal production is limited by nutrients, light or zooplankton grazing, etc. Initially, the possibilities of securing lake protection through prophylactic legal activities should be explored, followed by what the physician calls 'dieting', that is, curbing the load before it is assimilated by the water body. If there is no option other than in-lake therapy, the character of the lake needs to be considered: depth, stratification, trophic state, limiting nutrients, prevailing plankton communities, oxygen conditions, sediments, etc. The most promising techniques include the curbing of algae by nutrient limitation in mesotrophic and moderately eutrophic lakes, where a small change in nutrient supply may affect immediate reduction in excessive algal production. Conversely, a highly eutrophic or polytrophic lake will not respond until the nutrient availability decreased to the sigmoid part of the Chl $a$/P curve (see Fig. 16.3). Prognoses about the success of restoration are difficult because lakes remain relatively stable about their respective trophic levels. Case studies have shown that the response to distinct treatment sometimes involves a time lag of years. Hydraulic retention times, accumulated resources stored in sediments and the established biota are all factors which buffer the lake against anthropogenically managed changes in the income and export of resources.

An important factor influencing implementation is the likely cost of the treatment in relation to its restorative benefits. The economic benefits of rehabilitation are reflected in a growth in the national product or diminished losses, primarily in relation to improved raw water quality. Among them are:
• saving in treatment costs for the production of drinking water of acceptable quality;
• increased fishery yield, or reduced damage to yields and avoidance of losses, owing to the greater provision of fish habitat with adequate food and high self-purification capacity;
• increased yield of crops, reduced risk of drought effects, as a result of the provision of nutrient-rich, deep-level water, for irrigation;
• increased yields of crops through the enriching application of dredge spoil arising from lake management to agricultural soils;
• availability of natural bodies of water for recreational purposes, in place of concrete-lined swimming pools with a consequent financial saving and diversion of available resources of labour and energy to more productive purposes;
• ecological benefits stemming from the presence of healthy bodies of water in the environment in general, providing island habitats and stabilisation of the biocenosis with an abundance of different species, together with biological pest control by amphibia, birds and so on.

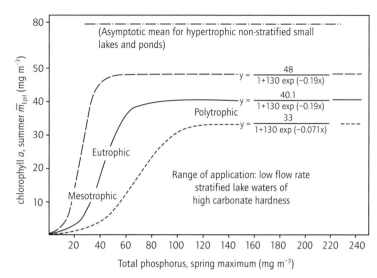

Fig. 16.3 Relation between phosphorus in early spring and chloropyll *a* in the epilimnion during the following summer. (From Standard TGL 27885/03, 1983.)

Apart from the direct economic consequences, there must be general public agreement on matters affecting the community as a whole, such as:
• improvement in living and working conditions especially in health and recreational opportunities for the population;
• scenically aesthetic benefits and better microclimatic conditions;
• maintenance of variety among the animal and plant kingdoms, including the protection of rare and endangered species.

In the interests of a correct rating of rehabilitation measures, as well as for water-quality management in the broadest sense, such considerations as the assessment of natural resources, the use of economic stimuli for upgrading quality and other economic controls for careful use of water resources should also be taken into account.

It is worthy of note that lake rehabilitation projects in highly developed, market-oriented countries are largely supported by the tourist industry. A clean lake possesses a market value in the form of high numbers of visitors, income to hoteliers and restaurant owners, high land prices, increasing and beneficial tax receipts to the local community. For the rehabilitation of the Puddingstone Reservoir by destratification, Whalis (1969) reported a cost/benefit ratio of 1:5 when the benefits extended to increased receipts from angling following the establishment of a population of game fish. In general, the recreation potential, or use-value, of a body of water with enhanced quality attributes is itself also enhanced. Public expenditure in order to attain an increased use-value to an equal or greater degree is fully justified.

## 16.11 REFERENCES

Bernhardt, H. & Clasen, J. (1985) Recent developments and perspectives of restoration for artificial basins used for water supply. In: *Proceedings of an International Congress on Lake Pollution and Recovery.* Associatione Nazionale di Ingegneria Sanitario, Milano, 213–27.

Bethune, S. (1996) *Biological Control of Salvinia molesta in the Eastern Caprivi.* Progress Report 1980–1995, Department of Water Affairs, Windhoek.

Carlson, R.E. (1977) A trophic state index for lakes. *Limnology and Oceanography*, **22**, 361–9.

Cooke, G.D., Welch, E.B., Peterson, S.A. & Newroth, P.R. (1993) *Restoration and Management of Lakes and Reservoirs* (2nd edition). Lewis Publishers, Boca Raton, 558 pp.

Fast, A.W., Overholtz, W.J. & Tubb, R.A. (1975) Hypolimnetic oxygenation using liquid oxygen. *Water Resources Research*, **11**, 294–9.

Fichtner, N. (1983) *Verfahren zur Nitrateliminierung im Gewässer. Acta hydrochimica et hydrobiologica*, **11**, 339–45.

Hammer, U.T. (1986) *Saline Lake Ecosystems of the world*. W. Junk, Dordrecht, 616 pp.

Heinonen, P. (1980) *Quantity and Composition of Phytoplankton in Finnish Inland Waters*. Publications of the Water Research Institute, No. 37, Vesihallitus National Board of Waters, Helsinki, 91 pp.

Hillbricht-Ilkowska, A. & Zdanowski, B. (1978) Effect of thermal effluents and retention time on lake functioning and ecological efficiencies in plankton communities. *Internationale Revue der gesamte Hydrobiologie*, **63**, 609–17.

Kerekes, J. (1975) Phosphorus supply in undisturbed lakes in Kejimkujik National Park, Nova Scotia (Canada). *Verhandlungen der internationalen Vereinigung für theoretische und angewandte Limnologie*, **19**, 349–57.

Kerekes, J. (1977) The index of lake basin permanence. *Internationale Revue der gesamte Hydrobiologie*, **62**, 291–3.

Klapper. H. (1991) *Control of Eutrophication in Inland Waters*. Ellis Horwood, London, 337 pp.

Klapper, H. (1992) Calcite covering of sediments as a possible way of curbing blue-green algae. In: Sutcliffe, D.W. & Jones, J.G. (eds), *Eutrophication: Research and Application to Water Supply*, Freshwater Biological Association, Ambleside, 107–11.

Klapper, H. (1998) Water quality problems in reservoirs of Rio de Janeiro, Minas Gerais and Sao Paulo. *Internationale Revue der gesamten Hydrobiologie*, **83**, 93–102.

Klapper, H. & Schultze, M. (1997) Sulfur acidic mining lakes in Germany-ways of controlling geogenic acidification. Proceedings of the 4th International Conference on Acid Rock Drainage, Vancouver, Canada, 1727–44.

Kudelska, D., Cydzik, D. & Soszka, H. (1981) Proprzycja systemu oceny jakosci jezior (= Design of lake quality evaluation systems). *Wiadomosci Ekologiczne*, **27**, 149–73.

Lueschow, L.A., Helm, J.M., Winter, D.R. & Karl, G.W. (1970) Trophic nature of selected Wisconsin lakes. *Proceedings of the Wisconsin Academy of Science, Arts and Letters*, **58**, 237–64.

Margalef, R. (1975) Typology of reservoirs. *Verhandlungen der internationalen Vereinigung für theoretische und angewandte Limnologie*, **19**, 1841–8.

McColl, R.H.S. (1972) Chemistry and trophic status of seven New Zealand lakes. *New Zealand Journal of Marine and Freshwater Research*, **6**, 399–447.

Müller, H.E. (1978) Belastung und Belastungsdynamik in See-Umland-Systemen. *Tagungsberichte und wissenschaftliche Abhandlungen*, **41**, 499–516.

Niemann, E. & Wegener, U. (1976) Verminderung des Stickstoff- und Phosphoreintrages in wasserwirtschaftliche Speicher mit Hilfe nitrophiler Uferstauden- und Verlandungsvegetation (Nitrophytenmethode). *Acta Hydrochimica et Hydrobiologica*, **4**, 269–75.

Oksijuk, O.P. & Stolberg, F.W. (1986) *Control of Water Quality in Canals*. Naukova Dumka, Kiev, 173 pp. (In Russian.)

Olem, H. (1991) *Liming Acidic Surface Waters*. Lewis Publishers, Chelsea, MI, 333 pp.

Patalas, K. (1960) Punktowa ocena pierwotnej productywnosci jezior okolic Wegorzewa. *Roczniki Nauk Rolniczych (Polish Agricultural Annual, Warsaw) Section B*, **77**, 299–325.

Patalas, K. (1968) Landschaft und Klima als Faktoren der Massenproduktion von Algen. *Fortschriftte der Wasserchemie*, **8**, 21–31.

Richardson, J.L. (1975) Morphometry and lacustrine productivity. *Limnology and Oceanography*, **20**, 661–7.

Ripl, W. (1978) *Oxidation of Lake Sediments with Nitrate—a Restoration Method for Former Recipients*. Editions of the Institute of Limnology, University of Lund, Lund, 151 pp.

Ryding, S.-O. & Rast, W. (1989) *The Control of Eutrophication of Lakes and Reservoirs*. UNESCO, Paris, 314 pp.

Schindler, D.W. (1971) A hypothesis to explain differences and similarities among lakes in the experimental lake area, northwestern Ontario. *Journal of the Fisheries Research Board of Canada*, **28**, 295–301.

Shannon, E.E. & Bresonik, P.L. (1972) Eutrophication analysis: a multivariate approach. *Journal of the sanitary Engineering Division, American Society of Civil Engineers*, **98**, 37–57.

Shapiro, J. (1979) The need for more biology in lake restoration. *Proceedings of the National Conference on Lake Restoration, Minneapolis 1978*. U.S. Environmental Protection Agency, Washington, DC, 161–7.

Sheldon, A.L. (1972) A quantitative approach to the classification of inland waters. In: Krutilla, J.V. (ed.), *Natural Environments*. John Hopkins University Press, Baltimore, 205–61.

Standard TGL 27885/01 (1982) *Nutzung und Schutz der Gewässer/Stehende Binnengewässer/Klassifizierung*. Verlag für Standardisierung, Leipzig.

Standard TGL 27885/02 (1983) *Nutzung und Schutz der Gewässer/ Stehende Binnengewässer/ Nährstoffelimination in Vorsperren*. Verlag für Standardisierung, Leipzig.

Standard TGL 27885/03 (1983) *Nutzung und Schutz der Gewässer/Stehende Binnengewässer/Wassergütebewirtschaftung Seen*. Verlag für Standardisierung, Leipzig.

Sylvester, H., Hutchins, M., Gould, L. & Hall, M.W. (1974) *A Quantitative Classification of Maine Lakes*. Environmental Studies Center, University of Maine, Orono, 62 pp.

Teerink, J.R. & Martin, C.V. (1969) Artificial destratification in reservoirs of the California State Water Project. *Journal of the American Waterworks Association*, **67**, 436–40.

Thornton, J.A. (1987) The German technical standard for the assessment of water quality and its application to Hartbeespoort Dam (South Africa). *Water, South Africa*, **13**, 87–93.

Thornton, J.A. & Nduku, W.K. (1982) Water chemistry and nutrient budgets. In: Thornton J.A. (ed.), *Lake McIlwaine, the Eutrophication and Recovery of a Tropical African Man Made Lake*. Monographiae Biologicae, No. 49, W. Junk, Den Haag.

Uhlmann, D. (1958) Die biologische Selbstreinigung in Abwasserteichen. *Verhandlungen der internationalen Vereinigung für theoretische und angewandte Limnologie*, **13**, 617–23.

Uhlmann, D. (1985) Anforderungen an die Gütebewirtschaftung von Oberflächengewässern in tropischen und subtropischen Klimaten. *Acta Hydrochimica et Hydrobiologica*, **13**, 507–25.

Uhlmann, D. & Hrbáček, J. (1976) Kriterien der Eutrophie stehender Gewässer. *Limnologica (Berlin)*, **10**, 245–53.

USEPA (1974) *An approach to a relative trophic index system for classifying lakes and reservoirs—national eutrophication survey*. Working Paper No. 24, Pacific Northwest Environment Research Laboratory, U.S. Environment Protection Agency, Corvallis, OR, 36 pp.

Uttormark, P.D. & Wall, J.P. (1975) *Lake Classification—a Trophic Characterisation of Wisconsin Lakes*. Report 660/3-75-033, U.S. Environment Protection Agency, Corvallis OR, 165 pp.

Vollenweider, R. (1968) *Scientfic Fundamentals of the Eutrophication of Lakes and Flowing Waters, with Particular Reference to Nitrogen and Phosphorus as Factors of Eutrophication*. Organization of Economic Cooperation and Development, Paris, 159 pp.

Whalis, M. (1969) Presentation on the Puddingstone Reservoir at a seminar on water destratification, 1968. Mimeographed.

Zafar, A.R. (1959) Taxonomy of lakes. *Hydrobiologia*, **13**, 287–99.

# 17 Biomanipulation in Shallow Lakes: Concepts, Case Studies and Perspectives

S. HARRY HOSPER, MARIE-LOUISE MEIJER, R.D. GULATI AND ELLEN VAN DONK

## 17.1 INTRODUCTION

During development of strategies for restoring lakes affected by eutrophication, emphasis has gradually broadened from solely controlling point sources of phosphorus inputs, to more comprehensive and ecosystem-based approaches (Benndorf 1988; Gulati et al. 1990; papers in de Bernardi & Giussani 1995; Gulati & van Donk 2002). Ecologists have consistently argued that the poor biotic structure and functioning of lakes with abundant algae, especially cyanobacteria, contributes to resistance of lakes to recovery, even after their external phosphorus loading has been reduced (Shapiro 1980; Moss 1983; van Liere & Gulati 1992). Early on, Shapiro et al. (1975) emphasised the necessity of treating lakes as ecosystems, rather than 'containers of algae and phosphorus'. Several of the studies cited above indicated that manipulation of the food chain from fish, through zooplankton, to the algae, could be a powerful tool for lake restoration (reviewed in Lammens et al. 1990a).

The importance of predation by planktivorous fish on large-bodied zooplankton, and its effects on grazing by zooplankton on algae, was originally emphasised by Hrbáček et al. (1961), and by Brooks & Dodson (1965). Later, top-down control of algal blooms through fish stock management was the subject of numerous studies (Hurlbert et al. 1972; Shapiro et al. 1975; Shapiro 1978, 1980; Andersson et al. 1978; Edmondson 1979, 1991; Fott et al. 1980; Benndorf et al. 1981, 1984; Lampert 1983; Moss 1983, 1987; Carpenter et al. 1985; McQueen et al. 1986; Lazzaro 1987; Lammens et al. 1990b; Drenner & Hambright 2002; papers in Kasprzak et al. 2002). Also, the role of filter-feeding zooplankton (Timms & Moss 1984; Gulati 1990) and of benthic organisms, especially the zebra mussel (*Dreissena polymorpha*; Reeders & bij de Vaate 1990; Noordhuis et al. 1992; Reeders et al. 1993), were investigated. Shapiro et al. (1975) introduced the term 'biomanipulation' for the biological management of eutrophic lakes, and defined this technique as using biological interactions within lakes in order to control algal abundance and species composition (Shapiro 1978).

Here, discussion of biomanipulation is restricted to the management of fish stocks, especially in shallow lakes, and particularly to the removal of planktivorous and benthivorous fish. Most biomanipulation research in the USA and Europe (Benndorf 1988; papers in Gulati et al. 1990; Carpenter & Kitchell 1993; De Bernardi & Giussani 1995; Gulati 1995) has been focused on relatively deep lakes (>5m), and on stocking with predatory fish, rather than removal. Our knowledge that fish play a major role in the ecosystem (Carpenter & Kitchell 1993; see papers cited in Moss 1998) has contributed significantly to the concept of biomanipulation, which Moss (1998) has even called 'the lynchpin of shallow lake restoration'. In his extensive review, Reynolds (1994) suggested that biomanipulation is likely to be more successful in small and shallow water bodies, wherein the ecosystem can be inhabited by benthic and peripheral macrophytes. Mehner et al. (2002) also argue in favour of restricting biomanipulation to lakes which meet a set of criteria required to ensure that a top-down control cascades from piscivorous fish to phytoplankton, so that biomanipulation measures are successful. They mention five such criteria: important among these are shallowness, the presence of macrophytes and the extent of reduction of phosphorus.

In theoretical terms, in shallow and eutrophic lakes, the success of biomanipulation measures will rest on the hypothesis of 'alternative stable states', suggesting that at moderate nutrient concentrations in shallow lakes two alternative stable states may exist: a turbid-water and a clear-water state (May 1977; Timms & Moss 1984; Scheffer 1998; Moss 1999). Major disturbance of the turbid-water system by substantial reduction of fish stocks, both planktivores and benthivores, will trigger a shift from the algal turbid-water state into the macrophyte clear-water state.

This chapter considers key processes in the functioning of shallow lake ecosystems in the temperate zones, based on the experiences gained from several field studies over the past 25 years or so. The phenomena of alternative stable states and the perspectives of biomanipulation (i.e. fish stock reduction) are central to much of this work. Results of long-term case studies are discussed, particularly in the light of investigations in The Netherlands (Hosper 1997; Meijer 2000; Gulati & van Donk 2002).

## 17.2 KEY PROCESSES IN SHALLOW LAKE ECOSYSTEMS

The lakes referred to here are, on average, 1 to 4 m deep, and polymictic (i.e. vertically mixed throughout the year). Typical features are intense water–sediment interaction and the potentially large impact of aquatic vegetation. Turbidity of the lake water is caused both by phytoplankton and by resuspended sediments. Phytoplankton abundance is the net result of production and loss processes. Production is controlled by external and internal nutrient loading, and by availability of sunlight. Loss factors are grazing by zooplankton and other filter feeders, non-grazing mortality and net sedimentation of algae.

Figure 17.1 shows a simplified food chain from phytoplankton and benthos up to predatory fish, fish-eating birds and fishing. Manipulations at the top of the food chain, such as reduction of planktivores, or stimulation of predatory fish, may cascade down and give rise to reduced phytoplankton abundance. Wind-induced waves, and the feeding activities of benthivorous fish, lead to resuspension of the finer sediment particles. Submerged macrophytes can play a key role in stabilising the clear-water state, for example, by providing a refuge for the zooplankton against fish predation, and by protecting the sediments against wave action and benthivorous fish feeding. The following aspects of lake ecosystem functioning relevant to the application of biomanipulation will be discussed: (i) zooplankton dynamics and the 'spring clear-water phase', (ii) the role of top predators in structuring the food chain, (iii) the role of submerged macrophytes, and (iv) the resuspension of sediments by benthivorous fish.

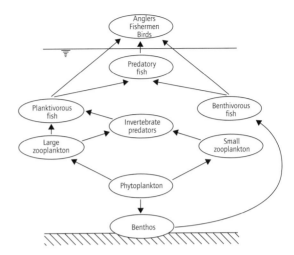

Fig. 17.1 Simplified food chain of a shallow lake ecosystem.

### 17.2.1 Zooplankton dynamics and the 'spring clear-water phase'

Planktonic algae are grazed by several genera of filter-feeding cladocera. The filtering capacities of *Daphnia* populations, however, are the only ones likely to develop the scale of consumption exploitable in practical biomanipulation (Reynolds 1994). In the temperate zones, larger zooplankton species such as *Daphnia galeata*, *D. pulex* and *D.*

*magna* are especially effective as grazers (Gulati 1989, 1990). In eutrophic lakes, grazing by zooplankton increases rapidly during spring, and may give rise to a clear-water phase with extremely low phytoplankton numbers (Gulati 1983; Lampert *et al.* 1986; Sommer *et al.* 1986). Such a clear-water phase may last for a period up to 1 month in May–June (Sommer *et al.* 1986) and evidently, for shallow lakes, clear water in that period of the year will be crucial in triggering development of submerged vegetation. A collapse in the grazer community is usually observed in summer, and subsequently algal biomass increases again (Sommer *et al.* 1986).

In many shallow lakes in The Netherlands which have become hypertrophic, for example, the Loosdrecht lakes, a distinct clear-water phase no longer occurs. This is because in most such lakes large-bodied daphnids are either virtually absent or are very sparse. As a result, these lakes fail to exhibit any increase in numbers of large-bodied *Daphnia* (Gulati 1990). Other similar shallow lakes (e.g., Tjeukemeer) may exhibit some increase in large-bodied *Daphnia* in spring, but their grazing is not sufficient to cause effective phytoplankton mortality, or therefore a major improvement in transparency (Lammens *et al.* 1990b). The question therefore arises as to which factors control the success of *Daphnia*.

The quality and quantity of food resources on the one hand, and of planktivorous fish and invertebrate planktivores on the other, are generally considered to be important in controlling the dynamics of *Daphnia* populations (Threlkeld 1987). Grazing effectiveness may be severely reduced by the type of algae. Moreover, as demonstrated by DeMott *et al.* (2001) in the Loosdrecht lakes, feeding inhibition in natural lake seston appears to increase markedly with increase in *Daphnia* size.

Exclusion of large-bodied *Daphnia* species in these lakes was attributed to interfering filaments of cyanobacteria. In a pioneering study, Gliwicz & Siedlar (1980) found algal food size to be the cause of food limitation (i.e. larger algae interfere with food collection so that cladocerans fail to control algal blooms; Gliwicz 1990). Larger algae, such as colonial or filamentous cyanobacteria (*Oscillatoria*, *Lyngbya* and *Aphanizomenon*), seem to depress the filtering rate mechanically (Bloem & Vijverberg 1984; Reynolds 1994; Gulati *et al.* 2001). At low filament concentrations in the seston, growth and reproduction of *Daphnia* are not affected. However, as filament concentrations increase, growth and reproduction tend to be reduced and even completely halted. Critical filament concentration depends both on the morphology and physiological state of the filaments, and on *Daphnia* body size (Gliwicz 1990).

All juvenile fish, and many adults, feed on zooplankton. Generally, in temperate lakes, planktivory will exhibit a conspicuous seasonal cycle, with:

• relatively low predation in the spring by the overwintering fish population;
• high predation in the summer, after the eggs of the fish hatch in June–July;
• low predation in the autumn after a part of the fish stock has died from food limitation.

As the food demand, and the production, of the young-of-the-year (YOY) fish is significantly higher than that of older fish, a more than proportional increase in planktivory may be expected during summer (Cryer *et al.* 1986; Barthelmes 1988). What is the effect of the planktivorous fish on the zooplankton community? Brooks & Dodson (1965) studied the zooplankton in Crystal Lake (USA) before and after the introduction of the planktivorous 'glut herring' (*Alosa aestivalis*). Ten years after *Alosa* had become abundant in the lake, the zooplankton was sampled again. Their modal size in the presence of *Alosa* was 0.285 mm, whereas in the absence of *Alosa* it was 0.785 mm. This seems to be clear evidence that predation by *Alosa* falls more heavily upon the larger zooplankton, eliminating those organisms greater than about 1 mm in length.

In their classic paper, these authors presented the 'size efficiency hypothesis', suggesting that: (i) larger zooplankton graze more efficiently and can also take larger particles than smaller zooplankton; (ii) when fish predation is low, small zooplankton will be eliminated by larger forms; and (iii), conversely, when fish predation is high, the most efficient grazers will be eliminated by the fish.

Although there is abundant evidence that planktivorous fish can change the species composition and size structure of zooplankton communities, the evidence that these predators actually regulate seasonal dynamics, and the spring peak of *Daphnia* in particular, is less clear. As accurate data on fish biomass or numbers are difficult to obtain, seasonal time-series of both *Daphnia* and fish stocks are very scarce. Mills & Forney (1987) presented detailed data for Oneida Lake (USA), which recorded strong inverse correlations between numbers of YOY yellow perch (*Perca flavescens*) and *Daphnia* during the summer, confirming the role of predation in that period of the year.

The clearest (if negative!) demonstration of the role of fish in the timing and magnitude of the spring peak of *Daphnia* is the aftermath of a massive natural fish-kill. Of this, the case study of Lake Mendota (Vanni et al. 1990; Kitchell 1992) offers an excellent example. In this lake, approximately 85% of the population of the main planktivore (Cisco; *Coregonus artedii*) perished in late summer of 1987 because of unusual high summer temperatures and consequent depletion of hypolimnetic oxygen. After fish mortality, the larger *Daphnia pulicaria* replaced the smaller *Daphnia galeata mendotae*, leading to greater grazing pressure on phytoplankton, and a much longer spring clear-water phase. Similar effects of catastrophic (winter) fish-kills and subsequent pronounced clear-water phases are described by Schindler & Comita (1972), De Bernardi & Giussani (1978) and Haertel & Jongsma (1982).

Invertebrate planktivores occupy a special position in the pelagic food web. Like zooplankton, they are vulnerable to predation by fish, yet they can also prey heavily on certain zooplankton (Lane 1978; Murtaugh 1981; Hanazato 1990; Mackay et al. 1990; Luecke et al. 1992). Many invertebrates, such as the crustaceans *Neomysis* spp., *Leptodora kindtii* and *Bythotrephes longimanus*, the cyclopoid copepod *Cyclops vicinus*, the larvae of the phantom midge *Chaoborus* and water mites, feed on *Daphnia*, but the impact of invertebrate predation on *Daphnia* dynamics, however, seems to be limited. Threlkeld (1987) concluded that, although correlations suggest that invertebrate predators may control *Daphnia*, experimental manipulations of invertebrate predators have not produced significant changes in timing or magnitude of seasonal peaks.

In conclusion, the observation that 'spring clearing' in many shallow lakes fails to develop may be explained by the presence of high stocks of planktivorous fish and high densities of filamentous cyanobacteria. Resuspension of sediments caused by abundant benthivorous fish (see below) or wind may be an additional factor preventing the spring clearing of the water.

### 17.2.2 The role of predatory fish

A clear consensus has now emerged that predatory fish play a major role in regulating stocks of fish. In addition, both fish-eating birds (e.g. cormorants, *Phalacrocorax carbo*) and fisheries can contribute significantly to the consumption or removal of fish. Here attention is focused on predatory fish, the most important of which, in The Netherlands, are northern pike (*Esox lucius*), pike-perch (*Stisostedion lucioperca*) and perch (*Perca fluviatilis*).

The first may have played an important regulatory role in structuring the food chain before shallow lakes and ponds in The Netherlands became hypertrophic (Grimm 1989, 1994; Grimm & Backx 1990). Survival of young pike appears to be strongly related to availability of refuges, which are usually provided by aquatic vegetation. Pike, being cannibalistic, need hiding places in order to reduce the risk of intraspecific predation and to provide cover for their own hunting.

Pike are able to swallow prey of a size up to two-thirds of their own length, irrespective of species, so that cannibalism can be quite common. Moreover, pike larvae attach themselves to plants for their first few days. In waters with abundant emergent and submerged vegetation, survival of young pike is relatively high, and their large numbers may regulate the abundance of young bream and roach. Hosper et al. (1987) suggested a conceptual model with a key role for pike (Fig. 17.2). The left-hand side of the circle represents the food chain where pike, as top predator, structure the food

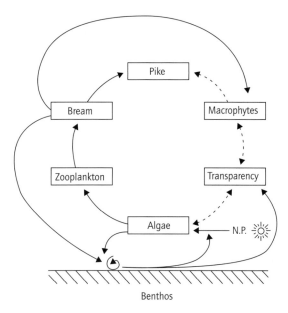

Fig. 17.2 Interactions in shallow, eutrophic lakes with a key role for the predatory fish pike. Severe eutrophication is leading to more algae, less light reaching the plants, fewer plants and fewer pike. This results in more bream, decrease in large zooplankton and even more algae, etc. Bream is feeding on zooplankton as well as on benthic organisms and stirs up the sediments, resulting in release of nutrients, increased turbidity of the water and mechanical disturbance of rooted macrophytes (from Hosper et al. 1987).

chain via the prey fish and the zooplankton, down to the algal biomass. The right-hand side indicates the indirect effects of algal biomass on the pike population, via transparency of the water, and the abundance of submerged macrophytes. Thus, pike are coupled to the algal biomass in two different ways, leading to development of both a positive (self-reinforcing) feedback loop (the reduction in transparency as a result of increased algal biomass causes the decline of the macrophytes) and indirectly a decline in the pike population, leading to even more algal biomass, etc. The model also shows the effects of the feeding strategies of bream, reinforcing and stabilising the turbid-water state (see below).

Considering the dependency of pike on littoral zones, and their restricted home range, it will be clear that the validity of the 'pike model' (Fig. 17.2) is limited to relatively small water bodies. In small lakes (<20 ha) with a high shoreline:area ratio, and high degree of 'patchiness' (open water and plant beds), optimal conditions for pike are easily met (Klinge et al. 1995). In such lakes, maximum pike biomass of 75 kg ha$^{-1}$ has been observed (Grimm 1989). However, in larger lakes (>30 ha), pike biomass will not exceed 25–35 kg ha$^{-1}$ (Klinge et al. 1995).

Although pikeperch are well adapted to turbid waters, they are originally a warm-water species; thus their growth in temperate lakes is usually suboptimal. Summer temperatures strictly control the success of annual recruitment (Willemsen 1980), and in cold summers especially the growth of young pikeperch is slow. In such years, bream, which grow relatively rapidly, soon become too large to be swallowed by pikeperch (van Densen & Grimm 1988).

Similarly, it is unlikely that perch can control recruitment of planktivorous fish in shallow, eutrophic lakes. Perch are fish from clear, vegetated waters of moderate productivity (Persson 1994). Young perch are planktivorous; during later stages of growth, only larger individuals (>25 cm) shift fully to piscivory (Grimm et al. 1992). In highly productive systems, recruitment of juvenile perch to piscivorous stages is limited by the presence of planktivorous cyprinids, such as roach. Conversely, in moderately productive systems (chlorophyll a concentration <5–10 mg m$^{-3}$; Persson 1994), recruitment limitation is less severe, allowing a larger proportion of perch to become piscivorous, thereby affecting cyprinid populations through predation (Persson 1994; Hargeby et al. 1994).

From scarce data on piscivorous perch production, Klinge et al. (1995) derived the maximum annual prey fish consumption by perch of 50 kg ha$^{-1}$, which can be produced by a total fish biomass of 60–80 kg ha$^{-1}$. Using the Hanson & Legget (1982) relationship for total phosphorus versus fish biomass, the authors suggested that top-down control of prey fish by perch can be expected only in lakes with concentrations as low as 20–40 g m$^{-3}$.

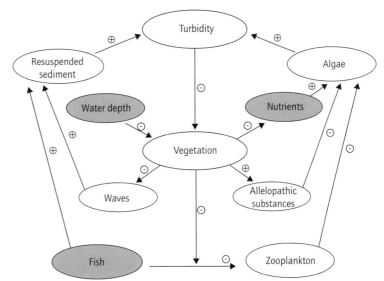

Fig. 17.3 Interactions in shallow, eutrophic lakes with a key role for the submerged vegetation. The qualitative effect of each route in the diagram can be determined by multiplying the signs along the way. In this way it can be seen that both the vegetated and the turbid state are self-reinforcing. The shaded boxes are possible steering variables for lake management and the effects of measures can be determined in a similar way (from Scheffer et al. 1993).

In conclusion, in shallow, eutrophic lakes (in The Netherlands) pike seem to be the best candidate for controlling the recruitment of major planktivores such as bream and roach. However, because of the dependency of pike on vegetated littoral zones, effective top-down control will be restricted to relatively small lakes and ponds (<10–20 ha) with sufficient macrovegetation. Pike predation in larger lakes with poor development of littoral vegetation is likely to be insufficient.

### 17.2.3 The role of submerged macrophytes

Thanks to biomanipulation research, our knowledge of the importance of macrophytes in lakes has increased rapidly over the past 15 years (Gulati & van Donk 2002). There is now increasing agreement that development of macrovegetation is crucial for maintaining clear water in lakes (Moss 1990; Jeppesen et al. 1990; Meijer et al. 1990; Jeppesen 1998; Scheffer 1998). Submerged macrophytes may reduce turbidity in many different ways, not just by providing a habitat for pike.

Meijer et al. (1990) and Scheffer et al. (1993) suggested a model for interactions in shallow lakes leading to turbidity, which demonstrates the significance of macrophytes (Fig. 17.3). The positive effect of vegetation on water clarity is the result of a number of mechanisms including:

1 aquatic plants provide a refuge against planktivorous fish for phytoplankton-grazing zooplankton (Timms & Moss 1984; Schriver et al. 1995; Stansfield et al. 1995);

2 structural complexity (provided by plants) promotes the piscivorous perch (Persson 1994) and pike (Grimm 1994) and deters the planktivorous–benthivorous bream (Lammens 1986), resulting in more top-down control of planktivores, and less fish-induced resuspension (Mehner et al. 2002);

3 vegetation reduces the availability of nutrients for phytoplankton, by uptake from the water and by promoting denitrification in the sediment (van Donk et al. 1990b, 1993; Gumbricht 1993); plants may release allelopathic substances which are toxic to algae (Wium-Andersen 1987; Jasser 1995; van Donk & van de Bund 2002; Mulderij et al. 2003);

4 wind-induced and fish-induced resuspension of sediments is reduced by vegetation (Jackson & Starrett 1959; James & Barko 1990), and shoreline erosion is prevented if littoral vegetation is well developed. In short, reduced sediment resuspen-

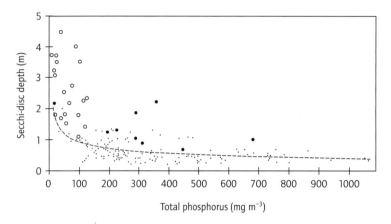

Fig. 17.4 Summer mean Secchi-disc depth in relation to lake water total phosphorus for shallow Danish lakes: (○) lakes with a high cover of submerged macrophytes and a surface area >3 ha; (●) lakes with a high cover of submerged macrophytes and a surface area <3 ha; and (·) lakes with only minor or unknown cover of submerged macrophytes. (From Jeppesen et al. 1990.)

sion and erosion lead to increased sedimentation. The increased sediment stability which follows will contribute to seasonally persistent clear-water patches. This has often been observed where *Chara* meadows develop after restoration measures, as in several Dutch lakes (Gulati & van Donk 2002).

Although the relative importance of these submechanisms is often difficult to assess, and may vary between lakes, analyses of extensive data sets support the view that, in freshwater lakes, there is an overall positive effect of vegetation on water transparency. Lakes with extensive submerged macrophyte cover tend to possess a higher transparency than those with the same nutrient status, but in which vegetation is sparse or absent (Fig. 17.4). Large lakes may develop subsystems consisting of shallow, clear-water areas with macrophytes, and deeper, turbid-water areas inhabited by algae (Scheffer et al. 1994; Scheffer 1998).

### 17.2.4 Resuspension of sediments by fish

In shallow, eutrophic lakes in The Netherlands, large bream and common carp (*Cyprinus carpio*) represent the benthivorous fishes. In particular, bream appear to be exceptionally successful in turbid waters with sediments rich in benthic organisms, and hardly any vegetation to hinder feeding. The reduced risk of predation and the fact that they can efficiently use zooplankton as well as benthos as a food source (Lammens 1986) are additional factors contributing to the success of bream.

In The Netherlands, almost all hypertrophic, shallow lakes are infested with bream, with stocks of around 100 kg ha$^{-1}$ in large bodies such as the Loosdrecht lakes (van Donk et al. 1990a) and Wolderwijd (Meijer et al. 1994b), rising to over 500 kg ha$^{-1}$ in smaller lakes and ponds (Meijer et al. 1990; Driessen et al. 1993). Large numbers of common carp may occur locally, mainly originating from artificial stocking. Larger bream (>20 cm) grub up sediments intensively in search of food (mainly midge larvae), each bream processing tens of litres of mud daily, often to a depth of several centimetres (Lammens 1986). This behaviour promotes nutrient release and turbidity of the overlying water (Meijer et al. 1990; Havens 1993; Breukelaar et al. 1994) and hampers plant growth through direct mechanical disturbance of the plant roots (Crivelli 1983; ten Winkel 1987).

Breukelaar et al. (1994) studied the effects of benthivorous bream and carp in 16 experimental ponds (0.1 ha, mean depth 1 m, sandy clay/clay sediment), stocked with bream (>25 cm) or carp (>40 cm) at densities varying from 0 to 500 kg ha$^{-1}$. It was concluded that, beginning from a bream or carp biomass of 100 kg ha$^{-1}$, significant effects on suspended solids, total phosphorus and total nitrogen concentrations may be expected. In the bream ponds, suspended solids roughly doubled, rising from 50–100 to 200 kg ha$^{-1}$, levelling off with further increases in bream biomass from 200 to 500 kg ha$^{-1}$. The resuspension effect of bream appeared to be twice as great as that of carp.

These results clearly indicate the relevance of fish-induced resuspension of fine sediments in small and shallow lakes. A clear-water phase in spring (see above), the effect of which is thought to trigger the submerged vegetation, may be completely masked by the effects of resuspension. Note that in deeper lakes, and in lakes with coarser sediment, the effects will be proportionally weaker. Owing to lower bream stocks in large lakes, fish-induced resuspension is apparently less important. However, low numbers of benthivorous bream could possibly loosen the top layer of the sediment, making it more susceptible to wind-induced resuspension in large lakes.

## 17.3 ALTERNATIVE STABLE STATES

From the conceptual models presented above (Figs 17.2 & 17.3), it can be inferred that simply because of the presence or absence of aquatic vegetation both the clear-water state and the turbid-water state of freshwater lakes are self-reinforcing. Different buffering mechanisms lead to a stable clear-water or a stable turbid-water state. Apparently, lake ecosystems exhibit resistance to change with increasing as well as decreasing nutrient loading. Resistance is defined here after Pimm (1991): 'Resistance measures the consequences when a variable is permanently changed. How much do other variables change as a consequence? If the consequent changes are small, the system is relatively resistant'.

Resistance of the ecosystem during eutrophication, as well as during the reverse process of oligotrophication, gives rise to the typical phenomenon of hysteresis (Fig. 17.5). Note that at the same nutrient loading, the lake 'under eutrophication' and 'under oligotrophication' may exhibit very different algal biomass.

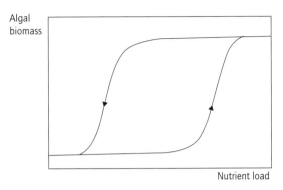

Fig. 17.5 Eutrophication and oligotrophication in relation to algal biomass, showing a typical hysteresis curve.

### 17.3.1 Theory

Theory indicates that natural multispecies assemblages of plants and animals are likely to possess several different equilibrium points (May 1977). If there is a unique stable state, the system will tend towards it from all initial conditions, and after any disturbance (like a marble seeking the bottom of the cup; see also Pahl-Wostl 2003, Volume 1, Chapter 17). If there are many stable alternatives, the state into which the system will settle depends on initial conditions. The system may return to this state following small perturbations, but large disturbances are likely to carry it to some new region of the dynamic landscape. If there is a unique stable state, historical events are unimportant, but if there are many alternative stable states, such events can be of overriding significance. The dynamic behaviour of ecosystems with two or more stable states is such that a continuous variation in a control variable can produce discontinuous effects (May 1977).

It will be obvious that questions of this kind are very important to the understanding and management of ecosystems. Scheffer (1990) explored the significance of multiple stable states for shallow, freshwater systems, and by the use of simple mathematical models demonstrated that over a certain range of nutrient concentrations several ecological relationships may give rise to the existence of

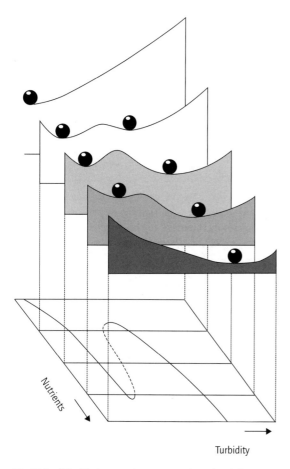

**Fig. 17.6** 'Marble in a cup' representation of stability at five different nutrient levels. The minima correspond to stable equilibria, maxima to unstable breakpoints (see text for explanation). (From Scheffer 1990.)

nutrient loading gradually changes the shape of the stability landscape, and gives rise to an alternative turbid equilibrium. However, if no major disturbances occur, the system will remain in its current state, responding only slightly to eutrophication. If nutrient loading is raised further, the stability of the clear state decreases and slight perturbations are enough to cause a shift to the turbid equilibrium.

At still higher nutrient loadings, the clear equilibrium disappears, causing an irreversible shift to a turbid state. Efforts to restore the system by reducing the nutrient loading will change the stability landscape again, but even at loadings at which the system formerly was clear there will hardly be any response. An alternative equilibrium may be present, but the stable state is sustained. Only a severe reduction of nutrient loading will result in a shift back to the clear state.

### 17.3.2 Evidence from the field

Lakes shifting between alternative stable states of clear and turbid waters are well known in fisheries science (Klapper 1992). Without any noticeable change in nutrient conditions, vegetated, clear water lakes with pike and tench (*Tinca tinca*) can change, from one year to another, into a lasting state of non-vegetated turbid water with pikeperch (Klapper 1992).

Shifts from turbid to clear, and back to turbid, were observed in the 2 m deep Sewekowersee (Germany). Klapper (1969) attributed these changes to weather conditions at the beginning of the growing season. The cold winter of 1962–63, followed by an extremely cold spring, produced low algal biomass and clear water at the beginning of the growing season. Submerged macrophytes took advantage of these favourable light conditions, colonising the whole lake and stabilising the clear-water state. Three years later, during the warm spring of 1966, the opposite circumstance occurred, and the lake reverted to a stable, turbid-water state with a transparency of 0.5 m.

As noted before, catastrophic fish-kills during cold winters (Schindler & Comita 1972; de Bernardi & Giussani 1978; Haertel & Jongsma 1982) may

alternative equilibria. His mathematical analysis supports the idea that shallow lakes may possess two alternative stable states: a clear, vegetated configuration with low fish stocks, and a turbid, unvegetated state with high densities of planktivorous and benthivorous fish.

The landscape of alternative stable states is visualised in Fig. 17.6 by means of the 'marble in a cup' analogy (see also Volume 1, Chapter 17). At low nutrient loading, the system exists as only one stable equilibrium: a clear-water state. Increase of

produce similar effects. Timms & Moss (1984) studied two linked, shallow freshwater basins with similar hydromorphological conditions and nutrient loading: Hudson's Bay and Hoveton Great Broad in the Norfolk Broads (UK). Hudson's Bay supported a large stand of water lilies with adjacent open, clear water; here, large-bodied Cladocera were abundant in the zooplankton. By contrast, Hoveton Great Broad produced algal blooms, and the water was turbid; zooplankton populations consisted of rotifers and small-bodied Cladocera. The authors concluded that these lakes seem to exemplify two alternative community states. They suggested that the relevant buffering mechanisms for the clear-water state included (i) nutrient competition between algae and macrophytes, (ii) secretion of algae-inhibiting metabolites by macrophytes, and especially (iii) the refuge function of macrophytes for zooplankton.

During the past few decades, the two moderately eutrophic lakes Tåkern and Krankesjön (Sweden) have shifted 'spontaneously' several times between a clear-water state with abundant submerged vegetation and a turbid-water state with high phytoplankton densities (Blindow et al. 1993). For both lakes, it was most likely that water-level fluctuations, affecting submerged macrophytes, caused these shifts, either through changes in light availability or through catastrophic events such as drying out or mechanical damage by ice movement (Blindow et al. 1993). Macrophyte control by herbicides during the 1960s caused a shift to the turbid-water state in the small Dutch lake IJzeren Man (Driessen et al. 1993).

### 17.3.3 Nutrient thresholds for a sustainable clear-water state

Scheffer's (1990) model (Fig. 17.6) suggests the existence of alternative stable states at moderate nutrient concentrations. At low nutrient concentrations, only the clear state will be stable. At high nutrient concentrations, the turbid state will be stable, and at intermediate concentrations, the clear-water and turbid-water states may exist as stable alternatives. The question now is how to quantify the appropriate nutrient concentrations for each state.

Results from multilake studies can help to find the total phosphorus (TP) concentration limits for the stable clear-water state. Stable clear water with Secchi-disc depth >1 m can be expected at TP <50 mg m$^{-3}$ (Fig. 17.4). If non-algal turbidity is high, lower TP concentrations will be required.

Indicating the upper and lower TP concentration limits for the existence of alternative stable states is more difficult. Algal-rich turbid water is associated with TP >100 mg m$^{-3}$ (Fig. 17.4). However, these high concentrations do not necessarily exclude the existence of an alternative clear-water state. Danish multilake studies have clearly shown the existence of vegetated clear-water lakes at high TP concentrations (Fig 17.4). This is particularly the case for small lakes (<3 ha), indicating the importance of a high shoreline:area ratio for the clearing effects of submerged vegetation.

For larger lakes (>3 ha), both clear and turbid waters are found at TP concentrations of 50–100 mg m$^{-3}$ (Fig. 17.4). In The Netherlands, several biomanipulation case studies in small lakes with high total phosphorus concentrations also indicated a more or less stable clear-water state (see next sections). Along with Jeppesen et al. (1990, 1991, 1997), it is tentatively concluded that alternative stable states may be expected at TP concentrations of 50–100 mg m$^{-3}$. In very small lakes, a sustainable clear-water state is possible at higher concentrations, owing to the stronger clearing effects of submerged vegetation.

## 17.4 MECHANISMS FOR BIOMANIPULATION IN SHALLOW LAKES

The idea of alternative stable states implies that, at moderate nutrient concentrations, one drastic intervention could tip the balance towards a new, stable clear-water equilibrium (Fig. 17.6). Essentially, biomanipulation measures include substantial reduction of planktivorous and benthivorous fish. By reduction of overwintering planktivorous fish stock (cf. the natural winter fish-kills), a *Daphnia*-mediated, spring clear-water phase is produced. Benthivore reduction, particularly in productive, clay-bottom lakes, further supports clearing of the lake.

Clear water during spring provides chances for submerged vegetation and a sustainable clear-water state. In addition to fish-stock reduction, stocking of pike fingerlings may help reduce YOY fish during summer. For assessing the perspectives of biomanipulation for lake restoration, three questions are relevant.
1 Is substantial fish stock reduction technically feasible?
2 Will the water clear up after the fish removal?
3 Will the new clear-water state be stable? (Fig. 17.7).

## 17.5 CASE STUDIES

A number of whole-lake case studies in The Netherlands have been evaluated for testing the applicability of biomanipulation for shallow lakes (Table 17.1). Note that the lakes studied mostly differ in surface area, total phosphorus concentration and original fish stock. The small lakes in particular contained a high fish stock and were rich in total phosphorus. Table 17.2 summarises the Secchi-disc depth effects in relation to total fish-stock reduction, the peaking of *Daphnia* and the reduction of benthivorous fish. The possible impact of inedible algae and of invertebrate predators (*Neomysis*, *Leptodora*) on the response of *Daphnia* is also shown.

Table 17.2 shows that where the winter fish stock was substantially reduced (>75% and no immigration) all lakes (except for Sondelerleijen) cleared up during the following spring. Clearing can be explained by *Daphnia* grazing (Zwemlust, Wolderwijd, Duinigermeer ), or by a combination of that factor and reduced benthivorous feeding (Bleiswijkse Zoom, Noorddiep, IJzeren Man). Despite effective total fish removal and reduction of benthivores, Sondelerleijen failed to develop a clear water phase. This lack of response could be explained by predation of *Neomysis integer* on *Daphnia* (Table 17.2). Breukeleveense plas and

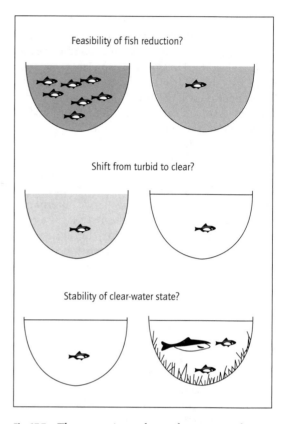

Fig. 17.7 Three questions relevant for assessing the chances for biomanipulation. Technical feasibility of substantial fish reduction is a prerequisite for the success of biomanipulation. Additionally, immigration of fish from adjacent lakes should be prevented, e.g. by the construction of fish barriers. Here, we will pass over this technical subject and focus attention on questions of clearing-up after fishing, the re-establishment of submerged vegetation and the stability of the clear-water state.

Klein Vogelenzang did not clear up either, and, apart from ineffective fishing, inedible algae and invertebrate predators (Breukeleveense plas) could explain the lack of response in these lakes, Klein Vogelenzang did not clear up in spring, but according to van der Vlugt *et al.* (1992) produced a *Daphnia* peak, and a clear water phase in the autumn, after the collapse of the cyanobacteria bloom. Apparently, food quality prevented the peaking of

**Table 17.1** Biomanipulation case studies

| Case study | Year | Area (ha) | Mean depth (m) | Total phosphorus (mg m$^{-3}$) | Fish stock before/after (kg ha$^{-1}$) | Fish reduction (%) | Fish immigration |
|---|---|---|---|---|---|---|---|
| Zwemlust (ZL)* | 1987 | 1.5 | 1.5 | 1000 | 800/10 | 98 | No |
| Bleiswijkse Zoom (BZ)† | 1987 | 3.1 | 1.1 | 250 | 760/120 | 84 | No |
| Noorddiep (ND)‡ | 1988 | 4.5 | 1.5 | 200 | 690/145 | 79 | No |
| Breukeleveense plas (BP)§ | 1989 | 180 | 1.5 | 100 | 150/60 | 60 | Yes |
| Klein Vogelenzang (KV)¶ | 1989 | 20 | 1.5 | 300 | 260/200 | 30 | Yes |
| IJzeren Man (IJM)** | 1990 | 11 | 1.8 | 250 | 710/120 | 83 | No |
| Sondelerleijen (SL)†† | 1992 | 25 | 1.5 | 300 | 500/75 | 85 | No |
| Wolderwijd (WW)‡‡ | 1991 | 2700 | 1.5 | 120 | 210/45 | 79 | No |
| Duinigermeer (DM)§§ | 1992 | 28 | 1.0 | 100 | 150/40 | 73 | No |

\*  Van Donk et al. (1989, 1990b, 1993), van Donk & Gulati (1995) and van de Bund & van Donk (2002). After complete fish removal, stocked with 10 kg ha$^{-1}$ rudd (Scardinius erythrophthalmus).
† Meijer et al. (1990, 1994a, 1995).
‡ Walker (1994), Meijer et al. (1994a, 1995) and van Berkum et al. (1996).
§ Van Donk et al. (1990a). After 60% fish reduction, fish immigrated from adjacent waters.
¶ Van der Vlugt et al. (1992). 60 kg ha$^{-1}$ removed in April 1989, 40 kg ha$^{-1}$ in December 1989. Fish immigrated from adjacent waters.
\*\* Driessen et al. (1993). After complete fish removal, stocked with 120 kg ha$^{-1}$ roach, rudd and other species.
†† Clewits (1994). Major fish removal took place in early 1992; also fish removals in early 1991 and 1993.
‡‡ Meijer et al. (1994b).
§§ Van Berkum et al. (1995) and Witteveen & Bos (1995).

**Table 17.2** Will the lake clear up after fish reduction? Secchi-disc depth effects in the following spring (May-June), in relation to total fish reduction, peaking of Daphnia and the reduction of benthivores. The possible effects of inedible algae and invertebrate predators (Neomysis, Leptodora) on the response of Daphnia are also shown. Criteria are from Hosper et al. (1992) and Hosper & Meijer (1993) or explained in the text. Data from references in Table 17.1 and Meijer (unpublished results)

| | Zwemlust | Bleiswijkse Zoom | Noorddiep | Breukeleveense plas | Klein Vogelenzang | Sondelerleijen | IJzeren Man | Wolderwijd | Duinigermeer |
|---|---|---|---|---|---|---|---|---|---|
| Clearing status of lake* | • | • | • | □ | □ | □ | • | • | • |
| Fish reduction† | • | • | • | □ | □ | • | • | • | • |
| Daphnia peak‡ | • | • | • | □ | □ | • | • | • | • |
| Inedible algae§ | • | • | • | □ | □ | • | • | • | • |
| Invertebrate predators¶ | • | • | • | □ | • | □ | ? | • | ? |
| Benthivore reduction** | □ | • | • | □ | □ | • | • | □ | □ |

\* •, significant increase in Secchi-disc depth (>bottom); □, no or minor increase.
† •, >75% and no immigration of fish; □, <75% or immigration of fish.
‡ •, peak of large-bodied Daphnia (>100 L$^{-1}$); □, no peak.
§ •, <50,000 mL$^{-1}$ filamentous cyanobacteria; □, >50,000 mL$^{-1}$.
¶ •, <100 m$^{-2}$ Neomysis, <5 L$^{-1}$ Leptodora; □, >100 m$^{-2}$ Neomysis, >5 L$^{-1}$ Leptodora.
\*\* •, >50% and reduction >150 kg ha$^{-1}$; □, <50% or <150 kg ha$^{-1}$.

*Daphnia* populations in the spring. Unfortunately, fishing was ineffective in both *Oscillatoria* lakes (Breukeleveense plas and Klein Vogelenzang).

Therefore, current case studies give no clue as to the question whether effective fish reduction can lead to removal of *Oscillatoria* blooms, through the grazing by *Daphnia*. As several alternative explanations are available for the failing biomanipulation in Breukeleveense plas and Klein Vogelenzang (with a relatively low fish reduction), the case studies do not allow firm conclusions to be drawn about the amount of fish it is necessary to remove for successful clearing of the lakes.

The next question refers to the stability of the clear-water state during the following summer, and those of subsequent years. Attention is focused on long-term results for Secchi-disc depth and submerged vegetation, during July–August (Fig. 17.8). That period (the summer) is the most critical for lake clarity because, at that time, young fish prey upon the zooplankton, and the benthivores are most active.

All lakes (except for Wolderwijd) remained clear throughout the summer, and exhibited a significant increase in submerged macrophytes in the first growing season after fish removal (Fig. 17.8). Within 1 or 2 years, 30–80% of lake area was covered with dense vegetation. Bleiswijkse Zoom and IJzeren Man returned to the turbid-water state after 7 and 5 years, respectively. Both small lakes (Noorddiep and Zwemlust) were still clear during summer, respectively 7 and 8 years after fish reduction. The clear state in Zwemlust was not stable in the long term. In 1999, another removal of zooplanktivorous fish (mainly rudd) produced similar effects as in 1987.

## 17.6 DISCUSSION AND CONCLUSIONS

Fish stocks in algal lakes tend to impose homeostasis on the system, preventing recovery of the lake. Biomanipulation, including substantial fish stock reduction, could trigger a shift from a stable turbid-water state to an alternative stable clear-water state. Case studies were evaluated for testing the applicability and perspectives of biomanipulation.

### 17.6.1 Case studies

As with natural winter fish-kills, biomanipulation aims at drastic reduction of planktivorous and benthivorous fish stocks. Its purpose is to trigger a shift to a stable clear-water state by means of a single, winter fishing. Nine case studies of shallow lakes in The Netherlands were evaluated with respect to the question of (i) clearing up right after fishing and (ii) stability of the clear-water state. Six out of the nine lakes cleared after reduction of the fish stock (Table 17.2). These lakes possess very different surface areas, total phosphorus concentrations and original fish stock, indicating that these factors are not critical to the mechanisms responsible for the clearing process. All lakes share in common that the total fish stock was reduced by more than about 75%. Conversely, >75% fish removal is no guarantee for clearing, as shown by Sondelerleijen. *Daphnia* population peaking, and the resultant grazing, is the major process in clearing in these lakes.

In some small lakes, removal of large quantities of benthivores contributed significantly to the clearing. In shallow Dutch lakes, invertebrate predators play a minor role in *Daphnia* dynamics during spring (Gulati *et al.* 1992; Boersma 1994). However, results from Sondelerleijen indicated that after fish reduction *Neomysis* may have been responsible for the absence of a *Daphnia* peak. In Breukeleveense plas, *Leptodora* could have been important, but for this lake alternative explanations (inedible algae, ineffective fish reduction) are also available. All lakes, except for the large Wolderwijd, exhibited a rapid response in submerged vegetation.

In Wolderwijd, the spring clear-water period only lasted for 6 weeks (Meijer & Hosper 1997) and the vegetated area remained practically unchanged. However, as shown by Meijer & Hosper (1997), during May–June 1991, this period triggered gradual colonisation of the lake by dense stands of Characeae, with clear water overlying, replacing thinner beds of *Potamogeton pectinatus*.

Biomanipulation in Shallow Lakes   475

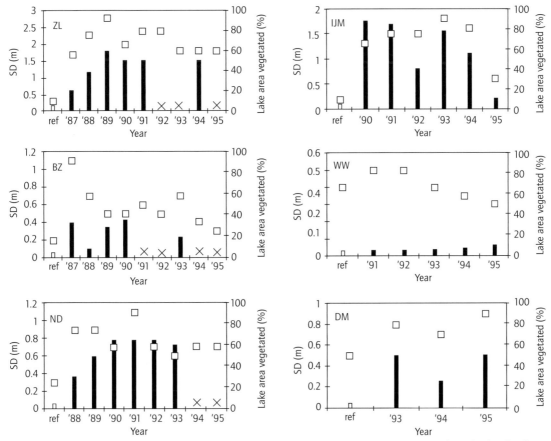

Fig. 17.8 Will the clear lake be stable? Effects of biomanipulation on submerged vegetation and Secchi-disc depth during July–August, in the years following the fish reduction. Bars, lake area vegetated; vegetation (%) of lake surface covered with dense macrophyte beds (covering 50–100% of the lake bottom); □, Secchi-disc depth; Ref, reference lake (no fish reduction) in year 1, or experimental lake in year 0. Vegetated lake area was derived from vegetation maps and qualitative information on coverage (personal communication with lake managers). Data from references in Table 17.1 and Meijer (unpublished results). Names of lakes in Table 17.1.

Bleiswijkse Zoom and IJzeren Man returned to the turbid-water state after, respectively, 7 and 5 years. For Bleiswijkse Zoom, repeated inputs of water from an adjacent lake, rich in total phosphorus and cyanobacteria, probably caused collapse of the submerged macrophytes (Meijer et al. 1994a). In IJzeren Man, an explosive development of filamentous epiphytic green algae (*Spirogyra* sp.) was observed during 1992 (Driessen et al. 1993). Such a bloom may also have induced the shift back to turbid water seen during 1995.

Both small lakes (Noorddiep and Zwemlust) were still clear during summer, respectively 7 and 8 years after fish reduction. In Noorddiep, both planktivores and benthivores seem to be controlled by a well-developed stock of piscivores (Walker 1994; Meijer et al. 1995). In this lake, macrophyte beds of varying density and patches of open water provide an optimal habitat for pike as well as perch (Walker 1994). Submerged macrophytes also act as a sink for nutrients from the water column (by uptake of nitrogen and phosphorus, and by promoting denitrification). Despite a high nitrogen loading of about $10\,g\,N\,m^{-2}\,yr^{-1}$, the summer phytoplankton in Zwemlust is nitrogen-limited (van Donk et al. 1993).

Additionally, allelopathic effects from *Ceratophyllum demersum* may have reduced the phytoplankton biomass in Zwemlust (van Donk & Gulati 1995). Here, total phosphorus (and soluble reactive phosphorus) loading is extremely high, and the fish stock developed to $300-400\,kg\,ha^{-1}$ (mainly planktivorous-herbivorous rudd, Meijer et al. 1995). The clear-water state in Zwemlust was not stable in the long term. In 1999, another removal of zooplanktivorous fish (mainly rudd) produced similar effects as in 1987, although macrophytes returned more slowly (van de Bund & van Donk 2002). In order to improve water quality in Zwemlust in the long term, reduction of nutrient loading is necessary. In the short term, repeated fish-stock reduction is a reasonable management strategy to keep the lake clear.

In conclusion, the Dutch case studies show that substantial fish-stock reduction (>75%) offers a likely chance for clearing small as well as large lakes. Only a 50% reduction was sufficient for clearing the Danish lake Vaengsø (15 ha, mean depth 1.2 m, TP = $100\,mg\,m^{-3}$) in 1987 (Lauridsen et al. 1994). The large and shallow Lake Christina (USA, 1600 ha, mean depth 1.5 m, TP = $80\,mg\,m^{-3}$) cleared up after a complete fish removal in 1987 (Hanson & Butler 1994). In Finjasjön (Sweden, 1100 ha, mean depth 3.0 m, TP = $200\,mg\,m^{-3}$), about 90% fish removal in 1994 resulted in clearing of the lake (Annadotter et al. 1999).

Rapid colonisation by submerged macrophytes, stabilising the clear-water state, can be demonstrated in relatively small lakes (<30 ha) in The Netherlands. In Vaengsø, recolonisation began 2 to 3 years after fish removal. Waterfowl grazing could explain this delay in response. A positive response in submerged vegetation was also observed in Lake Christina (Hanson & Butler 1994) and in Finjasjön (Annadotter et al. 1999). In spite of high nutrient loadings, biomanipulated small lakes may remain clear and covered with macrophytes for many years. However, high nutrient loadings lead, ultimately, to destabilisation and a reverse shift to the turbid-water state.

Our analysis of Dutch experimental lakes indicates that, in some cases (Bleiswijkse Zoom, Noorddiep, IJzeren Man), the removal of benthivores contributed significantly to clearing of the lakes, but that in all cleared lakes *Daphnia* grazers reached a pronounced peak following fishing. Changes in nutrient fluxes resulting from fishing are irrelevant to the Dutch case studies; all lakes studied are rich in nutrients, particularly during the period of spring clearing. In stating that the water body should be small (<4 ha) and that steps should be taken to promote macrophyte growth, Reynolds' (1994) checklist for successful biomanipulation seems too conservative. The Dutch results show that after clearing, and also in larger lakes (Duinigermeer, 28 ha), macrophytes rapidly recolonise the lake bed. Very large lakes (Wolderwijd, 2555 ha) need several consecutive spring clear-water years for successful re-establishment of macrovegetation.

### 17.6.2 Successes and failures of biomanipulation

There are more long-term failures on record than successes in lake studies using biomanipulation, mainly because 'bottom' effects (of nutrients) on the structure of the pelagic food web remain effective even in strongly top-down manipulated lakes (McQueen et al. 1986). Reduction of nutrient load from the catchment is an important prerequisite for successful biomanipulation (Benndorf 1988). Thus, for grazers to control phytoplankton biomass, phosphorus input must fall below a certain threshold (Carpenter et al. 2001).

De Melo et al. (1992) evaluated a large number of biomanipulation experiments with respect to the hypothesis of top-down control of algal biomass. These authors did not find any relationship between increased water clarity and zooplankton grazing. They therefore cast doubt on the 'trophic cascade theory' of Carpenter & Kitchell

(1992, 1993), mainly because of the weakening (or absence) of the top-down response at the zooplankton–phytoplankton level in 80% of the cases analysed. Alternative explanations include direct nutrient additions by fish, and by small-bodied zooplankton associated with increased planktivore biomasses, dead fish effects, and bioturbation by fish (De Melo et al. 1992).

On the other hand, Mehner et al. (2002) have attributed the failure of most biomanipulation work in deeper lakes to extremely high phosphorus loading, implying some bias in the analysis of De Melo et al. (1992), probably because the latter included more deep water bodies than shallow lakes in their analysis. In their analyses of data from 39 published studies, Drenner & Hambright (2002), however, found no support for Benndorf et al. (1981), and only limited evidence for 'the trophic cascade hypothesis' of Carpenter et al. (1985) and Carpenter & Kitchell (1988). In 22 of these studies, top-down effects on phytoplankton biomass were confounded by simultaneous reduction of nutrient loadings and numbers of planktivores. In the remaining cases, piscivore effects were generally wanting. However, lakes with both piscivores and planktivores contained relatively fewer phytoplankton than those with the planktivores alone, regardless of total phosphorus concentration.

In contrast, optimism is based on the consideration that biomanipulation is not only possible but also a relatively inexpensive and attractive method for management of eutrophic lakes, particularly as a follow-up measure to reduced nutrient loadings. Owing to differences in sediment–water interactions, the nutrient dynamics, and therefore the efficacy of restoration measures, are expected to differ markedly between shallow and deep lakes (Moss 1998). In addition, shallow lakes are more likely to be colonised by macrophytes, and therefore to experience a shift to a clear-water state at an earlier stage than deeper lakes.

## 17.7 PERSPECTIVES

Our knowledge of the functioning of the pelagic food chain in lake ecosystems has increased tremendously owing to whole-lake biomanipulation studies. However, a number of questions which are relevant to the application of biomanipulation as a shallow lake restoration strategy remain unanswered; first of all the extent to which the fish stock needs to be reduced in order to trigger a *Daphnia*-mediated clear-water phase, and then the role of filamentous cyanobacteria and invertebrate predators, which may prevent clearing of the lake by *Daphnia* grazers. In large lakes, and particularly in networks of interconnected lakes, fish stocks are difficult to control. Additionally, in large lakes, re-establishment of submerged vegetation takes more time. Winter fishing on a regular basis, rather than a single fishing operation, may therefore be promising. However, case studies are needed for further evaluation of the efficacy of repeated fishing for promoting *Daphnia*.

Other challenging questions for further research concern the role of infochemicals (kairomones) in steering food chain interactions (van Donk et al. 1999), the role of various macrophyte species in stabilising the clear-water state, and the possibilities for lake management of controlling species composition of aquatic vegetation in lakes. Finally, it should be stressed that the prudent lake manager, charged with the responsibility of reducing algal biomass in a particular lake, might be best advised to focus first on nutrient abatement, and then on biomanipulation (De Melo et al. 1992; Reynolds 1994; De Bernardi & Giussani 1995).

## 17.8 REFERENCES

Andersson, G., Berggren, H., Cronberg, G. & Gelin, C. (1978). Effects of planktivorous and benthivorous fish on organisms and water chemistry in eutrophic lakes. *Hydrobiologia*, **59**, 9–15.

Annadotter, H., Cronberg, G., Aagren, R., Lundstedt, B., Nilsson, P.A. & Strobeck, S. (1999). Multiple techniques for lake restoration. *Hydrobiologia*, **396**, 77–85.

Barthelmes, D. (1988). Fish predation and resource reaction: biomanipulation background data from fisheries research. *Limnologica*, **19**, 51–9.

Benndorf, J. (ed.) (1988). Biomanipulation. *Limnologica*, **19**, 1–110.

Benndorf, J., Uhlmann, D. & Pütz, K. (1981). Strategies for water quality management in reservoirs in the German Democratic Republic. *Water Quality Bulletin*, **6**, 68–73.

Benndorf, J., Kneschke, H., Kossatz, K. & Penz, E. (1984). Manipulation of the pelagic food web by stocking with predacious fishes. *Internationale Revue des gesamten Hydrobiologie*, **69**, 407–28.

Blindow, I., Andersson, G., Hargeby, A. & Hansson, S. (1993). Long-term pattern of alternative stable states in two shallow eutrophic lakes. *Freshwater Biology*, **30**, 159–67.

Bloem, J. & Vijverberg, J. (1984). Some observations on the diet and food selection of *Daphnia hyalina* (Cladocera) in an eutrophic lake. *Hydrobiological Bulletin*, **18**, 39–45.

Boersma, M. (1994). *On the seasonal dynamics of Daphnia species in a shallow eutrophic lake.* Ph.D. thesis, University of Amsterdam, Amsterdam.

Breukelaar, A.W., Lammens, E.H.R.R., Klein Breteler, J.G.P. & Tatrai, I. (1994). Effects of benthivorous bream (*Abramis brama*) and carp (*Cyprinus carpio*) on sediment resuspension and concentrations of nutrients and chlorophyll *a*. *Freshwater Biology*, **32**, 113–21.

Brooks, J.L. & Dodson, S.I. (1965). Predation, body size and composition of plankton. *Science*, **150**, 28–35.

Carpenter, S.R. & Kitchell, J.F. (1988). Consumer control of lake productivity. *BioScience*, **38**, 764–9.

Carpenter, S.R. & Kitchell, J.F. (1992) Trophic cascade and biomanipulation interface of research and management—a reply to the comment by De Melo et al. *Limnology and Oceanography*, **37**, 208–13.

Carpenter, S.R. & Kitchell, J.F. (eds) (1993). *The Trophic Cascade in Lakes*. Cambridge University Press, Cambridge. 386 pp.

Carpenter, S.R., Kitchell, J.F. & Hodgson, J.R. (1985). Cascading trophic interactions and lake productivity. *BioScience*, **35**, 634–9.

Carpenter, S.R., Cole J.J., Hodgson J.R., *et al.* (2001). Trophic cascades, nutrients, and lake productivity: whole-lake experiments. *Ecological Monograph*, **71**(2), 163–86.

Clewits, M. (1994). *Evaluatie van actief biologisch beheer in de Sondelerleijen*. Stageverslag Waterschap Friesland en Landbouwuniversiteit Wageningen.

Crivelli, A.J. (1983). The destruction of aquatic vegetation by carp. *Hydrobiologia*, **106**, 37–41.

Cryer, M., Peirson, G. & Townsend, C.R. (1986). Reciprocal interactions between roach (*Rutilus rutilus*) and zooplankton in a small lake: prey dynamics and growth and recruitment. *Limnology and Oceanography*, **31**, 1022–38.

De Bernardi, R. & Giussani, G. (1978). Effect of mass fish mortality on zooplankton structure and dynamics in a small Italian lake (Lago di Annone). *Verhandlungen der internationalen Vereinigung für theoretische und angewandte Limnologie*, **20**, 1045–8.

De Bernardi, R. & Giussani, G. (eds) (1995). *Guidelines of Lake Management*, Vol. 7, *Biomanipulation in Lakes and Reservoirs Management*. International Lake Environment Committee (United Nations Environmental Programme), Kusatsu, Shiga.

De Melo, R., France, R. & McQueen, D.J. (1992). Biomanipulation: Hit or myth?. *Limnology and Oceanography*, **37**, 192–207.

DeMott, W.R., Gulati, R.D. & van Donk, E. (2001). *Daphnia* food limitation in three hypertrophic Dutch lakes: Evidence for exclusion of large-bodied species by interfering filaments of cyanobacteria. *Limnology and Oceanography*, **46**, 2054–60.

Drenner, R.W. & Hambright, K.D. (2002). Piscivores, trophic cascades, and lake management. *The Scientific World*, **2**, 284–307

Driessen, O., Pex, B. & Tolkamp, H.H. (1993). Restoration of a lake: first results and problems. *Verhandlungen der internationalen Vereinigung für theoretische und angewandte Limnologie*, **25**, 617–20.

Edmondson, W.T. (1979). Lake Washington and predictability of limnological events. *Ergebnisse der Limnologie*, **13**, 234–41.

Edmondson, W.T. (1991). *The Uses of Ecology: Lake Washington and Beyond*. University of Washington Press, Seattle. 312 pp.

Fott, J., Pechar, L. & Pražáková, M. (1980). Fish as a factor controlling water quality in ponds. In: Barica, J. & Mur, L. (eds), *Hypertrophic Ecosystems*. Developments in Hydrobiology, No. 2. W.Junk, Den Haag, 255–61.

Gliwicz, Z.M. (1990). Why do cladocerans fail to control algal blooms? *Hydrobiologia*, **200/201**, 83–98.

Gliwicz, Z.M. & Siedlar, E. (1980). Food size limitation and algae interfering with food collection in *Daphnia*. *Archiv für Hydrobiologie*, **88**, 155–77.

Grimm, M.P. (1989). Northern pike (*Esox lucius* L.) and aquatic vegetation, tools in the management of fisheries and water quality in shallow waters. *Hydrobiology Bulletin*, **23**, 59–65.

Grimm, M.P. (1994). The characteristics of the optimum habitat of northern pike (*Esox lucius* L.). In: Cowx, I.G. (ed.), *Rehabilitation of Freshwater Fisheries*. Fishing

News Books, Blackwell Scientific Publications, Oxford, 235–43.

Grimm, M.P. & Backx, J.J.G.M. (1990). The restoration of shallow lakes, and the role of northern pike, aquatic vegetation and nutrient concentration. *Hydrobiologia*, **200/201**, 557–66.

Grimm, M.P., Jagtman, E. & Klinge, M. (1992). Fosfaatgehalten en de haalbaarheid van 'actief biologisch beheer', een visbiologisch perspectief. $H_2O$, **25**, 424–31.

Gulati, R.D. (1983). Zooplankton and its grazing as indicators of trophic status in Dutch lakes. *Environmental Monitoring and Assessment*, **3**, 343–54.

Gulati, R.D. (1989). Structure and feeding activities of the zooplankton community in Lake Zwemlust, in the two years after biomanipulation. *Hydrobiological Bulletin*, **23**, 35–48.

Gulati, R.D. (1990). Structural and grazing responses of zooplankton community to biomanipulation of some Dutch water bodies. *Hydrobiologia*, **200/201**, 99–118.

Gulati, R.D. (1995). Manipulation of fish population for lake recovery from eutrophication in temperate region. In: de Bernardi, R. & Giussani, G. (eds), *Guidelines of Lake Management*, Vol. 7, *Biomanipulation in Lakes and Reservoirs Management*. International Lake Environment Committee (United Nations Environmental Programme), Kusatsu, Shiga, 53–79.

Gulati, R.D. & van Donk, E. (2002). Lakes in the Netherlands, their origin, eutrophication and restoration: review of the state-of-the-art. *Hydrobiologia*, **478**, 73–106.

Gulati, R.D., Lammens, E.H.R.R., Meijer, M.-L. & van Donk, E. (eds) (1990). *Biomanipulation, Tool for Water Management*. Proceedings of an International Symposium held at Amsterdam, The Netherlands, 8–11 August 1989. Kluwer, Dordrecht, 628 pp.

Gulati, R.D., Ooms-Willems, A.L., van Tongeren, O.F.R., Postma, G. & Siewertsen, K. (1992). The dynamics and role of limnetic zooplankton in Loosdrecht lakes (The Netherlands). *Hydrobiologia*, **233**, 69–86.

Gulati, R.D., Bronkhorst, M. & Van Donk, E. (2001). Feeding in *Daphnia galeata* on *Oscillatoria limnetica* and on detritus derived from it. *Journal of Plankton Research*, **23**, 705–18.

Gumbricht, T. (1993). Nutrient removal processes in freshwater submersed macrophyte systems. *Ecological Engineering*, **2**, 1–30.

Haertel, L. & Jongsma, D. (1982). Effect of winterkill on the water quality of prairie lakes. *Proceedings of the South Dakota Academy of Sciences*, **61**, 134–51.

Hanazato, T. (1990). A comparison between predation effects on zooplankton communities by *Neomysis* and *Chaoborus*. *Hydrobiologia*, **198**, 33–40.

Hanson, M.A. & Butler, M.G. (1994). Responses to food web manipulation in a shallow waterfowl lake. *Hydrobiologia*, **279/280**, 457–66.

Hanson, J.M. & Legget, W.C. (1982). Empirical prediction of fish biomass and yield. *Canadian Journal of Fisheries and Aquatic Science*, **39**, 257–63.

Hargeby, A., Andersson, G., Blindow, I. & Johansson, S. (1994). Trophic web structure in a shallow eutrophic lake during a dominance shift from phytoplankton to submerged macrophytes. *Hydrobiologia*, **279/280**, 83–90.

Havens, K.E. (1993). Responses to experimental fish manipulations in a shallow hyper-eutrophic lake: the relative importance of nutrient recycling and trophic cascade. *Hydrobiologia*, **254**, 73–80.

Hosper, H. (1997). *Clearing lakes: an ecosystem approach to the restoration and management of shallow lakes in The Netherlands*. PhD thesis, Wageningen University.

Hosper, S.H. & Meijer, M.-L. (1993). Biomanipulation, will it work for your lake? A simple test for the assessment of chances for clear water, following drastic fish-stock reduction in shallow, eutrophic lakes. *Ecological Engineering*, **2**, 63–72.

Hosper, S.H., Jagtman, E. & Meijer, M.-L. (1987). Actief biologisch beheer, nieuwe mogelijkheden bij het herstel van meren en plassen. $H_2O$, **20**, 274–9.

Hosper, S.H., Meijer, M.-L. & Walker, P.A. (1992). *Handleiding Actief Biologisch Beheer*. Rijksinstituut voor integraal zoetwater beheer en afvalwaterbehandeling & Organisatie voor Verbetering van de binnenvisserij (RIZA/OVB), Lelystad, 102 pp.

Hrbáček, J., Dvořaková, M., Kořínek, V. & Procházková, L. (1961). Demonstration of the effect of the fish stock on the species composition of zooplankton and the intensity of metabolism of the whole plankton assemblage. *Verhandlungen der internationalen Vereinigung für theoretische und angewandte Limnologie*, **14**, 192–5.

Hurlbert, S.H., Zedler, J. & Fairbanks, D. (1972). Ecosystem alteration by mosquitofish (*Gambusia affinis*) predation. *Science*, **175**, 639–41.

Jackson, H.O. & Starrett, W.C. (1959). Turbidity and sedimentation at Lake Chautauqua. *Illinois Journal of Wildlife Management*, **23**, 157–68.

James, W.F. & Barko, J.W. (1990). Macrophyte influences on the zonation of sediment accretion and composition in a north-temperate reservoir. *Archiv für Hydrobiologie*, **120**, 129–42.

Jasser, I. (1995). The influence of macrophytes on a phytoplankton community in experimental conditions. *Hydrobiologia*, **306**, 21–32.

Jeppesen, E. (1998). *The ecology of shallow lakes—trophic interactions in the pelagial*. Doctorate dissertation, Technical Report No. 247, National Environmental Research Institute.

Jeppesen, E., Jensen, J.P., Kristensen, P., et al. (1990). Fish manipulation as a lake restoration tool in shallow, eutrophic, temperate lakes 2: threshold levels, long-term stability and conclusions. *Hydrobiologia*, **200/201**, 219–27.

Jeppesen, E., Kristensen, P., Jensen, J.P., Søndergaard, M., Mortensen, E. & Lauridsen, T. (1991). Recovery resilience following a reduction in external phosphorus loading of shallow, eutrophic Danish lakes: duration, regulating factors and methods for overcoming resilience. *Memorie dell'Istituto italiano di Idrobiologia*, **48**, 127–48.

Jeppesen, E., Jensen, J.P., Søndergaard, M., Lauridsen, T., Pedersen, L.J. & Jensen, L. (1997). Top-down control in freshwater lakes with special emphasis on the role of fish, submerged macrophytes and water depth. *Hydrobiologia*, **342**, 151–64.

Kasprzak, P., Benndorf, J., Mehner, T. & Koschel, R. (2002) Biomanipulation of lake ecosystems: an introduction. *Freshwater Biology*, **47**, 2277–81.

Kitchell, J.F. (ed.) (1992). *Food web Management, a Case Study of Lake Mendota*. Springer-Verlag, New York.

Klapper, H. (1969). Über die Wirkung einiger Primärfaktoren auf die Wasserbeschaffenheit von Seen. *Wissenschaftliche Zeitschrift der Universität Rostock, Mathematische Naturwissenschafliche Reihe*, **Heft 7**, 751–4.

Klapper, H. (1992). *Eutrophierung und Gewässerschutz*. Gustav Fischer Verlag Jena, Stuttgart, 277 pp.

Klinge, M., Grimm, M.P. & Hosper, S.H. (1995). Eutrophication and ecological rehabilitation of Dutch lakes: explanation and prediction by a new conceptual framework. *Water Science and Technology*, **31**, 207–18.

Lammens, E.H.R.R. (1986). *Interactions between fishes and the structure of fish communities in Dutch shallow, eutrophic lakes*. PhD thesis, Agricultural University of Wageningen,

Lammens, E.H.R R., Gulati, R.D., Meijer, M.L. & van Donk, E. (1990a). The first biomanipulation conference: a synthesis. *Hydrobiologia*, **200/201**, 619–27.

Lammens, E.H.R.R., van Densen, W.L.T. & Knijn, R. (1990b). The fish community structure in Tjeukemeer in relation to fishery and habitat utilisation. *Journal of Fish Biology*, **36**, 933–45.

Lampert, W. (1983). Biomanipulation—eine neue Chance zur Seensanierung? *Biologie in unserer Zeit*, **13**, 79–86.

Lampert, W., Fleckner, W., Rai, H. & Taylor, B.E. (1986). Phytoplankton control by grazing zooplankton. *Limnology and Oceanography*, **31**, 478–90.

Lane, P.A. (1978). Role of invertebrate predation in structuring zooplankton communities. *Verhandlungen der internationalen Vereinigung für theoretische und angewandte Limnologie*, **20**, 480–5.

Lauridsen, T.L., Jeppesen, E. & Sondergaard, M. (1994). Colonization and succesion of submerged macrophytes in shallow Lake Vaeng during the first five years following fish manipulation. *Hydrobiologia*, **275–6**, 233–42.

Lazzaro, X. (1987). A review of planktivorous fishes: their evolution, feeding behaviours, selectivities, and impacts. *Hydrobiologia*, **146**, 97–167.

Luecke C., Rudstam, L.G. & Allen, Y. (1992). Interannual patterns of planktivory 1987–89: an analysis of vertebrate and invertebrate predators. In: Kitchell, J.F. (ed.), *Food web Management, a Case Study of Lake Mendota*. Springer-Verlag, New York, 275–301.

Mackay, N.A., Carpenter, S.R., Soranno, P.A. & Vanni, M.J. (1990). The impact of two *Chaoborus* species on a zooplankton community. *Canadian Journal of Zoology*, **68**, 981–5.

May, R. (1977). Thresholds and breakpoints in ecosystems with a multiplicity of stable states. *Nature*, **269**, 471–7.

McQueen, D.J., Post, J.R. & Mills, E.L. (1986). Trophic relationships in freshwater pelagic ecosystems. *Canadian Journal of Fisheries and aquatic Science*, **43**, 1571–81.

Mehner, T., Benndorf, J., Kasprzak, P. & Koschel, R. (2002). Biomanipulation of lake ecosystems: successful applications and expanding complexity in the underlying science. *Freshwater Biology*, **47**, 2453–65.

Meijer, M.-L. (2000). *Biomanipulation in the Netherlands. 15 years of experience*. PhD Thesis, Wageningen University, 208 pp.

Meijer, M.-L. & Hosper, S.H. (1997). Effects of biomanipulation in the large and shallow lake Wolderwijd, the Netherlands. *Hydrobiologia*, **342/343**, 335–43.

Meijer, M.-L., de Haan, M.W., Breukelaar, A.W. & Buiteveld, H. (1990). Is reduction of the benthivorous fish an important cause of high transparency following biomanipulation in shallow lakes? *Hydrobiologia*, **200/201**, 303–15.

Meijer, M.-L., Jeppesen, E., Van Donk, E., et al. (1994a). Long-term responses to fish-stock reduction in small shallow lakes: interpretation of five-year results of four biomanipulation cases in The Netherlands and Denmark. *Hydrobiologia*, **275/276**, 457–66.

Meijer, M.-L., van Nes, E.H., Lammens, E.H.R.R., *et al.* (1994b). The consequences of a drastic fish stock reduction in the large and shallow lake Wolderwijd, The Netherlands. Can we understand what happened? *Hydrobiologia*, **275/276**, 31–42.

Meijer, M.-L., Lammens, E.H.R.R., Raat, A.J.P., Klein Breteler, J.P.G. & Grimm, M.P. (1995). Development of fish communities in lakes after biomanipulation. *Netherlands Journal of Aquatic Ecology*, **29**, 91–101.

Mills, E.L. & Forney, J.L. (1987). Trophic dynamics and development of freshwater pelagic food webs. In: Carpenter, S.R. (ed.), *Complex Interactions in Lake Communities*. Springer-Verlag, New York, 11–30.

Moss, B. (1983). The Norfolk Broadland: experiments in the restoration of a complex wetland. *Biological Reviews*, **58**, 521–61.

Moss, B. (1987). The art of lake restoration. *New Scientist*, **5 March**, 41–3.

Moss, B. (1990). Engineering and biological approaches to the restoration from eutrophication of shallow lakes in which aquatic plant communities are important components. *Hydrobiologia*, **200/201**, 367–78.

Moss, B. (1998). Shallow lakes, biomanipulation and eutrophication. *Scope Newsletter*, **29**, 1–45.

Moss, B. (1999). British Phycological Society presidential address 1999. From algal culture to ecosystem: from information to culture. *European Journal of Phycology*, **34**, 193–203.

Mulderij, G., Van Donk, E. & Roelofs, J.G.M. (2003). Differential sensitivity of green algae to allelopathic substances from Chara. *Hydrobiologia*, **491**, 261–71.

Murtaugh, P.A. (1981). Selective predation by *Neomysis mercedis* in Lake Washington. *Limnology and Oceanography*, **26**, 445–53.

Noordhuis, R.H., Reeders, H.H. & bij de Vaate, A. (1992). Filtering rate and pseudofaeces production in zebra mussels and their application in water quality management. *Limnologie Aktuell*, **4**, 101–14.

Pahl-Wostl, C. (2003) Self-regulation of limnetic ecosystems. In: O'Sullivan, P.E. & Reynolds, C.S. (eds), *The Lakes Handbook*, Vol. 1, *Limnology and Limnetic Ecology*. Blackwell, Oxford, 581–608.

Persson, L. (1994). Natural shifts in the structure of fish communities: mechanisms and constraints on perturbation sustenance. In: Cowx, I.G. (ed.), *Rehabilitation of Freshwater Fisheries*. Fishing News Books, Blackwell Scientific Publications, 421–34.

Pimm, S.L. (1991). *The Balance of Nature?* The University of Chicago Press, Chicago, 737 pp.

Reeders, H.H. & bij de Vaate, A. (1990). Zebra mussels (*Dreissena polymorpha*): a new perspective for water quality management. *Hydrobiologia*, **200/201**, 437–50.

Reeders, H.H., bij de Vaate, A. & Noordhuis, R. (1993). Potential of the zebra mussel (*Dreissena polymorpha*) for water quality management. In: Nalepa, T.F. & Schloesser, D.W. (eds), *Zebra Mussels, Biology, Impacts, and Control*. Lewis Publishers, Boca Raton, FL, 439–52.

Reynolds, C.S. (1994). The ecological basis for the successful biomanipulation of aquatic communities. *Archiv für Hydrobiologie*, **130**, 1–33.

Scheffer, M. (1990). Multiplicity of stable states in freshwater systems. *Hydrobiologia*, **200/201**, 475–86.

Scheffer, M. (1998). *Ecology of Shallow Lakes*. Chapman & Hall, London, 357 pp.

Scheffer, M., Hosper, S.H., Meijer, M.-L., Moss, B. & Jeppesen, E. (1993). Alternative equilibria in shallow lakes. *Trends in Ecology and Evolution*, **8**, 275–9.

Scheffer, M., van den Berg, M., Breukelaar, A., *et al.* (1994). Vegetated areas with clear water in turbid shallow lakes. *Aquatic Botany*, **49**, 193–6.

Schindler, D.W. & Comita, G.W. (1972). The dependence of primary production upon physical and chemical factors in a small, senescing lake, including the effects of complete winter oxygen depletion. *Archiv für Hydrobiologie*, **69**, 413–51.

Schriver, P., Bøgestrand, J., Jeppesen, E. & Søndergaard, M. (1995). Impact of submerged macrophytes on fish–zooplankton–phytoplankton interactions: large-scale enclosure experiments in a shallow eutrophic lake. *Freshwater Biology*, **33**, 255–70.

Shapiro, J. (1978). The need for more biology in lake restoration. *Lake Restoration: Proceedings of a National Conference*, 22–24 August, Minneapolis, EPA 440/5-79-001, 161–7.

Shapiro, J. (1980). The importance of trophic level interactions to the abundance and species composition of algae in lakes. In: Barica, J. & Mur, L.R. (eds), *Hypertrophic Ecosystems*. Developments in Hydrobiology, No. 2. W.Junk, Den Haag, 105–16.

Shapiro, J., Lamarra, V. & Lynch, M. (1975). Biomanipulation: an ecosystem approach to lake restoration. In: Brezonik, P.L. & Fox, J.L. (eds), *Proceedings of a Symposium on Water Quality Management through Biological Control*. University of Florida, Gainesville, 85–96.

Sommer, U., Gliwicz, Z.M., Lampert, W. & Duncan, A. (1986). The PEG-model of seasonal succession of planktonic events in freshwaters. *Archiv für Hydrobiologie*, **106**, 433–71.

Stansfield, J.H., Perrow, M.R., Tench, L.D., Jowitt, A.J.D. & Taylor, A.A.L. (1995). Do macrophytes act as refuges for grazing Cladocera against fish predation. *Water Science and Technology*, **32**, 217–20.

Ten Winkel, E.H. (1987). *Chironomid larvae and their foodweb relations in the littoral zone of Lake Maarsseveen*. PhD thesis, University of Amsterdam, Amsterdam.

Threlkeld, S.T. (1987). *Daphnia* population fluctuations: patterns and mechanisms. In: Peters, R.H. & de Bernardi, R. (eds), *'Daphnia'*. *Memorie dell'Istitutoi di Idrobiologia*, **45**, 367–88.

Timms, R.M. & Moss, B. (1984). Prevention of growth of potentially dense phytoplankton populations by zooplankton grazing, in the presence of zooplanktivorous fish, in a shallow wetland ecosystem. *Limnology and Oceanography*, **29**, 472–86.

Van Berkum, J.A., Klinge, M. & Grimm, M.P. (1995). Biomanipulation in the Duinigermeer. *Netherlands Journal of aquatic Ecology*, **29**, 81–90.

Van Berkum, J.A., Meijer, M.-L. & Kemper, J.H. (1996). Actief biologisch beheer in het Noorddiep. $H_2O$, **29**, 308–13.

Van Densen, W.L.T. & Grimm, M.P. (1988). Possibility for stock enhancement of pikeperch (*Stisostedion lucioperca* L.) in order to increase predation on planktivores. *Limnologica*, **19**, 45–9.

Van De Bund, W. & Van Donk, E. (2002). Short- and long-term effects of zooplanktivorous fish removal in Lake Zwemlust: a synthesis of 15 years of data. *Freshwater Biology*, **47**, 2380–7.

Van der Vlugt, J.C., Walker, P.A., van der Does, J. & Raat, A.J.P. (1992). Fisheries management as an additional lake restoration measure: biomanipulation scaling-up problems. *Hydrobiologia*, **233**, 213–24.

Van Donk, E., Gulati, R.D. & Grimm, M.P. (1989). Food-web manipulation in Lake Zwemlust: positive and negative effects during the first two years. *Hydrobiological Bulletin*, **23**, 19–34.

Van Donk, E., Grimm, M.P., Gulati, R.D., Heuts, P.G.M., de Kloet, W.A. & van Liere, E. (1990a). First attempt to apply whole-lake food-web manipulation on a large scale in the Netherlands. *Hydrobiologia*, **200/201**, 291–302.

Van Donk, E., Gulati, R.D., Grimm, M.P. & Klein Breteler, J.P.G. (1990b). Whole-lake food-web manipulation as a means to study community interactions in a small ecosystem. *Hydrobiologia*, **200/201**, 275–90.

Van Donk, E., Gulati, R.D., Iedema, A. & Meulemans, J.T. (1993). Macrophyte-related shifts in the nitrogen and phosphorus contents of the different trophic levels in a biomanipulated shallow lake. *Hydrobiologia*, **251**, 19–26.

Van Donk, E. & Gulati, R.D. (1995). Transition of a lake to turbid state six years after biomanipulation: mechanisms and pathways. *Water Science Technology*, **32**, 197–206.

Van Donk, E., Lürling, M. & Lampert, W. (1999). Consumer induced changes in phytoplankton: inducibility, costs, benefits and the impact on grazers. In: Harvell, J. & Tollrian, J. (eds), *Consequences of Inducible Defences for Population Biology*. Princeton University Press, Princeton, 89–103.

Van Donk, E. & van de Bund, W. (2002). Impact of submerged macrophytes including charophytes on phyto- and zooplankton communities: allelopathy versus other mechanisms. *Aquatic Botany*, **72**, 261–74.

Van Liere, L. & Gulati, R.D. (eds) (1992). Restoration and recovery of shallow eutrophic lake Ecosystems in The Netherlands. *Hydrobiologia*, **233**: 1–287.

Vanni, M.J., Luecke, C., Kitchell, J.F. & Magnuson, J. (1990). Effects of planktivorous fish mass mortality on the plankton community of Lake Mendota, Wisconsin: implications for biomanipulation. *Hydrobiologia*, **200/201**, 329–36.

Walker, P.A. (1994). Development of pike and perch populations after biomanipulation of fish stocks. In: Cowx, I.G. (ed.), *Rehabilitation of Freshwater Fisheries*. Fishing News Books, Blackwell Scientific Publications, Oxford, 376–89.

Willemsen, J. (1980). Fishery aspects of eutrophication. *Hydrobiological Bulletin*, **14**, 12–21.

Witteveen & Bos (Consulting Engineers) (1995). *Integrale eutrofiëringsbestrijding in Noordwest-Overijssel, actief biologisch beheer in het Duinigermeer*. Rapport 44.008, Deventer.

Wium-Andersen, S. (1987). Allelopathy among aquatic plants. *Ergebnisse der Limnologie*, **27**, 167–72.

# 18   Restoring Acidified Lakes: an Overview

LENNART HENRIKSON, ATLE HINDAR
AND INGEMAR ABRAHAMSSON

## 18.1   INTRODUCTION

Liming, i.e. adding neutralising agents to the water, is a way of restoring acidified lakes. Acid deposition has changed the natural water chemistry and, thus, the biological community in 50,000–100,000 lakes and watercourses in Europe and North America (Brodin 1995a). Aquatic biodiversity is affected directly and the possibility for human use of the natural resource may be prejudiced. The accumulation of heavy metals, such as mercury, in top carnivores, is a latent threat to animal species and human health.

Since the mid-1980s, the deposition of sulphur has been reduced in Europe, although further reductions are needed. Many European countries have agreed upon a reduction target for sulphur emissions of 70–80% by the year 2010 relative to 1980 (UN 1994); however, acidification will continue to be a problem in large base-poor catchments for many decades hence (Henriksen & Hindar 1993; Brodin 1995a). No large-scale reduction in nitrogen deposition has yet been achieved and, soon, acidification by nitrogen compounds will become a more crucial issue than sulphur deposition.

One way to counteract the acidification of freshwaters is to lime the water. Sweden and Norway have each chosen large-scale liming as a national strategy for preserving species threatened by acidification (e.g. Baalsrud et al. 1985; Hindar & Rosseland 1991; Henrikson & Brodin 1995a; Sandøy & Romundstad 1995; Svensson et al. 1995). Smaller-scale liming efforts, mainly as research or experimental projects, have been conducted in several other countries, e.g. Canada, Finland, the UK (Wales and Scotland) and USA. In Sweden, liming has also been used to reduce the content of mercury in fish in acidified lakes (Meili 1995).

In this chapter, we present an overview of the need for countermeasures and of the experiences of liming in various countries. Surveys on liming and its effects have been presented elsewhere, by Brocksen & Wisniewski (1988), Weatherley (1988), Olem (1991), Olem et al. (1991), Henrikson & Brodin (1995a) and Henrikson et al. (1995).

## 18.2   ACIDIFICATION STATUS AND CRITICAL LOAD

The composition and diversity of aquatic organisms in a given habitat is correlated with a set of water-quality parameters, such as pH, dissolved inorganic aluminium and calcium (Økland & Økland 1986; Wood & McDonald 1987; Brown & Sadler 1989; Rosseland et al. 1990; Herrmann et al. 1993; Brodin 1995a). Calcium may ameliorate the toxic effects of aluminium in acidic water. Dissolved organic matter forms complexes with aluminium and thereby reduces the inorganic concentrations (see Volume 1, Chapter 7). Iron and manganese have been shown to reach toxic concentrations in some cases (e.g. Nyberg et al. 1995).

Aquatic organisms are subjected to sublethal and lethal effects at different pH/aluminium-concentrations. Differences have been found between fish species, fish strains, invertebrate and phytoplankton species (e.g. Almer et al. 1978; Eriksson et al. 1983; Engblom & Lingdell 1984; Raddum & Fjellheim 1984; Brett 1989; Kroglund et al. 1992; Lien et al. 1992; Havas & Rosseland 1995). The critical tolerance of pH, or a combination of pH, aluminium or other components, varies from species to species. Regional differences in pH/aluminium-tolerance may be found, in response to regional differences in the natural water quality and climate.

Fish species composition, although related to the direct toxic effects of pH and aluminium, is well correlated to ANC (acid neutralising capacity) (Lien *et al.* 1992; Bulger *et al.* 1993). Lien *et al.* (1992) found that viable brown trout *Salmo trutta* populations corresponded in most cases to ANC values higher than $20\,\mu eq\,L^{-1}$. Exceedance of the critical loading limit is reflected by ANC scores below $20\,\mu eq\,L^{-1}$. A variable $ANC_{limit}$ has been introduced recently in order to take into account very low (even lower than $20\,\mu eq\,L^{-1}$) ANC occurring naturally in parts of Norway (Henriksen *et al.* 1995). Exceedances and thereby anticipated damage to aquatic life owing to acidification are manifest in large areas of Scandinavia (Henriksen *et al.* 1992).

Changes in the fish community owing to toxic effects change the predatory and competitive interactions, which in turn lead to further changes at the community level (e.g. Appelberg *et al.* 1993).

Biodiversity is the variation at genetic, species and ecosystem level, and the ecological processes within the ecosystem. All levels of aquatic biodiversity are affected by acidification. Unique genotypic populations of the Salmonidae family have been damaged (Appelberg & Degerman 1991; Bergquist 1991; Hesthagen & Hansen 1991; Snucins *et al.* 1995). The total number of species is lower in acidified than in non-acidified waters (e.g. Brodin 1995b), even though some tolerant species may be favoured in acidified waters, which they may successfully colonise (e.g. Henrikson & Oscarson 1981; B.-I. Henrikson 1988). Although biodiversity is affected at species level we only know of one species, the spring-spawning Cisco *Coregonus trybomi*, that may be threatened at the national and global scales (cf. Henrikson & Brodin 1995b). On regional and local scales, several species may be considered as threatened (e.g. European roach *Rutilus rutilus*, freshwater pearl mussel *Margaritifera margaritifera*), as are semi-aquatic species such as red-throated diver, *Gavia stellata* (Eriksson 1987; Brodin 1995b). Changes at the ecosystem level are manifested by altered community structure as well as ecological processes, e.g. retarded breakdown of leaves (e.g. Appelberg *et al.* 1993). If acid deposition decreases, the biodiversity increases again (e.g. Keller & Gunn 1995; Raddum & Fjellheim 1995).

## 18.3 LIMING: OBJECTIVES AND TARGETS

Yearly, approximately 300,000 t of fine-grained limestone is spread in lakes and streams in acidified areas, in order to raise the pH in surface waters. The annual cost is approximately US$40–50 million. In Sweden and Norway, a total of over 11,000 lakes and streams is treated on a continuing basis. Here, the liming is financed wholly or in large part by the respective governments.

When operational liming of acidified surface waters was commenced in the 1970s, the main driver was the concern for fish populations and recreational fisheries. Now, the aim and direction have been broadened and are focused on the preservation or recovery of biodiversity and on human health. In Sweden (cf. Svensson *et al.* 1995), the official aims are now: (i) an intent to detoxify the water adequately for the survival and re-establishment of the natural flora and fauna; (ii) to raise the pH above 6.0 and alkalinity above 0.1 meq $L^{-1}$. In Norway (Sandøy & Romundstad 1995), the intentions are (i) to improve conditions for recreational fishing and (ii) to preserve biological diversity. Specific chemical targets are varied according to the biological criteria to be satisfied. In Norwegian salmon rivers, the chemical target is set on an annual basis. Target pH is higher (pH 6.5) during February–June than during the rest of the year (pH 6.2–6.3), in order to take account of the extreme pH sensitivity of Atlantic salmon *Salmo salar* smolt.

## 18.4 LIMING AGENTS

Many kinds of deacidifying bases, such as carbonates, oxides, hydroxides and alkaline industrial wastes, have been used to neutralise acid waters

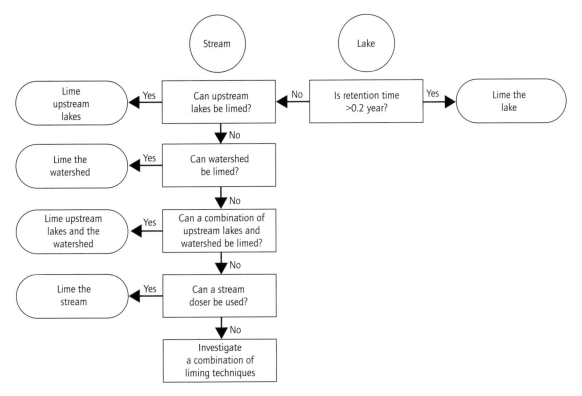

**Fig. 18.1** Flow diagram for decisions of liming strategy (SEPA 1988).

(Olem 1991; Dickson & Brodin 1995). Common agents include dolomite (CaMg(CO$_3$)$_2$), sodium carbonate (Na$_2$CO$_3$), olivine (Mg$_2$SiO$_4$) and hydrated lime (Ca(OH)$_2$). Calcium carbonate (CaCO$_3$) in the form of a dry, finely grained powder is the most widely used (>95%). It stabilises pH at intermediate values (pH 6–8), is cheap and is easy to handle, and contains few contaminants. The particle size is generally within the size range 0–0.2 mm, with 50% being finer than 50 μm. Yet finer-grained products (0–0.1 mm) are used where rapid dissolution is required.

## 18.5 LIMING STRATEGIES AND MODELS

Liming may be applied to lakes or to wetlands (Rosseland & Hindar 1988; Dickson & Brodin 1995; Henrikson et al. 1995). Often, a mix of different application methods is preferred for an optimal result. The methods can be seen as tools which need to be used in an appropriate way and in suitable combinations to achieve the chemical and biological targets (Fig. 18.1). Increased knowledge and experience may lead to changes in the liming strategy over time (cf. Alenäs et al. 1995). In devising a liming strategy for a given drainage system, it is important to consider the desirable chemical and biological effects in relation to the cost-effectiveness of the various methods. Dickson & Brodin (1995) argued that 'the optimal strategy is to utilise an agent at the lowest cost that achieves the desired effect, that lasts as long as possible and whose undesired side-effects are as limited and as brief as possible'.

Adequate water quality, with elevated pH and low concentrations of toxic inorganic aluminium-monomers, is necessary to achieve the biological targets. When pH increases, the aluminium species change and enter more stable forms. Temperature and the concentrations of ligand-forming compounds may affect the kinetics and products of these reactions (Lydersen 1990). For this reason, it is important to avoid zones of mixing between acid and buffered waters, which may be even more toxic than the acid water itself (Rosseland & Hindar 1991; Rosseland et al. 1992). Mixing zones are probably especially significant in salmon rivers because of the extreme sensitivity of Atlantic salmon smolts. Acid, aluminium-rich tributaries of some size should therefore be limed, although the biological targets in these particular rivers may not themselves be important.

Carbonates dissolve as a function of pH, dissolved carbon dioxide, particle size and the time and conditions available for dissolution to take place (Sverdrup 1985). Together with other factors, these variables are incorporated into mathematical models for the calculation of doses and the simulation of anticipated effects, including reacidification after the liming. Run-off conditions departing from average may well lead to longer or shorter duration of treatment. A model based on calculations of the critical load exceedence has been developed for dose calculations on a catchment or regional scale (Hindar et al. 1998). Both present and expected future sulphur and nitrogen deposition data are used and scenarios for future liming have been interpolated.

## 18.6 LIMING METHODS

### 18.6.1 *Lake liming*

Lake liming is the most common method for mitigating acidification (Olem et al.1991; Dickson & Brodin 1995). It is a cheap technique and simple to implement. The limestone is dispersed from boats or from pontoon vessels as a dry powder or powder mixed with lake water (Fig. 18.2). At lakes situated far from roads, dry power is applied from a helicopter. Also, application of powder onto the winter ice cover, using tracked vehicles or tractors, has been tried.

Fig. 18.2 Lake liming by ponton vessel. (Photograph by Lennart Henrikson.)

The limestone powder is applied evenly over the whole surface and/or along the shorelines of the lake. In lakes with short retention times the chemical effects of the effort can be extended by restricting the treatment to shallow bottoms. Doses of up to 5 t ha$^{-1}$ on shallow bottoms are not unusual (Dickson & Brodin 1995). The process can be divided into three phases: the initial dissolution, the long-term dissolution and the dilution phase. During the first phase, the dissolution rate exceeds the speed of dilution and the ANC and calcium concentrations improve with time. Almost all of the lime can be dissolved during this short (months) initial period, especially if the dose is small, the lake depth is great and the dissolution conditions are optimal, as in the case of the liming of Lake Nisser in Norway in 1996 (total amount added: 10,000 t; dose: 1.5 g m$^{-3}$; mean depth: 93 m).

During the long-term dissolution phase, the dilution exceeds the dissolution rate and the water volume is slowly reacidified. The dissolution can proceed for several years, depending on retention time, rate of deactivation and dispersal method. When dissolution ceases, the reacidification rate will increase and follow the course of dilution. The amount of lime dissolved usually exceeds 60% and up to 95–100% has been measured (Wilander et al. 1995; Hindar unpublished data).

Doses may be calculated on a theoretical basis, giving consideration to the concentrations of H$^+$, inorganic monomeric aluminium, dissolved or-

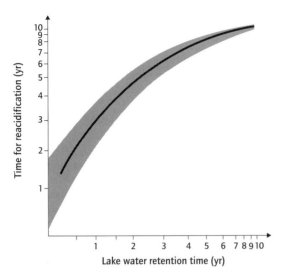

**Fig. 18.3** The duration of lime treatment in Swedish lakes (SEPA 1988).

ganic matter and the level of buffering they provide. A more direct approach is titration analyses. When liming a given site for the first time, the dose is usually calculated on the basis of acidity of the water, lake volume and the ANC to be achieved. Common doses are 10–45 g $CaCO_3\,m^{-3}$ lake volume, depending on water quality and retention time. The duration of the treatment is often equivalent to at least two retention times (Fig. 18.3). In Sweden, the median duration is 2.5 hydrological turnovers (Wilander et al. 1995). The doses required on subsequent occasions are estimated from the acidity of the tributaries (or acidity prior to liming) and the amount of water to be neutralised pending the next liming.

Lake liming can be inadequate in instances of severe reacidification of the littoral zone during the winter ice-cover and during snowmelt (Hasselrot et al. 1987; Abrahamsson 1993). The extent of impact depends on the retention time and mean depth of the lake and the acidity of the inlet stream water. Dimictic lakes with retention times of approximately 1 year or less are particularly sensitive to such acid events. To avoid negative effects on sensitive littoral species (Henrikson 1988), lake liming should be carried out in conjunction with other techniques, such as terrestrial liming or liming of inlet streams.

**Fig. 18.4** Lake liming by helicopter. (Photograph by Lennart Henrikson.)

### 18.6.2 Wetland liming

Wetland liming is becoming increasingly common, especially in Sweden. It is an efficient method for neutralising acidified streams and lakes with short turnover times (Degerman et al. 1995; Dickson & Brodin 1995). The lime is dispersed as dry powder from a helicopter (Fig. 18.4). The wetlands so treated are treeless mires having a peat substratum and a vegetation made up mainly of *Sphagnum* mosses.

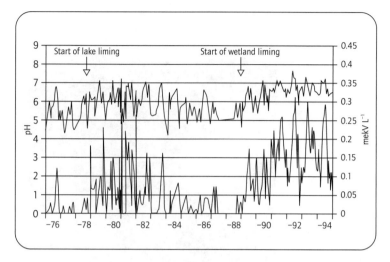

Fig. 18.5 Alkalinity and pH in the river Fageredsån situated in southwest Sweden. Since 1989 wetlands covering 1.4% of the catchment area have been limed. During the period 1978–1988 the river was unsuccessfully treated by lake and stream liming methods.

The lime doses in Sweden are 40–120 kg ha$^{-1}$ of catchment per year, depending on the acidity of the water and local run-off characteristics. Treatments are repeated yearly or every 2–3 years. The area treated needs to represent 1–4% of the catchment, if sufficient increase in pH and buffer capacity in the target water is to be achieved. The area-specific lime dosing is adjusted to the hydraulic load in each case. Typically, some 5–10 t ha$^{-1}$ yr$^{-1}$ are applied. Owing to the accumulated acidity in the peat layer, primary treatments need to be applied at higher doses — perhaps 20–30 t ha$^{-1}$ limed area. Figure 18.5 shows the impact of a wetland liming on the acid-base chemistry of the River Fageredsån in southwest Sweden. Since 1989, treeless mires have been treated yearly with 90 kg CaCO$_3$ ha$^{-1}$. The efforts have successfully kept the pH over 6.0 despite the high load of acid deposition in the region. Wetland liming is more expensive than lake liming, owing to the relative costs of helicopter and boat use.

The dissolution rate of lime powder (0–0.2 mm) is very fast in wetlands. Most of the lime dissolves within a few months after application and the level of dissolution can be almost as high as for lake liming (up to 70%: Dickson & Brodin 1995). The outflux is usually slow and depends on the hydraulic load of the treated wetland. In fens with medium hydraulic load, the lime effects can be detected for more than 10 years. In flooded wetlands and mud-bottoms with high hydraulic loads, the effects can pass within a year or two.

Wetland liming possesses several advantages compared with lake liming. Most important is terrestrial retention of aluminium and deacidification of melt-water (Gubala & Driscoll 1991; Dalziel et al. 1992; Hindar et al. 1996; Traaen et al. 1997). However, *Sphagnum* mosses, liverworts and lichens are usually eliminated from the treated fens whereas other mosses increase in abundance (Larsson 1995). Such undesired effects may encourage the use of forest soil liming. The positive effects from wetland liming may be retained (Hindar et al. 1996), with only minor drawbacks (Eilertsen et al. 1997; Hindar et al. 1997).

### 18.6.3 Dosers

Dosers for dry limestone powder or slurried powder are used in small streams as well as large rivers. The most advanced are equipped with automatic dosing control based on pH upstream or downstream and water flow (Hindar & Henriksen 1992) (Figs 18.6 & 18.7). Dosage capacities up to 2 t h$^{-1}$ and storage bins up to 80 m$^3$ are available on the market (Olem 1991). A doser needs to be located close to a road access suitable for bulk transports and on a length of fast flowing water that resists excessive sedimentation of lime powder on the streambed. In case of aluminium precipitation

Fig. 18.6 Principles of an electrically powered doser equipped with automatic dosing controlled by water level. Dry limestone powder is dispensed as a wet slurry after mixing with river water in a mixing tank. (Drawing by Kemira Ltd.)

Fig. 18.7 Large lime doser. (Photograph by Lennart Henrikson.)

downstream, dosers also need to be sited several hundred metres upstream of sensitive populations (Dickson & Brodin 1995). Common doses are 10–25 g $CaCO_3$ $m^{-3}$ stream water but they may be significantly larger if the doser is supposed to neutralise a long river stretch. When located in waters with high flow velocity, dosers may favour dissolution of 70–90% of the carbonate (Dickson & Brodin 1995).

Closedowns of dosers may be detrimental, especially to the extremely sensitive smolts of Atlantic salmon, if no other liming measures are adopted. A large survey in 1995 of Swedish dosers showed that one-third did not operate during episodes of high rainfall runoff (Svahnberg 1996). The closedowns were caused by both technical defects and lack of maintenance or supervision. The functional problems affected mainly older models and low-tech dosers operated by the water flow or by battery power. The survey argued that a doser must deliver 100–200 t of $CaCO_3$ $yr^{-1}$ to achieve the cost-effectiveness of lake (i.e. boat) or wetland (i.e. helicopter) liming.

### 18.6.4 Other techniques

Other techniques for liming streams have been practised for many years. Applying shell sand or coarsely ground limestone to streambeds may improve spawning conditions for salmonid fishes and raise pH in the spawning gravel. The lime utilisation is low, owing to the slow dissolution and covering of precipitated metal compounds, and the neutralisation effects are inversely correlated to the level of discharge (Dickson & Brodin 1995).

## 18.7 ECOLOGICAL EFFECTS OF LIMING

Generalising about the ecological effects after liming is difficult, partly because of a lack of representative data. In some instances, long-term biological changes have been detected over 16 years after liming (Appelberg 1995) but such rates may be related to delayed recolonisation of excluded species rather than the liming itself. With respect to chemical and biological targets, the effects may be considered as desirable, others as undesirable. The experiences of large-scale liming in Sweden and Norway, as well as the results of liming in USA and Scotland, indicate that the vast majority of chemical, biological and ecological changes have been beneficial (Baalsrud et al. 1985; Porcella 1989; Hindar 1992; Howells & Dalziel 1992; Henrikson & Brodin 1995b).

### 18.7.1 Desirable effects

The primary effects on water quality are increases in pH, alkalinity, ANC, calcium content and decreases in toxic metal species (cf. Wilander et al. 1995). Besides the elevated pH, the most important biological benefit is the decrease in the toxicity of aluminium and other metals such as iron and manganese.

In polyhumic lakes, the content of dissolved organic substances (water colour) decreases, whereas in waters of lower humic content, variations may be upward or downward. Such changes may affect the underwater light climate. Immediately after liming, turbidity is usually increased and there is a transient decrease in transparency. In the long term, the background transparency may well increase but enhanced phytoplankton production might cause a consequent diminution in water clarity.

Short-term studies indicate a transient elevation of the phosphorus content, probably emanating, in part, from the liming agent and through release of phosphorus bound in the sediments or in decaying plants in response to accelerated mineralisation. Liming may decrease dissolved phosphate concentrations owing to its adsorption by carbonate particles, as has been measured by Blomqvist et al. (1993). Long-term studies indicate unchanged or decreased contents. However, the accessibility of phosphorus to algae may increase. In the few studies of nitrogen, both increases and decreases have been documented.

The most desirable biological effects of liming come with enhanced fish reproduction, the increased density of sensitive species and the re-establishment of excluded species (Appelberg 1995; Bergquist 1995; Degerman et al. 1995; Larsson 1995). Another desirable effect is the decrease or elimination of species favoured by acidification, for example *Sphagnum* mosses and water bugs (Corixidae).

In Fig. 18.8, important factors and processes for the biological development after liming are summarised. The abiotic changes are the triggers for all other changes but, in the long run, the biotic mechanisms (competition, predation) determine the community structure. This will stabilise within the new abiotic limits set by the mix of liming techniques used. The community will be more complex, with increased numbers of species and more trophic levels (Appelberg 1995). Decomposition rates will become normalised (Gahnström 1995). Ecosystems severely damaged by acidification will experience the most profound community changes following liming (Henrikson et al. 1985; Degerman et al. 1995).

### 18.7.2 Undesirable effects

The overwhelmingly beneficial effects of liming must be weighed against the undesirable effects of liming that have been noted. These may be sub-divided among terrestrial and aquatic effects.

Direct damage to terrestrial vegetation after wetland or whole-catchment liming is probably the most serious of the undesirable effects. Terres-

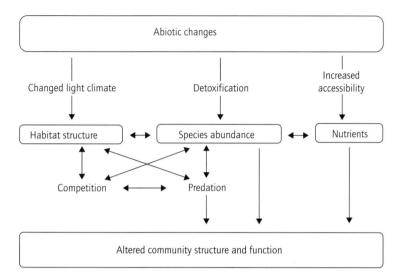

**Fig. 18.8** Schematic illustration of important factors and processes for the biological development after liming. The direct action of altered water quality will affect species abundance, nutrient conditions and habitat structure. The detoxification will facilitate the fish reproduction and populations increase for fish and other sensitive organisms and the recolonisation of eradicated species. This will affect the species abundance. The accessibility of the important plant nutrient phosphorus will increase when the aluminium content decreases or less toxic species of aluminium are formed, which will favour the production of algae. If the transparency is decreased, the habitat structure is changed, which may affect visual-dependent predators such as birds. The decrease of *Sphagnum* mat will change the habitat for benthic species. The primary changes caused by the new abiotic environment will release further biotic changes caused by competition and predation. The result is a 'new' structure of the organism community. As a consequence of changes in community structure also the functions within the ecosystems are changed. One example is increased breakdown of leaves owing to increased numbers of shredders.

trial liming, especially wetland liming, is now rather widely used as a liming technique in Sweden. Death of *Sphagnum* mosses (Mackenzie 1992; Hindar *et al.* 1995; Larsson 1995) changes mire surfaces in the short term and probably alters both the structure and function of the system after several decades of repeated liming. More frequent use of smaller doses, coarser liming material and introduction of Mg-containing dolomite might reduce these unwanted effects.

In some lakes in Norway and Sweden, increased expansion of *Juncus bulbosus* and *Myriophyllum alterniflorum* after liming has occurred (Brandrud & Roelofs 1995; Dickson *et al.* 1995; Larsson 1995; Roelofs *et al.* 1994). This may greatly influence littoral flora and, indirectly, the littoral fauna and, thus, the whole-lake ecology. Increased availability of inorganic carbon in the littoral sediments as a result of carbonate addition may stimulate the increased growth.

## 18.8 SOCIO-ECONOMIC EFFECTS

Cost-benefit analyses in Norway and Sweden indicate that liming may be profitable (e.g. Navrud 1990; Bengtsson & Bogelius 1995). In these studies, the cost of liming and fishery management has been related to benefit expressed as recreational value, i.e. the willingness of the public to pay for fishing. The people around a limed Finnish lake were willing to contribute to the costs of liming even if the lake alone was used for outdoor recreation (Iivonen *et al.* 1995). However, this kind of analysis only recognises economic issues and not the unaccounted value of (say) conserved biodiversity. Bengtsson & Bogelius (1995) also state that liming practices contribute to enhanced popular environmental awareness.

## 18.9 LIMING TO RECREATE PRIOR WATER QUALITY AND RESTORE ECOSYSTEMS

Successful liming operations usually lead to improved water quality and to increased numbers of the species susceptible to acidification, but do the ecosystems recover completely? Several difficulties must be considered. First of all, nobody knows the exact structure and function of the ecosystems before they were influenced by anthropogenic acidification. Second, all ecosystems experience a developmental process and a certain degree of interannual instability even if the external conditions remain more or less constant (cf. Brink *et al.* 1988).

Well aware of the uncertainties, we state that if the chemical target is met, liming will, in the long run, restore the ecosystems but will not make them identical to what may be the original ones (Fig. 18.9; cf. Degerman *et al.* 1995; Henrikson & Brodin 1995b; Lingdell & Englblom 1995; Wilander *et al.* 1995). That the ecosystems are not completely recovered is not peculiar to liming but to almost all other attempts to repair disturbed aquatic and terrestrial systems (cf. Brink *et al.* 1988; Cairns 1988). There may be several reasons why the limed waters differ from those unaffected by acidification (cf. Appelberg 1995; Degerman *et al.* 1995). The most common is probably that the chemical target is not met owing to insufficient planning or to severe practical constraints. Limed waters may become rather unstable if acidified waters are discharged continuously into lakes and streams. Toxic water and the environmental variability will affect the normal relations, favouring non-competitive, 'opportunistic' species. Also, changed nutrient availability and slow recolonisation may affect the outcome of liming. Improved liming strategy and complementary measures may improve the success of liming.

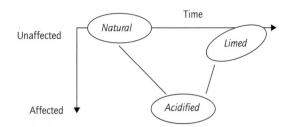

**Fig. 18.9** Generalisation of changes occurring in freshwater ecosystems influenced by acidification and liming. Acidification results in decreased ecosystem complexity. Successful liming entails a 'normalisation', i.e. the complexity and function of the ecosystem will be largely similar to unaffected ecosystems, yet the ecosystems will not become identical and liming alone might not bring back completely natural conditions. However, with time the similarity with unaffected waters will increase (Henrikson *et al.* 1995).

## 18.10 COMPLEMENTARY MEASURES

Complementary measures are a part of the Swedish liming programme. Two categories of 'biological restoration' can be defined: (i) measures aimed at facilitating natural recolonisation and re-establishment of species that have been eliminated; and (ii) direct re-introduction of species (Bergquist 1995). Examples include the elimination of migration obstacles, biotope reconstruction, and replenishment of individuals of species occurring in small numbers. Supply of nutrients, especially phosphorus, is another complementary measure in order to restore the productivity of limed waters. Hitherto, such measures have been taken only in a few research projects (Blomqvist *et al.* 1993) and more experience is needed before these measures can be operational.

## 18.11 RECOMMENDATIONS FOR LIMING OF LAKES AND WATERCOURSES

Experiences acquired from liming practices, particularly in Scandinavia, are summarised and listed below (cf. Henrikson & Brodin 1995b). These may be regarded as recommendations.

1 Accurate planning, taking the entire or at least a considerable part of the watershed into account, is essential for good results and to eliminate conflicts with other conservation interests, e.g. wetland protection.
2 Liming treatment in an early stage of acidification results in moderate ecological changes and is therefore recommended for protection of biological diversity. Limited variations in water quality should be targeted. This can be achieved by means of, for example:
  (a) using small doses of lime frequently rather than large doses separated by long intervals;
  (b) liming of littoral zones in lakes in order to alleviate acid surges;
  (c) repeated liming treatment before there is a risk of biological damage owing to reacidification.
3 Overdosage should generally be avoided except for rare cases when special circumstances are at hand.
4 Complementary measures to liming are often required, such as biological restoration and elimination of obstacles for the passage of fish and other animals.
5 A combination of different strategies such as direct lake liming, wetland liming and liming by means of dosers will generally bring about the best ecological results.
6 Wetland liming is only to be adopted when other strategies have turned out to be insufficient. Careful evaluations and comparison of the value of the wetland versus that of the surface water is needed, as limed wetland can be permanently damaged.
7 Otherwise accurate liming operations may fail to produce desirable results if the watershed is influenced by major land-use activities, such as clear-felling and forest-soil fertilisation.
8 Limestone is beyond comparison the most adequate and consequently mainly used neutralising agent for surface waters. Other neutralising agents can nevertheless be used.
9 Reacidification should not be allowed in limed surface waters, as there is an obvious risk to lose what has been achieved.

## 18.12 PERSPECTIVES ON LONG-TERM LIMING

As a consequence of deposition scenarios based upon the commitments of the UN sulphur protocol (UN 1994), liming in Norway will be reduced to one-third some time after year 2010 (Henriksen & Hindar 1993). Entries in Table 18.1 are based on the introduction of a variable $ANC_{limit}$ (Henriksen et al. 1995).

Decreased exceedance of the critical load will probably lead to reduced aluminium concentrations and to a decline in the frequency and magnitude of acid episodes. Target-pH for liming may therefore be lowered in this improved environment, thus reducing the liming costs further.

We are not yet able to estimate the exact timing of water quality improvements to given levels after future deposition reductions. Dynamic models such as MAGIC (model for acidification of groundwater in catchments) (Wright et al. 1988) have been used but lack of long data records during

Table 18.1 Effects of decreased exceedance of the critical load in southern and northern Norway according to the commitments of the UN (1994) protocol for sulphur deposition reductions. Figures are based on a variable $ANC_{limit}$, see Henriksen et al. (1995). (Data from A. Henriksen)

| Region/country | Area exceeded (%) | | Amount of lime (t yr$^{-1}$) | | Cost (million NKR) | |
|---|---|---|---|---|---|---|
| | Today | Year 2010 | Today | Year 2010 | Today | Year 2010 |
| South of Norway | 55.0 | 21.2 | 389,505 | 149,444 | 311.6 | 119.9 |
| North of Norway | 6.4 | 1.2 | 22,333 | 5615 | 23.3 | 4.3 |
| Norway | 29.7 | 10.8 | 411,838 | 155,059 | 334.9 | 124.2 |

recovery makes the forecasts uncertain. Owing to the time lags, liming operations may probably be longer lasting than recognised from sulphur deposition scenarios, i.e. several decades. In Sweden, liming of aquatic ecosystems will, most probably, be necessary for at least another 50 years (Henrikson & Brodin 1995b). This would remain true even if there were to be a major international reduction in acid emissions within the next few years.

Metals such as aluminium may polymerise and eventually be precipitated as humic–metal complexes after liming. Some have argued that easily available aluminium from such sediments could represent a threat to aquatic life during reacidification after liming and after closedowns of lime dosers. However, both the theoretical considerations and the monitoring data indicate that precipitated aluminium does not represent an additional source of any significance for aquatic life during these circumstances (Hindar & Lydersen 1995).

Wetland liming and whole-catchment liming may stimulate the decomposition of organic matter. A high carbon/nitrogen ratio in organic matter may reduce this stimulating effect, with increased decomposition becoming less pronounced under humic conditions and a low-temperature climate such as obtains in Scandinavia. Nevertheless, over a period of decades, liming will probably change the structure and function of mire surfaces and the humic layers of, for example, forest soils. Knowledge of the quality and speed of these changes is needed.

Long-lasting liming obviously results in accumulation of undissolved minerals on lake bottoms and stream beds. So far, research activities have focused on the suitability of limed sediments as habitat for invertebrates and as spawning sites for fish. The other ecological effects have not been investigated. Thus, in spite of increasing knowledge, it is important to point out the shortcomings and deficiencies in current scientific knowledge. We highlight the following (see also Henrikson & Brodin 1995b):

- long-term ecological effects, i.e. more than 20 years;
- concentrations and accessibility of nutrients;
- primary production;
- the structure and function of limed wetlands;
- recolonisation and re-introduction of species;
- strategies and methods to eliminate harmful effects of acidic episodes;
- improvements of liming techniques in order to reduce harmful effects in limed wetlands;
- measurable biological goals which can be quantified.

## 18.13 REFERENCES

Abrahamsson, I. (1993) Impact of overflows on acid-base chemistry in limed lakes. *Vatten*, **49**, 24–33.

Alenäs, I, Degerman, E. & Henrikson, L. (1995) Liming strategies and effects; the River Högvadsån case study. In: Henrikson, L. & Brodin, Y.-W. (eds), *Liming of Acidified Surface Waters; a Swedish Synthesis*. Springer-Verlag, Berlin, 363–74.

Almer, B., Dickson, W., Ekström, C. & Hörnström, E. (1978) Sulphur pollution and the aquatic ecosystem. In: Nriagu, O. I. (ed.), *Sulphur in the Environment*. Wiley, New York, 271–86.

Appelberg, M. (1995) The impact of liming on aquatic communities. In: Henrikson, L. & Brodin, Y.-W. (eds), *Liming of Acidified Surface Waters; a Swedish Synthesis*. Springer-Verlag, Berlin, 283–308.

Appelberg, M. & Degerman, E. (1991) Development and stability of fish assemblages after lime treatment. *Canadian Journal of Fisheries and Aquatic Sciences*, **30**, 545–54.

Appelberg, M., Henrikson, B.-I., Henrikson, L. & Svedäng, M. (1993) Biotic interactions within the littoral community of Swedish forest lakes during acidification. *Ambio*, **22**, 290–7.

Baalsrud, K., Hindar, A., Johannessen, M. & Matsow, D. (eds) (1985) *Liming of Acid Water*. Liming Project, Final Report, Department of the Environment and Directorate for Nature Management, Norway, Oslo, 147 pp.

Bengtsson, B. & Bogelius, A. (1995) Socio-economic consequences of aquatic liming. In: Henrikson, L. & Brodin, Y.-W. (eds), *Liming of Acidified Surface Waters; a Swedish Synthesis*. Springer-Verlag, Berlin, 423–58.

Bergquist, B. (1991) Extinction and natural recolonization of fish in acidified and limed lakes. *Nordic Journal of freshwater Research*, **66**, 50–62.

Bergquist, B. (1995) Supplementary measures to aquatic

liming. In: Henrikson, L. & Brodin, Y.-W. (eds), *Liming of Acidified Surface Waters; a Swedish Synthesis.* Springer-Verlag, Berlin, 399–422.

Blomqvist, P., Bell, R.T., Olofsson, H., Stensdotter, U. & Vrede, K. (1993) Pelagic ecosystem responses to nutrient additions in acidified and limed lakes in Sweden. *Ambio,* **22,** 283–9.

Brandrud, T.E. & Roelofs, J.G.M. (1995) Enhanced growth of the macrophyte *Juncus bulbosus* in S Norwegian limed lakes; a regional survey. *Water, Air, and Soil Pollution,* **85,** 913–8.

Brett, M.T. (1989) Zooplankton communities and acidification processes (a review). *Water, Air, and Soil Pollution,* **44,** 387–414.

Brink, P., Nilsson, L.M. & Swedin, U. (1988) Ecosystem redevelopment. *Ambio,* **17,** 84–9.

Brocksen, R.W. & Wisniewski, J. (eds) (1991) *Restoration of Aquatic and Terrestrial Systems.* Kluwer, Dordrecht, 501 pp.

Brodin, Y.-W. (1995a) Acidification of lakes and watercourses in a global perspective. In: Henrikson, L. & Brodin, Y.-W. (eds), *Liming of Acidified Surface Waters; a Swedish Synthesis.* Springer-Verlag, Berlin, 45–62.

Brodin, Y.-W. (1995b) Acidification of Swedish freshwaters. In: Henrikson, L. & Brodin, Y.-W. (eds), *Liming of Acidified Surface Waters; a Swedish Synthesis.* Springer-Verlag, Berlin, 63–80.

Brown, D.J.A. & Sadler, K. (1989) Fish survival in acid waters. In: Morris, R., Taylor, E.W., Brown, D.J.A. & Brown, J.A. (eds), *Acid Toxicity and Aquatic Animals.* Cambridge University Press, Cambridge, 31–44.

Bulger, A.J., Lien, L., Cosby, B.J. & Henriksen, A. (1993) Brown trout (*Salmo trutta*) status and chemistry from the Norwegian Thousand Lake survey; statistical analysis. *Canadian Journal of Fisheries and Aquatic Sciences,* **50,** 575–85.

Cairns, J., Jr. (1988) Restoration and the alternative, a research strategy. *Restoration Management Notes,* **6,** 65–7.

Dalziel, T.R.K., Proctor, M.V. & Paterson, K. (1992) Water quality of surface waters before and after liming. In: Howells, G. & Dalziel, T.R.K. (eds), *Restoring Acid Waters; Loch Fleet 1984–1990.* Elsevier Applied Science, London, 229–57.

Degerman, E., Henrikson, L., Herrmann, J. & Nyberg, P. (1995) The effects of liming on aquatic fauna. In: Henrikson, L. & Brodin, Y.-W. (eds), *Liming of Acidified Surface Waters; a Swedish Synthesis.* Springer-Verlag, Berlin, 221–82.

Dickson, W. & Brodin, Y.-W. (1995) Strategies and methods for freshwater liming. In: Henrikson, L. & Brodin, Y.-W. (eds), *Liming of Acidified Surface Waters; a Swedish Synthesis.* Springer-Verlag, Berlin, 81–124.

Dickson, W., Borg, H., Ekström, C., Hörnström, E. & Grönlund, T. (1995). Liming and reacidification on lake water; chemistry, phytoplankton and macrophytes. *Water, Air, and Soil Pollution,* **85,** 919–24.

Eilertsen, O., Stabbetorp, O.E., Aarrestad, P.A. & Bendiksen, E. (1997) Counteractions against acidification in forest ecosystems: vegetation dynamics in a forested catchment after dolomite application in Gjerstad, S Norway. *BIOGEOMON' 97, Journal of Conference Abstracts,* **2,** 167. (Cambridge Publications.)

Engblom, E. & Lingdell, P.-E. (1984) The mapping of short-term acidification with the help of biological pH indicators. *Report of the Institute of Freshwater Research, Drottningholm,* **61,** 60–8.

Eriksson, F., Hörnström, E., Mossberg, P. & Nyberg, P. (1983) Ecological effects of lime treatment of acidified lakes and rivers in Sweden. *Hydrobiologia,* **101,** 145–64.

Eriksson, M.O.G. (1987) Some effects of freshwater acidification on birds in Sweden. International Council for Bird Preservation, Cambridge, Technical Publication, 6, 1–8.

Gahnström, G. (1995) The effects of liming on microbial activity and the decomposition of organic material. In: Henrikson, L. & Brodin, Y.-W. (eds), *Liming of Acidified Surface Waters; a Swedish Synthesis.* Springer-Verlag, Berlin, 179–92.

Gubala, C.P. & Driscoll, C.T. (1991) Watershed liming as a strategy to mitigate acidic deposition in the Adirondack region of New York. In: Olem, H., Schreiber, R.K., Brocksen, R.W. & Porcella, D.B. (eds) (1991) *International Lake and Watershed Liming Practices.* The Terrene Institute, Washington, DC, 145–59.

Havas, M. & Rosseland, B.O. (1995) Response of zooplankton, benthos, and fish to acidification; an overview. *Water, Air, and Soil Pollution,* **85,** 51–62.

Hasselrot, B., Alenäs, I., Andersson, I. & Hultberg, H. (1987) Response of limed lakes to episodic acid events in southwestern Sweden. *Water, Air, and Soil Pollution,* **32,** 341–62.

Henriksen, A. & Hindar, A. (1993) Environmental measures in water; are we able to calculate the necessary liming amount for Norway? In: Romundstad, A.J. (ed.), *Liming in Lakes and Rivers.* Directorate for Nature Management, Oslo, 162–70. (In Norwegian.)

Henriksen, A., Kämäri, J., Posch, M. & Wilander, A. (1992) Critical loads for surface waters in the Nordic countries. *Ambio,* **21,** 356–63.

Henriksen, A., Posch, M., Hultberg, H. & Lien, L. (1995) Critical loads of acidity for surface waters—can the $ANC_{limit}$ be considered as a variable? *Water, Air, and Soil Pollution,* **85**, 2419–24.

Henrikson, B.-I. (1988) The absence of antipredator behaviour in the larvae of *Leucorrhinia dubia* (Odonata) and the consequences for their distribution. *Oikos,* **51**, 179–83.

Henrikson, L. (1988) Effects on water quality and benthos of acid water inflow to the limed Lake Gårdsjön. In: Dickson, W. (ed.), *Liming of Lake Gårdsjön; an Acidified Lake in SW Sweden.* Report 3426, Swedish Environmental Protection Agency, Stockholm, 309–27.

Henrikson, L. & Brodin, Y.-W. (eds) (1995a) *Liming of Acidified Surface Waters; a Swedish Synthesis.* Springer-Verlag, Berlin, 458 pp.

Henrikson, L. & Brodin, Y.-W. (1995b) Liming of surface waters in Sweden; a synthesis. In: Henrikson, L. & Brodin, Y.-W. (eds), *Liming of Acidified Surface Waters; a Swedish Synthesis.* Springer-Verlag, Berlin, 1–44.

Henrikson, L. & Oscarson, H.G. (1981) Corixids (Hemiptera-Heteroptera); the new top predators in acidified lakes. *Verhandlungen der internationalen Vereinigung für theoretische und angewandte Limnologie,* **21**, 1616–20.

Henrikson, L., Nyman, H.G., Oscarson, H.G. & Stenson, J.A.E. (1985) Changes in the zooplankton community after lime treatment of an acidified lake. *Verhandlungen der internationalen Vereinigung für theoretische und angewandte Limnologie,* **22**, 3008–13.

Henrikson, L., Hindar, A. & Thörnelöf, E. (1995) Freswater liming. *Water, Air, and Soil Pollution,* **85**, 131–42.

Herrmann, J., Degerman, E., Gerhardt, A., Johansson, C., Lingdell, P.-E. & Muniz, I.P. (1993) Acid-stress effects on stream biology. *Ambio,* **22**, 298–307.

Hesthagen, T. & Hansen, L.P. (1991) Estimates of the annual loss of Atlantic salmon, *Salmo salar* L., in Norway due to acidification. *Aquaculture and Fisheries Management,* **22**, 85–91.

Hindar, A. (1992) Is it possible to lime both economically and ecologically correct? In: *Liming Acid Water; Strategies and Effects.* Directorate for Nature Management, Oslo, 43–54. (In Norwegian.)

Hindar, A. & Henriksen, A. (1992) Acidification trends, liming strategy and effects of liming for Vikedalselva, a Norwegian salmon river. *Vatten,* **48**, 54–8.

Hindar, A. & Lydersen, E. (1995) *Does Precipitated/ Sedimented Aluminium Represent a Possible Environmental Problem after Liming?* Report O-92149, Norwegian Institute for Water Research, Oslo, 22 pp. (In Norwegian.)

Hindar, A. & Rosseland, B.O. (1991) Liming strategies for Norwegian lakes. In: Olem, H., Schreiber, R.K., Brocksen, R.W. & Porcella, D.B. (eds) (1991) *International Lake and Watershed Liming Practices.* The Terrene Institute, Washington, DC, 173–92.

Hindar, A, Nilsen, P., Skiple, A. & Høgberget, R. (1995) Counteractions against acidification in forests ecosystems. Effects on stream water quality after dolomite application to forest soil in Gjerstad, Norway. *Water, Air, and Soil Pollution,* **85**, 1027–32.

Hindar, A., Kroglund, F., Lydersen, E., Skiple, A. & Høgberget, R. (1996) Liming of wetlands in the acidified Røynelandsvatn catchment in southern Norway—effects on stream water chemistry. *Canadian Journal of Fisheries and aquatic Sciences,* **53**, 985–93.

Hindar, A., Norgaard, E., Nilsen, P., Skiple, A. & Høgberget, R. (1997) Effects on soil water and stream water quality after dolomite application to an acidified forested catchment in Gjerstad, Norway. *BIOGEOMON' 97, Journal of Conference Abstracts,* **2**, 198. (Cambridge Publications.)

Hindar, A., Henriksen, A., Sandøy, S. & Romundstad, A.J. (1998) Critical load concept to set restoration goals for liming acidified Norwegian waters. *Restoration Ecology,* **6**(4), 1–13.

Howells, G. & Dalziel, T.R.K. (1992) The Loch Fleet project and catchment liming in perspective. In: Howells, G. & Dalziel, T.R.K. (eds), *Restoring Acid Waters; Loch Fleet 1984–1990.* Elsevier Applied Science, London, 393–411.

Iivonen, P., Järvenpää, T., Lappalainen, A., Mannio, J. & Rask, M. (1995) Chemical, biological and socio-economic approaches to the liming of Lake Alinenjärvi in southern Finland. *Water, Air, and Soil Pollution,* **85**, 937–42.

Keller, W.B. & Gunn, J.M. (1995) Lake water quality improvements and recovering aquatic communities. In: Gunn, J.M.(ed.), *Restoration and Recovery of an Industrial Region.* Springer-Verlag, New York, 67–80.

Kroglund, F., Dalziel, T., Rosseland, B.O., Lien, L., Lydersen, E. & Bulger, A. (1992) *Restoring Endangered Fish in Stressed Habitats; ReFish Project 1988–1991.* Acid Rain Research Report No. 30, Norwegian Institute for Water Research, Oslo, 43 pp.

Larsson, S. (1995) The effects of liming on aquatic flora. In: Henrikson, L. & Brodin, Y.-W. (eds), *Liming of Acidified Surface Waters; a Swedish Synthesis.* Springer-Verlag, Berlin, 193–220.

Lien, L., Raddum, G.G. & Fjellheim, A. (1992) *Critical Loads for Surface Water; Invertebrates and Fish*. Acid Rain Research Report No. 21, Norwegian Institute for Water Research, Oslo. 36 pp.

Lingdell, P.-E. & Engblom, E. (1995) Liming restores the benthic community to 'pristine' state. *Water, Air, and Soil Pollution*, **85**, 955–60.

Lydersen, E. (1990) The solubility and hydrolysis of aqueous aluminium hydroxides in dilute fresh waters at different temperatures. *Nordic Journal of Hydrology*, **21**, 195–204.

Mackenzie, S. (1992) The impact of catchment liming on blanket bogs. In: Bragg, O.M., Hulme, P.D., Ingram, H.A.P. & Robertson, R.A. (eds) *Peatland Ecosystems and Man; an Impact Assessment*. University of Dundee, Dundee, 31–7.

Meili, M. (1995) Liming effects on mercury concentrations in fish. In: Henrikson, L. & Brodin, Y.-W. (eds), *Liming of Acidified Surface Waters; a Swedish Synthesis*. Springer-Verlag, Berlin, 383–98.

Navrud, S. (1990) *Cost-benefit Analysis of Liming of Streams. A Study in River Audna*. Directorate for Nature Management (Norway), Oslo, 51 pp. (In Norwegian.)

Nyberg, P., Andersson, P., Degerman, E., Borg, H. & Olofsson, E. (1995) Labile manganese; an overlooked reason for fish mortality in acidified streams? *Water, Air, and Soil Pollution*, **85**, 333–40.

Økland, J. & Økland, K.A. (1986) The effects of acid deposition in benthic animals in lakes and streams. *Experientia*, **42**, 471–86.

Olem, H. (1991) *Liming Acidic Surface Waters*. Lewis Publishers, Chelsea, MI, 331 pp.

Olem, H., Schreiber, R.K., Brocksen, R.W. & Porcella, D.B. (eds) (1991) *International Lake and Watershed Liming Practices*. The Terrene Institute, Washington, DC.

Porcella, D.B. (1989) Lake Acidification Mitigation Project (LAMP); an overview of an ecosystem perturbation experiment. *Canadian Journal of Fisheries and aquatic Sciences*, **46**, 246–8.

Raddum, G.G. & Fjellheim, A. (1984) Acidification and early warning organisms in freshwater in western Norway. *Verhandlungen der internationalen Vereinigung für theoretische und angewandte Limnologie*, **22**, 1973–80.

Raddum, G.G. & Fjellheim, A. (1995) Acidification in Norway—status and trends; biological monitoring—improvements in the invertebrate fauna. *Water, Air, and Soil Pollution*, **85**, 647–52.

Roelofs, J.G.M., Brandrud, T.E. & Smolders, A.J.P. (1994) Massive expansion of *Juncus bulbosus* L. after liming of acidified Norwegian lakes. *Aquatic Botany*, **48**, 187–202.

Rosseland, B.O. & Hindar, A. (1988) Liming of lakes, rivers and catchments in Norway. *Water, Air, and Soil Pollution*, **41**, 165–88.

Rosseland, B.O. & Hindar, A. (1991) Mixing zones; a fishery management. In: Olem, H., Schreiber, R.K., Brocksen, R.W. & Porcella, D.B. (eds) (1991) *International Lake and Watershed Liming Practices*. The Terrene Institute, Washington, DC, 161–73.

Rosseland, B.O., Eldhuset, T.D. & Staurnes, M. (1990) Environmental effects of aluminium. *Environment, Geochemistry and Health*, **12**, 17–27.

Rosseland, B.O., Blakar, I., Bulger, A., et al. (1992) The mixing zone between limed and acidic river waters; complex aluminium chemistry and extreme toxicity for salmonids. *Environmental Pollution*, **78**, 3–8.

Sandøy, S. & Romundstad, A.J. (1995) Liming of acidified lakes and rivers in Norway; an attempt to preserve and restore biological diversity in acidified regions. *Water, Air, and Soil Pollution*, **85**, 997–1002.

Snucins, J., Gunn, J.M. & Keller, W. (1995) Preservation of biodiversity; Aurora trout. In: Gunn, J.M. (ed.), *Restoration and Recovery of an Industrial Region*. Springer-Verlag, New York, 195–204.

Svahnberg, A. (1996) *Lime Dosers, a Method for Liming of Lakes and Streams*. Report No.4627, Swedish Environmental Protection Agency, Stockholm. (In Swedish.)

Svensson, T., Dickson, W., Hellberg, J., Moberg, G. & Munthe, N. (1995) The Swedish liming programme. *Water, Air, and Soil Pollution*, **85**, 1003–8.

Sverdrup, H.U. (1985) *Calcite dissolution kinetics and lake neutralization*. Thesis, Lund University Institute of Technolgy, Sweden, 170 pp.

SEPA (1988) *Liming of lakes and watercourses*. Official Guidelines 88:3, Swedish Environmetal Protection Agency, Stockholm. (In Swedish.)

Traaen, T.S., Frogner, T., Hindar, A., Kleiven, E., Lande, A. & Wright, R.F. (1997) Whole-catchment liming at Tjønnstrond, Norway; an 11-year record. *Water, Air, and Soil Pollution*, **94**, 163–80.

UN (1994) *Protocol to the 1979 Convention on Longerange Transboundary Air Pollution on Further Reduction of Sulphur Emissions*. Annex II, United Nations, Oslo, June, 106 pp.

Weatherley, N.S. (1988) Liming to mitigate acidification in freshwater ecosystems: a review of the biological consequences. *Water, Air, and Soil Pollution*, **39**, 421–37.

Wilander, A., Andersson, P., Borg, H. & Broberg, O. (1995) The effects of liming on water chemistry. In: Henrikson, L. & Brodin, Y.-W. (eds), *Liming of Acidified Surface Waters; a Swedish Synthesis*. Springer-Verlag, Berlin, 125–78.

Wood, C.M. & McDonald, D.G. (1987) The physiology of acid/aluminium stress in trout. *Annales du Société royale de Zoologie de la Belgique*, **117**(Supplement 1), 399–410.

Wright, R.F., Kämäri, J. & Forsius, M. (1988) Critical loads for sulfur; modelling time of response of water chemistry to changes in loading. In: Nilsson, J. & Grennfelt, P. (eds), *Critical Loads for Sulphur and Nitrogen*. UN-ECE and Nordic Council Ministers, Miljørapport 1988-15, Oslo, 201–24.

# Part V  Legal Frameworks

# 19 The Framework for Managing Lakes in the USA

## THOMAS DAVENPORT

## 19.1 INTRODUCTION

There are more than 41.6 million acres (168.4 × $10^3$ km$^2$) of lakes, reservoirs and ponds in the USA. There are over 100,000 lakes which exceed 100 surface acres (0.4 km$^2$) in the USA, not including Alaska, where there are several million. With over 90% of the USA population living within 50 miles of a lake (Holdren 1997), there is tremendous demand for, and effect on, these resources. Although serving as a recreational resource for millions of people, an estimated 80% of the 3700 urban lakes possess significant water-quality problems. Besides recreational use, some of these lakes may serve as drinking water supplies, provide flood control or support industrial uses.

Over 42% of the lake area of the USA was assessed for the 1998 National Water Quality Report. The States and Tribes determined that nutrient enrichment, metal pollution and siltation constitute the most common impacts on the surveyed lakes. These pollutants affected 45% of all the lakes surveyed and were the cause of 78% of all the water-quality problems identified (USEPA 2000). Agriculture was found to be the leading source of pollution among the surveyed lakes. According to the Report, agricultural pollution affected 14% of all lakes surveyed and it constituted 31% of all the water-quality problems identified.

In the USA, there is no one law which specifically protects lakes. Regulations are established to carry out legislation (laws) and then guidance is developed to assist the carrying out and/or compliance with the regulations and legislation. Protecting and managing water quality in lakes requires an effective management framework which integrates appropriate management and restoration programmes. This framework is usually based upon a combination of various provisions of Federal, State and local regulations and efforts. The two main federal laws governing lake-water quality issues are the Clean Water Act, as amended (www.USEPA.gov/region5/def/html/cwa.htm), and the Safe Drinking Water Act, as amended (www.USEPA.gov/safewater/sdwa/sdwa.html).

## 19.2 CLEAN WATER ACT

The Act (originally named the Water Quality Act of 1965), as amended, is the primary legislation in the USA directed to the restoration and protection of the quality of the Nation's water resources. In its various forms, the Act has been around for over 35 years, during which time it has been expanded and refocused several times. However, the overall goal of the Act has remained constant: to restore and maintain the chemical, physical and biological integrity of the Nation's waters and, where attainable, to achieve a level of water quality which provides for the protection and propagation of fish and shellfish, wildlife and recreation in and on the water.

The United States Environmental Protection Agency (USEPA) is responsible for enforcing the Act nationwide. However, the States and Tribes (henceforth, the term 'States' will refer to both 'States and Tribes') can assume authority, within their jurisdictions, for the management of parts of the Act, provided that they meet the appropriate legal and programme criteria. A number of provisions within the Act relate to the management of lakes: these include the water-quality standards, issue of permits, control of non-point sources, and the Clean Lakes Program.

Water-quality standards establish the foundation for most activities carried out under the Act. Figure 19.1 shows the relationship between water-quality management approaches and water-quality standards. The Water Quality Act of 1965 relied on violations of water-quality standards as the basis for pollution controls. This approach was not successful in reducing water pollution. Accordingly, the 1972 amendments (Federal Water Pollution Control Act) addressed this weakness and established the first comprehensive national framework for improving the quality of the Nation's waters. This framework established a dual system which set (i) water-quality standards for ambient waters and (ii) minimum requirements for abatement of point sources and area-wide management plans for the control and prevention of non-point sources.

### 19.2.1 Water-quality standards

Water-quality standards provide the benchmark tools to manage, protect and guide restoration of water resources (Fig. 19.1). In the code of Federal Regulations, under 40 CFR Part 131, USEPA issued regulations governing the development, review, revision and subsequent approval of water-quality standards by States. The system of standards is the basis for the strong Federal/State water-quality management partnership. In these cases, the States play the primary role in setting water-quality standards which serve as the foundation for their respective programmes for the management of water quality. Besides setting goals for protecting resources, water-quality standards provide the legal basis for wastewater-discharge permits and for controlling non-point source pollution (NALMS 1992). However, NALMS (North American Lake Management Society) often found that the water-quality standards or criteria adopted were addressed only to the regulation of point sources to rivers and streams (lotic waters).

The water-quality standard defines the quality goals for a water body in relation to its designated or desired uses, then sets standards and criteria for the proposed uses as well as protecting the existing uses through an anti-degradation policy. Section 304 of the Act requires USEPA to develop and publish ambient water-quality criteria which support the various designated uses. Then, under Section 303, the States are required to adopt the ambient standards and USEPA is required to either approve or disapprove the State standards proposed. State water-quality standards are required to be updated on a triennial cycle, with the approval of USEPA. The USEPA may promulgate a Federal standard under Section 303 if it considers a State standard to

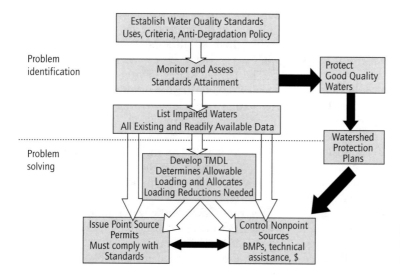

Fig. 19.1 Restoring polluted waters within a water-quality management framework.

be inconsistent with the Act. Generally, States adopt water-quality standards to protect six standard designated uses for surface water bodies, including lakes: aquatic life support, fish consumption, primary contact recreation (i.e., swimming and diving), secondary contact recreation (e.g., boating), drinking water supply and agricultural use. At a minimum, the designated use of a water resource must include secondary contact and the propagation of fish and wildlife.

Water-quality criteria are adopted as part of the water-quality standard. These criteria are numeric and narrative and are established according to the designated use(s) to be protected. Criteria set finite limits on particular pollutants or define the particular condition of a water resource to be attained. Numeric criteria are values assigned to measurable components in the water resource; narrative criteria are verbal descriptions of the desired condition of the resource.

State water-quality standards must contain an anti-degradation policy and a methodology for implementing the policy. These policies contain three tiers. The first requires all existing uses of a water to be maintained and protected. The second stipulates that the water quality which currently fulfils the minimum standards necessary to satisfy (protect) the existing uses is maintained, unless to do so interfered with an important social and economic development. The third tier contains waters of special ecological or recreational significance, categorised as Outstanding National Resources. The existing quality of these waters needs to be maintained, regardless of other social or economic considerations.

Although many water-quality standards can be effectively applied to protect lake quality, such applications are not universal. For nutrients, the inherent differences between lotic (flowing) and lentic (lake) resources necessitate the development of separate standards to apply to either type. The main difference is that lentic waters are more sensitive to nutrient loading, owing to the time of hydraulic residence, than are most lotic waters. Prolonged residence facilitates uptake and growth by plants and it encourages nutrient settling and recycling. The USEPA is working with the States to develop and adopt numeric criteria for nutrients which are tailored to reflect the different types of water bodies and the various eco-regions making up the country. The goal is to establish objective, scientifically sound bases for assessing nutrient overenrichment problems. The USEPA encourages States to modify or improve the basic approach of defining nutrient ecoregions and the natural and anthropogenic components influencing the standards to be set. For example, in the Upper Midwest region, one such refinement to the recommended approach has been adopted. Here, the relative importance of various environmental characteristics affecting nutrient concentrations is determined through the use of regression-tree analysis. The most significant characteristics are diagnosed and incorporated into the water-quality standards set. Nutrient standards appropriate to the area or region of a State are then defined only on the basis of the statistically most significant characteristics.

### 19.2.2 Pollution control

There are two major pollution control programmes under the Act, linked to the achievement of applicable lake-water quality standards. The two programmes are related to how dischargers are addressed through the Act: regulatory or voluntary. In either case, the Act distinguishes between regulatory and voluntary controls on discharges.

1 The National Pollutant Discharge Elimination System (NPDES). The Federal Government established the NPDES system to regulate discharges from point sources. The backbone of the system is the issue of NPDES permits which regulate the amount and/or concentration of pollutants which may be legally discharged into a water body. These regulatory discharge limits are set for each permit so as not to violate the quality of the receiving water. Section 502 of the Act defines point sources to include pipes, ditches and other discernible, confined and discrete conveyances from which pollutants are or may be discharged. Although concentrated animal feedlots over a certain size (presently the threshold is set at 1000 animal units) are considered to be point sources, most agricul-

tural storm water discharges are explicitly excluded. In the 1987 Amendments, the point-source category was expanded to include storm-water runoff from urban areas with over 100,000 population, as well as a small number of industrial operations. In 2003, a further expansion of the storm-water provisions included more municipalities of smaller size.

During the first 25 years of the Act, the Federal and State Governments focused entirely on controlling point sources of pollution, primarily the discharges from industrial and publicly owned waste-water treatment plants. Considerable progress has been made in remediating and preventing point-source discharges. After three decades of increasingly more stringent controls on point-source discharges, much of the remaining pollution impacting lakes and rivers derives from non-point sources.

2 Control of non-point sources. The 1987 Amendments established the first comprehensive national programme for non-point source pollution. Non-point sources are defined as water pollution sources excluded from the definition of a point source. Non-point sources are generally episodic in nature, originating from diffuse sources at divergent locations. This makes them inherently difficult to measure and manage. Non-point sources are not regulated at the Federal level, nor are they traditionally regulated at the State or local level. The 1987 Amendment (Section 319) required States to develop assessment reports of non-point sources, specifying the nature of the non-point sources and the locations which they threatened, the water-quality problems they caused or the uses they impaired. Based upon the reported findings, States were to develop comprehensive management programmes in order to address the problems thus identified and to prevent future ones. Once USEPA had approved both the assessment report and the management programme (or portions thereof), the State became eligible for financial assistance towards the implementation of the approved parts of the management programme.

In order to gain the approval of USEPA, State management programmes were specifically required to include a minimum six components. The most important of these for lake management are: (i) an identification of best management practices and measures for reducing loadings from non-point sources; and (ii) an identification of regulatory or non-regulatory programmes providing enforcement, technical assistance, financial assistance, education, training, technology transfer or demonstration projects designed to apply best management practices. Several in-lake management/restoration techniques which control non-point sources of pollution were eligible for funding if they were included on the approved list of best-management practices. Some examples of approved practices for the control of non-point sources and the benefit of in-lake restorations include dredging to remove phosphorus from the system and lake-shore stabilisation to prevent sediments from entering the lake. The overall State management framework included a provision for public involvement and management of whole watersheds. These are basic to the successful management of any lake.

Section 319 did not establish a scheme of permits and quantified discharge limits applicable to non-point sources. The basic mechanism for controlling pollution from non-point sources is the implementation of best-management practices throughout the area of jurisdiction or to whole watersheds. The problem with this arrangement is that, in most States, the programmes for the management of non-point sources are not founded on water-quality criteria. Most States have developed voluntary watershed-based programmes which focus on providing technical and financial assistance to polluters, as part of broad-based stewardship, rather than on the quality of the affected waters. Moreover, the stewardship approach distributes technical and financial resources equally, rather than in any prioritised manner. The absence of targeting the technical and financial resources towards the main pollutants of concern or their sources left the stewardship approach rather ineffective in achieving its intent.

In contrast, controls based on water quality require the identification of the pollutant(s) responsible for the quality impairment and they direct the focus of land management improvements to the critical source areas of the pollutant(s). In this way, pollutant sources and transport to the impaired (or threatened) water resources is effectively and directly addressed. The high level of treatment in the critical areas secures a greater reduction in pollutant delivery than does the broad-brush approach involving less intense land treatment over larger areas.

The Act encourages an approach to water quality based on the calculation of a total maximum daily load (TDML). If the limits on point-source discharges are insufficient to achieve and maintain satisfactory water quality in a particular water body, then that water is entered on the list of 'impaired' locations on the Section 303(d) list. The TMDLs are particularly targeted towards water bodies either failing to meet the quality standards or which are projected to do so even when remedial technology is in place. States are to establish TMDLs which will meet the water-quality standards for each listed water.

A TMDL is calculated as the sum of the waste loads from individual point sources plus the load allocations from non-point sources, as well as the natural background flux (40 CFR 130.2), subject to a margin of safety (CWA Section 303(d)(1)(C)). A TMDL can be described generically by the following equation:

$$TMDL = LC = \textstyle\sum WLA + \textstyle\sum LA = MOS$$

where LC is the loading capacity that a water body can receive without violating water-quality standards, WLA is the wasteload allocation, LA is the load allocation and MOS is the margin of safety. Once listed in these terms, the water bodies are prioritised for TMDL development, based upon a number of factors pertinent to the State. The water body remains on the 303(d) list until such time as it complies with the water-quality standards. An implementation plan needs to be developed, using a programme-neutral planning process currently being developed jointly by the United States department of Agriculture (USDA) and USEPA. Presently, the only enforceable component of the TMDL is the wasteload allocation (WLA), as this is covered by the NPDES scheme.

### 19.2.3 Watershed approaches

The implementation of the TMDL should include empirical measures and actions for both point and non-point sources of pollution. Depending upon the pollutant, one of the most efficient ways to implement a TMDL is through what is known as the Watershed Approach. In short, the Watershed Approach is designed to secure effective control and such restoration action which will mitigate or eliminate specific problems. Because watersheds are the hydrological units through which surface water flows, there are several linkages among water quality and habitat conditions, problems and solutions. The Watershed Approach is flexible and geographically customised; consequently, there is no single definition which encompasses all the programmes and activities which it might include. As it is a general, strategic approach, it is best defined by describing the common component steps for successful watershed management. The USEPA is recommending a four-phase process (Davenport 2002) which develops programme-neutral watershed plans, with appropriate support at each stage. The process comprises:

1 information collection and assessment;
2 development and decision-making;
3 implementation;
4 evaluation.

The process relies on an adaptive management philosophy to support pragmatic adjustments in the management, as new information becomes available. The combination of natural variability in the hydrological cycle and the uncertainty associated with off-site impacts of pollution controls requires that watershed management is sufficiently flexible to accommodate pragmatic modifications to the implementation in the light of progress achieved or new information derived from further studies. For the watershed approach to be successful, there needs to be active local involvement and opportunities for participation by all stakeholders interested in or affected by decisions about the management of the water resource. Active efforts to engage the public to participate at all stages of the scheme and having an effective project manager and organisation are essential ingredients to a successful watershed management strategy. With a targeted geographical focus, watershed management enables an efficient concentration of scarce resources which then builds upon the foundation of protection provided by the regulatory point source programmes and undertakes additional actions (controls and restoration) to address specific problems in each watershed.

### *19.2.4 The Clean Lakes Program*

The last main part of the framework of the Act is the Clean Lakes Program. This was established under Section 314 of the 1972 Amendments. Since then, the Clean Lakes Program has evolved into the most successful watershed programme in the nation's history. Four principles provide the foundation for this success: (i) local involvement and commitment; (ii) State management; (iii) matching funds; and (iv) good science. Recognising the importance of lakes to the American public and the need to protect and restore them, Congress encouraged the States to survey and classify their publicly owned lakes according to: (i) tropic condition; (ii) pollution problems; (iii) pollution control measures and restoration programmes; and (iv) lake restoration and eutrophication projects.

Because the science of lake restoration was, at the time, still in its infancy, the USEPA assigned a low priority for the Clean Lakes Program from 1972 through to 1975. The prevailing view was that aggressive implementation of the NPDES programme and the construction of municipal wastewater-treatment plants would be sufficient to protect and improve lake-water quality. In 1975, Congress appropriated $4 million to develop a programme to implement Section 314. However, USEPA was still uncertain about the feasibility and the appropriate scope of a national Clean Lakes Program because the existing information on lake restoration was still inconclusive and incomplete. From 1975 to 1979, the USEPA provided $35 million to support a research and development effort on lake restoration techniques and approaches (USEPA 1993). The first programme actions supported a series of demonstration projects designed to identify appropriate in-lake restoration technology. Projects funded by these grants proved that in-lake restoration technology existed to the extent that, when combined with appropriate watershed pollution control measures, quality could be restored to degraded lakes and that lake restoration could become an integral component of a national water-quality strategy.

In 1980, the focus of the Clean Lakes Program changed from research and development to an operational scheme of financial and technical assistance. At the same time, USEPA issued regulations for the administration of the national programme. These restricted the financial assistance available under the programme to State agencies. The restriction was designed to build State capacity and strengthen the role of the State agencies when working with local people to solve problems of poor lake quality. The idea was to build 50 strong State agency programmes, carrying statewide authority, rather than support a plethora of smaller, weaker lake management agencies with limited jurisdictional authority. The regulations established the basic structure of the programme, which is designed to take a given lake project from its planning phase (diagnostic/feasibility) through to implementation. Besides the regulations, USEPA issued the Clean Lakes Program Strategy to provide specific programme goals. Realising there would not be sufficient Federal resources for all lakes needing assistance, USEPA developed the Strategy to stimulate new approaches to lake management and to build public and political support. One goal of the strategy was to protect or restore at least one lake within 25 miles (40 km) of every major USA population centre to a water quality suitable for recreational use. In addition, the strategy focused particularly on projects which would: (i) maximise public benefits; (ii) integrate other USEPA/State programmes such as NPDES enforcement and wetlands' protection; (iii) emphasise watershed management; (iv) develop active State involvement and maintain a Federal/State partnership; (v) encourage active local participation; (vi) ensure capable and effective project management; (vii) foster project support from applicable local, State and Federal agencies and organisations; and (viii) conduct continuous programme and project evaluation.

The regulations established two types of co-operative agreement for financial assistance: Phase I Diagnostic/Feasibility Studies and Phase II Implementation Projects. Under Phase I funding, States would assess the condition of a lake and determine the causes of poor quality, then recommend the procedures necessary to restore and protect lake quality. Phase II funding was to help implement the lake restoration and protection measures identified in Phase I, or in a similar study. Restoration and protection measures might include control or reduction of non-point sources of pollution in the watershed, in-lake techniques to restore water quality, or a combination of the two. Point-source controls were ineligible for funding under Section 314, as they are covered in the NPDES programme.

The National Clean Lakes Program expanded as a result of the 1987 Amendments. The types of financial assistance available doubled and USEPA was authorised to treat qualified Native American Tribes as States, thereby making them eligible for financial assistance under Section 314. In addition, the amendments required a significant redirection of State responsibility. To be eligible for financial assistance under Section 314 of the Act, States were required to submit a report, once every 2 years, on the water quality and condition of each lake addressed. These biennial reports were to include: (i) a revised Lake Classification Report; (ii) a list of threatened and impaired lakes; and (iii) an assessment of the status and trends of lake-water quality. As noted earlier, the Amendments also added a new programme, Section 319, to address non-point sources of pollution. The integration of the 314 and 319 programmes became a key step in accelerating efforts to resolve many of the water-quality problems affecting the nation's lakes. Two new co-operative assistance programmes were established to support the enhanced efforts, in the form of Lake Water Quality Assessment grants and Phase III Post-Restoration Monitoring grants. Phase III was designed to increase the scientific base of knowledge on the longevity and effectiveness of restoration and protection measures carried out through Phase II projects through ongoing monitoring and assessment. Lake Water Quality Assessment grants were provided to States to support the sampling and analysis of lake-water quality, volunteer citizens' monitoring programmes, regional lake-water quality assessments, and other activities in support of States' Lakes programmes. The Lake Water Quality Assessment grants are the only grants in the Section 314 programmes which are not lake specific.

The greatest obstacles facing the Clean Lakes Program have been uncertainty of continued funding and agency support (USEPA 1993). Funding for the programme has fluctuated widely from year to year and has been zeroed out by both Congress and the Administration in seven of the last 11 years. Before 1992, the Program enjoyed tremendous Congressional support. This support was shown in several ways, including:

• Congressional 'add-ons' in each year from 1980 until 1991;
• Congressional refusal to allow USEPA to use Clean Lakes appropriation for anything other than to support the Clean Lakes Program;
• Expansion of the Clean Lakes Program in 1987;
• Requirement for a biennial Report to Congress on the Quality of the Nation's Lakes;
• Direction to the USEPA, through the appropriation process, to provide personnel to carry out and manage the Clean Lakes Program.

Funding uncertainty has discouraged State and local support for and participation in the programme. In general, States responded to this funding uncertainty by reducing the level of support (staff) and involvement in the programme. The present lack of specific funding for the Clean Lakes Program (Lake Water Quality Assessment grants) has led to the loss of dedicated staff positions in most of the States. This lack of resources is a result of competing budget requirements and a misconception that the Clean Lakes Program benefits primarily rich owners of lake-side homes and not the public in general. The lack of programme support is in itself a reflection of the success of the Clean Lakes Program. The four principles which provide the foundation for the Clean lakes Program have been embedded in many new initiatives (e.g., the

Watershed Approach, community-based environmental protection) and programmes (e.g., the Nonpoint Source Program, National Estuary Program). The increased eligibility of a number of traditional lake measures and activities under other programmes, such as those under Section 319, do not translate into a priority for lake management under any of these other funding sources. Instead everyone assumes that the other programmes will provide the support, so little or none is actually forthcoming. To overcome the misconception about the public benefits, the North American Lakes Management Society and other conservation/environmental organisations are actively promoting a better understanding of society's need for, and the benefits of, protecting, managing and restoring lakes. This grass-roots support and involvement is key to the future of the Clean Lakes Program.

## 19.3 SAFE DRINKING WATER ACT

Many lakes in the USA are designated as public drinking-water supplies. The Safe Drinking Water Act (SDWA) delegates authority to USEPA to establish and enforce standards for the protection of drinking-water supplies. Regulations addressing nitrates and turbidity are promulgated by USEPA under this authority. The SDWA regulations are geared toward ensuring that the public receives safe drinking water. With this focus, they have concentrated on the quality of the water after it has undergone treatment and is ready for distribution to the general public, rather than on what pollutants and in what quantity may be present in the raw water. The 1996 SDWA Amendments introduced provisions addressing source protection. Source-water protection is a community-based approach to protecting sources of drinking water from contamination. The approach is based on three principles, commonly known as the 'Three Rs': **R**estore the public's Right and Responsibility to protect their drinking water and their health through community-based pollution-prevention efforts and, thereby, **R**aise public confidence in the drinking-water supply and **R**educe the costs of providing safe drinking water, which is essential to the sustainable development of the community (USEPA 1995). Source-water protection needs to be viewed as the SDWA counterpart to watershed management for achieving the objectives of the Clean Water Act. The USEPA and the States are just beginning to put in place the guidance and the programmes required to carry out the source-water provisions of the 1996 Amendments.

## 19.4 STATE AND LOCAL EFFORTS

As noted earlier, the key to effective lake management, protection and restoration is the establishment of a management framework which combines State and local authorities and programmes in a single, comprehensive plan. The management framework must include natural resource-management programmes in the water-quality programmes mentioned earlier. Figure 19.2 shows a representative example of the number of programmes which might be invoked in managing a typical lake in the Midwest (Phillips 2002). Several States, such as Illinois and Minnesota, have established lakes' management programmes modelled on the Federal Clean Lakes method of helping local people to solve problems in their lakes. Other States have taken the initiative of establishing water-quality (eutrophication) standards to protect and manage their lake resources. According to NALMS (1992), State eutrophication criteria possess a variety of applications including:

1 enforcing and establishing NPDES or State permit limits;
2 goal-setting and prioritisation;
3 managing cumulative impact and watershed planning; and
4 reporting on attainment of beneficial uses and state performance audits.

Some States, such as Wisconsin, have enacted legislation to promote the establishment of lake districts. The law provides for special units of government, called lake protection and rehabilitation

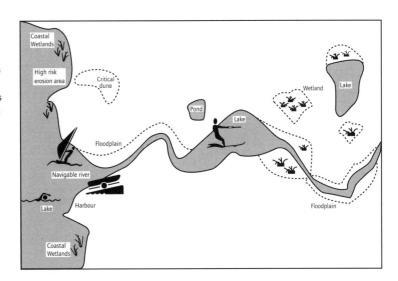

Fig. 19.2 Resource areas within a watershed. In general, wetland areas are protected by Section 404 of the Clean Water Act and other Wetlands Protection Acts, rivers and harbours by Section 10 of the U.S. Rivers and Harbors Act of 1899 and other river protection acts. Floodplains, shorelines, and inland lakes and ponds are also protected by statutes or ordinances. In addition, all resource areas are protected by erosion and sediment control legislation. (Phillips 2002: reproduced with permission of the Terrene Institute 1991.)

districts, to be formed to undertake lake management at the local level (UWEX/WDNR 1995).

The greatest strength and, simultaneously, the greatest weakness of these measures is local ordinance and zoning. Local support and involvement are important to effective lake management but local decisions are made mainly by combinations of elected, appointed and volunteer officials serving on land-use commissions for planning and zoning. The strength of the approach is that local people are solving their own problems for the good of the community. The weakness is that one vote is all that is needed to undo their good work. Several guidance documents are available (Dresen & Korth 1994; Terrene Institute 1995) which provide excellent examples of what needs to be in an ordinance to make it work. They cover such topics as: wetlands, wildlife corridors, buffer zones, on-site disposal systems, shoreline erosion and docks. In a survey of local units of government in Florida, Watkins et al. (1997) found that most local units were not using their authority to require buffer zones around lakes, neither were they applying density or commercial restrictions in the watersheds of their lakes. For any lake management framework to be effective, local support and leadership is needed; as noted earlier, it is the key to success. Public education is the foundation for local support and leadership.

## 19.5 REFERENCES

Davenport, T. (2002) *The Watershed Management Project Guide*. CRC Press–Lewis Publishers, Boca Raton, 312 pp.

Dresen, M.D. & Korth, R.M. (1994) *Life on the Edge... Owning Waterfront Property*. University of Wisconsin-Extension, Madison, 95 pp.

Holdren, C. (1997) NALMS looks at greater national focus, welcomes board members. *Lakeline*, **17** (2), North American Lake Management Society, Madison, 9 pp.

NALMS (1992). *Developing Eutrophication Standards for Lakes and Reservoirs*. Report to USEPA, the Lakes Standards Subcommittee of NALMS (North American Lake Management Society), Madison, 18 pp.

Phillips, N. (2002) *A Watershed Approach to Urban Runoff: Handbook for Decision Makers*. Terrene Institute, Alexandria, VA, 110 pp.

Terrene Institute (1995) *Local Ordinances: A User's Guide*. Terrene Institute, Alexandria, VA, 83 pp.

USEPA (1993) *A Commitment to Watershed Protection: a Review of the Clean Lakes Program.* USEPA-841-R-93–001, Office of Wetlands, Oceans and Watersheds, U.S. Environmental Protection Agency, Washington, DC, 52 pp.

USEPA (1995) *Source Water Protection: Protecting Drinking Water Across The Nation.* USEPA-813-F-95–005, Office of Water, U.S. Environmental Protection Agency, Washington, DC, 2 pp.

USEPA (2002). *National Water Quality Inventory.* Report. EPA841-R-02–001, Office of Water, U.S. Environmental Protection Agency, Washington, DC, 208 pp.

UWEX/WDNR (1995) *A Guide to Wisconsin's Lake Management Law* (9th edition). University of Wisconsin Extension and Wisconsin Department of Natural Resources, University of Wisconsin, Madison, 48 pp.

Watkins, C., Baird, R., Britt, M., McCann, K. & Medley, G. (1997) A survey of local government urban lake management practices in Florida (Defay). *Bulletin of the Florida Lake Management Society*, **1**, 1–6.

# 20 Nordic Lakes – Water Legislation with Respect to Lakes in Finland and Sweden

MARIANNE LINDSTRÖM

## 20.1 INTRODUCTION

Both Finland and Sweden possess large numbers of lakes. Close to 10% of Finland's total area consists of inland waters. There are 47 lakes over 100 km² and 56,000 lakes over 1 ha. The average lake area is 18 ha and the average depth is 7 m. Lakes are more numerous and larger in central and eastern Finland than in the southern and western parts. Finnish inland waters are, as a rule, complicated systems made up of lakes of various shapes and sizes, linked by rivers and brooks. The lakes are shallow and their water volumes are small (Wahlström et al. 1993). A dense network of lakes, rivers and wetlands makes Sweden similarly rich in water resources. Sweden contains about 80,000 lakes, which cover around 9% of the territory, and there are around 60,000 km of rivers. The buffering capacity of the lakes and that of the numerous wetlands (11% of the surface) guarantee an even supply of water, protecting the country from floods as well as droughts (OECD 1996).

Average annual precipitation and evapotranspiration in Finland amount, respectively, to 660 mm and 341 mm. Water abstraction in 1994 amounted to about 2.44 billion cubic metres, or 2.2% of the available resource. Industry consumed 67%, households around 17% and electricity production accounted for 15% of the surface water withdrawals. According to conventional criteria, water quality is good in most rivers and lakes. However, the humus content of water is naturally high and there are risks posed by acidification and eutrophication. Levels in about one-third of Finland's lakes have at some time been regulated, in order to meet human need (Wahlström et al. 1993; OECD 1997).

The annual precipitation in Sweden is 729 mm, the evapotranspiration is low at 128 mm yr$^{-1}$ and there is plenty of water in relation to the country's need. Demand for water amounts to about 3.6 billion cubic metres per year, or about 2% of available resources. Industry consumes 70%, household use about 15% and agriculture about 5% of the surface water withdrawals. Acidification from air pollutants affects many water bodies, particularly in the southern and southwestern areas. The quality of most inland waters is suitable for most purposes (OECD 1996).

Both Finland and Sweden have made major revisions to their environmental legislation in recent years (Table 20.1). International trends towards integrating legislation on preventing environmental pollution and the need to implement the environmental legislation (Table 20.1) of the European Union (EU) have contributed to this development.

In this review, I will give a short presentation of the development of water protection and development legislation in Finland and Sweden. I will also mention some of the problems these countries have experienced in implementing the EU Directives on environmental issues since joining the European Union. The impact of EU membership on the water protection laws of these Nordic countries will be discussed. In addition, I will point out some other current problems and discuss possible solutions.

## 20.2 THE EARLY DEVELOPMENT OF THE LEGISLATION

Sweden and Finland have possessed laws to govern the use of water since the thirteenth century. Finland was part of Sweden until 1809, which explains why the legislation is, to a degree, quite similar in the two countries. The numerous lakes

Table 20.1  Main European Union (EU) Directives and national legislation

| EU Directive | National legislation |
| --- | --- |
| Council Directive 76/464/EEC of 4 May 1976 on pollution caused by certain dangerous substances discharged into the aquatic environment of the Community | Finnish Water Act 19.5.1961/264 |
| Council Directive 85/337/EEC of 27 June 1985 on the assessment of the effects of certain public and private projects on the environment | Finnish Water Decree 6.4.1984/413 |
| Council Directive 91/271/EEC of 21 May 1991 concerning urban waste-water treatment | Finnish Environmental Protection Act, Ympäristönsuojelulaki 4.2.2000/86 |
| Council Directive 91/676/EEC of 12 December 1991 concerning protection of waters against pollution caused by nitrates from agricultural sources | Finnish Environmental Protection Decree, Ympäristönsuojeluasetus 18.2.2000/169 |
| Council Directive 96/61/EC of 24 September 1996 concerning integrated pollution prevention and control | Swedish Water Act, Vattenlagen (1983:291) |
| Council Directive 2000/60/EC of the of 23 October 2000 establishing a framework for Community action in the field of water policy | Swedish Environmental Code, Miljöbalken (1998:808) |

and rivers were important for transport, the floating of logs, fishery exploitation and, later, for the construction of mills. Under the old Germanic and Scandinavian laws, the ownership of land and the access to the use of waters led to private ownership of all watercourses. Today, this is still the basis of water rights in these countries. Legal protection of ownership has been exceptionally strong in Finland in comparison with the rest of Europe. Until 1997, Finland was unique in being the only country in Europe which lacked general prohibitions on building in coastal areas. However, from 1970, a land-use plan approved by the relevant authorities is normally needed for building for recreational purposes in shore areas (Vihervuori 1998).

The Water Rights Act in Finland was passed in 1902. It stipulated requirements on: the uses of waters in general; water supply plants; floating timber; land drainage by ditching; embankment protection; and water-level regulation. In the 1950s, a new water act was planned which covered issues of both construction and water-pollution control, because the Water Rights Act was no longer appropriate to regulate the demands of an industrialised society for water and hydropower. This new Water Act was adopted in 1961 (Hollo 1997).

The Swedish Water Act of 1918 was aimed at efficient exploitation of rivers and watercourses for hydropower plants. It was commonly felt that the Act favoured exploiters' interests at the cost of private and public interests. Originally, the Act regulated the conveyance of municipal wastewater in sewers but, after the reform of 1941, industrial wastewater was included within the scope of the Act. Both the Water Courts and the King (i.e. the Cabinet) were vested with competence to grant permits (Kuusiniemi 1997).

During the 1960s in Sweden, efforts to create legislation to regulate environmental pollution control led to the Environmental Protection Act of 1969, which covered all forms of environmental pollution, including air and water pollution, noise abatement and pollution from waste management. The Environmental Protection Act explicitly stipulated integrated pollution control. The Act applied to all environmentally hazardous activities, including the discharge of wastewater, solid matter and gas (Kuusiniemi 1997).

## 20.3  LEGISLATIVE POWERS

The legislative powers in both Finland and Sweden are vested in the national Parliament. In Finland, the Ministry of Justice prepared the legislation concerning water management and pollution protection at the governmental level. The legislation concerning environmental issues excluding waters has been part of the responsibility of the Ministry of the Environment, which was established in 1983. Water pollution control is now being connected to the legislation on environmental pollution protection in the Environmental Protection Act and thus the Ministry of the Environment will be charged with the main responsibility for the legislation concerning water pollution prevention. Other ministries also possess certain duties in the environmental field, related to topics such as fishery, chemicals and health (Lindström 1997a; Vihervuori 1998).

The Swedish Ministry of the Environment, which was established in 1987, formulates legislation and policy with respect to environmental protection, water conservation and pollution control. The legal environmental control in both countries is founded not only in parliamentary acts, but is also reflected in regulations at lower levels. In the legislative hierarchy, the Constitution is at the highest level, followed by the Acts issued by the Parliament. At lower levels, there are the Decrees and the agency enactments of central directives (Westerlund 1996; Kuusiniemi 1997).

### 20.3.1 Structure of the present environmental legislation

Until March 2000, Finnish environmental legislation was based on a number of individual acts, secondary regulations and instructions. Finland possesses a few general laws dealing with nature or the environment as a whole, e.g. the Environmental Impact Assessment Act (1994), the Environmental Damages Act (1994) and the rules on environmental crimes in the Penal Code (1995). If damage is caused through the illegal use of water, the general Damage Act is applied. Air and soil pollution, noise abatement and waste management were regulated by a set of sector-wise and general laws. This legislation was not integrated with water pollution control until year 2000.

It is required by the EC Directive on Integrated Pollution Prevention and Control (1996) that permitted procedures and permitted decisions concerning major industries and their emissions to air, water and soil are effectively integrated. Partly as a result of this directive and partly for domestic reasons, there has been intensive work to reform the legislation controlling pollution. The new Finnish Environment came into force in March 2000.

The main objectives of this Act are: to prevent the pollution of the environment; to safeguard a healthy, pleasant ecologically diverse and sustainable environment; to prevent the generation and the harmful effects of waste; to improve and integrate assessment of the impact of activities which pollute the environment; to improve citizens´ opportunities to influence decisions concerning the environment; to promote sustainable use of natural resources; and to combat climate change and otherwise support sustainable development.

The Swedish system of environmental law prior to 1999 was fragmented but not to the same extent as the Finnish system. Many laws not only overlapped but were, in some cases, also contradictory. Standards with regard to protection and compensation differed and some aspects were not regulated anywhere. The Natural Resources Act (1987) included general provisions on the use of resources, including water. More detailed measures concerning water protection were based on the Water Act (1983) and discharges to water were covered by the Environmental Protection Act (1969). Central parts of the environmental law, including the Environmental Protection Act, the Water Act and the Nature Conservation Act of 1964, are believed to have originated primarily from private law (Westerlund 1996).

Sweden has developed a new Environmental Code, which came into force in January 1999. The code amalgamates rules drawn from 15 existing Acts, among which are the Water Act, the Environmental Protection Act, the Natural Resources Act, the Dumping of Waste in Water (Prohibition) Act and the Environmental Damage Act. All new or extended development activities with an environmental impact will be examined within the terms of this code. The Code represents a major piece of legislation: it contains 33 chapters comprising almost 500 sections. However, it is only the fundamental rules which are included in the

Environmental Code. More detailed provisions will be laid down in decrees or Government decisions (Hepola 1997; Rubenson 1998).

### 20.3.2 The Water Acts and their development

The Finnish Water Act, which is a very complex and detailed code, is in many respects influenced by the fact that the water areas are privately owned (Vihervuori 1998). The Water Act contains 22 chapters. Its first chapter sets out the basic concepts and principles for the application of the Act including the definition of what can be done without a permit. The second chapter contains the general rules and principles for water management and chapters 3–9 contain provisions for particular categories of water use. Chapter 10 is concerned with water discharges and chapter 11 stipulates the criteria for compensation in cases of permitted activities and compulsory rights which enactment might compromise. The remaining chapters relate mainly to procedures (Hollo 1997).

The Finnish Water Act introduced two principal restrictions: from 2000, the closing off or altering of any water body is banned. The notion, 'body of water', applies to lakes, rivers, territorial seas and groundwaters. Largely following international conventions, Finland has developed additional legal provisions for the protection of coastal waters, of fisheries and for the conservation of natural wetlands and other natural water bodies (Vihervuori 2002).

The overall objective of the new Finnish Environment Protection Act is to prevent pollution of the environment and to improve and integrate assessment of the impact of activities leading to pollution. Most of the provisions on water-pollution prevention in the Water Act were transferred to the new act. The Government may now, by decree, introduce norms on quality or stipulations on monitoring and observation of the environment. The transposition of the EU Water Framework Directive will influence the monitoring programmes from a physical and chemical monitoring approach towards one that is more ecological.

The Swedish Water Act of 1983 also contained 22 chapters. In the first chapter, the principles governing the use and protection of waters and the application of the Act were set out. The second chapter contained the stipulations on water rights and the third one the permissibility and principles for management. Special stipulations for various resource management projects were described in the next four chapters, and the remainder dealt with the procedures (Strömberg 1984).

This Act was still based upon the old approach under which water areas were to be developed and wetland to be drained. There had been a presumption in the law in favour of development and this meant that nature conservation and environmental protection would influence decisions only if there was substantial evidence that a particular development would lead to the destruction of important environmental interests. In spite of that, the general principles of the Act paid more attention than before to these interests. The principles were rather vague and left a great deal of discretion to the permit authorities, the Water Courts. This legislation contained no water quality standards (Westerlund 1996; Kuusiniemi 1997).

In 1999, the Water Act was amalgamated along with several other acts, into the Swedish Environmental Code. All new or extended development activities with a potential environmental impact are to be examined under the Environmental Code. The examinations may be carried out by local authorities, by the county administrative boards, the environmental courts or by the Government. The Code includes a list of geographical areas which are considered to be of national interest for various purposes. The protective principle is absolute in these instances. For example, Vänern is one such area which is thus protected. An important new provision in the new Environmental Code is the possibility of introducing standards of environmental quality (Rubenson 1998).

## 20.4 ADMINISTRATION

### 20.4.1 Administration at the central level

In Finland, the State administration of environmental matters was somewhat fragmented prior to the 1980s. This has often been used as an explanation for the slow progress in the development of the environmental legislation. After a lively political debate, the Ministry of the Environment was created in 1983. The environmental administration is now headed by the Ministry of the Environment and partly also by the Ministry of Agriculture and Forestry. The latter is responsible for the development of water systems, flood control, land drainage, water supply and sewerage, as well as for operations and maintenance linked to these activities (Lindström 1997a; Vihervuori 1998).

A general trend in the development of Finnish administration since the beginning of the 1990s has been to concentrate the politically relevant management tasks to ministries, and the traditionally strong National Boards have been transformed into expert bodies. At the same time the practical administration has been delegated to provincial authorities and municipalities.

The Swedish Ministry of the Environment deals with most of the issues with environmental implications. Environmental protection is, in general, supervised by the Swedish National Environmental Protection Agency. This agency advises, for example, the regional agencies (County Administrations) and local environmental agencies (Westerlund 1996).

### 20.4.2 Administration at regional and local levels

In 1997, Finland was divided into five provinces. These were not allocated environmental legislation. Instead, these responsibilities were passed to the regional environmental administration, consisting of the 13 regional Environmental Centres which had been created in 1995 from the former Finnish Water and Environment Districts and the Environmental Subdivisions of the Provincial Offices. The regional Environmental Centres now act as regulatory and supervisory authorities in most environmental matters. From March 2000, the Environmental Permit Authorities act as permit authorities, having taken over this task from the Water Courts. At the local level, municipal Environmental Boards take decisions in minor water and other environmental cases.

In Sweden, there are five regional Environmental Courts, which are mainly based on the former Water Courts as they existed until 1999. The Environmental Courts act as permitting or appellate agencies but not as supervisory authorities. Sweden is divided into 24 counties, their head agencies being the county administrations. These agencies play a very important role in more direct environmental control. The County Administrative Boards grant permits under the Environmental Code and they also supervise, inspect and enforce permits issued by themselves or by the Environmental Courts.

At the local level, Finland comprises 455 and Sweden 285 municipalities which have been granted increasing powers to legislate on environmental matters. This could prove to be wrong as regards environmental protection in general, because most impacts cross municipal borders. In addition, some of the polluting activities are owned by the municipalities themselves (Westerlund 1996), who, in both Finland and Sweden, play the primary role in land-use planning.

## 20.5 PERMIT AUTHORITIES

Since March 2000, the Finnish permit authorities are the three Environmental Permit Authorities, the 13 regional environment centres and the municipality environmental authorities. The chairman of the Environmental Permit Authority is a judge and the members are experts in technology or ecology. The presenters of a case are often lawyers. The permit authorities deal *ex officio* with compensation issues as an integrated part of the permit procedure. The first appellate instance is the Vasa Administrative Court. An appeal against a decision of this Court may be made to the Supreme Administrative Court. Matters of Private law and criminal law in Finland are handled through the general jurisdiction, which consists of local courts, regional courts and the Supreme Court (Vihervuori 2002).

In Sweden, the former Water Courts were reorganised as Environmental Courts in 1999. These now deal with both water and other environmental issues. Ordinary permit authorities under the Environmental Code are the Environmental Courts, the County Administrative Boards and the environmental authorities in the municipalities. The Environmental Courts are staffed by personnel consisting of lawyers and experts in different fields, in technical or environmental matters. The competence of the Environmental Courts, the County Administrative Boards and municipal authorities are divided roughly on the basis of the environmental significance of the installation, so that the activities with the greatest impact on the environment are handled in the courts. In Sweden appeals against the decisions of the municipal authorities are lodged in the County Administrative Boards and appeals against other decisions are lodged in the Supreme Environmental Court.

### 20.5.1 The Finnish permit system

One of the most important instruments in environmental legislation is the permit system. In Finland the permit procedure according to the new Environment Protection Act does not differ much from that under the former Water Act. One difference is that the procedure is an administrative one and that there is a list of those polluting activities which need a permit. Permit applications are submitted to the competent permit authority. The application documents include detailed information on the polluting activities and possible impacts on the environment and on various interests, both public and private. The documents are made public by public notification and parties especially concerned by the matter are separately notified. The legislation is based on the 'polluter pays' principle in the sense that the polluter pays all pollution abatement costs. A typical feature of the licensing procedure in Finland is the case-by-case deliberation of applicants and the tailoring of permit conditions. Thus environmental, technical and economic aspects can be taken into account in the choice of solutions for each case (Hollo 1997; Lindström 1997a).

In granting a permit for the discharge of wastewater, the competent authority issues the necessary discharge regulations on the quantity and the composition of the wastewater. It also assumes the duty to take measures to build protective and treatment facilities at the sites of emission. The permit also includes regulations on how to monitor the formation, treatment, discharge and effects of wastewaters. In granting a permit for conducting water for use as a liquid or for abstracting groundwater, the permit shall also include regulations on how the activity should be operated. General regulations or guidelines on maximal concentrations of different emissions or on the state of the aquatic environment are missing (Kuusiniemi 1995). Those concerned possess the right to appeal, and a review of permit conditions can be made (Hollo 1997; Lindström 1997a).

Under the present Finnish Water Act, the activities covered by the term 'construction' in the context of water use is very wide. It may refer to the building of hydroelectric power plants, dams, quays, jetties, waterways, timber floating facilities, cables, water intake pipes, sewage outlets, embankments, bridges, road banks, reservoirs and fish farms, as well as to the damming up or impounding of water, lowering water surface, filling of watered areas with earth, extraction of sand from the seabed, dredging and clearing of rapids. The drainage projects are divided into two legal categories: ditching, which normally is free of permit control, and alteration or lowering of the water level, which normally needs a permit of the Environmental Permit Authority.

There is no list setting out the types of water utilisation which require a permit. The rules specifying where and when a permit is required are complicated. The necessity of a permit may depend on the impact on the water bodies or the water use. In practise, the rules elaborated by the former Water Court are sufficiently clear for operators to be able to determine in advance whether a permit will be needed. The system of the Finnish Water Act is based on restrictions on closing off or altering a body of water. These bans are not exactly absolute prohibitions, but rather they may be described as provisions for determining the eventual necessity of a permit case by case. The permit judgement rules are somewhat complicated. The permits for construction projects normally include a number of various technical provisions and also provisions by which the harmful impacts on public and private interests are reduced or eliminated. Any damage, harm or loss caused by the measures must be compensated for (Hollo 1997).

The ban on altering water bodies is the most common judicial reason for the necessity of a construction permit. If a permit is granted, the activities are no longer forbidden and the damages are assessed *ex officio* by the Water Court. Compensations for latent damages can be claimed later on. Normally the permit consideration depends on a weighing of interests; the benefits from a project need to be significant compared with the various damages, inconveniences and loss of benefits it may cause. As an alternative to the weighing of interests, and irrespective of its outcome, a permit may also be granted when the project is considered necessary for the public good, e.g. public water supply interests, interests of public recreation, waterway interests and public energy supply interests (Vihervuori 1998).

Damages caused by illegal pollution, including permit violations, are in Finland regulated by the Act on Compensation for Environmental Damage (1994) and, if damage is caused by other illegal water use, by the general Damages Act (Vihervuori 2002).

### 20.5.2 The Swedish permit system

The Swedish Environmental Code requires the issue of a permit before certain activities affecting the aquatic environment may be started, altered or modified. The party exercising the activity is liable to prove that the general rules of the Environmental Code are complied with. Among these general rules are the application of the precautionary principle, best available technique, the localisation principle, the resource management and ecocycle principles (Westerlund 1999).

The Environmental Code sets the conditions requiring the application for a permit. Section 9 specifies the provisions on environmentally hazardous activities requiring health protection. 'Environmentally hazardous activity' means all uses of land, buildings or fixed installations which involve emissions to land, atmosphere or water. The same applies to such uses which risk nuisance to human health or to the environment (Westerlund 1999).

In certain parts of Sweden, the Government may issue regulations or prohibitions against the emission of, for example, wastewater. This applies if the activity is likely to result in an area of water becoming polluted or infected. For instance, the provisions empower the prohibition of emissions to a lake which is important to the supply of drinking water or which contains rare or particularly valuable species of fauna and flora (Rubenson 1998).

The Government may also issue rules concerning prohibitions, protective measures, limitations and other precautionary measures. The powers granted will be used particularly to integrate EC legislation in Swedish law and satisfy other international obligations.

The environmentally hazardous activities require a permit either from the Environmental Court (A list) or the county administrative board (B list). The C list includes environmentally hazardous activities which are subject to a duty to give a notification. Such notification must be given to the county administrative board or to the municipality (Westerlund 1999).

An important new provision in the Environmental Code is the possibility to introduce environmental quality norms. Regulations may be issued for the whole of Sweden or for certain geographical areas on the quality of land, water, air or the environment in general (Chapter 5 of the Code) (Westerlund 1999).

For water management projects, a permit is required from an environmental court (Chapter 11 in the Code). The main rule is that a permit is always needed for a water undertaking. The exemptions apply if it is clear that neither public nor private interest is harmed by the effects of the water undertaking on the water conditions. Nor is a permit needed for specially listed water undertakings, for example, wells for single or double family dwellings (Rubenson 1998). In this context, 'water undertaking' refers to a number of various measures involving water. This list has not been changed in the Code. Examples of water undertakings include:

**1** construction, alteration, storage of water; demolition of dams and other installations in watercourses, lakes or other water areas; filling or piling works in water areas; extraction of water; excavation, exploiting or dredging in water areas; or other measures in water areas having the intention to alter the depth or the location of water;
**2** extraction for groundwater and making installations for this purpose;
**3** harnessing of water to increase the quantity of groundwater and undertaking installations and measures for this purpose;
**4** measures to drain land, to lower a lake, or to protect land against water, provided that the aim of the measures is to permanently improve the suitability of certain property for particular uses (Strömberg 1984; Kuusiniemi 1997).

A water undertaking may be conducted only if the advantages from the general and individual viewpoint exceed the expense and damage caused by the undertaking.

## 20.6 ENFORCEMENT, CONTROL AND MONITORING

Enforcement and control is based on monitoring. An environmental monitoring system is an important instrument for following up environmental work with regard to stated goals. The present state of the environment is the decisive starting point for all proposed environmental measures and political decisions.

In Finland, the water monitoring system consists of nationwide and regional monitoring carried out by authorities, as well as mandatory monitoring carried out at the polluter's expense by an independent laboratory or the laboratory of the polluting enterprise. At the national level, the water-quality monitoring programme applies to all lake basins greater than 100 km$^2$ in area and to some locally important lakes as well (Antikainen et al. 1996). The regional environment centres are responsible for ensuring that permit conditions are met and that the monitoring is properly carried out. This is checked out four to six times a year (Lindström 1997b).

The Swedish Environmental Protection Agency (SNV) started a National Environment Monitoring Programme in 1978, which was reviewed in 1990. Its main purpose was to monitor long-term and large-scale changes in the environment and thus identify problems which might call for more research or the adoption of some countermeasures. The monitoring programme was also to collect environmental data from areas relatively unaffected by pollution and to determine the transport pathways within and among air, soil and water. Environmental monitoring of lakes is also organised regionally in the form of Regionally Integrated Water Quality Programmes. Surveys of the quality of the waters in Swedish lakes and rivers are carried out every fifth year (Fejes 1995).

In many places water quality is also monitored by the so-called water pollution protection societies, which are cooperatives of licence holders discharging into a given water body. It is not practically possible, especially owing to the lack of administrative resources, for authorities alone to exercise supervision and control over environmentally hazardous activities. In both Finland and Sweden operators themselves partly measure and submit reports to the authorities. This self-control is an essential part of enforcement of the regulations and of the permit provisions. Every permit is supplemented by an individual control programme, which is exercised by the operator but with periodical reports to and inspections by the supervising authority (Westerlund 1996; Lindström 1997b).

## 20.7 WATER QUALITY POLICY

In the 1970s and 1980s, environmental issues were at the forefront of the political agenda in both Finland and Sweden. Environmental policies during those years were distinguished by an almost 'case-by-case' approach, involving, with strong public support, the introduction of cleaner industrial technologies. Sweden took the initiative in hosting the first UN Conference for the Environment in Stockholm in 1972, with the aims of identifying the most urgent environmental problems and obtaining agreements on the actions needed to deal with them. In the latter part of the 1980s, Finland and Sweden also became strongly involved in global issues (OECD 1996, 1997).

In both Finland and Sweden, the promotion of sustainable development has been comprehensively adopted as the goal of broad co-operation among government, the private sector, interest groups and NGOs, the scientific community, the education system and the media (Swedish Ministry of the Environment and Natural Resources 1994; Finnish National Commission on Sustainable Development 1995). In developing environmental measures, the countries often refer to the principles of the Rio Declaration and Agenda 21, the Framework Convention on Climate Change and the Convention on Biological Diversity. How these principles are going to be implemented in practise is not yet altogether clear.

The Finnish national objectives for decreasing industrial and municipal discharges into waters as well as for reducing nutrient losses from agriculture has been defined in the Decision of the Council of State on the Water Protection Programme, operative until 2005. According to this programme, the purpose of water protection in Finland is to improve the state of water bodies and to remedy harmful effects — specifically:
• the goal of industrial water protection is to reduce emissions and to minimise the harmful impact of products during their whole life cycle;
• the load from municipal sewage will be reduced in areas in which it constitutes a problem;
• new techniques will be developed for tackling problems of water protection caused by fish farming;
• the objective in water protection related to arable farming and livestock breeding is to reduce the load of surface water and groundwater enough to halt deterioration in water quality and to improve the state of polluted waters;
• environmentally sound forest management will restrict the harmful effects on water caused by forest drainage, felling, tillage and fertilisation.
The objectives are to reduce load and adverse change and to conserve biodiversity (Lindström 1997c).

Most of Sweden's pollution reduction objectives are derived from agreements concerning the Baltic and North Seas, rather than national objectives aimed at improving inland water quality. There are no formal objectives for water quality in lakes and rivers. The Parliament has in 1990 approved targets relating, *inter alia*, to discharges of nitrogen, mercury, cadmium and lead. Under the eutrophication theme, a few objectives refer specifically to water:
• measures should be taken to ensure the survival of vigorous, balanced populations of naturally occurring species in marine and aquatic areas;
• pollution shall not limit the use of water, either from lakes and watercourses or from groundwater sources;
• waterborne emissions of nitrogen from human activities shall be halved between 1985 and 1995 (Swedish Ministry of the Environment and Natural Resources 1994).

Sweden's performance in terms of municipal sewage treatment is among the best in the OECD and Swedish industry has significantly reduced its pollutant discharges. However, some problems still occur, among which are: acidification; nitrogen leaching from agricultural land; appropriate use and disposal of sewage sludge; and the high levels of mercury in pike in Swedish lakes. Acidification is one of the most serious problems in Sweden together with coastal and Baltic Sea pollution. The Environmental Protection Act, together with the special Sulphur Act, has provided useful instruments in diminishing the discharge of sulphur into the atmosphere. However, the framework of the act provides for yet more stringent standards to be applied (OECD 1996; Westerlund 1996).

In the special legislation on agriculture, stricter standards have been introduced aiming at reducing nitrate run-off. So far, this has not resulted in any considerable improvement of water quality. There is reason to assume that measures at the sources will not be sufficient but that additional hydrological measures (recreating wetland, etc.) are necessary to increase the denitrification of water before it reaches the sea (Westerlund 1996).

## 20.8 IMPLEMENTING EC DIRECTIVES

Finland has been active in incorporating EC environmental legislation into its legal system and the delays in implementing directives are few (OECD 1997). The directives on industrial and municipal wastewaters (91/271/EEC), groundwater and waters used for drinking (80/778/EEC) have been implemented in Finland by decisions of the Council of State based on the Water Act. The new legislation came into force in June 1994. The implementation of the directive on the quality of surface waters used as drinking water has resulted in an extended inventory of raw water quality and the revision of monitoring programmes. Many of the analyses required by the directive are considered as outdated. The Environmental Impact Assessment (EIA) directive (85/337/EEC) has been fully implemented in Finland (Silvo 1997).

The Dangerous Substances directive (76/464/EEC) and its daughter directives have been implemented by introducing a ban against discharging certain harmful substances into waters or sewers. The directive on Urban Wastewater Treatment (91/271/EEC) sets minimum requirements for the treatment of urban wastewater, monitoring wastewater and the pretreatment of industrial wastewater discharged into public sewers. The implementation of the directive has not caused major difficulties. Some treatment plants are required to be complemented with a biological unit, and nitrogen removal must be improved in major plants along the southern coast of Finland. Protection of surface and groundwater from pollution by nitrates from agricultural sources is regulated by the directive (91/676/EEC). The Nitrates directive was implemented in 1999 but it requires improved control of nutrient loadings from agriculture, forestry and fish farming. The Fish Water directive (78/659/EEC) is now also implemented in the Water Act. The Directive on Integrated Pollution Prevention and Control (96/61/EC) has led to a major reform of the wastewater permit system aiming at a single environmental permit procedure for industrial plants and other polluting activities. The principles of applying the best available techniques have already been incorporated into the Water Act, the Air Pollution Control Act and the Waste Act. Membership of the European Union sped up legislation on the environmental impact assessment procedure. Finland has been exempted from implementing the shellfish water directive (72/923/EEC) owing to natural conditions (Silvo 1997).

Sweden has also been active in complying with EC directives. The directive on surface water for drinking purposes has only resulted in an agency regulation prohibiting the use of surface water where the quality standards are not met, but there is no legal requirement to improve the quality where it falls below standards. The fish water directive or the shellfish water directive are not reflected in Swedish law, neither is the EIA directive. The so-called Seveso directive is not implemented. The Swedish Environmental Protection Act explicitly requires integrated pollution control. Therefore the EC IPPC Directive does not require any major structural changes in the Swedish legislation on pollution control (Westerlund 1996).

The water legislation of the European Union aims particularly at curbing emissions from industry and municipalities as well as safeguarding the quality of drinking and bathing waters. Legislation has also been proposed for reducing nitrate discharge from agriculture. There are a number of activities affecting the status of waters, which are not presently within the scope of EC water legislation, for example, water construction, flood control, water abstraction and such sectors important in Finland as forestry, peat extraction and fish farming. The renewal of the EU water policy and legislation started in the European Commission and in the Council in 1995. The new water policy framework directive will replace several existing water directives and form the basis for general water protection and management within the EU (Silvo 1997).

## 20.9 CURRENT DEVELOPMENTS

In general, the Finnish Water Act has proved successful in securing the use and protection of natural waters. Substantial progress has been made in reducing discharges from industry and municipalities. In industry, especially the pulp and paper industry, process changes and pollution control have led to large reductions in discharges of phosphorus, biochemical oxygen demand (BOD), organochlorines and heavy metals. Municipal sewage collection and treatment have also improved over the past 10 years, bringing reductions in the discharges of organic matter and phosphorus (Lindström 1997c; OECD 1997). How to deal with pollution from farm land seems to be the outstanding challenge for the near future, even though the Nitrates directive is in force.

The Swedish legal control of waters, especially of lakes, has been effective for the major point sources. All major industries possess treatment equipment, often of a fairly high standard. When applications for expansion are made, some of the equipment will be exchanged for even newer and more efficient equipment. Pollution from farm land and to some extent from forest land seems to pose very serious problems in Sweden, threatening not just surface water but groundwater as well. This is not due to a lack of relevant rules but to lack of implementation and enforcement (OECD 1996; Westerlund 1996).

Both in Finland and Sweden the case-by-case approach to licensing discharges will be complemented by a greater emphasis on setting quality objectives for water and aquatic ecosystems. In Finland, the EU directive on integrated pollution prevention and control (IPPC Directive 1996) is being implemented in March 2000.

The purpose of the Swedish Environmental Code is not just to provide a synthesis of existing environmental laws, but also to present a uniform approach to environmental questions. The purpose of the Code is: to guarantee existing and future generations a healthy, pleasant environment to live in; to preserve biological diversity and maintain good living conditions for all plant and animal species occurring naturally in Sweden and its surrounding waters; to preserve a living landscape and to make it possible for society to develop in a sustainable manner in the long term. The Code covers three closely interconnected fields, i.e., protection, care and improvement, and management (Westerlund 1996).

The fundamental principles by which all environmental work should be governed are defined in the Code. The focus is firstly at preventing damage to the environment in the first place. The very risk of damage is taken into account (the precautionary principle, the principle of prevention, the principle of best available technology and the substitution principle). That which is extracted from nature will be used and finally disposed of without any damage to nature (the cyclical principle). A person potentially or actually causing damage or nuisance shall pay the cost of its prevention or rectification (the 'polluter pays' principle). The Code is, to a large extent, a framework legislation and the authorities are empowered to issue more detailed provisions necessary to achieve the aims of the Code (Westerlund 1996).

Both in Finland and Sweden the legal control of lakes has been very effective for the major point sources, whereas it appears that measures should be taken to reduce the nitrogen leaching from agricultural land (SEPA 1997). The case-by-case approach to licensing discharges should in the near future be complemented by a greater emphasis on setting quality objectives for water and aquatic ecosystems. In both countries, owing to international trends, major changes of the environmental legislation have been enacted recently and the system of permits has been renewed.

## 20.10 REFERENCES

Antikainen, S., Puupponen, M., Vuoristo, H. & Seuna, P. (1996) Country paper of Finland. Surface monitoring networks in Finland. In: Federal Institute of Hydrology (ed.), *Optimizing Freshwater Data Monitoring Networks including Links with Modelling.* EurAqua, Koblenz, 51–9.

Fejes, J. (1995) Country Paper of Sweden. Freshwater monitoring in Sweden. In: Federal Institute of Hydrology (ed.), *Optimizing Freshwater Data Monitoring Networks including Links with Modelling.* EurAqua, Koblenz, 191–7.

Finnish National Commission on Sustainable Development (1995) Objectives and means. In: M. Wilkki (ed.),

Finnish Action for Sustainable Development. Ministry of the Environment, Forssa, 11–15.

Hepola, M. (1997) Piirteitä Ruotsin ympäristökaarihankkeesta (Features in the Swedish proposal for an environmental code). In: Ympäristöjuridiikka 2–3/1997. Suomen Ympäristöoikeustieteen Seura r.y., Helsinki, 4–13.

Hollo, E. (1997) Legal and institutional solutions in Finland. In: Lindström, M. (ed.), Water Legislation in Selected Countries. A Comparative Study made for the South-African Water Law Review. Finnish Environment Institute, Helsinki, 68–102.

Kuusiniemi, K. (1995) Environmental standards and pollution control law. In: Hollo, E.J. & Marttinen, K. (eds), North European Environmental Law. Hakapaino Oy, Helsinki, 239–61.

Kuusiniemi, K. (1997) Legal and institutional solutions in Sweden In: Lindström, M. (ed.), Water Legislation in Selected Countries. A Comparative Study made for the South-African Water Law Review. Finnish Environment Institute, Helsinki, 142–60.

Lindström, M. (1997a) Administration and Legislation of Water Issues. In: Ministry of the Environment (ed.), Environmental Policies in Finland. Ministry of the Environment, Helsinki, 28–30.

Lindström, M. (1997b) Monitoring and assessment in Finland. In: Lindström, M. (ed.), Water Legislation in Selected Countries. A Comparative Study made for the South-African Water Law Review. Finnish Environment Institute, Helsinki, 269–78.

Lindström, M. (1997c) Water quality policy in Finland. In: Lindström, M. (ed.), Water Legislation in Selected Countries. A Comparative Study made for the South-African Water Law Review. Finnish Environment Institute, Helsinki, 235–53.

OECD (1996) Environmental Performance Reviews; Sweden. Organization for Economic Co-operation and Development, Paris, 17–19, 33–46.

OECD (1997) Environmental Performance Reviews; Finland, Organization for Economic Co-operation and Development, Paris, 45–61.

Rubenson, S. (1998) Miljöbalken; Den nya miljörätten. Norstedts Tryckeri AB, Stockholm, 96–107.

SEPA (1997) Progress so far—a review taken by the Swedish authorities to achieve nine environmental objectives In: Swedish EPA Report 4701. Swedish Environmental Protection Agency, Stockholm, 33–8.

Silvo, K. (1997) The effects of EU-membership on Finnish water management. In: Ministry of the Environment (ed.), Environmental Policies in Finland. Ministry of the Environment, Helsinki, 23–4.

Strömberg, R. (1984) Vattenlagen med kommentarer. LiberFörlag, Stockholm, 16–18.

Swedish Ministry of the Environment and Natural Resources (1994) Towards Sustainable Development in Sweden. Implementation of the Resolutions of the United Nations Conference on Environment and Development. Ministry of the Environment and Natural Resources, Stockholm, 55–63.

Vihervuori, P. (1998) Environmental law in Finland. In: Blanpain, R., Boes, M. & Billiet, C. (eds), International Encyclopaedia of Laws, Environmental Law. Kluwer Law International, The Hague, 28–43, 72–103, 173–85.

Vihervuori, P. (2002) Public environmental law in Finland. In: Seerden, R., Heldeweg, M. & Deketelaire, K. (eds), Public Environmental Law in the European Union and the UnitedStates. A Comparative Analysis. Kluwer Law International, The Hague, 127–70.

Wahlström, E., Hallanaro, E.-L. & Reinikainen, T. (1993) The State of the Finnish Environment. Environment Data Centre and Ministry of the Environment, Forssa, 119–23.

Westerlund, S. (1996) Public environmental law in Sweden. In: Seerden, R. & Heldeweg, M. (eds), Comparative Environmental Law in Europe. MAKLU Uitgevers, Antwerpen, 367–93.

Westerlund, S. (1999) Miljöbalken med kommentarer. In: Institutet för miljörätt i Uppsala (ed.), Miljörättslig Tidskrift 1999:1. Åmyra Förlag, Uppsala, 164–81.

# 21 The Problem of Rehabilitating Lakes and Wetlands in Developing Countries: the Case Example of East Africa

F.W.B. BUGENYI

## 21.1 INTRODUCTION AND BACKGROUND

It is generally accepted that the future well being of the human population, especially in Africa, will depend upon a much wiser balance between exploitation and conservation of all natural resources, in order to achieve appropriate sustainable production systems which avoid environmental degradation. Water, in lakes, wetlands, rivers and reservoirs, is one of the most precious of all resources. Most of the water on the planet is unavailable to human consumption. More than 97% is saline, making up the oceans. Of the world's inland waters, 69.9% is frozen and most of the rest is stored underground in natural reservoirs which are replenished only very slowly. The balance of available freshwater amounts to just 0.26% of the total (UNEP 1994), hence the great importance attached to this increasingly scarce resource. Water is an absolute necessity for humans and for all plants and animals on Earth, on which humans depend. Freshwater provides us with food (as fish, which provides up to one-quarter of the protein intake in developing countries; UNEP 1994), transport facilities and energy (as hydropower). It is crucial to domestic, industrial and agricultural processes. Water acts as a hope and a habitat for significant elements of the biological diversity (one-third of all fish species, for example, live in lakes, reservoirs, rivers and wetlands). Lakes and reservoirs play an important role in flood control and they are also important culturally and aesthetically. Besides, an adequate water supply is one of the factors necessary for economic development, so playing an important role in development of society.

Wetlands also furnish a wide range of ecological and hydrological services, ranging from flood control to stormwater transport and water purification. Wetlands in general and wetland ecotones (Holland 1988) in particular have played a significant role in the development and sustenance of human populations. Today these systems deliver many practical services, including crucial habitat-control of nutrients (and pollutants) and water flow, as well as being sources of the predators which control agricultural pests. For decades, ecologists and wildlife managers, geographers and other scientists, and land and water managers have all been interested in the spatial changes among biological communities. More recently, their attention has widened to the ecological processes which apply across landscapes but which are mediated through the transitional wetland ecotone zones (Holland 1988). These have been regarded as integral components of both the landscape and the water (lake, river and reservoir).

Because water and its resources are so important to life, people have always chosen, whenever they could, to live reasonably close to it. This is the very reason why so many of our lakes and rivers are now dangerously threatened by many forms of pollution and ecological perturbations, which compromise their future as useful resources, and why the well being of the African Great Lakes is now so threatened. Increasing pressures come *from* population growth, catchment development, overfishing and the introduction of alien species. It is well known that the impact of the introduced Nile perch (*Lates niloticus*) on the fish population of Lakes Victoria and Kyoga, for example (Ogutu-Ohwayo 1985; Ogutu-Ohwayo *et al.*, 1996), as well as the invasion of floating weed, such as water hyacinth (*Eichornia crassipes*; Twongo 1992), have been devastating. Some exotic species have been introduced with good intentions. Nile perch was introduced, it is said, to provide table fish, in replacement of non-table fish, the native haplochromines. Others, such as the waterweed, were introduced inadvertently from cultivation for its beautiful flower and the ease of growing it in home aquaria. The major impacts of its release have arisen through the disruption of traditional community socio-economic structures and further environmental degradation.

The East African region, with its many freshwater systems (Fig. 21.1), is rich in aquatic resources, particularly fish (Lowe-McConnell 1996). There is an impressive array of faunal and floral diversity, and the fisheries production is remarkable. Human activities, however, including misguided agricultural practices, urbanisation, domestic and industrial waste disposal, have affected water quality, biological diversity and fish stocks. Environmental degradation and water-resource depletion have been increasing in the region in the past decades. Wetlands (many of which are fish habitats) are rapidly being degraded in many parts of the world by industrial, urban and agricultural pollution, landfill, and the damming and diversion of rivers. Clearance of wetland vegetation, sedimentation and the exploration and extraction of minerals and soils all exert further tolls on the wetland resources (Chapman *et al.* 2001).

With regard to environmental damage for fisheries, only some states or countries in Africa seem sufficiently aware of the problems and are attempting to solve them. In the case of the fisheries, where damaging fishing practices persist, there are often considerable difficulties in enforcing regulations and laws and fishermen are reluctant to change their practices.

Several major problems have emerged to confront policy makers, as the complexity of management has become increasingly understood. They often include a lack of adequate information relating to key biological and physico-chemical data, such as the assimilative capacity of the resources to economic parameters. These are compounded by uncertainties about the economic valuation of prospective fisheries against the loss of biological diversity, often because there is a lack of suitable expertise to deal with such problems (FAO 1997).

The increasing globalisation of the world economy, the ever-continuing data gathering and the attendant information explosion all affect the future of world water resources. As a consequence, those responsible for formulating and implementing national policies in the fisheries and other water resources sectors find that the nature and scope of their task are changing. Mainstream environmental concerns are especially challenging in developing countries. Many of them, however, have embarked upon the 'National Environmental Action Plan' (NEAP) strategies in order to try and address environmental concerns (World Bank 1996; see also Fig. 21.2).

Fisheries have played an important role in many tropical or sub-Saharan African countries as a major contributor to animal protein supplies, a foreign exchange earner and a generator of rural employment. An estimated eight million people are directly employed in the sector (FAO 1997). Total production by the countries in the region amounted to 3.9 million tonnes in 1994. Fish consumption did decline recently from an average per capita supply of about 9 kg per person per year in 1990 to less than 7 kg per person per year in 1994 (live weight equivalent). The overall trade balance of the sub-Saharan region has been positive (in value terms) for the past decade, even though the region plays only a marginal role in international trade.

Since the mid-1980s, many countries have sought to develop medium-term, sectorally tailored systems in support of governmental macro-economic policies but these plans have often not come to fruition. The main obstacles to good fisheries management planning in the region are low government budgets, the weak institutional base and the lack of political will to implement management policies and measures. Structural adjustments and related budget cuts have also allowed fewer subsidies to be granted (if at all) than previously.

Awareness of harm to the aquatic environment from human activities has led to the political and legislative authorities of industrially developed

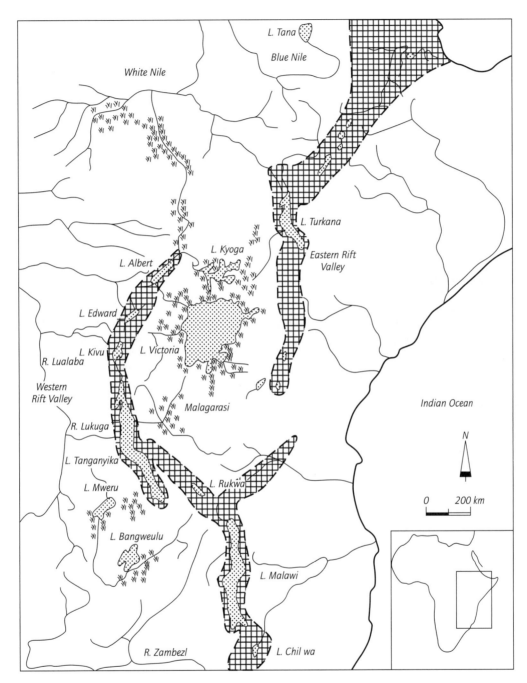

Fig. 21.1 East African Great Lakes and other freshwater systems in the region.

countries to introduce or review regulations to protect the environment. In recent years, many countries in Africa have experienced remarkable population growth, accompanied by an intense urbanisation, increased industrial activities and a greater exploitation of cultivatable land (for eastern Africa, see Tables 21.1 & 21.2; UNECA 1995). These transformations have brought about a huge increase in the amount of waste discharged and a wide diversification in the types of pollutants which reach river waters and consequently lakes and reservoirs. These produce undesirable effects on fish and on the potential for fishery exploitation. A few reviews exist on the state of pollution and other environmental state aspects of African inland/freshwaters (e.g. Symoens *et al.* 1981; Livingstone & Melack 1984; Hecky & Bugenyi 1992; Balirwa 1995; Cohen *et al.* 1996; Johnson & Odada 1996). A need has long been recognised for improved legislation and training for more African experts in order to address the problems of aquatic resources. A case example of the problems affecting the 'Lake Victoria basin' is detailed below.

## 21.2 HUMAN PERTURBATION AND PROBLEMS IN THE LAKE VICTORIA BASIN

### 21.2.1 General introduction

Lake Victoria, with a surface area of 68,800 km$^2$, is the largest in Africa and the second largest in the world (only Lake Superior is larger). Its catchment area of 195,000 km$^2$ extends to Burundi and Rwanda. The lake is relatively shallow (maximum depth about 80 m, mean depth about 40 m) and touches the Equator in its northern reaches. Its shoreline is long and convoluted (3500 km), enclosing innumerable small, shallow wetlands. These differ greatly from one another and from the environment of the main lake.

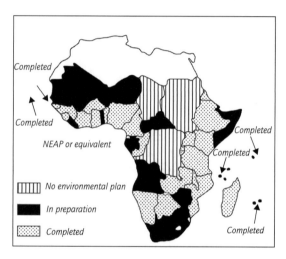

**Fig. 21.2** Status of National Environment Action Plans (NEAPs) or equivalents (1996) in sub-Saharan Africa.

**Table 21.1** Composite water balance of Lake Victoria (averaged over 1946–1970)

| Parameter | Input* (milliards m$^3$) | Losses* (milliards m$^3$) | Output* (milliards m$^3$) |
|---|---|---|---|
| Rainfall on land | 208.57 | | |
| Evapotranspiration and losses | | 189.77 | |
| Runoff | | | 18.80 |
| Rainfall over lake | 114.31 | | |
| Evaporation | | 99.57 | |
| Rainfall – evaporation | | | 14.74 |
| Initial storage | 0 | | |
| Change of storage | | 4.73 | |
| Output | | | −4.73 |
| Total | 322.81 | 294.07 | 28 |

* 1 km$^3$ = 10$^9$ m$^3$ = 1 milliard.

At the level corresponding with the above dimensions, lake volume is 2700 km$^3$, although the maximum storage capacity is 3400 km$^3$ (Table 21.1; UNDP/WMO 1974). Owing to the influence of the lake and the wide variation in topographical features in the basin—ranging from low lying swamps and flat-topped hills, to the mountains of Bufumbira, Elgon, and Kenyan highlands—the climate over the basin varies from moist subhumid to semi-arid. Twenty-one significant riverine inflows to Lake Victoria include the Kagera (the largest), Nzoia, Yala, Sondu, Awach Kaboun, Sio and Mara. The only outflow is the Victoria Nile, at Jinja. The lake level and the outflowing discharge undergo conspicuous fluctuations, with serious socio-economic implications for the basin and the infrastructure around the lake, as well as, of course, for the hydrology of the Nile.

Owing to its relative shallowness, the lake is seasonally ventilated during wind-mixing episodes. Seasonality at this equatorial location is also imposed through cyclical variations in the inflow; affluent rivers at the southern end of the lake swelling with the annual summer rains and the bi-annual equinoxial rains affecting those at the northern end. However, the greatest input is supplied by rain falling directly onto the surface of the lake (even for the other two Great Lakes, Tanganyika and Malawi). The tropical location and consequent high temperatures lead to low oxygen saturations, and simultaneously high rates of

**Table 21.2** Percentage of population residing in urban areas (1980–2020)

| Country | 1980 | 1990 | 2000 | 2010 | 2020 |
|---|---|---|---|---|---|
| Burundi | 4.3 | 6.3 | 9.0 | 12.9 | 18.1 |
| Ethiopia | 10.5 | 12.3 | 14.9 | 19.5 | 26.2 |
| Kenya | 16.1 | 23.6 | 31.8 | 39.7 | 47.6 |
| Rwanda | 4.7 | 5.6 | 6.7 | 8.9 | 12.6 |
| Uganda | 8.8 | 11.2 | 14.2 | 18.8 | 25.1 |
| Tanzania | 14.8 | 20.8 | 28.2 | 36.3 | 44.3 |

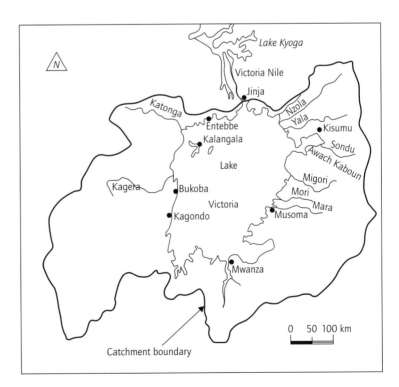

**Fig. 21.3** Location map of Lake Victoria basin.

consumption. Thus, oxygen concentrations are especially sensitive to the oxidative demands of biogenic materials. Recent studies of the eutrophication of this lake (Hecky 1993; Hecky et al. 1994) indicate that bottom waters (below about 20 m over much of the lake) now become quickly deoxygenated and remain so for most of the year.

The large size of the lake, coupled with high evaporation rates, makes for long pollutant retention times (this is even more true for other tropical Great Lakes: Tanganyika and Malawi), particularly in the hypolimnion (Hecky & Bugenyi 1992; Bootsma & Hecky 1993). Extensive cover of water hyacinth also impedes oxidative destruction of pollutants.

Human pressures also threaten the condition of the extensive wetland vegetation around Lake Victoria (Bugenyi & Balirwa 1989; Bugenyi 1991; Balirwa 1995). Until recently, wetlands were hardly considered to be important natural resources in their own right, only as potential agricultural land if they could be drained and reclaimed. It is only recently that their inherent value has been fully recognised (see O'Sullivan, this Volume, Chapter 1).

As is true of other African countries, constitutional changes in the three nations bordering Lake Victoria (Kenya, Tanzania, Uganda) have opened the way for the democratic opportunities for local practices and institutions to enter the administrative and legal frameworks which underpin natural resource management. This, however, needs to be seen as a 'trade liberalisation' which is gaining ground in the region, and which has become increasingly central to the economic policies of developing countries. It is clear that for resources which are shared internationally (transboundary resources), harmonisation of policies is essential to the management of those resources. There are instances in which 'transboundary programmes' have helped to develop solutions to regional environmental issues (World Bank 1996), hence the establishment of the *Lake Victoria Fisheries Organisation* (LVFO) and the formation of the *Lake Victoria Environment Management Project* (LVEMP) by the three countries bordering the lake, to be funded by the World Bank/GEF (Global Environmental Facility) and the three Governments concerned.

### 21.2.2 Major threats to the lake

It has been difficult to ascribe for certain any of the problems afflicting Lake Victoria to any single or unique causal agent. In some cases, this has been because of the inadequacies in data (e.g. Kaufman 1992; Kling 1992) but, more often, it is clear that there are interactions among complex and multiple forces for change. Despite the fact that Lake Victoria is large and is a most complex ecosystem, it is far from being immutable: indeed, it can be said to have been severely altered by human interference. The story of change of the lake as a consequence of human activities is real and the rate of change seems faster than the scientific literature can track (Cohen et al. 1996), as the lake ecosystem suffers the confluent impacts of anthropogenic and other forces.

Like any natural perturbation, the historical context of human activity in the East African lake basins is revealing. The human population of the region of East Africa is high, growing rapidly and is heavily concentrated near the lakes. For example, most of the people in the Lake Victoria basin live in small villages and towns. Urban areas accommodate only a small (albeit growing) fraction of the population (Table 21.2), which is predicted to more than double in the next 20 years. The dense rural populations are often matched by equally high cattle populations. The greatest concentrations are within a short distance of the water (Bootsma & Hecky 1993). Thus, the cumulative impact of human activity is powerfully focused upon the lake, having many ramifications for patterns of resource consumption (notably increased demands for potable water and fresh fish) and of waste discharges (mostly direct to the lake). Population densities around the sub-Saharan lakes, especially in eastern Africa (Table 21.3), are not just high, they are growing at some of the fastest rates in the world, averaging more than 3.1% per year during the 1970s and 1980s (doubling time, approximately 20 years). The environmental indicators shown in Table 21.3 should be considered seriously.

Table 21.3 Total area, total population and the environmental related basic indicators. (From UNECA, 1995)

| Country | Total area ('000 ha) | Land area ('000 ha) | Percentage water area | Total population ('000) | Population density (persons/1000 ha of land area) | Average/annual population growth |
|---|---|---|---|---|---|---|
| Burundi | 2783 | 2568 | 7.8 | 6026 | 2347 | 3.00 |
| Kenya | 58,037 | 56,969 | 1.8 | 26,391 | 463 | 3.59 |
| Rwanda | 2634 | 2467 | 6.3 | 7554 | 3062 | 2.59 |
| Uganda | 23,588 | 19,965 | 15.4 | 19,940 | 999 | 3.42 |
| Tanzania | 94,509 | 88,359 | 6.2 | 28,019 | 317 | 2.96 |

Rapid population growth has resulted in an equally rapid conversion of most of the watershed areas from forest and savannah woodland (including those areas subject to traditional slash-and-burn shifting cultivation) to agricultural rangeland. For example, lowland deforestation of the mesic forests and woodlands of both Uganda and Burundi is now virtually complete, but for some small isolated patches. Even these are subject to severe encroachment. The conversion has occurred rapidly, within the past 50 years. The history of this change provides an important background to understanding other problems for the lake (Ogutu-Ohwayo et al., 1996). As a result of these intensive, non-shifting, agricultural practices, soil erosion and loss of soil fertility are rife. The farmers are more likely to be persuaded towards the use of fertilisers in preference to traditional mulching and, similarly, to abandon the kind of mixed cropping which required tilling land by hand. Both trends possess serious implications for the export of nutrients from croplands through soil erosion and their subsequent discharge to lakes (Hullsworth 1987).

The use of fuel wood (and particularly charcoal) as a primary energy source among rural people in the Great Lakes region is a further contributory factor to accelerating rates of deforestation and nutrient loading to the lakes (including through windborne particle transport). Presently, 70% of total energy consumption in sub-Saharan Africa comes from fuel wood, a figure which has remained remarkably constant in recent years (Davidson 1992). Clearly, high population densities and growth rates are at the crux of the problems facing the watersheds of the African Great Lakes in general, and that of Lake Victoria in particular. All potential solutions rest upon the ability of governments to halt these demographic trends and to legislate for rational land and water usage, and to be able to gain the acceptance of the legislation and the compliance of the people.

### 21.2.3 The eutrophication of the lake and related problems

Eutrophication is a form of environmental degradation that may be best defined as 'a process whereby water bodies become progressively enriched with the plant nutrients (especially nitrogen and phosphorus) with the resulting excessive production of the plant (or more often the algal) biomass'. The increased accumulation rates of nitrogen, since the beginning of the 1920s, and of phosphorus, from the 1950s, were probably the result of changes to the watershed and airshed of Lake Victoria (Hecky 1993). Clearing and burning of forests and savannah (Bugenyi & Balirwa 1989; Simons 1989) for planting and grazing to sustain the rapidly increasing population released nitrogen as the natural vegetation was removed (Bayley et al. 1992). This trend intensified until exposed soil and widespread burning led to the mobilisation of terrestrial phosphorus through erosion by water and wind. These events have supported more photosynthetic (chlorophyll-mediated) production in the lake in the 1990s (Hecky & Bugenyi 1992; Hecky 1993; Mugidde 1993) than occurred in the 1960s (Talling 1965). The sedimentation and decomposition of the additional organic product consumes proportionately more oxygen and causes deoxygenation of the lower and bottom depths (Hecky et al. 1994).

Deoxygenation is likely to have contributed to the observed decline of haplochromine (in addition to the Nile perch predation) stocks in the lake. 'Anoxia' forces the demersal populations to shallower waters where they are exposed to predation by Nile perch (Kudhongania & Cordone 1974; Witte 1992). Recent changes in fish communities may be, in part, the consequence of physicochemical changes but alterations in the fish communities, as a result of overfishing, by using the wrong fishing methods and gear, could also have exacerbated the low-oxygen conditions by altering nutrient flows and distribution within the lake.

Water hyacinth has been present in Lake Victoria since the mid-1980s (Twongo 1992). Originating in Venezuela, the weed has now spread to 50 countries around the world. Wherever it has invaded, it has presented a persistent and expensive problem. The spread of the infestation imposed a wide range of direct costs on the lake community. The rate of spread of water hyacinth around the shores of the lake had been extremely rapid, favoured by the vast size of the lake and the convoluted shorelines, particularly on its northern, western, and southern margins. Assessing the impact of proliferation of water hyacinth in lakes in East Africa requires baseline data on water quality, biodiversity and socio-economic activities. Among the impacts are: the disruption of fishing activities and transportation; and interference with source waters for domestic supply and hydroelectric power.

Diseases such as schistosomiasis, whose vectors flourish in shallow water environments, do recur in these waters infested by the weed. Twongo et al. (1995) gave preliminary measurements under small hyacinth mats in sheltered environments showing depressed oxygen concentrations and lowered pH values, as well as restricted water mixing and gas exchange. All contribute to a decline in local biodiversity. The uptake of nutrients would be accelerated in certain weed-affected locations, however, owing to its rapid growth rate.

## 21.3 WETLAND DEGRADATION AND ATTEMPTS TO CORRECT IT

With respect to the entire African continent, eastern Africa is estimated to contain about one-third of all wetland areas (Fig. 21.4). The value of wetlands to human society is often poorly appreciated but it is, perhaps, more important even than the biological diversity their presence brings, and is especially significant in eastern Africa. People are using wetland resources in a number of ways, some of which are environmentally benign but others are deleterious. Uses include cultivation, hunting, livestock grazing, gathering of building materials, supplying cultural crafts, collection of natural/traditional medicine, and as valuable fishing grounds. The wetlands also function as 'filtering and regulating' zones for the movement of nutrients, pollutants, soil and sediments, and 'damping' water surges contributing to floods. Thus, great pressure is exerted on these rather fragile systems. 'Drainage' and 'reclamation' (Chapman et al. 2001) pose the greatest threats to wetlands.

There are other factors which affect sound wetland management. The question of 'ownership' of land and the right to exploit a given wetland very often hinder its proper and sustainable management. Attempts to legislate new tenure systems in

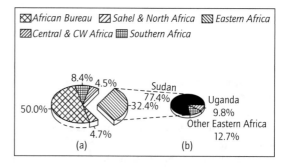

Fig. 21.4 Comparison of Africa's wetland distribution (source: World Resources Institute, *Natural Resources Indicators for Africa – 1992*, 30 September 1992).

post-colonial Africa have focused on establishing mechanisms for ownership of freehold property. The objectives appear to be narrowly restricted to providing a legal framework for encouraging economic development and market production. The economic argument for providing a clear title to land and full hereditary rights is that this will encourage long-term investment in the productive capacity of the land or wetland system. The most disadvantaged in the above are the women.

## 21.4 DEVELOPMENT OF NATIONAL AND REGIONAL POLICIES FOR THE SUSTAINABLE USE OF NATURAL RESOURCES

After the countries bordering Lake Victoria had gained independence, about 40 years now, most basic aspects of the colonial policies and laws governing natural resources remained intact and continued to operate for some time. The failure immediately to develop homegrown laws for the governance of use of natural resources is now exacting an expensive premium on the environment. For the protection and management of the environment, many statutes have been enacted (more than 60 in Uganda) governing various aspects of natural resources. However, the implementation and enforcement of these laws and policies present problems. Local involvement in decision making is often lacking in formulating and enacting legislation considered to be the prerogative of central government. Laws often pay too little attention to important environmental consequences to biological diversity. Reactive legislation in response to crisis situations is always less satisfactory than considered anticipatory instruments. Poor implementation and enforcement of the laws and policies often follows weaknesses in the logistic, financial and human resources allocated. Poor co-ordination of laws and policies tailored to sectoral needs also favours compartmentalised application.

A number of sectoral institutions operate within the various ministries of the three countries bordering the lake which are directly concerned with protecting the environment and its natural resources. The necessary ingredient to sound decision-making in the wise management of resources is the availability of accurate, reliable, up-to-date and timely environmental data and information. In many cases, this is lacking, for any of the following reasons: there are too few trained personnel to collect the data; there is too little money to buy the equipment and other facilities to collect the data, or to pay others to collect it; and no widespread structure and proper pathway in the system for the dissemination of even the little information which may become available.

In spite of these problems, it has been possible for Uganda to develop a 'national policy for the conservation and management of wetland resources'. In September 1986, the Uganda Government issued administrative guidelines to curtail the devastation of wetland resources. A ban on large-scale drainage scheme was imposed until such time that a more elaborate, scientifically proven and socially harmonious policy could be put in place, and which aimed at environmentally sound management and rational use of all resources. The national wetlands policy set five goals:

1 to establish the principles by which wetlands can be optimally used both now and in future;
2 to end practices reducing wetland productivity;
3 to maintain the biological diversity of the wetlands;
4 to maintain wetland functions and values;
5 to integrate wetland concerns into the planning and decision making of other sectors.

It is beneficial to develop national policies which apply to one country's natural resources. However, there are many internationally shared resources, such as those of water and air, which cut across national boundaries. An activity affecting the water of one country will spill over to the next downstream country. Resource-sharing nations need to harmonise the management measures for the good of the resource. With respect to Lake Victoria, the three bordering countries have done exactly this. An organisation, Lake Victoria Fisheries Organisation, LVFO, was established, among many other functions, to co-ordinate management measures of the fisheries in the three countries. The countries acknowledge the need for data on which to base, formulate and implement harmonised policies for the proper management of Lake Victoria, its fisheries and its other resources.

Lake Victoria is an international water body which is of both great economic worth to the three countries and of great scientific and cultural significance to the global community, mainly in respect of its unique waterborne biodiversity. It is suffering severely from three of the four major global environmental concerns highlighted in the 'GEF Operational Strategy for International Waters': degradation of water quality owing to pollution from land-based activities; introductions of non-indigenous species; and excessive exploitation of living resources. It is also facing their typical consequences—the potential irreversible environmental damage, hardship to poor people and serious health implications. The World Bank/IDA and GEF offered funding for the LVEMP for the three countries. The assistance is designed to help the three countries to: develop a better understanding of the lake functions; learn how the actions of their populations affect the lake environment; and work out and implement jointly (under LVFO) a comprehensive approach to managing the lake ecosystem to achieve global environment benefits.

## 21.5 ACKNOWLEDGEMENTS

I thank Drs J.S. Balirwa and T. Twongo and the editors for their useful comments, Mr. S. Sowobi for the artwork and Ms F. Bazannya for processing the manuscript.

## 21.6 REFERENCES

Balirwa, J.S. (1995) The Lake Victoria environment: its fisheries and its wetlands—a review. *Journal of Wetland Ecology and Management*, **3**, 209–24.

Bayley, S.E., Schindler, D.W., Beaty, K.G., Parker, B.R. & Stainton, M.P. (1992) Effects of multiple fires on nutrient yields from streams draining boreal forest and fen watersheds. *Canadian Journal of Fisheries and Aquatic Sciences*, **49**, 584–96.

Bootsma, H. & Hecky, R.E. (1993) Conservation of African Great Lakes: a limnological perspective. *Conservation Biology*, **7**, 644–56.

Bugenyi, F.W.B. (1991) Ecotones in a changing environment: management of adjacent wetlands for fisheries production in the tropics. *Verhandlungen der internationale Vereinigung fur theoretische und angewandte Limnologie*, **24**, 2547–51.

Bugenyi, F.W.B. & Balirwa, J.S. (1989) Human intervention in natural processes of the Lake Victoria ecosystem—a problem. In: Solanki, J. & Herodek, S. (eds), *Conservation and Management of Lakes*. Symposia Biologica Hungaria, Vol. 38. Akademiai Kiado, Budapest, 311–40.

Chapman, L.J., Balirwa, J.S., Bugenyi, F.W.B., Chapman, C.A. & Crisman, T.L. (2001) Wetlands of East Africa: biodiversity, exploitation, and policy perspectives. In Gopal, B., Junk, W.J. & Davis, J.A. (eds), *Biodiversity in Wetlands: Assessment, Function and Conservation*, Vol. 2. Backhuys Publishers, Leiden, 101–31.

Cohen, A.S., Kaufman, L. & Ogutu-Ohwayo, R. (1996) Anthropogenic threats, impacts and conservation strategies in the African Great Lakes: a review. In: Johnson, T.C. & Odada, E.O. (eds), *The Limnology, Climatology and Paleoclimatology of the East African Lakes*. Gordon and Breach, Toronto, 574–624.

Davidson, O. (1992) Energy issues in sub-Saharan Africa: Future directions. *Annual Review of Energy and the Environment*, **17**, 359–404.

FAO (1997) *The State of the World Fisheries and Aquaculture*. Food and Agriculture Organisation, Rome, 126 pp.

Hecky, R.E. (1993) The eutrophication of Lake Victoria. *Verhandlungen der Internationale Vereinigung fur theoretische und angewandte Limnologie*, **25**, 39–48.

Hecky, R.E. & Bugenyi, F.W.B. (1992) Hydrology and chemistry of the African Great Lakes and water quality issues: problems and solutions. *Mitteilungen Internationale Vereingung fuer Theoretische und Angewandte Limnologie*, **23**, 45–54.

Hecky, R.E., Bugenyi, F.W.B., Ochumba, P., Gophen, M. Mugidde, R. & Kaufman, L. (1994) De-oxygenation of the deep water of Lake Victoria. *Limnology and Oceanography*, **39**, 1476–80.

Holland, M.M. (1988) SCOPE/MAB Technical Consultation on Landscape Boundaries: Report of a SCOPE/MAB Workshop on Ecotones. *Biology International* (Special Issue), **17**, 47–106.

Hullsworth, G.E. (1987) *Anatomy, Physiology and Psychology of Erosion*. Wiley, New York, 176 pp.

Johnson, T.C. & Odada, E.O. (eds) (1996) *The Limnology, Climatology and Paleoclimatology of the East African Lakes*. Gordon and Breach Publishers, Toronto, 664 pp.

Kaufman, L. (1992) Catastrophic change in a species rich freshwater ecosystem. *Bioscience*, **42**, 846–58.

Kling, G. (1992) Limnological change in Lake Victoria; contrasts in structure and function between temperate and tropical great lake ecosystem. *Conference on the Biodiversity, Fisheries, and the Future of Lake Victoria*, Jinja, Uganda, 17–20 August, Abstracts, 35–6.

Kudhongania, A.W. & Cordone, A.J. (1974) Bathospatial distribution pattern and biomass estimate of the major demersal fishes in Lake Victoria. *African Journal of Tropical Hydrobiology and Fisheries*, **3**, 15–31.

Livingstone, D.A. & Melack, J.M. (1984) Some lakes of sub-Saharan Africa. In: Taub, F.B. (ed.), *Lakes and Reservoirs*. Elsevier, Amsterdam, 467–97.

Lowe-McConnell, R. (1996) Fish communities in the African Great Lakes. *Environmental Biology of Fishes*, **45**, 219–35.

Mugidde, R. (1993) The increase in phytoplankton primary production and biomass in Lake Victoria (Uganda). *Verhandlungen der internationale Vereinigung fur Theoretische und angewandte Limnologie*, **25**, 846–9.

Ogutu-Ohwayo, R. (1985) The effects of production of Nile Perch introduced into Lake Kyoga (Uganda) in relation to the fisheries of Lakes Kyoga and Victoria. *FAO Fishery Report*, **335**, 18–39.

Ogutu-Ohwayo, R., Hecky, R.E., Bugenyi, F.W.B., *et al.* (1996) Some causes and consequences of rapid ecosystem changes in a large tropical lake—Lake Victoria. *Proceedings of the joint Victoria Falls Conference on Aquatic Systems and International Symposium on Exploring the Great Lakes of the World (GLOW) Food-web Dynamics, Health and Integrity*, Victoria Falls, Zimbabwe, 15–19 July.

Simons, M. (1989) High ozone and acid rain levels found over African rain forest: *New York Times*, **138**(47), 1.

Symoens, J.J., Burgis, M. & Gaudet, J.J. (1981) *The Ecology and Utilisation of African Inland Waters/Ecologie et utilisation des eaux continentales africaines*. UNEP Report, Proceedings Series 1, United Nations Environment Programme, Nairobi, Kenya. 191 pp.

Talling, J.F. (1965) The photosynthetic activity of phytoplankton in East African Lakes. *Internationale Revue der gesamten Hydrobiologie*, **50**, 1–32.

Twongo, T. (1992) The spread of water hyacinth on Lakes Victoria and Kyoga and some implications for aquatic biodiversity and fisheries. *Conference on the Biodiversity, Fisheries and the Future of Lake Victoria*, Jinja, Uganda, 17–20 August, Abstracts, 42.

Twongo, T., Bugenyi, F.W.B. & Wanda, F. (1995) The potential for further proliferation of water hyacinth in Lakes Victoria, Kyoga and Kwania and some urgent aspects for research. *African Journal of Tropical Hydrobiology and Fisheries*, **6**, 1–10.

UNDP/WMO (1974) *HydromeTechnological Survey of the Catchments of Lakes Victoria, Kyoga and Albert*, Vol. 1, *Meteorology and Hydrology of the Basin, Part II*. UNDP/WMO, RAF 66-025 Technical Report 1, United Nations Development Programme and World Meteorological Organisation, Geneva, 925 pp.

UNECA (1995) *African Compendium of Environment Statistics*. Report, United Nations Economic Commission for Africa, Addis Ababa.

UNEP (1994) *The Pollution of Lakes and Reservoirs*. Environment Library No 12, United Nations Environment Programme, Nairobi, 35 pp.

Witte, F. (1992) The destruction of an endemic flock: Qualitative data on the decline haplochromine cichlids of Lake Victoria. *Environmental Biology of Fishes*, **34**, 1–28.

World Bank (1996) *Towards Environmentally Sustainable Development in sub-Saharan Africa. A World Bank Agenda*. Report, World Bank, Washington, DC, 142 pp.

# 22 South Africa – Towards Protecting our Lakes

## G.I. COWAN

## 22.1 INTRODUCTION

South Africa is a country which finds itself on the bridge between the first and third world, having both a strong and competitive industrial component as well as a large rural population dependent on a subsistence economy. With the change in government from the apartheid system to a democratic system, South African legislators have been provided with an exceptional opportunity to revise all our laws, many of which were inadequate and confusing at best. Not the least requiring scrutiny are those relating to water, wetlands and biodiversity, and their sustainable and equitable utilisation. Under development in terms of our constitution (Constitution of the Republic of South Africa: Act No. 108 of 1996), South Africa's legislation and its administration are in an interesting and dynamic state of change (see Appendix). It is in this context that this paper considers some of the major approaches being used to protect our lakes.

South Africa is not well endowed with lakes, and those which do occur are all relatively small. They are classified as one of five natural wetland types. Wetlands in general have been inadequately protected in our legislation, particularly as regards their ecological functions. Our lakes are not used for navigation or as transportation corridors, therefore the controls developed, and which are developing for lakes, are related to: recreation (e.g. for small boats, fishing, water sports), resource utilisation (subsistence, traditional, artisinal, commercial), water exploitation (e.g. pumping, opening of estuary mouths), religious (sacred lake Fundudzi) and pollution control. Artificial lakes are not discussed here, even though they are the most numerous and largest water bodies in South Africa. However, many of the threats and controls are equally applicable to these water bodies as to natural lakes.

## 22.2 TYPES OF LAKES FOUND IN SOUTH AFRICA

Lakes or lacustrine wetlands have been defined as areas of permanent water with little flow (Barbier et al. 1997). Although Dugan (1990) includes seasonal freshwater lakes (>8 ha), ponds and pans (<8 ha) in a classification of wetlands in South Africa, Hart (1995) excludes the pans from lakes, which he defines as reasonably extensive and relatively deep open waters, although Shaw (1988) considers endorheic pans to be a lake type. In this paper, endorheic pans have been excluded (based primarily on salinity and depth), although the continuum is recognised, particularly given the classification used in South Africa (Cowan 1997).

Hart (1995) emphasises the fact that South Africa is deficient in terms of natural freshwater lakes, and notes that besides Lake Fundudzi, which was created by a landslide blockage of the Mutale River, South Africa's only significant natural standing waters are its coastal lakes. Most of these were formed by Holocene sea-level rise (Allanson et al. 1990; Orme 1990).

Coastal lakes have been classified into three major regional groups (Cowan 1995; Hart 1995): subtropical lakes on the northeastern seaboard (subtropical coastal plain wetland region); warm temperate lakes on the Cape south coast (southern coast temperate wetland region); and a disparate group in the southwestern Cape experiencing a

Mediterranean climate (western coastal slope—Mediterranean wetland region). Besides the permanent freshwater lakes identified by Hart (1995), a number of largely freshwater permanent and temporary pans, mostly part of floodplain systems, are listed as lacustrine wetlands (Fig. 22.1). Even though some of these systems may tend to become saline as they dry out and could be categorised as pans, they have been included here. Some 41 permanent freshwater lakes, 16 permanent freshwater ponds or pans, nine seasonal freshwater lakes and three seasonal freshwater ponds or pans are included in a directory of South Africa's wetlands (Cowan 1997).

Most of the major lake systems have been included in sites designated by South Africa as Wetlands of International Importance in terms of the Ramsar Convention (de Hoop, de Mond, Kosi System, Lake Sibaya, Ndumo, St Lucia System, Verlorenvlei and Wilderness Lakes; Fig. 22.2), all of which have been afforded legal protection (Cowan & Marneweck 1996). Lake Fundudzi forms the focus of a rich heritage of ceremonial rituals and myths amongst the local Venda people and is protected by tribal custom (van der Waal 1996).

## 22.3 THREATS TO SOUTH AFRICA'S LAKES

The major threats to South Africa's lakes include: agricultural and silvicultural developments in their catchments, leading to reduced flows into the systems, nutrient loading and herbicide/biocide pollution; industrial developments with corresponding increases in pollutants; direct demands of an expanding population (e.g. recreational demands and overexploitation of resources); inappropriate riparian developments including reclamation, leading to changes in hydrological regimes and sediment loading (Hart 1995).

Major threats can be considered regionally or topically. Regionally, the following issues can be identified:

**1** The southwest Cape and Cape Flats—eutrophication from formal agriculture and from informal settlements; marina developments; wetland infilling accompanying both purposeful 'reclamation' and sediment infilling resulting from erosion.

**2** The Wilderness lakes (collectively)—catchment development for agriculture, silviculture or residential reasons combined with invasion by alien invasive plants, leading to altered sediment

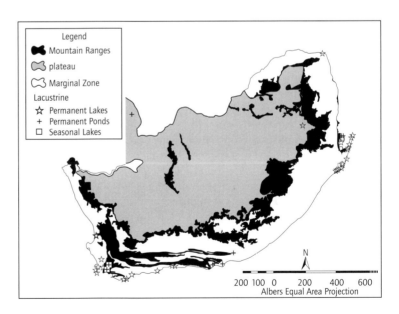

**Fig. 22.1** Distribution of South Africa's lacustrine wetlands, found primarily along the northeastern, southern and southwestern seaboard (after Cowan, 1997).

Fig. 22.2 Two views of the Wilderness Lakes, a coastal lake system in the southern Cape.

and nutrient loadings, and unnatural hydrological characteristics.

3 Natal/KwaZulu — the manifest consequences of human population increase; air pollution stemming from industrial development centred at Richards Bay; clearing of swamp forest for informal settlements and cash crops; silviculture development of former coastal grasslands leading to a lowering of the water table; bunding of coastal lakes for increased water supply, and increased water abstractions reducing throughflow and enhancing macrophyte encroachment of drainage channels, in addition to interrupting natural faunal migrations; increasing sedimentation.

Topically, the following threats, in no order of priority, can be listed:

1 Agricultural/silvicultural developments in the catchment, leading to increased water abstraction/lowering of the water table, to reduced inflows into the coastal lakes and, thus, to alterations in plant and dependent animal communities and consequent changes in ecological structure and function; to sediment mobilisation, with consequent changes in underwater light climate and increased sedimentation (infilling) rates; to enhanced nutrient loading and corresponding increases in primary production by a community of lowered species diversity, with consequent alterations of foodweb structures; to herbicide/biocide pollution, with its obvious detrimental effects on ecological structure and function.

2 Industrial developments, with corresponding increases in pollutants, such as heavy metals, and alterations in the quantity and nature of airborne substances which have a deleterious impact on lake ecosystems.

3 Direct environmental consequences of expanding human population pressure, resulting in: demands to remove macrophytes for enhanced recreational boating and swimming possibilities; heavier exploitation of natural stocks, fish in particular; increased power-boating and associated wake-washing of littoral communities, and disturbance of normal feeding and breeding activities of fish and wildfowl, etc.

4 Breaching of estuary mouths at suboptimal levels, minimising scour and thus resulting in infilling with coastal sand or riverine silt, thus exacerbating the consequences of greater catchment-mobilised sediment inputs.

## 22.4 PROTECTION MEASURES

A single legislative system for wetland management in general, or lakes specifically, does not exist in South Africa. The provisions in various acts outlined below have been enforced by a wide range of authorities, with differing objectives in mind, including the Department of Environmental Affairs and Tourism, the Department of Water Affairs and Forestry, the Department of Agriculture, the National Parks Board and the provincial nature conservation departments. These acts have been replaced, or are in the process of being replaced, by new legislation addressing the previous shortcomings in both social and environmental terms.

The Lake Areas Development Act (No. 39 of 1975) does not specifically indicate the purpose of establishment of Lake Areas, nor does it define them. The Minister of Environment Affairs may, by notice in the Government Gazette, declare any land comprising or adjoining a tidal lagoon, a tidal river or any part thereof, or any other land comprising or adjoining a natural lake or river or any part thereof which is within the immediate vicinity of a tidal lagoon or a tidal river, to be a Lake Area. This Act may serve to protect the natural habitats in and around lakes. All existing Lake Areas have been included into national parks or parts thereof and are controlled by the South African National Parks.

The Environment Conservation Act (No. 73 of 1983) provides for the protection and controlled utilisation of the environment and the determination of policy to protect, use sustainably and rehabilitate the environment. Protection of the natural environment can be achieved under this act through the declaration of protected natural environments, special nature reserves, limited development areas or by restriction of certain activities or developments which may harm the environment. To date, neither the protected natural environment nor the special nature reserve clauses have been used to protect lakes. The Department of Environmental Affairs and Tourism is primarily responsible for this act, and the responsibility for certain sections has been devolved to provincial level.

The objective of the National Parks Act (No. 57 of 1976) is the establishment of national parks for the preservation and study of wild animal, marine and plant life and objects of geological, archaeological, historical, ethnological, oceanographic, educational and other scientific interest and objects relating to the life, events or history of the park, which is retained in its natural state for the benefit of visitors. National Parks are the responsibility of the South African National Parks, an autonomous parastatal organisation.

The Water Act (No. 54 of 1956) dealt only with inland running water. It differentiated between public and private water and streams. In principle, private water was defined as that which is not capable of common use for irrigation purposes. Although several provisions of the Act regulated the use of private water, in principle the owner of the land on which it is found enjoyed sole and exclusive use and enjoyment of it. Public water was defined as water from a natural stream following a known and defined channel which can be used for irrigation of two or more pieces of riparian land. The control and use of this water was regulated by the Act so that it could not be privately owned.

In terms of the Forest Act (No. 122 of 1984) the Minister of Water Affairs and Forestry may prohibit afforestation or reafforestation (i.e. plantation forestry with alien species) of certain land to protect any natural water source. The prohibition, in practice, depends on the discretion of the forest officer, who will evaluate the status and hydrological value of the water source and will determine how far from that water source the land will be afforested. This provision, *inter alia*, protects riverine habitats from destruction and prevents soil erosion.

## 22.5 NEW INITIATIVES

Since South Africa's first democratic election in 1994, a number of major initiatives to rectify the injustices of the apartheid legal system have been set in motion, the most important of which is the development and enactment of the constitution (Constitution of the Republic of South Africa: Act No. 108 of 1996). All legislation is required to be in line with this act. In terms of the environment, Section 24 of the Bill of Rights is specifically relevant:

> 24. *Everyone has the right-*
> *(a) to an environment that is not harmful to their health or wellbeing; and*
> *(b) to have the environment protected, for the benefit of present and future generations, through reasonable legislative and other measures that-*

(i) *prevent pollution and ecological degradation;*
(ii) *promote conservation;*
(iii) *secure ecologically sustainable development and use of natural resources while promoting justifiable economic and social development.*

Although this section provides a sound base on which to build measures for the protection of wetlands in general, and lakes in particular, in terms of Schedule 4, the constitution complicates the issue by assigning legislative competence, at national and provincial levels, for the following functional areas:

- environment;
- nature conservation, excluding national parks, national botanic gardens and marine resources;
- pollution control;
- regional planning and development.

In terms of administration, this means that the number of agencies responsible for developing and implementing legislation in these areas has become inordinately large for a country with a limited capacity. It is quite likely that numerous loopholes in the legislation and its implementation will be identified, unless a major co-ordination effort is driven from the National Ministry of Environment.

The mechanism for such co-ordination existed in terms of the Environmental Conservation Act (No. 73 of 1989), where the statutory Committee for Environmental Coordination is established. The relevant sections of this act were replaced with the introduction of the National Environmental Management Act (No. 107) in 1998. A number of draft policies relating to the environment have been or are in the process of being accepted by this committee, before being published as national policy, the most important of which are:

**1** the White Paper on the Conservation and Sustainable Use of South Africa's Biological Diversity, which gives effect to South Africa's obligations in terms of the Convention on Biological Diversity, with special reference to the three objectives of the convention (i.e. the conservation of biodiversity, the sustainable use of biological resources and the fair and equitable sharing of benefits arising from the use of genetic resources);
**2** the draft National Policy and Strategy for Wetland Conservation in South Africa, which gives effect to South Africa's obligations in terms of the Convention on Wetlands of International Importance especially as Waterfowl Habitat;
**3** the draft National Policy on Pollution and Solid Waste Disposal.

The National Environmental Management Act was promulgated as framework legislation providing for co-operative environmental governance. After the publication of the White Paper, it was decided that further legislation was required to address the conservation of South Africa's biological diversity. As part of a suite of legislation within the framework of the amended National Environmental Management Act, two new acts were passed by parliament during 2003, the Biodiversity Act and the Protected Areas Act.

The National Environmental Management: Biodiversity Act (No. 10 of 2004) aims to provide for the management and conservation of South Africa's biological diversity, the protection of both ecosystems and species in need of protection and the sustainable use of indigenous biological resources. In terms of its relevance to protecting our lakes, the Act gives effect to relevant international agreements which South Africa is a party to (notably the Ramsar Convention on Wetlands and the Convention on Biodiversity), as well as providing for co-operative governance, which is important when there are a number of agencies at different tiers of government responsible for the management of ecosystems. Importantly, the act identifies the state as the trustee of South Africa's biological diversity. The act makes provision for the listing of ecosystems, along with the identification of threatening processes leading to the development of environmental management plans to manage the processes and to mitigate the effects of such processes on the relevant ecosystem. Controls on the introduction and spread of alien and invasive species are also addressed in this act.

The National Environmental Management:

Protected Areas Act (No. 57 of 2003) provides for the establishment of a representative system of ecologically viable protected areas as part of the strategy to manage and conserve South Africa's biological diversity. Included in the objectives of establishing these protected areas is the need to ensure sustained environmental goods and services and to manage the interrelationship between natural biodiversity, human settlement and economic development. This act repeals the Lakes Area Act, the National Parks Act and relevant sections of the Environment Conservation Act. It also adds weight to the World Heritage Convention Act (No. 49 of 1999), which has been used in order to consolidate the Kosi Lake system, Lake Sibaya and the St Lucia system under the Greater St Lucia Wetland Park Authority.

Water, as opposed to environment, is firmly entrenched as a national legislative competency only, under the administration of the Department of Water Affairs and Forestry. A major review of both the policy and legislation relating to water was completed in 1998, where many of the issues relating to access to water and the differences between private and public water were addressed. One of the major steps forward in this policy is the determination of 'the reserve', where 'the quantity, quality and reliability of water required to maintain the ecological functions on which humans depend shall be reserved so that human use of water does not individually or cumulatively compromise the long term sustainability of aquatic and associated ecosystems'.

The National Water Act (No. 36 of 1998), as a means of implementing the policy, took a major step forward, providing for fundamental reform of the law relating to water resources in South Africa. For the first time, water resources are regarded as part of a unitary water cycle; therefore the legislation moves away from its predecessor which dealt only with inland running water, and which differentiated between public and private water. The new act acknowledges the national government's overall responsibility in the management of South Africa's water resources. Sustainable use of water for the benefit of all users is set as the aim for such management, and, to achieve this, the act recognises that it is necessary to protect the quality of our water resources.

The purpose of the act is to ensure that the nation's water resources are protected, used, developed, conserved, managed and controlled in ways which take in to account a range of factors including, importantly, protecting aquatic and associated ecosystems and their biological diversity as well as reducing and preventing pollution and degradation of water resources. Lakes, in being defined as part of a watercourse, now enjoy the same level of attention as any other element in the water cycle which provides those water resources.

As with the suite of legislation relating to the environment, sustainability and equity are identified as central guiding principles in the protection, use, development, conservation, management and control of water resources. These guiding principles recognise the basic human needs of present and future generations, the need to protect water resources, and to share and promote social and economic development through the use of water.

The Act requires a national water resource strategy which will set out, *inter alia*, strategies, objectives and plans relating to the protection, use, development, conservation, management and control of water resources, and provide for at least the requirements of the reserve, which consists of two parts—the basic human-needs reserve and the ecological reserve. The basic human-needs reserve provides for the essential needs of individuals served by the water resource in question and includes water for drinking, for food preparation and for personal hygiene. The ecological reserve relates to the water required to protect the aquatic ecosystems of the water resource. The reserve refers to both the quantity and quality of the water in the resource and will vary depending on the class of the resource. The strategy will promote the management of catchments and take into account the class of water resources and quality objectives. The classification system is used to determine the class and resource quality objectives of all or part of the water resources to be considered significant.

## 22.6 CONCLUSION

Wetland conservation, including the conservation of lake systems, has been subject to a plethora of confusing legislation and administration, not the least of which has been the differences between water law and environmental law (Fuggle & Rabie 1992; Hanks & Glavovic 1992; O'Keefe et al. 1992; Rabie & Day 1992; Department of Water Affairs and Forestry 1996, 1997). With the change to a democratic system, South Africa has been afforded a unique opportunity to rectify these problems. New policies and legislation are aimed at rectifying these problems, the most exciting being the move towards the catchment approach in the National Water Act and the ecosystem approach in the National Environmental Management: Protected Areas Act and Biodiversity Act.

This series of new acts has taken us a long way down the road towards the integration of our legislation and towards meeting the needs of the citizens of South Africa. The use of 'common language' has been a hallmark of these new bills and acts, making them more easily understood by the South African public. However, administration of this legislation remains in parochial departmental territories. The challenge now is to ensure that the spirit of co-operative governance is applied.

## 22.7 APPENDIX

Constitution of the Republic of South Africa (No. 108 of 1996)
Environment Conservation Act (No. 73 of 1983)
Forest Act (No. 122 of 1984)
Lake Areas Development Act (No. 39 of 1975)
National Environmental Management Act (No. 107 of 1998)
National Environmental Management: Biodiversity Act (No. 10 of 2004)
National Environmental Management: Protected Areas Act (No. 57 of 2003)
National Parks Act (No. 57 of 1976)
National Water Act (No. 36 of 1998)
Water Act (No. 54 of 1956)
World Heritage Convention Act (No. 49 of 1999)

## 22.8 REFERENCES

Allanson, B.R., Hart, R.C., O'Keefe, J.H. & Robarts, R.D. (1990) *Inland Waters of Southern Africa: an Ecological Perspective.* Monographiae Biologicae, Vol. 64, W. Junk, Den Haag, 458 pp.

Barbier, E.B., Acreman, M. & Knowler, D. (1997) *Economic Valuation of Wetlands: a Guide for Policy Makers and Planners.* Ramsar Convention Bureau, Gland, 127 pp.

Cowan, G.I. (1995) Wetland regions of South Africa. In: Cowan, G.I. (ed.), *Wetlands of South Africa.* Department of Environmental Affairs and Tourism, Pretoria, 21–31.

Cowan, G.I. (1997) *The development of a national policy and strategy for wetland conservation in South Africa.* Unpublished PhD thesis, University of Pretoria, 181 pp.

Cowan, G.I. & Marneweck, G.C. (1996) *South African National Report to the Ramsar Convention 1996.* Department of Environmental Affairs and Tourism, Pretoria, 44 pp.

Department of Water Affairs and Forestry (1996) *The Philosophy and Practice of Integrated Catchment Management: Implications for Water Resource Management in South Africa.* Water Research Commission, Pretoria, 140 pp.

Department of Water Affairs and Forestry (1997) *White Paper on a National Water Policy for South Africa.* Government Printer, Pretoria, 37 pp.

Dugan, P. (ed.) (1990) *Wetland Conservation: a Review of Current Issues and Required Action.* IUCN, Gland, 96 pp.

Fuggle, R.F. & Rabie, M.A. (eds) (1992) *Environmental Management in South Africa.* Juta, Cape Town, 823 pp.

Hanks, J. & Glavovic, P.D. (1992) Protected areas. In: Fuggle, R.F. & Rabie, M.A. (eds), *Environmental Management in South Africa.* Juta, Cape Town, 690–714.

Hart, R.C. (1995) South African coastal lakes. In: Cowan, G.I. (ed.), *Wetlands of South Africa.* Department of Environmental Affairs and Tourism, Pretoria, 103–30.

O'Keefe, J.H., Uys, M. & Bruton, M.N. (1992) Freshwater systems. In: Fuggle, R.F. & Rabie, M.A. (eds), *Environmental Management in South Africa.* Juta, Cape Town, 227–315.

Orme, A.R. (1990) Wetland morphology, hydrodynamics and sedimentation. In: M. Williams, M. (ed.), *Wetlands: a Threatened Landscape.* Special Publications

Series No. 25, Institute of British Geographers, London, 42–94.

Rabie, M.A. & Day, J.A. (1992) Rivers. In: Fuggle, R.F. & Rabie, M.A. (eds), *Environmental Management in South Africa*. Juta, Cape Town, 647–68.

Shaw, P.A. (1988) Lakes and pans. In: Moon, B.P. & Dardis, G.F. (eds), *The Geomorphology of Southern Africa*. Southern Book Publishers, Johannesburg, 221–40.

Van der Waal, B.C.W. (1996) *Fundudzi, a Sacred Lake Profaned*. Inaugural address. University of Venda, Thoyanadou, 8 pp.

# Index

Note: page numbers in *italics* refer to figures, those in **bold** refer to tables and boxes.

acclimatisation works, Aral Sea 188
acid neutralising capacity (ANC) 484, 486–7
acid rain 451
acidification 15, 19, *20*, 50–1
 bioproductivity 451
 European alpine lakes 171, 173–4
 Fennoscandian lakes 127, 130, 135–6, 139, *140*
 geogenic lakes 452
 models of lakes/reservoirs 407–10
  fish 410
 remediation 451
 status 483–4
 Sweden 519
acids, precipitation 50
*Acipenser fulvescens* (sturgeon) 72
aeolian action 210
aeration 456
 hypolimnetic 457
African Great Lakes 523, *525*
ageing of lakes 43, *355*
 natural 25
agriculture
 Aral Sea 187, 188
 Fennoscandian lakes 124–5
 forest conversion 250–1
 Lake Baikal 185
 Lake Balkhash 195, 196, 197
 Lake Victoria basin 529
 land use 335
 nutrient loading 18
 production increase 42
 reservoirs 311–12
 shallow tropical lakes 289
 soil 47–50
alewife 74, 79, 81–3
algae
 barriers 454
 biomass indicator 368
 competition with macrophytes 265

 control 458
 floodplain lakes of tropical/subtropical South America 246–7
 Laurentian Great Lakes 76
  production 70–1
 macrophyte influence on growth 266–7
 nutrients
  limitation 458
  stoichiometry 359–62
 saline lakes 214
 self-shading 428
 synthesis phase 359
 top-sown control of biomass 476–7
algae shields/booms 454
algal blooms 54, 55
 algae shields/booms 454
 *Daphnia* grazing 474
 European alpine lakes 171–2
 Fennoscandian lakes 127, 133–4, 136–7, 139
 likelihood 362
 polymictic lakes 439
 tropical shallow lakes 290–1
 turbidity shift 475
 urban reservoirs 243
alkalinity 409
 European alpine lakes 174
allelopathy 267
*Alosa aestivalis* (glut herring) 464
*Alosa pseudoharengus* (alewife) 74
alpine lakes, European 159–74
 acidification 173–4
 area 160–1
 benthos 170–1
 chemistry 165–6
 depth 161, *163*
 eutrophication 171–3
 fish 170–1
 groups 159

 human impact 171–4
 locations 161
 morphometry **162–3**
 physics 163–5
 phytoplankton 166, 167–9, 171–2, 174
 plankton 166–7
 primary production 167–9, *170*
alternative stable states
 biomanipulation 469–71
 evidence 470–1
 hypothesis 463, 469–70
 shallow lakes 469–71
aluminium salts 264, 455
 liming 493–4
 toxicity decrease 490
Aman, Lake (Malaysia) 284
Amazon, hydrological regimes 249
amphipods 80
anemotrophy 207
*Anguilla rostrata* (American eel) 80
animal wastes 339
anthropogenic interventions 41–2
 North American Great Lakes 65
aquaculture, shallow tropical lakes 288
aquatic–terrestrial transition zone 242
Aral Sea (Kazakhstan) 186–91, 229, *230*, 231–3
 area 231
 bathymetry *230*, 231
 biomass 232
 biota composition changes 232
 canal construction 188–9
 characteristics 186–7
  1990s 188–9
 chemistry 189
 composition 187
 crisis 187–8
  consequences 190
 dam construction 191

Aral Sea (cont'd)
  ecological value 218
  fauna 187, 188
  fish 187, 188, 190, 232
  hydrological regimes 189–90
  measures to save 190–1
  morphometry 231–2
  phytoplankton 189–90, 232
  regional water supply 191
  salinity 220
  water level 188, 189, 190
  water source 204
  zooplankton 190
argillotrophy 207
arid environments 200–36
  aseasonal regimes 203
  categories 201
  climatic variability 202–3
  distribution *202*, **203**
  extent 201–3
  indices 201–2
  lakes 203–15
    aesthetic value 217
    atmospheric changes 222
    bed changes 221
    biological processes 208
    biota 220–1
    case studies 223–36
    catchment activities 220
    categories 206
    chemistry 207–8
    climate changes 222
    cultural value 217–18
    distribution 204–5
    ecological value 218
    economic use/value 215–16
    educational value 216–17
    fish 208, 220–1
    impacts 218–22
    morphometric features **204**
    pollution 220
    recreational value 218
    salinisation 219
    scientific value 216
    therapeutic value 218
    total area 205
    types 205–15
    use 215–18
    value 215–18
    water diversions 219–20
    water permanency 206

  nature 201–3
  salinisation 451
  spatial variability 203
  temporal variability 202, 203
  unpredictability 202
    water-body occurrence 202, **203**
*Artemia* (brine shrimp) 221
Aswan High Dam (Egypt/Sudan) 203, 204, 208, 225–6
Athabasca, Lake (North America) 65, 66
atmosphere 44–5
  changes and dryland lakes 222
atmospheric deposition, Fennoscandian lakes 130, 138–9
autotrophs 29–30

bacteria 319
Baikal, Lake (Russia) 179–85
  characteristics 179–80, *181*, 182
  ecological problems 182–4
  fauna 182
  flora 182
  phytoplankton 182
  protection measures 184–5
  self-purification potential 182
  water quality 184–5
Balbina reservoir (Brazil) 244, *245*
Balkhash, Lake (Kazakhstan) 191–7
  biology 194–5
  characteristics 191–2
  chemistry 192, 194–5
  dam construction 197
  development 195
  economic use 195
  environmental problems 195–6
  hydrology 192–4, 195–6
  measures to improve 196–7
  morphometrics 192–4
  pollution 196
  reservoir construction 195
  salinity 213
Baltic Sea pollution 519
Baltic Shield *see* Fennoscandian lakes
Barra Bonita Reservoir (Brazil) 243
base cations
  carriage in rivers 49
  immobilisation 47
  input 38

  leaching rate 50
  loss during human settlement 41
Batata, Lake (Brazil) 247–8
bathing waters 58
bauxite tailings 247–8
benthic fauna, Aswan High Dam 226
benthic processes 56–7
benthivores
  reduction 472, 474
  removal 476
benthos, European alpine lakes 170–1
bilharzia 457, 458, 530
Bill of Rights (South Africa) 537–8
biochemical oxygen demand (BOD) 318
biochemical oxygen demand–dissolved oxygen models 410–11
biodiversity 484
  *see also* species diversity
biogeochemical modelling 427
biological control 458
biological processor properties of water 30
biological productivity 443
biology of lakes 14
biomanipulation
  alternative stable states 469–71
  conditions 420
  failures 476–7
  Fennoscandian lakes 134–5
  fish 462
  macrophyte regrowth 271
  matter production regulation 454
  modelling 420–1, *422*, 423–4
    structural dynamic 424–30
    success/failure 430–4
  phosphorus 423–4
  shallow lakes 271, 462–77
    case studies 472, **473**, 474–6
    mechanisms 471–2
    tropical 293–4
  structural dynamic models 431
  submerged macrophytes 467–8
  successes 476–7
  urban reservoirs 243
biomelioration 454
biota, abundance/composition variation 19

birds
  Coongie Lakes 227
  eutrophic lakes 283
  fish-eating 465
  herbivorous 271, 272
  Lake Balkhash 195
  Lake Eyre 236
  saline lakes 214, 217, 221
  shallow lakes 271–2
bog formation 39–40
boundary layers 28, 31
  atmosphere and water 51
bream 465, 466
  sediment resuspension 468–9
bream state 420–1, *422*
buffering 319
burbot 70
burning, deforestation 251
*Bythotrephes longimanus* (spiny waterflea) 83

Calado, Lake (Brazil) 248–50
calcite
  flushing 455
  precipitation in European alpine lakes 165
calcium loss in rivers 49
carbon 56
  anaerobic oxidation 46
  limiting nutrient 357–8
  organic in Fennoscandian lakes 122, *123*
carbon dioxide 46
  concrete dissolution 307
  partial pressure in soil 49
  reservoir emissions 245
carbonates 486
  European alpine lakes 165
carbonic acid
  equilibrium changes 51–2
  partial pressure 52
carbon:nitrogen:phosphorus ratio 360
carp 80
  grass 42
  sediment resuspension 468–9
carrying capacity 397–8, 438
catastrophe theory 429, 431
catchment area 45–51
  activities affecting dryland lakes 220

definition for export coefficient approach 334
development in South Africa 535–6
liming 490–1, 494
nutrient attenuation 341
phosphorus exports 338
reservoirs 313
sewage diversion 453
shallow tropical lakes 285
categories of lakes **15**
catfish 80
cathedral view 6, 9
cation discharges 50
Chad, Lake (West African Sahara) 204, 223–4
*Chaoborus* (phantom midge) 294
char, arctic 170–1
chemical efficiency 28, 43
chemical oxidative demand (COD) 318
chemical potentials 28
chemical processor properties of water 28, 29, 30
chemical properties, Lake Washington 98, **100**
chemistry of lakes 14
chemocline, saline lakes 213
Chilwa, Lake (Malawi) 281–2
chironomids, European alpine lakes 169, 170
chloride, Lake Balkhash 194
chlorophyll
  concentration and phosphorus load 366–7
  determinations 443
  phosphorus relationship 458, *459*
  Secchi-disc depth 368
chub 79
circulation, artificial 456–7
cisco 72, 484
cladocerans
  exotic 83
  macrophyte beds 268
  phytoplankton grazing 267
classification of lakes 19
clay particles 312, 317
Clean Lakes Program (USA) 501, 506–8
  funding uncertainty 507

Clean Water Act (USA) 501–8
  watershed approaches 505–6
clear water
  sustainable state 471, 474, 476
  *see also* spring clear-water phase
climate 44–5
  changes
    deforestation 251
    dryland lakes 222
    reservoir impacts 312
  development 38–9
  variability in arid environments 202–3
coastal lakes 534–5
Cocibolca, Lake (Central America) 293
colour of water, reservoirs 319
community stability 52
Compensation for Environmental Damage Act (Finland) 517
Comprehensive Everglades Restoration Programme (CERP) 292
condensation 25
  heating effect 30
  influence on microclimate 38
  spatiotemporal separation from evapotranspiration 38–9
consumers, dissipative ecological unit 33
Convention on Biological Diversity 538
Coongie Lakes (Australia) 226–7, *228, 229, 230*
coots 271, 272
copper contamination 404
*Coregonus* (cisco) 72, 484
*Coregonus artedii* (lake herring) 79
*Coregonus clupeaformis* (lake whitefish) 79
County and Parish Summaries of Agricultural Returns (UK) 335, 336
critical load 18, 483–4
  European alpine lakes 173
cyanobacteria
  floc blankets 320
  Lake Washington 104, *105*, 106
  reservoirs 306, 317

cyanobacteria (cont'd)
  shallow tropical lakes 290, 291
  see also algal blooms
*Cyprinus carpio* (carp) 80

dams/dam construction
  Amazonian reservoirs 243–4
  Aswan High Dam 225
  downstream flow 206, 244
  dryland lakes 221–2
  regulation of water level 54
dangerous substances directive (EU) 520
*Daphnia*
  algal bloom grazing 474
  fish role in spring peak 465
  invertebrate planktivores 465
  Lake Washington 107, *108*, 109–10
  spring clear-water phase 472
  spring populations 463–5
DDT
  ecotoxicological model 404
  Laurentian Great Lakes 84, 85, *86*, 89
decision support systems 323
decomposers
  dissipative ecological unit 33
  respiration 281
decomposition processes 55–6
deflation 206, 210
deforestation
  burning 251
  climate changes 251
  floodplain lakes of tropical/subtropical South America 248–51
  Lake Victoria basin 529
  nutrient loading impact 249–50
degradation of waters, Laurentian Great Lakes 71–2
denitrification 57
deoxygenation, Lake Victoria 530
depopulation, Fennoscandian lakes 144–5
desertification, Lake Balkhash 196
destabilisation 44
destratification 456–7
desulphurisation 57
detergents, phosphorus levels 76

detritus
  decomposition in hypolimnion 411–12
  dissipative ecological unit 33
  littoral zone 52
  reservoirs 305
developing countries, rehabilitation of lakes 523–32
diatoms
  European alpine lakes 166–7, 171, 174
  Fennoscandian lakes 143, 144–5
  fossil assemblages 285
  models of lakes/reservoirs 401
  reservoirs 320
  transfer function approach 347
dieldrin, Laurentian Great Lakes 84
dimictic lakes 487
discharge
  accelerated 44–5
  quantified limits 504
disease
  Amazonian reservoirs 244
  Aswan High Dam 226
  Lake Victoria 530
  reservoirs 312, 313
  undesirable organism removal 457
dispersive processes 32
dissipative ecological unit (DEU) 32–3, 35, 36
  components 32, *33*
  littoral 52, 53
  rivers 52
dissipative processes 25
dissolved organic carbon (DOC) 174
dissolved organic matter (DOM), Fennoscandian lakes 120–2, 125–6
disturbance of lakes, physical 348
drainage
  arheic 211
  endorheic 206, 211–12, 223
  exorheic 211
  Fennoscandian lake region 127
drainage basin area 363
draining, dryland lakes 222
dredging 455–6
*Dreissena polymorpha* (zebra mussel) 83–4, 137
drinking water 453, 508

droughts, reservoir use 302
drying out
  Aral Sea 188, 189
  Lake Eyre 233, *234*, 235
dunes, marginal 210
dynamic model of toxic substance in a trophic level 403–4
dynamic prescriptive models, complex 322

East African region 524–32
echotechnology 415
ecological engineering 415
ecological models 367, 386
  fifth/next generation 424
ecological–economic models 388
ecology, changes caused by major impact 19
ecoregions 321
ecosystems
  buffer 453
  change in prevailing conditions 426
  disturbance pulses 43–4
  long-term stability 349
  natural functionality 25
  optimisation 36
  restoration with liming 492
  sustainability 41
ecotechnological measures, eutrophication control 452–8
ecotoxicological models 401–7
  dynamic model of toxic substance in a trophic level 403–4
  effect components 405
  food chain/web dynamic 403
  fugacity models 405–6
  population dynamics 404–5
  risk assessment component 405–6
  static models of mass flows of toxic substances 403
  uncertainty 406
eel, American 80
efficiency 33–4
*Eichhornia crassipes* (water hyacinth) 284–5
electron acceptors 57
endemicity
  Aral Sea biota 232
  temporary lakes 210–11

energetics 28
  landscape 28–30
energy
  dissipation 27–8, 35, 38
    malfunction 46
  distribution 36
  dynamics of lakes 30–1
  flow density lowering 34–5
  free relative to environment 426
  generation 5
  transport 31
energy potential
  dissipation 28, *29*
  ETR model 27–8
energy–transport–reaction (ETR)
    conceptual model 26, 27–36
  energy potential 27–8
  landscape energetics 28–30
  matter transport 31–2
  reaction 32–6
  sustainability 43
entropy 28
environment
  organism interactions 359
  spatial patterns 37
  temporal patterns 37
Environment Conservation Act
    (South Africa) 537, 538, 539
Environment Protection Act
    (Finland) 513, 514, 516
Environmental Code (Sweden) 513,
    514, 517
environmental degradation,
    Laurentian Great Lakes 66
environmental management models
    390–1
environmental models 386, *387*,
    388, 390–1
Environmental Protection Act
    (Sweden) 512, 513, 520
Environmental Protection Agency
    (EPA, USA) 501, 502–3
  Clean Lakes Program 506, 507
environmental value, quantifiable
    measure 21
environmentally hazardous
    activities 517
epilimnion 441
  photosynthesis 411
epiphytes 267
episodic lakes 206

Erie, Lake (North America) 65, 66
  areas of concern 85
  fisheries 80
  OECD phosphorus load lake
    response models 378
  oxygen depletion 71
  perturbations 71
  phosphorus targets 77
  population 68
*Esox lucius* (pike) 465
European Union (EU) directives
    345
  Finland 520
euryoecious taxa 222
eutrophic lakes 356
  recovery
    delayed 332–3
    evaluation 332
  shallow tropical 283, 289–92
eutrophication 15, 19, *20*, 56
  anthropogenic 438
  assessment schemes 448
  beginning 18
  bream state 421, *422*
  chemical background 448, 451–2
  climatic conditions 448, 451–2
  control strategies 452–8
  curative treatments 455–8
  definition 354, 356
  depth increase 455
  dietary measures 453–4
  European alpine lakes 171–3
  Fennoscandian lakes 124, 133–4,
    139–40, 143
  fish stocks 370
  floodplain lakes of
    tropical/subtropical South
    America 246
  indices 448
  Lake Glumsø 398–401
  Lake Victoria 529–30
  Lake Washington 104, *105*, 106
  Laurentian Great Lakes 71, 74
  legal regulations 452–3
  limiting-nutrient concept 357
  macrophytes in shallow lakes
    264–5
  management 362
    lag period 367
    phosphorus load–lake response
      model 375–9

matter
  chemical control 455
  hydromechanical control
    methods 455
  loss 26
  production regulation 454–5
models of lakes/reservoirs 388,
    393–4, 395–401
  prognoses 401
  natural 438
  nitrogen 356–7
  nutrient loads 356–7
  phosphorus 356–7
    lag period in response to
      reduction 367
  phytoplankton response 430
  pike state 421, *422*
  process 354, 356–9
  prophylaxis 452–3
  rehabilitation methods 455–8
  reservoirs 308
  reversibility 332
  shallow temperate lakes 261–72
  shallow tropical lakes 279–94,
    284–5
    control strategies 292
  South African lakes 535
  state-changed scheme 347
  symptoms 356
  therapy 454–5
  trophic status **355**, 358–9, *361*
  urban reservoirs 243
  water quality **355**
evaporation 25
  influence on microclimate 38
  reduced 44–5
evaporation basins 219
evaporation–condensation cycle
    30
evaporative structures 44
evapotranspiration 30
  spatiotemporal separation from
    condensation 38–9
exergy 425–6, 433
  calculations 427
  computing 426
  flow 425
  indices 432, *433*
exotic species invasions
  Laurentian Great Lakes 80, 83–4,
    89

exotic species invasions (cont'd)
    see also mussel, zebra; water hyacinth
expert systems 323
export coefficient approach to prediction of nutrient loading 331–49
    calibration 343
    catchment definition 334
    current uses 345–9
    data
        collection 334–7
        computing 341–2
    derivation 337–43
    distance decay 341–2
    hindcast values 347–8
    land use 334–6
    limitations 341–3
    morphometric data 334
    rehabilitation 344
    restoration 344–5
    results
        evaluation 343–4
        using 344–5
    site selection 333–4
    validation 343
Eyre, Lake (Australia) 228, 233, 234, 235–6
    biota 235–6
    dry phase fauna 236
    drying out 233, 234, 235
    filling patterns 233, 234, 235

fauna
    Aswan High Dam 226
    Lake Eyre 235–6
Fennoscandian lakes 117–46
    acidification 135–6
    agriculture 124–5
    atmospheric deposition 130
    biomanipulation 134–5
    characteristics **121**
    depopulation effects 144–5
    environment monitoring 117–20
    eutrophication 124
    Finland 123–35
    fish 135
    forestry 125–7
    human impact 123–4
    humus content 120–2, *123*
    hydropower 129

ice cover 123
industrialisation 127–8
landscape types **122**
light attenuation 122
limnological constraints 120–3
methylmercury 135–6
natural conditions 120
nutrient loading 125, 130–5
organic carbon 122, *123*
peatland management 125–7
phytoplankton 129, 132–3, 136–7
sewage treatment 127–8, 134
Sweden 135–8
thermal stratification 122
water quality 124–5, 136, 137, 138–9, 140–1
    monitoring 117–20
    see also Nordic lakes legislation
fens 39–40
fertilisation, internal 455
fertilisers, phosphorus 338–9
filling patterns, Lake Eyre 233, *234*, 235
filters, reservoirs 319, 320
Finland
    EU directives 520
    lakes 123–35
    legislation 511–21
        administration 514–15
        current developments 520–1
        structure 513
    permit authorities 516–17
    water monitoring 518
    water quality policy 518–19
fish
    acidification
        effects 51
        models 410
    Aral Sea 187, 188, 190, 232
    Aswan High Dam 226
    biomanipulation
        removal 472, 474, *475*, 475–6
        role 462
    carnivorous 432, 433
    community changes 484
    Coongie Lakes 227
    culture 4–5
    *Daphnia* spring peak 465
    dryland lakes 208, 220–1
    European alpine lakes 170–1, 172
    Fennoscandian lakes 135

floodplain lakes of tropical/subtropical South America 246–7
kills in cold winters 470–1
Lake Baikal 182, 184
Lake Balkhash 195
Lake Eyre 236
Lake Victoria 530
Lake Washington 100
Laurentian Great Lakes 70, 72
    management programmes 78–83
liming effects 490
models 412–14
    without age structure 413–14
phytophagous 458
piscivorous 269, 270, 465–7
planktivorous 267, 268, 269, 270, 433
    biomanipulation 462
    introduction 454
    reduction 471–2
    zooplankton impact 464–5
predatory 269, 270, 465–7
production in lakes 412–14
saline lakes 214
sediment resuspension 467, 468–9
shallow lakes 267–9
species composition 484
stocking of shallow lakes 270, 462
stocks with eutrophication 370
structural dynamic model 428
threatened species 484
toxic effects 484
yield and phosphorus load 370
young-of-the-year production 464
    reduction 472
zooplankton interactions 267–8, 397
    macrophyte influence 268–9
    zooplankton predation 267
    removal 476
fisheries 58
    commercial 5
        Laurentian Great Lakes 72, 73, 74, 79–81
        modelling 412–13
    East African region 524
    Lake Chad 224
    management 89

shallow tropical lakes 286–7, 288, 289
stocking programmes in Laurentian Great Lakes 81–3
fishing, nutrient fluxes 476
flamingos 214, 217, 221
flood water attenuation 46
flooding, Coongie Lakes 227
floodplain lakes, tropical/subtropical South America 241–52
Amazonian reservoir environmental impacts 243–5
bauxite tailings impacts 247–8
deforestation 248–51
nitrogen limitation 245–7
phosphorus limitation 245–7
floodplains
discharge 47
shallow tropical lakes 283
flushing 333
rapid rate 363
rates 363–4
food chain
dynamic models 403
hazardous substance accumulation in Lake Baikal 184
foragers, Stone Age 40–1
forest, tropical, inundation 307
Forest Act (South Africa) 537
forest soil liming 488
forestry
Fennoscandian lakes 125–7
Lake Baikal 185
Framework Directive on Water Resources (EU) 345
freezing period, European alpine lakes 163–4
fuel wood use 529
fugacity models 405–6
*Fulica atra* (coot) 271

gas ebullition 307
general lake water quality index (GLWQI) 320–1
geographical information systems (GIS) 322
George, Lake (Uganda) 281, 283
gibbsite equilibrium 409

globalisation of world economy 524
Glumsø, Lake, eutrophication 398–401
Great Bear Lake (North America) 65, 66
population 69
Great Lakes Water Quality Agreement (GLWQA) 74–6
Great Slave Lake (North America) 65, 66
population 69
green revolution 42
greenhouse gases, fluxes 245
groundwater 46
abstraction 55
Fennoscandian lakes 127
levels 47
pumping 42
runoff 37
throughflow 37
gymnasium argument 6

habitat disturbance 18, 19
halocline 412
halotolerant biota 215
Havel River (Germany) 54–5
heavy metals
Lake Balkhash 196
reservoirs 312, 319
herbicides 457–8
herring
glut 464
lake 79, 80
hindcasting 17–18, 21
*Hippopotamus amphibius* (hippopotamus) 283
Hjälmaren (Sweden) 137
Hoveton Great Broad (UK), alternative stable states 471
Hudson's Bay (UK), alternative stable states 471
human impact
European alpine lakes 171–4
Fennoscandian lakes 123–4
floodplain lakes of tropical/subtropical South America 242, 249–51
Lake Victoria basin 526–30
reservoirs 311–12
human settlement 40–1

hummock-and-hollow systems 39–40
humus
Fennoscandian lakes 120–2, *123*
water-storing 37
Huron, Lake (North America)
areas of concern 85
fisheries 80
stocking programmes 83
perturbations 71
phosphorus targets 75–6, 76–7
population 68
hydraulic resistance times, shallow tropical lakes 285
hydraulics of lakes
disturbances 55–6
Lake Baikal 180
hydrogen carbonate ions 408, 410
hydrogen sulphide 55–6, 307
mixing events 451
hydrography
determination of body of water 439–41
European alpine lakes 159–60
response 47
hydrological regimes
Amazon 249
Aral Sea 189–90
dryland lakes 220, 221–2
Lake Balkhash 192–4, 195–6
shallow tropical lakes 288
hydropower
Amazonian reservoirs 243–4
Fennoscandian lakes 129
Lake Balkhash 195
hypolimnion 439–41
aeration 412
detritus decomposition 411–12
oxygen 56
hysteresis reaction 429, 430, 433

ice cover
Fennoscandian lakes 123
reacidification 486–7
*Ictalurus* (catfish) 80
industrial waste, Lake Baikal 182, 183–4
industrialisation 41–2
Fennoscandian lakes 127–8
Lake Balkhash 195
river impact 55

inherent value of lakes 9
instrumental value 5–7, 17
Integrated Pollution Prevention and
    Control Directive (EU) 520
intermittent lakes 206
internal disturbances in lakes 55–7
intrinsic value of lakes 7–8
    objective 8–9
    problems with 12–13
    subjective 8–9
invertebrates
    benthic of Laurentian Great Lakes
        78
        palaeolimnology 70
    grazers 267
    Lake Eyre 235–6
    planktivores 465
iron
    shallow tropical lakes 286
    toxicity decrease 490
iron hydroxide 452
iron/phosphorus ratio 263
irrigation
    Aral Sea 187, 188
    Aswan High Dam 226
    dryland lakes 215–16
    Lake Balkhash 197
    reservoirs 209
    salinisation 308

Kenya 528
Kitkajärvi (Russia) 141–2
Kovda River drainage area (Russia)
    141–3
Krankesjön (Sweden) 471

laboratory argument 6
Ladoga, Lake (Russia) 143–4
Laguna de Bay (Philippines) 287–9
lake basins, origins 36
Lake Victoria Fisheries Organisation
    (LVFO) 531–2
Lake Water Quality Assessment
    grants (US) 507
Lakes Areas Development Act
    (South Africa) 537
lamprey, sea 72, 78
    control 80–1
land development, Lake Washington
    103, 106, *107*
land ownership in East Africa 530–1

land use
    categories 337
    export coefficient approach
        334–6, 339, 341
    region identification 346
    spatial distribution 45–50
landscape
    catchment area 45–51
    destabilisation 42
    discharge *48*
    energetics 28–30
    Fennoscandian lakes **122**
    functionality 46
    hydromorphic 36–7
    overheating 45
    processes *48*
    stabilisation of natural function
        36–40
    thermostatic function 30
    water balance interference 46–7,
        49
land–water ecosystem 25
    characterisation 26
    holistic concept 26–7
    reed colonisation 53
Lappajärvi (Finland) 130–1
*Lates niloticus* (Nile perch) 523
Laurentian Great Lakes (North
    America) 65–90
    areas of concern **88**, 84, 85, 87
    degradation
        early evidence 70–2, 73, 74
        environmental 66
    European colonisation 66–9
    exotic species invasions 80, 83–4
    fisheries 72, 73, 74
        management 89
        stocking programmes 81–3
    mining operations 68–9
    OECD phosphorus load lake
        response models 378
    phosphorus 71
        control programmes 74–8
    physical features **67**
    pollutants 66, **88**, 85, 87
    population **67**, 68, 69
    research
        programmes/publications
        89–90
    sea lamprey control 80–1
    toxic substances 84–5, *86*

Lavijärvi (Finland/Russia) 144–5
levée banks 221
level of lakes, fluctuation 46
Liebig's Law of the Minimum 357
light
    littoral zone deposits 53
    shallow lakes 263
light attenuation
    European alpine lakes 164,
        *165*
    Fennoscandian lakes 122
limestone powder 486
    dosers 488–9, *489*
liming 451, 483
    abiotic changes 490
    agents 484–5
    aluminium salts 493–4
    catchment area 490–1, 494
    complementary measures 492
    dose 486–7, 488
    dosers 488–9, *489*
    ecological effects 490–91
    Fennoscandian lakes 136
    forest soil 488
    lakes 486–7
    littoral zone 491
    long-term 493–4
    methods 485–90, *489*
    models 485–6
    objectives 484
    phosphorus 490
    recommendations 492–3
    restoration programmes 492
    socio-economic effects 491–2
    strategies 485–6
    streams 488–90
    targets 484
    undesirable effects 490–1
    vegetation 490–1
    water quality 492
    wetlands 487–8, 490–1, 494
limit cycles 433
limiting-nutrient concept 357–8
liquid–solid boundary 31
littoral zone
    functionality damage 53–4
    liming effects 490–1
    reacidification 486–7
    retentive structure
        disturbance/removal 54
    riparian 52–3

shallow tropical lakes 282–3, 289
structural robustness 52–3
livestock numbers 336–7
loading capacity 505
*Lota lota* (burbot) 70
lunettes 210

macroclimate 38
vegetation succession 39
macrophytes
algal growth influence 266–7
Aswan High Dam 226
biofilters 454
competition with algae 265
dryland lakes 221
European alpine lakes 170
fish species impact 467
fish–zooplankton interactions 268–9
floodplain lakes of tropical/subtropical South America 246, 247
Lake Chad 224
Lake Washington 98, 100
littoral zone 52
decrease 55
mechanical control methods 457
nutrient availability 266
recolonisation 476
reservoirs 209
saline lakes 214
sedimentation 266
shallow lakes 264–7
restoration 271–2
submerged 467–8
clear-water state stabilisation 474, 476
synthesis phase 359
MAGIC (model for acidification of groundwater in catchments) 493–4
magnesium loss in rivers 49
Mälaren (Sweden) 136–7
management optimisation models 323
manganese
shallow tropical lakes 286
toxicity decrease 490
Manitoba, Lake (North America) 65, 66
margin of safety 505

matter
loading pathways 44
loss 25–6
reduction 26
short-distance cycles 38
transport 31–2, 43
mercury
Fennoscandian lakes 138
*see also* methylmercury
meromixis 14
Lake Eyre 235
metabolism, organismic 43
methane 307
reservoir emissions 245
methanogens 57
methylmercury, Fennoscandian lakes 135–6
Michaelis–Menten equations 396
Michigan, Lake (North America) 65, 66
areas of concern 85
fisheries 79, 80
stocking programmes 82
perturbations 71
phosphorus targets 76
population 68
microbial community, Lake Eyre 235
microbial degradation 46
microclimate 38
midge, phantom 294
mineral content
Lake Baikal 180, 182, 183–4
Lake Balkhash 194
reservoirs 318–19
mineral resources of saline lakes 216, 221
mining industry 42
Laurentian Great Lakes 68–9
mires 39–40
mixolimnion, saline lakes 213
Mjøsa (Norway) 139–40
models of lakes/reservoirs 386–434
accuracy 393
acidification 407–10
base 391
biomanipulation 420–1, *422*, 423–4
complexity 391–2, 393–4
conceptual diagram use 392–3
control functions 387–8

data
collection 388
requirements 393
decision makers 390
decision-making process 390
development 393
ecological–economic models 388
economics 393
ecotoxicological models 401–7
environmental 386, *387*, 388, 390–1
eutrophication 388, 393–4, 395–401
prognoses 401
submodels 396
fish production modelling 412–14
independent nutrient cycles 396
management tool 387–91
methodology 391–5
nitrogen 408
oxygen 410–12
phytoplankton 401
grazing by zooplankton 397–8
point sources 389–90
predictive capability 388–9
procedure 393
sediment–water submodel 396–7
simple 389, 392
usage 393
stratification of lakes 412
structurally dynamic 424–30
sulphur 408
thermocline 412
validation 389
water–sediment exchange submodel 397
models of wetlands 414–16, *417*, 418, *419*, 420
monimolimnion, saline lakes 213
montane regions, high 448, 451
*Morone americana* (white perch) 80
mosaic-cycle structures 38
Müritz district (Germany) *450*
mussel
quagga 83
zebra 80, 83–4
biomanipulation 462
Fennoscandian lakes 137
*Myoxocephalus thompsoni* (sculpin) 72

Naivasha, Lake (Kenya), eutrophication 287
Napindan Hydraulic Control Structure (NHCS) 288, 289
Nasser, Lake *see* Aswan High Dam (Egypt/Sudan)
National Environmental Management Act (South Africa) 538
National Environmental Management: Biodiveristy Act (South Africa) 538
National Environmental Management: Protected Areas Act (South Africa) 539
National Eutrophication Survey (EPA, USA) 360
National Parks Act (South Africa) 537
National Pollutant Discharge Elimination System (NPDES) 503, 506–7
National Water Act (South Africa) 539
Native American Tribes status 507
naturalness 13–14, **15, 16**, 15
 destruction 21
 future 17, 18, 21
 matrix 19, *20*
 original 17, 18–19
 past 18
nature conservation 453
navigation 58
nematodes, European alpine lakes 169
*Neomysis* (mysid) 109, 110
Netherlands shallow lake biomanipulation 472, **473**, 474–6
Niger River delta, eutrophic lakes 287
nitrate
 bauxite tailings impact 247
 European alpine lakes 173–4
 eutrophic reduction 57
 Lake Balkhash 194
 Laurentian Great Lakes 76
 outflow concentration 418, *419*, 420
Nitrates Directive, EU (1991) 345, 520

nitrogen
 animal wastes 339
 balance 418, *419*, 420
  agricultural regions 415
 bauxite tailings impact 247
 concentration in reservoirs 318
 cycling 416
 deforestation effects 251
 eutrophication 356–7
 export coefficients 347
 Fennoscandian lakes 125–6, *129*, 136, 137, 138, 139
 floodplain lakes of tropical/subtropical South America 250, 251
 limitation in reservoirs/floodplain lakes 245–7
 limiting nutrient 357–8
 loading 51, 56
 models of lakes/reservoirs 408
 phytoplankton growth 359
 productivity of lakes 360–1
 removal from wetland 415–16, *417*, 418, *419*, 420
 shallow tropical lakes 289–90, 290–1
 structural dynamic model 428
 transport 38
 wetland model 415–16, *417*, 418, *419*, 420
nitrogen fixation
 floodplain lakes of tropical/subtropical South America 246
 reservoirs 312
 shallow tropical lakes 290
Nordic lakes legislation 511–21
 administration 514–15
 control 518
 early development 511–12
 enforcement 518
 legislative powers 512–14
 monitoring 518
 permit authorities 515–18
 *see also* Fennoscandian lakes
Norfolk Broads (UK)
 alternative stable states 471
 coot–macrophyte populations 271, 272
 macrophytes 264–5, 265
 sediments 264

North American Great Lakes 65–90
 before colonial period 69, 70
 drainage patterns 65, 66
 *see also* Laurentian Great Lakes (North America)
Norway, lakes 138–40
Nubia, Lake *see* Aswan High Dam (Egypt/Sudan)
nutrient cycles, independent 396
nutrient loading 25
 agriculture 18
 deforestation impact 249–50
 European alpine lakes 171, 172–3
 export coefficient approach to prediction 331–49
 Fennoscandian lakes 125, 130–5, 144
 Lake Chad 224
 shallow lakes 271
 tropical 280
nutrient loads
 eutrophication 356–7
 nutrient concentration relationship 362–6
 trophic status 362–6
nutrient status 4
 macrophyte influence on fish–zooplankton interactions 269
nutrients
 attenuation in catchments 341
 availability and macrophyte effects 266
 biological elimination 454–5
 concentration 362–6
 cycling in shallow tropical lakes 281
 enrichment of shallow lakes 284
 fluxes following fishing 476
 limiting 357–8
 productivity of lakes 360–1
 redissolution processes 443
 runoff reduction 453
 thresholds for sustainable clear-water state 471
 total maximum daily loads (TMDL) 291–2
 wetlands 282

oil products in Lake Balkhash 196
Okeechobee Lake (Florida, USA) 289–92
  restoration 291–2
oligochaetes
  European alpine lakes 169
  shallow tropical lakes 290, 291
*Onchorhynchus nerka* (sockeye salmon) 100
Onega Lake (Russia) 144
Ontario, Lake (North America) 65, 66
  areas of concern 85
  fisheries 80
    stocking programmes 82
  perturbations 71
  phosphorus targets 77–8
  population 68
ontogeny of lakes 13, 14
  processes characterising 14, **16**
Opi, Lake (Nigeria) 281
organic contaminants
  persistent in Laurentian Great Lakes 85, 86
  toxic 319
organic matter concentration in reservoirs 318
Organisation for Economic Cooperation and Development (OECD) Eutrophication Programme 331–2, 333, 354, 370–80
  phosphorus load lake response models 372, **373**, 375–9
  predictive capability 375–8
  regional projects 371–2
  trophic boundary systems 372, **373**, 374–5
organisms
  properties 427
  removal of harmful 457
organochlorines, Fennoscandian lakes 128, 138
origin of lakes 36–7
orthophosphate 357
*Oscillatoria* (cyanobacteria)
  European alpine lakes 171, *172*
  Fennoscandian lakes 139
  Lake Washington 104, *105*, 106, 110

Netherlands shallow lake biomanipulation 474
oscillaxanthin 106
*Osmerus mordax* (rainbow smelt) 72, 74
overfishing, Laurentian Great Lakes 72, 78
oxygen
  Aral Sea 189
  areal hypolimnetic depletion 368–70
  Aswan High Dam 225–6
  concentration
    European alpine lakes 165–6
    reservoirs 317
  deficiency 456
  deficit in reservoirs 307
  depletion in Laurentian Great Lakes 71
  hypolimnion 56
  injection 457
  Lake Baikal 182
  Lake Chad 224
  macrophytes 266
  models 410–12
  photosynthetic production 410–11
  reservoir outflow 311
  sediments 411
    depletion 56–7
ozone, stratospheric 174

Pääjärvi (Finland) 131–2
Paanajärvi (Russia) 141, 142–3
Päijänne (Finland) 132–3
palaeoecology 40
palaeolimnology 18–19, 21, 40
  Aral Sea 231
  Fennoscandian lakes 119, 130, 144
  Lake Erie 77
  Lake Washington 103
  Laurentian Great Lakes 70
  pre-eutrophical states 344
pans 209, 210
papyrus wetlands 282
Paranoá Lake (Brazil) 243
PCBs
  Lake Baikal 183, 184
  Laurentian Great Lakes 66, 84, 85, 86, 89

peat bogs 39–40
peatland management of Fennoscandian lakes 125–7
Pechsee (Germany) 45
*Perca flavescens* (yellow perch) 80
*Perca fluviatilis* (perch) 465
perch 465, 466
  Nile 523, 524
  white 80
  yellow 80
periphyton
  acidification effects 51
  habitat removal 54
  littoral zone 52
  reservoirs 320
permaculture 4–5
permanent lakes 206–8
permissible uses of water 446, **447**
  use-limiting factors 448
permit schemes 504
perturbations in Laurentian Great Lakes 71
*Petromyzon marinus* (sea lamprey) 72
pH of lakes 30, 31, 409
  acidification 50–1
  Coongie Lakes 227
  critical tolerance 483–4
  equilibrium with hydrogen carbonate ions 410
  European alpine lakes 174
  reservoirs 319
pH/aluminium concentrations 483–4
phosphorus
  animal wastes 339
  biomanipulation modelling 423–4
  boundary loading 362
  chlorophyll relationship 458, *459*
  concentrations 14
  condensed 357
  deforestation effects 251
  European alpine lakes 172
  eutrophication 56, 356–7
    control programmes 361–2
    indices 448
  exchangeable 397
  export coefficient approach 334–7, 337–9

phosphorus (cont'd)
  Fennoscandian lakes  125–6, 129, 134, 136, 137–8
  floodplain lakes of tropical/subtropical South America  250, 251
  flushing corrected average annual inflow concentration  366, 367
  inflow concentration  378–9
  inorganic  338–9, 357
    loss  339–40
  interstitial  397
  Laurentian Great Lakes  71
    control programmes  74–8, 87
    summer concentrations  77
  liming  490
  limitation in reservoirs/floodplain lakes  245–7
  limiting nutrient  357–8
  limits for clear-water state  471
  load
    areal hypolimnetic oxygen depletion  368–70
    assimilation  363
    chlorophyll concentration  366–7
    critical  364
    fish yield  370
    lake response models  377
    Secchi-disc depth  367–8
    total  364
  loading
    concept  354–80
    deep lakes  477
  macrophyte effects  266
  mobilisation from sediments  262–4
  molybdate-reactive  318
  North American Great Lakes  69
  organic  338–9
    losses  340
  permissible loading  362–3
    boundary  364–6
  phytoplankton growth  359
  point-source  441
  productivity of lakes  360–1
  release from sediment  57
  reservoirs  304–5, 305–6, 314–15
    concentration  317–18
  retention in sediment  266

  shallow tropical lakes  286, 289–90, 291
  structural dynamic models  431–2
  total  349, 471
  transport  38
  urban reservoirs  243
phosphorus elimination plants  453–4
photosynthesis  31
  epilimnion  411
  Lake Baikal  182
  Lake Victoria  529–30
  oxygen production  410–11
  rate in European alpine lakes  169
  shallow tropical lakes  279
photosynthesis–respiration reaction  359
physical processor properties of water  30
physics of lakes  14
phytoplankton
  Aral Sea  189–90, 232
  biomass composition  359
  chemical control  457–8
  Coongie Lakes  227, 229
  European alpine lakes  166, 167–9, 171–2, 174
  eutrophication  430
  Fennoscandian lakes  129, 132–3, 136–7, 143
  grazing by zooplankton  267
    models  397–8
  growth  359
  high pH  428
  Lake Baikal  182
  Lake Balkhash  194, 196
  Lake Chad  224
  Lake Eyre  235
  Lake Washington  98, 101, 104, 105, 106
  models of lakes/reservoirs  401
  reservoirs  209, 305, 306, 310
    evaluation  319–20
    multiple-mediated feedback control  315
    tropical  312
  respiration  411
  shallow temperate lakes  261–2, 263, 264, 265–6
    persistence  267–8

  shallow tropical lakes  279–80, 284, 287–8, 290, 291
  respiration  293
  structural dynamic model  428, 429, 433
  top-down effects on biomass  477
Piburger See (Austria)  55
Pielinen lake (Finland)  10
pike  269, 465–7
pike state  420–1, 422
pikeperch  465, 466
*Pistia stratiotes*  312–13
  herbicides  457–8
plankton
  acidification effects  51
  biomass categories  448, **451**
  European alpine lakes  166–7
  Lake Baikal  184
  littoral zone functionality damage  54
  palaeolimnology of Laurentian Great Lakes  70
  screening  454
  *see also* phytoplankton; zooplankton
playas  210, 212
Ploetzensee (Germany)  56
polar regions  448, 451
pollutants/pollution  19, 20, 441
  anthropogenic  441
  assessment  345
  Baltic Sea  519
  control in USA  503–5
  damages  517
  dryland lakes  220
  emissions from heating surfaces  45
  Fennoscandian lakes  125–6, 138, 139, 141, 144
  identification  504–5
  Lake Baikal  182–4, 185
  Lake Balkhash  196
  Lake Victoria  528
  Laurentian Great Lakes  66, **88**, 85, 87, 89
  loading indices  **442**
  non-point sources  504
  point sources  502, 503–4
  protection societies  518
  reservoirs  308, 311, 312
  sediments  45

shallow tropical lakes 288–9
subtropical reservoirs 243
total maximum daily load 505
polymictic lakes 439–41
transient 224
polymictic reservoirs 305
population dynamics
Africa 526
Lake Victoria 529
models 404–5
South Africa 535–6
population-equivalent values 441
potassium
Fennoscandian lakes 125–6
loss in rivers 49
precipitation 38, 44–5
acids 50
water current generation 28
pre-reservoirs 454
primary production
dissipative ecological unit 32–3
European alpine lakes 167–9, 170
floodplain lakes of tropical/subtropical South America 246
process water 47
process-oriented approach 36
productivity of lakes 360–1
protective zones 453
protons
activity 31
discharges 50
pulp and paper industry
Fennoscandian lakes 128, 129, 132, 144
Lake Baikal 182, 183, 185
pumping, reservoir circulation 308–9

quantitative structure analysis research (QSAR) 401

rainfall, Aral Sea 229, 231
RAINS model 408, 409
rainwater pH prediction 408–9
Ramsar Convention on Wetlands 538
recreational use of lakes 4, 42
banks 46
dryland 218

redox potential of lakes 30, 31
changes 348
increase 455
reedbeds
decline 54–5
Lake Chad 224
temporal changes 53
regional development plans 453
rehabilitation
developing countries 523–32
legal regulations 452–3
methods 455–8
rating 459
reproductive (r) strategy 34
reservoirs 300–23
ageing 306–7
agriculture 311–12
Amazonian 243–4
anoxia 315–16
cascades 308
catchments 313
classification 307–8
deep 302, 304–5, **306**, 310
tropical 305
density stratification 310
depth 302, *303*
detritus 305
dryland 208–9, 222
ecological differences from natural lakes 301–3
ecosystem properties **314**
eutrophication 308
fluvial catchments 309–11
greenhouse gas fluxes 245
human impact 311–12
hydrodynamics 315
inflow treatment 453–4
influence on rivers 309–11
interflow 310
latitude 302–3
management 313–16
mathematical models 321–3
matter retention 304–5
models 386–434
morphological zonation 304
multiple systems 308–9
new 315–16
outflow
location 304
management 315, **318**

oxygen 311
temperature 310–11
oxygen deficit 307
phosphorus 304–5, 305–6, 314–15
phytoplankton 209, 305, 306, 310
evaluation 319–20
multiple-mediated feedback control 315
tropical 312
pollution 308, 311, 312
productivity phenomena 306–7
retention time 302, 304, 305, 306
self-purification capacity 309
shallow 302, 305–6, 310
stratification 304
subtropical 242–3
temperate 311–12
thermal stratification 312, 315
trophic upsurge 306–7
tropical 312–13, 451
urban 243
use impact 301–9
vegetation decay 307
water column management 313–15
water mixing 302
water quality *310*, 311–13, 316–21
evaluation 317–20
indices 320–1
monitoring 316–17
watersheds 302, 309
zooplankton 306, 320
resistance times, shallow tropical lakes 285
resources 4
allocation 36
internationally shared 531
sustainable use 531–2
respiration 31
phytoplankton 411
restoration programmes 332
acidified lakes 483–94
assessment 458–9
community impact 459
costs 458
export coefficient approach 344–5
liming 492
retention time 441, 458
dimictic lakes 487
reservoirs 302, 304, 305, 306
riparian buffer zones 345

riparian ecotone 52–5
river banks
  damage 53–4
  modifications 47
  recreational use 46
  structural robustness 52–3
  see also littoral zone
rivers
  allogenic 204
  engineering 51, 52
  hydrological pattern disruption 55
  modifications 47
  morphology damage 51
  reservoir influence 309–11
roach 269
Rodó, Lake (Uruguay) 294
running waters 51–2
runoff, discharge uniformity 46
Russia, northwest, lakes 140–5

Safe Drinking Water Act (US) 508
Saimaa lake complex (Finland) 133
salinas 212
saline lakes 206, 211–15
  aesthetic value 217
  athalassic 212
  biota 213–15
  deep 213
  mixing events 451
  shallow 212–13
  stratification 412
  thalassic 212
  tropical 451
  see also Eyre, Lake (Australia)
salinisation 308
  Aral Sea 188, 189, 190
  arid environments 451
  Aswan High Dam 226
  dryland lakes 219
  Lake Balkhash 196
salinity
  Aral Sea 220, 232
  Aswan High Dam 226
  biota restriction 213–14
  classification 215
  Coongie Lakes 227
  determination 443, **445**, 446
  dryland lakes 219–20
  Lake Chad 224
  Lake Eyre 235

reservoirs 318–19
saline lakes 213
*Salmo salar* (Atlantic salmon) 70
*Salmo trutta forma fario* (brown trout) 170
salmon
  Atlantic 70, 72
  Pacific 80
  sockeye 100
  stocking programmes 82
salt
  content determination 443, **445**, 446
  dryland lakes 207–8
  input to Lake Balkhash 196
salt pond management 221
salterns 212
saltwater intrusion, tropical shallow lakes 288, 289
*Salvelinus alpinus* (arctic char) 170–1
*Salvelinus namaycush* (lake trout) 70
*Salvinia* 294, 312–13
  herbicides 457–8
São Paulo State (Brazil) 242–3
*Schistosoma haematobium* (bilharzia) 457
sculpin 72
seasonality, shallow tropical lakes 282
Secchi-disc depth 443
  chlorophyll 368
  phosphorus load 367–8
  prediction 376
sediment memory 333
sedimentation
  accelerated 25
  littoral zone functionality damage 54
  macrophytes 266
  rate 43
    coefficient 366
sediments
  allochthonous materials 56
  autochthonous materials 56
  bauxite tailings impact 248
  chemical conditioning 455
  deposition 40
  disturbance in shallow tropical lakes 281–2

electron acceptors 57
Fennoscandian lakes 119
Lake Washington 101
limed 494
oxygen 411
  demand 411
  depletion 56–7
phosphorus release 57
pollution 45
records 344
removal 264
reservoirs 208
resuspension 467–8
  fish 468–9
sealing 455
shallow lakes 262–4
  tropical 286
sediment–water submodel 396–7
self-optimisation 52
self-organisation 35–6, 43–4
sewage/sewage treatment 42
  advanced 453
  consented discharges 337
  diversion from catchment 453
  Fennoscandian lakes 127–8, 134
  Lake Washington 103–4, 106
  reservoirs 311
  shallow tropical lakes 286
  sludge 42
  Sweden 519
  urban reservoirs 243
shallow lakes
  alternative stable states 469–71
  biomanipulation 271, 462–77
    case studies 472, **473**, 474–6
    mechanisms 471–2
  ecosystem key processes 463–9
  zooplankton dynamics 463–5
  see also temperate shallow lakes; tropical shallow lakes
shallowness of lakes 207, 261–2
shell sand 489–90
shorelines, reservoirs 209
shrimp, brine 221
silica, Laurentian Great Lakes 70
silo argument 6
silt
  Aswan High Dam 226
  reservoirs 311–12, 317
siltation of shallow tropical lakes 289

size efficiency hypothesis 464
smelt, rainbow 72
snowmelt 487
Søbygaard, Lake 427–9
soil
  agriculture 47–50
  base saturation 409
  buffer capacity 408
  carbon dioxide partial pressure 49
  chemistry 407, 408
  development 38
  erosion 312
  hydrological conductivity 418, 420
  leaching 41
    processes 42
  pH 408, 409
  processes 407, 408
soil water
  chemical composition 37
  process intensity 46
solar energy pulse, daily 36
solar radiation
  attenuation in European alpine lakes 164, 165
  shallow tropical lakes 279
solute transport in Amazon basin 249
South Africa 534–40
  coastal lakes 534–5
  lake types 534–5
  new initiatives for lakes 537–9
  protection of lakes 536–7
  threats to lakes 535–6
  water resources 539
South America, floodplain lakes/reservoirs 241–52
spatial state classification scheme for lakes 346–7
species diversity 43, 52
  Aral Sea 232
  Coongie Lakes 227
  European alpine lakes 171
  maintenance 36
  North American Great Lakes 69, 70
  saline lakes 215
  temporary lakes 210
species growth/survival 427
spring clear-water phase 472, 474
  sediment resuspension 469
  zooplankton 463–5

static bodies of water
  classification **440**, 446, 448, **449**, *450*
  depth increase 455
  permissible use assignment 446, **447**
  use-limiting factors 448
static calculation models, simple 321–2
static models of mass flows of toxic substances 403
STELLA software 416, *417*
*Stizostedion* (walleye) 72
*Stizostedion lucioperca* (pikeperch) 465
stratification of lakes 441
  disturbances 55–6
  models 412
  saline lakes 412
  *see also* thermal stratification
streams
  liming 488–90
  pasture 250
structural dynamic models 424–30
  biomanipulation 431
  parameter shifts 429
  phosphorus 431–2
  phytoplankton 428, 429, 433
structure analysis research (SAR) 401
sturgeon 72
sulphate
  European alpine lakes 173
  eutrophic reduction 57
  geogenic acidic lakes 452
  microbially-mediated reduction 452
sulphide ions 57
sulphur 55–6
  deposition 483
  Fennoscandian lakes 139
  models of lakes/reservoirs 408
sulphur dioxide, industrial emissions 50
Superior, Lake (North America)
  areas of concern 85
  fisheries 79
  phosphorus targets 76
  population 68
surface waters, classification 345
survival *(K)* strategy 34

sustainability 25, 43
Sweden
  acidification 519
  environmental monitoring 518
  EU directives 520
  lakes 135–8
  legislation 511–21
    administration 515
    current developments 520–1
    structure 513–14
  permit system 517–18
  pollution reduction objectives 519
  sewage treatment 519
  water quality policy 518–19
synthetic aperture radar (SAR) images 244, *245*
system efficiency 33–4
systemic value of lakes 11–12

Tåkern (Sweden) 471
Tanzania 528
teleological concepts 12
temperate shallow lakes 261–72
  fish 267–9
  macrophytes 264–7
  multiple stable states 265–6
  physico-chemical environment 266
  phytoplankton 265–6
  rehabilitation 270–1, 272
  restoration 272
  sediments 262–4
  switch mechanisms 266, 270–1, 433
temperature
  Aral Sea 231
  Aswan High Dam 225
  Coongie Lakes 227
  European alpine lakes 163–4
  Lake Baikal 180
  Lake Balkhash 193–4
  Lake Eyre 235
  permanent lakes 207
  reservoir outflow 310–11
  shallow tropical lakes 279
  temporary lakes 206, 209–11
  thereimictic lakes 207
  thermal efficiency 25, 28, 34
  thermal stratification
    Aral Sea 231

thermal stratification (cont'd)
  Coongie Lakes 227
  Fennoscandian lakes 122
  Lake Eyre 235
  reservoirs 312, 315
  shallow tropical lakes 281
  subtropical reservoirs 243
thermocline
  models 412
  saline lakes 213, 412
Titiwangsa, Lake (Malaysia) 284
topographic criteria of body of water 439–41
toxaphene 66, 84
toxic substances, Laurentian Great Lakes 84–5, 86
transparency 367–8
  alternative stable states 470
  Aswan High Dam 225
  European alpine lakes 164–5
  Lake Baikal 180
  Lake Balkhash 193
  reservoirs 317
transport, water-borne 42
trichopterans, European alpine lakes 169
trophic boundary system 372, **373**, 374–5
trophic cascade theory 476–7
trophic categories of lakes 332
trophic criteria determination 441, **442**, 443, **444**
trophic index number 448
trophic probability diagrams 374–5
trophic state index 448
trophic status
  assessment 438–9
  classification 372, **373**, 374, *375*, 376
  eutrophication **355**, *361*
  historical records 344
  nutrient loads 362–6
  tropical reservoirs 451
trophic upsurge, reservoirs 306–7
tropical lakes
  fisheries 289
  study 279
tropical reservoirs 305
tropical shallow lakes 279–80, 279–94
  biomanipulation 293–4

  case studies 285–92
  catchment processes 285
  depth role 280–2
  eutrophic 283
  eutrophication 284–5
  features 279–80
  fisheries 286–7, 288
  littoral zone 282–3
  mixing depth 280
  phosphorus 291
  phytoplankton 279–80, 284, 287–8, 290, 291
    respiration 293
  pollution 288–9
  restoration 292–4
  saltwater intrusion 288, 289
  seasonality 282
  vegetation 285–6, 287–8
  water column mixing 280, 281, 285
  water level 280, 282–3
  water quality 287, 292
trout
  brown 170
  lake 70, 80
    stocking programmes 81, 82
Trummen, Lake (Sweden) 38
  phosphorus-rich sediment removal 264
Tucuruí dam 244, 245
turbid bream equilibrium 421
turbidity 367–8
  algal blooms 475
  alternative stable states 470
  Aswan High Dam 225
  liming 490
  saline lakes 212
  shallow tropical lakes 281–2, 288
turbulence, Lake Baikal 180

Uganda 528, 531
ultraviolet light penetration
  dryland lakes 222
  European alpine lakes 165, 169, 174
undisturbed lakes 19
Urban Waste Water Treatment (UWWT) EU directive 345
urbanisation, Lake Washington 103–4

USA
  framework for managing lakes 501–9
  lake area 501
  local/State efforts 508–9
utilisation of lakes 57–8

values of lakes 5–13, 21
Vänern (Sweden) 138, 514
Vättern (Sweden) 137–8
vegetation
  biomass 433
  climax stage 37–8
  decomposition of submerged 244, 307
  development 38
  floating 312–13
  liming effects 490–1
  littoral zone 53, 282
  nutrient availability for phytoplankton 467
  progressive zones 53
  recolonisation 37
  regressive zones 53
  removal by humans 41
  sediment effects 467–8
  shallow tropical lakes 285–6, 287–8
  submerged 433
  succession 38, 39
  terrestrial 490–1
  wetland 282
Vesijärvi (Finland) 133–5
Victoria, Lake (East Africa) 526–32
  bordering nations 528, 531
  eutrophication 529–30
  location 527
  seasonality 527
  size 526–7, 528
  sustainable use of natural resources 531–2
  threats to 528–9
  transboundary programmes 528, 531–2
vleis 209, 210

Wahnbach Reservoir (Germany) 315
Waigani Lake (Papua New Guinea) 285–7
walleye 72, 79

Washington, Lake (USA) 96–113
  alkalinity 106, *107*
  chemical properties 98, **100**, 106, *107*
  contained material movement 111–12
  density flow 111–12
  development 100–1, **102**
  eutrophication 104, *105*, 106
  features 96, *97*, 98, *99*
  fish 100
  formation 100–1, **102**
  horizontal variations 110–11
  hydrology 101
  land development 103, 106, *107*
  level lowering 101
  macrophytes 98, 100
  past condition changes 101–3
  pH variation 106, *107*
  phytoplankton 98, 101, 104, *105*, 106
  present condition 98, 100, 101, **102**
  recovery 104, *105*, 106
  sediments 101
  ship canal 112
    building 101
  urbanisation effects 103–4
  water movement 111–12
  watershed 96, *99*
  zooplankton 98, *107*, *108*, 109–10
waste treatment plants, Laurentian Great Lakes 75–6
wasteload allocation 505
wastewater
  discharge 516
  disposal 42
  diversion 400
  treated 399–400
  urban 243
water
  abstraction from Lake Balkhash 195–6
  currents 28
  diversions 219–20
  exchange 455
  processor properties 28–30
  transformation 28, *29*
Water Act (Finland) 514, 516, 520–1
Water Act (South Africa) 537
Water Act (Sweden) 512, 514

water bodies
  altering 517
  artificial 300–23
water budget, European alpine lakes 159–60
water courses, engineering 42
water cycle 44–5
  disturbance 46
  fast 363
  long-range 45
  long-wave 38, 39
  short-wave 38, 39
Water Framework Directive (EU) 514
water hardness/softness
  European alpine lakes 165
  Lake Baikal 180
  reservoirs 318–19
water hyacinth 284–5, *286*, 312–13
  African Great Lakes 523–4
  biological control 287
  herbicides 457–8
  introduction 294
  Lake Victoria 530
water legislation (EU) 520
water level
  Amazonian reservoirs 244
  Aral Sea 188, 189, 190, 231
  astatic 207
  dryland lakes 219–20
  Fennoscandian lakes 128–9
  floodplain lakes of tropical/subtropical South America 242
  Lake Baikal 179
  Lake Balkhash 197
  Lake Chad *224*
  Lake Washington 101
  regulation of lakes 54
  reservoirs 311
  shallow tropical lakes 280, 282–3
water management projects 517–18
Water Management Sector Group (WMSG) 370
water quality 4
  classification 446, 448
  control standards 453
  eutrophication **355**
  Fennoscandian lakes 124–5, 136, 137, 138–9, 140–1
  Finnish policies 518–19

Lake Baikal 184–5
  liming 492
  management 458–9
  mass development of algae 439
  microbiological 312
  monitoring in Fennoscandian lakes 117–20, 130–5, 518
  objectives 333
  permissible use assignment 446, **447**
  reservoirs *310*, 311–13, 316–21
    evaluation 317–20
    indices 320–1
    monitoring 316–17
  shallow tropical lakes 287, 292
  Swedish policies 518–19
  USA standards 501–3
Water Quality Objectives (WQO) 346
water regime, vegetation development 37
water resources, South Africa 539
Water Resources Act (UK, 1991) 346
Water Rights Act (Finland) 512
water sports, non-bathing 4
waterflea, spiny 83
water–sediment exchange submodel 397
watersheds
  clean water approaches 505–6
  export coefficients 337
  management 313
  reservoirs 302, 309
weathering processes 409
wetlands
  degradation in East Africa 530–1
  liming 487–8, 490–1, 494
  modelling 414–16, *417*, 418, *419*, 420
  papyrus 282
  rehabilitation in developing countries 523–32
  South Africa 535
    protection 536–7
  Ugandan management 531
whitefish, lake 79, 80
wilderness 6, 9
Wilderness lakes (South Africa) 535–6
wind action 207
  saline lakes 212

Winnipeg, Lake (North America) 65, 66
  population 68
Winnipegosis, Lake (North America) 65, 66
Wisconsin Lake Model Spreadsheet (WILMS) 379
World Heritage Conservation Act (South Africa) 539

Xolotlán, Lake (Central America) 293

zooplankton
  Aral Sea 190, 232
  carrying capacity 397–8
  Coongie Lakes 227
  dynamics in shallow lakes 463–5
  European alpine lakes 167, 174
  Fennoscandian lakes 139–40
  fish interactions 267–8, 397
    macrophyte influence 268–9
  grazing 430, 432, 462
  herbivory 315
  Lake Baikal 182, 184
  Lake Balkhash 194–5, 196
  Lake Washington 98, 107, *108*, 109–10
  phytoplankton grazing 267, 397–8
  planktivorous fish impact 464–5
  reservoirs 316, 320
  saline lakes 214
  shallow tropical lakes 279–80
  spring clear-water phase 463–5